CHILTON'S GUIDE TO
AUTOMATIC TRANSMISSION REPAIR – 1980-84 DOMESTIC CARS & TRUCKS

President	Dean F. Morgantini, S.A.E.
Vice President–Finance	Barry L. Beck
Vice President–Sales	Glenn D. Potere
Executive Editor	Kevin M. G. Maher
Production Manager	Ben Greisler, S.A.E.
Project Managers	Michael Abraham, George B. Heinrich III, S.A.E., Will Kessler, A.S.E., S.A.E., Richard Schwartz
Schematics Editor	Christopher G. Ritchie

CHILTON™ Automotive Books

PUBLISHED BY **W. G. NICHOLS, INC.**

Manufactured in USA
© 1988 Chilton Book Company
1020 Andrew Drive
West Chester, PA 19380
ISBN 0-8019-7890-4
Library of Congress Catalog Card No. 88-48012
7890123456 7654321098

ACKNOWLEDGEMENTS

Chilton Book Company expresses appreciation to the following firms for their cooperation and technical assistance:

ATRA—Automatic Transmission Rebuilders Association, Ventura, California
    Special Thanks to: Mr. Gene Lewis—Executive Director
                    Mr. John Maloney—National President
                    Mr. Robert D. Cherrnay—Technical Director
                    Mr. Michael Abell—Service Engineering
                    Mr. C. W. Smith—Technical Staff
ASC—Automotive Service Councils, Inc.® Elmhurst, Illinois
    Special Thanks to: Mr. Del Wright—Chairman of the Board
                    Mr. John F. Mullins, Jr.—Vice Chairman of the Board
                    Mr. George W. Merwin, III—President
American Motors Corporation and Regie Nationale des Usines Renault, Detroit, Michigan
American Honda Motor Company, Gardena, California
American Isuzu Motors, Inc., Whittier, California
Audi, Division of Volkswagen of America Incorporated, Englewood Cliffs, New Jersey
Brandywine Transmission Service, Inc., Wilmington, Delaware
Borg Warner Transmission Service Center, Ramsey, New Jersey
    Special Thanks to: Mr. Michael LePore
Chrysler Corporation, Detroit, Michigan
Detroit Diesel Allison, Division of General Motors Corporation, Indianapolis, Indiana
Ford Motor Company, Dearborn, Michigan
Fuji Heavy Industries, Ltd., Tokyo, Japan
General Motors Corporation, Flint, Michigan
Hydra-Matic, Division of General Motors Corporation, Ypsilanti, Michigan
Jaguar, Rover, Triumph Motor Company, Inc., Leonia, New Jersey
Japanese Automatic Transmission Company, Tokyo, Japan
Lee's Auto Service, Trainer, Chester, Pennsylvania
Mazda Motors of America, Incorporated, Montvale, New Jersey
Mercedes Benz of North America, Incorporated, Montvale, New Jersey
Nissan Motor Corporation of USA, Carson, California
Ralph's Garage, Chester, Pennsylvania
Seuro Transmissions, Incorporated, Pittsburgh, Pennsylvania
Subaru of America, Incorporated, Pennsauken, New Jersey
Toyota Motor Sales USA, Incorporated, Torrence, California
Transmissions By Lucille, Pittsburgh, Pennsylvania
Volkswagen of America, Incorporated, Englewood Cliffs, New Jersey
Volvo of America Corporation, Rockleigh, New Jersey
ZF of North America, Inc., Chicago, Illinois

CONTENTS

DOMESTIC CARS, LIGHT TRUCKS & VANS TRANSMISSIONS AND TRANSAXLES

PART NUMBERS

Part numbers listed in this reference are not recomendations by Chilton for any product by brand name. They are references that can be used with interchange manuals and aftermarket supplier catalogs to locate each brand supplier's discrete part number.

Although information in this manual is based on industry sources and is complete as possible at the time of publication, the possibilty exists that some car manufacturers made later changes which could not be included here. While striving for total accuracy, Chilton Book Company cannot assume responsibility for any errors, changes or omissions that may occur in the compilation of this data.

There are six major sections in this manual. And each is appropriately numbered for easy location:

1. General Information
2. Chrysler Corporation
3. Ford Motor Company
4. General Motors Corporation
5. Modifications & Changes

Further, each transmission/transaxle section is indexed and organized by the following functions:

Applications
General Description
Modifications
Trouble Diagnosis
On Car Services
Removal and Installation Procedures
Bench Overhaul
Specifications
Special Tools

Graphic symbols throughout the manual aid in the location of sections and speed the pinpointing of information.

METRIC NOTICE

Certain parts are dimensioned in the metric system. Many fasteners are metric and should not be replaced with a customary inch fastener.

It is important to note that during any maintenance procedure or repair, the metric fastener should be salvaged for reassembly. If the fastener is not reusable, the equivalent fastener should be used.

A mismatched or incorrect fastener can result in component damage or possibly, personal injury.

SAFETY NOTICE

Proper service and repair procedures are vital to the safe, reliable operation of all motor vehicles, as well as the personal safety of those performing repairs. This manual outlines procedures for servicing and repairing vehicles using safe, effective methods. The procedures contain many NOTES, CAUTIONS and WARNINGS which should be followed along with standard safety procedures to eliminate the possiblity of personal injury or improper service which could damage the vehicle of compromise its safety.

It is important to note that repair procedures and techniques, tools and parts for servicing motor vehicles, as well as the skill and experience of the individual performing the work, vary widely. It is not possible to anticipate all of the conceivable ways or conditions under which vehicles may be serviced, or to provide cautions to all of the possible hazards that may result. Standard and accepted safety precautions and equipment should be used when handling toxic or flammable fluids, and safety goggles or other protection should be used when handling toxic or flammable fluids, and safety goggles or other protection should be used during cutting, grinding, chiseling, prying or any process than can cause material removal or projectiles.

Some procedures require the use of tools especialy designed for a specific purpose. Before substituting another tool or procedure, you must be completely satisfied that neither your personal safety, nor the performance of the vehicle will be endangered.

INDEX

GENERAL INFORMATION

Introduction

This concise, but comprehensive service manual places emphasis on diagnosing, troubleshooting, adjustments, testing, disassembly and assembly of the automatic transmission.

This manual will consist of the following major sections:
1. General Information section
2. Domestic Car Automatic Transmission section
3. Domestic Truck Automatic Transmission section
4. Modification section
5. Correction section.

Within the Automatic Transmission sections, the following information is included:

1. Transmission Application Chart.
2. General description—to include:
   a. Model and type.
   b. Capacities.
   c. Fluid specifications.
   d. Checking fluid level.
3. Transmission modifications
4. Trouble diagnosis—to include:
   a. Hydraulic system operation
   b. Oil pressure test.
   c. Air pressure test.
   d. Stall test.
   e. Control pressure specifications.
   f. Shift speed specifications (when available).
5. On Car services—to include:
   a. Adjustments.
   b. Removal and installation.
6. Transmission/Transaxle removal and installation.
7. Bench overhaul—to include:
   a. Transmission/transaxle disassembly.
   b. Internal component disassembly and assembly.
   c. Transmission/transaxle assembly.
8. Specifications.
9. Factory recommended tools.

Metric Fasteners And Inch System

Metric bolt sizes and bolt pitch are more commonly used for all fasteners on the automatic transmissions/transaxles now being manufactured. The metric bolt sizes and thread pitches are very close to the dimensions of the similar inch system fasteners and for this reason, replacement fasteners must have the same measurement and strength as those removed.

Do not attempt to interchange metric fasteners for inch system fasteners. Mismatched and incorrect fasteners can result in dam-

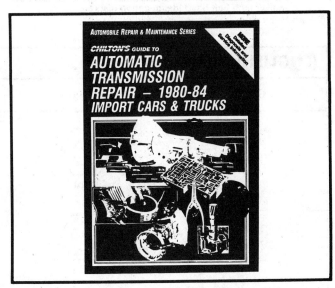

Companion manual—"Chilton's Guide to Automatic Transmission Repair, 1980–84 Import Cars & Trucks"

GENERAI INFORMATION

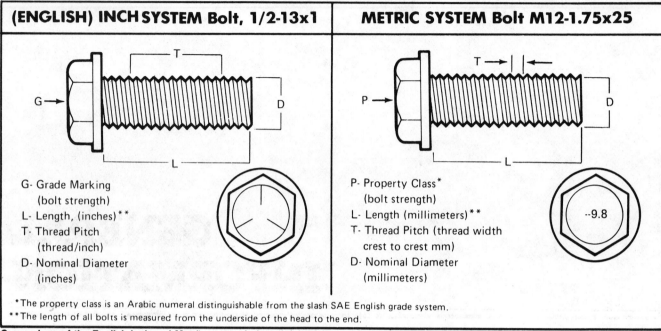

(ENGLISH) INCH SYSTEM Bolt, 1/2-13x1

G- Grade Marking
   (bolt strength)
L- Length, (inches)**
T- Thread Pitch
   (thread/inch)
D- Nominal Diameter
   (inches)

METRIC SYSTEM Bolt M12-1.75x25

P- Property Class*
   (bolt strength)
L- Length (millimeters)**
T- Thread Pitch (thread width
   crest to crest mm)
D- Nominal Diameter
   (millimeters)

*The property class is an Arabic numeral distinguishable from the slash SAE English grade system.
**The length of all bolts is measured from the underside of the head to the end.

Comparison of the English Inch and Metric system bolt and thread nomenclature(© Ford Motor Co.)

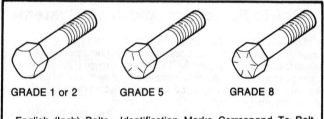

GRADE 1 or 2   GRADE 5   GRADE 8

English (Inch) Bolts—Identification Marks Correspond To Bolt Strength—Increasing Number Of Slashes Represent Increasing Strength.

Typical English Inch bolt head identification marks (© Ford Motor Co.)

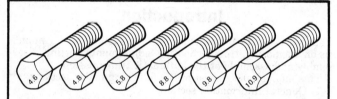

Metric Bolts—Identification Class Numbers Correspond To Bolt Strength—Increasing Numbers Represent Increasing Strength. Common Metric Fastener Bolt Strength Property Are 9.8 And 10.9 With The Class Identification Embossed On The Bolt Head.

Typical Metric bolt head identification marks (© Ford Motor Co.)

(ENGLISH) INCH SYSTEM

Grade	Identification
Hex Nut Grade 5	3 Dots
Hex Nut Grade 8	6 Dots

Increasing dots represent increasing strength.

METRIC SYSTEM

Class	Identification
Hex Nut Property Class 9	Arabic 9
Hex Nut Property Class 10	Arabic 10

May also have blue finish or paint daub on hex flat.
Increasing numbers represent increasing strength.

Comparison of English Inch and Metric hex nut strength identification marks(© Ford Motor Co.)

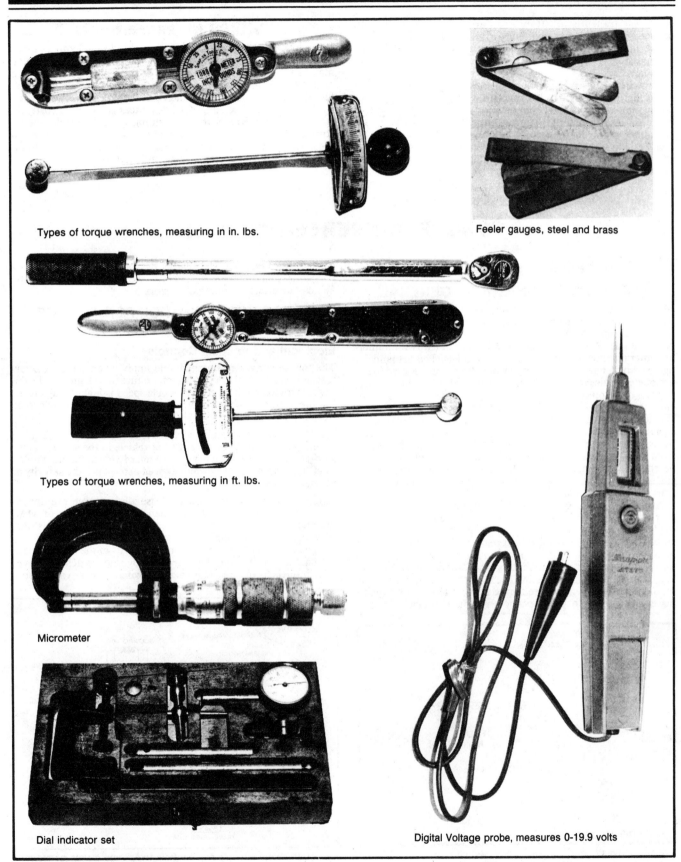

Types of torque wrenches, measuring in in. lbs.

Feeler gauges, steel and brass

Types of torque wrenches, measuring in ft. lbs.

Micrometer

Dial indicator set

Digital Voltage probe, measures 0-19.9 volts

Typical precision measuring tools

age to the transmission/transaxle unit through malfunction, breakage or possible personal injury. Care should be exercised to re-use the fasteners in their same locations as removed, when ever possible. If any doubt exists in the re-use of fasteners, install new ones.

To avoid stripped threads and to prevent metal warpage, the use of the torque wrench becomes more important, as the gear box assembly and internal components are being manufactured from light weight material. The torque conversion charts should be understood by the repairman, to properly service the requirements of the torquing procedures. When in doubt, refer to the specifications for the transmission/transaxle being serviced or overhauled.

Critical Measurements

With the increase use of transaxles and the close tolerances needed throughout the drive train, more emphasis is placed upon making the critical bearing and gear measurements correctly and being assured that correct preload and turning torque exists before the unit is re-installed in the vehicle. Should a comeback occur because of the lack of proper clearances or torque, a costly rebuild can result. Rather than rebuilding a unit by "feel", the repairman must rely upon precise measuring tools, such as the dial indicator, micrometers, torque wrenches and feeler gauges to insure that correct specifications are adhered to.

TORQUE CONVERTER CLUTCH

Principles of Operation

ELECTRONIC CONTROLS—GENERAL

Many changes in the design and operation of the transmission/transaxles have occurred since the publishing of our first Automatic Transmission Manual. New transaxles and transmissions have been developed, manufactured and are in use, with numerous internal changes made in existing models. The demand for lighter, smaller and more fuel efficient vehicles has resulted in the use of electronics to control both the engine spark and fuel delivery at a more precise time and quantity, to achieve the fuel efficient results that are required by law. Certain transmission/transaxle assemblies are a part of the electronic controls, by sending signals of vehicle speed and throttle opening to an on-board computer, which in turn computes these signals, along with others from the engine assembly, to determine if spark occurence should be changed or the delivery fo fuel should be increased or decreased. The computed signals are then sent to the respective controls and/or sensors as required.

Automatic transmissions with microcomputers to determine gear selections are now in use. Sensors are used for engine and road speeds, engine load, gear selector lever position, kick down switch and a status of the driving program to send signals to the microcomputer to determine the optimum gear selection, according to a preset program. The shifting is accomplished by solenoid valves in the hydraulic system. The electronics also control the modulated hydraulic pressure during shifting, along with regulating engine torque to provide smooth shifts between gear ratio

changes. This type of system can be designed for different driving programs, such as giving the operator the choice of operating the vehicle for either economy or performace.

ELECTRICAL CONTROL FOR TORQUE CONVERTER CLUTCH

Electrical and Vacuum Controls

The torque converter clutch should apply when the engine has reached near normal operating temperature in order to handle the slight extra load and when the vehicle speed is high enough to allow the operation of the clutch to be smooth and the vehicle to be free of engine pulses.

NOTE: When the converter clutch is coupled to the engine, the engine pulses can be felt through the vehicle in the same manner as if equipped with a clutch and standard transmission. Engine condition, engine load and engine speed determines the severity of the pulsations.

The converter clutch should release when torque multiplication is needed in the converter, when coming to a stop, or when the mechanical connection would affect exhaust emissions during a coasting condition.

The electrical control components consists of the brake release switch, the low vacuum switch and the governor switch. Some vehicle models have a thermal vacuum switch, a relay valve and a

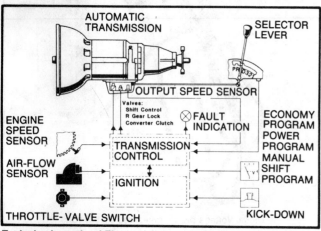

Typical schematic of Electronic gear selection with microcomputer control(© ZF of North America, Inc.)

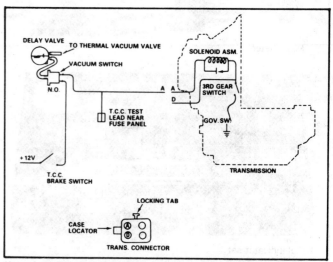

Use of electrical and vacuum controls to operate torque converter clutch(© General Motors Corp.)

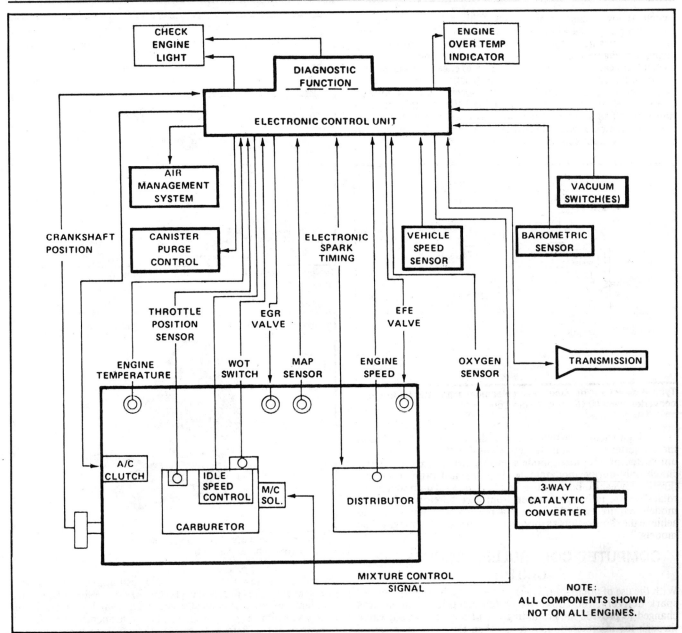

Typical schematic of a computer command control system(© General Motors Corp.)

delay valve. Diesel engines use a high vacuum switch in addition to certain above listed components. These various components control the flow of current to the apply valve solenoid. By controlling the current flow, these components activate or deactivate the solenoid, which in turn engages or disengages the transmission converter clutch, depending upon the driving conditions as mentioned previously. The components have the two basic circuits, electrical and vacuum.

ELECTRICAL CURRENT FLOW

All of the components in the electrical circuit must be closed or grounded before the solenoid can open the hydraulic circuit to engage the converter clutch. The circuit begins at the fuse panel and flows to the brake switch and as long as the brake pedal is not depressed, the current will flow to the low vacuum switch on the gasoline engines and to the high vacuum switch on the diesel engines. These two switches open or close the circuit path to the so-

lenoid, dependent upon the engine or pump vacuum. If the low vacuum switch is closed (high vacuum switch on diesel engines), the current continues to flow to the transmission case connector, into the solenoid and to the governor pressure switch. When the vehicle speed is approximately 35-50 mph, the governor switch grounds to activate the solenoid. The solenoid, in turn, opens a hydraulic circuit to the converter clutch assembly, engaging the unit.

It should be noted that external vacuum controls include the thermal vacuum valve, the relay valve, the delay valve, the low vacuum switch and a high vacuum switch (used on diesel engines). Keep in mind that all of the electrical or vacuum components may not be used on all engines, at the same time.

VACUUM FLOW

The vacuum relay valve works with the thermal vacuum valve to keep the engine vacuum from reaching the low vacuum valve switch at low engine temperatures. This action prevents the clutch from engaging while the engine is still warming up. The delay valve slows down the response of the low vacuum switch to changes in engine vacuum. This action prevents the low vacuum switch from causing the converter clutch to engage and disengage too rapidly. The low vacuum switch deactivates the converter clutch when engine vacuum drops to a specific low level during moderate acceleration just before a part-throttle transmission downshift. The low vacuum switch also deactivates the clutch while the vehicle is coasting because it receives no vacuum from its ported vacuum source.

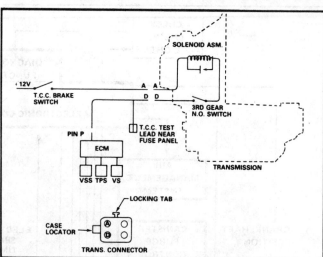

Typical Computer Command Control schematic
(© General Motors Corp.)

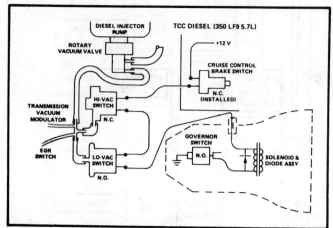

Typical diesel engine vacuum and electrical schematic for torque converter clutch(© General Motors Corp.)

The high vacuum switch, when on diesel engines, deactivates the converter clutch while the vehicle is coasting. The low vacuum switch on the diesel models only deactivates the converter clutch only during moderate acceleration, just prior to a part-throttle downshift. Because the diesel engine's vacuum source is a rotary pump, rather than taken from a carburetor port, diesel models require both the high and the low vacuum switch to achieve the same results as the low vacuum switch on the gasoline models.

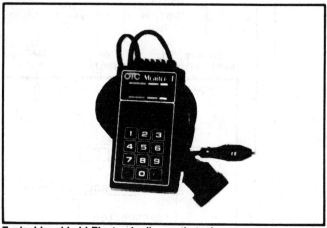

Typical hand held Electronic diagnostic tool
(© OTC Tools and Equipment)

COMPUTER CONTROLLED CONVERTER CLUTCH

With the use of micro-computers governoring the engine fuel and spark delivery, the converter clutch electronic control was changed to provide the grounding circuit for the solenoid valve through the micro-computer, rather than the governor pressure switch. Sensors are used in place of the formerly used switches and send signals back to the micro-computer to indicate if the engine is in its proper mode to accept the mechanical lock-up of the converter clutch.

Normally a coolant sensor, a throttle position sensor, an engine vacuum sensor and a vehicle speed sensor are used to signal the micro-computer when the converter clutch can be applied. Should a sensor indicate the need for the converter clutch to be deactivated, the grounding circuit to the transmission solenoid valve would be interrupted and the converter clutch would be released.

Diagnostic Precautions

We have entered into the age of electronics and with it, the need for the diagnostician to increase the skills needed for this particular field. By learning the basic components, their operation, the testing procedures and by applying a "common sense" approach to the diagnosis procedures, repairs can be made accurately and quickly. Avoid the "short-cut" and "parts replacement" ap-

proach, but follow the recommended testing and repair procedures. Much time effort and costs can be saved. When diagnosing problems of an electronically equipped vehicle, consider the changes in vehicle speed sensing and throttle opening sensing that are encountered. Before electronics, the governor sensed vehicle speed and by allowing more governor pressure to be produced, overcame throttle and main line pressures to move the shifting valves. Now, the vehicle speed is sensed from a small electric producing generator, mounted on or near the speedometer or cable and is activated by the rotation of the speedometer cable. This signal is then directed to the micro-computer for computation on when the converter clutch should be engaged or disengaged.

The governor operation still remains the same as to the shifting and producing of governor pressure, but its importance of being the vehicle speed sensor has now diminished. A throttle positioner switch has been added to the carburetor to more precisely identify the position of the throttle plates in regards to vehicle speed and load. The throttle cable or modulator are still used to control the throttle pressure within the transmission/transaxle assemblies. Added pressure switches are mounted to the various passages of the valve bodies to sense gear changes, by allowing electrical current to either pass through or be blocked by the switches. The repairman must be attentive when overhauling or replacing the valve body, to be sure the correct pressure switch is installed in the correct hydraulic passage, because the switches have either normally closed (N.C.) contacts or normally open (N.O.) contacts and would be identified on the electrical schematic as such. To interchange switches would cause malfunctions and possible costly repairs to both the transmission and to the electronic system components.

The many types of automatic transmissions that employ the electronic controlled converter clutch, do not follow the same electronic or hydraulic routings nor have the same sensors, oil pressure switches and vacuum sensing applications. Therefore, before diagnosing a fault in the converter clutch assembly, the electric, the electronic or hydraulic circuits, identify the automatic transmission as to its model designation and code.

NOTE: Refer to the proper diagnostic outline in the individual automatic transmission outline or to the General Motors Turbo Hydra-Matic Converter Clutch Diagnostic outline, at the beginning of the General Motors Automatic Transmission Sections.

HYDRAULIC CONVERTER CLUTCH OPERATION

Numerous automatic transmissions rely upon hydraulic pressures to sense, determine when and to apply the converter clutch assembly. This type of automatic transmission unit is considered to be a self-contained unit with only the shift linkage, throttle cable or modulator valve being external. Specific valves, located within the valve body or oil pump housing, are caused to be moved when a sequence of events occur within the unit. For example, to engage the converter clutch, most all automatic transmissions require the gear ratio to be in the top gear before the converter clutch control valves can be placed in operation. The governor and throttle pressures must maintain specific fluid pressures at various points within the hydraulic circuits to aid in the engagement or disengagement of the converter clutch. In addition, check valves must properly seal and move to exhaust pressured fluid at the correct time to avoid "shudders" or "chuckles" during the initial application and engagement of the converter clutch.

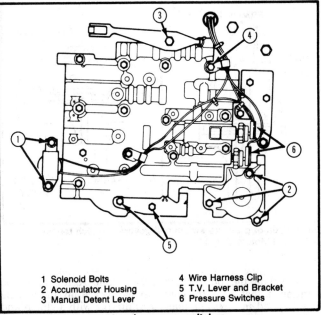

1 Solenoid Bolts
2 Accumulator Housing
3 Manual Detent Lever
4 Wire Harness Clip
5 T.V. Lever and Bracket
6 Pressure Switches

Valve body components and pressure switches
(© General Motors Corp.)

CENTRIFUGAL TORQUE CONVERTER CLUTCH OPERATION

A torque converter has been in use that mechanically locks up centrifugally without the use of electronics or hydraulic pressure. At specific input shaft speeds, brake-like shoes move outward from the rim of the turbine assembly, to engage the converter housing, locking the converter unit mechanically together for a 1:1 ratio. Slight slippage can occur at the low end of the rpm scale,

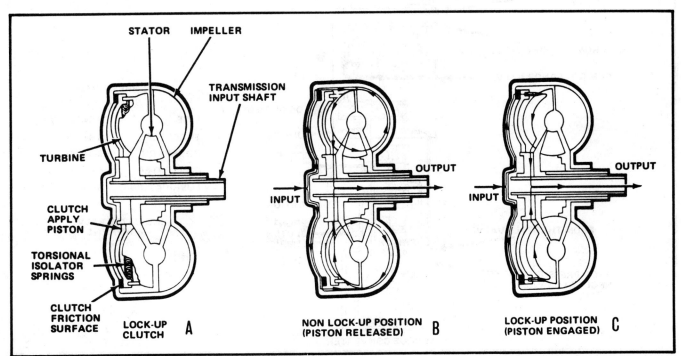

Power direction through lock-up converter clutch, typical(© Chrysler Corp.)

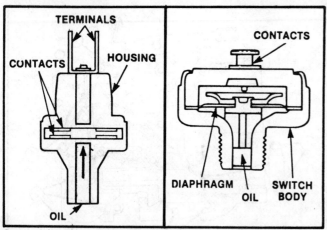

Comparison of pressure switch and governor switch terminals (© General Motors Corp.)

but the greater the rpm, the tighter the lock-up. Again, it must be mentioned, that when the converter has locked-up, the vehicle may respond in the same manner as driving with a clutch and standard transmission. This is considered normal and does not indicate converter clutch or transmission problems. Keep in mind if engines are in need of tune-ups or repairs, the lock-up "shudder" or "chuckle" feeling may be greater.

MECHANICAL CONVERTER LOCK-UP OPERATION

An other type of converter lock-up is the Ford Motor Company's AOD Automatic Overdrive transmission, which uses a direct drive input shaft splined to the damper assembly of the torque converter cover to the direct clutch, bypassing the torque converter reduction components. A second shaft encloses the direct drive input shaft and is coupled between the converter turbine and the reverse clutch or forward clutch, depending upon their applied phase. With this type of unit, when in third gear, the input shaft torque is split, 30% hydraulic and 70% mechanical. When in the overdrive or fourth gear, the input torque is completely mechanical and the transmission is locked mechanically to the engine.

CONFIRMING LOCK-UP OF TORQUE CONVERTER

To confirm the lock-up of the torque converter has occurred, check the engine rpm with a tachometer while the vehicle is being driven. If the torque converter is locked-up, the engine rpm will decrease approximately 200-400 rpm, at the time of lock-up.

Overdrive Units

With need for greater fuel economy, the automatic transmission/transaxles were among the many vehicle components that have been modified to aid in this quest. Internal changes have been made and in some cases, additions of a fourth gear to provide the

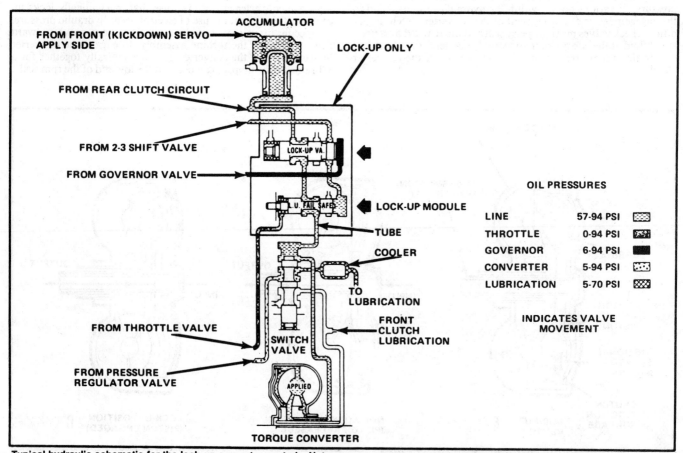

OIL PRESSURES		
LINE	57-94 PSI	
THROTTLE	0-94 PSI	
GOVERNOR	6-94 PSI	
CONVERTER	5-94 PSI	
LUBRICATION	5-70 PSI	

INDICATES VALVE MOVEMENT

Typical hydraulic schematic for the lock-up converter controls. Note governor pressure reaction area on the lock-up valve (© Chrysler Corp.)

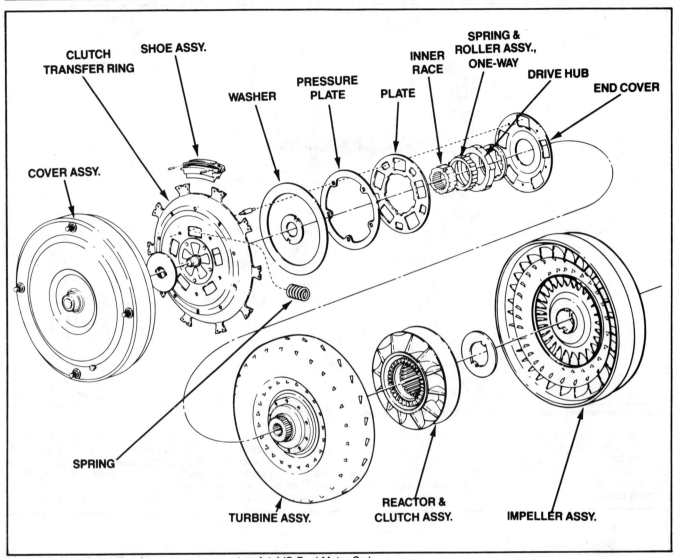

CLUTCH
TRANSFER RING

SHOE ASSY.

COVER ASSY.

WASHER

PRESSURE
PLATE

PLATE

INNER
RACE

SPRING &
ROLLER ASSY.,
ONE-WAY

DRIVE HUB

END COVER

SPRING

TURBINE ASSY.

REACTOR &
CLUTCH ASSY.

IMPELLER ASSY.

Exploded view of centrifugal type torque converter clutch(© Ford Motor Co.)

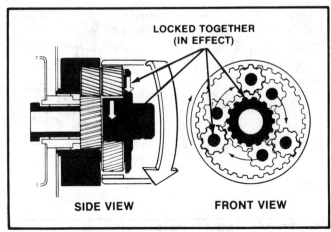

LOCKED TOGETHER
(IN EFFECT)

SIDE VIEW

FRONT VIEW

Planetary gear rotation in direct drive(© Ford Motor Co.)

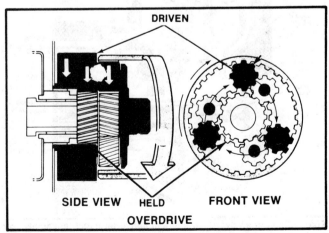

DRIVEN

SIDE VIEW

HELD

FRONT VIEW

OVERDRIVE

Planetary gear rotation in overdrive(© Ford Motor Co.)

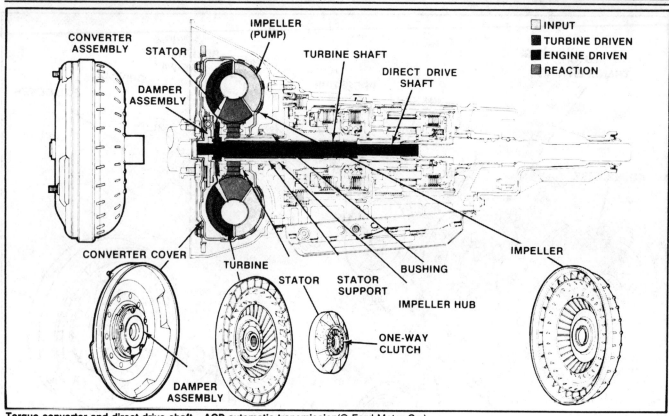

Torque converter and direct drive shaft—AOD automatic transmission(© Ford Motor Co.)

overdirect or overdrive gear ratio. The reasoning for adding the overdrive capability is that an overdrive ratio enables the output speed of the transmission/transaxle to be greater than the input speed, allowing the vehicle to maintain a given road speed with less engine speed. This results in better fuel economy and a slower running engine.

The overdrive unit usually consists of an overdrive planetary gear set, a roller one-way clutch assembly and two friction clutch assemblies, one as an internal clutch pack and the second for a brake clutch pack. The overdrive carrier is splined to the turbine shaft, which in turn, is splined into the converter turbine.

Another type of overdrive assembly is a separation of the over-

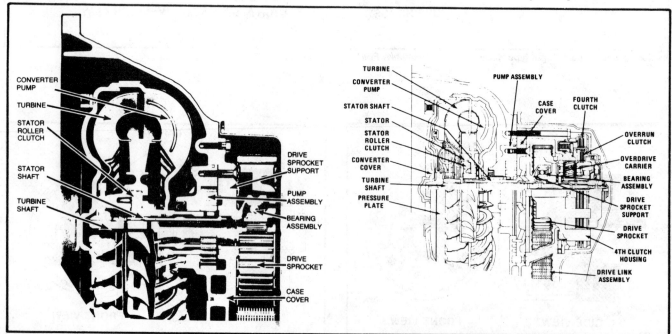

The addition of an overdrive unit and converter clutch assembly to an existing transaxle, changes its top end operation
(© General Motors Corp.)

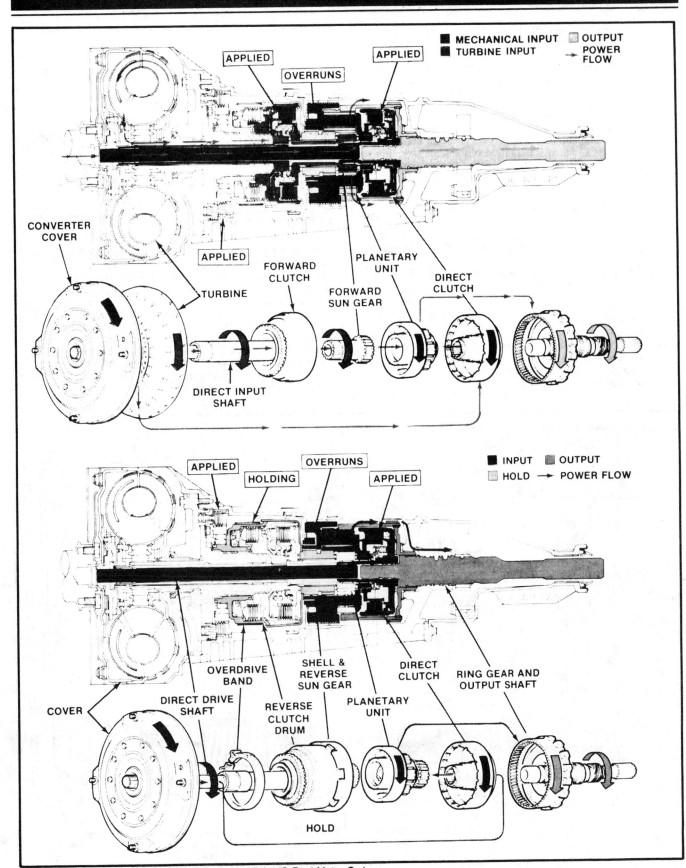

MECHANICAL INPUT **OUTPUT**
TURBINE INPUT → **POWER FLOW**

APPLIED

OVERRUNS

APPLIED

CONVERTER COVER

APPLIED

TURBINE

FORWARD CLUTCH

PLANETARY UNIT

DIRECT CLUTCH

FORWARD SUN GEAR

DIRECT INPUT SHAFT

APPLIED

HOLDING

OVERRUNS

APPLIED

INPUT **OUTPUT**
HOLD → **POWER FLOW**

OVERDRIVE BAND

SHELL & REVERSE SUN GEAR

DIRECT CLUTCH

RING GEAR AND OUTPUT SHAFT

COVER

DIRECT DRIVE SHAFT

REVERSE CLUTCH DRUM

PLANETARY UNIT

HOLD

Comparison of direct drive and overdrive power flows(© Ford Motor Co.)

drive components by having them at various points along the gear train assembly and also utilizing them for other gear ranges. Instead of having a brake clutch pack, an overdrive band is used to lock the planetary sun gear. In this type of transmission, the converter cover drives the direct drive shaft clockwise at engine speed, which in turn drives the direct clutch. The direct clutch then drives the planetary carrier assembly at engine speed in a clockwise direction. The pinion gears of the planetary gear assembly "walk around" the stationary reverse sun gear, again in a clockwise rotation. The ring gear and output shaft are therefore driven at a faster speed by the rotation of the planetary pinions. Because the input is 100% mechanical drive, the converter can be classified as a lock-up converter in the overdrive position.

Planetary Gears

Our aim is not to discuss the basics of gearing, but to stress the importance of the planetary gear set in the operation of the automatic transmission/transaxle assemblies. The advantages of planetary gear sets are as follows;

a. The location of the gear components makes the holding of the various members or the locking of the unit relatively easy.

b. Planetary gears are always in constant mesh, making quick gear changes without power flow interruption.

c. Planetary gears are strong and sturdy. The gears can handle larger torque loads as it is passed through the the planetary gear set, because the torque load is distributed over several planet pinion gears, allowing more tooth contact area to handle the power flow.

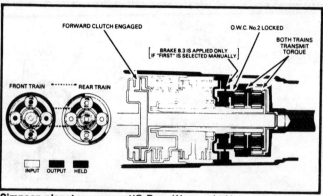

Simpson planetary gear set(© Borg Warner, Ltd.)

d. The planetary gear set is compact in size and easily adapted to different sized gear boxes.

In the automatic transmission/transaxle, the planetary gears must provide Neutral, Reduction, Direct Drive, Overdrive and Reverse. To accomplish this, certain gear or gears are held, resulting in the desired gear ratio or change of direction.

Two or more planetary gear sets are used in the three and four speed automatic units, providing the different gear ratios in reduction or to provide a separate overdrive ratio.

Two types of planetary gear sets are used, the Simpson and the Ravigneaux. The Simpson gear set is two planetary gear sets sharing a common sun gear shaft and output shaft. The Ravigneaux

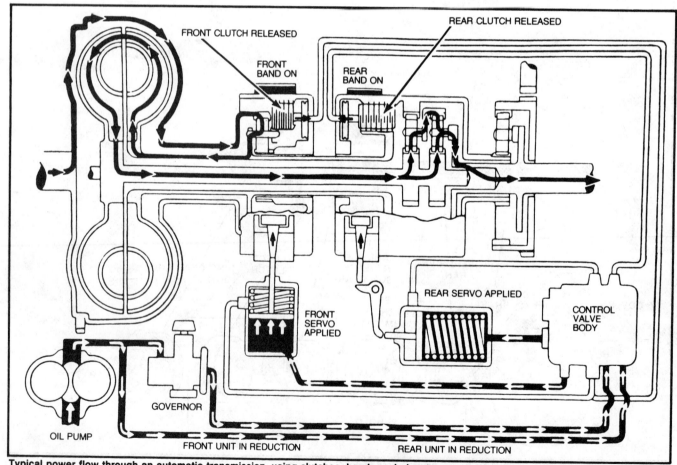

Typical power flow through an automatic transmission, using clutches, bands and planetary gear sets, with servo application illustrated

gear set utilizes a dual pinion planetary carrier, two sun gears and a ring gear. The pinion gears consists of short pinions (primary) and long pinions (secondary). Most automatic gear boxes use the Simpson type planetary gear assemblies.

HOLDING OR LOCKING-UP COMPONENTS OF THE PLANETARY GEARS

The holding or locking-up of the planetary gears is accomplished by hydraulic pressure, directed to a specific component, by the opening or closing of fluid passages in the transmission/transaxle assembly by spool type valves, either operated manually or automatically. The holding components are either clutch packs, internal or external, bands or overrunning "one-way" clutch units. Depending upon the design of the transmission/transaxle assembly, would dictate the holding of a specific part of the planetary gear unit by the holding components. It is important for the repairman to refer to the clutch and band application chart to determine the holding components in a particular gear ratio.

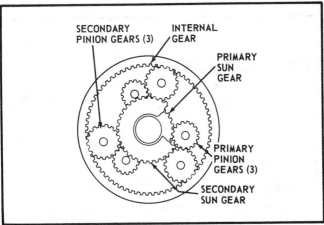

Ravigneaux planetary gear set(© General Motors Corp.)

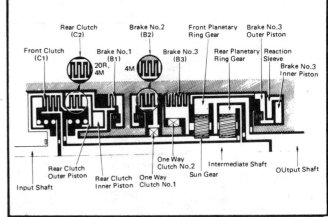

Typical component arrangement using clutch brakes instead of bands(© Toyota Motor Co.)

DIAGNOSING AUTOMATIC TRANSMISSION/TRANSAXLE MALFUNCTIONS

Diagnosing automatic transmission problems is simplified following a definite procedure and understanding the basic operation of the individual transmission that is being inspected or serviced. Do not attempt to "short-cut" the procedure or take for granted that another technician has performed the adjustments or the critical checks. It may be an easy task to locate a defective or burned-out unit, but the technician must be skilled in locating the primary reason for the unit failure and must repair the malfunction to avoid having the same failure occur again.

Each automatic transmission manufacturer has developed a diagnostic procedure for their individual transmissions. Although the operation of the units are basically the same, many differences will appear in the construction, method of unit application and the hydraulic control systems.

The same model transmissions can be installed in different makes of vehicles and are designed to operate under different load stresses, engine applications and road conditions. Each make of vehicle will have specific adjustments or use certain outside manual controls to operate the individual unit, but may not interchange with another transmission/vehicle application from the same manufacturer.

The identification of the transmission is most important so that the proper preliminary inspections and adjustments may be done and if in need of a major overhaul, the correct parts may be obtained and installed to avoid costly delays.

CUSTOMER EXPLANATION OF MALFUNCTION

The customer should be approached and questioned in a professional, but friendly and courteous manner as to the malfunction that could exist in the gearbox. By evaluating the answers, a pattern could emerge as to why this problem exists, what to do to correct it and what should be done to prevent a recurrence. Should the vehicle be towed because of apparent transmission failure, the cause should be determined before the unit is re-

Obtaining customer explanation of malfunction(© Ford Motor Co.)

moved to be certain a stalled engine, a broken or worn drive line component, broken drive plate or lack of fluid could be the cause. Again, question the owner/driver as to what happened, when and where the malfunction took place, such as engine flare-up, starting from a stop and/or on a hill. From the answers given, usually the correct diagnosis can be determined. Physical inspection of the vehicle components is the next step, to verify that either the outer components are at fault or the unit must be removed for overhaul.

SYSTEMATIC DIAGNOSIS

Transmission/transaxle manufacturers have compiled diagnostic aids for the use of technicians when diagnosing malfunctions through oil pressure tests or road test procedures. Diagnostic symptom charts, operational shift speed charts, oil pressure specifications, clutch and band application charts and oil flow schematics are some of the aids available.

Numerous manufacturers and re-manufacturers require a diag-

AUTOMATIC TRANSMISSION	CUSTOMER QUESTIONNAIRE

1. How long have you had the condition?   R. O. _____

   ☐ Since car was new
   ☐ Recently (when?) _____
   ☐ Came on gradually   ☐ Suddenly

2. Describe the condition?

	P-R-N-D-2-1 SELECTOR POSITION(S)	CHECK AS APPROPRIATE WHICH GEAR?		
		HIGH	INTERMEDIATE	LOW
☐ Slow Engagement				
☐ Rough Engagement				
☐ Slip				
☐ No Drive				
☐ No Upshift				
☐ No Downshift				
☐ Slip During Shift				
☐ Wrong Shift Speed(s)				
☐ Rough Shift				
☐ Mushy Shift				
☐ Erratic Shift				
☐ Engine "runaway or "buzzy"				
☐ No Kickdown				
☐ Starts in high gear in D				
☐ Starts in intermediate gear in D				
☐ Oil leak (where?)				

3. Which of the following cause or affect the condition?

   ☐ Transmission cold   ☐ Engine at fast (cold) idle
   ☐ After warm-up   ☐ Normal idle
   ☐ High speed   ☐ Wet road
   ☐ Cruising speed   ☐ Dry road
   ☐ Low Speed   ☐ Braking
   ☐ Accelerating   ☐ Coasting down

4. Does the engine need a tune-up?

   ☐ Yes   ☐ No   ☐ When was last tune-up? _____

5. Describe any strange noises

   ☐ Rumble   ☐ Squeak
   ☐ Knock   ☐ Grind
   ☐ Chatter   ☐ Hiss
   ☐ Snap or pop   ☐ Scrape
   ☐ Buzz   ☐ Other (describe) _____
   ☐ Whine

Ford Parts and Service Division
Training and Publications Department

SERVICE ADVISOR HELPER

Customer questionnaire, published by Ford Motor Co., for use by their Dealer body diagnostic personnel. Typical of other manufacturers (© Ford Motor Co.)

nosis check sheet be filled out by the diagnostician, pertaining to the operation, fluid level, oil pressures (idling and at various speeds), verification of adjustments and possible causes and the needed correction of the malfunctions. In certain cases, authorization must be obtained before repairs can be done, with the diagnostic check sheet accompanying the request for payment or warranty claim, along with the return of defective parts.

It is a good policy to use the diagnostic check sheet for the evaluation of all transmission/transaxles diagnosis and include the completed check sheet in the owners service file, should future reference be needed.

Many times, a rebuilt unit is exchanged for the defective unit, saving down time for the owner and vehicle. However, if the diagnostic check sheet would accompany the removed unit to the rebuilder, more attention could be directed to verifying and repairing the malfunctioning components to avoid costly comebacks of the rebuilt unit, at a later date. Most large volume rebuilders employ the use of dynamometers, as do the new unit manufacturers, to verify proper build-up of the unit and its correct operation before it is put in service.

GENERAL DIAGNOSIS

Should the diagnostician not use a pre-printed check sheet for the diagnosing of the malfunctioning unit, a sequence for diagnosis of the gear box is needed to proceed in an orderly manner. A suggested sequence is as follows:
1. Inspect and correct the fluid level.
2. Inspect and adjust the throttle or kick-down linkage.
3. Inspect and adjust the manual linkage.
4. Install one or more oil pressure gauges to the transmission as instructed in the individual transmission sections.

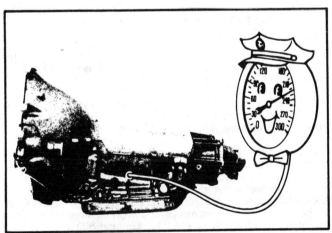

Use the oil pressure gauge(© General Motors Corp.)

5. Road test the vehicle (with owner if possible).

NOTE: During the road test, use all the selector ranges while noting any differences in operation or changes in oil pressures, so that the unit or hydraulic circuit can be isolated that is involved in the malfunction.

Engine Performance

When engine performance has declined due to the need of an engine tune-up or a system malfunction, the operation of the transmission is greatly affected. Rough or slipping shift and overheating of the transmission and fluid can occur, which can develop into serious internal transmission problems. Complete the adjustments or repairs to the engine before the road test is attempted or transmission adjustments made.

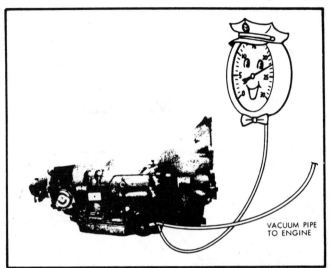

Use the vacuum gauge as required(© General Motors Corp.)

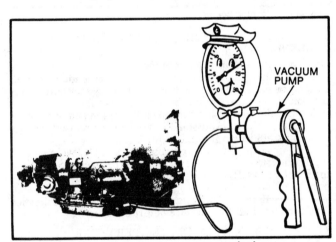

Using the hand operated vacuum pump as required (© General Motors Corp.)

Inspection of the Fluid Level

Most automatic transmissions are designed to operate with the fluid level between the ADD or ONE PINT and FULL marks on the dipstick indicator, with the fluid at normal operating temperature. The normal operating temperature is attained by operating the engine-transmission assembly for at least 8 to 15 miles of driving or its equivalent. The fluid temperature should be in the range of 150° to 200°F when normal operating temperature is attained.

NOTE: If the vehicle has been operated for long periods at high speed or in extended city traffic during hot weather, an accurate fluid level check cannot be made until the fluid cools, normally 30 minutes after the vehicle has been parked, due to fluid heat in excess of 200° F.

The transmission fluid can be checked during two ranges of temperature.
1. Transmission at normal operating temperature.
2. Transmission at room temperature.
During the checking procedure and adding of fluid to the transmission, it is most important not to overfill the reservoir to avoid foaming and loss of fluid through the breather, which can cause slippage and transmission failure.

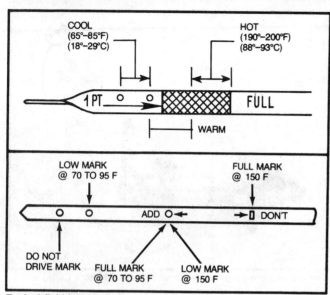

Typical fluid level indicators

TRANSMISSION AT NORMAL OPERATING TEMPERATURE

(150° to 200° F.—Dipstick hot to the touch)

1. With the vehicle on a level surface, engine idling, wheels blocked or parking brake applied, move the gear selector lever through all the ranges to fill the passages with fluid.
2. Place the selector lever in the Park position and remove the dipstick from the transmission. Wipe clean and reinsert the dipstick to its full length into the dipstick tube.
3. Remove the dipstick and observe the fluid level mark on the dipstick stem. The fluid level should be between the ADD and the FULL marks. If necessary, add fluid through the filler tube to bring the fluid level to its proper height.
4. Reinstall the dipstick and be sure it is sealed to the dipstick filler tube to avoid the entrance of dirt or water.

TRANSMISSION AT ROOM TEMPERATURE

(65° to 95° F.—Dipstick cool to touch)

─────────────── CAUTION ───────────────

The automatic transmissions are sometimes overfilled because the fluid level is checked when the transmission has not been operated and the fluid is cold and contracted. As the transmission is warmed to normal operating temperature, the fluid level can change as much as ¼ inch.

1. With the vehicle on a level surface, engine idling, wheels blocked or parking brake applied, move the selector lever through all the ranges to fill the passages with fluid.
2. Place the selector lever in the Park position and remove the dipstick from the transmission. Wipe clean and re-insert it back into the dipstick tube.
3. Remove the dipstick and observe the fluid level mark on the dipstick stem. The fluid should be directly below the FULL indicator.

NOTE: Most dipsticks will have either one mark or two marks, such as dimples or holes in the stem of the dipstick, to indicate the cold level, while others may be marked HOT or COLD levels.

4. Add enough fluid, as necessary, to the transmission, but do not overfill.

─────────────── CAUTION ───────────────

This operation is most critical, due to the expansion of the fluid under heat.

FLUID TYPE SPECIFICATIONS

The automatic transmission fluid is used for numerous functions such as a power-transmitting fluid in the torque converter, a hydraulic fluid in the hydraulic control system, a lubricating agent for the gears, bearings and bushings, a friction-controlling fluid for the bands and clutches and a heat transfer medium to carry the heat to an air or cooling fan arrangement.

Because of the varied automatic transmission designs, different frictional characteristics of the fluids are required so that one fluid cannot assure freedom from chatter or squawking from the bands and clutches. Operating temperatures have increased sharply in many new transmissions and the transmission drain intervals have been extended or eliminated completely. It is therefore most important to install the proper automatic transmission fluid into the automatic transmission designed for its use.

Types of Automatic Transmission Fluid

DEXRON® II

This fluid supersedes the Dexron® type fluid and meets a more severe set of performance requirements, such as improved high temperature oxidation resistance and low temperature fluidity. The Dexron® II is recommended for use in all General Motors, Chrysler, American Motors and certain imported vehicles automatic transmissions. This fluid can replace all Dexron® fluid with a B- number designation.

Container Identification number—D-XXXXX

DEXRON® II—SERIES D

This fluid was developed and is used in place of the regular Dexron® II fluids.

The container identification is with a "D" prefix to the qualification number on the top of the container.

TYPE F

Ford Motor Company began developing its own specifications for automatic transmission fluid in 1959 and again updated its specifications in 1967, requiring fluid with different frictional characteristics and identified as Type F fluid.

Beginning with the 1977 model year, a new Type CJ fluid was specified for use with the C-6 and newly introduced Jatco model PLA-A transmissions. This new fluid is not interchangeable with the Type F fluid.

Prior to 1967, all Ford automatic transmissions use fluids in containers marked with qualification number IP-XXXXXX, meeting Ford specification number ESW-M2C33-D.

Fluids in containers marked with qualification number 2P-XXXXXX meets Ford specification ESW-M2C33-F and is used in the Ford automatic transmissions manufactured since 1967, except the 1977 and later C-6, the Automatic Overdrive, and Jatco models PLA-A, PLA-A1, PLA-A2 transmissions.

The container identification number for the new fluid is Ford part number D7AZ-19582-A and carries a qualification number ESP-M2C138-CJ.

TYPE CJ

─────────────── CAUTION ───────────────

The CJ fluid is NOT compatible with clutch friction material of other Ford transmissions and must only be used in the 1977 and later C-6, the Automatic Overdrive and Jatco models PLA-A, PLA-A1 and PLA-A2 automatic transmissions.
Do not mix or interchange the fluids through refills or topping off as the Type F and the Type CJ fluids are not compatible.

A technical bulletin has been issued by Ford Motor Company, dated 1978, advising the compatibility of Dexron® II, series D fluid with the CJ fluid. It can be substituted or mixed, if necessary, in the 1977 and later C-6, the Automatic Overdrive and the Jatco PLA-A, PLA-A1 and PLA-A2 automatic transmissions.

With approved internal modifications, CJ or Dexron® II, series D automatic transmission fluid can be used in the past models of the C-4 transmissions. To insure the proper fluid is installed or added, a mylar label is affixed or is available to be affixed to the dipstick handle with the proper fluid designation on the label.

TYPE H
With the introduction of the C-5 automatic Transmission, Ford Motor Company developed a new type fluid, designated "H", meeting Ford's specification ESP-M2C166-H. This fluid contains a special detergent which retains in suspension, particles generated during normal transmission operation. This suspension of particles results in a dark discoloration of the fluid and does not indicate need for service. It should be noted that the use of other fluids in the C-5 automatic transmission could result in a shuddering condition.

TYPE G
The type G fluid is an improvement over the type F fluid and meets Ford Motor Company specification of M2C-33G.

Type G fluid has the capability of reducing oxidization at higher transmission operating temperatures. Should an automatic transmission be filled with type G fluid, type F fluid can be used to top off the level. However, the more type F fluid that is mixed with the type G fluid, will proportionally reduce the maximum working temperature of the type G fluid.

FLUID CONDITION
During the checking of the fluid level, the fluid condition should be inspected for color and odor. The normal color of the fluid is deep red or orange-red and should not be a burned brown or black color. If the fluid color should turn to a green/brown shade at an early stage of transmission operation and have an offensive odor, but not a burned odor, the fluid condition is considered normal and not a positive sign of required maintenance or transmission failure.

With the use of absorbent white paper, wipe the dipstick and examine the stain for black, brown or metallic specks, indicating clutch, band or bushing failure, and for gum or varnish on the dipstick or bubbles in the fluid, indicating either water or anti-freeze in the fluid.

Should there be evidence of water, anti-freeze or specks of residue in the fluid, the oil pan should be removed and the sediment inspected. If the fluid is contaminated or excessive solids are found in the removed oil pan, the transmission should be disassembled, completely cleaned and overhauled. In addition to the cleaning of the transmission, the converter and transmission cooling system should be cleaned and tested.

Fluid Overfill Problems
When the automatic transmission is overfilled with fluid, the rotation of the internal units can cause the fluid to become aerated. This aeration of the fluid causes air to be picked up by the oil pump and causes loss of control and lubrication pressures. The fluid can also be forced from the transmission assembly through the air vent, due to the aerated condition.

Fluid Underfill Problems
When the fluid is low in the transmission, slippage and loss of unit engagement can result, due to the fluid not being picked up by the pump. This condition is evident when first starting after the vehicle has been sitting and cooled to room temperature, in cold weather, making a turn or driving up a hill. This condition should be corrected promptly to avoid costly transmission repairs.

Throttle Valve and Kickdown Control Inspection
Inspect the throttle valve and kickdown controls for proper operation, prior to the road test. Refer to the individual transmission section for procedures.

THROTTLE VALVE CONTROLS
The throttle valve can be controlled by linkage, cable or engine vacuum.

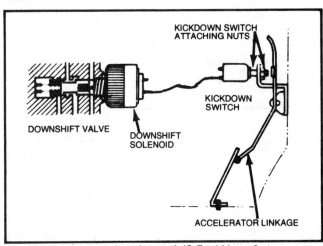

Typical kickdown switch and controls(© Ford Motor Co.)

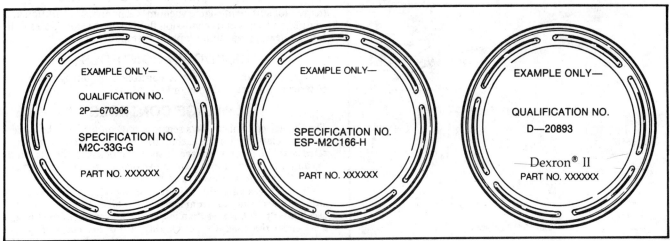

Comparison of types H, G and Dexron® II identifying codes found on container tops—typical

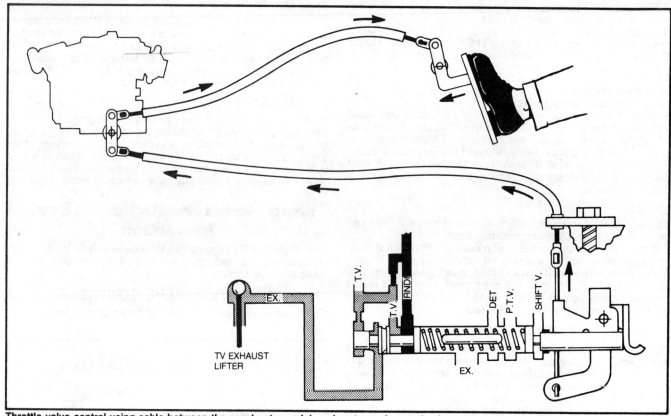

Throttle valve control using cable between the accelerator pedal, carburetor and control valve (© Borg Warner, Ltd.)

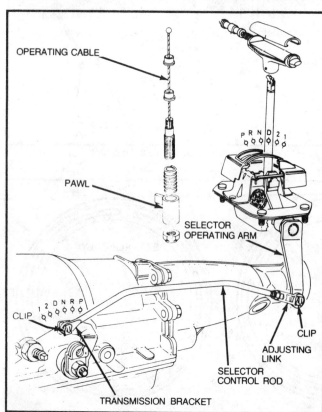

Manual Control linkage phasing—typical(© Ford Motor Co.)

LINKAGE CONTROL

Inspect the linkage for abnormal bends, looseness at the bellcrank connections and linkage travel at the wide open throttle stop. Be sure the linkage operates without binding and returns to the closed position upon release of the accelerator.

CABLE CONTROL

Inspect the cable for sharp bends or crimps, secured retainers, freedom of cable movement throughout the full throttle position and the return to the closed throttle position without binding or sticking, and connection of the throttle return spring.

ENGINE VACUUM CONTROLS

Inspect for sufficient engine vacuum, vacuum hose condition and routing, and signs of transmission fluid in the vacuum hoses indicating a leaking vacuum diaphragm (modulator).

KICKDOWN CONTROLS

The transmission kickdown is controlled by linkage, cable or electrical switches and solenoid.

LINKAGE CONTROLS

The linkage control can be a separate rod connected to and operating in relation with the carburetor throttle valves, or incorporated with the throttle linkage. Inspect for looseness, bends, binding and movement into the kickdown detent upon the movement of the throttle to the wide open stop.

NOTE: It is a advisable to inspect for the wide open throttle position at the carburetor from inside the vehicle, by depressing the accelerator pedal, rather than inspecting movement of the linkage from under the hood. Carpet matting, dirt or looseness of the accelerator can prevent the opening of the throttle to operate the kickdown linkage.

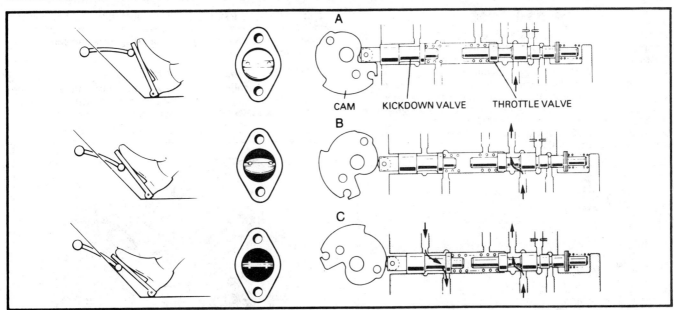

CAM   KICKDOWN VALVE   THROTTLE VALVE

Typical throttle and kickdown valve operation through three possible operating positions of the throttle plates and accelerator pedal (© Borg Warner, Ltd.)

CABLE CONTROLS

The kickdown cable control can be a separate cable or used with the throttle valve control cable. It operates the kickdown valve at the wide open throttle position. Inspect for kinks and bends on bracket retention of the cable. Inspect for freedom of movement of the cable and see that the cable drops into the kickdown detent when the accelerator pedal is fully depressed and the throttle valves are fully open.

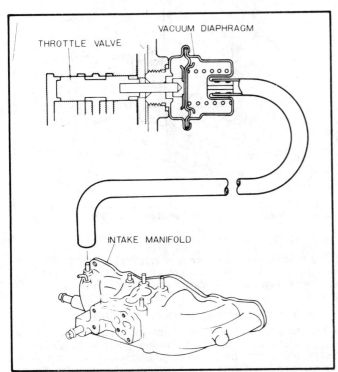

THROTTLE VALVE   VACUUM DIAPHRAGM

INTAKE MANIFOLD

Vacuum diaphragm operated throttle valve(© Toyo Kogyo Co. Ltd.)

ELECTRICAL CONTROLS

The electrical kickdown controls consist of a switch, located on the accelerator linkage and a solenoid control, either mounted externally on the transmission case or mounted internally on the control valve body, in such a position as to operate the kickdown valve upon demand of the vehicle operator. Inspect the switch for proper operation which should allow electrical current to pass through, upon closing of the switch contacts by depressing the accelerator linkage. Inspect the wire connector at the transmission case or the terminals of the externally mounted solenoid for current with the switch contacts closed. With current present at the solenoid, either externally or internally mounted, a clicking noise should be heard, indicating that the solenoid is operating.

Manual Linkage Control Inspection

The manual linkage adjustment is one of the most critical, yet the most overlooked, adjustment on the automatic transmission. The controlling quadrant, either steering column or console mounted, must be properly phased with the manual control valve detent. Should the manual valve be out of adjustment or position, hydraulic leakage can occur within the control valve assembly and can result in delay of unit engagement and/or slippage of the clutches and bands upon application. The partial opening of apply passages can also occur for clutches and bands not in applying sequence, and result in dragging of the individual units or bands during transmission operation.

Inspect the selector lever and quadrant, the linkage or cable control for looseness, excessive wear or binding. Inspect the engine/transmission assembly for excessive lift during engine torque application, due to loose or broken engine mounts, which can pull the manual valve out of position in the control valve detent.

─── CAUTION ───

The neutral start switch should be inspected for operation in Park and Neutral positions, after any adjustments are made to the manual linkage or cable.

Road Test

Prior to driving the vehicle on a road test, have the vehicle operator explain the malfunction of the transmission as fully and as accurate as possible. Because the operator may not have the same technical knowledge as the diagnostician, ask questions concerning the malfunction in a manner that the operator can understand. It may be necessary to have the operator drive the vehicle on the road test and to identify the problem. The diagnostician can observe the manner in which the transmission is being operated and can point out constructive driving habits to the operator to improve operation reliability.

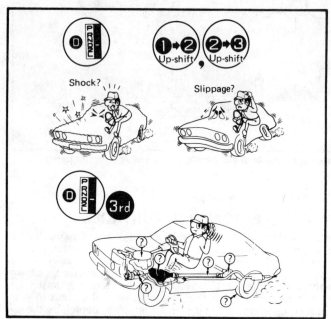

Road Test vehicle to determine malfunctions(© Toyota Motor Co.)

Many times, an actual transmission malfunction can occur without the operator's knowledge, due to slight slippages occurring and increasing in duration while the vehicle is being driven. Had the operator realized that a malfunction existed, minor adjustments possibly could have been done to avoid costly repairs.

As noted previously in this section, be aware of the engine's performance. For example, if a vacuum modulator valve is used to control the throttle pressure, an engine performing poorly cannot send the proper vacuum signals to the transmission for proper application of the throttle pressure and control pressure, in the operation of the bands and clutches. Slippages and changes in shift points can occur.

Perform the road test with the customer, whenever possible, to determine the cause of the malfunction

During the road test, the converter operation must be considered. Related converter malfunctions affecting the road test are as follows, with the converter operation, diagnosis and repairs discussed later in the General Section.

STATOR ASSEMBLY FREE WHEELS

When the stator roller clutch freewheels in both directions, the vehicle will have poor acceleration from a standstill. At speeds above approximately 45 MPH, the vehicle will act normally. A

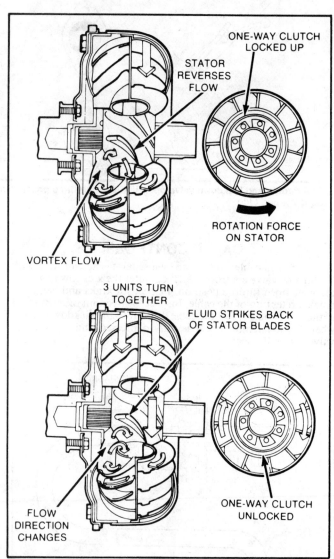

Stator operation in lock-up and freewheeling modes

check to make on the engine is to accelerate to a high RPM in neutral. If the engine responds properly, this is an indication that the engine is operating satisfactorily and the problem may be with the stator.

STATOR ASSEMBLY REMAINS LOCKED UP

When the stator remains locked up, the engine RPM and the vehicle speed will be restricted at higher speeds, although the vehicle will accelerate from a standstill normally. Engine overheating may be noticed and visual inspection of the converter may reveal a blue color, resulting from converter overheating.

Clutch and Band Application

During the road test, operate the transmission/transaxle in each gear position and observe the shifts for signs of any slippage, variation, sponginess or harshness. Note the speeds at which the upshifts and downshifts occur. If slippage and engine flare-up occurs in any gears, clutch, band or overrunning clutch problems are indicated and depending upon the degree of wear, a major overhaul may be indicated.

The clutch and band application chart in each transmission/transaxle section provides a basis for road test analysis to determine the internal units applied or released in a specific gear ratio.

NOTE: Certain transmission/transaxles use brake clutches in place of bands and are usually indicated as B-1 and B-2 on the unit application chart. These components are diagnosed in the same manner as one would diagnose a band equipped gearbox.

EXAMPLES

Using the Borg Warner Model 66 Clutch and Band application chart as a guide, the following conditions can be determined.

1. A customer complaint is a slippage in third speed and reverse. By referring to the clutch and band application chart, the commonly applied component is the rear clutch unit, applied in both third speed and reverse. By having a starting point, place the gear selector in the reverse position and if the slippage is present and slippage is verified on the road in third speed, all indications would point to rear clutch failure.

Broken input shaft can operate intermittently at low torque applications, until worn smooth as illustrated

2. A customer complaint is a flare-up of engine speed during the 1-2 shift, D position, with a delayed application or sponginess in the second gear. By referring to the clutch and band application chart, the application of the components can be located during each shift and gear ratio level. With the front clutch unit applied in each of the gear ranges and the front band applied only in the second speed, the complaint can then be pinpointed to either the front band being out of adjustment, worn out or a servo apply problem. Because the front clutch is applied in the three forward speeds and no apparent slippage or flare-up occurs in the first or third speeds, the front clutch unit is not at fault.

Many times, malfunctions occur within a unit that are not listed on a diagnosis chart. Such malfunctions could be breakage of servo or clutch return springs, stripped serrated teeth from mated components or broken power transfer shaft, just to name a few. When a malfunction of this type occurs, it can be difficult to diagnose, but by applying the information supplied in a clutch and band application chart, the use of an oil pressure gauge, when necessary, and using a "common sense" approach, the malfunction can be determined.

SUMMARY OF ROAD TEST

This process of elimination is used to locate the unit that is malfunctioning and to confirm the proper operation of the transmission. Although the slipping unit can be defined, the actual cause of the malfunction cannot be determined by the band and clutch application charts. It is necessary to perform hydraulic and air pressure tests to determine if a hydraulic or mechanical component failure is the cause of the malfunction.

Pressure Tests

OIL PRESSURE GAUGE

The oil pressure gauge is the primary tool used by the diagnostician to determine the source of malfunctions of the automatic transmissions.

Oil pressure gauges are available with different rates, ranging from 0-100, 0-300 and 0-500 PSI that are used to measure the pressures in the various hydraulic circuits of the automatic transmissions. The high-rated pressure gauges (0-300, 0-500 PSI) are used to measure the control line pressures while the low-rated gauge (0-100 PSI) is used to measure the governor, lubrication and throttle pressures on certain automatic transmissions.

The gauges may be an individual unit with a 4- to 10-foot hose attached, or may be part of a console unit with other gauges, normally engine tachometer and vacuum. The diagnostician's preference dictates the type used.

CLUTCH AND BAND APPLICATION CHART
Borg Warner Model 66

Gear	Front Clutch	Rear Clutch	Front Band	Rear Band	One-Way Clutch
Drive 1st	Applied	—	—	—	Holding
Drive 2nd	Applied	—	Applied	—	—
Drive 3rd	Applied	Applied	—	—	—
1—Low	Applied	—	—	Applied	—
2—1st	Applied	—	—	—	Holding
2—2nd	Applied	—	Applied	—	—
Reverse	—	Applied	—	Applied	—

NOTE: Rear band is released in "N", but applied in "P" for constructional reasons only.

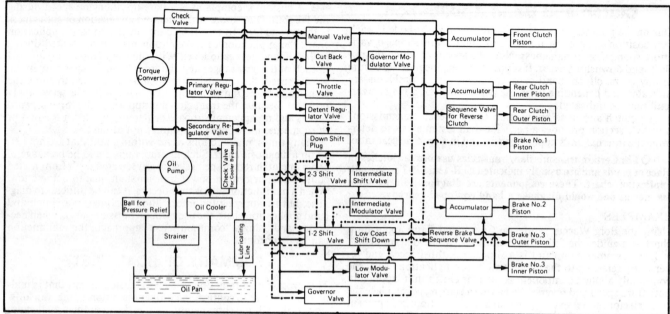

Typical hydraulic control system(© Toyota Motor Co.)

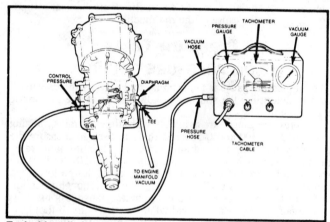

Typical installation of oil pressure and vacuum gauges on transmission(© Ford Motor Co.)

CONTROL PRESSURE TESTING

The methods of obtaining control pressure readings when conducting a pressure test, vary from manufacturer to manufacturer when vacuum modulators are used to modulate the throttle pressures. Since engine vacuum is the controlling factor as the vehicle is being driven, the amount of vacuum must be controlled when testing. The use of a motorized vacuum pump, a hand held mechanically operated vacuum pump or an air bleed valve may be recommended by the manufacturer.

Before any vacuum tests are performed on the modulator, the engine, being the primary vacuum source, must be checked to ascertain that vacuum is available to the modulator. The hose at the modulator should be disconnected and a vacuum gauge attached to the hose. With the brakes set and the engine idling at normal operating temperature, the engine vacuum reading should be in

To measure the hydraulic pressures, select a gauge rated over the control pressure specifications and install the hose fitting into the control pressure line tap, located either along the side or the rear of the transmission case. Refer to the individual automatic transmission sections for the correct locations, since many transmission cases have more than one pressure tap to check pressures other than the control pressure.

The pressure gauges can be used during a road test, but care must be exercised in routing the hose from under the vehicle to avoid dragging or being entangled with objects on the roadway.

The gauge should be positioned so the dial is visible to the diagnostician near the speedometer area. If a console of gauges is used, the console is normally mounted on a door window or on the vehicle dash.

— CAUTION —

During the road test, traffic safety must be exercised. It is advisable to have a helper to assist in the reading or recording of the results during the road test.

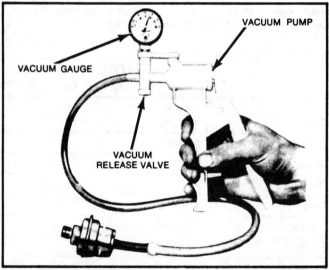

Typical vacuum pump used to test vacuum diaphragms on and off the transmission(© Ford Motor Co.)

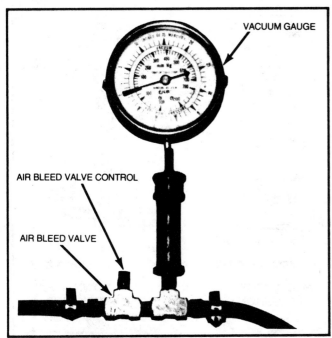

VACUUM GAUGE

AIR BLEED VALVE CONTROL

AIR BLEED VALVE

Air bleed with vacuum gauge attached

the 17 to 20 in. Hg range. Should the vacuum be low, check the vacuum reading at the engine and compare the two readings. If the vacuum reading is lower at the modulator end of the hose than at the engine, look for leaking or defective hoses or lines. If both readings are low, the engine is not producing sufficient vacuum to properly operate the modulator at road speeds or under specific road conditions, which could result in transmission malfunctions and premature internal wear. To correct this condition, the reason for the low engine vacuum would have to be determined and repaired.

--------- CAUTION ---------

To install a rebuilt unit under this type of vacuum condition would invite operational problems.

An air bleed valve was placed in the vacuum line from the engine to the modulator to control the amount of vacuum reacting on the diaphragm, simply by allowing more or less atmospheric air to enter the vacuum line, controlled by a screw valve. This type of testing has been replaced by the use of the hand held mechanical vacuum pump, in most cases.

Testing Control Pressure—Typical

To illustrate a typical control pressure test, the following explanation is given concerning the accompanying chart.

NOTE: **Refer to the individual transmission sections for correct specifications.**

TRANSMISSION MALFUNCTION RELATED TO OIL PRESSURE

15-20" vacuum applied to modulator							0" vacuum to modulator	
③ Drive Brakes Applied 1000 RPM	③ Reverse Brakes Applied 1000 RPM	③ Super or Lo Brakes Applied 1000 RPM	Neutral Brakes Applied 1000 RPM	③ Drive 1000 RPM Brakes on Detent* Activated	Drive Idle	① Drive 30 MPH Closed Throttle	Drive—from 1000 to 3000 RPM Wheels free to move	Pressure Test Conditions
60-90	85-150	85-110	55-70	90-110	60-85	55-70	Pressure drop of 10 PSI or more	Normal Results Note2
							DROP	Malfunction in Control Valve Assembly
							NO DROP	Malfunction in Governor or Governor Feed System
┌ALL PRESSURES HIGH WITH LESS THAN 35 PSI BETWEEN PRESSURE READINGS					┐		—	Malfunction in Detent System
┌ALL PRESSURES HIGH WITH MORE THAN 35 PSI BETWEEN PRESSURE READINGS					┐		—	Malfunction in Modulator
Low						—	Low to Normal	— |Oil Leak in Feed System to the Direct Clutch
Low		Low to Normal		Low to Normal		—	Low to Normal	— |Oil Leak in Feed System to the Forward Clutch
			Low				—	Detent System

A blank space = Normal pressure
A dash (—) in space = Pressure reading has no meaning
① Coast for 30 mph—read before reaching 20 mph

② If high line pressures are experienced see "High Line Pressures" note.

③ Cable pulled or blocked thru detent position or downshift switch closed by hand

NOTE: It is assumed the oil pressure gauge, the vacuum gauge, the vacuum pump, and the tachometer are attached to the engine/transmission assembly, while having the brakes locked and the wheels chocked. The fluid temperature should be at normal operating temperature.

1. Start the engine and allow to idle.

2. Apply 15-20 inches of vacuum to the vacuum modulator with the vacuum hand pump.

3. Place the selector lever in the Drive position and increase the engine speed to 1000 RPM. The pressure reading should be 60-90 PSI, as indicated on the chart.

4. Taking another example from the chart, place the selector lever in the Super or Low position and increase the engine speed to 1000 RPM. As indicated on the chart, the pressure should read 85-110 PSI.

5. Referring to the chart column to test pressures with the detent activated, the following must be performed. Place the selector lever in the Drive position and increase the engine speed to 1000 RPM. With the aid of a helper, if necessary, pull the detent cable through the detent, or if the transmission is equipped with an electrical downshift switch, close the switch by hand. The pressure reading should be 90-110 PSI, as indicated on the chart.

6. A pressure test conditions column is included in the chart to assist the diagnostician in determining possible causes of transmission malfunctions from the hydraulic system.

CAUTION

Do not use the above pressure readings when actually testing the control pressures. The above readings are only used as a guide for the chart explanation. Refer to the individual transmission section for the correct specifications and pressure readings.

HIGH CONTROL PRESSURE

If a condition of high control pressure exists, the general causes can be categorized as follows.
1. Vacuum leakage or low vacuum
2. Vacuum modulator damaged
3. Pump pressure excessive
4. Control valve assembly
5. Throttle linkage or cable misadjusted

LOW CONTROL PRESSURE

If a condition of low control pressure exists, the general causes can be categorized as follows.
1. Transmission fluid low
2. Vacuum modulator defective
3. Filter assembly blocked or air leakage
4. Oil pump defective
5. Hydraulic circuit leakage
6. Control valve body

NO CONTROL PRESSURE RISE

If a control pressure rise does not occur as the vacuum drops or the throttle valve linkage/cable is moved, the mechanical connection between the vacuum modulator or throttle valve linkage/cable and the throttle valve should be inspected. Possible broken or disconnected parts are at fault.

Vacuum Modulator

TESTING

A defective vacuum modulator can cause one or more of the following conditions.
   a. Engine burning transmission fluid
   b. Transmission overheating
   c. Harsh upshifts
   d. Delayed shifts
   e. Soft up and down shifts

Whenever a vacuum modulator is suspected of malfunctioning, a vacuum check should be made of the vacuum supply.

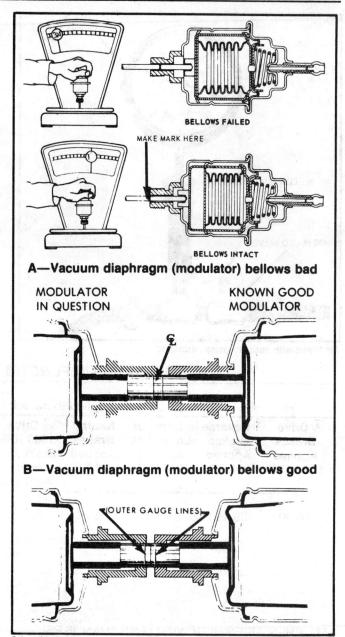

A—Vacuum diaphragm (modulator) bellows bad

B—Vacuum diaphragm (modulator) bellows good

Other methods used to test vacuum diaphragms

1. Disconnect the vacuum line at the transmission vacuum modulator connector pipe.

2. With the engine running, the vacuum gauge should show an acceptable level of vacuum for the altitude at which the test is being performed.

3. If the vacuum reading is low, check for broken, split or crimped hoses and for proper engine operation.

4. If the vacuum reading is acceptable, accelerate the engine quickly. The vacuum should drop off and return immediately upon release of the accelerator. If the gauge does not register a change in the vacuum reading, indications are that the vacuum lines are plugged, restricted or connected to a reservoir supply.

5. Correct the vacuum supply as required.

When the vacuum supply is found to be sufficient, the vacuum modulator must be inspected and this can be accomplished on or off the vehicle.

On Vehicle Tests

1. Remove the vacuum line and attach a vacuum pump to the modulator connector pipe.
2. Apply 18 inches of vacuum to the modulator. The vacuum should remain at 18 inches without leaking down.
3. If the vacuum reading drops sharply or will not remain at 18 inches, the diaphragm is leaking and the unit must be replaced.
4. If transmission fluid is present on the vacuum side of the diaphragm or in the vacuum hose, the diaphragm is leaking and the unit must be replaced.

NOTE: Gasoline or water vapors may settle on the vacuum side of the diaphragm. Do not diagnose as transmission fluid.

Off Vehicle Tests

1. Remove the vacuum modulator from the transmission.
2. Attach a vacuum pump to the vacuum modulator connector pipe and apply 18 inches of vacuum.
3. The vacuum should hold at 18 inches, if the diaphragm is good and will drop to zero if the diaphragm is leaking.
4. With the control rod in the transmission side of the vacuum modulator, apply vacuum to the connector pipe. The rod should move inward with light finger pressure applied to the end of the rod. When the vacuum is released, the rod will move outward by pressure from the internal spring.

Stall Speed Tests

The stall speed test is performed to evaluate the condition of the transmission as well as the condition of the engine.

The stall speed is the maximum speed at which the engine can drive the torque converter impeller while the turbine is held stationary. Since the stall speed is dependent upon the engine and torque converter characteristics, it can vary with the condition of the engine as well as the condition of the automatic transmission, so it is most important to have a properly performing engine before a stall speed test is attempted, thereby eliminating the engine from any malfunction that may be present.

EFFECT OF ALTITUDE ON ENGINE VACUUM

Elevation in Feet	Number of Engine Cylinders		
	FOUR	SIX	EIGHT
Zero to 1000	18 to 20	19 to 21	21 to 22
1000 to 2000	17 to 19	18 to 20	19 to 21
2000 to 3000	16 to 18	17 to 19	18 to 20
3000 to 4000	15 to 17	16 to 18	17 to 19
4000 to 5000	14 to 16	15 to 17	16 to 18
5000 to 6000	13 to 15	14 to 16	15 to 17

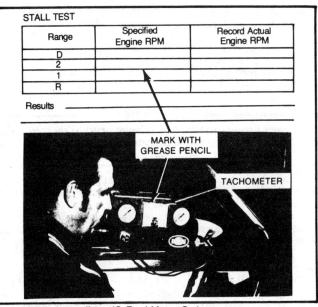

Preparation for stall test(© Ford Motor Co.)

Because engines perform differently between high and low altitudes, the stall speeds given in specification charts are for vehicles tested at sea level and cannot be considered representative of stall speed tests performed at higher altitudes. Unless specific stall speed tests specification charts are available for use at higher altitudes, representative stall speeds can be determined by testing several vehicles known to be operating properly, averaging the results and recording the necessary specifications for future reference.

PERFORMING THE STALL SPEED TESTS

1. Attach a tachometer to the engine and an oil pressure gauge to the transmission control pressure tap. Position the tachometer and oil pressure gauge so the operator can read the dials.
2. Mark the specified maximum engine RPM on the tachometer clear cover with a grease pencil so that the operator can immediately check the stall speed to see if it is over or under specifications.
3. Start the engine and bring to normal operating temperature.
4. Check the transmission fluid level and correct as necessary.
5. Apply the parking brake and chock the wheels.

— **CAUTION** —
Do not allow anyone in front of the vehicle during the preparation or during the stall speed test.

SAMPLE STALL TEST DIAGNOSIS CHART

Selector Lever Range	Specified Engine RPM	Actual Engine RPM	Control Pressure PSI	Holding Members Applied
D (Drive)				
1				
2				
R (Reverse)				

NOTE: The range identifications are to be taken from the selector quadrant of the vehicle being tested. Before stall test, fill in the specified engine RPM and the holding members applied columns from the clutch and band application and specification charts of the automatic transmission being tested.

6. Apply the service brakes and place the selector lever in the "D" position and depress the accelerator pedal to the wide open throttle position.

7. Do not hold the throttle open any longer than necessary to obtain the maximum engine speed reading and never over five (5) seconds for each test. The stall will occur when no more increase in engine RPM at wide open throttle is noted.

――――――― CAUTION ―――――――

If the engine speed exceeds the maximum limits of the stall speed specifications, release the accelerator immediately as internal transmission slippage is indicated.

8. Shift the selector lever into the Neutral position and operate the engine from 1000 to 1500 RPM for at least 30 seconds to two minutes to cool the transmission fluid.

9. If necessary, the stall speed test can be performed in other forward and reverse gear positions. Observe the transmission fluid cooling procedure between each test as outlined in step 8.

RESULTS OF THE STALL SPEED TESTS

The stall speed RPM will indicate possible problems. If the engine RPM is high, internal transmission slippage is indicated. If the engine RPM is low, engine or converter problems can exist.

The transmission will not upshift during the stall test and by knowing what internal transmission members are applied in each test range, an indication of a unit failure can be pinpointed.

It is recommended a chart be prepared to assist the diagnostician in the determination of the stall speed results in comparison to the specified engine RPM, and should include the range and holding member applied.

LOW ENGINE RPM ON STALL SPEED TEST

The low engine RPM stall speed indicates either the engine is not performing properly or the converter stator one-way clutch is not holding. By road testing the vehicle, the determination as to the defect can be made.

If the stator is not locked by the one-way clutch, the performance of the vehicle will be poor up to approximately 30-35 MPH. If the engine is in need of repairs or adjustments, the performance of the vehicle will be poor at all speeds.

HIGH ENGINE RPM ON STALL SPEED TEST

When the engine RPM is higher than specifications, internal transmission unit slippage is indicated. By following the holding member application chart, the defective unit can be pinpointed.

It must be noted that a transmission using a one-way overrunning clutch while in the "D" position, first gear, and having a band applied in the "2" position, first gear (for vehicle braking purposes while going downhill), may have the band slipping unnoticed during the stall test, because the overrunning clutch will hold. To determine this, a road test must be performed to place the transmission in a range where the band is in use without the overrunning clutch.

NORMAL ENGINE RPM ON STALL SPEED TEST

When the engine RPM is within the specified ranges, the holding members of the transmission are considered to be operating properly.

A point of interest is if the converter oneway clutch (overrunning) is seized and locks the stator from turning either way, the engine RPM will be normal during the test, but the converter will be in reduction at all times and the vehicle will probably not exceed a speed of 50-60 miles per hour. If this condition is suspected, examine the fluid and the converter exterior for signs of overheating, since an extreme amount of heat is generated when the converter remains in constant reduction.

Transmission Noises

During the stall speed test and the road test, the diagnostician must be alert to any abnormal noises from the transmission area or any excessive movement of the engine/transmission assembly during torque application or transmission shifting.

――――――― CAUTION ―――――――

Before attempting to diagnose automatic transmission noises, be sure the noises do not orginate from the engine components, such as the water pump, alternator, air conditioner compressor, power steering or the air injection pump. Isolate these components by removing the proper drive belt and operate the engine. Do not operate the engine longer than two minutes at a time to avoid overheating.

1. Whining or siren type noises—Can be considered normal if occurring during a stall speed test, due to the fluid flow through the converter.

2. Whining noise (continual with vehicle stationary)—If the noise increases and decreases with the engine speed, the following defects could be present.
  a. Oil level low
  b. Air leakage into pump (defective gasket, "O"-ring or porosity of a part)
  c. Pump gears damaged or worn
  d. Pump gears assembled backward
  e. Pump crescent interference

3. Buzzing noise—This type of noise is normally the result of a pressure regulator valve vibrating or a sealing ring broken or worn out and will usually come and go, depending upon engine/transmission speed.

4. Rattling noise (constant)—Usually occurring at low engine speed and resulting from the vanes stripped from the impeller or turbine face or internal interference of the converter parts.

5. Rattling noise (intermittent)—Reflects a broken flywheel or flex plate and usually occurs at low engine speed with the transmission in gear. Placing the transmission in "N" or "P" will change the rattling noise or stop it for a short time.

6. Gear noise (one gear range)—This type of noise will normally indicate a defective planetary gear unit. Upon shifting into another gear range, the noise will cease. If the noise carries over to the next gear range, but at a different pitch, defective thrust bearings or bushings are indicated.

7. Engine vibration or excessive movement—Can be caused by transmission filler or cooler lines vibrating due to broken or disconnected brackets. If excessive engine/transmission movement is noted, look for broken engine/transmission mounts.

――――――― CAUTION ―――――――

When necessary to support an engine equipped with metal safety tabs on the mounts, be sure the metal tabs are not in contact with the mount bracket after the engine/transmission assembly is again supported by the mounts. A severe vibration can result.

8. Squeal at low vehicle speeds—Can result from a speedometer driven gear seal, a front pump seal or rear extension seal being dry.

The above list of noises can be used as a guide. Noises other than the ones listed can occur around or within the transmission assembly. A logical and common sense approach will normally result in the source of the noise being detected.

Air Pressure Tests

The automatic transmission have many hidden passages and hydraulic units that are controlled by internal fluid pressures, supplied through tubes, shafts and valve movements.

The air pressure test are used to confirm the findings of the fluid pressure tests and to further pinpoint the malfunctioning area. The air pressure test can also confirm the hydraulic unit operation after repairs have been made.

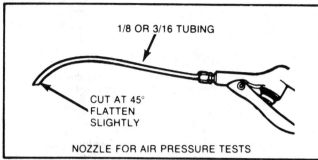

Type of air nozzle to be used in the air pressure tests
(© Ford Motor Co.)

To perform the air pressure test, the control valve body must be removed from the transmission case, exposing the case passages. By referring to the individual transmission section, identify each passage before any attempt is made to proceed with the air pressure test.

───── CAUTION ─────

It is a good practice to protect the diagnostician's face and body with glasses and protective clothing and the surrounding area from the oil spray that will occur when air pressure is applied to the various passages.

NOTE: **The air pressure should be controlled to approximately 25 psi and the air should be clean and dry.**

When the passages have been identified and the air pressure applied to a designated passage, reaction can be seen, heard and felt in the various units. Should air pressure be applied to a clutch apply passage, the piston movement can be felt and a soft dull thud should be heard. Some movement of the unit assembly can be seen.

When air pressure is applied to a servo apply passage, the servo rod or arm will move and tighten the band around the drum. Upon release of the air pressure, spring tension should release the servo piston.

When air pressure is applied to the governor supply passage, a whistle, click or buzzing noise may be heard.

When failures have occurred within the transmission assembly and the air pressure tests are made, the following problems may exist.

1. No clutch piston movement
2. Hissing noise and excessive fluid spray
3. Excessive unit movement
4. No servo band apply or release

Torque Converter

The torque converter is a simple, but yet complex torque multiplication unit, designed and applied to specific engine/transmis-

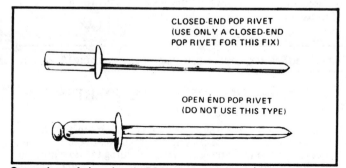

Comparison of closed end "POP" rivet and open end "POP" rivet

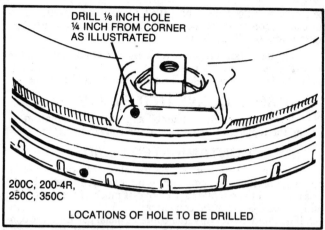

Location of hole to be drilled in the converters of models THM 325, 200C, 200-4R, 250C and 350C transmissions
(© General Motors Corp.)

sion/transaxle applications. Manufacturers apply different ratings for their application to a specific engine. An example is the use of the "K" factor method by several of the vehicle manufacturers to indicate the performance curve of a given engine size. The larger the "K" factor, the smaller the engine size, which gives more engine break-away torque at lower speeds.
Example:

$$K = \frac{RPM}{\sqrt{Torque\ (Nm)}}$$

$$K = \frac{2200}{\sqrt{100}}$$

$$K = \frac{2200}{10}$$

$$K = 220$$

NOTE: **Torque of 100 N•m is usually constant for the finding of the "K" factor.**

NOTE: **Performance cars are rated differently.**

Regardless of the type of rating used, the correct torque converter must be coupled to the specific engine/transmission/transaxle assembly to achieve the desired operational efficiency. Converter rebuilders are supplying quality rebuilt units to the trade, that in most cases, equal the performance of the new converter units. In cases where problems occur, it is generally the use of a converter assembly that does not match the desired performance curve of the engine/transmission/transaxle assembly.

DRAINING AND FLUSHING THE CONVERTER

When the converter has become filled with contaminated fluid that can be flushed out and the converter reused, the different manufacturers recommend procedures that should be followed for their respective converters. All recommend the use of a commercial flushing machine, if available. Certain converters will have drain plugs that can be removed to allow the fluid to drain. Other manufacturers recommend the drilling of drain holes and the use of a rivet plug to close the hole after the fluid has been drained and the converter flushed. With the use of lock-up converters, the drilling of a drain hole in the converter shell must be done correctly or the lock-up mechanism can be damaged. Various after-market suppliers have drill and tap guide kits available to drill a hole in the converter, tap threads in the shell and install a threaded plug or a kit with a drill, pop rivets and instructions on the location of the hole to be drilled. Certain manufacturers require that a contaminated converter be replaced without attempting to drain and flush the unit.

NOTE: Be certain the oil cooler is flushed of all contaminates before the vehicle is put back in service. Should a question arise as to the efficiency of the cooler, a replacement should be installed.

TORQUE CONVERTER EVALUATION

The following is a "rule of thumb" as to the determination regarding the replacement or usage of the converter.

CONTAMINATED FLUID

1. If the fluid in the converter is discolored but does not contain metal particles, the converter is not damaged internally and does not need to be replaced. Remove as much of the discolored fluid as possible from the converter.

2. If the fluid in the converter contains metal particles, the converter is damaged internally and must be replaced.

3. If the fluid contamination was due to burned clutch plates, overheated fluid or engine coolant leakage, the unit should be flushed. The degree of contamination would have to be decided by the repairman in regards to either the replacement or flushing of the unit.

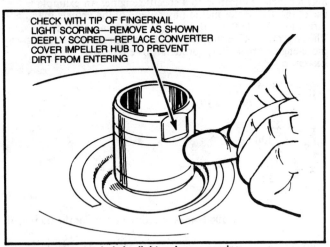

CHECK WITH TIP OF FINGERNAIL
LIGHT SCORING—REMOVE AS SHOWN
DEEPLY SCORED—REPLACE CONVERTER
COVER IMPELLER HUB TO PREVENT
DIRT FROM ENTERING

Checking converter hub for light or heavy scoring
(© Ford Motor Co.)

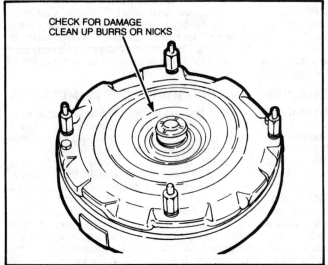

CHECK FOR DAMAGE
CLEAN UP BURRS OR NICKS

Checking converter cover for nicks or burrs(© Ford Motor Co.)

4. If the pump gears or cover show signs of damage or are broken, the converter will contain metal particles and must be replaced.

STRIPPED CONVERTER BOLT RETAINERS

1. Inspect for the cause, such as damaged bolt threads. Repair the stripped bolt retainers, using Heli-coils or its equivalent.

FLUID LEAKAGE

1. Inspect the converter hub surface for roughness, scoring or wear that could damage the seal or bushing. If the roughness can be felt with a fingernail, the front seal could be damaged. Repair the hub surface with fine crocus cloth, if possible, and replace the front seal.

2. Inspect the inside of the bell housing. If fluid is present, leakage is indicated and the converter should be leak tested. If leaks are found in the converter, the unit should be replaced.

CAUTION
Do not attempt to re-weld the converter.

CONVERTER NOISE OR SLIPPAGE

1. Check for loose or missing flywheel to converter bolts, a cracked flywheel, a broken converter pilot, or other engine parts that may be vibrating. Correct as required.

NOTE: Most converter noises occur under light throttle in the "D" position and with the brakes applied.

2. Inspect the converter for excessive end play by the use of the proper checking tools. Replace the converter if the turbine end play exceeds the specifications (usually 0.050 in.).

3. Inspect the converter for damages to the internal roller bearings, thrust races and roller clutch. The thrust roller bearing and thrust races can be checked by viewing them when looking into the converter neck or feeling through the opening to make sure they are not cracked, broken or mispositioned.

4. Inspect the stator clutch by either inserting a protected finger into the splined inner race of the roller clutch or using special tools designed for the purpose, and trying to turn the race in either direction. The inner race should turn freely in a clockwise direction, but not turn or be very difficult to turn in a counterclockwise direction. The converter must be replaced if the roller bearings, thrust races or roller clutch are damaged.

CONVERTER VIBRATION

1. Isolate the cause of the vibration by disconnecting other engine driven parts one at a time. If the converter is determined to be the cause of the vibration, check for loss of balance weights and should they be missing, replace the converter. If the weights are in place, relocate the converter 120 degrees at a time to cancel out engine and converter unbalanced conditions. Washers may be used on the converter to flywheel bolts to isolate an area of unbalance.

CAUTION
Be sure sufficient clearance is available before starting the engine.

INSPECTION OF THE CONVERTER INTERNAL PARTS

The average automatic transmission repair shop can and should inspect the converter assembly for internal wear before any attempt is made to reuse the unit after rebuilding the transmission unit. Special converter checking tools are needed and can be obtained through various tool supply channels.

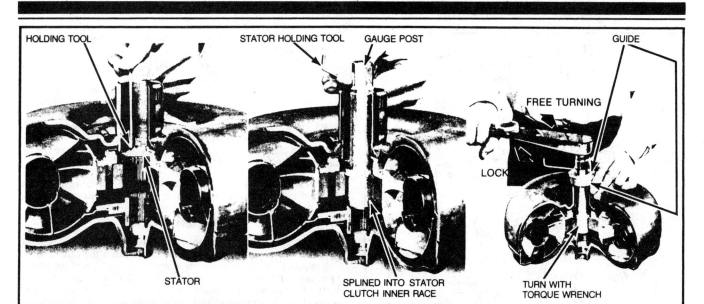

Preparing tool to inspect converter stator overrunning clutch operation(© Ford Motor Co.)

CHECKING CONVERTER END PLAY (STATOR AND TURBINE)

1. Place the converter on a flat surface with the flywheel side down and the converter hub opening up.
2. Insert the special end play checking tool into the drive hub opening until the tool bottoms.

NOTE: Certain end play checking tools have a dual purpose, to check the stator and turbine end play and to check the stator one-way clutch operation. The dual purpose tool will have an expandable sleeve (collet) on the end, along with splines to engage the internal splines of the stator one-way clutch inner race.

3. Install the cover or guide plate over the converter hub and tighten the screw nut firmly to expand the split sleeve (collet) in the turbine hub.
4. Attach a dial indicator tool on the tool screw and position the indicator tip or button on the converter hub or the cover. Zero the dial indicator.
5. Lift the screw upward as far as it will go, carrying the dial indicator with it. Read the measurement from the indicator dial. The reading represents the converter end play.
6. Refer to the individual automatic transmission sections for the permissible converter end play.
7. Remove the tools from the converter assembly. Do not leave the split sleeve (collet) in the turbine hub.

STATOR TO IMPELLER INTERFERENCE CHECK

1. Place the transmission oil pump assembly on a flat surface with the stator splines up.
2. Carefully install the converter on oil pump and engage the stator splines.
3. Hold the pump assembly and turn the converter counterclockwise.
4. The converter should turn freely with no interference. If a slight rubbing noise is heard, this is considered normal, but if a binding or loud scraping noise is heard, the converter should be replaced.

STATOR-TO-TURBINE INTERFERENCE CHECK

1. Place the converter assembly on a flat surface with the flywheel side down and the converter hub opening up.

2. Install the oil pump on the converter hub. Install the input shaft into the converter and engage the turbine hub splines.
3. While holding the oil pump and converter, rotate the input shaft back and forth.
4. The input shaft should turn freely with only a slight rubbing noise. If a binding or loud scraping noise is heard, the converter should be replaced.

STATOR ONE-WAY CLUTCH CHECK

Because the stator one-way clutch must hold the stator for torque multiplication at low speed and free-wheel at high speeds during the coupling phase, the stator assembly must be checked while the converter is out of the vehicle.

1. Have the converter on a flat surface with the flywheel side down and the hub opening up.
2. Install the stator race holding tool into the converter hub opening and insert the end into the groove in the stator to prevent the stator from turning.
3. Place the special tool post, without the screw and split sleeve (collet), into the converter hub opening. Engage the splines of the tool post with the splines of the stator race. Install the cover or guide plate to hold the tool post in place.
4. With a torque wrench, turn the tool post in a clockwise manner. The stator should turn freely.
5. Turn the tool post with the torque wrench in a counterclockwise rotation and the lock-up clutch should lock up with a 10 ft.-lb. pull.
6. If the lock-up clutch does not lock up, the one-way clutch is defective and the converter should be replaced.

Certain vehicle manufacturers do not recommend the use of special tools or the use of the pump cover stator shaft as a testing device for the stator one-way clutch unit. Their recommendations are to insert a finger into the converter hub opening and contact the splined inner race of the one-way clutch. An attempt should be made to rotate the stator inner race in a clockwise direction and the race should turn freely. By turning the inner race in a counter-clockwise rotation, it should either lock-up or turn with great difficulty.

CAUTION

Care should be exercised to remove any metal burrs from the converter hub before placing a finger into the opening. Personal injury could result.

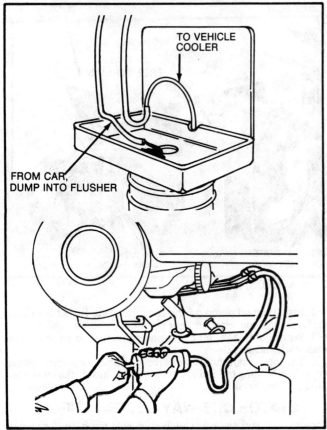

TO VEHICLE COOLER

FROM CAR, DUMP INTO FLUSHER

Two methods of flushing transmission cooler and lines

CONVERTER LOCK-UP CLUTCH AND PISTON

Unless a direct malfunction occurs and/or fluid contamination exists from the converter clutch unit, it is extremely difficult to diagnose an internal wear problem. The diagnostician should exercise professional expertise when the determination is made to replace or reuse the converter unit, relating to mileage, type of operation and wear of related parts within the gearbox.

Flushing the Fluid Cooler and Lines

Much reference has been made to the importance of flushing the transmission/transaxle fluid coolers and lines during an overhaul procedure. With the increased use of converter clutch units and the necessary changes to the internal fluid routings, the passage of contaminated fluid, sludge or metal particles to the fluid cooler is more predominate. In most cases, the fluid returning from the fluid cooler is directed to the lubrication system and should the system be deprived of lubricating fluid due to blockage, premature unit failure will occur.

GENERAL FLUSHING PROCEDURES

Two methods of flushing the fluid cooling system can be used.
   a. Disconnect both fluid lines from the cooler and the transmission/transaxle and flush each line and cooler separately.
   b. Disconnect both fluid lines from the transmission /transaxle assemblies, leaving the lines attached to the cooler. Add a length of hose to the return line and place in a container. Flush both lines and the cooler at the same time.
When flushing the cooling components, use a commercial flushing fluid or its equivalent. Reverse flush the lines and cooler with the flushing fluid and pulsating air pressure. Continue the flushing process until clean flushing fluid appears. Remove the flushing fluid by the addition of transmission fluid through the lines and cooler.

COOLER FLOW

To check the fluid flow through the cooler, place the return line to the transmission/transaxle in a clean container of approximately one quart capacity. Overfill the transmission/transaxle by one quart of fluid, start the engine with the shift in the neutral position. Run the engine for exactly twenty (20) seconds. If the cooler flow is less than one quart in the twenty (20) seconds, have the radiator fluid cooler reconditioned or replaced.

   NOTE: Commercial flushing machines are available that flush the fluid cooling system and measure the rate of flow.

Special Tools

There are an unlimited amount of special tools and accessories available to the transmission rebuilder to lessen the time and effort required in performing the diagnosing and overhaul of the automatic transmission/transaxles. Specific tools are necessary during the disassembly and assembly of each unit and its sub-assemblies. Certain tools can be fabricated, but it becomes the responsibility of the repair shop operator to obtain commercially manufactured tools to insure quality rebuilding and to avoid costly "come backs."
   The commercial labor saving tools range from puller sets, bushing and seal installer sets, compression tools and presses (both mechanically and hydraulically operated), holding fixtures, oil pump aligning tools (a necessity on most automatic transmissions) to work bench arrangements, degreaser tanks, steam cleaners, converter flushing machines, transmission jacks and lifts, to name a few. For specific information concerning the various tools, a parts and tool supplier should be consulted.
   In addition to the special tools, a complete tool chest with the necessary hand tools should be available to the repairman.

BASIC MEASURING TOOLS

The use of the basic measuring tools has become more critical in the rebuilding process. The increased use of front drive transaxles, in which both the automatic transmission and the final drive gears are located, has required the rebuilder to adhere to specifications and tolerances more closely than ever before.
   Bearings must be torqued or adjusted to specific preloads in order to meet the rotating torque drag specifications. The end play and backlash of the varied shafts and gears must be measured to avoid excessive tightness or looseness. Critical tensioning bolts must be torqued to their specifications to avoid warpage of components and proper mating of others.
   Dial indicators must be protected and used as a delicate measuring instrument. A mutilated or un-calibrated dial indicator invites premature unit failure and destruction. Torque wrenches are available in many forms, some cheaply made and others, accurate and durable under constant use. To obtain accurate readings and properly applied torque, recalibration should be applied to the torque wrenches periodically, regardless of the type used. Micrometers are used as precise measuring tools and should be properly stored when not in use. Instructions on the recalibration of the micrometers and a test bar usually accompany the tool when it is purchased.
   Other measuring tools are available to the rebuilder and each in their own way, must be protected when not in use to avoid causing mis-measuring in the fitting of a component to the unit. A good example of poorly cared-for tools is the lowly feeler gauge blades. Many times a bent and wrinkled blade is used to measure clearances that, if incorrect, can cause the failure of a rebuilt unit. Why risk the failure of a $1000.00 unit because of a $2.98 feeler gauge? Good tools and the knowledge of how to use them reflects upon the longevity of the rebuilt unit.

STANDARD TORQUE SPECIFICATIONS
AND CAPSCREW MARKINGS

Newton/Metre has been designated as the world standard for measuring torque and will gradually replace the foot-pound and kilogram-meter torque measuring standard. Torquing tools are still being manufactured with foot-pounds and kilogram-meter scales, along with the new Newton-Metre standard. To assist the repairman, foot-pounds, kilogram-meter and Newton-Metre are listed in the following charts, and should be followed as applicable.

U. S. BOLTS

SAE Grade Number	1 or 2			5			6 or 7			8		
Capscrew Head Markings Manufacturer's marks may vary. Three-line markings on heads shown below, for example. Indicate SAE Grade 5.												
Usage	Used Frequently			Used Frequently			Used at Times			Used at Times		
Quality of Material	Indeterminate			Minimum Commercial			Medium Commercial			Best Commercial		
Capacity Body Size (inches)—(Thread)	Torque			Torque			Torque			Torque		
	Ft-Lb	kgm	Nm	Ft-Lb	kgm	Nm	Ft-Lb	kgm	Nm	Ft-Lb	kgm	Nm
1/4-20	5	0.6915	6.7791	8	1.1064	10.8465	10	1.3630	13.5582	12	1.6596	16.2698
-28	6	0.8298	8.1349	10	1.3830	13.5582				14	1.9362	18.9815
5/16-18	11	1.5213	14.9140	17	2.3511	23.0489	19	2.6277	25.7605	24	3.3192	32.5396
-24	13	1.7979	17.6256	19	2.6277	25.7605				27	3.7341	36.6071
3/8-16	18	2.4894	24.4047	31	4.2873	42.0304	34	4.7022	46.0978	44	6.0852	59.6560
-24	20	2.7660	27.1164	35	4.8405	47.4536				49	6.7767	66.4351
7/16-14	28	3.8132	37.9629	49	6.7767	66.4351	55	7.6065	74.5700	70	9.6810	94.9073
-20	30	4.1490	40.6745	55	7.6065	74.5700				18	10.7874	105.7538
1/2-13	39	5.3937	52.8769	75	10.3725	101.6863	85	11.7555	115.2445	105	14.5215	142.3609
-20	41	5.6703	55.5885	85	11.7555	115.2445				120	16.5860	162.6960
9/16-12	51	7.0533	69.1467	110	15.2130	149.1380	120	16.5960	162.6960	155	21.4365	210.1490
-18	55	7.6065	74.5700	120	16.5960	162.6960				170	23.5110	230.4860
5/8-11	83	11.4789	112.5329	150	20.7450	203.3700	167	23.0961	226.4186	210	29.0430	284.7180
-18	95	13.1385	128.8027	170	23.5110	230.4860				240	33.1920	325.3920
3/4-10	105	14.5215	142.3609	270	37.3410	366.0660	280	38.7240	379.6240	375	51.8625	508.4250
-16	115	15.9045	155.9170	295	40.7985	399.9610				420	58.0860	568.4360
7/8-9	160	22.1280	216.9280	395	54.6285	535.5410	440	60.8520	596.5520	605	83.6715	820.2590
-14	175	24.2025	237.2650	435	60.1605	589.7730				675	93.3525	915.1650
1-8	236	32.5005	318.6130	590	81.5970	799.9220	660	91.2780	894.8280	910	125.8530	1233.7780
-14	250	34.5750	338.9500	660	91.2780	849.8280				990	136.9170	1342.2420

SUGGESTED TORQUE FOR COATED BOLTS AND NUTS

Metric Sizes		6&6.3	8	10	12	14	16	20
Nuts and All Metal Bolts	N·m	0.4	0.8	1.4	2.2	3.0	4.2	7.0
	In. Lbs.	4.0	7.0	12	18	25	35	57
Adhesive or Nylon Coated Bolts	N·m	0.4	0.6	1.2	1.6	2.4	3.4	5.6
	In. Lbs.	4.0	5.0	10	14	20	28	46

Inch Sizes		1/4	5/16	3/8	7/16	1/2	9/16	5/8	3/4
Nuts and All Metal Bolts	N·m	0.4	0.6	1.4	1.8	2.4	3.2	4.2	6.2
	In. Lbs.	4.0	5.0	12	15	20	27	35	51
Adhesive or Nylon Coated Bolts	N·m	0.4	0.6	1.0	1.4	1.8	2.6	3.4	5.2
	In. Lbs.	4.0	5.0	9.0	12	15	22	28	43

METRIC BOLTS

Description	Torque ft-lbs. (Nm)			
Thread for general purposes (size x pitch) (mm)	Head mark 4		Head mark 7	
6 x 1.0	2.2 to 2.9	(3.0 to 3.9)	3.6 to 5.8	(4.9 to 7.8)
8 x 1.25	5.8 to 8.7	(7.9 to 12)	9.4 to 14	(13 to 19)
10 x 1.25	12 to 17	(16 to 23)	20 to 29	(27 to 39)
12 x 1.25	21 to 32	(29 to 43)	35 to 53	(47 to 72)
14 x 1.5	35 to 52	(48 to 70)	57 to 85	(77 to 110)
16 x 1.5	51 to 77	(67 to 100)	90 to 120	(130 to 160)
18 x 1.5	74 to 110	(100 to 150)	130 to 170	(180 to 230)
20 x 1.5	110 to 140	(150 to 190)	190 to 240	(160 to 320)
22 x 1.5	150 to 190	(200 to 260)	250 to 320	(340 to 430)
24 x 1.5	190 to 240	(260 to 320)	310 to 410	(420 to 550)

CAUTION: Bolts threaded into aluminum require much less torque.

DECIMAL AND METRIC EQUIVALENTS

Fractions	Decimal In.	Metric mm.	Fractions	Decimal In.	Metric mm.
1/64	.015625	.397	33/64	.515625	13.097
1/32	.03125	.794	17/32	.53125	13.494
3/64	.046875	1.191	35/64	.546875	13.891
1/16	.0625	1.588	9/16	.5625	14.288
5/64	.078125	1.984	37/64	.578125	14.684
3/32	.09375	2.381	19/32	.59375	15.081
7/64	.109375	2.778	39/64	.609375	15.478
1/8	.125	3.175	5/8	.625	15.875
9/64	.140625	3.572	41/64	.640625	16.272
5/32	.15625	3.969	21/32	.65625	16.669
11/64	.171875	4.366	43/64	.671875	17.066
3/16	.1875	4.763	11/16	.6875	17.463
13/64	.203125	5.159	45/64	.703125	17.859
7/32	.21875	5.556	23/32	.71875	18.256
15/64	.234375	5.953	47/64	.734375	18.653
1/4	.250	6.35	3/4	.750	19.05
17/64	.265625	6.747	49/64	.765625	19.447
9/32	.28125	7.144	25/32	.78125	19.844
19/64	.296875	7.54	51/64	.796875	20.241
5/16	.3125	7.938	13/16	.8125	20.638
21/64	.328125	8.334	53/64	.828125	21.034
11/32	.34375	8.731	27/32	.84375	21.431
23/64	.359375	9.128	55/64	.859375	21.828
3/8	.375	9.525	7/8	.875	22.225
25/64	.390625	9.922	57/64	.890625	22.622
13/32	.40625	10.319	29/32	.90625	23.019
27/64	.421875	10.716	59/64	.921875	23.416
7/16	.4375	11.113	15/16	.9375	23.813
29/64	.453125	11.509	61/64	.953125	24.209
15/32	.46875	11.906	31/32	.96875	24.606
31/64	.484375	12.303	63/64	.984375	25.003
1/2	.500	12.7	1	1.00	25.4

Work Area

The size of the work area depends upon the space available within the service shop to perform the rebuilding operation by having the necessary benches, tools, cleaners and lifts arranged to provide the most logical and efficient approach to the removal, disassembly, assembly and installation of the automatic transmission. Regardless of the manner in which the work area is arranged, it should be well lighted, ventilated and clean.

Precautions to Observe When Handling Solvents

All solvents are toxic or irritating to the skin to some degree. The amount of toxicity or irritation normally depends upon the skin exposure to the solvent. It is a good practice to avoid skin contact with solvent by using rubber gloves and parts drainers when cleaning the transmission parts.

——————— CAUTION ———————

Do not, under any circumstances, wash grease from hands or arms by dipping into the solvent tank and air drying with compressed air. Blood poison can result.

Drive Line Service

Drive line vibrations can affect the operation and longevity of the automatic transmission/transaxles and should be diagnosed during the road test and inspected during the unit removal phase. The drive shafts are designed for specific applications and the disregard for the correct application can result in drive shaft failure with extremely violent and hazardous consequences. A replace-

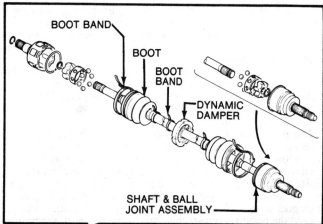

Typical front drive axle assembly using CV joint components on both ends of shaft(© Toyo Kogyo Co. Ltd.)

ment shaft assembly must always be of the same design and material specifications as the original to assure proper operation.

Natural drive line vibrations are created by the fluctuations in the speed of the drive shaft as the drive line angle is changed during a single revolution of the shaft. With the increased use of the front drive transaxles, the drive shafts and universal joints must transfer the driving power to the front wheels and at the same time, compensate for steering action on turns. Special universal joints were developed, one a constant velocity (CV) or double offset type, and a second type known as the tripod joint. The constant velocity joint uses rolling balls in curved grooves to obtain

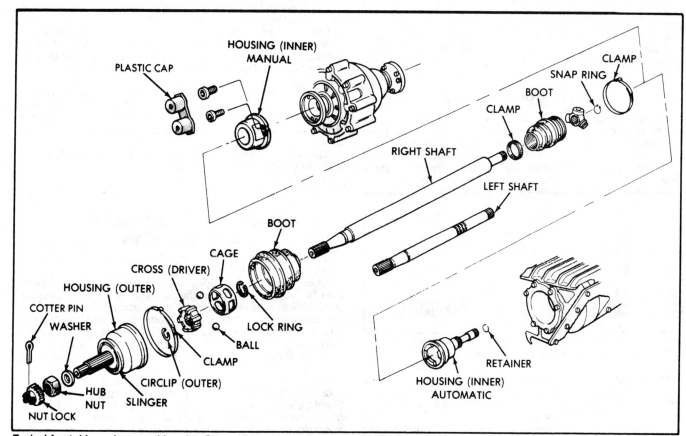

Typical front drive axle assembly using CV and Tri-pot joint components(© Chrysler Corp.)

uniform motion. As the joint rotates in the driving or steering motion, the balls, which are in driving contact between the two halfs of the joint coupling, remain in a plane which bisects the angle between the two shafts, thus cancelling out the fluctuations of speed in the drive shaft.

The tripod type uses a three legged spider, with needle bearing and balls incased in a three grooved housing. With the spider attached to the driveshaft, the joint assembly is free to roll back and forth in the housing grooves as the shaft length varies in normal drive line operation.

The front driveshafts are normally of two different lengths from the transaxle to the drive wheels, due to the location of the engine/transaxle mounting in the vehicle. Care should be exercised when removing or replacing the driveshafts, as to their locations (mark if necessary), removal procedures and handling so as not to damage the boots covering the universal joints, or if equipped, with boots covering the transaxle driveshaft opening. Should the boots become torn or otherwise damaged, premature failure of the universal joint would result due to the loss of lubricant and entrance of contaminates.

ATTACHMENT OF THE DRIVESHAFT TO THE TRANSAXLE

The attachment of the driveshafts to the transaxle is accomplished in a number of ways and if not familiar with the particular shaft attachment, do not pry or hammer until the correct procedure is known.

The shafts can be attached by one of the following methods:

1. Driveshaft flange to transaxle stub shaft flange, bolted together. Mark flanges and remove bolts.
2. Circlips inside differential housing. Remove differential cover, compress circlips and push axle shafts outward.
3. Spring loaded circlip mounted in groove on axle shaft and mating with a groove in the differential gear splines. Is usually pried or taped from differential gear with care.
4. Universal joint housing, axle shaft flange or axle shaft stub end pinned to either the differential stub shaft or differential gear flange with a roll pin. Mark the two components and drive the pin from the units.

BOOT REPLACEMENT

The most common repairs to the front driveshafts are boot replacement and boot retaining ring replacement. Many automatic transmission repair shops are requested to perform this type of repairs for their customers. EOM and after-market replacement boots are available, with special tools used to crimp and tighten

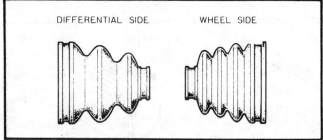

DIFFERENTIAL SIDE   WHEEL SIDE

Look for boot differences between joint locations. Typical of one manufacturer's drive axle(© Toyo Kogyo Co. Ltd.)

the retaining rings. Most boot replacement procedures require the removal of the driveshafts. A boot kit is available that provides a split boot that can be installed without driveshaft removal. The boot is then sealed with a special adhesive along its length and the procedure finished with the installation of the boot retaining rings.

DRIVE LINE DIAGNOSIS—FRONT WHEEL DRIVE

Clicking Noise In Turns

1. Worn or damaged outboard joint. (Check for cut or damaged seals).

"Clunk" When Accelerating From "Coast" To "Drive"

1. Worn or damaged inboard joint

Shudder Or Vibration During Acceleration

1. Excessive joint angle
   a. Excessive toe-in.
   b. Incorrect spring heights.
2. Worn or damaged inboard or outboard joints.
3. Sticking inboard joint assembly (Double Offset Design).
4. Sticking spider assembly (Tri-Pot Design).

Vibration At Highway Speeds

1. Out of balance front wheels or tires.
2. Out of round front tires.

Towing

Proper towing is important for the safe and reliable transfer of an inoperative vehicle from one point to another.

The basic towing instructions and procedures are general in nature and the towing procedures may not apply to the same make, model or vehicle throughout the model years without procedure changes, modification to equipment or the use of auxiliary equipment designed for a specific purpose.

It is important to minimize the risk of personal injury, to avoid damage to the towed vehicle or to render it unsafe while being towed. There are many conceivable methods of towing with possible hazardous consequences for each method. Therefore, it is the responsibility of the tow truck operator to determine the correct connection of the towing apparatus in a safe and secure manner.

Towing manuals are available from varied sources, explaining the vehicle manufacturer's recommended lifting procedures, towing speeds and towing distances. The operating instructions for the towing truck should be understood and followed by the operator.

The disabled vehicle should never be pushed or pulled on a highway because of safety reasons.

Automatic Transmission Identification

The need to identify automatic transmissions occurs when transmission units are obtained; for example, through bulk buying for overhaul and storage as replacement units. To assist the repairman in coupling the proper transmission to the vehicle/engine combination, a listing is given of automatic transmission codes and their corresponding vehicle model and engine usage. The torque converter identification is difficult, as most replacement converters are rebuilt and distributed to suppliers for resale. The transmission model and serial number should be considered when converter replacement is required.

The following listings contain the transmission models or codes for the most commonly used automatic transmissions. Should a model or code be needed for a non-listed vehicle or transmission, refer to the individual transmission section and, if available, the model or code will be noted along with the model and code location.

AMERICAN CAR MANUFACTURER'S BODY CODES

AMERICAN MOTORS CORPORATION

Year	Series	Model
1980-82	01	Concord
1980-83	30	Eagle
1980-83	40	Spirit
1980	40	AMX
1981-83	50	SX-4, Kammback
1980	60	Pacer

Chrysler Corporation

Year	Series	Model
1980-82	L	M-Horizon/TC3, Z-Omni/024
1984	L	M-Horizon/turismo, Z-Omni/Charger
1980	F	H-Volare, N-Aspen
1980-81	M	B-Caravelle, F-LeBaron, G-Diplomat
1982	M	V-Dodge 400, C-LeBaron, G-Diplomat
1984	M	G-Diplomat/Gran Fury, F-New Yorker/Fifth Avenue
1980-84	J	X-Mirida, S-Cordoba
1980-81	R	J-Gran Fury, E-Saint Regis, T-Newport/New Yorker
1982	R	B-Gran Fury, F-New Yorker
1981-84	Y	Y-Imperial
1981-84	K	D-Aries, P-Reliant
1984	CV	V-Dodge 400, C-LeBaron
1984	E	E-Chrysler E class, New Yorker, Dodge 600, 600ES

Ford Motor Company

Year	Series	Model
1980-82		LTD, Marquis, Continental
1984	L	LTD, Marquis, Continental
1980-82		Thunderbird, Cougar
1984	S	Thunderbird, Cougar
1980-82		Town Car, Mark VI
1984	Panther	LTD Crown Victoria, Grand Marquis, Town Car, Mark VI
1980-82		Fairmont, Zephyr, Mustang, Capri

Ford Motor Company

Year	Series	Model
1984	Fox	Fairmont, Zephyr, Mustang, Capri
1980		Pinto, Bobcat
1980-82		Granada, Monarch①
1981-82		Escort, Lynx, EXP, LN7
1984	Erika	Escort, Lynx, EXP, LN7
1984	Topaz	Tempo, Topaz

① 1980 only

General Motors Corporation

Year	Series	Model
1980-81	A	Century, Regal, Malibu, El Camino, Monte Carlo, Cutlass, LeMans, Grand AM, Safari
1982-84	A	Century, Celebrity, Ciera, 6000
1980-81	B	LeSabre, Estate Wagon, Impala, Caprice, Delta 88, Catalina, Bonneville
1982-84	B	LeSabre, Impala, Caprice, Delta 88
1980-84	C	Electra, Limited, Fleetwood Cp.① Deville, Ninety-Eight
1980-84	E	Riviera, Eldorado, Toronado
1980	H	Skyhawk, Monza, Starfire, Sunbird
1980-84	X	Skylark, Citation, Omega, Phoenix
1982-84	J	Skyhawk, Cimarron, Firenza, Cavalier, 2000
1980-82	D	Fleetwood Sedan, Limousine
1984	D	Fleetwood, Limousine
1980-84	K	Seville
1980-84	Z	Commercial Chassis
1980-84	F	Camaro, Firebird
1980-84	T	Chevette
1982-84	T	1000
1980-84	Y	Corvette
1982-84	G	Regal, Malibu, El Camino, Monte Carlo, Cutlass, Bonneville, Grand Prix

① 1984 Fleetwood Brougham

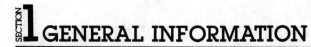

GENERAL MOTORS CORPORATION
Automatic Transmission/Transaxle Listing

Year	Model	Engine	Trans. code
	THM 125 (M34,MD9)		
1980-81	Skylark, Citation, Omega	151 eng.	PZ
	Phoenix	173 eng.	CT,CV
1982-84	Century, Celebrity, Citation, Ciera, Omega, Phoenix, 6000	151 eng.	PL,PW,PI,PD,PZ
		173 eng.	CE,CL,CT,CW,CC
	Century, Ciera	3.0L eng.	BL,BF
	Century, Celebrity, Ciera 6000	260 eng. (diesel)	OP
	Cavalier, Cimarron, Firenza, Skyhawk, J2000	112 eng. (1.8L)	PG,P3,C1,CU,CF,C3,HV,CA,CJ,PE, PJ
	Cavalier, Cimarron, Firenza, Skyhawk	122 eng. (2.0L)	CB,CA,CF
	Fiero	151 eng.	PF
	THM 180 (MD2,MD3)		
1980-81	Chevette	1.6L eng.	Trans. Code unavailable
	THM 180 C		
1982	Chevette, T1000	1.6L eng.	VQ
1983	Chevette, T1000	1.6L eng.	JY,TN
1984	Chevette, T1000	1.6L eng.	TP
	THM 200 (MV9, M29)		
1980	Chevette	1.6L eng.	CN
	Monza, Starfire, Sunbird	151 eng.	PB,PY
	El Camino, Grand Am, LeMans, Malibu, Monte Carlo	229 eng.	CA,CK
	Century, Grand Am, Grand Prix, LeMans, Cutlass, Regal	231 eng.	BZ
	Bonneville, Catalina, Century, Grand Am, Grand Prix, LeMans Regal	260 eng.	PG
	Delta 88	260 eng.	OW
	Caprice, Impala	267 eng.	CE
	Century, Grand AM, Grand Prix, LeMans, LeSabre, Regal	301 eng.	PW
	Firebird	301 eng.	PD
	Cutlass, El Camino, Malibu, Monte Carlo	305 eng.	CC
	Bonneville, Catalina, LeSabre,	350 eng. (Gas)	BA
	Delta 88	350 eng. (Gas)	OS
	Bonneville, Catalina, Cutlass, Delta 88, Electra, LeSabre, Olds. 98	350 eng. (Diesel)	OT
	DeVille, Fleetwood	350 eng. (Diesel)	AS
1981	Chevette	1.6L eng.	CN
	Chevette	1.8L eng. (Diesel)	CY
	Starfire	151 eng.	PB,PY
	El Camino, Grand Prix, LeMans, Malibu, Monte Carlo	229 eng.	CA

GENERAL MOTORS CORPORATION
Automatic Transmission/Transaxle Listing

Year	Model	Engine	Trans. code
1981	Century, Cutlass, El Camino. Grand Prix, LeMans, Malibu, Monte Carlo, Regal	231 eng.	BZ
	Delta 88	260 eng.	OW
	Bonneville, Catalina, Century, Grand Prix, LeMans, Regal	260 eng.	DW,PG
	Firebird	260 eng.	PF
	Camaro, El Camino, Malibu, Monte Carlo	267 eng.	CE
	Cutlass	267 eng.	WA
	Grand Prix, LeMans	301 eng.	PD,PE
	Cutlass, El Camino, Malibu, Monte Carlo	305 eng.	CC
	LeSabre	307 eng.	OG
	Bonneville, Catalina, Caprice, Century, Cutlass, Delta 88, Electra, Grand Prix, Impala, LeMans, LeSabre, Olds. 98, Regal	350 eng. (Diesel)	OT
	DeVille, Fleetwood	350 eng. (Diesel)	AS
1982-83	Chevette, T1000	1.8L eng. (Diesel)	JY,CY
	Cutlass, El Camino, Malibu, Monte Carlo, Regal	260 eng. (Diesel)	OR
1984	Chevette, T1000	1.8L eng. (Diesel)	JY
	Bonneville, Caprice, Cutlass, Delta 88, El Camino, Grand Prix, Impala, LeSabre, Monte Carlo, Parisienne, Regal	231 eng.	BH
	Cutlass	260 eng. (Diesel)	OR
	Cutlass, Delta 88, LeSabre	307 eng.	OI
	Bonneville, Caprice, Cutlass, Delta 88, El Camino, Grand Prix, Impala, LeSabre, Monte Carlo, Parisienne	350 eng. (Diesel)	OU
	THM 200-4R (MW9)		
1981	DeVille, Fleetwood, LeSabre	252 eng.	BM,BY
	Caprice, Impala	305 eng.	CU
	Bonneville, Catalina, Delta 88, Electra, Olds. 98	307 eng.	OG
1982-83	Regal	231 eng.	BR
	DeVille, Electra, Fleetwood LeSabre, Olds. 98	252 eng. (V-6)	BY
	Regal	252 eng. (V-6)	BT
	DeVille, Fleetwood	252 eng. (V-8)	AA,AP
	Caprice, Impala	267 eng.	CQ
	Caprice, Impala	305 eng.	CR

GENERAL MOTORS CORPORATION
Automatic Transmission/Transaxle Listing

Year	Model	Engine	Trans. code
1982-83	Cutlass	307 eng.	OZ
	Electra, LeSabre	307 eng.	OG
	Caprice, Delta 88, Deville, Electra, Fleetwood, Impala, LeSabre, Olds. 98,	350 eng. (Diesel)	OM
	THM 250 (M31)		
1980	Caprice, Impala	229 eng.	WK
	Firebird	260 eng.	MC
	Caprice, Impala, El Camino, Malibu, Monte Carlo	267 eng.	WH,XL
	Bonneville, Catalina, Century, Grand AM, Grand Prix, LeMans, LeSabre, Regal	301 eng.	MD,TB
	Cutlass, El Camino, Malibu, Monte Carlo	305 eng.	WL
	Delta 88	307 eng.	TT
1981	Camaro, Caprice, Impala	229 eng.	TA
	Cutlass, El Camino, Grand Prix LeMans, Malibu, Monte Carlo	231 eng.	XX
	El Camino, Malibu, Monte Carlo	267 eng.	XS
	El Camino, Malibu, Monte Carlo	305 eng.	XK
	Cutlass, Delta 88, LeSabre	307 eng.	XL,XA
1982-84	Caprice, El Camino, Impala, Malibu, Monte Carlo	229 eng.	XP,XE
	Bonneville, Cutlass, El Camino, Grand Prix, Malibu, Monte Carlo Regal	231 eng.	WK
	Bonneville, El Camino, Grand Prix, Malibu, Monte Carlo, Caprice	305 eng.	XK
	Cutlass	307 eng.	XN
	Delta 88, LeSabre	307 eng.	XL
	El Camino, Monte Carlo	229 eng.	CH
	Regal	231 eng.	BQ
	LeSabre, Electra, Regal	252 eng. (V6)	BY,BT
	Fleetwood	252 eng. (V8)	AA,AP
	Cutlass, Regal	260 eng. (Diesel)	OF,OY
	Cutlass, Regal	305 eng.	HG
	Bonneville, El Camino, Grand Prix, Monte Carlo	305 eng.	CQ,CR
	Delta 88, Electra, LeSabre, Olds. 98	307 eng.	OG,OJ,OZ
	Bonneville, Cutlass, Delta 88, El Camino, Electra, Grand Prix, LeSabre, Monte Carlo, Olds. 98, Parisienne, Regal	350 eng. (Diesel)	OM

GENERAL MOTORS CORPORATION
Automatic Transmission/Transaxle Listing

Year	Model	Engine	Trans. code
		THM 325 (M32)	
1980-81	Riviera	231 eng.	BJ
	Riviera, Toronado	252 eng.	BE
	Eldorado, Seville	252 eng.	AG
	Riviera, Toronado	307 eng.	OH
	Riviera, Toronado	350 eng. (Gas)	OJ
	Eldorado, Seville	350 eng. (Gas)	AJ
	Riviera, Toronado	350 eng. (Diesel)	OK
	Eldorado, Seville	350 eng. (Diesel)	AK
	Eldorado, Seville	368 eng.	AF
		THM 325 4SP. (M57)	
1982-84	Riviera	231 eng.	BJ
	Eldorado, Seville	252 eng. (V-6)	AM
	Eldorado, Seville	252 eng. (V-8)	AB,AJ,AE
	Riviera, Toronado	252 eng.	BE
	Riviera, Toronado	307 eng.	OJ
	Eldorado, Seville	350 eng. (Diesel)	AL
	Riviera, Toronado	350 eng. (Diesel)	OK
		THM 350	
1980	Caprice, Impala	229 eng.	JA,XY
	Bonneville, Catalina, Caprice, Century, Cutlass, Delta 88, El Camino, Grand Am, Grand Prix, Impala, LeMans, Malibu, Monte Carlo	231 eng.	KJ,KT,KC,KS,KH
	Camaro, Firebird	231 eng.	KF
	Skyhawk, Starfire, Sunbird	231 eng.	KA
	DeVille, Electra, Fleetwood LeSabre	252 eng.	KD
	Cutlass	260 eng.	LC,LD
	Camaro, El Camino, Malibu, Monte Carlo	267 eng.	JJ,JN
	Bonneville, Catalina, Grand Am, Grand Prix, LeMans	301 eng.	MS
	Camaro, Firebird	301 eng.	MJ,MT
	Bonneville, Caprice, Catalina, Century, Cutlass, El Camino, Grand Am, Grand Prix, Impala, LeMans, Malibu, Monte Carlo, Regal	305 eng.	JE,JK,LJ
	Camaro, Firebird	305 eng.	JK,JD
	Corvette	305 eng.	JC
	Bonneville, Catalina, Delta 88, LeSabre	350 eng. (Gas)	KN,LA,TV
	Camaro	350 eng. (Gas)	JL

GENERAL MOTORS CORPORATION
Automatic Transmission/Transaxle Listing

Year	Model	Engine	Trans. code
		THM 350	
1980	Caprice, El Camino, Impala, Malibu, Monte Carlo	350 eng. (Diesel)	JS
1981	Bonneville, Catalina, Caprice, Century, Cutlass, Delta 88, El Camino, Grand Prix, LeMans, LeSabre, Impala, Malibu, Monte Carlo, Regal	231 eng.	KD,KT
	Camaro, Firebird	231 eng.	KY
	Electra, LeSabre	252 eng.	KK
	Cutlass, Grand Prix, LeMans	260 eng.	LC
	Firebird	301 eng.	MC
	Grand Prix, LeMans	301 eng.	MA
	Camaro	350 eng. (Gas)	JC
	Corvette	350 eng. (Gas)	JD
	Bonneville, Catalina, Caprice, Delta 88, Impala, LeSabre	350 eng. (Diesel)	LA,LD
1982	Caprice, El Camino, Impala, Malibu, Monte Carlo	229 eng.	WP
	Caprice, El Camino, Impala, Malibu, Monte Carlo	267 eng.	WC
	Bonneville, Cutlass, El Camino, Grand Prix, Malibu, Monte Carlo, Regal	350 eng. (Diesel)	WX
	Caprice, Delta 88, Electra, Impala, LeSabre, Olds. 98	350 eng. (Diesel)	WT
		THM 350C—(Lock-up Converter)	
1980	Caprice, Impala	267 eng.	TZ
	El Camino, Malibu, Monte Carlo	267 eng.	XN
	Bonneville, Catalina, LeSabre	301 eng.	WB
	Corvette	305 eng.	TW
	Caprice, Cutlass, El Camino, Impala, Malibu, Monte Carlo	305 eng.	WD
	Delta 88	307 eng.	WA
	All Models	350R eng. (Gas)	TY
	Bonneville, Catalina	350X eng. (Gas)	TV
	Bonneville, Catalina, Caprice, Delta 88, Impala, LeSabre	350 eng. (Diesel)	WC
1981	El Camino, Malibu, Monte Carlo	229 eng.	WP
	Caprice, Cutlass, El Camino, Impala, Malibu, Monte Carlo	267 eng.	WC
	Camaro, Caprice, El Camino, Impala, Malibu, Monte Carlo	305 eng.	WD
	Caprice, El Camino, Impala, Malibu, Monte Carlo	350 eng. (Gas)	WE
	Caprice, Impala	350 eng. (Diesel)	WS,WW

GENERAL MOTORS CORPORATION
Automatic Transmission/Transaxle Listing

Year	Model	Engine	Trans. code
THM 350C—(Lock-up Converter)			
1982	Bonneville, Caprice, El Camino Cutlass, Delta 88, Impala, Electra, Grand Prix, Malibu, Monte Carlo, Regal	231 eng.	KA
	Bonneville, Electra, Grand Prix, Regal	252 eng.	KE,KK
	Cutlass, Delta 88	260 eng.	LA
	Bonneville, El Camino, Grand Prix, Malibu, Monte Carlo	350 eng. (Diesel)	LB
	Caprice, Delta 88, Electra, Impala, Regal	350 eng. (Diesel)	LD
1983	Bonneville, Caprice, Cutlass, Delta 88, Electra, Impala Grand Prix, Regal	231 eng.	KA
	Electra	252 eng.	KE,KK
	Bonneville, El Camino, Grand Prix, Malibu, Monte Carlo	305 eng.	WD,WE
	Caprice, Delta 88, Impala	350 eng. (Diesel)	LD,LJ
	Bonneville, Cutlass, El Camino, Grand Prix, Malibu, Monte Carlo	350 eng. (Diesel)	LB
	Electra, Regal	350 eng. (Diesel)	LB,LJ
1984	El Camino, Monte Carlo	305 eng.	WS
	Impala, Caprice	305 eng.	WW
THM 400 (M40)			
1980	Electra, LeSabre	350 eng.	BB,OB
	Fleetwood, DeVille	368 eng.	AB,AD,AE
	Limo, Comm. Ch.	368 eng.	AD,AN
1981-84	Fleetwood, DeVille	368 eng.	AE
	Limo	368 eng.	AN
	Comm. Ch.	368 eng.	AD
THM 700 (MD8)			
1982	Caprice, Impala	267 eng.	Y4,YL
	Caprice, Impala	305 eng.	Y3,YK
	Corvette	350 eng. (Gas)	YA
1983	Caprice, Impala	305 eng.	YK
1984	Camaro, Firebird	151 eng.	PQ
	Camaro, Firebird	173 eng.	YF
	Camaro, Firebird	305 eng.	YG,YP
	Caprice, Impala, Parisienne	305 eng.	YK
	Corvette	350 eng.	Y9,YW
THM 440-T4 (ME9)			
1984	Celebrity, Century, Ciera, 6000	260 eng. (Diesel)	OB,OV

CHRYSLER CORPORATION
Transmission Listing

Year	Assy. No.	Engine Cu. In.	Type Trans.	Other Information
				AMERICAN MOTORS CORPORATION CARS
1980	3235770	2.5 Litre	A-904	Standard—Non-Lockup (4 Cyl)
	3237269	258	A-904	Standard—Lockup
	3236581	258	A-998	4-Wheel Drive—Non-Lockup
	3238455	258	A-998	Export Non-Lockup
	3238767	282	A-998	Mexico—Lockup
	3235220	258/304	A-999	Jeep CJ 4-Wheel Drive—Non-Lockup
	5359402	258/360	A-727	Jeep SR 4-Wheel Drive—Non-Lockup
1981	3238777	2.5 Litre	A-904	Wide Ratio—Non Lockup
	3238772	2.5 Litre	A-904	4x4 Wide Ratio—Non Lockup
	3238778	2.5 Litre	A-904	CJ-7 4x4 Wide Ratio—Non Lockup
	3240107	258	A-904	Standard—Lockup
	3238771	258	A-998	4x4 Eagle—Lockup
	3240296	258/282	A-998	VAM—Lockup
	3238773	258/304	A-999	CJ-7 4x4 Lockup (2.73 Axle)
	3239816	258/304	A-999	CJ-7 4x4 Lockup (3.31 Axle)
	3240219	258	A-999	CJ-7 4x4 Export—Non Lockup
	3238774	360	A-727	Sr Jeep 4x4 Lockup (2.73 Axle)
	3239817	258/360	A-727	Sr Jeep 4x4 Lockup (3.31-3.73 Axle)
	3240255	258/360	A-727	Sr Jeep 4x4 Export—Non Lockup
1982	3238772	2.5 Litre	A-904	4x4—Non-Lockup
	3238777	2.5 Litre	A-904	Non-Lockup
	—	2.5 Litre	A-904	CJ7 4x4—Non-Lockup
	5567640	2.5 Litre	A-904	AM General-Post Office Trucks
	3240229	258	A-904	Lock-up
	3241099	258	A-998	4x4 Eagle-Lockup
	3241100	6 & V8	A-998	VAM—Non-Lockup
	3240231	258	A-999	CJ7 4x4—Lockup
	3240231	258	A-999	SR Jeep 4x4—Lockup (2.73 Axle)
	—	258/304	A-999	CJ7 4x4—Lockup (3.31 Axle)
	3241098	258	A-999	CJ7 4x4 Export—Non-Lockup
	3238774	360	A-727	SR Jeep 4x4—Lockup (2.73 Axle)
	3239817	258/360	A-727	SR Jeep 4x4—Lockup (3.31/3.73 Axle)
	3240255	258/360	A-727	SR Jeep 4x4 Export—Non-Lockup
1983-84	8933000864	2.5 Litre	A-904	4x4—Non Lockup
	8933000863	2.5 Litre	A-904	Non Lockup
	8923000028	2.5 Litre	A-904	AM General—Post Office Truck
	8953000847	2.46 Litre	A-904	XJ 4x4—Lockup
	3240229	258	A-904	Lockup
	8933000916	258	A-998	4x4 Eagle—Lockup
	3241100	6 & 8	A-998	VAM—Non Lockup
	8933000913	258	A-999	CJ7-8 4x4—Lockup
				SJ 4x4—Lockup (2.73 Axle)
	8933000917	258	A-999	CJ 4x4 Export—Non Lockup
	8933000915	360	A-727	SJ 4x4—Lockup (2.73 Axle)
	8933000914	258/360	A-727	SJ 4x4—Lock (3.31/3.73 Axle)
	8933000918	258/360	A-727	SJ 4x4 Export—Non Lockup

CHRYSLER CORPORATION
Transmission Listing

Year	Assy. No.	Engine Cu. In.	Type Trans.	Other Information
CHRYSLER CORPORATION CARS				
1980	4130951	225	A-904	Standard—Lockup
	4130953	225	A-904	Wide Ratio Gear Set—Lockup
	4202095	225	A-904	Wide Ratio Gear Set—Non-Lockup
	4130952	225	A-904	Heavy Duty—Lockup
	4202094	225	A-904	Heavy Duty—Non-Lockup
	4130955	318	A-998	Standard—Lockup
	4130956	318	A-998	Wide Ratio Gear Set—Lockup
	4130957	360	A-999	Standard—Lockup
	4202084	1.6 Litre	A-904	MMC (Colt-Arrow)—Non-Lockup
	4202085	2.0 Litre	A-904	MMC (Colt-Arrow) Non-Lockup
	4202086	2.6 Litre	A-904	MMC (Colt-Arrow) Non-Lockup
	4202064	318	A-727	Standard—Lockup
	4130976	360	A-727	Hi-Performance—Lockup
	4058376	360	A-727	Hi-Performance—Non-Lockup
	5224442	1.7 Litre	A-404	Federal & California—3.48 Axle
1981	4202662	225	A-904	Wide Ratio—Non Lockup
	4202663	225	A-904	Wide Ratio—Lockup
	4202664	225	A-904	Heavy Duty—Wide Ratio—Non Lockup
	4058383	225	A-904T	Heavy Duty—Wide Ratio—Lockup
	4058398	318	A-999	Wide Ratio—Lockup
	4202675	318	A-999	Wide Ratio—Lockup
	4202729	318	A-999	"Imperial"-Wide Ratio—Lockup
	4202572	1.6 Litre	A-904	MMC (Arrow-Colt)—Non Lockup
	4202573	2.0 Litre	A-904	MMC (Arrow-Colt)—Non Lockup
	4202574	2.6 Litre	A-904	MMC (Arrow-Colt)—Non Lockup
	4202571	318	A-727	Hi-Performance—Lockup
	4058393	360	A-727	Export—Non-Lockup
1982	4202662	225	A-904	Non-Lockup
	4202663	225	A-904	Lockup
	4202664	225	A-904	Heavy Duty—Non Lockup
	4058383	225	A-904T	Heavy Duty—Lockup Also used in truck
	4058398	318	A-999	Lockup—Also used in truck
	4202675	318	A-999	Lockup
	4202729	318	A-999	"Imperial" lockup
	4269051	1.6 Litre	A-904	MMC Non-Lockup
	4269052	2.0 Litre	A-904	MMC Non-Lockup
	4269053	2.6 Litre	A-904	MMC Non-Lockup
	4202571	318 HP	A-727	Hi-Performance Lockup
	4058393	360	A-727	Export—Non-Lockup
1983	4202662	225	A-904	Non-Lockup
	4202663	225	A-904	Lockup
	4202664	225	A-904	Heavy Duty—Non Lockup
	4058398	318	A-999	Lockup-Also used in Truck
	4202675	318	A-999	Lockup
	4202729	318	A-999	"Imperial" lockup
	4269932	1.6 Litre	A-904	MMC Non Lockup
	4269933	2.0 Litre	A-904	MMC Non Lockup
	4269934	2.6 Litre	A-904	MMC Non-Lockup

CHRYSLER CORPORATION
Transmission Listing

Year	Assy. No.	Engine Cu. In.	Type Trans.	Other Information
1983	4202898	2.6 Litre	A-904	MMC 4x4 Non Lockup
	4202571	318 HP	A-727	Hi-Performance Lockup
	—	360	A-727	Export—Non Lockup
1984	4295512	2.2 Litre	Transaxle	2.78 Overall Ratio
	4295763	2.2 Litre	Transaxle	3.22 Overall Ratio
	4295513	2.2 Litre (E.F.I)	Transaxle	3.02 Overall Ratio
	4329827	2.2 Litre (Turbo)	Transaxle	3.02 Overall Ratio
	4295515	2.6 Litre	Transaxle	3.02 Overall Ratio
	4295517	2.6 Litre	Transaxle	3.22 Overall Ratio
	4295887	318	A904	Lock-up 2.24 Axle
	4329436	318	A904	Lock-up 2.26 Axle
	4058398	318	A904	Lock-up 2.94 Axle
	4329631	318	A904	Non Lock-up 2.94 Axle

DODGE TRUCK TRANSMISSIONS

Year	Assy. No.	Engine Cu. In.	Type Trans.	Other Information
1980	4058376	360	A-727	Hi-Perf. Long Extension—Non-Lockup
	4058371	225	A-727	Long Extension—Lockup
	4058355	225	A-727	4-Wheel Drive—Lockup
	4058351	318/360	A-727	Short Extension—Lockup
	4058375	318/360	A-727	Short Extension—Non-Lockup
	4058373	318/360	A-727	Long Extension—Lockup
	4058374	318/360	A-727	Long Heavy Duty Ext.—Lockup
	4058372	318/360	A-727	Long Heavy Duty Ext.—Non-Lockup
	4058356	318/360	A-727	4-Wheel Drive-Lockup
	4058358	318/360	A-727	4-Wheel Drive—Non-Lockup
	4058336	446	A-727	Medium Extension—Non-Lockup
1981	4058383	225	A-904T	Wide Ratio—Lockup
	4058398	318	A-999	Wide Ratio—Lockup
	4058384	225	A-727	Long Ext.—Lockup
	4058385	225	A-727	4x4—Lockup
	4058388	318/360	A-727	Short Ext.—Lockup
	4058389	318/360	A-727	Short Ext.—Non-Lockup
	4058392	318/360	A-727	Long Ext.—Lockup
	4058394	318/360	A-727	HD Long Ext.—Lockup
	4058395	318/360	A-727	HD Long Ext.—Non Lockup
	4058396	318/360	A-727	4x4—Lockup
	4058397	318/360	A-727	4x4—Non-Lockup
1982	4058383	225	A-904T	Lockup
	4058398	318	A-999	Lockup
	4058384	225	A-727	Long Ext.—Lockup
	4058385	225	A-727	4x4—Lockup
	4058388	318/360	A-727	Short Ext.—Lockup
	4058389	318/360	A-727	Short Ext.—Non-Lockup
	4058394	318/360	A-727	HD Long Ext.—Lockup
	4058395	318/360	A-727	HD Long Ext.—Non-Lockup
	4058396	318/360	A-727	4x4—Lockup
	4058397	318/360	A-727	4x4—Non-Lockup

CHRYSLER CORPORATION
Transmission Listing

Year	Assy. No.	Engine Cu. In.	Type Trans.	Other Information
			DODGE TRUCK TRANSMISSIONS	
1983	4058383	225	A-904T	Lockup
	4058398	318	A-999	Lockup
	4058384	225	A-727	Long Ext.—Lockup
	—	225	A-727	4x4—Lockup
	—	318/360	A-727	Short Ext.—Lockup
	4058389	318/360	A-727	Short Ext.—Non-Lockup
	—	318/360	A-727	HD Long Ext.—Lockup
	4058395	318/360	A-727	HD Long Ext.—Non Lockup
	—	318/360	A-727	4x4—Lockup
	4058397	318/360	A-727	4x4—Non Lockup
1984	4058384	225	A-727	Long Ext.—Lockup
	4295941	225	A-727	Long Ext.—Non Lockup
	4329438	318	A-727	Long Ext.
	4329468	360	A-727	Long Ext.
	4329482	318/360	A-727	Short Ext.
	4329458	318	A-727	4x4
	4329488	360	A-727	4x4
	4058383	225	A-904	Long Ext.
	4058398	318	A-999	Long Ext.—Lockup

Other Transmission Usage

Year	Assy. No.	Engine Cu. In.	Type Trans.	Other Information
			EXPORT	
1980	4193341	—	A-727	Aston-Martin—Non-Lockup
	4193390	—	A-727	Aston-Martin—Lockup
	4058377	225/Diesel	A-727	United Kingdon Export Short Ext.
			AM GENERAL TRANSMISSION	
	5565798	2.0L	A-904	Standard—Non-Lockup (4 Cyl)
			INTERNATIONAL HARVESTER CORP. TRANSMISSION	
	492448-C91	304/345	A-727	Scout 4-Wheel Drive—Non-Lockup
			MARINE AND INDUSTRIAL TRANSMISSIONS	
	4142312	225	A-727	Short Extension—Non-Lockup
	4142313	225	A-727	Medium Extension—Non-Lockup
	4142321	318/360	A-727	Medium Extension—Non-Lockup
	4142362	Diesel	A-727	Short Extension—Non-Lockup
	4142363	Diesel	A-727	Medium Extension—Non-Lockup
	4142364	Diesel	A-727	Long Extension—Non-Lockup
			EXPORT	
1981	4025739	—	A-727	Aston-Martin Lockup
	4058387	225/Diesel	A-727	UK Export—Short Ext.—Non Lockup
			MARINE & INDUSTRIAL TRANSMISSIONS	
	4142312	225	A-727	Short Ext.—Non Lockup
	4142313	225	A-727	Medium Ext.—Non Lockup
	4142321	318/360	A-727	Medium Ext.—Non Lockup
			MARINE AND INDUSTRIAL TRANSMISSIONS	
	4142362	Diesel	A-727	Short Ext.—Non Lockup
	4142363	Diesel	A-727	Medium Ext.—Non Lockup
	4142364	Diesel	A-727	Long Ext.—Non Lockup

CHRYSLER CORPORATION
Transmission Listing

Other Transmission Usage

EXPORT				
1982	4058387	225/Diesel	A-727	UK Export—Short Non-Lockup
	4025739	—	A-727	Aston Martin—Lockup
	4202717	—	A-727	Roadmaster Rail—Medium Non-Lockup
	3836023	—	A-727	Land Rover 4x4—Non-Lockup
	3836024	Diesel	A-727	IVECO—Medium Ext—Non-Lockup
MARINE AND INDUSTRIAL TRANSMISSIONS				
	4142312	225	A-727	Short Ext—Non-Lockup
	4142313	225	A-727	Medium Ext—Non-Lockup
	4142321	318/360	A-727	Medium Ext—Non-Lockup
	4142362	Diesel	A-727	Short Ext—Non-Lockup
	4142363	Diesel	A-727	Medium Ext—Non-Lockup
	4142364	Diesel	A-727	Long Ext—Non-Lockup
EXPORT				
1983	4058387	225/Diesel	A-727	UK Export-Short Non-Lockup
	4025739	—	A-727	Aston Martin-Lockup
	4202717	—	A-727	Roadmaster Rail—Med. Non-Lockup
	3836023	—	A-727	Land Rover 4x4—Non Lockup
	3836024	Diesel	A-727	IVECO—Med. Ext. Non-Lockup
	3836040	—	A-727	Maserati—Lockup
MARINE AND INDUSTRIAL TRANSMISSIONS				
	4142312	225	A-727	Short Ext—Non Lockup
	4142313	225	A-727	Med. Ext—Non Lockup
	4142321	318/360	A-727	Med. Ext—Non Lockup
	4142362	Diesel	A-727	Short Ext—Non Lockup
	4142363	Diesel	A-727	Med. Ext—Non Lockup
	4142364	Diesel	A-727	Long Ext—Non Lockup

FORD MOTOR CO.
Automatic Transmission/Transaxle Listing

Year	Model	Engine	Trans-code
C-3 TRANS.			
1980	Pinto—Bobcat	2.3L eng.	80DT-AA,-AB
	Mustang—Capri	2.3L eng. wo/turbo	800T-CA,-CB
		2.3L eng. w/turbo	800T-CDA,-CDB
		200 eng.	80DT-EA,-EB
	Fairmont—Zephyr	2.3L eng. wo/turbo	80DT-CA,-CB,-HA,-HB
		2.3L eng. w/turbo	80DT-CDA,-CDB
		200 eng.	80DT-EA,-EB,-LA,-LB,-CEB
	Thunderbird—Cougar	200 eng.	80DT-LB
1981	Mustang—Capri	2.3L eng.	81DT-CFA,-CHA
		200 eng.	81DT-DAA,-DAB,-DEA,-DEB
	Cougar-XR7—Granada Fairmont—Zephyr	2.3L eng.	81DT-CFA,-CHA,-CKA,-CMA
		200 eng.	81DT-DAA,-DAB,-DDA,-DDB,-DEA-DEB,-DGA,-DGB
	Thunderbird	200 eng.	81DT-DDA,-DDB,-DGA,-DGB

FORD MOTOR CO.
Automatic Transmission/Transaxle Listing

Year	Model	Engine	Trans-code
1982	Mustang—Capri	2.3L eng. 200 eng.	82DT-AAA,-ACA 82DT-BBA
	Cougar-XR7—Granada	2.3L eng.	82DT-AAA,-ABA,-ACA,-ADA,-ARA
	Fairmont—Zephyr	2.3L eng. 200 eng.	82DT-AAA,-ABA,-ACA,-ADA 82DT-BAA,-BBA
1983	Mustang—Capri	2.3L eng.	83DT-AAB
	LTD—Marquis	2.3L eng.	83DT-AAB,-ABB,-AGB,-AHB
	Fairmont—Zephyr	2.3L eng. 200 eng.	83DT-AAB,-ABB 83DT-BAA,-BBA
	Cougar-XR7—Thunderbird	232 eng. 255 eng.	PKA-BH1,2,3 PKA-AH5,6,7,8
	Lincoln Continental	232 eng. 302 eng.	PKA-BF1,2,3,4 PKA-BD1,2,3,4
	Lincoln Towncar—Mark VI	302 eng.	PKA-M8-M13,14,15,16-BC1,2,3
1983	Ford—Mercury	302 eng. 351 eng.	PKA-AG17-AU17-AY12-BB12 PKA-C25-AS17
	Cougar-XR7—Thunderbird	232 eng. 302 eng.	PKA-BR-BT PKA-K
	Lincoln Continental	302 eng.	PKA-BD12
	Lincoln Towncar—Mark VI	302 eng.	PKA-M25-BC5
	Marquis—LTD	232 eng.	PKA-BR-BT
1984	Ford—Mercury	302 eng. 351 eng.	PKA-AG23,24-AU23,24 -AY18,19-BB18,19 PKA-C31,32-AS,23
	Mustang—Capri	232 eng. 302 eng.	PKA-BZ,1-CD,1 PKA-BW,1
	Cougar-XR7—LTD Marquis—Thunderbird	232 eng.	PKA-BT6,7-CB6,7
	Cougar-XR7—Thunderbird	302 eng.	PKA-K6,7
	Lincoln Continental	302 eng.	PKA-BD18
	Lincoln Towncar	302 eng.	PKA-M31-BC12
	Lincoln Mark VII	302 eng.	PKA-BV
	ATX TRANS.		
1981	Escort—Lynx	1.6L eng.	PMA-A1,2
	LN7—EXP	1.6L eng.	PMA-K,1
1982	Escort—Lynx—LN7—EXP	1.6L eng. 1.6L eng.(H.O.)	PMA-A3-K2 PMA-R
1983	Escort—Lynx—LN7—EXP	1.6L eng. 1.6L eng.w/E.F.I. 1.6L eng.(H.O.)	PMA-K3,PMB-A1 PMA-P PMA-R1
1984	Tempo—Topaz Tempo, Topaz	2.3L eng. 2.3L eng.(Canada)	PMA-N PMA-AA
	Escort—Lynx—EXP	1.6L eng.w/E.F.I. 1.6L eng.(w/H.O.)	PMA-U1,2-PMB,D PMA-V3-PMB-C2
	Tempo—Topaz	2.3L eng.	PMA-N-N1-N2

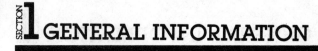
FORD MOTOR CO.
Automatic Transmission/Transaxle Listing

Year	Model	Engine	Trans-code
		C-3 TRANS.	
1984	Mustang—Capri	140 eng.	84DT-AAA,-ACA,-AJA
	LTD—Marquis	140 eng.	83DT-AGB,-AHB 84DT-ABA,-ACA,-ADA,-AKA
	Thunderbird—Cougar XR7	140 eng.	84DT-AEA,-AFA
		C-4 TRANS.	
1980	Ford—Mercury	302 eng.	PEE-DZ3,4,5,6,7,-EA3,4,5,6,7-EM3,4-FC1,2-FE1,2,3,4,5
	Mustang—Capri	2.3L eng.	PEJ-AC1,2,3
		200 eng.	PEB-P4,5,6,7
		255 eng.	PEM-B1,2,3,4-E1,2,3,4-N1,2,3,4
	Granada—Monarch	250 eng.	PEL-A1,2,3-B1,2,3,-C1-D1
		255 eng.	PEM-J1,2,3,4-K1,2,3,4-P1-R1
		302 eng.	PEE-CW5,6,7,8,9-FP1,2,3,4-FR1,2,3,4
	Pinto—Bobcat	2.3L eng.	PEJ-Z1,2,3
	Fairmont—Zephyr	2.3L eng.	PEJ-AC1,2,3-AD1,2,3
		200 eng.	PEB-N6,7,8,9,-P4,5,6,7-S3,4-T1,2,3,4-U1,2,3
		255 eng.	PEM-C1,2,3,4-D1,2,3,4-E1,2,3,4-G1-H1,2,3,4-M1,2,3,4-L1,2,3,4-N1,2,3,4
	Cougar—Thunderbird	200 eng.	PEB-T3,4
		255 eng.	PEM-D1,2,3,4-L1,2,3,4
		302 eng.	PEE-F1,2,3,4-FN1,2,3,4
	Versailles	302 eng.	PEE-EY2,3,4,5,6-FV1,2,3,4
1981	Mustang—Capri	2.3L eng.	PEJ-AC3-AC4
		200 eng.	PEB-P8,P9,P10-A1-B1
		255 eng.	PEM-E5-E6-W-W1-AD-AD1-AK-AK1
	Fairmont—Zephyr	2.3L eng.	PEJ-AC3-AC4-AD3-AD4
		200 eng.	PEB-N10-N11—P8,P9,P10,U4-U5-Z-Z1
		255 eng.	PEM-C5-C6-D5-D6-E5-E6-AC-AC1-AD-AD1-AL-AL1-AM-AM1-AN-AN1-AN2
	Cougar-XR7	2.3L eng.	PEJ-AC3-AC4-AD3-AD4
		200 eng.	PEB-N10-N11-P8,P9,P10-Z-Z1 PEN-A-A1-B-B1
		255 eng.	PEM-C5-C6-D6-D5-E5-E6-AC-AC1-AD-AD1-AE-AE1-AL-AL1-AM-AM1-AN-AN1-AN2
	Granada	2.3L eng.	PEJ-AC3-AC4-AD3-AD4
		200 eng.	PEB-N10-N11-P8-P9-P10 PEN-A-A1-B-B1
		255 eng.	PEM-C5-C6-E5-E6-AC-AC1-AD-AD1-AL-AL1-AM-AM1-AN-AN1-AN2
	Thunderbird	200 eng.	PEB-A-Z1-D5-D6-AC-AC1-AE-AE1-AN2

FORD MOTOR CO.
Automatic Transmission/Transaxle Listing

Year	Model	Engine	Trans-code
		C-5 TRANS.	
1982	Mustang—Capri	200 eng.	PEN-G-W
		255 eng.	PEM-AM3-AP
	Cougar-XR7—Granada	200 eng.	PEB-Z2-C-G-J-K-P-S
		232 eng.	PEP-B-D-E-F-G-H-N-P
	Fairmont—Zephyr	200 eng.	PEN-C-G-P-V-W
		255 eng.	PEM-AL3
1983	Mustang—Capri	232 eng.	PEP-B1-R
	Cougar-XR7—Thunderbird	232 eng.	PEP-V-W
	Fairmont—Zephyr	200 eng.	PEN-G1-P1-AA-AB-BA-CA
	Marquis—LTD	200 eng.	PEN-S1-U-Y-Z
		232 eng.	PEP-R-W
1984	Mustang—Capri	232 eng.	PEP-AF
	Cougar-XR7—Thunderbird	232 eng.	PEP-AD-AE
	LTD—Marquis	232 eng.	PEP-AC-AE-Z
		C-6 TRANS.	
1980	Ford—Mercury	302 eng.	PDG-BH5-CU5
		351 eng.	PGD-DD-DE-DG
		FMX TRANS.	
1980	Ford—Mercury	302 eng.	PHB-BH2-BH3
		351 eng.	PHB-BK-BK1-BP-BP1-BT-BT1-BU-BU1
1981	Ford—Mercury	302 eng.	PHB-BH2,3-BN
		JATCO TRANS.	
1980	Granada—Monarch	250 eng.	PLA-A2-A3
		A.O.T. TRANS.	
1980	Ford—Mercury	302 eng.	PKA-E1,2,3,4,5,6-W1,2,3
		351 eng.	PKA-C1,2,3,4,5,6-R1,2,3,4,5,6 T1,2,3,4,5,6-Z1,2,3,4,5,6
	Cougar-XR7—Thunderbird	302 eng.	PKA-Y1,2,3,4,5,6
	Lincoln—Mark VI	302 eng.	PKA-M1m2m3m4m5m6
		351 eng.	PKA-D1,2,3,4,5,6-U1,2,3,4,5,6
1981	Ford—Mercury	255 eng.	PKA-AF-AF5-AT-AT5
		302 eng.	PKA-E6-AG-AG5-AG50-AL-AV-AU5
		351 eng.	PKA-C6-C8-C13-R8-T8-Z8-AR-AS-AS5-AV
	Cougar-XR7—Thunderbird	255 eng.	PKA-AH-AH5
		302 eng.	PKA-Y8
	Lincoln—Mark VI	302 eng.	PKA-=M8-M13
1982	Ford	255 eng.	PKA-AF5,6-AT5
		302 eng.	PKA-AG-AG5-AU5-AU6-AY,1-BB,1
		351 eng.	PKA-C13,14-AS5,6
	Mercury	255 eng.	PKA-AF,5,6,7,8,-AT,5,6,7,8
		302 eng.	PKA-AG-AG5,6,7,8-AU-AU5,6,7,8-AY1,2,3-BB1,2,3
		351 eng.	PKA-C13,14,15,16-AS-AS5,6,7,8

INDEX

CHRYSLER CORPORATION
A404 • A413 • A415 • A470

 **APPLICATIONS**

Chrysler Corporation Transaxles

MODELS A-404, A-413 and A-470

1980—Omni/024, Horizon/TC3
1981—Omni/024, Horizon/TC3, Aries, Reliant
1982—LeBaron, Aries, Reliant, 400, Rampage,
Horizon/TC3/Turismo/Miser/E-Type
Omni/024/Miser/Charger/E-Type
1983—LeBaron, Aries, 600, Reliant, E-Class, New Yorker,
400, Scamp, Rampage, Horizon/Turismo, Omni/Charger
1984—Laser, Voyager, LeBaron, New Yorker/E-Class,
Reliant, Horizon/Turismo, Daytona, Caravan, 600,
Aries, Omni/Charger

Vehicle Line Letter Code Identification

References will be made to the vehicle models by a letter code,
throughout the section. An explanation, by year, follows:

1980
Z   Omni/024
M   Horizon/TC3

1981
Z   Omni/024
M   Horizon/TC3

P   Reliant
D   Aries

1982
C   LeBaron
     LeBaron Medallion
     Town and Country

D   Aries
     Aries Custom
     Aries S.E.
M   TC3 Miser
     Horizon Miser
     TC3
     Horizon
     TC3 Turismo
     E-Type
P   Reliant
     Reliant Custom
     Reliant S.E.
V   400
     400 LS
Z   024 Miser
     Omni Miser
     024
     Rampage
     Omni
     024 Charger 2.2
     E-Type

1983
C   LeBaron
     Town and Country
D   Aries
     Aries Custom
     Aries S.E.
E   600
     600 ES

J,L Caravelle (Canada)
M   Scamp
     Horizon Custom
     Horizon
     Turismo
     Scamp 2.2
P   Reliant
     Reliant Custom
     Reliant S.E.
T   E-Class
     New Yorker
V   400
Z   Rampage Sport
     Omni Custom
     Rampage
     Omni
     Charger

1984
C   LeBaron
D   Aries
E   600
J,L Caravelle (Canada)
M   Horizon
P   Reliant
T   New Yorker and E-Class
Z   Omni
V   Daytona (Dodge)
C   Laser (Chrysler)

CHRYSLER CORPORATION
Transaxles Models A-404, A-413, A-415 and A-470

Year	Assy. No.	Engine	Trans Type	Over-all Ratio	Other Information
1980	5224442	1.7L	A-404	3.48	Fed. & Calif. Emissions
1981	4207088	1.7L	A-404	3.48	L-Car
	4269522	2.2L	A-413	2.78	K-Car
	4269523	2.2L	A-413	2.78	L-Car
	5224475	2.6L	A-470	2.78	K-Car
1982	4207296	2.6L	A-470	2.78	K-Car
	4207293	1.7L	A-404	3.48	L-Car
	4207294	2.2L	A-413	2.78	C,V,E,T, L & K Car
	4269544	2.2L	A-413	3.22	C,V,E,T, L & K Hi-Alt.
1983	4269686	2.6L	A-470	2.78	K-Car
	4295231	1.7L	A-404	3.50	L-Car
	4269683	2.2L	A-413	2.78	L & K Car
	4269685	2.2L	A-413	3.22	K-Car/Mex
	4269684	2.2L	A-413	3.02	L & K Car
	4269791	2.6L	A-470	3.02	K-Car
1984	4329551	1.6L	A-415	3.02	M,Z
	4295512	2.2L	A-413	2.78	M,Z,P,D
	4329552	2.2L	A-413	3.02	—
	4329553	2.2L	A-413	3.22	—
	4329827	2.2L	A-413	3.02	V,C,ET,VC
	4295763	2.2L	A-413	3.22	H,K
	4295513	2.2L	A-413	3.02	M,Z,P,D, VC,ET
	4295515	2.6L	A-470	3.02	P,D,V,C,E,T
	4295517	2.6L	A-470	3.22	H,K,P,D,C

ENGINE TO TRANSAXLE APPLICATION

Engine Liter	Transaxle Models
1.6	A-415
1.7	A-404
2.2	A-413
2.6	A-470

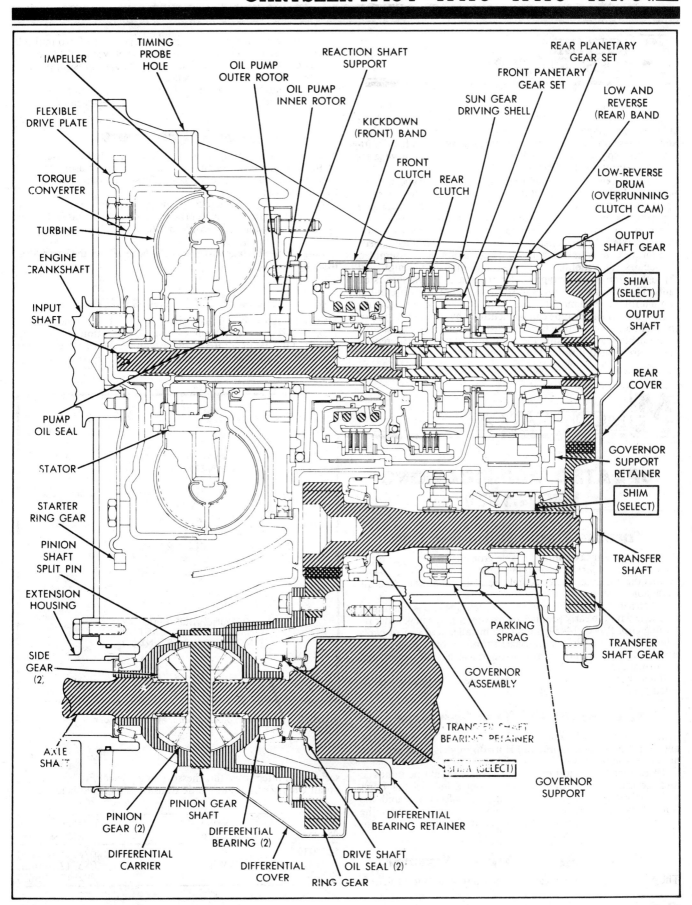

Cross section of TorqueFlite transaxle (©Chrysler Corp.)

GENERAL DESCRIPTION

The Chrysler Corporation TorqueFlite Transaxles, Models A-404, A-413, A-415 and A-470, are used on the front wheel drive vehicles, manufactured by Chrysler Corporation in the United States. These transaxles combine a torque converter, a fully automatic three speed transmission, differential and final drive gearing into a front wheel drive unit.

The transaxle is assembled with metric fasteners and numerous special tools are required during its overhaul. The transaxle operation requirements are different for each vehicle/engine combination, with different internal parts used in the specific application. It is important to refer to the parts number stamped on the transaxle oil pan flange when obtaining replacement parts. The torque converter is a sealed unit and cannot be disassembled for repairs. The cooling of the transaxle fluid is accomplished through an oil-to-water type cooler, located in the radiator side tank. The transaxle sump was separated from the differential sump until the 1983 models, when both were combined as a common sump. The later type can be identified by the elimination of the fill hole and plug from the cover, which is not interchangeable with earlier models. Both the later transaxle and differential sumps are vented through the dipstick.

The internal mechanical components consists of two multiple disc clutches, an overrunning clutch, two apply servos, two bands and two planetary gear sets. The hydraulic system is supplied by an engine driven oil pump, operating valves within a single valve body to control the shifting, along with a governor assembly, mounted on the transfer shaft of the transaxle assembly.

Metric Fasteners

The metric fastener dimensions are very close to the dimensions of the familiar inch system fasteners. For this reason, replacement fasteners must have the same measurement and strength as those removed.

Do not attempt to interchange metric fasteners for inch system fasteners. Mismatched or incorrect fasteners can result in damage to the transmission unit through malfunctions, breakage or possible personal injury.

Care should be taken to re-use the fasteners in the same locations as removed.

Fluid Specifications

Use only automatic transmission fluids of the type marked Dexron® II or its equivalent. Chrysler Corporation does not recommend the use of additives, other than the use of a dye to aid in the determination of fluid leaks.

MODIFICATIONS

UPDATED MODIFICATIONS FROM 1978

Throttle Cable Retaining Clip

A new wider transaxle throttle cable grommet retaining clip has recently entered production on the subject models. This new clip retains the grommet more securely, resulting in a more accurate cable adjustment.

When servicing a vehicle where the retaining clip is missing or a vehicle with the early production narrow clip, the new wider retaining clip, P/N 5214484, should be installed and proper cable adjustment made. Proper transaxle throttle cable adjustment is *necessary* to ensure proper automatic transaxle shift quality and performance.

If a new clip is not available, a replacement can be easily fabricated from .021" sheet steel stock.

Driveshaft Oil Seal Leakage

Transmission fluid leakage from the driveshaft oil seal may be encountered on some Omni and Horizon models equipped with an automatic transaxle.

To correct this condition, the driveshaft should be removed and the inner driveshaft seal area inspected for nicks or burrs which may cause seal failure and leakage. If defects are found, the inner driveshaft section should be replaced. In all instances, the driveshaft oil seal, blue in color, should be replaced with an improved seal, green in color, P/N 5205591.

"Select Fit" #3 Thrust Washer

The No. 3 thrust washer located on the front of the output shaft is

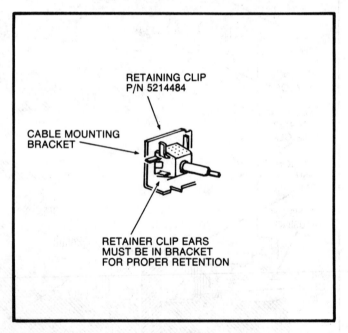

Installation of cable mounting bracket retaining clip
(©Chrysler Corp.)

serviced in three sizes (thicknesses) for a "Select Fit" installation on all Chrysler built automatic transaxles, as listed below:

Washer Part No.	Color Code	Washer Thickness
5224039	Blue	.078" (1.98 mm)
5224040	White	.085" (2.16 mm)
5224041	Yellow	.092" (2.34 mm)

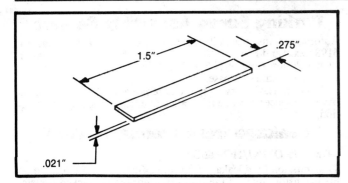

Fabrication of retaining clip (©Chrysler Corp.)

The function of No. 3 thrust washer is to set input shaft end play. The procedure for measuring input shaft end play is as follows:

Measuring Input Shaft End Play

Measure input shaft end play prior to disassembly; which will indicate whether a thrust washer change is required. If, during repair, the input and/or output shafts are replaced, it will be necessary to assemble the transaxle and measure "Input Shaft End Play"; this will allow you to select the proper No. 3 thrust washer.

1. Attach a dial indicator to the transaxle bell housing with its plunger seated against the end of the input shaft.

Move the input shaft in until it bottoms and zero the dial indicator. Pulling on the input shaft, measure the amount of input shaft end play. Input shaft end play should be .030" to .106".

2. Note the amount of end play. When assembling the transaxle, select the No. 3 thrust washer that will provide the minimum amount of specified end play.

EXAMPLES: 1. End play as measured   .037 inch
Original Thrust Washer   .078 inch (Blue)
Corrected End Play   In Specification

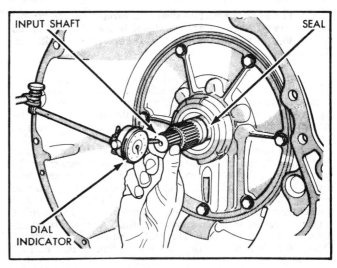

Measuring input shaft seal (©Chrysler Corp.)

2. End play as measured   .027 inch
Original Thrust Washer   .085 inch (White)
Replacement Thrust Washer .078 inch (Blue)
Corrected End Play   .034 inch

Capacity

CHECKING FLUID LEVEL

1981-82

Place the selector in "PARK" and allow the engine to idle. The fluid should be at operating temperature (180°). Check the dipstick. The fluid level is correct if it is between the "FULL" and "ADD" marks. Do not overfill.

CAPACITY CHART

Year	Transaxle	Engine (Liter)	U.S. Quarts	Liters	Imperial Quarts
1981-82	A-404	1.7	7.3	6.9	6.2
	A-413	2.2	7.5	7.1	6.3
	A-470	2.6	8.5	8.1	7.0

NOTES: 1. Differential capacity—1.2 U.S. quarts, 1.1 liter, 1.0 Imperial quart.
2. Replacement fluid volume is approximately 3.0 U.S. quarts, 2.8 liter, 2.5 Imperial quarts.

Year	Transaxle	Engine (Liter)	U.S. Quarts	Liters	Imperial Quarts
1983-84	A-404	1.7	8.4	7.9	7.0
	A-413	2.2 Except Fleet	8.9	8.4	7.4
	A-415	1.6	8.9	8.4	7.4
	A-470	2.6 Except Fleet	8.9	8.4	7.4
	A-413	2.2 Fleet only	9.2	8.7	7.6
	A-470	2.6 Fleet only	9.2	8.7	7.6

NOTES: 1. Differential sump included with transaxle sump.
2. Replacement fluid volume is approximately 4.0 U.S. quarts, 3.8 liters, 3.7 Imperial quarts.

NOTE: The differential oil sump is separate from the "transmission sump." Special emphasis should be placed on filling and maintaining the differential oil level to the fill hole in the differential cover.

It is necessary to treat these units separately when performing lubricant changes or oil level inspections. The transaxle assembly does not have the conventional filler tube, nor a drain plug in the pan or converter. The transaxle is filled through a die cast filler (and dipstick) hole in the case. The differential assembly contains both a fill and drain plug.

Both the automatic transaxle and differential assembly require Dexron® automatic transmission fluid. Although they share the same housing, they are internally sealed from each other. Both should have their fluid levels checked every six months in normal service. Although in normal service no changes of fluid and filter are necessary the transaxle should be serviced every 15,000 miles if used for severe service. At that time, fluid, filter and band adjustment should be done. The refill capacity is 3 quarts of Dexron® adding, if necessary, to bring the fluid to the "FULL" mark when warmed up.

1983 AND LATER

Place the vehicle on a level surface with the engine/transaxle at or near normal operating temperature (minimum operating time of six minutes) to stablize the oil level between the transaxle and the differential. Apply the parking brake and momentarily place the gear selector in each gear position, ending in the P (Park) position.

Remove the dipstick, wipe clean and check the fluid to determine if it is hot or warm. If the fluid is approximately 180°F (82°C) and cannot be held comfortably between the fingers, it would be considered at normal operating temperature. If the fluid is between 85 to 125°F (29 to 52 C) and can be held comfortably between the fingers, it is considered to be warm.

Re-insert the dipstick to its seat on the cap seal, remove the dipstick and note the oil level indication. If the fluid temperature is considered hot, the reading should be in the crosshatched area marked 'OK' (Between the two dots in the dipstick). If the fluid temperature is considered warm, the fluid level should be between the bottom of the dipstick and the lower dot.

If the fluid level indicates low, add sufficient fluid to bring the level to within the marks indicated for the appropriate fluid temperature.

The Automatic Transaxle fluid and filter should be changed, and the bands adjusted if the vehicle is used under severe usage, at 15,000 miles (24,000 km). Severe usage is defined as more than 50% operation in heavy city traffic during hot weather (above 90°F (32°C). A band adjustment and filter change should be made at the time of the transaxle fluid change, regardless of vehicle usage.

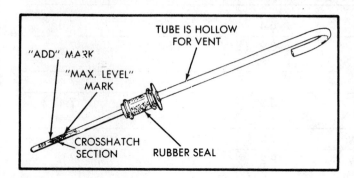

Dipstick assembly with internal vent (©Chrysler Corp.)

Parking Sprag Assembly Service

Automatic transaxles built between August 17, 1978, (Serial No. 6198-xxxx) and February 28, 1979, (Serial No. 6424-xxxx) have a Planet Pinion Ring, P/N 3681949, between the park sprag pawl and the transmission case.

When transaxle service requires removal of the parking sprag pawl, the spacer must be reinstalled to ensure proper pawl operation.

Leakage from Transaxle Vent

1978-1979 OMNI/HORIZON

Transmission fluid leakage from the differential vent, located on the extension, may be the result of a porous or cracked transfer bearing and oil seal retainer, a leaking oil seal or a nicked or broken "O" ring(s).

If vent leakage exists, it will be necessary to remove the transfer shaft bearing and oil seal retainer to allow complete inspection of the retainer for defects. If the retainer is found to be defective, it should be replaced with P/N 5222120. To improve sealing of the retainer, a second "O" ring has been added to the retainer on transaxles with a starting transaxle serial number of 6324-xxxx.

If either the transfer shaft or transfer shaft bearing and oil seal retainer are replaced, the procedure for setting transfer shaft end play contained in the transaxle assembly procedure must be followed.

The transfer shaft bearing cup should be reused whenever the transfer shaft bearing retainer is replaced, unless the cup is damaged during removal. The Transfer Shaft Oil Seal, P/N 5222015, and Transfer Shaft Bearing Retainer "O" Ring(s), P/N 6500169, must always be replaced.

Automatic Transaxle Gear and Bearing Noise Diagnosis and Repair

ALL 1978 THROUGH 1980 OMNI AND HORIZON EQUIPPED WITH AUTOMATIC TRANSAXLE

Abnormal gear and bearing noises must be distinguished to facilitate proper repair of the A-404 automatic transaxle for these conditions.

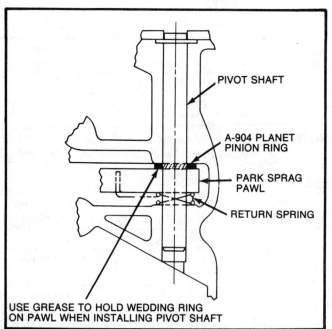

Use of planet pinion ring between park sprag pawl and case (©Chrysler Corp.)

Before attempting any repair for a noise condition on an A-404 automatic transaxle, a thorough road test should be performed to determine whether such repairs are warranted or unnecessary.

NOTE: The automatic transaxle by its design characteristics (a multi-angle gear drive unit), being located in front of the driver, develops higher levels of gear noise than conventional rear wheel drive units.

GEAR NOISE DIAGNOSIS

Planetary Gear Noise:

Low frequency gear whine which occurs in low and second gear but not in direct drive (high gear). This is similar in sound and pitch to rear wheel drive automatic transmission planetary gear noise.

Transfer Gear Noise:

High frequency gear whine most noticeable in direct drive or on coast below 25 mph. This noise may occur at any car speed.

Final Drive Gear Noise (Differential Gear):

Low frequency gear whine most noticeable above 40 mph in direct drive (high gear) or coast (the audible frequency is usually ⅓ that of transfer gear noise).

BEARING NOISE DIAGNOSIS

Bearing noise is a low frequency, rumbling type noise which is apparent throughout the entire speed range. Bearing noise is not torque sensitive, whereas gear noise usually changes with torque and speed.

NOTE: Do not mistake front wheel bearing noise for transaxle bearing noise. Transaxle bearing noise is not sensitive to steering direction, whereas front wheel bearing noise is sensitive to changes in steering direction.

Once the source of gear or bearing noise has been determined, the gear set or bearing must be replaced properly to assure correction of the noise condition.

NOTE: When gear set replacement is required, the replacement gears should be replaced in a matched set. This will ensure proper gear tooth contact and reduction in gear noise.
When bearing replacement is required, replace both the bearing cup and cone. The replacement of both the cup and cone will ensure maximum bearing life.

SERVICE COMPONENTS FOR CORRECTION OF ABNORMAL GEAR AND BEARING NOISE

Planetary Gear Set Components

Replace all components shown for planetary gear noise

Description	Part Number
Rear Annulus Gear	5222047
Rear Annulus Bearing Cone	5224300
Rear Annulus Bearing Cup	5224299
Rear Carrier Assembly	5205740
Sun Gear Assembly	5205733
Front Carrier Assembly	5222495
Front Annulus Assembly (Service Component Parts)	
Snap Rings (2)	5205725
Gear	5212201
Support Assembly	5212203

Transfer Gear Set Components

Replace all components shown for transfer gear set noise.

Description	Part Number
Transfer and Output Gear Set	4131048
Output Shaft Gear Bearing Cone	5224300
Output Shaft Gear Bearing Cup	5224299
Transfer Shaft Gear Bearing Cone	5224301
Transfer Shaft Gear Bearing Cup	5222259

Final Drive Gear Set Components

Replace all components shown for final drive gear noise.

Description	Part Number	
	3.48 (1978/1979 FEDERAL) (1980 ALL)	**3.67** (1978/1979 CALIF)
Ring Gear and Transfer Shaft Set	4131049 (54T gear-19T shaft)	4131050 (54T gear-18T shaft)
Transfer Shaft Bearing Cone	5224301	5224301
Transfer Shaft Bearing Cup	5222259	5222259

─── **CAUTION** ───

Ring Gear Service Package, P/N 4131050, has an identification groove on the O.D. of the gear teeth. Extreme care must be exercised to use this ring gear with the 18-tooth transfer shaft only.

NOTE: Improperly seated bearing cups and cones are subject to low mileage failures.

When servicing the transaxle for gear or bearing noise, particular attention should be given to the following points:
 a. All bearing adjustments, except transfer shaft bearing, must be made with no other gear train component interference or in gear mesh.
 b. Used (original) bearings may lose up to 50% of the original drag torque after break-in.
 c. When replacement of either the output or transfer gear is required, both gears should be replaced in a matched set.

After servicing the transaxle and installing the unit back in the vehicle, remember the transmission and differential have separate oil sumps and fill each sump to the proper specification. Check and adjust shift linkage and throttle cable adjustments.

Driveline Vibration— Automatic Transaxle

1979 & 1980 OMNI/HORIZON MODELS
Driveline vibration sensed at speeds from 25 to 35 mph, primarily felt through the steering column, may be caused by excessive transfer shaft runout.
If abnormal driveline vibration is present, use the following procedure to verify the condition.
 a. Road test vehicle. This condition cannot be effectively diagnosed on a hoist.
 b. Verify that condition occurs between 25 to 35 mph and in both drive and coast. This condition is extremey speed sensitive and is most noticeably transferred through the steering column.
Tire balance or C/V joint malfunction can also cause driveline vibration.

a. TIRE BALANCE—most often causes vibration above 40 mph

b. C/V JOINT—most noticeable in direct gear in drive, **not** coast, and in long smooth straight away areas.

When transfer shaft runout condition is incurred, it will be necessary to replace the following parts.

Description	Part Number	
	3.48 (1978/1979 FEDERAL) (1980 ALL)	**3.67** (1978/1979 CALIF)
Ring Gear and Transfer Shaft Set	4131049 (54T gear-19T shaft)	4131050 (54T gear-18T shaft)
Transfer Shaft Bearing Cone	5224301	5224301
Transfer Shaft Bearing Cup	5222259	5222259

————— CAUTION —————

Ring Gear Service Package, P/N 4131050, has an identification groove on the O.D. of the gear teeth. Extreme care must be exercised to use this ring gear with the 18-tooth transfer shaft only.

When servicing the transaxle for gear or bearing noise, particular attention should be given to the following points:

a. All bearing adjustments, except transfer shaft bearing, must be made with no other gear train components interference or in gear mesh.

b. Used (original) bearings may lose up to 50% of the original drag torque after break-in.

c. When replacement of either the output or transfer gear is required, both gears must be replaced in a matched set.

After servicing the transaxle and installing the unit back in the vehicle, remember the transmission and differential have separate oil sumps and fill each sump to the proper specification. Check and adjust shift linkage and throttle cable adjustment.

Transfer Shaft Bearing Service A-404, 413 & 470 Automatic Transaxle

OMNI, HORIZON, ARIES, RELIANT

Transfer shaft bearing and bearing cup retainer service in many instances may be difficult, due to the pressed fit of the transfer shaft bearing cup in the bearing retainer.

To improve service of the bearing and its retainer, a service package has been released. This package, PN #4205899, consists of:

1—Transfer Shaft Bearing Cone
1—Transfer Shaft Bearing Cup
1—Transfer Shaft Retainer
1—Transfer Shaft Oil Seal
2—Transfer Shaft Retainer 'O' Ring

This package will have the bearing cup pressed into the retainer. Use the package in its entirety, **Do Not Use Individual Parts.**

Simplifying the service of the transaxle should provide easier and better repair.

Automatic Transaxle Kickdown Band Adjustment & Application

1981 & 1982 OMNI/HORIZON, ARIES/RELIANT

The kickdown band adjustment for automatic transaxle assemblies released for the 1982 subject model vehicles differs from the 1981 model assemblies. The adjustment is changed from 2 turns for the 1981 assembly to 2¾ turns for the 1982 assembly.

Since 1981 and 1982 transaxle assemblies may have been interchanged on a limited number of late built 1981 and early built 1982 subject model vehicles, it is important that the transaxle part number be observed prior to adjusting the kickdown on these vehicles.

NOTE: Remember that kickdown band adjustment is accomplished by tightening the kickdown band adjustment screw to 72 inch pounds (8 N•m), then backing the screw out 2 or 2¾ turns depending on model year, then tightening the lock nut.

The list below gives the model year, transaxle part number, engine application, and kickdown band adjustment for the transaxles involved.

Model Year	Transaxle Part• No. & Model	Engine Application	Kickdown Band Adjustment
81	4269534-A-413	2.2L	2 turns
81	4269535-A-470	2.6L	2 turns
82	4207294-A-413	2.2L	2¾ turns
82	4207296-A-470	2.6L	2¾ turns
82	4269544-A-413 (Hi Alt.)	2.2L	2¾ turns

Automatic Transaxle—Transfer Gear Noise Reduction

1981 OMNI, HORIZON, ARIES, RELIANT; 1982 OMNI, HORIZON, ARIES, RELIANT, LEBARON & DODGE 400

On some 1981 and 1982 vehicles equipped with automatic transaxle, light gear noise may be noticeable between 40 and 60 miles per hour (MPH).

To properly determine this condition a road test must be performed. The gear noise will be high pitch in sound.

To reduce this noise a Rubber Shift Cable Sleeve, PN 4269881, must be installed on the gear shift selector cable between the selector cable mounting on the transaxle and the firewall using the following procedure:

1. Open hood and locate gear selector cable.
2. Unravel the sleeve from the center and wrap it around the gear shift selector cable.

NOTE: For best performance the sleeve should be centered between the firewall and the mounting bracket.

Speedometer pinion gear failure

1981 OMNI, HORIZON, ARIES, RELIANT; 1982 OMNI, HORIZON, ARIES, RELIANT, LEBARON, DODGE 400

Check the routing of the speedometer cable or cables. Make certain the cable is not misrouted or improperly bound by any component that may shorten its length or restrict its travel.

NOTE: Speedometer cable travel is the movement of the cable during engine rock (movement when shifting gears or accelerating).

Inspect the cable for damage due to improper routing and replace if necessary. Proper cable routing should prevent pinion gear failures in all applications. See the attached illustrations for proper cable routing.

On vehicles equipped with the 2.6 litre engine and speed control that have experienced speedometer pinion gear failure, the following procedure should be used:

REPAIR PROCEDURE

VEHICLES EQUIPPED WITH 2.6 LITRE ENGINE & SPEED CONTROL

Speedometer pinion gear failures on these vehicles may be due to engine movement pulling on the speedometer cable. In this application the cable length may be shortened due to routing. This results in side loading of the speedometer adapter, deflecting the drive pinion gear away from the drive shaft and failing of the pinion gear.

To correct this condition and insure unrestricted cable travel, install a new 5″ longer Speedometer Cable (PN 4047761).

Transaxle Oil Leak at Oil Pan Bolt

ALL 1981-1982 OMNI, HORIZON, RELIANT, ARIES; RAMPAGE, & LEBARON

Oil leak at transaxle oil pan which appears to be a result of improper application of RTV sealer at the oil pan.

The leak may be due to porosity of the case between the governor pressure port and the nearest oil pan tapped hole. The porosity may allow oil to leak from the governor circuit, into the oil pan bolt hole, and out around the bolt. Use the following procedure to correct.

Parts Required: RTV Sealer—PN 4026070
1. Remove the oil pan bolt from tapped hole.
2. Check for transmission fluid presence in the hole and on the bolt threads.
3. Thoroughly clean the bolt and the hole of transmission fluid. Use Mopar Brake and Carburetor Parts Cleaner, PN 3879889, or equivalent to clean the parts. Blow dry with compressed air.
4. Apply RTV sealer to the bolt threads, reinstall and torque bolt to 165 inch pounds (19 N•m).

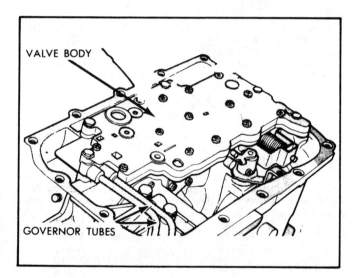

Location of possible fluid leakage at oil pan tapped bolt hole (©Chrysler Corp.)

Shift Cable Adjustment

1981 & 1982 OMNI, HORIZON, ARIES, RELIANT, LEBARON & DODGE 400 EQUIPPED WITH AUTOMATIC TRANSMISSION

Transmission cable adjustment procedures have been clarified for subject model vehicles equipped with automatic transmissions.

Note that the following adjustment methods vary depending on the type of shift (column versus console shift) and/or the model vehicle being serviced.

Adjustment Procedure

1. Place the gear shift lever in "P" (Park).

2. Loosen the cable retaining clamp on the cable mounting bracket of the automatic transmission.
3. Apply loading as follows while retightening the cable retaining clamp bolt to 90 inch pounds (10 N•m).
All models equipped with column shift—apply a 10 pound load in the forward direction on the cable housing isolator.
Aries, Reliant, LeBaron and Dodge 400 models equipped with console shift—apply a 10 pound minimum load in the forward direction on the console shift lever knob.
Omni and Horizon models equipped with console shift—apply a 10 pound load in the forward direction on the transmission lever (at the transmission).
4. Check adjustment by moving the shift lever through all positions while attempting to key start. Engine should start only in the Park and Neutral positions.

TRANSAXLE MANUFACTURING RUNNING CHANGES

1980 Changes

OVER-ALL RATIO

The optional 3.67 over-all ratio for California cars was dropped this year.

THROTTLE VALVE SPRING

The A404 throttle valve spring was replaced by the current A904/A727 throttle valve spring. This change will produce lower vehicle speed upshifts, improve fuel economy and reduce part throttle downshift sensitivity.

VALVE BODY AND TRANSFER PLATE ASSEMBLY

A new valve body and transfer plate assembly was introduced as a running change in preparation for major hydraulic changes that took place in the 1981 Model Year. The parts involved are: Valve body casting, transfer plate casting, steel plate, torque converter switch valve, throttle valve, regulator valve cover, and an additional cover on top of the valve body. These individual parts are not interchangeable with previous style parts, however, the complete valve body and transfer plate assembly can be used on all prior model transaxles.

OUTPUT SHAFT BEARINGS

The two output shaft bearings were moved six millimeters further apart in the over running clutch race. This improves the alignment between the input and output shafts, thus reducing output plug wear and rear clutch disc spline wear. The parts involved in this change are: over running clutch race, output shims, output shaft, transfer gear, and end cover. The transaxle assembly part number with these changes is 4207442 for the 1.7 liter engine application. The parts are not interchangeable with the previous style parts.

1981 Changes

In 1981 two new models were introduced into the transaxle family—one for the 2.2 liter Chrysler-built engine and one for the 2.6 liter MMC engine. The two new models each have a new case configuration, making a total of four transaxle cases.

WIDE RATIO

All models have a wide ratio planetary gear set (similar to the A-904). The gear ratios are 2.69 (low) and 1.55 (second) as compared to the previous standard ratios of 2.48 and 1.48. The wide ratio transaxle offers fuel economy improvements with little loss of performance because it is used with lower final drive ratios.

The wide ratio gear set consists of a new larger front annulus gear, support, snap ring, front carrier, stepped diameter sun gear, carrier thrust washers, and drive shell thrust plate, along with rear clutch discs and plates which have larger inside diameters. These parts are unique to the wide ratio and are not interchangeable with previous model transaxles.

DIFFERENTIAL BEARING RETAINER

The differential bearing retainer was changed from aluminum to steel. This adds stability and reduces deflection of the retainer, which is required for the higher torque engines. The steel retainer requires the use of shorter bolts, and maybe used on all prior model transaxles.

LOW-REVERSE ADJUSTABLE LEVER

A two piece low-reverse lever with an adjustment screw and a new narrow strut is used in transaxles for the 2.2 and 2.6 liter engines. The adjustment on this lever is 3½ turns backed off from 41 inch pounds. This lever cannot be used in prior model transaxles.

KICKDOWN LEVER

A 3.00 to 1 ratio kickdown lever instead of the 2.76 to 1 ratio is used in transaxles for the 2.2 and 2.6 liter engines. The adjustment for the 3.00 lever is 2 turns backed off from 72 inch pounds. A new configuration 2.76 ratio lever is used with the new servo. This lever may be used on all previous models.

TRANSFER GEAR SET

A new transfer gear set having a lower ratio was introduced for use with the 2.2 and 2.6 liter engines. The gear set consists of a 57 tooth output gear and a 52 tooth transfer shaft gear.

FINAL DRIVE GEARS

A new final drive gear set with a 20-61 tooth combination was introduced for use with the 2.6 liter engine. An 18-55 tooth combination set previously used in 1978 is used with the 2.2 liter engine.

REAR CLUTCH

All 1981 transaxles have a 3 disc rear clutch assembly instead of a 2 disc assembly as in prior models. The 3 disc assembly consists of a new retainer which is longer and a new waved snap ring, belleville spring, discs and plates which have a larger inside diameter to fit the wide ratio front annulus gear (mentioned above). These parts are not interchangeable with the previous style parts.

SUN GEAR DRIVE SHELL

The drive shell configuration was slightly modified to clear the longer rear clutch retainer. This part may be used on previous models.

FRONT CLUTCH

A redesigned front clutch assembly was introduced, consisting of a new retainer, smaller outside diameter piston, outer seal (same as A-904), thicker separator plates and a thinner reaction plate. A two disc front clutch with two separator plates between the discs is used in transaxles for the 1.7 liter and Simca engines. A three disc front clutch with one separator plate between each disc is used in transaxles for the 2.2 and 2.6 liter engines. These new front clutch parts are not interchangeable with the previous style parts.

VALVE BODY

The valve body was redesigned to improve shift quality, reduce internal leakage and to provide lower line pressures. Several of the changes were made as running 1980 model year changes as previously stated. The parts involved in the 1981 changes are: valve body, steel plate, shuttle valve and plug, regulator valve, springs and the additional bypass valve. None of these parts or the complete valve body and transfer plate assembly are interchangeable with prior model parts.

KICKDOWN PISTON

The diameter of the kickdown piston was increased. The large diameter is the same as the A-904 and uses the same seal ring; the small diameter is the same as the A-998 and uses the same seal ring. The inner piston, o-ring, snap ring and rod are A-904 parts. A new kickdown guide was also required. The case piston bore is larger, therefore making none of these parts interchangeable with the previous style parts. This change was made to accommodate the lower line pressures and to improve shift quality with the larger engines.

GOVERNOR

The size of the governor valves and governor body bores were redesigned to accommodate the lower line pressures and wide ratio which require different shift points. The parts are not interchangeable with the previous style parts.

LOW-REVERSE SERVO

The low-reverse servo was modified to give more stroke to improve neutral to reverse shift quality. The servo stem length was modified, cushion spring wire diameter was reduced, and the spring retainer replaced with the A-904 MMC spring retainer. These parts, as a group, can be used on prior models.

DIFFERENTIAL COVER AND FILL PLUG

The differential cover fill plug thread boss was changed from a welded-on boss to one extruded from the sheet metal cover. A new length plug and washer assembly is used. These parts reduce plug leaks and can be used on all prior model transaxles.

GOVERNOR

The governor for the 1.7 liter engine application was modified to reduce the wide open throttle shift speeds. The secondary valve was replaced by the Simca valve which was a hole through its center and the spring load decreased. This governor can be used in all wide ratio transaxles for the 1.7 liter engine.

1982 Changes

KICKDOWN PISTON

The length of the bore for the inner piston was increased by 2.5mm. The purpose was to improve shift quality throughout the operating temperature range. With this servo, the Kickdown Band adjustments are:

Engine	Trans. Part No.	Turns	Trans Model
2.2L	4207294	2¾	A-413
2.6L	4207296	2¾	A-470
1.7L	4207293	3	A-404
2.2L	4269544 (Hi Alt.)	2¾	A-413

This piston may be used in 1981 model transaxles with the above kickdown band adjustments.

VALVE BODY AND TRANSFER PLATE

The regulator valve line pressure plug and sleeve were eliminated and the regulator valve throttle pressure plug replaced with a longer plug having a different diameter. With this system, a new regulator valve throttle plug spring is required. The transfer plate cast passages were modified in the area of the regulator valve. The 1-2 and 2-3 shift valve springs were changed to the springs used in 1980 Model Year. These individual parts are not interchangeable with previous style parts, however, the complete valve body and transfer plate assembly may be used on 1981 Model Year transaxles.

TRANSFER GEAR SET

A new transfer gear set was introduced in 1982 for high altitude and Mexican applications. The tooth combination is a 53 tooth output shaft gear and a 56 tooth transfer shaft gear.

GOVERNOR VALVE

The primary governor valve used in the high altitude and Mexican applications was made lighter by making a stepped diameter hole in it, this increases the shift speeds.

TEFLON SEAL RINGS

Teflon seal rings for the Kickdown and Accumulator pistons and the Input Shaft were incorporated. They may be used in previous model transaxles.

DIFFERENTIAL ASSEMBLY

The 18-55 tooth combination was replaced with the 20-61 tooth combination.

FRONT PUMP SEAL

A Vamac (black) front pump seal replaced the Silicone (orange) seal. This seal may be used on all prior model transaxles.

DIFFERENTIAL COVER AND PLUG

The differential cover and plug was changed from the threaded plug to a push-in rubber plug, similar to conventional rear axles. This cover and plug can be use on all prior model transaxles.

EXTENSION

The extension speedo bore was re-designed giving a machined pilot bore for the and of the speedometer pinion. This extension can be used on all prior model transaxles.

MAIN OIL PAN BOLT

A nylon sealing patched bolt was incorporated for use in the main oil pan screw hole which is next to the governor pressure circuit. This bolt reduces leakage, due to porosity, between the governor circuit and the tapped hole thus reducing external leaks.

1983 Changes
COMMON TRANSMISSION AND DIFFERENTIAL OIL SUMP

This change makes a common reservoir of oil for the differential and the transmission. The purpose of "Common Sump" is to reduce the possibility of transfer shaft bearing failures due to low differential oil level. The intent is to signal the customer via changes in transmission performance (slipping) when the oil level is too low. If the oil level is not corrected, a rear clutch failure will occur before a more serious and costly transfer shaft or differential bearing failure occurs. The possibility of leaks is reduced since the differential cover plug and extension vent are eliminated.

Following is a description of the parts that changed:

CASE—There is a machined passage between the transmission and differential cavities. This case is not interchangeable with previous models.

TRANSFER SHAFT BEARING RETAINER—One O-ring groove and the transfer shaft seal has been eliminated. The outside and inside diameter of the hub where the seal went is reduced in size to 17.8mm. This part is not interchangeable with previous model parts.

PUMP HOUSING GASKET—The gasket was modified by adding a torque converter circuit bleed hole which feeds oil to the differential cavity under all conditions, and a cut out to match the case passage hole. This gasket may be used on all previous model transaxles.

PUMP HOUSING—The casting was changed to accommodate the revised gasket and to make a space for oil to flow between the differential and transmission cavities.

EXTENSION—The machined vent hole and the vent were eliminated. This extension is not interchangeable with previous models.

DIFFERENTIAL COVER—The fill hole and fill plug were eliminated. This cover cannot be used on previous models.

DIFFERENTIAL THRUST WASHERS—The thrust washers in the differential assembly were changed from free spinning washers to tabbed washers to reduce the possibility of dirt contamination to the transmission as the washers were. Differential carriers with these washers may be used in all previous model transaxles.

VALVE BODY OIL FILTER—A new style body filter which has the filter material in a partially enclosed plastic and steel case is used. This style filter does not have to be completely immersed in oil to avoid sucking air. With common sump the filter at times may be tipped out of the oil reservoir and exposed to air within the transmission.

A gasket is used between the filter and the transfer plate.

This new filter requires a new deeper main oil pan.

This filter can be used on all prior model transaxles but must be accompanied by the pan and gasket.

DIFFERENTIAL OIL RETAINER—A new different oil retainer with a revised shape and smaller inside diameter which improves lubrication to the differential half shaft journal is used. The left side hub of the differential carrier was reduced in length by 1mm to accommodate this retainer. The oil retainer cannot be used in previous model transaxles.

OIL DIPSTICK—A new length oil dipstick is required for the common sump transaxles. The "FULL" mark is 101mm from the washer face, 1982 models had the "FULL" mark 90mm from the washer face.

COOLER LINE FITTING

The cooler line fitting was redesigned in order to reduce cooler line leakage. The new fitting has a nippled extension on it to accommodate a rubber hose and clamp instead of the inverted flare tube joint. This fitting may be used on all previous model transaxles with the proper rubber hoses.

DIFFERENTIAL GEAR SETS

The 19-54 tooth combination gear set was replaced by a higher capacity 21-60 tooth combination gear set. This gear set may be used to replace the 19-54 set on prior models.

REAR SERVO SEAL

The material of the rear servo seal was changed to Viton to improve its resistance to high temperatures. This seal may be used in all prior model transaxles.

1983 Model transaxles may be used to service all prior models, it is important that the 1983 Models have the correct dipstick.

TRANSAXLE GOVERNOR REPLACEMENT
A-404, A-413, A-415 and A-470

Model Year	Transmission P/N	Gov. Body	Primary Valve	Secondary Valve	Spring
1978	5212256 5222187	4269501	5222457 or 5224390	5222458 or 5224391	5222459-Plain
1979	5224023 5224196	4269501	5222457 or 5224390	5222458 or 5224391	5222459-Plain
1980	5224442 4207442	4269501	5222457 or 5224390	5222458 or 5224391	5222459-Plain
1981	5224442 4207442	4269501	5222457 or 5224390	5222458 or 5224391	5222459-Plain

NOTE: Governor repair package 4186092 can be used to service the above transmissions. The size of the governor valves and governor body bores were redesigned for use with wide ratio and low line pressure transmissions introduced in 1981. The new parts *shown below* are not interchangeable with previous parts.

Model Year	Transmission P/N	Gov. Body	Primary Valve	Secondary Valve	Spring
1981	4207088 4207215	4207055	4207186-Solid	4207183-Hole	4207193-Blue
1981	4269522-4269523 4269534-4269535 5224473-5224474 5224475	4207055	4207186-Solid	4207185-Solid	4207193-Blue
1982	4207293	4207055	4207186-Solid	4207183-Hole	4207193-Blue
1982	4207294 4207296	4207055	4207186-Solid	4207185-Solid	4207193-Blue
1982	4269544 4269651	4207055	4269641-2 Dia Hole	4207183-Hole	4207194-Orange

TROUBLE DIAGNOSIS

CLUTCH AND BAND APPLICATION CHART
A-404, A-413, A-415 and A-470 TorqueFlite Transaxles

Lever Position	Clutches			Bands	
	Front	Rear	Over-running	(Kickdown) Front	(Low-Rev.) Rear
P—PARK	—	—	—	—	—
R—REVERSE	X	—	—	—	X
N—NEUTRAL	—	—	—	—	—
D—DRIVE					
First	—	X	X	—	—
Second	—	X	—	X	—
Direct	X	X	—	—	—
2—SECOND					
First	—	X	X	—	—
Second	—	X	—	X	—
1—LOW (First)	—	X	—	—	X

—Not applicable

CHILTON'S THREE "C's" TRANSAXLE DIAGNOSIS CHARTS
A-404, A-413, A-415 and A-470 TorqueFlite Transaxles

Condition	Cause	Correction
Harsh engagement from Neutral to D	a) Engine idle speed too high b) Valve body malfunction c) Hydraulic pressure too high d) Worn or faulty rear clutch	a) Adjust to specification b) Clean or overhaul c) Adjust to specifications d) Overhaul rear clutch
Harsh engagement from Neutral to R	a) Low/reverse band mis-adjusted (A-413, A-470) b) Engine idle speed too high c) Low/reverse band worn out d) Low/reverse servo, band or linkage malfunction e) Hydraulic pressure too high f) Worn or faulty rear clutch	a) Adjust to specifications b) Adjust to specifications c) Overhaul d) Overhaul e) Adjust to specifications f) Overhaul rear clutch
Delayed engagement from Neutral to D	a) Hydraulic pressure too low b) Valve body malfunction c) Malfunction in low/reverse servo, band or linkage d) Low fluid level e) Manual linkage adjustment f) Oil filter clogged g) Faulty oil pump h) Bad input shaft seals i) Idle speed too low j) Bad reaction shaft support seals k) Bad front clutch l) Bad rear clutch	a) Adjust to specification b) Clean or overhaul c) Overhaul d) Add as required e) Adjust as required f) Change filter and fluid g) Overhaul pump h) Replace seal rings i) Adjust to specifications j) Replace seal rings k) Overhaul l) Overhaul
Delayed engagement from Neutral to R	a) Low/reverse band mis-adjusted (A-413, A-470) b) Hydraulic pressures too low c) Low/reverse band worn out d) Valve body malfunction or leakage e) Low/reverse servo, band or linkage malfunction f) Low fluid level g) Manual linkage adjustment h) Faulty oil pump i) Worn or broken input shaft seal rings j) Aerated fluid k) Idle speed too low l) Worn or broken reaction shaft support seal rings m) Worn or faulty front clutch n) Worn or faulty rear clutch o) Oil filter clogged	a) Adjust to specifications b) Adjust to specifications c) Overhaul d) Clean or overhaul e) Overhaul f) Add as required g) Adjust as required h) Overhaul pump i) Replace seal rings j) Check for overfilling k) Adjust to specifications l) Replace support seal rings m) Overhaul n) Overhaul o) Change filter and fluid
Runaway upshift	a) Hydraulic pressure too low b) Valve body malfunction c) Low fluid level d) Oil filter clogged e) Aerated fluid f) Manual linkage adjustment g) Bad reaction shaft support seals h) Malfunction in kickdown servo, band or linkage i) Bad front clutch	a) Adjust to specifications b) Clean or overhaul c) Add as required d) Change filter and fluid e) Check for overfilling f) Adjust as required g) Replace seal rings h) Overhaul i) Repair as needed

2-15

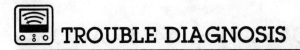

CHILTON'S THREE "C's" TRANSAXLE DIAGNOSIS CHARTS
A-404, A-413, A-415 and A-470 TorqueFlite Transaxles

Condition	Cause	Correction
No upshift	a) Hydraulic pressure too low	a) Adjust to specifications
	b) Valve body malfunction	b) Clean or overhaul
	c) Low fluid level	c) Add as required
	d) Manual linkage adjustment	d) Adjust as required
	e) Incorrect throttle linkage adjustment	e) Adjust as required
	f) Bad seals on governor support	f) Replace seals
	g) Bad reaction shaft support seals	g) Replace seal rings
	h) Governor malfunction	h) Service or replace unit
	i) Malfunction in kickdown servo, band or linkage	i) Overhaul
	j) Bad front clutch	j) Overhaul
3-2 Kickdown runaway	a) Hydraulic pressure too low	a) Adjust to specifications
	b) Valve body malfunction	b) Clean or overhaul
	c) Low fluid level	c) Add as required
	d) Aerated fluid	d) Check for overfilling
	e) Incorrect throttle linkage adjustment	e) Adjust as required
	f) Kickdown band out of adjustment	f) Adjust to specifications
	g) Bad reaction shaft support seals	g) Replace seal rings
	h) Malfunction in kickdown servo, band or linkage	h) Overhaul
	i) Bad front clutch	i) Overhaul
No kickdown or normal downshift	a) Valve body malfunction	a) Clean or overhaul
	b) Incorrect throttle linkage adjustment	b) Adjust as required
	c) Governor malfunction	c) Service or replace unit
	d) Malfunction in kickdown servo, band or linkage	d) Service or replace parts as required
Shifts erratic	a) Hydraulic pressure too low	a) Adjust to specifications
	b) Valve body malfunction	b) Clean or overhaul
	c) Low fluid level	c) Add as required
	d) Manual linkage adjustment	d) Adjust as required
	e) Oil filter clogged	e) Change filter and fluid
	f) Faulty oil pump	f) Overhaul oil pump
	g) Aerated fluid	g) Check for overfilling
	h) Incorrect throttle linkage adjustment	h) Adjust as required
	i) Bad seals on governor support	i) Replace seals
	j) Bad reaction shaft support seals	j) Replace seal rings
	k) Governor malfunction	k) Service or replace unit
	l) Malfunction in kickdown servo, band or linkage	l) Overhaul
	m) Bad front clutch	m) Overhaul
Slips in forward drive positions	a) Hydraulic pressure too low	a) Adjust to specifications
	b) Valve body malfunction	b) Clean or overhaul
	c) Low fluid level	c) Add as required
	d) Manual linkage adjustment	d) Adjust as required
	e) Oil filter clogged	e) Change filter and fluid
	f) Faulty oil pump	f) Overhaul pump
	g) Bad input shaft seals	g) Replace seal rings
	h) Aerated fluid	h) Check for overfilling
	i) Incorrect throttle linkage adjustment	i) Adjust as required

CHILTON'S THREE "C's" TRANSAXLE DIAGNOSIS CHARTS
A-404, A-413, A-470 TorqueFlite Transaxles

Condition	Cause	Correction
Slips in forward drive positions	j) Overrunning clutch not holding k) Bad rear clutch	j) Overhaul or replace k) Overhaul
Slips in reverse only	a) Hydraulic pressure too low b) Low/reverse band out of adjustment c) Valve body malfunction d) Malfunction in low/reverse servo, band or linkage e) Low fluid level f) Manual linkage adjustment g) Faulty oil pump h) Aerated fluid i) Bad reaction shaft support seals J) Bad front clutch	a) Adjust as required b) Adjust to specifications c) Clean or overhaul d) Service or replace parts as required e) Add as required f) Adjust as required g) Overhaul pump h) Check for overfilling i) Replace seal rings j) Overhaul
Slips in all positions	a) Hydraulic pressure too low b) Valve body malfunction c) Low fluid level d) Oil filter clogged e) Faulty oil pump f) Bad input shaft seals g) Aerated fluid	a) Adjust as required b) Clean or overhaul c) Add as required d) Change fluid and filter e) Overhaul pump f) Replace seal rings g) Check for overfilling
No drive in any position	a) Hydraulic pressure too low b) Valve body malfunction c) Low fluid level d) Oil filter clogged e) Faulty oil pump f) Planetary gear sets broken or seized	a) Adjust to specifications b) Clean or overhaul c) Add as required d) Change filter and fluid e) Overhaul pump f) Replace affected parts
No drive in forward drive positions	a) Hydraulic pressure too low b) Valve body malfunction c) Low fluid level d) Bad input shaft seals e) Overrunning clutch not holding f) Bad rear clutch g) Planetary gear sets broken or seized	a) Adjust to specifications b) Clean or overhaul c) Add as required d) Replace seal rings e) Overhaul or replace f) Overhaul g) Replace affected parts
No drive in reverse	a) Hydraulic pressure too low b) Low/reverse band out of adjustment c) Valve body malfunction d) Malfunction in low/reverse servo, band or linkage e) Manual linkage adjustment f) Bad input shaft seals g) Bad front clutch h) Bad rear clutch i) Planetary gear sets broken or seized	a) Adjust to specifications b) Adjust to specifications c) Clean or overhaul d) Overhaul e) Adjust as required f) Replace seal rings g) Overhaul h) Overhaul i) Replace affected parts
Drives in neutral	a) Valve body malfunction b) Manual linkage adjustment c) Insufficient clutch plate clearance d) Bad rear clutch e) Rear clutch dragging	a) Clean or overhaul b) Adjust as required c) Overhaul clutch pack d) Overhaul e) Overhaul

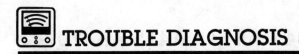

CHILTON'S THREE "C's" TRANSAXLE DIAGNOSIS CHARTS
A-404, A-413, A-415 and A-470 TorqueFlite Transaxles

Condition	Cause	Correction
Drags or locks	a) Stuck lock-up valve b) Low/reverse band out of adjustment c) Kickdown band adjustment too tight d) Planetary gear sets broken or seized e) Overrunning clutch broken or seized	a) Clean or overhaul b) Adjust to specifications c) Adjust to specifications d) Replace affected parts e) Overhaul or replace
Grating, scraping or growling noise	a) Low/reverse band out of adjustment b) Kickdown band out of adjustment c) Output shaft bearing or bushing bad d) Planetary gear sets broken or seized e) Overrunning clutch broken or seized	a) Adjust to specifications b) Adjust to specifications c) Replace d) Replace affected parts e) Overhaul or replace
Buzzing noise	a) Valve body malfunction b) Low fluid level c) Aerated fluid d) Overrunning clutch inner race damaged	a) Clean or overhaul b) Add as required c) Check for overfilling d) Overhaul or replace
Hard to fill, oil blows out filler tube	a) Oil filler clogged b) Aerated fluid c) High fluid level d) Breather clogged	a) Change filter and fluid b) Check for overfilling c) Bad converter check valve d) Clean, change fluid
Transmission overheats	a) Engine idle speed too high b) Hydraulic pressure too low c) Low fluid level d) Manual linkage adjustment e) Faulty oil pump f) Kickdown band adjustment too tight g) Faulty cooling system h) Insufficient clutch plate clearance	a) Adjust to specifications b) Adjust to specifications c) Add as required d) Adjust as required e) Overhaul pump f) Adjust to specifications g) Service vehicle's cooling system h) Overhaul clutch pack
Harsh upshift	a) Hydraulic pressure too low b) Incorrect throttle linkage adjustment c) Kickdown band out of adjustment d) Hydraulic pressure too high	a) Adjust to specifications b) Adjust as required c) Adjust to specifications d) Adjust to specifications
Delayed upshift	a) Incorrect throttle linkage adjustment b) Kickdown band out of adjustment c) Bad seals on governor support d) Bad reaction shaft support seals	a) Adjust as required b) Adjust as required c) Replace seals d) Replace seal rings

CHILTON'S THREE "C's" TRANSAXLE DIAGNOSIS CHARTS
A-404, A-413, A-415 and A-470 TorqueFlite Transaxles

Condition	Cause	Correction
Delayed upshift	e) Governor malfunction f) Malfunction in kickdown servo, band or linkage g) Bad front clutch	e) Service or replace unit f) Overhaul g) Overhaul

DIAGNOSIS TEST

Automatic Transaxle problems are caused by five general conditions. These are:

1. Poor engine performance (out of tune, improperly maintained).
2. Incorrect fluid levels (too low or too high).
3. Improper linkage, band or control pressure adjustments.
4. Malfunctions in the hydraulic system.
5. Actual break down of mechanical parts.

Two procedures can be followed in checking out a vehicle with transaxle problems. If the vehicle can be driven, follow this procedure:

1. Check the level of the transaxle fluid. Note the condition, color and appearance of the fluid. (Air bubbles, metal particles, burnt smell, etc.)
2. If the complaint was that the shifts were delayed, erratic or harsh, adjust the throttle and shift linkages before road testing. Check for obvious faults like broken linkage, etc.
3. If the complaint was that the acceleration is slow, or sluggish, or if an unusual amount of throttle is needed to keep up road speed, then a stall test should be performed.
4. Road test vehicle, preferably with the vehicle operator doing the driving. Observe for any malfunction, noting speed, load, any unusual noises or vibrations, gear range, etc.
5. Perform hydraulic pressure test.
6. Perform air pressure test of clutch and band operation.

If the vehicle cannot be driven, follow the procedure shown below.

1. Check the level of the transaxle fluid. Note the condition, color and appearance of the fluid. (Air bubbles, metal particles, burnt smell, etc.)
2. Check for broken or disconnected throttle linkage.
3. Check for broken cooler lines, and loose or missing pressure port plugs, thus causing massive fluid loss.
4. Raise the car, start engine, shift into gear and check:

(a.) If drive shafts turns but not the wheels, then the problem is in the differential or axle shaft, and not in the transaxle.

(b.) If drive shafts do not turn, and transaxle is noisy, immediately stop engine, remove the pan and check for debris lying in the pan. If debris is not found, then the transaxle must be removed. Check for broken drive plate, attaching bolts, broken converter hub shafts or oil pump.

(c.) If drive shafts do not turn and transaxle is not noisy, perform a hydraulic pressure test to determine if the problem is caused by a hydraulic or mechanical component.

FLUID LEVEL AND CONDITION

Since the torque converter fills in both the Park and Neutral positions, place the selector in Park and allow the engine to idle. The fluid should be at operating temperature (180°). Check the dipstick. The fluid level is correct if it is between the "Full" and

"Add" marks. Do not overfill, or the fluid will be churned into foam and cause the same symptoms of low fluid level and can lead to transmission overheating, and fluid being forced from the vent. Check the condition of the fluid, examining the fluid for metal bits or friction material particles. A milky appearance indicates water contamination, possibly from the cooling system. Check for a burned smell. If there is any doubt about the condition of the fluid, drain out a sample for a better check.

MANUAL LINKAGE

A quick way to check the manual linkage adjustment is to check the operation of the neutral safety switch. If the starter operates in both Park and Neutral, the manual linkage is properly adjusted. If not, either the neutral switch is bad or the linkage needs adjustment. See the section "On Car Services" for the procedure.

THROTTLE CABLE

The throttle cable adjustment is critical to the proper operation of the transmission, because it controls a valve in the valve body which determines shift speed, shift quality and part throttle downshift sensitivity. If the setting is to short, early shifts and slippage between shifts may happen. If the setting is too long, shifts may be delayed and the part throttle downshifts will be sensitive.

Diagnostic Test Sequence

The following order should be used when trouble shooting a TorqueFlite transaxle.

1. Road Test
2. Hydraulic Pressure Test
3. Stall Test
4. Air Pressure Test
5. Check diagnosis charts for probable cause of malfunction.

ROAD TEST

Before road testing, check for obvious faults, such as low fluid level, disconnected linkage, bad fluid leaks, etc.

During the road test, observe the engine performance. An out of tune engine will often cause symptoms mistaken for transmission difficulties. It is a good idea to let the vehicle operator drive so that the service technician is free to record any problems. The transaxle should be shifted to each position to check for slipping or any variation in shifting. Take note whether the shifts feel spongy or are too harsh. Record the speed at which the downshifts and upshifts occur. Listen for engine speed flare up or slippage. In most cases the clutch or band that is slipping can be determined by checking how the transaxle operates in each shift position.

For example, the rear clutch is applied in both "D" first gear and "1" first gear positions, but the overrunning clutch is applied "D" first and the low and reverse band is applied in "1" first. If the transaxle slips in "D" first gear but does not slip in "1" first gear, then the overrunning clutch must be the unit that is slipping.

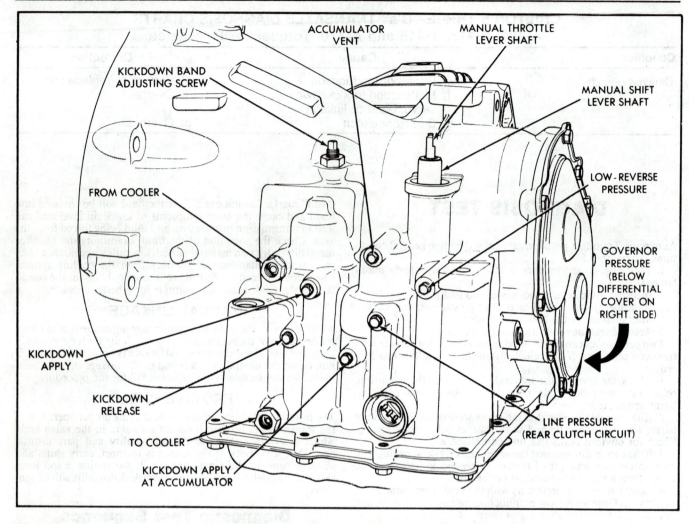

Location of pressure point taps, manual control and band adjustment screw, left side of transaxle (©Chrysler Corp.)

In the same way, if the transaxle slips in any two forward gears, then the rear clutch is the slipping unit. Using this same procedure, notice that the rear clutch and front clutch are applied in "D" third gear. So if the slippage is in the third gear, then either the front clutch or rear clutch is slipping. By shifting to another gear which does not use one of those units, the unit that is slipping can be determined. This process of elimination can be used to detect any unit which slips as well as to confirm proper operation of good units.

Although a road test analysis can usually diagnose the slipping units, the actual cause usually cannot be determined. Any of the above conditions can be caused by leaking hydraulic circuits or sticking valves. So before attempting to disassemble the transaxle, a hydraulic pressure test should be done.

CONTROL PRESSURE TEST

Before starting on a pressure test, make sure the fluid level is correct and the linkage is adjusted properly. Fluid should be at operating temperature (approx. 180°F).

Test Procedure

1. Hook up engine tachometer and route wires so that it can be read under the car.

2. Raise vehicle on hoist so that the front wheels can turn. It may be helpful to disconnect the throttle valve and shift controls so that they can be shifted from under the vehicle. Two size gauges are needed, one 150 psi and the other a 300 or 400 psi. The higher pressure gauge is required for the "reverse test."

TEST ONE (Selector in "1")
The purpose of this test is to check pump output, the pressure regulation and also to check on the condition of the rear clutch and servo hydraulic circuits.

1. Hook up gauges to the "line pressure" port and "low-reverse pressure" port (rear servo).
2. Adjust engine speed to 1000 rpm.
3. Shift into "1" position (selector lever on transaxle all the way forward).
4. Read pressures on both gauges as the throttle lever on the transaxle is moved from the full forward position to full rearward position.
5. Line pressure should read 60 to 66 psi (1981 and later—52 to 58 psi), with the throttle lever forward and it should gradually increase as the lever is moved rearward, to 97 to 103 psi (1981 and later—80 to 88 psi).
6. Rear servo pressure should read the same as the line pressure to within 3 psi.

TEST TWO (Selector in "2")

The purpose of this test is to check pump output, the pressure regulation and also to check on the condition of the rear clutch and lubrication hydraulic circuits.

1. Hook up gauge to "line pressure" port and, with a "tee" fitting, hook into the lower cooling line to read "lubrication" pressure.
2. Adjust engine speed to 1000 rpm.
3. Shift into "2" position. (This is one "detent" rearward from the full forward position).
4. Read pressures on both gauges as the throttle lever on the transaxle is moved from the full forward position to full rearward position.
5. Line pressure should read 60 to 66 psi (1981 and later—52 to 58 psi), with the throttle lever forward and it should gradually increase as the lever is moved rearward, to 97 to 103 psi (1981 and later—80 to 88 psi).
6. Lubrication pressure should read 10 to 25 psi with the lever forward and 10 to 35 psi with the lever rearward.

TEST THREE (Selector in "D")

The purpose of this test is to check pump output, the pressure regulation and also to check on the condition of the rear clutch and front clutch hydraulic circuits.

1. Attach gauges to the "line pressure" port and "kickdown *release*" port (front servo *release*).
2. Adjust engine speed to 1600 rpm.
3. Shift into "D" position. (This is two "detents" rearward from the full forward position of the selector lever.)
4. Read pressures on both gauges as the throttle lever on the transaxle is moved from the full forward position to the full rearward position.
5. Line pressure should read 60 to 66 psi (1981 and later—52 to 58 psi), with the throttle lever forward and it should gradually increase as the lever is moved rearward.
6. The "kickdown release" port is pressurized only in direct drive and should be the same as the line pressure within 3 psi, up to the downshift point.

TEST FOUR (Selector in "Reverse")

The purpose of this test is to check pump output, pressure regulation and the condition of the front clutch and rear hydraulic circuits. Also, at this time, a check can be made for leakage into the rear servo due to case porosity, cracks, valve body or case warpage which can cause reverse band burn out.

1. Attach the 300 psi gauge to the "low-reverse pressure" port (rear servo).
2. Adjust engine speed to 1600 rpm.
3. Shift into "R" position. (This is four "detents" rearward from the full forward position of the selector lever.)
4. Read pressure on the gauge. It should read 176 to 180 psi (1981 and later—160 to 180 psi), with the throttle lever forward and it should gradually increase as the lever is moved rearward to 270 to 280 psi (1981 and later—250 to 300 psi).
5. Move the selector lever on the transaxle to "D" position. The rear servo pressure should drop to zero since even a little pressure can cause the band to apply and burn up.

Analyzing the Pressure Test

1. If the pressure readings from minimum to maximum are correct, the pump and pressure regulator are working properly.
2. If there is low pressure in "D 1, 2" but correct pressure in "R," then there is leakage in the rear clutch circuit.
3. If there is low pressure in "D and R" but correct pressure in "1," then there is leakage in the front clutch circuit.
4. If there is low pressure in "R and 1" but correct pressure in "2," then there is leakage in the rear servo circuit.
5. If there is low line pressure in all positions then there could be a band pump, stuck pressure regulator valve or clogged filter.

GOVERNOR PRESSURE

The governor pressure only needs to be tested if the transaxle shifts at the wrong vehicle speeds, when the throttle cable adjustment has been verified correct.

1. Connect a gauge that will read 0 to 150 psi to the "governor pressure" port which is on the lower side of the transaxle case, below the differential cover.
2. With the engine running and in third gear, read the pressures and compare the speeds shown in the chart.

If the pressures are wrong at a given speed, the governor valves are probably sticking. The pressure should respond smoothly to any change in rpm and should drop to 0 to 3 psi when the vehicle is stopped. If there is high pressure at standstill (more than 3 psi) then the transaxle will be prevented from downshifting.

THROTTLE PRESSURE

The TorqueFlite Transaxle has no provision for testing throttle pressure with a gauge. The only time incorrect throttle pressure should be suspected is if the part throttle upshift speeds are either too slow in coming or occur too early in relation to vehicle speeds. Engine "runaway" on either upshifts or downshifts can also be an indicator of incorrect (low) throttle pressure setting. The throttle pressure really should not be adjusted until the throttle cable has been checked and adjustment has been verified to be right.

STALL TEST

The stall test involves determining the maximum engine speed obtainable at full throttle in "D" position. This test checks on the torque converter's stator clutch and the clutches' holding ability. The transaxle oil level should be checked and temperatures brought to normal operating levels.

——————— CAUTION ———————

Never allow anyone to stand in front of the car when performing a stall test. Both the parking and service brakes must be fully applied during the test.

NOTE: Do not hold the throttle open any longer than necessary and never longer than five seconds at a time. If more than one stall test is required, operate the engine at 1,000 rpm in neutral for at least 20 seconds to cool the transmission fluid between runs.

1. After checking the fluid level, blocking wheels and setting the brakes, use the tachometer (previously hooked up for the pressure test) to record the maximum engine rpm by opening the throttle completely in drive, "D."
2. If engine speed exceeds 2000-2400 rpm, release accelerator immediately. This indicates that the transaxle is slipping.
3. Shift transaxle into Neutral; operate engine for at least 20 seconds at 1,000 rpm to cool the transmission fluid before shutting off engine.

STALL SPEED TOO HIGH

If the stall speed is more than 200 rpm above specifications, clutch slippage is the likely problem. The hydraulic pressure tests and air pressure will help pinpoint the problem unit.

STALL SPEED TOO LOW

Low stall speeds with a properly tuned engine would indicate a problem in the torque converter stator clutch. This condition should be confirmed by road testing prior to converter replacement.

If the stall speed is 250 to 350 rpm below what it should be and the vehicle runs properly on the highway, but has poor acceleration through the gears, then the stator over-running clutch is slipping, and the converter must be replaced.

If both the stall speed and acceleration are normal, but an unusually large amount of throttle is needed to keep up highway speeds, then the stator clutch has seized and the converter must be replaced.

AUTOMATIC SHIFT SPEEDS AND GOVERNOR PRESSURE CHART
(Approximate Miles and Kilometers Per Hour)

1980	—Carline— Federal (M,Z)		—Carline— California (M,Z)	
Engine (liter)	1.7		1.7	
Axle ratio	3.48		3.67	
Throttle Minimum	MPH	Km/h	MPH	Km/h
1-2 Upshift	8-15	13-24	7-14	11-23
2-3 Upshift	11-21	18-34	10-19	16-31
3-1 Downshift	8-15	13-24	7-14	11-23
Throttle Wide Open				
1-2 Upshift	32-44	51-71	30-41	48-66
2-3 Upshift	52-65	84-105	48-60	77-97
Kickdown Limit				
3-2 WOT Downshift	48-61	77-98	44-57	71-92
3-2 Part Throttle Downshift	35-50	56-80	32-46	50-74
3-1 WOT Downshift	30-37	48-60	28-34	45-55
Governor Pressure①				
15 psi	21-23	34-37	19-21	31-34
40 psi	35-41	56-66	32-38	51-61
60 psi	50-56	80-90	46-52	74-84

1981	—Carline— M and Z		—Carline— M, Z, P and D	
Engine (liter)	1.7		2.2 and 2.6	
Overall Top Gear Ratio	3.48		2.78	
Throttle Minimum	MPH	Km/h	MPH	Km/h
1-2 Upshift	9-14	14-23	9-13	14-21
2-3 Upshift	15-18	24-29	14-17	23-27
3-1 Downshift	9-13	14-21	8-12	13-19
Throttle Wide Open				
1-2 Upshift	27-39	43-63	33-45	53-72
2-3 Upshift	51-62	82-100	57-69	92-111
Kickdown limit				
3-2 WOT Downshift	48-59	77-95	54-66	87-106
3-2 Part Throttle Downshift	36-47	58-76	42-53	68-85
3-1 WOT Downshift	27-35	43-56	29-37	47-60
Governor Pressure①				
15 psi	22-23	35-37	21-23	34-37
60 psi	52-59	83-95	59-66	95-106

1982	—Carline— M and Z		—Carline— M,Z,P,D,C,V		—Carline— M,Z,P,D,C,V High Altitude	
Engine (Liter)	1.7L.		2.2 and 2.6L.		2.2L.	
Overall Top Gear Ratio	3.48		2.78		3.22	
Throttle Minimum	MPH	km/h	MPH	km/h	MPH	km/h
1-2 Upshift	11-15	18-24	10-14	16-23	11-15	18-24
2-3 Upshift	16-21	26-34	15-20	24-32	16-21	26-34
3-1 Downshift	11-14	18-23	10-13	16-21	16-22	26-35
Throttle Wide Open						
1-2 Upshift	33-39	53-63	37-44	60-71	33-38	53-61
2-3 Upshift	55-64	89-103	61-71	98-114	62-73	100-117

AUTOMATIC SHIFT SPEEDS AND GOVERNOR PRESSURE CHART
(Approximate Miles and Kilometers Per Hour)

1982	—Carline— M and Z		—Carline— M,Z,P,D,C,V		—Carline— M,Z,P,D,C,V High Altitude	
Kickdown Limit						
3-2 WOT Downshift	51-60	82-97	57-66	92-106	56-66	90-106
3-2 Part Throttle Downshift	28-32	45-51	26-30	42-48	29-33	47-53
3-1 WOT Downshift	30-35	48-56	32-38	51-61	31-36	50-58
Governor Pressure①						
15 psi	23-26	37-42	22-24	35-39	24-27	39-43
50 psi	54-61	87-98	61-68	98-109	61-68	98-109

1983	—Carline— M and Z		—Carline— M,Z,P,D,C,V,L		—Carline— M,Z,P,D,C,V,L,J,E,T High Altitude	
Engine (Liter)	1.7L.		2.2 and 2.6L.		2.2 and 2.6L.	
Overall Top Gear Ratio	3.50		2.78		3.02	
Throttle Minimum	MPH	km/h	MPH	km/h	MPH	km/h
1-2 Upshift	11-15	18-24	10-14	16-23	12-15	20-25
2-3 Upshift	16-21	26-34	15-20	24-32	17-22	28-35
3-1 Downshift	11-14	18-23	10-13	16-21	11-15	20-25
Throttle Wide Open						
1-2 Upshift	33-39	53-63	37-44	60-71	35-39	57-62
2-3 Upshift	55-64	89-103	63-71	101-114	60-70	92-113
Kickdown Limit						
3-2 WOT Downshift	51-60	82-97	58-66	93-106	54-64	87-103
3-2 Part Throttle Downshift	28-32	45-51	43-54	70-87	39-49	62-79
3-1 WOT Downshift	30-35	48-56	31-39	50-62	33-36	53-58
Governor Pressure①						
15 psi	23-26	37-42	22-24	35-39	26-29	42-47
50 psi	54-61	87-98	61-68	98-109	59-66	95-106

1984	—Carline— M,Z,P,D		—Carline— H,K,P,D,C		—Carline— M,Z,P,D,C,H,K,E,T		—Carline— C,V	
Engine (Liter)	2.2L		2.6L		2.2 and 1.6L		2.2L	
Overall Top Gear Ratio	2.78		3.22		3.02 (except turbocharged)		3.02 (turbocharged)	
Throttle Minimum	MPH	km/h	MPH	km/h	MPH	km/h	MPH	km/h
1-2 Upshift	13-17	21-27	13-16	21-26	13-17	21-27	15-19	24-31
2-3 Upshift	17-21	27-34	17-21	27-34	18-22	29-35	21-25	34-40
3-2 Downshift	13-16	21-26	12-15	19-24	13-16	21-26	15-19	24-31
Throttle Wide Open								
1-2 Upshift	35-42	56-68	34-42	55-68	36-44	58-71	38-42	61-68
2-3 Upshift	61-68	98-109	59-66	95-106	63-71	101-114	70-80	113-129
Kickdown Limit								
3-2 WOT Downshift	56-64	90-103	55-62	89-100	58-66	93-106	64-74	103-119
3-2 Part Throttle Downshift	44-52	71-84	44-51	71-82	46-54	74-87	47-55	76-89
3-1 WOT Downshift	31-38	50-61	30-37	48-60	32-39	51-63	37-40	60-64
Governor Pressure								
15 psi	23-25	37-40	24-27	39-43	26-29	42-47	28-31	45-50
50 psi	59-65	95-105	57-63	92-101	61-68	98-109	69-76	111-122

NOTE: Changes in tire size will cause shift points to occur at corresponding higher or lower vehicle speeds.
Km/h. = Kilometers per hour
① Governor pressure should be from zero to 3 psi at stand still or downshift may not occur.

NOTE: A siren-like noise, or a whining is normal on some converters during the stall test due to the fluid flow. However, loud metallic noises or banging from loose parts indicate a defective converter. To make sure, operate the vehicle on a lift at light throttle in both "D" and then "N," listening under the transaxle bell housing to confirm the area from which the noise originates.

AIR PRESSURE TESTS

Even though all fluid pressures are correct and the hydraulic pressure test checks out, it is still possible to have a "no drive" condition, due to inoperative clutches or bands. By testing with air pressure instead of hydraulic pressure, the defective unit can be pinpointed.

NOTE: Compressed air should be limited to 30 psi and must be free from dirt and moisture.

1. After the car is safely supported, remove the oil pan and carefully remove the valve body.
2. Locate the front clutch "apply" passage and apply air pressure. Listen for a dull "thud" and/or place fingertips on the front clutch housing to feel piston movement, confirming that the front clutch is operating. Look for excessive oil leaking.
3. Locate the rear clutch "apply" passage and apply air pressure. As before, listen for a dull "thud" and/or place fingertips on the rear clutch housing to feel piston movement confirming that the rear clutch is operating, and again looking for excessive oil leaks.
4. Locate the kickdown servo "on" passage (front servo apply) and apply air pressure. This tests the kickdown servo and operation is indicated by the front band tightening. The spring on the servo piston should release the band.
5. Locate the low-reverse servo "apply" passage (rear servo) and apply air pressure. This tests the low and reverse servo and operation is indicated by the rear band tightening. The spring on the servo piston should release the band.

If, after the air pressure tests, correct operation of the clutches and servos are confirmed, and the complaint was no upshift or erratic shifts, then the problem has been narrowed down to the valve body.

FLUID LEAKAGE DIAGNOSIS

Fluid leakage can be deceptive as to its origin, in and around the transaxle and converter areas, due to engine oil leakages. Factory fill fluid for the automatic transaxles is red in color and should easily be distinguished from engine oil. However, with contaminants in the automatic transaxle fluid, its color appearance can change to that of engine oil. Therefore, the leaking fluid should be examined closely and its point of origin be determined before repairs are started.

1. The following leaks may be corrected without removing the transaxle assembly:
  a. Manual lever shaft oil seal.
  b. Pressure gauge plugs.
  c. Neutral start switch.
  d. Oil pan sealer.
  e. Oil cooler fittings or lines.
  f. Extension housing to case bolts.
  g. Speedometer adapter "O" ring.
  h. Front band adjusting screw.
  i. Extension housing axle seal.
  j. Differential bearing retainer axle seal.
  k. Rear end cover sealer.
  l. Extension housing "O" ring.
  m. Differential bearing retainer sealer.
2. The following leaks require the removal of the transaxle and torque converter assemblies for repairs.
  a. Transaxle fluid leaking from the lower edge of the converter housing, caused by the front pump oil seal.
  b. Pump to case seal.
  c. Torque converter weld.
  d. Cracked or porous transaxle case.

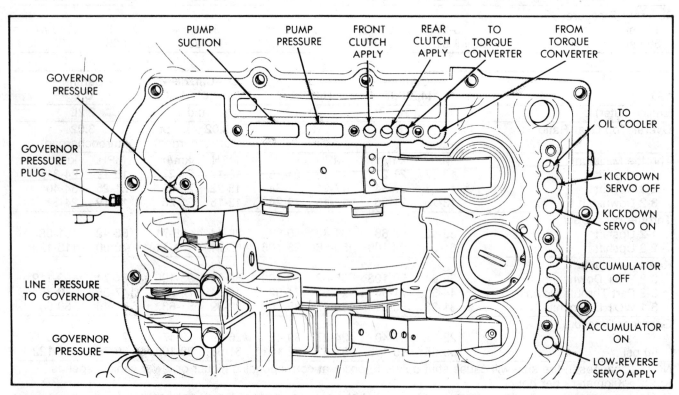

Air pressure test locations in transaxle case (©Chrysler Corp.)

OIL COOLERS AND TUBES
REVERSE FLUSHING

When a transaxle failure has contaminated the fluid, the oil cooler(s) should be flushed and the torque converter replaced with an exchange unit to insure that metal particles or sludged oil are not later transferred back into the reconditioned or replaced transaxle assembly.

Procedure

1. Place a length of hose over the end of the lower oil cooler tube (from cooler) and insert the other end of the hose into a waste oil container.
2. Apply compressed air into the upper oil cooler hose in short, sharp blasts.
3. Pump approximately one quart of automatic transaxle fluid into the upper oil cooler hose (to cooler).
4. Repeat the short, sharp blasts of air into the lower cooler line. Repeat if necessary.
5. If the reverse flushing of the cooler system fails to clear all the obstructions from the system, the cooler and/or radiator assembly must be replaced.

NOTE: The fluid flow through the cooler should be, as a rule of thumb, one quart in 20 seconds, the transaxle in neutral and the engine at curb idle.

ABNORMAL NOISE DIAGNOSIS

1. Examine fluid level and condition.
2. Road test vehicle to determine if an abnormal noise exists.
3. Identify the type of noise, the driving ranges and conditions when the noise occurs.

Gear Noise

1. Check for correct location of rubber insulator sleeve on the center of the shift cable.
2. Planetary gear noise—Necessary to remove the transaxle assembly and replace the planetary gear set.
3. Transfer Shaft Gear noise—Necessary to remove the transaxle assembly and replace the output and transfer shaft gears.
4. Differential gear noise—Necessary to remove the transaxle assembly and replace the transfer shaft and ring gears and/or the differential carrier gears.

Grinding Noise

1. Remove the transaxle/converter assembly.
2. Disassemble, clean and inspect all parts. Clean the valve body, install all new seals, rings and gaskets.
3. Replace all worn or defective parts.

Knock, Scrape or Clicking Noise

1. Remove the converter dust shield and inspect for loose or cracked converter drive plate.
2. Inspect for contact of the starter drive with the starter ring gear.

Whine or Buzzing Noise

1. Determine source of noise, from either the transaxle or converter.
2. If the transaxle has the buzzing or whining noise, remove all three pans and inspect for debris indicating worn or failed parts. If no debris, check valve body.
3. If the converter has the buzzing or whining noise, replace the converter assembly.

TORQUEFLITE TRANSAXLE STALL SPEED CHART

Year	Engine Liter	Transaxle Type	Converter Diameter	Stall R.P.M.
1980-81	1.7	A-404	9½ inches (241 millimetres)	2250-2450
	2.2	A-413	9½ inches (241 millimetres)	2190-2410
	2.6	A-470	9½ inches (241 millimetres)	2400-2630
1982-83	1.6	A-415	9½ inches (241 millimetres)	2250-2450
	1.7	A-404	9½ inches (241 millimetres)	2300-2500
	2.2	A-413	9½ inches (241 millimetres)	2200-2410
	2.6	A-470	9½ inches (241 millimetres)	2400-2630
1984	1.6	A-415	9½ inches (241 millimetres)	2250-2450
	2.2	A-413	9½ inches (241 millimetres)	2200-2400
	2.2 EFI	A-413	9½ inches (241 millimetres)	2280-2480
	2.2 EFI (turbocharged)	A-413	9½ inches (241 millimetres)	3020-3220
	2.6	A-470	9½ inches (241 millimetres)	2400-2600

ON CAR SERVICES

ADJUSTMENTS
Gearshift Linkage

If it should be necessary to take the linkage apart, the plastic grommets that are used as retainers should be replaced with new ones. Use a prying tool (large screwdriver) to force rod from grommet, then cut away old grommet. Use pliers to snap new grommet into the lever and to snap the rod into the grommet.

1980 MODELS

1. Make sure that the adjustable swivel block is free to slide on the shift cable. Disassemble and clean or repair, to assure free action, as necessary.
2. With all linkage assembled and free and the adjustable swivel lock bolt loose, place the gearshift lever in Park, then move the shift lever on the transaxle all the way to the rear detent position.
3. Tighten the swivel lock bolt to 90 in. lbs. (10N•m).
4. To check operation, key start must work only when the gear shift lever is in Park or Neutral.

1981 AND LATER MODELS

1. Place the gear shift lever in the "P" (Park) position.
2. Loosen the cable retaining clamp on the cable mounting bracket of the automatic transaxle.
3. Apply the loading as follows while tightening the cable retaining clamp bolt to 90 in. lbs. (10N•m).
   a. All models equipped with column shift—Apply a 10 pound load in the forward direction on the cable housing isolator.
   b. All models except Omni and Horizon equipped with console shift—Apply a 10 pound load minimum, in the forward direction on the console shift lever knob.
   c. Omni and Horizon Models equipped with console shift—Apply a 10 pound load in the forward direction on the transaxle lever, at the transaxle.
4. Check the adjustment by moving the shift lever through all positions while attempting to start the engine with the key. The engine should only start with the transaxle shift lever in the Park or Neutral positions.

Throttle Linkage

Bring engine to operating temperature and be sure the carburetor is off fast idle. Check idle speed with tachometer. Disconnect choke if necessary to keep carburetor off fast idle.

1. Loosen the adjusment bracket lock screw. The bracket must be free to slide on its slot. If necessary, disassemble and clean or repair. Lube with a good quality light grease.
2. a. 1980-82—Hold the transaxle throttle lever firmly rearward against its internal stop, and then tighten the adjusting lock screw to 105 in. lbs. (12N•m). This automatically removes cable backlash.
   b. 1983 and Later—Slide the bracket to the left (towards the engine) to the limit of its travel. Release the bracket and move the throttle lever fully to the right against its internal stop and tighten the adjusting bracket lock screw to 105 in. lbs. (12N•m).
3. Connect choke if it was disconnected. Test for freedom of movement by pushing the lever forward and slowly release it to confirm that it will return fully rearward.

Band Adjustment
KICKDOWN (FRONT) BAND

The kickdown band (front band) has its adjusting screw located on the top front (left side) of the transaxle case. Adjustment is as follows:

1. Loosen the lock nut and back off about five turns.
2. Tighten the adjusting screw to 72 in. lbs. (8N•m):
3. Back off the adjusting screw 3 turns from 72 in. lbs. on the A-404 and A-415 models. Back off the adjusting screw 2¾ turns from 72 in. lbs. on the 1982 and later A-413 and A-470 transaxles. 1981 A-413 and A-470 models require the adjusting screw to be backed off 2 turns.
4. Hold the adjusting screw in this position and tighten the locknut to 35 ft. lbs. (47 N•m).

LOW-REVERSE BAND
A-404

The Low-reverse band (rear) is not adjustable. If excessive band wear is suspected, the following procedure should be used to verify wear.

1. Remove the transaxle oil pan and pressurize the low-reverse servo with 30 psi of air pressure.
2. With the low-reverse servo pressurized, measure the gap between the band ends. If the gap is less than 0.080 inch (2.0 mm) the band has worn excessively and should be replaced.

A-413 and A-470

The low-reverse band is adjustable. However, before adjustments are done, the procedure outlined to check the low-reverse band for the A-404 transaxle should be done to verify that the proper end gap exists on these transaxle models. To adjust the band, follow the outlined procedure.

1. Loosen and back off the lock nut approximately 5 turns.
2. With an inch pound torque wrench, tighten the adjusting screw to 41 inch pound (5N•m) torque.
3. Back off the adjusting screw 3½ turns from the 41 in. lbs. (5N•m) torque.
4. While holding the adjusting screw, tighten the lock nut to 10 ft. lbs. (14N•m) torque.

SERVICES
Fluid Change and/or Oil Pan Removal

1. Raise the vehicle and support safely. Loosen, but do not remove the oil pan bolts. Gently pull one corner of the oil pan downward so the fluid will drain into a container with a large opening.

—————— CAUTION ——————
If the fluid is hot, do not allow it to touch the person. Burns can result.

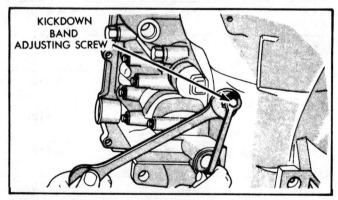

Adjusting kickdown band (©Chrysler Corp.)

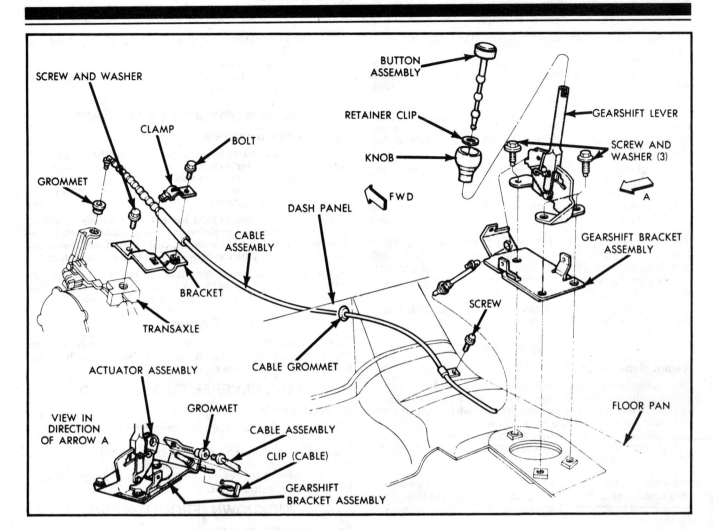

Console model shift linkage and cable assembly (©Chrysler Corp.)

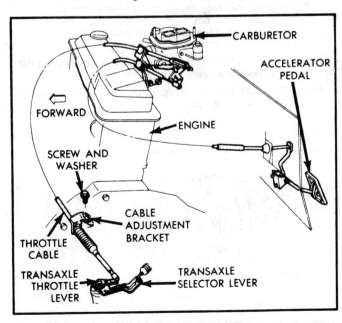

Throttle linkage assembly (©Chrysler Corp.)

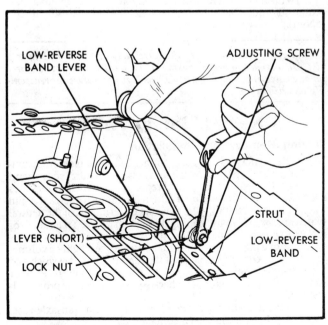

Adjusting low-reverse band, A-413 and A-470 models
(©Chrysler Corp.)

2. When the fluid has drained, remove the bolts and the oil pan.

3. Carefully inspect the filter and pan bottom for a heavy concentration of friction material or metal particles. A small accumulation can be considered as normal, but a heavy build-up indicates damaged or worn parts.

NOTE: Filter replacement and band adjustments are recommended whenever the fluid is changed. The oil filter screws should be torqued to 35 in. lbs., (7N•m) through 1980, and 40 in. lbs. (5N•m), 1981 and later.

4. Check the oil pan carefully for distortion, straightening the flanges with a block of wood and mallet, as required.

5. Apply a bead of RTV sealant on the pan flanges and install to the transaxle case. Install the retaining bolts and torque to 150 in. lbs. (16N•m) through 1980, and 165 in. lbs. (19N•m), 1981 and later.

6. Fill with four quarts of Dexron R II. Idle the engine for at least six minutes, moving the selector lever through each position, ending in Park. Check the fluid level and correct as required.

7. Be sure the dipstick is properly seated to prevent water and dirt from entering the transaxle fill tube.

NOTE: The A-404, A-413 and A-470 transaxles do not have drain plugs on the torque converters.

Valve Body

Removal and Intallation

To remove the valve body, first drain the fluid and remove the pan as outlined under "Oil Pan R&R." The oil filter is removed next which requires the proper size Torx® drive bit. After the pan and filter are removed, proceed as follows:

1. Unscrew the neutral safety switch and remove.
2. Remove the "E" clip that connects the parking rod to its lever. Remove parking rod.
3. Remove the seven valve body attaching bolts.
4. Remove the valve body along with the governor tubes being very careful not to bend or force the tubes. Place the valve body where it will be protected from damage if other transaxle work is to be done.
5. To install the valve body assembly, reverse the removal procedure. Torque the valve body attaching bolts to 40 in. lbs. (5N•m).

─────────── CAUTION ───────────

Do not clamp any portion of the valve body or transfer plate in a vise. Distortion can occur causing sticking valves. Do not use force on any valve or spring. Identify all springs and valves for reassembly identification.

Neutral Safety Switch

Testing, Removal and Installation

The neutral safety switch used on this transaxle includes provision for the backup lamp switch function. The neutral start circuit is through the center pin on the three terminal switch. It provides a ground for the starter solenoid circuit through the center pin when in Park or Neutral. The two outside terminals of the neutral switch are for the circuit feeding the backup lamps. To test and/or replace the switch, follow this procedure:

1. Remove the wiring connector and with a test light or voltmeter, check for continuity between the center pin of the switch and the transaxle case. There should be continuity only in Park or Neutral. If switch tests "bad" check gearshift cable for proper adjustment.
2. If switch needs replacement, unscrew from transaxle case. Expect fluid to run out; keep container close.
3. Move selector lever to Park and Neutral to see if the lever in the transaxle is centered in the switch opening.

4. Install switch (making sure seal is on switch) and torque to 24 ft. lbs. (33N•m).

5. Add fluid as necessary.

6. To retest, check for continuity between the two outside pins, which should only be in Reverse.

Speedometer Pinion Gear

Removal and Installation

1. Remove the locking bolt assembly securing the speedometer pinion adapter in the extension housing. Carefully work the adapter/pinion assembly out of the extension housing.

2. Remove the retaining spring from the assembly and separate the pinion from the adapter.

NOTE: If transmission fluid has entered the cable assembly, install a new speedometer pinion and seal assembly. If the transmission fluid is leaking between the cable and adapter, the small "O" ring must be replaced between the cable and the adapter.

3. Install the adapter on the cable, the pinion on the adapter with a new large "O" ring and install the retainer on the pinion and adapter. Be certain the retainer is properly seated.

Servos

To remove the servos, first drain the fluid and remove the pan and valve body as previously outlined. With valve body removed, proceed as follows:

LOW-REVERSE (REAR) SERVO

Removal and Installation

1. Remove the snap ring, the servo retainer and the return spring.

2. The low-reverse servo assembly can be pulled from its bore, and if necessary, have the lip seal renewed.

3. Assembly is the reverse, being sure to lube the seal with Dexron® II or petroleum jelly.

KICKDOWN (FRONT) SERVO

Removal and Installation

1. Remove the snap ring, using two screwdrivers to pry it from its groove, being careful not to damage the case.

2. Remove the rod guide, spring and piston rod.

3. Remove the kickdown piston.

NOTE: There may be minor variations due to different applications, but the basic procedure is the same.

Assembly is the reverse, being sure to lube the seals with Dexron® II or petroleum jelly.

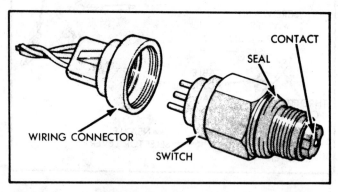

Neutral start and back-up lamp switch (©Chrysler Corp.)

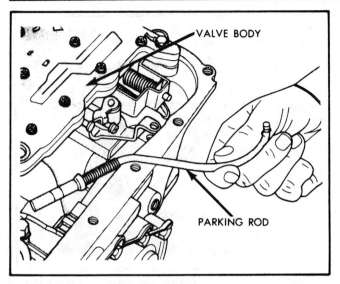

Removing park rod from valve body (©Chrysler Corp.)

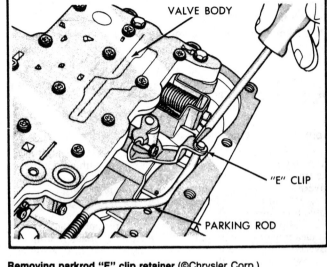

Removing parkrod "E" clip retainer (©Chrysler Corp.)

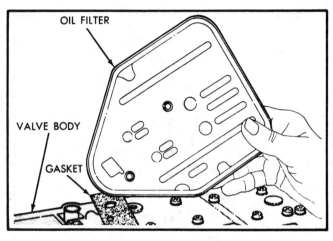

Oil filter removal (©Chrysler Corp.)

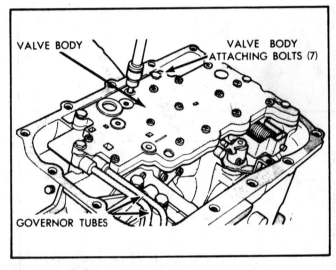

Valve body attaching bolts and governor tube location (©Chrysler Corp.)

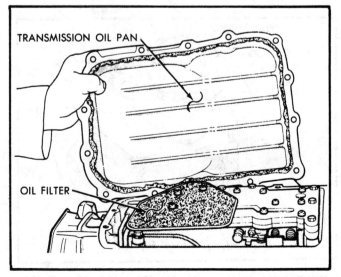

Transaxle oil pan removal and filter location (©Chrysler Corp.)

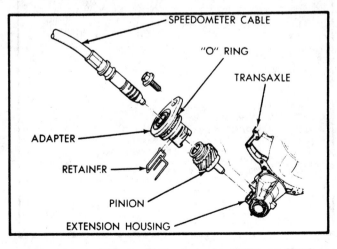

Exploded view of speedometer drive assembly (©Chrysler Corp.)

Accumulator

Removal and Installation

The accumulator is easily removed in much the same manner as the servos.

1. Remove the snap ring and accumulator plug.
2. Remove the spring and accumulator piston.
3. Assembly is the reverse, being sure to lube the seals with Dexron® II or petroleum jelly.

Governor

Removal and Installation

The governor assembly is a special offset design. No shaft runs through it and so it can be removed for service without removing the transaxle gear cover, transfer gear or governor support. The governor may be serviced by the following procedure:

1. As outlined previously, remove the transaxle oil pan and valve body assembly.
2. Unbolt the governor from the governor support and remove for cleaning or reconditioning. When cleaning or assembling the governor, make sure that the valves move freely in their bores.
3. When installing the governor, torque the bolts to 60 in. lbs. (7 N•m).

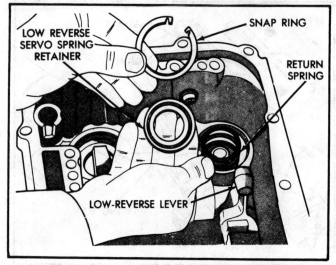

Low-reverse servo retainer, snap ring and spring removal or installation (©Chrysler Corp.)

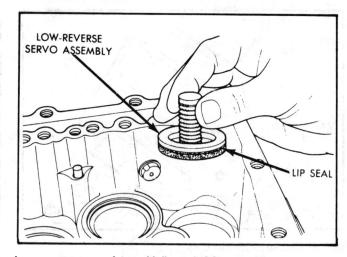

Low-reverse servo piston with lip seal (©Chrysler Corp.)

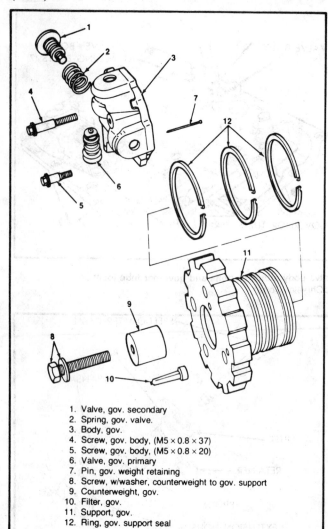

1. Valve, gov. secondary
2. Spring, gov. valve.
3. Body, gov.
4. Screw, gov. body, (M5 × 0.8 × 37)
5. Screw, gov. body, (M5 × 0.8 × 20)
6. Valve, gov. primary
7. Pin, gov. weight retaining
8. Screw, w/washer, counterweight to gov. support
9. Counterweight, gov.
10. Filter, gov.
11. Support, gov.
12. Ring, gov. support seal

Exploded view of governor assembly (©Chrysler Corp.)

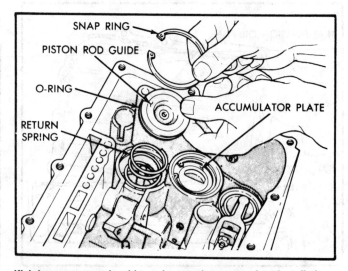

Kickdown servo rod guide and snap ring removal or installation (©Chrysler Corp.)

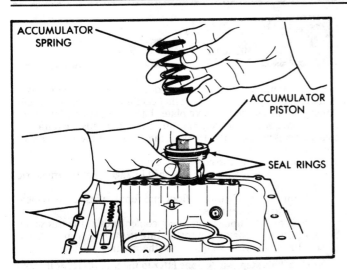

Accumulator piston assembly removal or installation
(©Chrysler Corp.)

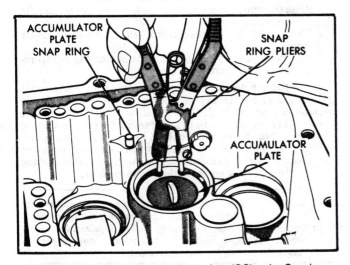

Removal or installation of accumulator plate (©Chrysler Corp.)

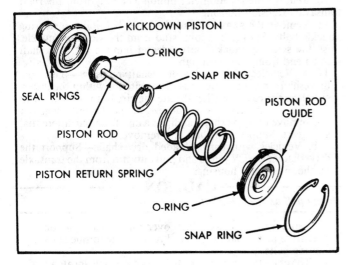

Exploded view of controlled load kickdown servo assembly
(©Chrysler Corp.)

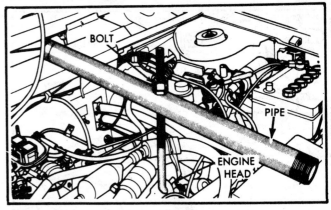

Typical engine support fixture (©Chrysler Corp.)

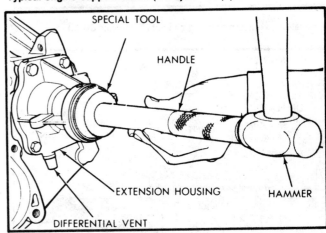

Installation of extension housing oil seal using special driver tool
(©Chrysler Corp.)

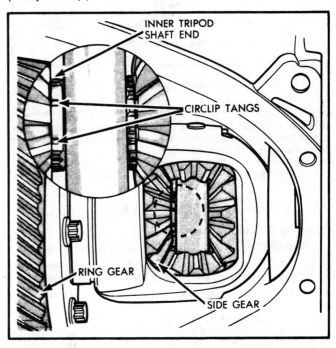

Circlip location within the differential carrier assembly
(©Chrysler Corp.)

REMOVAL & INSTALLATION

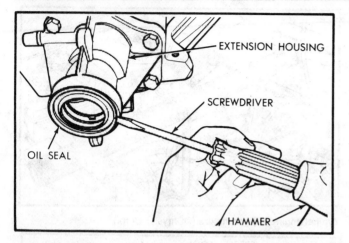

Removal of extension housing oil seal (©Chrysler Corp.)

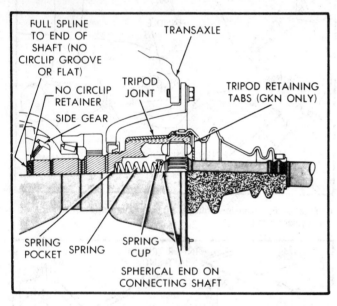

Spring loaded drive shaft assembly (©Chrysler Corp.)

─────────────── CAUTION ───────────────
Dirt and sand can cause misalignment, resulting in speedometer pinion gear damage. Be certain all parts and mating surfaces are clean and free of foreign material.
───

4. Install the locking bolt assembly and torque to 60 in. lbs. (7N•m).

DRIVESHAFT TRANSAXLE OIL SEALS

Removal and Installation

1. Remove the drive shaft assembly from the transaxle.

NOTE: Refer to the transaxle Removal and Installation section for drive shaft removal, handling and installation information.

2. Remove the oil seal with a seal remover tool or its equivalent.
3. Inspect the seal seat and bore.
4. Lubricate the seal lip with transmission fluid or petroleum jelly, seat it squarely in its bore and install the seal to its seat with a seal installer tool.
5. Reinstall the drive shaft in the reverse order of its removal.

 REMOVAL & INSTALLATION

While the removal of the transaxle does not require the removal of the engine, it should be noted that care must be used to prevent damage to the converter drive plate. The drive plate will not support any weight, so the transaxle and converter must be removed as an assembly. Do not let any of the weight of the transaxle or converter rest.

1. Disconnect the negative battery cable.
2. Disconnect the throttle and shift linkage from the transaxle levers.
3. With the vehicle on the floor and foot brakes applied, loosen the hub and wheel nuts.
4. Remove the upper and lower cooler lines at the transaxle.
5. Position an engine support fixture to the engine and tighten to equalize the engine weight from the mounts.

NOTE: This step can be done later in the procedure, at the discretion of the repairman.

6. Remove the upper bolts of the bell housing.
7. Raise the vehicle, remove the wheel/tire assemblies and the hub nut with washer from the drive shaft.
8. From under the left front fender, remove the inner fender splash panel.
9. Vehicles equipped with driveshafts having the circlip retainers, remove the differential cover.

NOTE: Driveshafts were changed from a circlip retainer to a spring loaded Tripod (inner joint) during the 1982 model year. To determine if a spring loaded joint/shaft is used in the vehicle, place a pry bar between the right side transaxle extension housing and the face of the tripod joint housing. Pry the joint housing outward (into the rubber boot). If the joint housing can be moved at least ½ inch from the extension housing, the driveshaft is a spring loaded model and does not have the circlip retainer.

─────────────── CAUTION ───────────────
Mishandling of the driveshaft assemblies, such as allowing the assemblies to dangle unsupported, pulling or pushing the ends can result in pinched rubber boots or damage to the CV joints. Boot sealing is vital to retaining special joint lubricants and to prevent contaminents from entering the joint areas.
───

10. Remove the speedometer pinion assembly from the right transaxle extension housing.
11. Remove the sway bar and the lower ball joint to steering knuckle bolts. Pry the ball joint stud from the steering knuckle. Push the steering knuckle outward and remove the driveshaft splined end from the wheel hub.
12. a. Vehicles with circlip driveshaft retainers—Turn the driveshafts to expose the ends of the circlips within the differential carrier case opening. Rotate the circlip to coincide with the flats of the driveshafts, squeeze the circlip ends together and remove the driveshaft/circlip as a unit from the differential side gears. Support the shaft and remove.
   b. Vehicles with spring loaded driveshafts—Support the driveshaft at the CV joints and pull outward from the transaxle on the inner joint housing.

─────────────── CAUTION ───────────────
Do not pull on the driveshaft.
───

13. Remove the converter dust cover, mark the torque converter and drive plate relationship and remove the torque converter mounting bolts.
14. To rotate the engine, remove the access plug located in the right fender splash panel, and use a suitable socket/ratchet assembly to turn the engine crankshaft.

15. Remove the connector from the neutral starter switch. Certain models may require the removal of the lower cooler pipe at this time, if not accomplished earlier,

16. Remove the engine mount bracket from the front cross member.

17. Remove the front mount insulator through bolt and front engine mount bolts.

NOTE: It will be necessary to adjust the engine support to obtain the zero clearance needed to relieve the pressure from the mounts and bolts.

18. Place the removal jack under the transaxle, remove the left engine mount and the long bolt through the mount.

19. Remove the lower bell housing bolts. Pry against the engine and lower transaxle, being careful of the torque converter.

Installation of the transaxle is a reversal of the procedure but remember to fill the differential with Dexron® II automatic

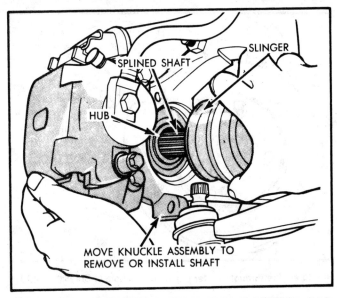

Separating or installing splined shaft into hub assembly (©Chrysler Corp.)

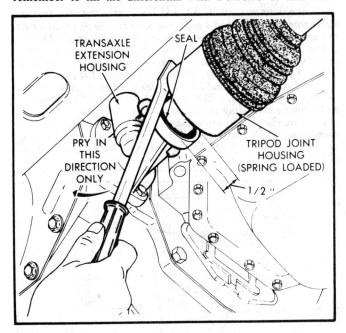

Identifying spring loaded drive shaft (©Chrysler Corp.)

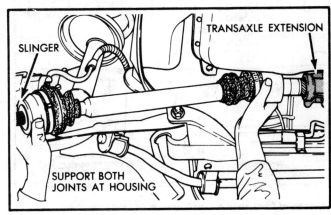

Supporting of drive shaft during removal or installation (©Chrysler Corp.)

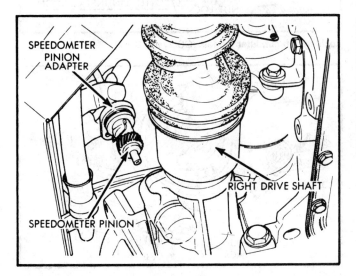

Speedometer drive pinion adapter removal (©Chrysler Corp.)

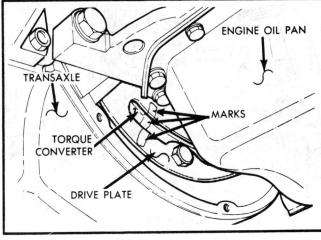

Mark torque converter and drive plate before separation (©Chrysler Corp.)

2-33

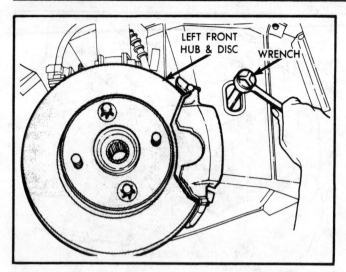

Rotate engine through fender panel access hole (©Chrysler Corp.)

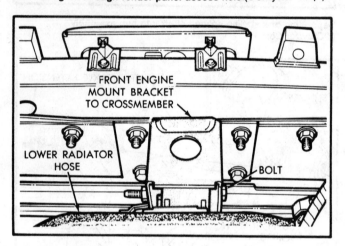

Location of front engine mount bracket to cross member (©Chrysler Corp.)

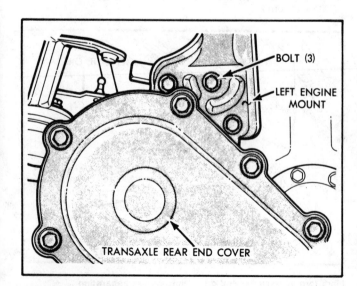

Left engine mount location (©Chrysler Corp.)

transmission fluid, if equipped with separate sump, before lowering the car. Also pay particular attention to the proper positioning of the "circlips" on the driveshaft inboard ends, with the "tangs" laying on the flattened end of the shaft. Holding the joint housing, a quick, firm push will complete the lock-up of the "circlips" on the axle side gear.

CAUTION

Make sure the clips are positioned correctly.

When reinstalling the lower ball joint to the steering knuckle, torque the clamp bolt to 50 foot-pounds (68N•m). The differential cover is not installed with a gasket, but with RTV sealant in a ribbon about ⅛" wide. The screws are torqued to 165 inch-pounds (19N•m). If the inner drive shaft boots appear collapsed or deformed, slip a small round rod under the boot to vent some air into it since a vacuum may have formed when it was pushed home.

CAUTION

Early transaxle units had a unique locking feature on the hub nuts to maintain preload and prevent the nut from backing off. This required that the nut be staked in place. The hub nuts are not reusable. Install washer and a new hub nut. Apply brakes and torque to 200 foot-pounds (271N•m). Stake nut into place but do not use a sharp chisel. The tool should be ⁷/₁₆" wide and have a radius ground on the end of about ¹/₁₆". The completed stake should be about ³/₁₆" to ¼" long, conforming closely to the axle shaft slot.

Later transaxle units use a more conventional lock and cotter pin to maintain the bearing preload. Again, a new nut should be used on the hub, and apply the brakes and torque to 180 foot-pounds (245N•m). Install the nut lock and a new cotter pin.

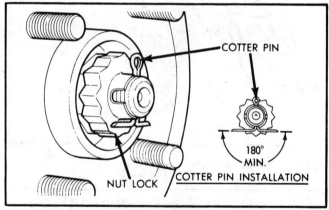

Locking hub nut, later models (©Chrysler Corp.)

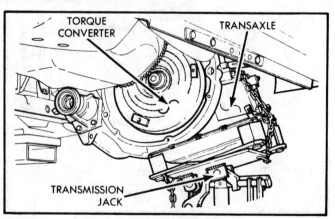

Lowering or raising transaxle assembly (©Chrysler Corp.)

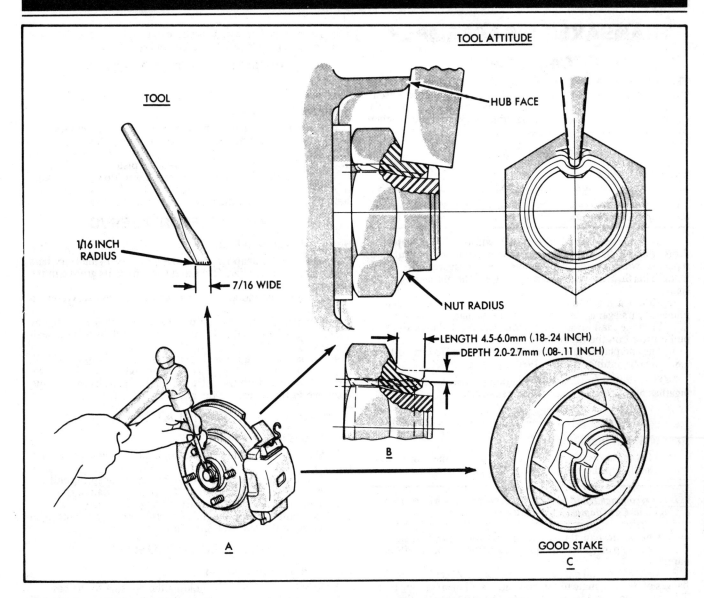

Staking hub nut, early models (©Chrysler Corp.)

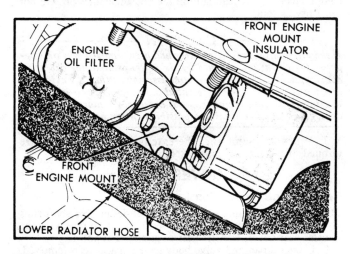

Engine mount bracket removal or installation (©Chrysler Corp.)

BENCH OVERHAUL

Before removing any of the transaxle subassemblies for bench overhaul, the unit should be cleaned. Cleanliness during disassembly and assembly is necessary to avoid further transaxle trouble after assembly. Before removing any of the transaxle subassemblies, plug all the openings and clean the outside of the transaxle thoroughly. Steam cleaning or car wash type high pressure equipment is preferable. Parts should be washed in a cleaning solvent, then dried with compressed air. *Do not wipe parts with shop towels.* The lint and fibers will find their way into the valve body and other parts and cause problems later. The case assembly was accurately machined and care must be used to avoid damage. Pay attention to the torque values to avoid case distortion.

2–35

TRANSAXLE DISASSEMBLY

Converter

The torque converter is removed by sliding the unit out of the transaxle input and reaction shaft. If the converter is to be reused, set it aside so it will not be damaged. Since the units are welded and have no drain plugs, converters subject to burnt fluid or other contamination should be replaced.

INPUT SHAFT

End Play Check

------ CAUTION ------

This measurement is considered critical, before disassembly and at the completion of the assembly procedures.

Measuring the input shaft end play will usually indicate if a thrust washer change is required, unless major components are replaced, which would then require new measurements to be made. This thrust washer is located between the input and output shafts.

1. Mount a dial indicator in such a way that the indicator plunger bears against the end of the input shaft.
2. Push the shaft inward to its limit, zero the dial indicator and pull outward on the input shaft.
3. The end play should be between 0.007 to 0.073 inch (0.18 to 1.85 mm) beginning with the 1980 transaxle models.

NOTE: Both removal and installation procedures are outlined together for easier reference by the repairman.

Oil Pan

Removal and Installation

1. With the transaxle held securely, remove the pan to case bolts. Gently tap the pan loose from the case.

------ CAUTION ------

It is not recommended to insert a tool between the pan and case as a prying tool. Case damage can result.

2. Check the pan flange for distortion, straightening the flanges of the pan with a straight block of wood and a soft faced hammer.
3. To install the oil pan, apply a bead of RTV sealer around the flanges of the pan, mate to the case and install the retaining bolts. Torque the bolts to 150 inch-pounds (16N•m) through 1980, and 165 inch-pounds (19N•m) 1981 and later.

Oil Filter

Removal and Installation

1. Three Torx R type screws are used to hold the filter to the valve body. The screws must be removed and replaced with the use of the proper tool bit.
2. When re-installing or replacing the filter assembly, torque the screws to 35 inch-pounds (4N•m) through 1980, and 40 inch-pounds (5N•m), 1981 and later.

Valve Body

Removal and Installation

1. After oil pan and filter removal, take the "E" clip off the park rod and pull the park rod free of its support.
2. Remove the seven valve body attaching bolts.
3. Remove the valve body along with the governor tubes being very careful not to bend or force the tubes. Place the valve body where it will be protected from damage if other transaxle work is to be done.

4. Reverse the removal procedure to install. Torque the valve body attaching bolts to 105 inch-pounds (12N•m).

LOW/REVERSE (REAR) SERVO

Removal and Installation

1. With the valve body removed, disconnect the snap ring, remove the servo retainer and the servo return spring.
2. The low/reverse servo piston can be removed from the bore in the case. Remove the lip seal from the piston.
3. To assemble, install a new lip seal, lubricate with transmission fluid or petroleum jelly and install into the piston bore. Install the return spring, the servo retainer and the snap ring.

KICKDOWN (FRONT) SERVO

Removal and Installation

1. Remove the snap ring with the aid of two small pry bars (1980) or snap ring pliers (1981 and later) from its groove in the case bore.
2. Remove the rod guide, spring and piston rod/servo piston assembly.
3. Separate the piston rod from the servo piston assembly by the removal of the small snap ring. Remove and discard the sealing rings and "O" rings.
4. To assemble and install, place new sealing rings and "O" rings on the piston rod guide, the piston rod and the kickdown piston. Lubricate with transmission fluid or petroleum jelly and assemble in the reverse of the removal procedure.

ACCUMULATOR

Removal and Installation

1. Slightly depress the accumulator plate and remove the snap ring.
2. Remove the accumulator plate, accumulator spring and piston from the case bore. Remove "O" ring and sealing rings.
3. To assemble, install new sealing rings and a new "O" ring. Lubricate with transmission fluid or petroleum jelly and install in the reverse of the removal procedure.

EXTENSION HOUSING

Removal and Installation

1. Remove the four bolts holding the housing to the case.
2. Twist the housing to break it free, turning clockwise and counterclockwise, repeating until the housing comes off. Do not use a prying tool.

------ CAUTION ------

If the differential bearing retainer had been removed (the unit opposite the extension, retained by six bolts) prior to the extension housing, hold on to the differential assembly. When the extension housing is removed, the differential assembly will roll out of the housing (providing the differential cover was removed) and fall, damaging the differential and possibly causing personal injury.

3. When reassembling, the extension housing bolts are torqued to 250 inch-pounds (28N•m).

GOVERNOR

Removal and Installation

The governor assembly is a special offset design. No shaft runs through it so it can be removed for service without removing the transfer gear cover, transfer gear or governor support.

1. If the oil pan and valve body have already been removed, just unbolt the governor from the governor support and remove.

NOTE: Since the transfer gears need pullers to remove them, and a special fixture to hold the gear when the transfer shaft nut is removed, (it is torqued to 200 foot-pounds or 270N•m) the above method is the best way to remove the governor for service if there is no need to open up the transfer gear end of the transaxle

2. At reassembly, the governor-to-support screws are torqued to 60 inch-pounds (7N•m).

NOTE: The following assemblies are outlined for removal only. The assembly is outlined during the unit re-assembly.

OIL PUMP

Removal

1. Tighten the front band adjusting screw until the band is tight on the front clutch retainer. This prevents the front clutch retainer from coming out with the pump, which might cause unnecessary damage to the clutches.

2. Remove the seven pump attaching bolts.

—— CAUTION ——

Because the oil pump bolts are metric, conventional non-metric threaded slide hammers cannot be used to pull the pump. Adapters are available to attach to the slide hammer threads, but in no case should non-metric tools be forced into the metric threads.

3. Pull oil pump, remove the gasket, and loosen the kickdown band adjusting screw.

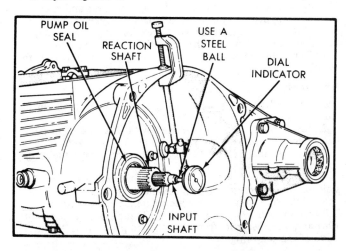

Measuring input shaft end play (©Chrysler Corp.)

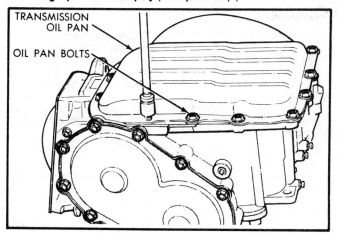

Removing oil pan retaining bolts (©Chrysler Corp.)

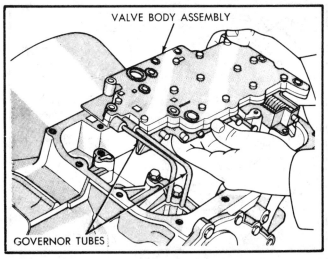

Removal of valve body and governor tubes (©Chrysler Corp.)

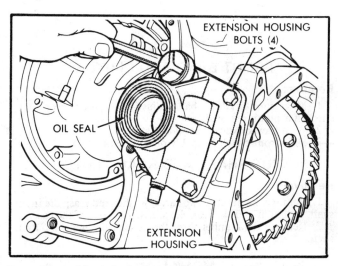

Removing retaining bolts from extension housing (©Chrysler Corp.)

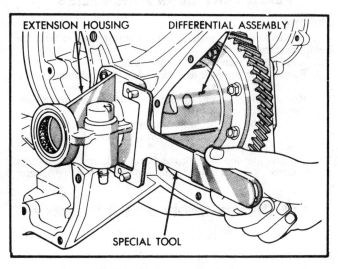

Using special tool to slowly rotate the extension housing during removal or installation (©Chrysler Corp.)

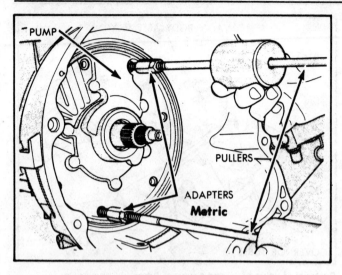

Removal of oil pump with metric pullers (©Chrysler Corp.)

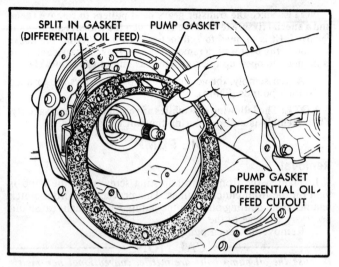

Oil pump gasket removal or installation (©Chrysler Corp.)

Front Unit

FRONT BAND AND CLUTCH

Removal

1. Make sure kickdown band adjusting screw has been loosened after pump removal.
2. Withdraw kickdown band and strut.
3. Slide front clutch assembly out of the case.

INPUT SHAFT AND REAR CLUTCH

Removal

1. Remove the number two thrust washer and grasp the input shaft, pulling it and the rear clutch assembly out of the case.
2. Remove the number three thrust washer from the output shaft.

Rear Unit

PLANETARY GEAR ASSEMBLIES

Removal

1. Remove the front planetary gear snap ring and remove the front planetary gear.
2. Remove the number six thrust washer.
3. Remove the sun gear driving shell and the numbers seven and eight thrust washers.
4. Remove the number nine thrust washer and the rear planetary gear assembly.
5. Remove the number ten thrust washers, the overrunning clutch cam and clutch rollers and springs. There should be eight rollers and eight springs.
6. Remove the low-reverse band and strut and the number eleven thrust washer. (A-413 and A-470—loosen low/reverse band.)

Unit Disassembly and Assembly

FRONT CLUTCH

Disassembly

1. Remove the large, waved snap ring and lift out the thick steel plate.

2. Take particular note of the order of the clutch plates. Through 1980, there are two driving discs and three clutch plates. Note that there are two clutch plates together. 1981 and later uses three driving discs and three clutch plates, stacked alternately.
3. With a compressor, relieve the tension on the snap ring, remove the ring and release the compressor.
4. Take out the spring retainer, spring and piston from the clutch retainer.

Assembly

1. After all parts have been cleaned, renew the seals on the clutch piston.
2. Install the piston; lube the seals with Dexron® II or petroleum jelly to keep from damaging them.
3. Reverse the disassembly procedure and install the return spring, retainer and snap ring.
4. Install the new clutches; make sure the order is correct. Install the thick steel (reaction) plate and snap ring.
5. Make sure front clutch plate clearance is .067-.106 in (1.7-2.7mm) through 1980, and 0.87 to 0.133 inch (2.22 to 3.37mm), 1981 and later. Measure from the reaction plate to the "farthest" wave of the snap ring.

REAR CLUTCH

Disassembly

1. Carefully pry the large snap ring from its groove.
2. Remove the thick steel (reaction) plate and the clutches, taking note of the order.
3. Carefully pry the large, waved snap ring from its groove so that the piston spring and piston can be removed from the rear clutch retainer.
4. Remove the snap ring in the rear clutch retainer if the input shaft is to be removed. The shaft will have to be pressed out.

Assembly

1. After all parts have been cleaned, renew the seals on the clutch piston.
2. Install the piston; lube the seals with Dexron® II or petroleum jelly to keep from damaging them.
3. Reverse the disassembly procedure and install the return spring, retainer and snap ring.
4. Install the new clutches; make sure the order is correct. Install the thick steel (reaction) plate and snap ring, selective. (Refer to specification.)

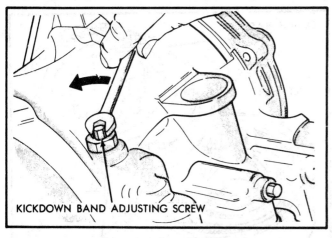

Loosening kickdown band adjusting screw (©Chrysler Corp.)

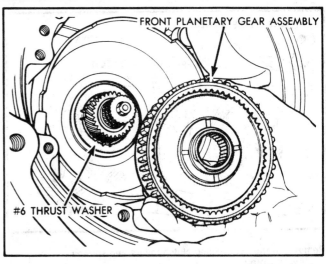

Removal of front planetary assembly (©Chrysler Corp.)

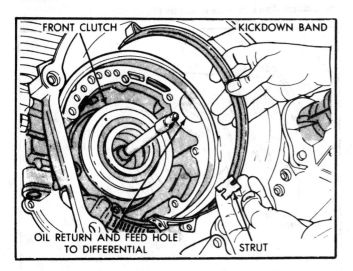

Removal of kickdown band and strut (©Chrysler Corp.)

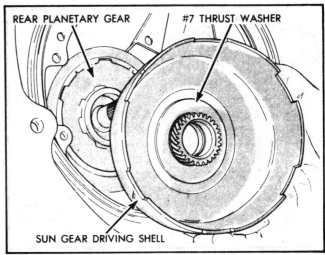

Removal of drive shell (©Chrysler Corp.)

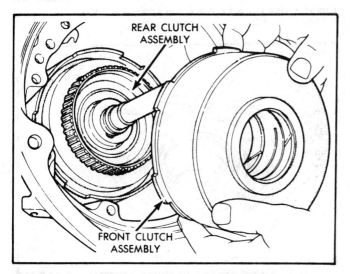

Removal of front clutch assembly (©Chrysler Corp.)

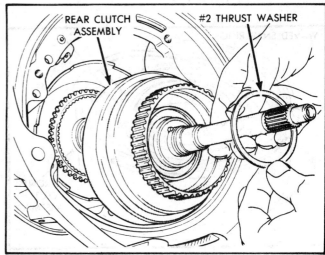

Removal of rear clutch assembly (©Chrysler Corp.)

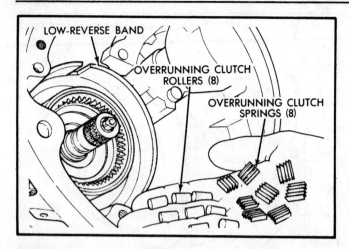

Removal of the overrunning clutch rollers and springs (©Chrysler Corp.)

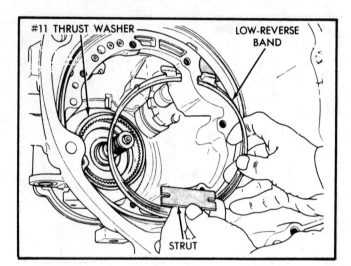

Low-reverse band and strut removal or installation (©Chrysler Corp.)

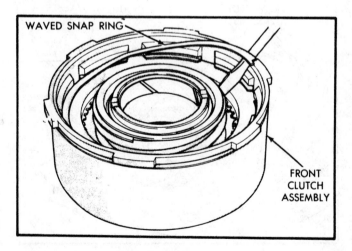

Removal of front clutch waved washer (©Chrysler Corp.)

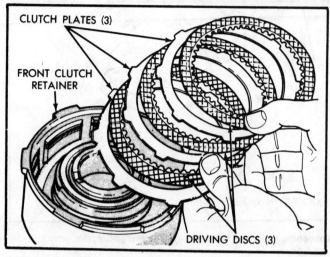

Front clutch disc and plate locations (©Chrysler Corp.)

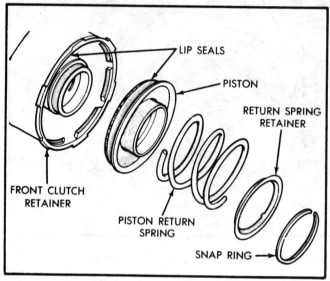

Exploded view of front clutch return spring and piston (©Chrysler Corp.)

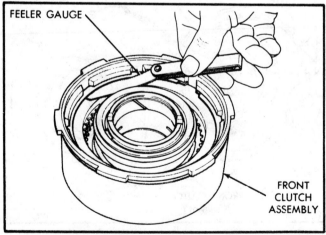

Measuring front clutch plate clearance (©Chrysler Corp.)

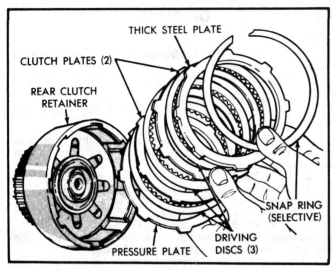

Rear clutch disc and plate location (©Chrysler Corp.)

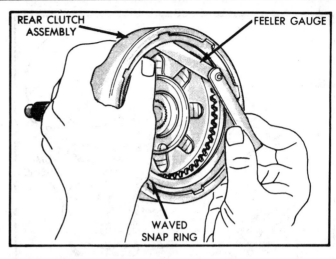

Measuring rear clutch plate clearance (©Chrysler Corp.)

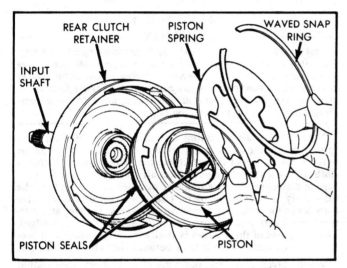

Removal of rear clutch waved snap ring, piston and spring (©Chrysler Corp.)

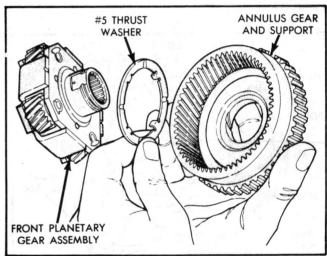

Separation of front planetary gear assembly (©Chrysler Corp.)

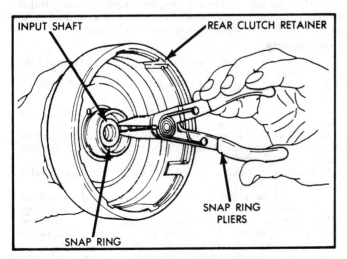

Removal of input shaft, if necessary (©Chrysler Corp.)

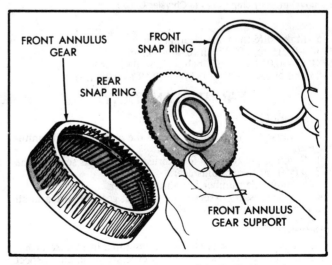

Separation of front annulus gear support from the front annulus gear (©Chrysler Corp.)

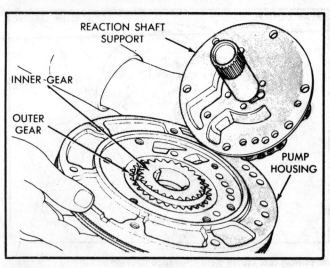

Separation of the reaction shaft support from the pump housing (©Chrysler Corp.)

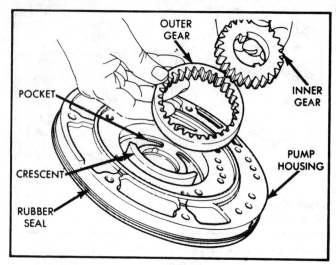

Removal of the pump gears (©Chrysler Corp.)

5. Make sure the plate clearance is:
1980—0.016 to 0.030 inch (0.40 to 0.94mm)
1981-82—0.018 to 0.037 inch (0.46 to 0.95mm)
1983 and later—0.023-0.037 inch (0.58 to 0.95mm)

FRONT PLANETARY

Disassembly

1. Remove the snap ring that holds the planetary to the annulus gear.
2. Remove the number four thrust washer and separate the planetary from the annulus gear.
3. Remove the number five thrust washer.
4. If it is necessary to service the annulus gear support, a small screwdriver can be used to remove the snap ring.

Assembly

After all parts have been cleaned, reassemble in reverse order. Check the thrust washers carefully for excessive wear and inspect the planetary gear carrier for cracks. Check the pinions for broken or worn teeth.

OIL PUMP

Disassembly

1. Remove the six bolts holding the reaction shaft support to the oil pump housing.
2. Remove the reaction shaft support and the inner and outer gears.
3. Clean all parts well and inspect for damage.

Assembly

1. If the inner and outer gears are still serviceable, reinstall in the oil pump body. The clearance must be checked and be with these specifications:
   a. Outer Gear to Pocket, 0.0018 to 0.0056 inch.
   b. Outer Gear I.D to Crescent, 0.0059 to 0.012 inch.
   c. Outer Gear Side Clearance, 0.001 to 0.002 inch.
   d. Inner Gear O.D. to Crescent, 0.0063 to 0.0124 inch.
   e. Inner Gear Side Clearance, 0.001 to 0.002 inch.
The side clearance is checked by laying a straight edge across the face and pump body and inserting a feeler gauge.
2. Install reaction shaft support and torque the six bolts to 250 inch-pounds (28N•m).
3. Renew seals.

VALVE BODY
— CAUTION —

Do not clamp any part of the valve body or transfer plate in a vise. Any slight distortion of the body or plate will cause sticking valves or excessive leakage or both. When removing and installing valves or plugs, slide them in or out very carefully. Do not use force to remove or install valve. Clean all parts well, blow dry with compressed air. Do not dry with shop towels.

NOTE: When disassembling the valve body, identify all valve springs with a tag for assembly reference later. Check parts for burns or nicks. Slight imperfections can be removed with crocus cloth. Using a straightedge, inspect all mating surfaces for warpage or distortion. Be sure all metering holes are open in both the valve body and separator plate. Use a penlight to inspect valve body bores for scratches, burrs, pits or scores. Remove slight irregularities with crocus cloth. Do not round off the sharp edges. The sharpness of these edges is vitally important because it prevents foreign matter from lodging between the valve and the bore.

When valves, plugs and bores are clean and dry, they should fall freely in the bores. Valve bodies and their bores do not change dimensionally with use. Therefore, a valve body that functioned properly when the vehicle was new, will operate properly if it is correctly and thoroughly cleaned. There should be no need to replace a valve body unless it is damaged in handling.

Disassembly

1. Remove the detent spring-attaching screw and remove the detent spring.
2. Remove the valve body screws.
3. Remove the transfer and separator plates.
4. Take note of the location of the 8 steel balls and remove them.
5. Remove the "E"-clip from the throttle valve shaft and remove the washer and oil seal. Lift the manual valve lever from the throttle valve lever assembly.
6. Remove the throttle valve lever and then the manual valve from the valve body.
7. Remove the screws holding the pressure regulator and adjuster assembly to the valve body. Use caution since the springs will be under tension. Do not alter the settings of the throttle pressure adjusting screws. Remove the line pressure valve, switch valve, and kickdown valves and their springs.
8. The governor plugs can be removed after removing their respective end covers.

9. Pressure regulator valve plugs, shift valves and shuttle valve can all be removed after their end covers have been removed. Be careful that the parts do not become mixed. Label parts if necessary.

Assembly

1. To reassemble, reverse the above sequence. Make certain parts are clean.

2. All screws on the valve body are to be torqued to 40 inch-pounds (4.5N•m) with the exception of the transfer plate-to-case screws. These are tightened to 105 inch-pounds (12N•m). Verify the correct position of the check balls.

Transfer and Output Shaft Service

To remove the output shaft, its bearing and select shim, it is recommended that the transfer shaft assembly be removed first. It should be noted that the planetary gear sets must be removed to accurately check the output shaft bearing turning torque. To remove the transfer shaft so that the output shaft can be serviced, follow this procedure.

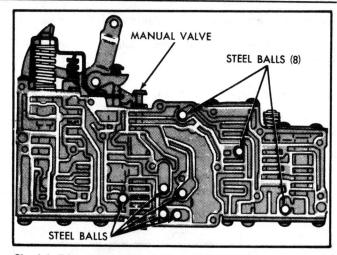

Check ball locations—1981 and later (©Chrysler Corp.)

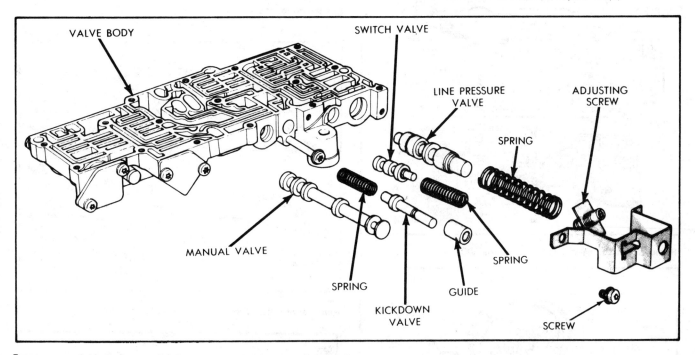

Pressure regulator and manual control valves—1980 (©Chrysler Corp.).

TRANSFER SHAFT

Disassembly

1. Remove the transfer gear cover. There are ten bolts. Note that RTV sealer is to be used on reassembly.

2. Because the nut on the transfer gear shaft has been torqued to 200 foot-pounds (270N•m) a special fixture is needed to hold the gear. A piece of plate or angle iron, drilled to suit the gear and bolted into place should hold the gear while sufficient force is applied to the 30mm socket and wrench to break free the nut. Remove it and the washer.

3. A gear puller is needed to pull the transfer shaft gear from its splined shaft. Note the select shim and bearing. Remove governor support retainer, the low/reverse band anchor pin and the governor assembly.

NOTE: Pullers and/or press will be required to change and in-

stall bearings or bearing cups. They are to be replaced in sets only.

4. With snap ring pliers, reach in alongside the transfer shaft and remove the snap ring at the head of the shaft. The transfer shaft may need the help of a slide hammer puller to help it out, due to a retainer with external seals.

5. Inspect bearings for excessive wear. Renew seals in the retainer.

Transfer Shaft, Output Shaft and Differential Bearing

PRECAUTIONS

Take extreme care when removing and installing bearing cups and cones. Use only an arbor press for installation, a hammer may not properly align the bearing cup or cone. Burrs or nicks on the bearing seat will give a false end play reading while gauging. Improperly seated bearing cup and cones are subject to low mileage failure. Bearing cups and cones should be replaced if they show signs of:

    a. Pitting
    b. Heat Distress

If distress is seen on either the cup or bearing rollers, both the cup and cone must be replaced. Bearing end play and drag torque specifications *must be maintained* to avoid premature bearing failures.

    a. All bearing adjustments, except transfer shaft bearing, must be made with no other gear train component interference or in gear mesh.

    b. Used (original) bearings may loose up to 50% of the original drag torque after break-in.

    c. When replacement of either the output or transfer gear is required, both gears should be replaced in a matched set.

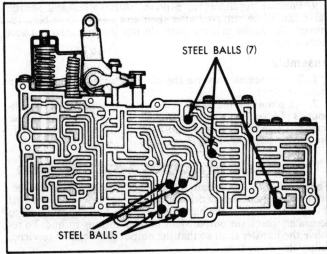

Check ball locations—1980 (©Chrysler Corp.)

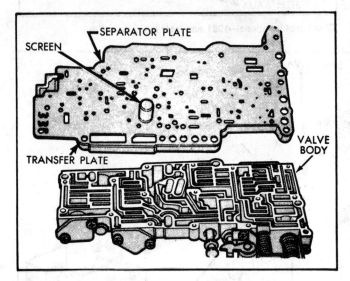

Removal of transfer and separator plates (©Chrysler Corp.)

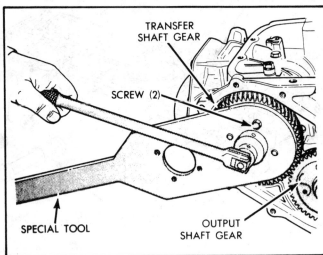

Removal of transfer shaft gear nut (©Chrysler Corp.)

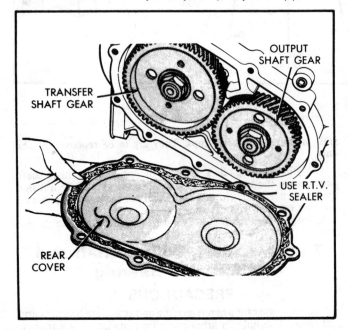

Rear cover removal or installation (©Chrysler Corp.)

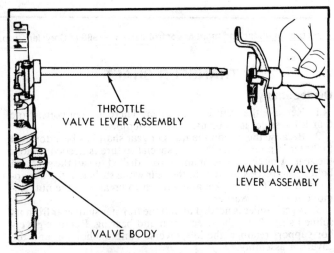

Separation of manual control lever from throttle valve lever after removal of "E" clip retainer (©Chrysler Corp.)

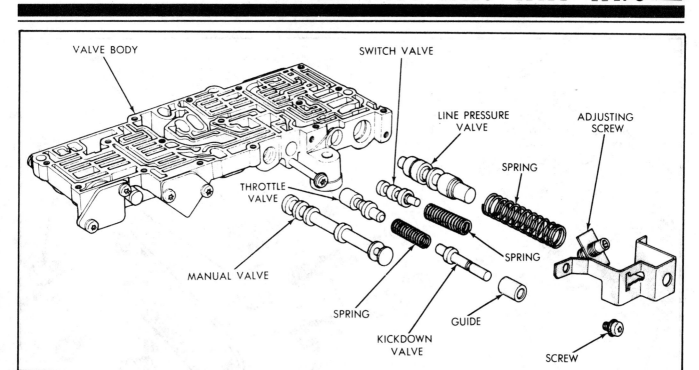

Pressure regulator and manual control valve—1981 and later (©Chrysler Corp.)

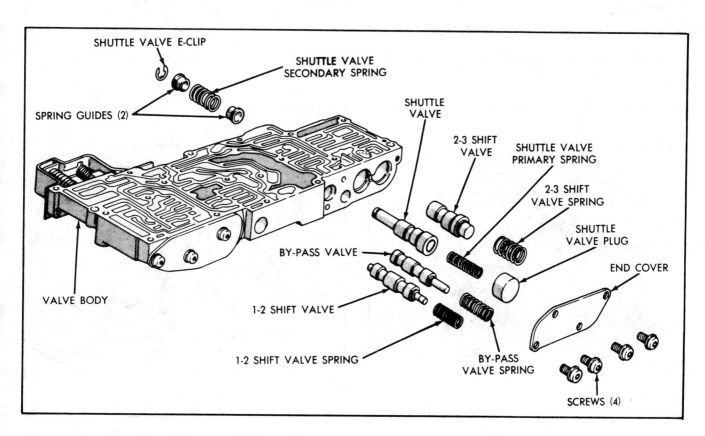

Shift valve and shuttle valve—1981 and later (©Chrysler Corp.)

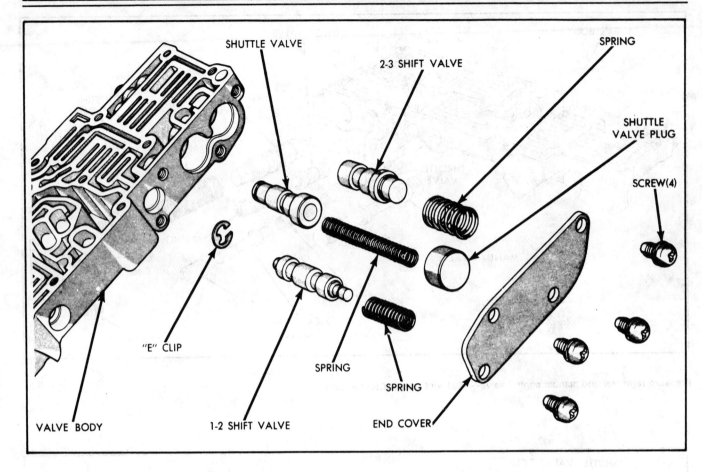

Shift valve and shuttle valve—1980 (©Chrysler Corp.)

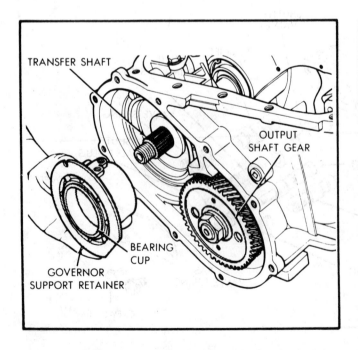

Removal of governor support retainer (©Chrysler Corp.)

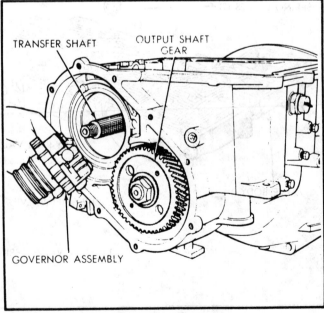

Removal of governor assembly (©Chrysler Corp.)

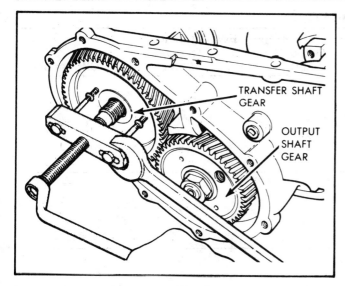

Use of puller to remove transfer shaft gear (©Chrysler Corp.)

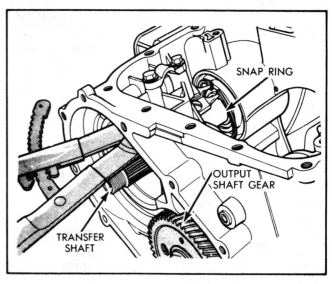

Removal of transfer shaft snap ring (©Chrysler Corp.)

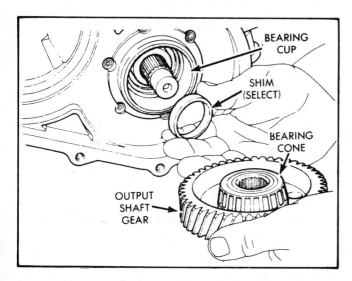

Output shaft gear and selective shims (©Chrysler Corp.)

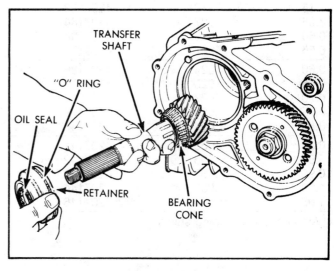

Removal of transfer shaft assembly (©Chrysler Corp.)

TRANSFER SHAFT BEARINGS

1. If the retaining nut and washer are to be removed, use the gear holding tool. Use a gear pulling tool to remove the gear from the shaft.

2. Install a 0.090 inch (2.29 mm) and 0.055 inch (1.39 mm) gauging shims on the transfer shaft behind the governor support.

3. Install transfer shaft gear and bearing assembly and torque nut to 200 foot-pounds (271 N•m).

NOTE: A few drops of Automatic Transmission Fluid applied to the bearing rollers will ensure proper seating and rolling resistance.

4. To measure bearing end play:
   a. Mount a steel ball with grease on the end of the transfer shaft.
   b. Push down on the gear while rotating back and forth to ensure proper seating of the bearing rollers.

   c. Using a dial indicator mounted to the transaxle case, measure transfer shaft end play by raising and lowering gear.

5. Refer to the Transfer Shaft Bearing Shim Chart for the required shim combination to obtain the proper bearing setting.

6. Use gear holding tool to remove the retaining nut and washer. Remove the transfer gear using a gear puller.

7. Remove the two gauging shims and install the correct shim combination. Install the transfer gear and bearing assembly.

8. Install the retaining nut and washer and torque to 200 foot-pounds (271 N•m).

9. Measure bearing end play as outlined in step 4. End play should be between 0.002 and 0.010 inch (0.05 mm and 0.25 mm).

NOTE: If end play is too high, install a 0.002 inch (0.05 mm) thinner shim combination. If end play is too low, install a 0.002 inch (0.05 mm) thicker shim combination. Repeat until the proper end play is obtained.

BEARING SHIM CHART

Shim Thickness			Bearing Usage		
mm	Inch	Part Number	Output Shaft	Transfer Shaft	Differential
0.94	.037	4207166	x	x	—
0.99	.039	4207167	x	x	—
1.04	.041	4207168	x	x	—
1.09	.043	4207169	x	x	—
1.14	.045	4207170	x	x	—
1.19	.047	4207171	x	x	—
1.14	.049	4207172	x	x	—
1.29	.051	4207173	x	x	—
1.34	.053	4207174	x①	x	—
1.39	.055	4207175	x	x①	—
1.84	.072	4207176	x	x	—
2.29	.090	4207177	x	x①	—
6.65	.262	4207159	x	—	—
7.15	.281	4207160	x	—	—
7.65	.301	4207161	x①	—	—
12.65	.498	4207162	x	—	—
13.15	.518	4207163	x	—	—
13.65	.537	4207164	x①	—	—
0.50	.020	4207134	—	—	x①
0.55	.022	4207135	—	—	x
0.60	.024	4207136	—	—	x
0.65	.026	4207137	—	—	x
0.70	.027	4207138	—	—	x
0.75	.029	4207139	—	—	x
0.80	.031	4207140	—	—	x
0.85	.033	4207141	—	—	x
0.90	.035	4207142	—	—	x
0.95	.037	4207143	—	—	x
1.00	.039	4207144	—	—	x
1.05	.041	4207145	—	—	x

①Also used as gauging shims

OUTPUT SHAFT

Disassembly

1. Using the same holding tool as was used on the transfer shaft gear, hold the output shaft gear securely and remove the nut and washer from the shaft.
2. With a puller, remove the output shaft gear from the shaft. Take note of the selective shim.
3. Remove the output shaft and annulus gear assembly. The shaft is a press fit in the annulus gear and must be pressed in and out.

SHIM THICKNESS

TRANSFER SHAFT

Shim thickness needs only to be determined if any of the following parts are replaced: (a) transaxle case, (b) transfer shaft, (c) transfer shaft gear, (d) transfer shaft bearings, (e) governor support retainer, (f) transfer shaft bearing retainer, (g) retainer snap ring and (h) governor support.

OUTPUT SHAFT

Shim thickness need only to be determined if any of the following parts are replaced: (a) transaxle case, (b) output shaft, (c) rear planetary annulus gear, (d) output shaft gear, (e) rear annulus and output shaft gear bearing cones, and (f) overrunning clutch race cups.

The output shaft bearing turning torque is 3 to 8 inch-pounds if the proper shim has been installed.

Differential Service

Disassembly

1. As described previously, remove the transfer shaft assembly.
2. Remove the ten differential cover bolts. Note that RTV sealer is to be used on reassembly.
3. With a 13mm socket, remove the six differential bearing retainer bolts.
4. Gently rotate the bearing retainer back and forth and pull out of the case.

TRANSFER BEARING SHIM CHART

End Play (with 2.29mm and 1.39mm gauging shims installed)		Required Shim Combination	Total Thickness	
mm	Inch	mm	mm	Inch
0	0	2.29+1.39	3.68	.145
.05	.002	2.29+1.39	3.68	.145
.10	.004	2.29+1.39	3.68	.145
.15	.006	2.29+1.39	3.68	.145
.20	.008	2.29+1.34	3.63	.143
.25	.010	2.29+1.29	3.58	.141
.30	.012	2.29+1.24	3.53	.139
.35	.014	2.29+1.19	3.48	.137
.40	.016	2.29+1.14	3.43	.135
.45	.018	2.29+1.09	3.38	.133
.50	.020	2.29+1.04	3.33	.131
.55	.022	2.29+ .99	3.28	.129
.60	.024	1.84+1.39	3.23	.127
.65	.026	1.84+1.34	3.18	.125
.70	.028	1.84+1.29	3.13	.123
.75	.030	1.84+1.24	3.08	.121
.80	.032	1.84+1.19	3.03	.119
.85	.034	1.84+1.14	2.98	.117
.90	.036	1.84+1.09	2.93	.115
.95	.038	1.84+1.04	2.88	.113
1.00	.040	1.84+ .99	2.83	.111
1.05	.042	1.39+1.39	2.78	.109
1.10	.044	1.39+1.34	2.73	.107
1.15	.046	1.39+1.29	2.68	.105
1.20	.048	1.39+1.24	2.63	.103
1.25	.049	1.39+1.19	2.58	.101
1.30	.050	1.39+1.14	2.53	.099
1.35	.052	1.39+1.09	2.48	.097
1.40	.055	1.39+1.04	2.43	.095
1.45	.057	1.39+ .99	2.38	.093
1.50	.059	.94+1.39	2.33	.091
1.55	.061	.94+1.34	2.28	.089
1.60	.063	.94+1.29	2.23	.087

5. With a 13mm socket, remove the four extension housing bolts.

6. Gently rotate the extension housing back and forth and pull out of the case.

─────────── CAUTION ───────────

When the extension housing is pulled out, the differential assembly will roll out of the case. To prevent damage, hold on to the differential assembly when removing extension housing.

NOTE: Pullers and/or press will be required to change and install bearings or bearing cups. They are to be replaced in sets only.

7. To remove pinion shaft, first withdraw the roll pin. Use a screw extractor like an "Easy-Out" and pull the pin from the case. Drive the pinion shaft out with a brass drift and hammer.

8. Remove the pinion gears and side gears as well as the four thrust washers by rotating the pinion gears to the opening in the differential case.

9. If the ring gear is to be removed, take note that a 10mm, 12 point socket will be required. There are eight bolts. Upon reassembly, they are to be torqued to 70 foot-pounds (95 N•m).

NOTE: Immerse the ring gear in boiling water for 15 minutes before installing ring gear onto the differential case.

10. Renew the seal in the differential bearing retainer.

NOTE: Shim thickness need only be determined if any of the following parts are replaced: a. transaxle case, b. differential carrier, c. differential bearing retainer, d. extension housing, or e. differential bearings.

Refer to the "Bearing Adjustment Outline" to determine the proper shim thickness for correct bearing preload and proper bearing turning torque.

Assembly

1. Reverse the above procedure to reassemble the differential. Torque the ring gear bolts to 70 foot-pounds (95 N•m). Be sure to install the pinion shaft with the proper end out to receive the roll pin. Inspect bearings and cups. Renew the seal in extension housing before assembling. Extension housing bolts are torqued to 250 inch-pounds (28 N•m), the bearing retainer bolts are also torqued to 250 inch-pounds (28 N•m).

NOTE: Install the differential cover without sealer, temporarily, since the axle (drive shaft) circlips should be checked after installation for proper seating. If the circlips fit satisfactorily, then the cover, with RTV sealer, can be fitted and the bolts torqued to 165 inch-pounds (19 N•m).

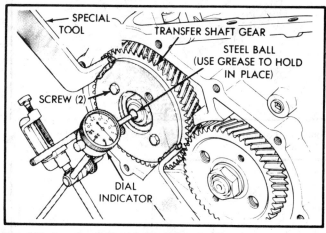

Checking transfer shaft end play (©Chrysler Corp.)

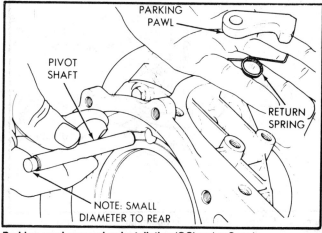

Parking pawl removal or installation (©Chrysler Corp.)

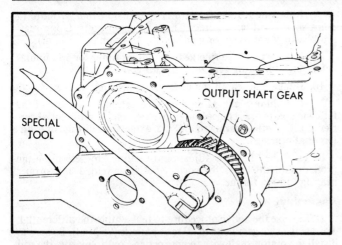

Removing output shaft retaining nut and washer (©Chrysler Corp.)

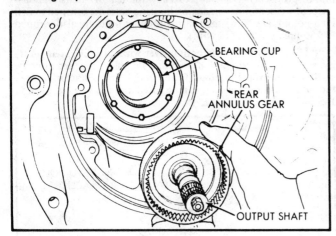

Removal of output shaft and rear annulus gear assembly (©Chrysler Corp.)

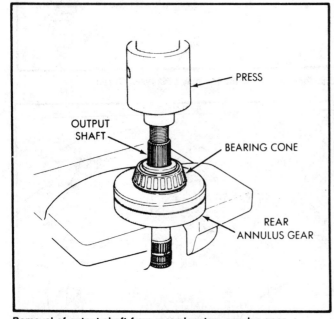

Removal of output shaft from rear planetary annulus gear (©Chrysler Corp.)

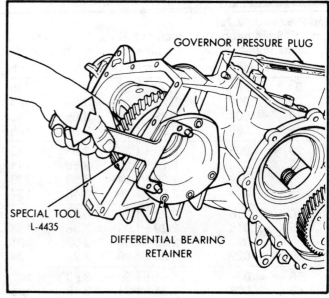

Rotate bearing retainer to remove (©Chrysler Corp.)

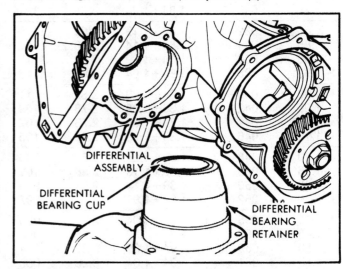

Differential bearing retainer (©Chrysler Corp.)

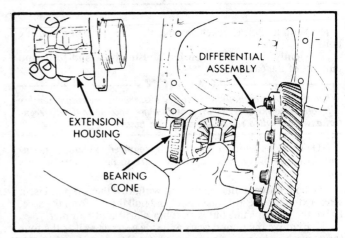

Holding differential carrier when removing the extension housing and bearing retainer (©Chrysler Corp.)

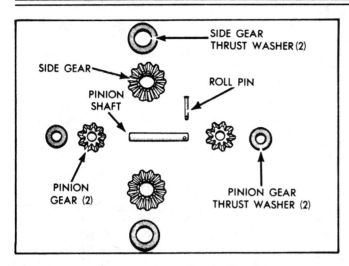

Exploded view of pinion gear set (©Chrysler Corp.)

OUTPUT SHAFT BEARINGS

1980

1. If the retaining nut and washer are to be removed, use the gear holding tool. Use a gear puller to remove the gear from the shaft.

2. With output shaft gear removed, install a 0.301 inch (7.65 mm) gauging shim on the planetary rear annulus gear hub and a 0.053 inch (1.34 mm) gauging shim on the output shaft using lubriplate to hold the shims in place. The 7.65 mm shim has a larger inside diameter and must be installed first. The 1.34 mm shim pilots on the output shaft.

3. Install output shaft gear and bearing assembly, torque the retaining nut to 200 foot-pounds (271 N•m).

NOTE: A few drops of Automatic Transmission Fluid applied to the bearing rollers will ensure proper seating and rolling resistance.

4. To measure bearing end play:
   a. Mount a steel ball with lubriplate on the end of the output shaft.
   b. Push down on the gear while rotating back and forth to ensure seating of the bearing rollers.
   c. Using a dial indicator, mounted to the transaxle case, measure output shaft end play by moving the gear up and down.

5. Once bearing end play has been determined, refer to the Output Shaft Bearing Shim Chart for the required shim combination to obtain proper bearing setting.

NOTE: a. The 6.65 mm (.262 inch) 7.15 mm (.281 inch) or 7.65 mm (.301 inch) shim is always installed first. These shims have a lubrication hole which is necessary for proper bearing lubrication.

b. Shims thinner than 6.65 mm listed in the chart are common to both the transfer and output shaft bearings.

6. Remove the retaining nut and washer. Remove the output shaft gear.

7. Remove the two gauging shims and install the proper shim combination, making sure to install the 6.65, 7.15, or 7.65 mm shim first. Use lubriplate to hold the shims in place. Install the output gear and bearing assembly.

8. Install the retaining nut and washer and torque to 200 foot-pounds (271 N•m).

9. Using an inch-pound torque wrench, check the turning torque. The torque should be between 3 to 8 inch-pounds.

NOTE: If the turning torque is too high, install a 0.002 inch (0.050 mm) thicker shim. If the turning torque is too low, install a

0.002 inch (0.050 mm) thinner shim. Repeat until the proper turning torque of 3 to 8 inch-pounds is obtained.

OUTPUT SHAFT BEARING SHIM CHART
1980

End Play (with 7.65mm and 1.34mm gauging shims installed)		Required Shim Combination	Total Thickness	
mm	Inch	mm	mm	Inch
0	0	7.65 + 1.34	8.99	.354
.05	.002	7.65 + 1.24	8.89	.350
.10	.004	7.65 + 1.19	8.84	.348
.15	.006	7.65 + 1.14	8.79	.346
.20	.008	7.65 + 1.09	8.74	.344
.25	.010	7.65 + 1.04	8.69	.342
.30	.012	7.65 + .99	8.64	.340
.35	.014	7.65 + .94	8.59	.338
.40	.016	7.15 + 1.39	8.54	.336
.45	.018	7.15 + 1.34	8.49	.334
.50	.020	7.15 + 1.29	8.44	.332
.55	.022	7.15 + 1.24	8.39	.330
.60	.024	7.15 + 1.19	8.34	.328
.65	.026	7.15 + 1.14	8.29	.326
.70	.028	7.15 + 1.09	8.24	.324
.75	.030	7.15 + 1.04	8.19	.322
.80	.032	7.15 + .99	8.14	.320
.85	.034	7.15 + .94	8.09	.318
.90	.036	6.65 + 1.39	8.04	.316
.95	.038	6.65 + 1.34	7.99	.314
1.00	.040	6.65 + 1.29	7.94	.312
1.05	.042	6.65 + 1.24	7.89	.311
1.10	.044	6.65 + 1.19	7.84	.309
1.15	.046	6.65 + 1.14	7.79	.307
1.20	.048	6.65 + 1.09	7.74	.305
1.25	.049	6.65 + 1.04	7.69	.303
1.30	.051	6.65 + .99	7.64	.301
1.35	.053	6.65 + .94	7.59	.299

Note: Average Conversion .05mm = .002 inch

1981 AND LATER

With output shaft gear removed.

1. Install a 0.537 inch (13.65 mm) and a 0.053 inch (1.34 mm) gauging shims on the planetary rear annulus gear hub using grease to hold the shims in place. The 13.65 mm shim has a larger inside diameter and must be installed over the output shaft first. The 1.34 mm shim pilots on the output shaft.

2. Install output shaft gear and bearing assembly, torque to 200 ft. lbs. (271 N•m).

3. To measure bearing end play:
   a. Mount a steel ball with grease into the end of the output shaft.
   b. Push and pull the gear while rotating back and forth to insure seating of the bearing rollers.
   c. Using a dial indicator, mounted to the transaxle case, measure output shaft end play by raising and lowering gear.

4. Once bearing end play has been determined, refer to the output shaft bearing shim chart for the required shim combination to obtain proper bearing setting.

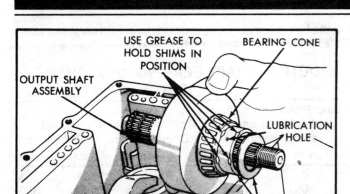

Installing output shaft assembly with selective shim held in place with grease (©Chrysler Corp.)

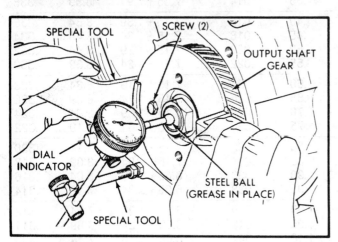

Checking output shaft and end play (©Chrysler Corp.)

a. The 0.498 inch (12.65 mm), 0.518 inch (13.15 mm), 0.537 inch (13.65 mm) shims are always installed first. These shims have lubrication slots which are necessary for proper bearing lubrication.

b. Shims thinner than 12.65 mm listed in the chart are common to both the transfer shaft and output shaft bearings.

5. Remove the retaining nut and washer. Remove the output shaft gear.

6. Remove the two gauging shims and install the proper shim combination, making sure to install the 12.65, 13.15, or 13.65 mm shim first. Use grease to hold the shims in place. Install the output shaft gear and bearing assembly.

7. Install the retaining nut and washer and torque to 200 ft. lbs. (271 N•m).

8. Using an inch-pound torque wrench, check the turning torque. The torque should be between 3 and 8 inch-pounds.

NOTE: If the turning torque is too high, install a 0.002 inch (0.050 mm) thicker shim. If the turning torque is too low, install a 0.002 inch (0.050 mm) thinner shim. Repeat until the proper turning torque is 3 to 8 inch pounds.

DIFFERENTIAL BEARINGS

NOTE: The use of special tools are noted in the outline and those that have been approved by Chrysler Corporation, or their equivalents, should be used when making the following critical measurements.

OUTPUT SHAFT BEARING SHIM CHART
1981 and Later

End Play (with 13.65mm and 1.34mm gauging shims installed)		Required Shim Combination	Total Thickness	
mm	Inch	mm	mm	Inch
.0	.0	13.65+1.34	14.99	.590
.05	.002	13.65+1.24	14.89	.586
.10	.004	13.65+1.19	14.84	.584
.15	.006	13.65+1.14	14.79	.582
.20	.008	13.65+1.09	14.74	.580
.25	.010	13.65+1.04	14.69	.578
.30	.012	13.65+ .99	14.64	.576
.35	.014	13.65+ .94	14.59	.574
.40	.016	13.15+1.39	14.54	.572
.45	.018	13.15+1.34	14.49	.570
.50	.020	13.15+1.29	14.44	.568
.55	.022	13.15+1.24	14.39	.566
.60	.024	13.15+1.19	14.34	.564
.65	.026	13.15+1.14	14.29	.562
.70	.028	13.15+1.09	14.24	.560
.75	.030	13.15+1.04	14.19	.558
.80	.032	13.15+ .99	14.14	.556
.85	.034	13.15+ .94	14.09	.554
.90	.036	12.65+1.39	14.04	.552
.95	.038	12.65+1.34	13.99	.550
1.00	.040	12.65+1.29	13.94	.548
1.05	.042	12.65+1.24	13.89	.547
1.10	.044	12.65+1.19	13.84	.545
1.15	.046	12.65+1.14	13.79	.543
1.20	.048	12.65+1.09	13.74	.541
1.25	.049	12.65+1.04	13.69	.539
1.30	.051	12.65+ .99	13.64	.537
1.35	.053	12.65+ .94	13.59	.535

Average Conversion .05mm = .002 inch

1. Remove the bearing cup from the differential bearing retainer and remove the existing shim, from under the cup. The original shim should not be re-used.

2. Install a 0.020 inch (0.50 mm) gauging shim, and reinstall the bearing cup into the retainer. Use an arbor press to install the cup.

NOTE: A few drops of Automatic Transmission Fluid applied to the bearing rollers will ensure proper seating and rolling resistance.

3. Install the bearing retainer into the case and the torque bolts to 250 inch-pounds (28 N•m).

4. Position the transaxle assembly vertically on the support stand and install Tool L-4436 or its equivalent, into the extension.

5. Rotate the differential assembly at least on full revolution to ensure the tapered roller bearings are fully seated.

6. Attach a dial indicator to the case and zero the indicator on the flat end of Tool L-4436 or its equivalent.

7. Place a large screwdriver to each side of the ring gear and

lift with enough force to take up the clearance between the bearings. Check the dial indicator for the amount of end play. Caution should be used not to damage the transmission case and/or differential cover sealing surface.

8. Once the end play has been determined, refer to the differential bearing shim chart for the required shim combination to obtain the proper bearing setting.

9. Remove the differential bearing retainer. Remove the bearing cup and the 0.020 inch (0.50 mm) gauging shim.

10. Install the proper shim combination under the bearing cup. Make sure the oil baffle is installed properly in the bearing retainer below the bearing shim and cup.

11. Install the differential retainer. Make sure to seal the retainer to the housing with RTV sealer and torque the bolts to 250 inch-pounds (28 N•m).

12. Using special Tool L-4436 or its equivalent, and an inch-pound torque wrench, check the turning torque of the differential. The turning torque should be between 5 and 18 inch-pounds.

NOTE: If the turning torque is too high, install a 0.002 inch (0.050 mm) thinner shim. If the turning torque is too low, install a 0.002 inch (0.050 mm) thicker shim. Repeat until 5 to 18 inch-pounds turning torque is obtained.

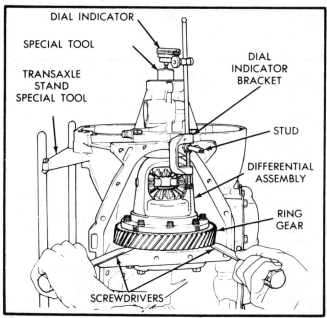

Checking differential bearing end play (©Chrysler Corp.)

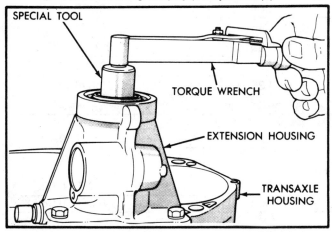

Checking differential bearing turning torque (©Chrysler Corp.)

DIFFERENTIAL BEARING SHIM CHART

End Play (with .50mm gauging shim installed)		Required Shim Combination		Total Thickness	
mm	Inch	mm		mm	Inch
0	0	.50		.50	.020
.05	.002	.75		.75	.030
.10	.004	.80		.80	.032
.15	.006	.85		.85	.034
.20	.008	.90		.90	.035
.25	.010	.95		.95	.037
.30	.012	1.00		1.00	.039
.35	.014	1.05		1.05	.041
.40	.016	.50+	.60	1.10	.043
.45	.018	.50+	.65	1.15	.045
.50	.020	.50+	.70	1.20	.047
.55	.022	.50+	.75	1.25	.049
.60	.024	.50+	.80	1.30	.051
.65	.026	.50+	.85	1.35	.053
.70	.027	.50+	.90	1.40	.055
.75	.029	.50+	.95	1.45	.057
.80	.031	.50+1.00		1.50	.059
.85	.033	.50+1.05		1.55	.061
.90	.035	1.00+	.60	1.60	.063
.95	.037	1.00+	.65	1.65	.065
1.00	.039	1.00+	.70	1.70	.067
1.05	.041	1.00+	.75	1.75	.069
1.10	.043	1.00+	.80	1.80	.071
1.15	.045	1.00+	.85	1.85	.073
1.20	.047	1.00+	.90	1.90	.075
1.25	.049	1.00+	.95	1.95	.077
1.30	.051	1.00+1.00		2.00	.079
1.35	.053	1.00+1.05		2.05	.081
1.40	.055	1.05+1.05		2.10	.083

Transaxle Assembly

NOTE: When assembling this unit, use only automatic transmission fluid or petroleum jelly to lubricate the transmission components.

With the transmission case thoroughly cleaned and the various sub assemblies overhauled and assembled, the assembly procedure is as follows:

1. Install the low-reverse band (rear) and strut assembly into the case along with the #11 thrust washer.

2. Install the overrunning clutch rollers and springs into the cam assembly and install into the case along with the #10 thrust washer.

3. Install the #9 thrust washer and the rear planetary gear assembly. Use petroleum jelly to retain the thrust washer if necessary.

4. Install the sun gear driving shell and the #7 thrust washer onto the rear planetary gear.

5. The #6 thrust washer is installed followed by the front planetary gear carrier and carefully set into the driving shell. Install the front planetary gear snap ring.

6. Install the #3 thrust washer. Install the #2 thrust washer onto the rear clutch assembly and install into the case.

7. Install the front clutch assembly onto the rear clutch assembly.

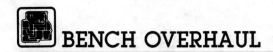

8. Install the kickdown band and strut assembly. Snug down the adjusting screw temporarily to help hold the band in place.

9. Lube the pump outer seal and install into the case. Use a new gasket. Torque the bolts to specifications.

10. Install the overhauled valve body along with the governor tubes into the case. Install the valve body attaching screws and torque to specifications. Adjust bands as required.

11. Install the parking rod and retain with the "E" clip.

12. Install a new oil filter

13. Install the oil pan using RTV sealer.

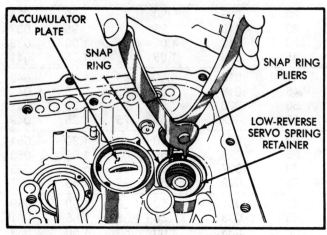

Removal or installation of low-reverse servo snap ring with location of accumulator assembly noted (©Chrysler Corp.)

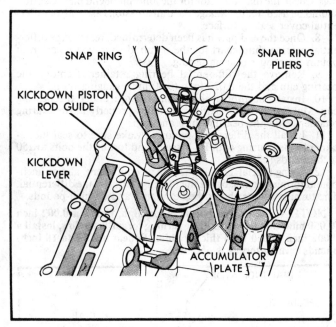

Kickdown piston rod guide snap ring removal or installation (©Chrysler Corp.)

TORQUEFLITE AUTOMATIC TRANSAXLE SPECIFICATIONS
A-404, A-413, A-415 and A-470

Pump clearances		Millimeter	Inch
Outer gear to pocket		0.045 -0.141	0.0018 -0.0056
Outer gear I.D. to crescent		0.150 -0.306	0.0059 -0.012
Outer gear side clearance		0.025 -0.050	0.001 -0.002
Inner gear O.D. to crescent		0.160 -0.316	0.0063 -0.0124
Inner gear side clearance		0.025 -0.050	0.001 -0.002
End play		**Millimeter**	**Inch**
Input shaft		0.18 -1.85	0.007 -0.073
Front clutch retainer		0.76 -2.69	0.030 -0.106
Front carrier		0.89 -1.45	0.007 -0.057
Front annulus gear		0.09 -0.50	0.0035 -0.020
Planet pinion		0.15 -0.59	0.006 -0.023
Reverse drum		0.76 -3.36	0.030 -0.132
Clutch clearance and selective snap rings		**Millimeter**	**Inch**
Front clutch (non-adjustable measured from reaction plate to "farthest" wave) (1980)		1.7 -2.7	0.067 -0.106
Front clutch (non-adjustable measured from reaction plate to "farthest" wave) (1981 and Later)	three disc	1.14 -2.45	0.045 -0.096
	two disc	0.86 -2.03	0.034 -0.080
Rear clutch (all are three disc) Adjustable		0.40-0.94	0.016-0.037

TORQUEFLITE AUTOMATIC TRANSAXLE SPECIFICATIONS
A-404, A-413, A-415 and A-470

Clutch clearance and selective snap rings		Millimeter	Inch
Sun gear drive shell, steel	Nos. 7, 8	0.85 -0.91	0.033 -0.036
1981 and later			
Front Carrier, Steel Backed Bronze	Nos. 5, 6	1.22 -1.28	0.048 -0.050
Sun Gear (Front)	No. 7	0.85 -0.91	0.033 -0.036
Sun Gear (Rear)	No. 8	0.85 -0.91	0.033 -0.036
Rear Carrier, Steel Backed Bronze	Nos. 9, 10	1.22 -1.28	0.048 -0.050
All			
Rev. drum, phenolic	No. 11	1.55 -1.60	0.061 -0.063
Tapered roller bearing settings		**Millimeter**	**Inch**
Output shaft		0.0-0.07 preload	0.0 -0.0028
Transfer shaft		0.05-0.25 end play	0.002-0.010
Differential		0.15-0.29 preload	0.006-0.012
Selective Snap Rings (3)		1.52-1.57	0.060-0.062
		1.93-1.98	0.076-0.078
		2.36-2.41	0.093-0.095

Band Adjustment:

Kickdown, Backed off from 8 N•m (72 in. lbs.)		A-404 3 Turns, A-413 & A-470 1981 —2 turns 1982 and later—2¾ turns
Low-Reverse	A-404	non-adjustable
	A-413 & A-470	3½ Turns backed off from 5 N•m (41 in. lbs.)

Thrust washers		Millimeter	Inch
All			
Reaction shaft support (phenolic)	No. 1	1.55 -1.60	0.061 -0.063
Rear clutch retainer (phenolic)	No. 2	1.55 -1.60	0.061 -0.063
Output shaft, steel backed bronze	No. 3	1.55 -1.65	0.061 -0.065
Front annulus, steel backed bronze	No. 4	2.95 -3.05	0.116 -0.120
1980			
Carrier, steel backed bronze	Nos. 5, 6, 9, 10	1.22 -1.28	0.048 -0.050

TORQUE SPECIFICATIONS
A-404, A-413, A-415 and A-470 Automatic Transaxle

	Qty	Torque Newton-meters	Torque Inch-Pounds
BOLT, SCREW OR NUT			
Bolt—Bell Housing Cover	3	12	105
Bolt—Flex Plate to Crank (A-404)	6	68	50①②③
Bolt—Flex Plate to Torque Converter (A-404)	3	54	40①④
Screw Assy. Transaxle to Cyl. Block	3	95	70①
Screw Assy. Lower Bell Housing Cover	3	12	105
Screw Assy. Manual Control Lever	1	12	105
Screw Assy. Speedometer to Extension	1	7	60
Connector, Cooler Hose to Radiator	2	12	110
Bolt—Starter to Transaxle Bell Housing	3	54	40①

TORQUE SPECIFICATIONS
A-404, A-413, A-415 and A-470 Automatic Transaxle

	Qty	Torque Newton-meters	Torque Inch-Pounds
BOLT, SCREW OR NUT			
Bolt—Throttle Cable to Transaxle Case	1	12	105
Bolt—Throttle Lever to Transaxle Shaft	1	12	105
Bolt—Manual Cable to Transaxle Case	1	28	250
Bolt—Front Motor Mount	2	54	40①
Bolt—Left Motor Mount	3	54	40①
CASE			
Connector Assembly, Cooler Line	2	28	250
Plug, Pressure Check	7	5	45
Switch, Neutral Safety	1	34	25①
DIFFERENTIAL AREA			
Ring Gear Screw	8	95	70①
Bolt, Extension to Case	4	28	250
Bolt, Differential Bearing Retainer to Case	6	28	250
Screw Assy., Differential Cover to Case	10	19	165
TRANSFER & OUTPUT SHAFT AREAS			
Nut, Output Shaft	1	271	200①
Nut. Transfer Shaft	1	271	200①
Bolt, Gov to Support	2	7	60
Bolt, Gov to Support	1	7	60
Screw Assy., Governor Counterweight	1	28	250
Screw Assy., Rear Cover to Case	10	19	165
Plug, Reverse Band Shaft	1	7	60
PUMP & KICKDOWN BAND AREAS			
Bolt, Reaction Shaft Assembly	6	28	250
Bolt Assy., Pump to Case	7	31	275
Nut, Kickdown Band Adjustment Lock	1	47	35①
VALVE BODY & SPRAG AREAS			
Bolt, Sprag Retainer to Transfer Case	2	28	250
Screw Assy., Valve Body	16	5	40
Screw Assy., Transfer Plate	16	5	40
Screw Assy., Filter	2	5	40
Screw, Transfer Plate to Case	7	12	105
Screw Assy., Oil Pan to Case	14	19	165
Nut, Reverse Band Adjusting Lock	1	14	120

① Foot pounds
② A-413 = 88 N•m 65 ft. lbs.
③ A-470 = 136 N•m 100 ft. lbs.
④ A-413 and A-470 = 54 N•m 40 ft. lbs.

SPECIAL TOOLS

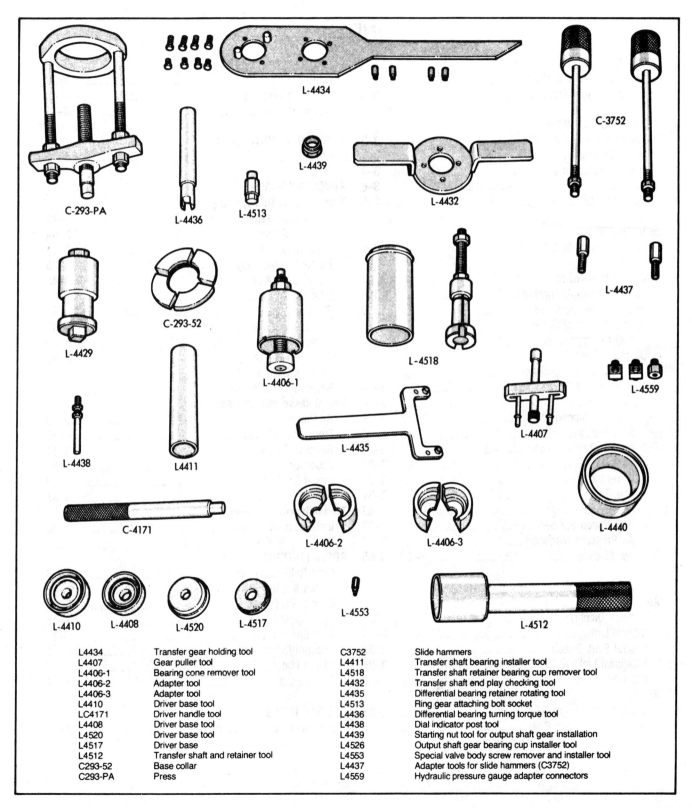

L4434	Transfer gear holding tool	C3752	Slide hammers
L4407	Gear puller tool	L4411	Transfer shaft bearing installer tool
L4406-1	Bearing cone remover tool	L4518	Transfer shaft retainer bearing cup remover tool
L4406-2	Adapter tool	L4432	Transfer shaft end play checking tool
L4406-3	Adapter tool	L4435	Differential bearing retainer rotating tool
L4410	Driver base tool	L4513	Ring gear attaching bolt socket
LC4171	Driver handle tool	L4436	Differential bearing turning torque tool
L4408	Driver base tool	L4438	Dial indicator post tool
L4520	Driver base tool	L4439	Starting nut tool for output shaft gear installation
L4517	Driver base	L4526	Output shaft gear bearing cup installer tool
L4512	Transfer shaft and retainer tool	L4553	Special valve body screw remover and installer tool
C293-52	Base collar	L4437	Adapter tools for slide hammers (C3752)
C293-PA	Press	L4559	Hydraulic pressure gauge adapter connectors

INDEX

FORD MOTOR COMPANY
C-5

APPLICATIONS

C-5 AUTOMATIC
TRANSMISSION (CODE C)

1982

Capri, Cougar, XR-7, Granada—200/232 CID Engines
Fairmont, Zephyr, Mustang—200/255 CID Engines
Thunderbird—200 CID Engine
F100

1983

Fairmont, Zephyr—200 CID Engine
Capri, Cougar, XR-7, Mustang, Thunderbird—232 CID Engine
LTD, Marquis—200/232 CID Engines
F100, Bronco II, Ranger

1984

LTD, Marquis, Cougar XR-7, Thunderbird, F150, Bronco II, Ranger

GENERAL
DESCRIPTION

TRANSMISSION AND
CONVERTER IDENTIFICATION

Transmission

The C-5 Automatic Transmission is a fully automatic unit with three forward and one reverse speeds. A lock-up converter is used in most models, to provide mechanical coupling of the engine to the rear wheels. Simpson type planetary gears are used for reduction.

The C-5 Automatic Transmission resembles the C-4 Automatic Transmission, both internally and externally. The C-5 unit has replaced the C-4 unit in production.

The major differences between the two units are in the hydraulic systems, where several new valves have been incorporated along with a new timing valve body. The converter relief valve has been moved from the pump assembly reactor support to the timing valve body, thereby causing the oil pump assemblies not to be interchangeable.

With the exception of an added tool to install the reverse and high clutch piston, most all special tools needed to service the C-4 unit will also service the C-5 unit.

The C-5 Automatic Transmission can be identified by the code letter C under the transmission section of the Vehicle Certification label, found on the driver's door lock post. A model identification tag is located under the lower intermediate servo cover bolt, with the transmission model, build date code and the assembly part number, containing the prefix and suffix.

The first line on the tag shows the transmission model prefix and suffix. A number appearing after the suffix indicates that internal parts have been changed after initial production start up. For example, a PEE-FL model transmission that has been changed internally would read PEE-FL1. Both transmissions are basically the same, but some service parts in the PEE-FL1 transmission are slightly different than the PEE-FL transmission.

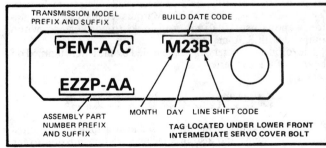

Identification plate explanation (©Ford Motor Company)

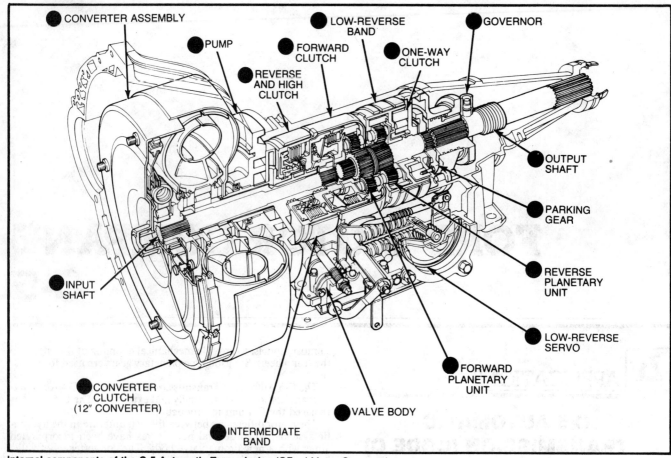

Internal components of the C-5 Automatic Transmission (©Ford Motor Company)

Therefore, it is important that the codes on the transmission identification tag be checked when ordering parts or making inquiries about the transmission.

Converter

A twelve inch torque converter is used in most models, incorporating a centrifugally operated lock-up clutch mechanism, negating the use of electrical or hydraulic lock-up components.

With the centrifugal lock-up feature, the torque converter is changed to a more efficient mechanical coupling as the speed of the input shaft is increased. The lock-up mechanism is designed to engage at various predetermined speeds of the converter, depending upon the vehicle model and driving conditions. With the centrifugal clutch lock-up, a mechanical connection exists between the engine and the rear wheels, resulting in improved driveline efficiency and fuel economy. The torque converter cannot be disassembled for field service. 1983—Fairmont Futura, Zephyr for altitude use and LTD, Marquis sedans (Calif) and station wagons use 10¼ inch non-lock-up converters, when equipped with 3.3L engines.

The torque converter can be identified by the letter stamped on the face of the converter.

METRIC FASTENERS

Metric bolts and fasteners may be used in attaching the transmission to the engine and also in attaching the transmission to the chassis crossmember mount.

The metric fastener dimensions are very close to the dimensions of the familiar inch system fasteners, and for this reason, replacement fasteners must have the same measurement and strength as those removed.

─────── WARNING ───────

Do not attempt to inter-change metric fasteners for inch system fasteners. Mismatched or incorrect fasteners can result in damage to the transmission unit through malfunctions or breakage and possible personal injury.

FLUID SPECIFICATIONS

Type H fluid, meeting Ford Motor Company's specifications ESP-M2C166-H, is used in the C-5 Automatic Transmission. To avoid filling or topping off the fluid level with the wrong type fluid, this information is stamped on the dipstick blade.

Checking the Fluid Level

The C-5 Automatic Transmission is designed to operate with the fluid level between the arrows at the hot low mark and the hot full mark.

**TRANSMISSION AT NORMAL
OPERATING TEMPERATURE
(150°-170° F. DIPSTICK HOT TO THE TOUCH)**

1. With the vehicle on a level surface, engine idling, wheels blocked, foot brakes applied, move the transmission gear selector

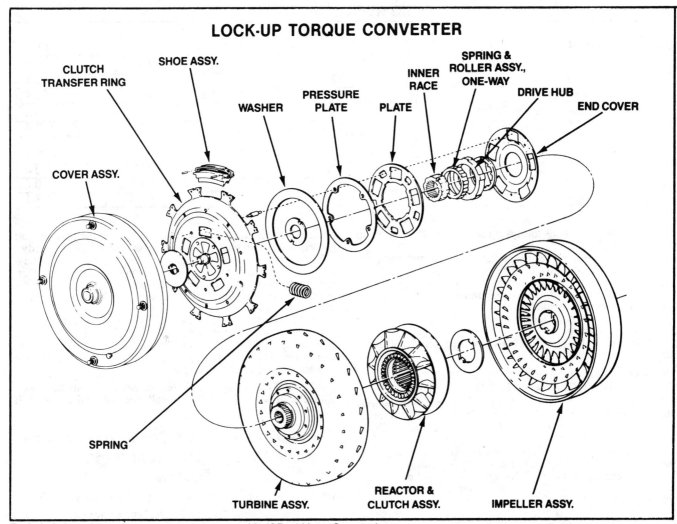

LOCK-UP TORQUE CONVERTER

Exploded view of torque converter lock-up assembly (©Ford Motor Company)

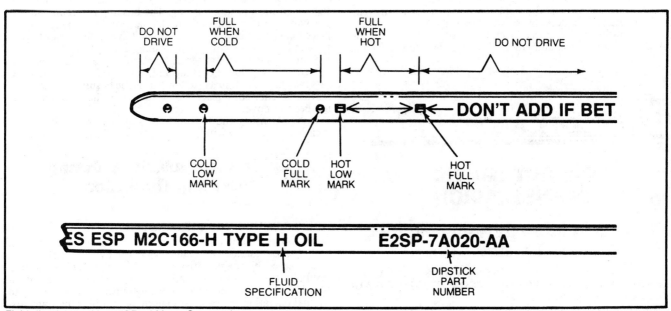

Fluid dipstick markings (©Ford Motor Company)

FLUID REFILL CAPACITY

Year	Model	Converter Size (inches)	Engine	U.S. Quarts	Liters
1982	Mustang, Capri, Fairmont, Zephyr	12	3.3, 4.2L	11	10.4
	Granada, Cougar, Thunderbird, XR-7	12	3.3, 3.8L	11	10.4
	F100, F150	12	3.8, 4.2L	11	10.4
1983-84	Fairmont, Futura, Zephyr (less Altitude)	12	3.3L	11	10.4
	LTD, Marquis (Sdn-less Calif.)	12	3.3L	10.3	9.8
	Fairmont Futura, Zephyr (Altitude), LTD, Marquis (Sdn-Calif.), (Stat. Wgn.)	10¼	3.3L	7.5	7.1
	Mustang, Capri, Thunderbird, XR-7	12	3.8L	11	10.4
	F100, F150	12	3.8L	11	10.4
	Ranger	10¼	2.3L	7.5	7.1

through the gear positions to engage each gear and to fill the oil passages with fluid.

2. Place the selector lever in the Park position and apply the parking brake. Do not turn off the engine. Allow to idle.

3. Clean the dipstick area of dirt and remove the dipstick from the filler tube. Wipe the dipstick clean, and reinsert it in the filler tube making sure it is seated firmly.

4. Remove the dipstick for the second time from the filler tube and check the fluid level.

5. If necessary adjust the fluid to its proper level and replace the dipstick firmly in the tube.

TRANSMISSION AT ROOM TEMPERATURE (70°-95° F. DIPSTICK COOL TO THE TOUCH)

1. With the vehicle on a level surface, engine idling, wheels blocked, foot brakes applied, move the transmission selector lever through the gear positions to engage each gear and to fill the oil passages with fluid.

2. Place the selector lever in the Park position and apply the parking brake. *Do not* turn off engine. Allow to idle.

3. Clean the dipstick area of dirt and remove the dipstick from the filler tube. Wipe the dipstick clean, reinsert it back into the filler tube and seat it firmly.

4. Again remove the dipstick from the filler tube and check the fluid level as indicated on the dipstick. The level should be between the middle and the top holes on the dipstick.

5. If necessary, add enough fluid to bring the level between the middle hole and the top hole on the dipstick.

6. When the fluid level is correct, fully seat the dipstick in the filler tube to avoid contamination of the fluid by entrance of dirt and water.

M MODIFICATIONS

C-5 AUTOMATIC TRANSMISSION

Rear Output Shaft Retaining Ring Deleted

1982 FORD MODELS, 1982 LINCOLN-MERCURY MODELS AND LIGHT TRUCKS EQUIPPED WITH THE C-5 AUTOMATIC TRANSMISSION

C-5 transmissions built after March 8, 1982 will not have a rear output shaft retaining ring. A design change to the parking gear (wider) and a shoulder will retain the governor collector body.

The deleted rear output shaft retaining ring was used to hold the governor collector body in position on the output shaft. Current production, after March 8, 1982, transmissions have a wider parking gear and a shoulder on the output shaft for governor collector body retention.

For both past and current model service replacement, the new output shaft will be used, and the old park gear and the retaining ring will be retained for service. The new output shaft has the retaining ring groove and the old park gear and retaining ring will be packaged together for service.

NOTE: When disassembling the transmission, hold the governor collector body when lifting the collector body and output shaft out of the case.

Service Procedure for Installation of New Design Extension Housing Bushing

1982 GRANADA, COUGAR AND 1983 RANGER

A new design bushing used on Rangers and some Granadas/Cougars.

Ranger applications with the C5 transmission have a new design extension housing bushing. The bushing is teflon/lead impregnated with no fluid drain-back hole. For service requiring replacement of the bushing, align the split line to the top of the extension housing when the new bushing is installed.

Some initial build Ranger C5 applications may have the passenger car type bronze bushing and some 1982½ Granada/Cougars may have the Ranger (teflon/lead) type bushing. In the event service is required, replace the bushing with the same type that has been removed.

If the extension and bushing assembly requires replacement, replace the assembly with the part cataloged for that application.

Upshift Irregularities During Warm-up Operation

ALL C-5 EQUIPPED VEHICLES (1982)

Some C-5 automatic transmissions, built prior to September 10, 1981 (transmission build date code-J10) may exhibit shift concerns caused by irregularities in the D2 valve. This condition results in a 1-3 shift or second gear starts, at all throttle openings. The condition may go away as the transmission warms up to normal operating temperatures and may return as the transmission cools down.

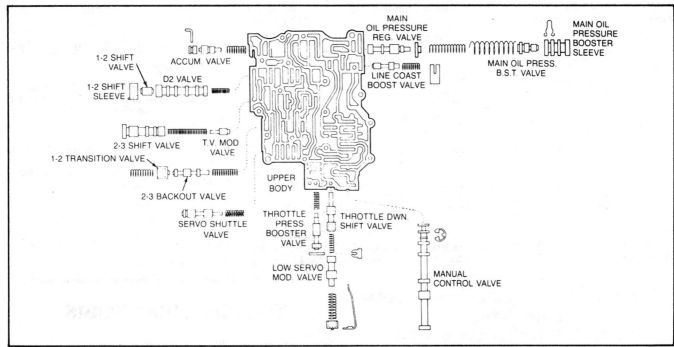

Exploded view of upper valve body (©Ford Motor Company)

Similar concerns, i.e., delayed shifts, no 1-2 upshift, second gear starts or 1-3 shifts may be also caused by chips, burrs in either the governor assembly or the main control. Service using the following procedures:

1. First determine whether there are metal particles or other material (such as clutch material, etc.) in the transmission fluid. If the fluid is found to have metal particles or materials, clean and service the transmission in the usual manner.

2. If the above fluid condition is not evident, road test the vehicle to determine if line pressure cutback is occuring. Attach a pressure gauge to the line pressure port of the case. Cutback should occur between 10 mph and 16 mph. If line pressure cutback is not observed, then the governor is not operating properly.

3. If line pressure cutback is observed, then the governor is operating properly and the main control is the probable source. Remove the main control and inspect the D2 valve and valve bore for observable causes (i.e., chips or burrs). If no obvious cause is observed, then D2 valve bore irregularity is the probable cause and the main control must be replaced.

Start-up Shudder and Vibration (10-15 MPH)

1983 T-BIRD, COUGAR W/3.8L ENGINE AND C-5 TRANSMISSION

This article outlines a procedure to service a start-up shudder condition at 10-15 mph on Thunderbird and Cougar vehicles built prior to January 20, 1983. The shudder may be felt in the steering column and is caused in part by the engine angle being near the high end of specification ($3\frac{1}{2}° \pm \frac{1}{2}°$).

To service this condition, use the following procedure:

1. Inspect for a spacer approximately ¼" thick between the No. 3 crossmember and the rear engine mount.

2. If the spacer is not found, replace the No. 3 crossmember, color coded white, with crossmember (Part No. EOSZ-6A023-A) color coded purple.

Intermediate Band Adjustment Screw Change—C-4 and C-5 Transmissions

1981-82 FAIRMONT, GRANADA, MUSTANG, T-BIRD, ZEPHYR, COUGAR, XR-7, CAPRI, F-SERIES AND E-SERIES LIGHT TRUCKS

During 1981 model C4 transmission production, a change of intermediate band adjustment screw and nut was incorporated—from coarse to fine thread.

The adjustment specifications are **different** for the intermediate band on C4 transmissions depending on the thread of the adjustment screw and nut.

Prior to any intermediate band adjustment or service, always examine the threads of the adjustment screw to determine the type of threads—fine or coarse.

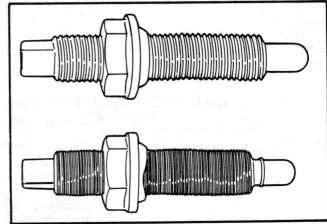

Comparison of coarse and fine threaded band adjusting bolts (©Ford Motor Company)

NOTE: This change affects the C-5 Automatic Transmission during its production stages.

The following chart denotes Intermediate Band Adjustment Specifications:

Transmission Type	Pitch on Thread Adjustment Screw & Nut	Intermediate Band Adjustment Specification
C4	Fine pitch thread (C4 trans W/C5 case)	Back-off 3 turns. Locknut torque 35-45 lbs.-ft.
C4	Coarse pitch thread	Back-off 1¾ turns. Locknut torque 35-45 lbs.-ft.
C5	Fine pitch thread only	Back-off 4¼ turns. Locknut torque 35-45 lbs.-ft.

NOTE: Enough of the thread can be seen from the outside of the transmission case so the screw does not have to be removed for visual inspection.

Fluid Leakage At 90 Degree Cooler Line Fitting To Transmission Case—C-5 Transmission

ALL 1982 MODELS SO EQUIPPED

Loss of transmission fluid at the cooler line fitting-to-case on units built prior to September 10, 1981 (transmission build date indicated by J-10) may be attributed to an undersize threaded fitting.

To service, the fitting must be inspected. If the 90 degree fitting does not appear tightly secured to the transmission case, the 90 degree fitting is to be replaced. DO NOT attempt to teflon tape the fitting threads.

Correct Fluid Usage in C-5 Transmissions

The C-5 Automatic Transmission uses a special fluid (Ford specification ESP-M2C166-H) Motorcraft part no. XT-4-H in quart cans only. Use only Type H fluid as specified on the dipstick. Use of any other fluid may cause a transmission shudder condition.

The H-type fluid used with the C-5 automatic transmission contains a special detergent which retains in suspension particles generated during normal transmission use.

This characteristic may result in a dark coloration of the fluid and does not by itself indicate need for service.

TROUBLE DIAGNOSIS

In order to properly diagnose transmission problems and avoid making second repairs for the same problem, all of the available information and knowledge must be used. Included is a knowledge of the components of the transmission and their function. Also, test procedures and their accompanying specifications charts aid in finding solutions to problems. Further answers are found by road testing vehicles and comparing all of the results of the above to the Ford C-5 Diagnostic Chart. The diagnostic chart gives condition, cause and correction for most possible trouble conditions in the Ford C-5 transmissions.

CLUTCH AND BAND APPLICATION CHART
C-5 Automatic Transmission

Gear/Range	Reverse and High Clutch	Forward Clutch	Intermediate Band	Low-Reverse Band	One-Way Clutch
Park/Neutral	—	—	—	—	—
Drive 1st	—	Applied	—	—	Holding
Drive 2nd	—	Applied	Applied	—	—
Drive 3rd	Applied	Applied	—	—	—
Manual 1	—	Applied	—	Applied	—
Manual 2	—	Applied	Applied	—	—
Reverse	Applied	—	—	Applied	—

CHILTON'S THREE "C's" TRANSMISSION DIAGNOSIS CHARTS
C-5 Automatic Transmission

Condition	Cause	Correction
Slow initial engagement	a) Improper fluid level b) Damaged or improperly adjusted linkage c) Contaminated fluid d) Improper clutch and band application, or oil control pressure	a) Add fluid as required b) Repair or adjust linkage c) Perform fluid level check d) Perform control pressure test

CHILTON'S THREE "C's" TRANSMISSION DIAGNOSIS CHARTS
C-5 Automatic Transmission

Condition	Cause	Correction
Rough initial engagement in either forward or reverse	a) Improper fluid level b) High engine idle c) Looseness in the driveshaft, U-joints or engine mounts d) Incorrect linkage adjustment e) Improper clutch or band application, or oil control pressure f) Sticking or dirty valve body	a) Perform fluid level check b) Adjust idle to specifications c) Repair as required d) Repair or adjust linkage e) Perform control pressure test f) Clean, repair or replace valve body
Harsh engagements—(warm engine)	a) Improper fluid level b) TV linkage misadjusted long disconnected/sticking/damaged return spring disconnected c) Engine curb idle too high d) Valve body bolts—loose/too tight e) Valve body dirty/sticking valves	a) Perform fluid level check b) Adjust linkage c) Check engine curb idle d) Tighten to specification e) Determine source of contamination. Service as required
No/delayed forward engagement (reverse OK)	a) Improper fluid level b) Manual linkage—misadjusted damaged c) Low main control pressure d) Forward clutch assembly burnt/damaged e) Valve body bolts—loose/too tight f) Valve body dirty/sticking valves g) Transmission filter plugged h) Pump damaged	a) Perform fluid level check b) Check and adjust or service as required c) Control pressure test, note results d) Perform air pressure test e) Tighten to specification f) Determine source of contamination. Service as required g) Replace filter h) Visually inspect pump gears. Replace pump if necessary
No/delayed reverse engagement (forward OK)	a) Improper fluid level b) Manual linkage misadjusted/damaged c) Low main control pressure in reverse d) Reverse clutch assembly burnt/worn e) Valve body bolts loose/too tight f) Valve body dirty/sticking valves g) Transmission filter plugged h) Pump damaged	a) Perform fluid level check b) Check and adjust or service as required c) Control pressure test d) Perform air pressure test e) Tighten to specification f) Determine source of contamination. Service as required g) Replace filter h) Visually inspect pump gears. Replace if necessary
No/delayed reverse engagement and/or no engine braking in manual low (1)	a) Improper fluid level b) Linkage out of adjustment c) Low reverse band servo piston burnt/worn d) Bands out of adjustment	a) Perform fluid level check b) Service or adjust linkage c) Perform air pressure test d) Adjust reverse band

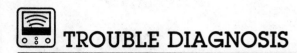

CHILTON'S THREE "C's" TRANSMISSION DIAGNOSIS CHARTS
C-5 Automatic Transmission

Condition	Cause	Correction
No/delayed reverse engagement and/or no engine braking in manual low (1)	e) Polished, glazed band of drum	e) Service or replace as required
	f) Planetary low one way clutch damaged	f) Replace
No engine braking in manual second gear	a) Improper fluid level	a) Perform fluid level check
	b) Linkage out of adjustment	b) Service or adjust linkage
	c) Intermediate band out of adjustment	c) Adjust intermediate band
	d) Improper band or clutch application, or oil pressure control system	d) Perform conrol pressure test
	e) Intermediate servo leaking	e) Perform air pressure test of intermediate service for leakage. Service as required
	f) Intermediate one way clutch damaged	f) Replace
	g) Polished or glazed band or drum	g) Service or replace as required
Forward engagement slips/shudders/chatters	a) Improper fluid level	a) Perform fluid level check
	b) Manual linkage misadjusted/damaged	b) Check and adjust or service as required
	c) Low main control pressure	c) Control pressure test
	d) Valve body bolts—loose/too tight	d) Tighten to specification
	e) Valve body dirty/sticking valves	e) Determine source of contamination. Service as required
	f) Forward clutch piston ball check not seating	f) Replace forward clutch cylinder. Service transmission as required
	g) Forward clutch piston seal cut/worn	g) Replace seal and service clutch as required
	h) Contamination blocking forward clutch feed hole	h) Determine source of contamination. Servicd as required
	i) Low one way clutch (planetary) damaged	i) Determine cause of condition. Service as required
No drive in any gear	a) Improper fluid level	a) Perform fluid level check
	b) Damaged or improperly adjusted linkage	b) Repair or adjust linkage
	c) Improper clutch or band application, or oil pressure	c) Perform control pressure test
	d) Internal leakage	d) Check and repair as required
	e) Valve body loose	e) Tighten to specification
	f) Damaged or worn clutches	f) Perform air pressure test
	g) Sticking or dirty valve body	g) Clean, repair or replace valve body
No drive forward, reverse OK	a) Improper fluid level	a) Perform fluid level check
	b) Damaged or improperly adjusted linkage	b) Repair or adjust linkage
	c) Improper clutch or band application, or oil pressure control system	c) Perform control pressure test
	d) Damaged or worn forward clutch or governor	d) Perform air pressure test

CHILTON'S THREE "C's" TRANSMISSION DIAGNOSIS CHARTS
C-5 Automatic Transmission

Condition	Cause	Correction
No drive forward, reverse OK	e) Valve body loose f) Dirty or sticking valve body	e) Tighten to specification f) Clean, repair or replace valve body
No drive, slips or chatters in first gear in D. All other gears normal	a) Damaged or worn (planetary) one-way clutch	a) Service or replace planetary one-way clutch
No drive, slips or chatters in second gear	a) Improper fluid level b) Damaged or improperly adjusted linkage c) Intermediate band out of adjustment d) Improper band or clutch application e) Damaged or worn servo and/or internal leaks f) Dirty or sticking valve body g) Polished, glazed intermediate band or drum	a) Perform fluid level check b) Service or adjust linkage c) Adjust intermediate band d) Perform control pressure test e) Perform air pressure test f) Clean, service or replace valve body g) Replace or service as required
Starts up in 2nd or 3rd	a) Improper fluid level b) Damaged or improperly adjusted linkage c) Improper band and/or clutch application, or oil pressure control system d) Damaged or worn governor. Sticking governor e) Valve body loose f) Dirty or sticking valve body g) Cross leaks between valve body and case mating surface	a) Perform fluid level check b) Service or adjust linkage c) Perform control pressure test d) Peform governor check. Replace or service governor, clean screen e) Tighten to specification f) Clean, service or replace valve body g) Service or replace valve body and/or case as required
Reverse shudders/chatters/slips	a) Improper fluid level Low main control pressure in reverse c) Reverse servo bore damaged d) Low (planetary) one-way clutch damaged e) Reverse clutch drum, bushing damaged f) Reverse clutch stator support seal rings, ring grooves worn/damaged g) Reverse clutch piston seal cut/worn h) Reverse band out of adjustment or damaged i) Looseness in the driveshaft, U-joints or engine mounts	a) Perform fluid level check b) Control pressure test c) Determine cause of condition. Service as required d) Determine cause of condition. Service as required e) Determine cause of condition. Service as required f) Determine cause of condition. Service as required g) Determine cause of condition. Service as required h) Adjust reverse band. Service as required i) Service as required
Shift points incorrect	a) Improper fluid level b) Improper vacuum hose routing or leaks c) Improper operation of EGR system	a) Perform fluid level check b) Correct hose routing c) Repair or replace as required

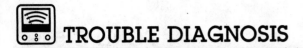

CHILTON'S THREE "C's" TRANSMISSION DIAGNOSIS CHARTS
C-5 Automatic Transmission

Condition	Cause	Correction
Shift points incorrect	d) Linkage out of adjustment	d) Repair or adjust linkage
	e) Improper speedometer gear installed	e) Replace gear
	f) Improper clutch or band application, or oil pressure control system	f) Perform shift test and control pressure test
	g) Damaged or worn governor	g) Repair or replace governor, clean screen
	f) Dirty or sticking valve body	f) Clean, repair or replace valve body
No upshift at any speed in D	a) Improper fluid level	a) Perform fluid level check
	b) Vacuum leak to diaphragm unit	b) Repair vacuum line or hose
	c) Linkage out of adjustment	c) Repair or adjust linkage
	d) Improper band or clutch application, or oil pressure control system	d) Perform control pressure test
	e) Damaged or worn governor	e) Repair or replace governor, clean screen
	f) Dirty or sticking valve body	f) Clean, repair or replace valve body
Shifts 1-3 in D, all upshifts harsh/delayed or no upshifts	a) Improper fluid level	a) Perform fluid level check
	b) Intermediate band out of adjustment	b) Adjust band
	c) Damaged intermediate servo and/or internal leaks	c) Perform air pressure test. Repair front servo and/or internal leaks
	d) Polished, glazed band or drum	d) Repair or replace band or drum
	e) Improper band or clutch application, or oil pressure control system	e) Perform control pressure test
	f) Dirty of sticking valve body	f) Clean, repair or replace valve body. (Refer to Modification Section)
	g) Manual linkage—misadjusted, damaged	g) Check and adjust or service as required
	h) Governor sticking	h) Perform governor test. Service as required
	i) Main control pressure too high	i) Control pressure test. Service as required
	j) TV control rod incorrect	j) Change TV control rod
	k) Valve body bolts—loose/too tight	k) Tighten to specification
	l) Vacuum leak to diaphragm unit	l) Check vacuum lines to diaphragm unit. Service as necessary. Perform vacuum supply and diaphragm tests.
	m) Vacuum diaphragm bent, sticking, leaks	m) Check diaphragm unit. Service as necessary
Mushy/early all upshifts pile up/upshifts	a) Improper fluid level	a) Perform fluid level check
	b) Low main control pressure	b) Control pressure test. Note results
	c) Valve body bolts loose/too tight	c) Tighten to specification
	d) Valve body valve or throttle control valve sticking	d) Determine source of contamination. Service as required

CHILTON'S THREE "C's" TRANSMISSION DIAGNOSIS CHARTS
C-5 Automatic Transmission

Condition	Cause	Correction
Mushy/early all upshifts pile up/ upshifts	e) Governor valve sticking	e) Perform governor test. Repair as required.
	f) TV control rod too short	f) Install correct TV control rod
No 1-2 upshift	a) Improper fluid level	a) Perform fluid level check
	b) Kickdown linkage misadjusted	b) Adjust linkage
	c) Manual linkage—misadjusted/ damaged	c) Check and adjust or service as required
	d) Governor valve sticking. Intermediate band out of adjustment	d) Perform governor test. Service as required. Adjust intermediate band
	e) Vacuum leak to diaphragm unit	e) Check vacuum lines to diaphragm unit. Service as required
	f) Vacuum diaphragm bent, sticking, leaks	f) Check diaphragm unit. Service as necessary
	g) Valve body bolts—loose/too tight	g) Tighten to specification
	h) Valve body dirty/sticking valves	h) Determine source of contamination. Service as required
	i) Intermediate clutch band and/or servo assembly burnt	i) Perform air pressure test
Rough/harsh/delayed 1-2 upshift	a) Improper fluid level.	a) Perform fluid level check
	b) Poor engine performance	b) Tune engine
	c) Intermediate band out of adjustment	c) Adjust intermediate band
	d) Main control pressure too high	d) Control pressure test. Note results
	e) Governor valve sticking	e) Perform governor test. Service as required.
	f) Damaged intermediate servo	f) Air pressure check intermediate servo
	g) Engine vacuum leak	g) Check engine vacuum lines. Service as necessary. Check vacuum diaphragm unit. Service as necessary. Perform vacuum supply and diaphragm tests
	h) Valve body bolts—loose/too tight	h) Tighten to specifications
	i) Valve body dirty/sticking valves	i) Determine source of contamination. Service as required
	j) Vacuum leak to diaphragm unit	j) Check vacuum lines to diaphragm unit. Service as required
	k) Vacuum diaphragm bent, sticking, leaks	k) Check diaphragm unit. Service as necessary
Mushy 1-2 shift	a) Improper fluid level	a) Perform fluid level check
	b) Incorrect engine performance	b) Tune adjust engine idle as required
	c) Improper linkage adjustment	c) Repair or adjust linkage
	d) Intermediate band out of adjustment	d) Adjust intermediate band

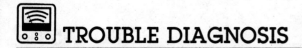

TROUBLE DIAGNOSIS

CHILTON'S THREE "C's" TRANSMISSION DIAGNOSIS CHARTS
C-5 Automatic Transmission

Condition	Cause	Correction
Mushy 1-2 shift	e) Improper band or clutch application, or oil pressure control system f) Damaged high clutch and/or intermediate servo or band g) Polished, glazed band or drum h) Dirty or sticking valve body i) Governor valve sticking	e) Perform control pressure test f) Perform air pressure test. Repair as required g) Repair or replace as required h) Clean, repair or replace valve body i) Test governor, clean or repair
No 2-3 upshift	a) Low fluid level b) TV linkage misadjusted (long)/sticking/damaged c) Low main control pressure to direct clutch d) Valve body bolts—loose/too tight. e) Valve body dirty/sticking valves f) Direct clutch or reverse/high clutch assembly burnt/worn	a) Perform fluid level check b) Adjust linkage. Service as required c) Control pressure test. Note results d) Tighten to specification e) Determine source of contamination, then service as required f) Stall test. Determine cause of condition. Service as required
Harsh/delayed 2-3 upshift	a) Low fluid level b) Incorrect engine performance c) Engine vacuum leak d) Damaged or worn intermediate servo release and high clutch piston check ball e) Valve body bolts—loose/too tight f) Valve body dirty/sticking valves g) Vacuum diaphragm or TV control rod bent, sticking, leaks	a) Perform fluid level check b) Check engine tune-up c) Check engine vacuum lines. Service as necessary. Check vacuum diaphragm unit. Service as necessary. Perform vacuum supply and diaphragm tests d) Air pressure test the intermediate servo. Apply and release the high clutch piston check ball. Service as required e) Tighten to specification f) Determine source of condition. Service as required g) Check diaphragm and rod. Replace as necessary
Soft/early/mushy 2-3 upshift	a) Improper fluid level b) Valve body bolts loose/too tight c) Valve body dirty/sticking valves d) Vacuum diaphragm or TV control rod bent, sticking, leaks.	a) Perform fluid level check b) Tighten to specification. c) Determine source of contamination. Service as required d) Check diaphragm and rod. Replace as necessary
Engine over-speeds on 2-3 shift	a) Improper fluid level b) Linkage out of adjustment c) Improper band or clutch application, or oil pressure control system d) Damaged or worn high clutch and/or intermediate servo	a) Perform fluid level check b) Service or adjust linkage c) Perform control pressure test d) Perform air pressure test. Service as required

CHILTON'S THREE "C's" TRANSMISSION DIAGNOSIS CHARTS
C-5 Automatic Transmission

Condition	Cause	Correction
Engine over-speeds on 2-3 shift	e) Dirty or sticking valve body	e) Clean, service or replace valve body
Erratic shifts	a) Improper fluid level	a) Perform fluid level check
	b) Poor engine performance	b) Check engine tune-up
	c) Valve body bolts—loose/too tight	c) Tighten to specification
	d) Valve body dirty/sticking valves	d) Line pressure test, note results. Determine source of contamination. Service as required
	e) Governor valve stuck	e) Perform governor test. Service as required
	f) Output shaft collector body seal ring damaged	f) Service as required
Rough 3-1 shift at closed throttle in D	a) Improper fluid level	a) Perform fluid level check
	b) Incorrect engine idle or performance	b) Tune, and adjust engine idle
	c) Improper linkage adjustment	c) Repair or adjust linkage
	d) Improper clutch or band application, or oil pressure control system	d) Perform control pressure test
	e) Improper governor operation	e) Perform governor test. Repair as required
	f) Dirty or sticking valve body	f) Clean, repair or replace valve body
No forced downshifts	a) Improper fluid level	a) Perform fluid level check
	b) Linkage out of adjustment	b) Repair or adjust linkage
	c) Improper clutch or band application, or oil pressure control system	c) Perform control pressure test
	d) Damaged internal kickdown linkage	d) Repair internal kickdown linkage
	e) Dirty or sticking valve body	e) Clean, repair or replace valve body
No 3-1 shift in D	a) Improper fluid level	a) Perform fluid level check
	b) Incorrect engine idle or performance	b) Tune, and adjust engine idle
	c) Damaged governor	c) Perform governor check. Repair as required
	d) Dirty or sticking valve body	d) Clean, repair or replace valve body
Runaway engine on 3-2 downshift	a) Improper fluid level	a) Perform fluid check
	b) Linkage out of adjustment	b) Repair or adjust linkage
	c) Intermediate band out of adjustment	c) Adjust intermediate band
	d) Improper band or clutch application, or oil pressure control system	d) Perform control pressure test
	e) Damaged or worn intermediate servo	e) Air pressure test check the intermediate servo. Repair servo and/or seals
	f) Polished, glazed band or drum	f) Repair or replace as required
	g) Dirty or sticking valve body	g) Clean, repair or replae valve body

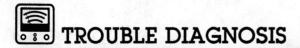

CHILTON'S THREE "C's" TRANSMISSION DIAGNOSIS CHARTS
C-5 Automatic Transmission

Condition	Cause	Correction
Engine over-speeds on 3-2 downshift	a) Improper fluid level	a) Perform fluid level check
	b) Linkage out of adjustment	b) Service or adjust linkage
	c) Intermediate band out of adjustment	c) Adjust intermediate band
	d) Improper band or clutch application, and one way clutch, or oil pressure control system	d) Perform control pressure test. Service clutch
	e) Damaged or worn intermediate servo	e) Air pressure test check the intermediate servo. Service servo and/or seals
	f) Polished, glazed band or drum	f) Service or replace as required
	g) Dirty of sticking valve body	g) Clean, service or replace valve body
Shift efforts high	a) Manual shift linkage damaged/misadjusted	a) Check and adjust or service as required
	b) Inner manual leve nut loose	b) Tighten nut to specification
	c) Manual lever retainer pin damaged	c) Adjust linkage and install new pin
Poor vehicle acceleration	a) Poor engine performance	a) Check engine tune-up
	b) Torque converter one-way clutch locked up	b) Replace torque converter
No engine braking in manual first gear	a) Improper fluid level	a) Perform fluid level check
	b) Linkage out of adjustment	b) Repair or adjust linkage
	c) Band out of adjustment	c) Adjust reverse band
	d) Oil pressure control system	d) Perform control pressure test
	e) Damaged or worn reverse servo	e) Perform air pressure test of reverse servo. Repair reverse clutch or rear servo as required
	f) Polished, glazed band or drum	f) Repair or replace as required
No engine braking in manual second gear	a) Improper fluid level	a) Perform fluid level check
	b) Linkage out of adjustment	b) Repair or adjust linkage
	c) Intermediate band out of adjustment	c) Adjust intermediate band
	d) Improper band or clutch application, or oil pressure control system	d) Perform control pressure test
	e) Intermediate servo leaking	e) Perform air pressure test of intermediate servo for leakage. Repair as required
	f) Polished or glazed band or drum	f) Repair or replace as required
Transmission noisy, valve resonance NOTE: Gauges may aggravate any hydraulic resonance. Remove gauge and check for resonance level	a) Improper fluid level	a) Perform fluid level check
	b) Linkage out of adjustment	b) Repair or adjust linkage
	c) Improper band or clutch application, or oil pressure control system	c) Perform control pressure test
	d) Cooler lines grounding	d) Free up cooler lines
	e) Dirty sticking valve body	e) Clean, repair or replace valve body
	f) Internal leakage or pump cavitation	f) Repair as required

CHILTON'S THREE "C's" TRANSMISSION DIAGNOSIS CHARTS
C-5 Automatic Transmission

Condition	Cause	Correction
Transmission overheats	a) Improper fluid level	a) Perform fluid level check
	b) Incorrect engine idle, or performance	b) Tune, or adjust engine idle
	c) Improper clutch or band application, or oil pressure control system	c) Perform control pressure test
	d) Restriction in cooler or lines	d) Repair restriction
	e) Seized one-way clutch	e) Replace one-way clutch
	f) Dirty or sticking valve body	f) Clean, repair or replace valve body
Transmission fluid leaks	a) Improper fluid level	a) Perform fluid level check
	b) Leakage at gasket, seals, etc.	b) Remove all traces of lube on exposed surfaces of transmission. Check the vent for free breathing. Operate transmission at normal temperatures and inspect for leakage. Repair as required
	c) Vacuum diaphragm unit leaking	c) Replace diaphragm

HYDRAULIC CONTROL SYSTEM

In order to diagnose transmission trouble the hydraulic control circuits must be traced. The main parts of the hydraulic control system are the oil pump, valve body, governor, and the servo systems togehter with the fluid passages connecting the units. The clutches and bands control the planetary gear units which determine the gear ratio of the transmission.

Major Components

The C-5 Automatic transmission oil pump operates constantly whenever the engine is operating and at engine speed, driven by the converter pump cover, which is attached to the engine flywheel or flex plate, providing fluid pressure to the hydraulic system.

The major components of the hydraulic control system are as follows:

1. Reservoir or sump—The oil pan containing a supply of automatic transmission fluid for use in the hydraulic control system.

2. Screens—Both a pan and a pump inlet screen protects the hydraulic system from dirt or other foreign material that maybe carried by the fluid.

3. Oil pump—Provides fluid pressure to the hydraulic system whenever the engine is operating.

4. Main line oil pressure regulator valve—Regulates the main control line pressure in the hydraulic system and also supplys converter, cooler and lubrication systems with pressured fluid.

5. Main pressure booster valve and sleeve—Causes fluid pressure to be boosted depending upon the range or gear and causes the line pressure to increase or decrease in relation to the engine load.

6. Fluid cooler—Located in the radiator and removes heat generated in the torque converter and transmission by having the fluid flow through the cooler core.

7. Intermediate servo accumulator—Cushions or smooths the 1-2 upshift.

8. Converter—Couples the engine to the transmission gear train input shaft.

9. Drain-back valve—Prevents the lubrication fluid from draining back into the sump after the engine is stopped.

10. Converter pressure relief valve—Prevents excessive pressure build-up in the converter unit during operation.

11. Reverse and High clutch—Applied by hydraulic pressure to couple the input shaft to the sun gear.

12. Forward clutch—Applied by hydraulic pressure to couple the input shaft to the forward ring gear.

13. Intermediate servo—Applies hydraulically to actuate the intermediate band.

14. Low-reverse servo—Applies hydraulically to actuate the low-reverse band.

15. Throttle valve—Regulates throttle pressure (T.V.) as an engine load signal to the hydraulic system.

16. Governor—Provides a road speed signal to the hydraulic control system.

17. 1-2/3-2 timing control valve—Routes servo release fluid to the accumulator on a 1-2 upshift and joins the servo release fluid pressure with the high clutch on a 3-2 downshift.

18. 3-2 timing valve—Regulates the reverse and high clutch pressure for a smooth 3-2 downshift.

19. Cutback valve—Provides a reduction or cut-back in line pressure as the road speed of the vehicle increases.

20. Manual valve—Directs the line pressure to the various passages to apply or block off passages during the application of clutches and band servos, depending upon the shift selector position in the valve body.

21. 1-2 shift valve and sleeve—Acted on by governor pressure to control the Drive 2 valve.

22. Drive 2 valve—Control the 1-2 upshift and the 3-1 or 2-1 downshift.

23. 2-3 Shift valve—Controls the 3-2 downshift and the 2-3 upshift.

24. Throttle pressure modulator—Modulates the T.V. or boosted T.V. pressure to cause a lower T.V. signal to the Drive 2 valve.

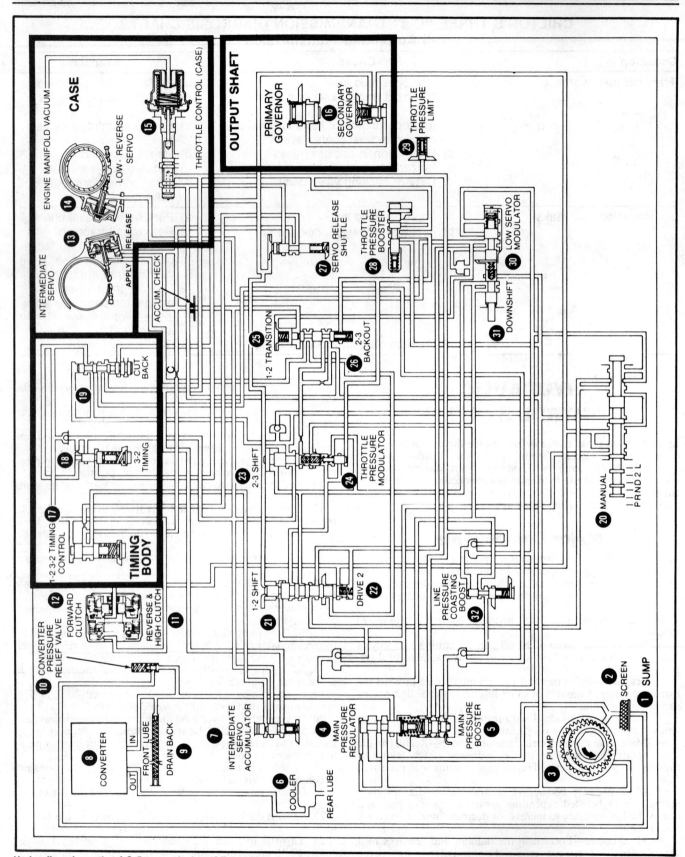

Hydraulic schematic of C-5 transmission (©Ford Motor Company)

25. 1-2 Transition valve—Prevents band to band tie-up on a manual 1-2 shift.

26. 2-3 Back-out valve—Prevents clutch to band tie-up on a closed throttle 2-3 upshift.

27. Servo release shuttle valve—Selects servo release orifices on a 3-2 downshift, depending upon road speed.

28. Throttle pressure booster valve—Provides a boosted T.V. pressure in a high engine load mode.

29. Throttle pressure limit valve—Regulates maximum T.V. pressures.

30. Low servo modulator valve—Modulates the reverse servo apply pressure in a manual shift to the first gear.

31. Downshift valve—Causes through detent kickdown shifts to occur at higher road speed than the torque demand downshifts. It also provides maximum delay on wide open throttle (WOT) upshifts.

32. Line pressure coasting boost valve—Causes a boost in the line pressure when in ranges 1 and 2.

Diagnosis Tests

GENERAL DIAGNOSTIC SEQUENCE

Before starting any test procedures, a selected sequence should be followed in the diagnosis of C-5 automatic transmission malfunctions. A suggested sequence is as follows;

1. Inspect the fluid level and correct as required.
2. Check the freedom of movement of the downshift linkage and adjust as required.
3. Check the manual linkage synchronization and adjust as required.

4. Inspect the vacuum routings to the modulator and be sure sufficient engine vacuum is available.

5. Install a 400 psi oil pressure gauge to the main line pressure port on the transmission case. Should this arrangement be used on a road test, route and secure the hose so as not to drag or be caught during the test.

Preform the pressure test in all gears and N/P positions. Record all results.

6. Perform the road test over a predetermined route to verify shift speeds and engine performance. Refer to the pressure gauge during all shifts for irregularities in the pressure readings. With the aid of a helper, record all readings for reference.

7. During the road test, governor operation can be noted and the shift speeds recorded as the throttle valves are moved through various positions. Should further testing of the governor system be needed, this can be accomplished when the vehicle is returned to the service center, by a shift test.

8. Should a verification of engine performance or initial gear engagement be needed, a stall test can be done to aid in pinpointing a malfunction.

9. Adjust transmission bands should the malfunction indicate loose or slipping bands.

10. Perform a case air pressure test, should the malfunction indicate internal transmission pressure leakage.

FLUID LEVEL AND CONDITION

The fluid level should be checked with the transmission at normal operating temperature, however, the cold fluid check can be made, if necessary (refer to the fluid level checking and fluid specification outline at the beginning of this section). Should top-

RANGE	CHECK FOR	CONDITION (OK OR NOT OK)
1	Engagement	
	Should be no 1-2 upshift	
	Engine braking in low gear	
	Shifts 3-2 and then 2-1 coming out of D at cruise	
	No slipping	
2	Engagement/Starts in second gear	
	Should be no 2-3 upshift	
	Shifts 3-2 coming out of D at cruise	
	No slipping	
D	Engagement/Starts in first gear	
	Upshifts not mushy or harsh	
	Upshifts and downshifts at specified speeds	
	● Minimum Throttle 1-2	
	● Minimum Throttle 2-3	
	● Minimum Throttle 3-2	
	● Minimum Throttle 2-1	
	● To-detent (heavy throttle) 1-2	
	● To-detent (heavy throttle) 2-3	
	● To-detent (heavy throttle) 3-2	
	● Through-detent (W.O.T.) 1-2	
	● Through-detent (W.O.T.) 2-3	
	● Through-detent (W.O.T.) 3-2	
	● Through-detent (W.O.T.) 3-1 or 2-1	
R	Engagement	
	Back up without slip	

Typical road test diagnosis sequence (©Ford Motor Company)

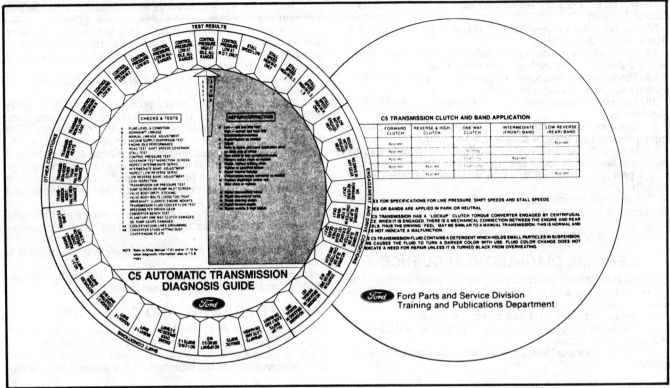

Slide wheel diagnostic tool, available from Ford Motor Company to assist in diagnosing C-5 automatic transmission (©Ford Motor Company)

ping off of the fluid level be necessary, add only type F automatic transmission fluid as specified on the dip stick. Should the fluid level be low, be sure to inspect the transmission for signs of leakage.

The fluid condition should be observed during the checking operation. The fluid should be clean and not discolored from contaminates.

NOTE: Type H fluid contains a detergent which retains particles in suspension, which are generated in normal automatic transmission operation. This characteristic does not in its self indicate a transmission malfunction, although the fluid may be dark in color.

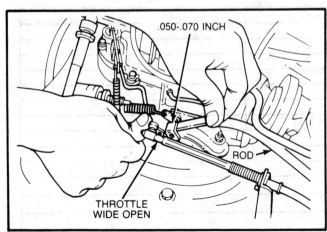

Downshift rod adjustment (©Ford Motor Company)

By smelling the fluid, a burnt or rotten egg odor indicates a major transmission failure with overhaul required. Should burned flakes, solid residue or varnish in the fluid or on the dipstick be evident, overhaul of the unit is indicated.

DOWNSHIFT LINKAGE CHECK

Check the linkage for binding and for proper operation. The rod must not move until the throttle is almost wide open. With the throttle at the wide open position, the downshift linkage should have some slight travel (0.050-0.070 inch). Should binding of the rod, a sticking valve or rod misalignment be evident, the condition must be corrected before the vehicle is road tested.

NOTE: Refer to the "On-Car Adjustment" outline for proper adjustment procedures.

MANUAL LINKAGE CHECK

Before starting the engine, move the shift lever through each gear range, feeling the detents in the transmission. The detents and the shift selector should be syncronized. Place the shift selector in the D position and against its stop. Try to move the selector lever to the manual 2 position without raising the lever. If there is free movement to the D stop or if the lever stop is up on the manual 2 stop, an adjustment is required.

─────────────── **CAUTION** ───────────────

Do not roadtest the vehicle until the adjustments have been completed. Refer to the "On-Car Adjustment" outline for proper adjustment procedures.

Vacuum Diaphragm

The modulated throttle system, which adjusts throttle pressure for the control of the shift valves, is operated by engine manifold

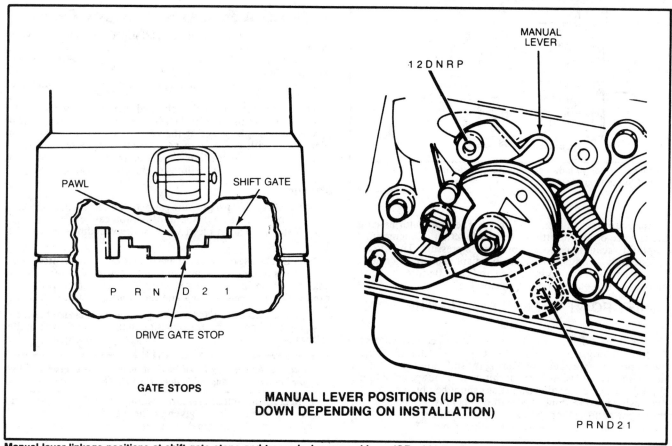

PAWL

SHIFT GATE

P R N D 2 1

DRIVE GATE STOP

GATE STOPS

MANUAL LEVER

1 2 D N R P

MANUAL LEVER POSITIONS (UP OR DOWN DEPENDING ON INSTALLATION)

P R N D 2 1

Manual lever linkage positions at shift gate stops and transmission manual lever (©Ford Motor Company)

vacuum through a vacuum diaphragm and must be inspected whenever a transmission defect is apparent.

MANIFOLD VACUUM

Testing

1. With the engine idling, remove the vacuum supply hose from the modulator nipple and check the hose end for the presence of engine vacuum with an appropriate gauge.

2. If vacuum is present, accelerate the engine and allow it to return to idle. A drop in vacuum should be noted during acceleration and a return to normal vacuum at idle.

3. If manifold vacuum is not present, check for breaks or restrictions in the vacuum lines and repair.

VACUUM DIAPHRAGM

Testing

1. Apply at least 18 in. Hg. to the modulator vacuum nipple and observe the vacuum reading. The vacuum should hold.

2. If the vacuum does not hold, the diaphragm is leaking and the modulator assembly must be replaced.

NOTE: A leaking diaphragm causes harsh gear engagements and delayed or no up-shifts due to maximum throttle pressure developed.

3. Remove the vacuum diaphragm from the transmission and attach it to a good vacuum source set at 18 in. of vacuum. If the vacuum holds at 18 in., then the diaphragm is not leaking.

4. Check for operation of the diaphragm return spring by holding a finger over the end of the control rod and removing the vac-

uum source. When the vacuum source is removed the spring should push out on the rod. If it does not push out, replace the diaphragm unit.

ALTITUDE COMPENSATING MODULATOR

To control shift spacing and shift timing where engine performance is greatly affected by changes in altitudes, and altitude

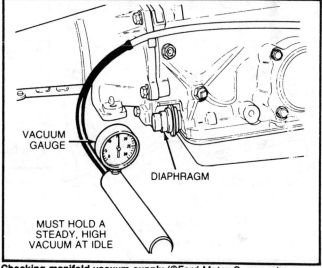

VACUUM GAUGE

DIAPHRAGM

MUST HOLD A STEADY, HIGH VACUUM AT IDLE

Checking manifold vacuum supply (©Ford Motor Company)

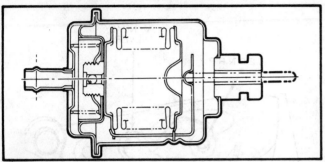

Cross-section of altitude compensating modulator
(©Ford Motor Company)

compensating modulator unit is used. The modulator assembly exerts force on the throttle valve as a function of engine intake manifold vacuum and local atmospheric pressure. The modulator assembly is composed of a flexible diaphragm, an aneroid bellows assembly, a calibration spring and case. Be sure the correct modulator is replaced in the transmission.

SHIFT POINT CHECKS DURING THE ROAD TEST

To determine if the governor pressure and shift control valves are functioning properly, a road test of the vehicle should be made over a predetermined course. During the shift point check operation, if the transmission does not shift within the specified limits, slippage occurs or certain gear ratios cannot be obtained, further diagnosing must be done. The shift points must be checked with the engine at normal operating temperature to avoid fast idle operation of the engine to affect the shift timing.

Control Pressure System Tests

Control pressure tests should be performed whenever slippage, delay or harshness is felt in the shifting of the transmission. Throttle and modulator pressure changes can cause these problems also, but are generated from the control pressures and therefore reflect any problems arising from the control pressure system.

The control pressure is first checked in all ranges without any throttle pressure input, and then is checked as the throttle pressure is increased by lowering the vacuum supply to the vacuum diaphragm, with the use of a vacuum bleed valve or stall test.

The control pressure tests should define differences between mechanical or hydraulic failures of the transmission.

Testing

1. Install a 0-400 psi pressure gauge to the control pressure tap, located on the left side of the transmission case.
2. Install a vacuum bleed valve in the vacuum line near the vacuum gauge.

NOTE: The use of a vacuum bleed valve or hand operated vacuum tester is recommended and will enable the repairman to set the engine vacuum to the required specifications without danger of overheating the transmission and fluid during a stall test.

3. Block wheels and apply both parking and service brakes.
4. Operate the engine/transmission in the ranges shown on the following charts and at the manifold vacuum specified.

CAUTION
When using the vacuum bleed method, operate the engine at 1000 rpm for the 10 inch and wide open throttle (WOT) tests.

5. Record the actual pressure readings in each test and compare to the specification as given on the specifications charts. Refer to Vacuum Diaphragm Adjustment, On-Car Services section.

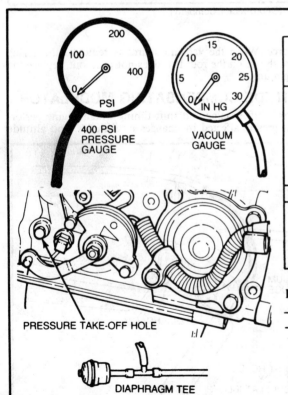

PRESSURE TAKE-OFF HOLE

DIAPHRAGM TEE

Control pressure test preparation (©Ford Motor Company)

Engine RPM	Manifold Vacuum In-Hg	Throttle	Range	PSI Record Actual	PSI Record Spec.
Idle	Above 12	Closed	P		
			N		
			D		
			2		
			1		
			R		
As Required	10	As Required	D,2,1		
As Required	Below 3	Wide Open	D		
			2		
			1		
			R		

Results: _____

_____

SPECIFICATIONS AND RESULTS

PRECAUTIONS BEFORE IDLE TESTS

1. Be sure manifold vacuum is above 15 in. Hg. If lower and unable to raise it, check for vacuum leaks and repair
2. Be sure the manifold vacuum changes with throttle plate opening. Check by accelerating quickly and releasing the accelerator and observing the vacuum reading.

PRECAUTIONS IF STALL TEST IS USED ON PRESSURE RISE TEST (REFER TO STALL TEST PROCEDURES)

1. Don't operate engine/transmission at stall for longer than 5 seconds per test.
2. Operate the engine between 1000 and 1200 rpm at the end of a test for approximately one to two minutes for cooling.
3. Release the accelerator immediately in case of slippage or spin-up of the transmission to avoid more damage to the unit.

CONTROL PRESSURE TESTS

Condition	Cause	Correction
Pressure low at idle in all ranges	a) Low transmission fluid level	a) Repair leaks and adjust fluid to proper level
	b) Restricted intake screen or filter	b) Change transmission fluid and screen
	c) Loose oil tubes	c) Re-install or replace tubes as needed
	d) Loose valve body or regulator to case bolts	d) Torque bolts or replace as needed
	e) Excessive leakage in front pump	e) Replace seals
	f) Leak in case	f) Replace case
	g) Control valves or regulator valve sticking	g) Clean valve body and replace fluid and screen
Pressure high at idle in all ranges	a) EGR system	a) Clean EGR system and replace valve if needed
	b) Improper operation of vacuum diaphragm (vacuum modulator)	b) Adjust or replace vacuum diaphragm
	c) Vacuum line	c) Repair or replace the vacuum line
	d) Throttle valves or control rod	d) Repair or replace as needed
	e) Regulator boost valves sticking	e) Clean or replace valve body
Pressure OK at idle in all ranges but low at 10 in. of vacuum	a) Vacuum diaphragm (modulator)	a) Adjust or replace vacuum diaphragm (modulator)
	b) Control rod or throttle valve sticking	b) Clean, repair or replace as needed
Pressure OK at 10 in. vacuum but low at 1 in. of vacuum	a) Excessive oil pressure leakage	a) Replace oil seals as needed
	b) Low oil pump capacity	b) Repair or replace oil pump
	c) Restricted oil pan screen	c) Change transmission fluid and screen
Pressure low in drive	a) Forward clutch	a) Replace forward clutch
Pressure low in 2	a) Forward clutch or intermediate servo	a) Replace forward clutch or intermediate servo
Pressure low in 1	a) Forward clutch and/reverse clutch or servo	a) Replace forward clutch or reverse clutch, and repair or replace servo
Pressure low in R	a) High clutch and/reverse clutch or servo	a) Replace high clutch or reverse clutch, and repair or replace servo
Pressure low in P/N	a) Valve body	a) Clean or replace valve body
Pressure high or low in all test conditions	a) Modulator assembly and/or diaphragm control rod	a) Replace modulator and/or diaphragm control rod

Automatic Shift Point Tests

Testing

The shift test can be performed in the repair shop to check the shift valve operation, governor operation, shift delay pressures, throttle boost and downshift valve action. The following procedure can be used for a guide.

1. Raise the vehicle and support safely so the drive wheels are clear of the floor. Install an oil pressure gauge.

2. Disconnect and plug the vacuum line to the modulator valve. Attach either a hand operated vacuum pump or an electrically operated vacuum pump hose to the nipple on the modulator valve.

3. Assuming the vacuum diaphragm is intact within the modulator valve, apply 18 inches of vacuum to the diaphragm. With the engine operating, place the transmission in the DRIVE mode and make minimum throttle shifts of the 1-2 and 2-3 gearing. As the shift is made, the speedometer needle and the oil pressure gauge needle will make a momentary movement, the speedometer needle moving up, while the oil pressure gauge needle moves downward. Both needles will stabilize as the shift is completed. As the shift is being made, the drive line will bump, due to the interruption of torque. If the shifts are within specifications, the 1-2, 2-3 and governor valves are operating satisfactory.

4. If the shift points are not in specifications, a governor test should be made to isolate the problem.

—————— CAUTION ——————

After each test, operate the engine in neutral at 1000 rpm to cool the transmission for a period of one to two minutes.

5. To check the shift delay pressures and the throttle boost, decrease the vacuum at the modulator diaphragm to 0-2 inches of vacuum and make the 1-2 shift. If the shift point raises to specifications, the throttle boost and shift delay valves are operating properly.

6. To check the downshift valve action, keep the vacuum at 0-2 inches at the modulator diaphragm. Position the downshift linkage in the wide open throttle position (WOD) and repeat the 1-2 shift test. The speed of the shift point should be higher.

—————— CAUTION ——————

Never exceed 60 mph speedometer speed during any of the above tests.

CONTROL PRESSURE SPECIFICATIONS—1982

Transmission Model*	Range	10" Vacuum
PEN-C, G, J, K, PEM-AL, AM	D	90-101
	2,1	123-136
	R	151-168
PEP-E, F, G, H, P, N	D	87-97
	2,1	119-132
	R	145-162
PEP-B, D	D	86-99
	2,1	120-132
	R	143-165

CONTROL PRESSURE SPECIFICATIONS—1983

Transmission Type	Transmission Model	Range	Idle 15" & Above	Idle 10"	WOT Stall Thru Detent
C5	PEN-G,P,S,U,Y,Z	D	55-70	94-107	162-174
		2,1	107-109	100-112	162-174
		R	91-117	156-178	271-291
		P,N	55-70	94-107	162-174
C5	PEN-AA,AB	D	54-72	92-109	161-176
		2,1	107-119	100-112	161-176
		R	90-121	154-182	268-293
		P,N	54-72	92-109	161-176
C5	PEP-V	D	64-68	86-97	160-169
		2,1	105-114	100-109	160-169
		R	76-96	144-162	267-281
		P,N	64-68	86-97	160-169
C5	PEP-R	D	67-80	99-110	157-165
		2,1	102-112	99-110	157-165
		R	113-133	166-184	261-275
		P,N	67-80	99-110	157-165
C5	PEP-B @	D	66-81	97-111	153-165
		2,1	102-112	103-111	153-165
		R	110-134	162-174	256-274
		P,N	66-81	97-111	153-165
C5	PEJ-AE,AG	D	60-64	83-94	157-165
		2,1	101-111	97-106	157-165
		R	70-90	139-156	262-275
		P,N	60-64	83-94	157-165

CONTROL PRESSURE SPECIFICATIONS—1983

Transmission Type	Transmission Model	Range	Idle 15″ & Above	Idle 10″	WOT Stall Thru Detent
C5	PEJ-AF,AH @	D	60-64	82-96	155-167
		2,1	102-110	97-106	155-167
		R	69-94	137-160	259-278
		P,N	60-94	82-96	155-167
C5	PEJ-AJ	D	57-61	83-93	157-165
		2,1	101-111	97-106	157-165
		R	70-90	139-156	262-275
		P,N	57-61	83-93	157-165
C5	PEJ-AK @	D	57-61	82-96	155-167
		2,1	102-110	97-106	155-167
		R	69-94	137-147	259-279
		P,N	57-61	83-93	157-167
C5	PEA-CR	D	64-68	83-94	163-172
		2,1	109-117	104-114	163-172
		R	79-85	139-157	272-287
		P,N	64-68	83-94	163-172

@ Absolute barometric pressure (ABP) 29.25

SHIFT SPEEDS—ACTUAL M.P.H.
C5 Automatic Transmission

1983 3.8L MUSTANG/CAPRI/LTD/MARQUIS/THUNDERBIRD/COUGAR PEP-R, PEP-W

Throttle	Range	Shift	OPS—R.P.M.	1	2	3
Closed (Above 17″ Vacuum)	D	1-2	413-456	12-13	11-13	10-11
	D	2-3	580-761	17-22	15-20	14-18
	D	3-1	331-366	10-11	9-10	8-9
	1	2-1	1074-1271	31-37	29-34	25-30
To Detent (Torque Demand)	D	1-2	907-1187	26-35	24-32	21-28
	D	2-3	1600-1852	47-54	43-49	38-44
	D	3-2	1459-1614	43-47	39-43	34-38
Through Detent (W.O.T.)	D	1-2	1480-1718	43-50	39-46	35-40
	D	2-3	2621-2901	76-85	70-77	62-68
	D	3-2	2359-2516	69-73	63-67	56-59
	D	3-1 2-1	1061-1292	31-38	28-34	25-30

Axle Ratio	Tire Size	Use Column No.
2.47:1	P185/75R14	1
	P195/75R14	1
	205/70HR14	1
	220/55R390	1
2.73:1	P185/75R14	2
	P195/75R14	2
	205/70HR14	2
	220/55R390	2
3.08:1	P185/75R14	3
	P195/75R14	3

1983 3.8L MUSTANG/CAPRI—ALTITUDE
PEP-B1

Throttle	Range	Shift	OPS—R.P.M.	1
Closed (Above 17″ Vacuum)	D	1-2	411-457	11-12
	D	2-3	580-761	15-20
	D	3-1	331-366	9-10
	1	2-1	1074-1269	28-33
To Detent (Torque Demand)	D	1-2	845-1185	22-31
	D	2-3	1527-1850	40-49
	D	3-2	1436-1610	38-42
Through Detent (W.O.T.)	D	1-2	1464-1716	39-45
	D	2-3	2600-2898	68-76
	D	3-2	2338-2513	62-66
	D	3-1 2-1	1046-1290	28-34

Axle Ratio	Tire Size	Use Column No.
2.73:1	P185/75R14	1
	P195/75R14	1
	205/70HR14	1
	220/55R390	1

1983 3.8L T'BIRD/COUGAR
PEP-V

Throttle	Range	Shift	OPS—R.P.M.	1	2
Closed (Above 17″ Vacuum)	D	1-2	421-467	12-14	12-14
	D	2-3	586-807	17-23	17-24
	D	3-1	331-366	10-11	10-11
	1	2-1	1091-1313	32-38	32-39
To Detent (Torque Demand)	D	1-2	831-1193	24-35	25-35
	D	2-3	1496-1845	44-54	44-55
	D	3-2	1330-1542	39-45	40-46
Through Detent (W.O.T.)	D	1-2	1506-1769	44-51	45-53
	D	2-3	2655-2971	77-87	79-89
	D	3-2	2387-2582	69-75	71-77
	D	3-1 2-1	1080-1332	31-39	32-40

Axle Ratio	Tire Size	Use Column No.
2.47:1	P185/75R14	1
	P195/75R14	2
	P205/70R14	1
	220/55R390	1
	205/60R390	1

1983 3.3L FAIRMONT/ZEPHYR-SEDAN, POLICE, TAXI, LTD/MARQUIS
PEN-P, PEN-G, PEN-S, PEN-U, PEN-Y, PEN-Z, PEN-CA

Throttle	Range	Shift	OPS—R.P.M.	1	2
Closed (Above 17″ Vacuum)	D	1-2	399-446	10-11	11-12
	D	2-3	585-847	15-22	15-22
	D	3-1	331-366	8-9	9-10
	1	2-1	1091-1317	28-34	29-35
To Detent (Torque Demand)	D	1-2	492-964	13-25	13-25
	D	2-3	1107-1551	29-40	29-41
	D	3-2	1040-1515	27-39	27-40
Through Detent (W.O.T.)	D	1-2	1514-1775	39-46	40-47
	D	2-3	2683-3008	69-78	71-80
	D	3-2	2426-2633	63-68	64-70
	D	3-1 2-1	1094-1344	28-35	29-35

Axle Ratio	Tire Size	Use Column No.
2.73	P175/75R14	1
	190/65R390	
2.73	P185/75R14	2
	P195/75R14	2
	P205/70R14	2
	195/70R14	2

1983 3.3L FAIRMONT/ZEPHYR-SEDAN, POLICE, TAXI
PEN-AA, AB, BA

Throttle	Range	Shift	OPS—R.P.M.	1	2
Closed (Above 17″ Vacuum)	D	1-2	398-449	10-12	10-12
	D	2-3	582-854	15-22	15-23
	D	3-1	331-366	8-9	9-10
	1	2-1	1091-1317	28-34	29-35
To Detent (Torque Demand)	D	1-2	490-994	13-26	13-26
	D	2-3	1085-1589	28-41	29-42
	D	3-2	978-1545	35-40	26-41
Through Detent (W.O.T.)	D	1-2	1506-1781	39-46	40-47
	D	2-3	2673-3016	69-78	71-80
	D	3-2	2416-2641	62-68	64-70
	D	3-1 2-1	1087-1348	28-35	29-36

Axle Ratio	Tire Size	Use Column No.
2.73	P175/75R14 190/65R390	1
2.73	P185/75R14	2
	P195/75R14	2
	P205/70R14	2
	195/70HR14	2

1983 3.8L F100 4 x 2
PEA-CR1

Throttle	Range	Shift	OPS—R.P.M.	1	2	3
Closed (Above 17″ Vacuum)	D	1-2	409-444	11-12	12-13	13-14
	D	2-3	655-864	19-24	19-25	20-27
	D	3-1	331-366	9-10	10-11	10-11
	1	2-1	1104-1303	31-37	32-38	34-40
To Detent (Torque Demand)	D	1-2	490-910	14-26	14-27	15-28
	D	2-3	1079-1489	30-42	31-45	33-46
	D	3-2	1017-1498	29-42	30-44	31-46
Through Detent (W.O.T.)	D	1-2	1513-1751	43-49	44-51	46-54
	D	2-3	2682-2975	76-84	78-87	82-91
	D	3-2	2435-2610	69-74	71-76	75-80
	D	3-1 2-1	1100-1329	31-38	32-39	34-41

Axle Ratio	Tire Size	Use Column No.
2.73:1	P195/75R15SL	1
	P215/75R15SL	2
	P235/75R15XL	3

1983 2.3L RANGER 4 x 2
PEJ-AE1

Throttle	Range	Shift	OPS—R.P.M.	1	2	3	4	5	6
Closed (Above 17″ Vacuum)	D	1-2	405-440	8-9	8-9	8-9	8-9	8-9	8-9
	D	2-3	880-1116	18-23	18-24	19-24	17-21	17-22	18-22
	D	3-1	331-366	6-7	7-8	7-8	6-7	6-7	6-7
	1	2-1	1478-1724	30-35	31-37	32-37	28-33	29-34	30-35
To Detent (Torque Demand)	D	1-2	831-1441	17-30	18-30	18-31	16-28	16-28	17-29
	D	2-3	1477-2004	30-41	31-43	32-44	28-38	29-39	30-40
	D	3-2	1458-1740	30-36	31-37	32-38	28-33	29-34	29-35
Through Detent (W.O.T.)	D	1-2	1977-2283	40-47	42-49	43-50	38-44	39-45	40-46
	D	2-3	3071-3438	63-70	65-73	67-75	59-66	66-68	62-69
	D	3-2	2743-2965	56-60	58-63	60-65	53-57	54-58	55-60
	D	3-1 2-1	1485-1768	30-36	32-38	32-39	28-34	29-35	30-36

Axle Ratio	Tire Size	Use Column No.
3.45	185/75R14SL	1
	195/75R14SL	2
	P205/75R14SL	3
	P205/75R14XL	3
3.73	185/75R14SL	4
	195/75R14SL	5
	P205/75R14SL	6
	P205/75R14XL	

1984 3.8L LTD/MARQUIS—50 STATES

Throttle	Range	Shift	OPS—R.P.M.	1	2
Closed	D	1-2	421-467	11-12	11-13
(Above	D	2-3	586-807	15-21	16-22
17″	D	3-1	331-366	9-10	9-10
Vacuum)	1	2-1	1091-1313	29-35	29-35
To Detent	D	1-2	612-1063	16-28	16-29
(Torque	D	2-3	1240-1659	33-44	33-45
Demand)	D	3-2	1139-1542	30-41	31-42
Through	D	1-2	1506-1769	40-47	41-48
Detent	D	2-3	2655-2971	70-78	72-80
(W.O.T.)	D	3-2	2387-2582	63-68	64-70
	D	3-1 2-1	1080-1332	28-35	29-36

Axle Ratio	Tire Size	Use Column No.
2.73:1	P185/75R14	1
	P195/75R14	2

1984 3.8L THUNDERBIRD/COUGAR—50 STATES

Throttle	Range	Shift	OPS—R.P.M.	1	2
Closed	D	1-2	421-467	11-13	11-12
(Above	D	2-3	586-807	16-22	15-21
17″	D	3-1	331-366	9-10	9-10
Vacuum)	1	2-1	1091-1313	29-35	29-35
To Detent	D	1-2	831-1193	22-32	22-32
(Torque	D	2-3	1496-1845	40-50	40-49
Demand)	D	3-2	1330-1542	36-42	35-41
Through	D	1-2	1506-1769	41-48	40-47
Detent	D	2-3	2655-2971	72-80	70-78
(W.O.T.)	D	3-2	2387-2582	64-70	63-68
	D	3-1 2-1	1080-1332	29-36	28-35

Axle Ratio	Tire Size	Use Column No.
2.73:1	P195/75R14	1
	P205/70R14	1
	205/70HR14	1
	220/55R390	2

1984 3.8L LTD/MARQUIS, THUNDERBIRD/COUGAR—UNIQUE CANADA

Throttle	Range	Shift	OPS—R.P.M.	1	2
Closed	D	1-2	399-446	9-10	9-11
(Above	D	2-3	500-744	12-17	12-18
17″	D	3-1	331-366	8-9	8-9
Vacuum)	1	2-1	1091-1317	25-31	26-31
To Detent	D	1-2	829-1192	19-28	20-28
(Torque	D	2-3	1493-1844	35-43	36-44
Demand)	D	3-2	1354-1564	32-36	32-37
Through	D	1-2	1514-1775	35-41	36-42
Detent	D	2-3	2666-2981	62-70	64-71
(W.O.T.)	D	3-2	2407-2602	56-61	57-62
	D	3-1 2-1	1094-1344	26-31	26-32

Axle Ratio	Tire Size	Use Column No.
3.08:1	P185/75R14	1
	P195/75R14	2
	P205/70R14	1
	205/70HR14	1
	220/55R390	1

1984 3.8L MUSTANG/CAPRI—50 STATES/ALTITUDE

Throttle	Range	Shift	OPS—R.P.M.	1	2
Closed	D	1-2	413-456	11-12	11-12
(Above	D	2-3	580-761	15-20	16-20
17"	D	3-1	331-366	9-10	9-10
Vacuum)	1	2-1	1074-1271	28-33	29-34
To Detent	D	1-2	907-1187	24-31	24-32
(Torque	D	2-3	1600-1852	42-49	43-50
Demand)	D	3-2	1459-1614	38-42	40-43
Through	D	1-2	1480-1718	39-45	40-46
Detent	D	2-3	2621-2901	69-76	71-78
(W.O.T.)	D	3-2	2359-2516	62-66	64-68
	D	3-1 2-1	1061-1292	28-34	29-35

Axle Ratio	Tire Size	Use Column No.
2.73:1	P185/75R14	1
	P195/75R14	2
	205/70HR14	2
	220/55R390	1

1984 3.8L MUSTANG/CAPRI—ALTITUDE

Throttle	Range	Shift	OPS—R.P.M.	1	2
Closed	D	1-2	411-457	11-12	11-12
(Above	D	2-3	580-761	15-20	16-20
17"	D	3-1	331-366	9-11	9-10
Vacuum)	1	2-1	1074-1269	28-33	29-34
To Detent	D	1-2	845-1185	22-31	23-32
(Torque	D	2-3	1527-1850	40-49	41-49
Demand)	D	3-2	1436-1610	38-42	39-43
Through	D	1-2	1464-1716	38-45	39-46
Detent	D	2-3	2600-2898	68-76	70-78
(W.O.T.)	D	3-2	2338-2513	61-66	63-68
	D	3-1 2-1	1046-1290	27-34	28-35

Axle Ratio	Tire Size	Use Column No.
2.73:1	P185/75R14	1
	P195/75R14	2
	205/70HR14	2
	220/55R390	1

1984 3.8L MUSTANG/CAPRI—CANADA

Throttle	Range	Shift	OPS—R.P.M.	1	2
Closed	D	1-2	421-467	11-12	11-13
(Above	D	2-3	586-807	15-21	16-22
17"	D	3-1	331-366	9-10	9-10
Vacuum)	1	2-1	1091-1313	29-35	29-35
To Detent	D	1-2	925-1193	24-31	25-32
(Torque	D	2-3	1613-1845	42-49	43-50
Demand)	D	3-2	1330-1542	35-41	36-41
Through	D	1-2	1506-1764	40-47	41-48
Detent	D	2-3	2655-2971	70-78	72-80
(W.O.T.)	D	3-2	2387-2582	63-68	64-70
	D	3-1 2-1	1080-1332	28-35	29-36

Axle Ratio	Tire Size	Use Column No.
2.73:1	P185/75R14	1
	P195/75R14	2
	205/70HR14	2
	220/55R390	1

1984 4.9L F-150—50 STATES
3.08 Axle Ratio

Throttle	Range	Shift	OPS—R.P.M.	1	2	3
Closed	D	1-2	405-449	10-12	11-12	10-11
(Above	D	2-3	574-776	15-20	16-21	14-19
17"	D	3-1	331-366	8-9	9-10	8-9
Vacuum)	1	2-1	1004-1192	26-31	27-32	25-30
To Detent	D	1-2	724-995	19-26	20-27	18-25
(Torque	D	2-3	1341-1536	35-40	36-42	34-38
Demand)	D	3-2	1070-1259	28-33	29-34	27-31
Through	D	1-2	1373-1598	35-41	37-43	34-40
Detent	D	2-3	2413-2681	62-69	65-73	60-67
(W.O.T.)	D	3-2	2170-2324	56-60	59-63	54-58
	D	3-1 2-1	998-1218	26-31	27-33	25-30

Axle Ratio	Tire Size	Use Column No.
3.08:1	P215/75R15SL	1
	P235/75R15XL	2
	P195/75R15SL	3

1984 5.0L F-150—49 STATES/CANADA
3.55 Axle Ratio

Throttle	Range	Shift	OPS—R.P.M.	1	2	3
Closed	D	1-2	413-456	11-12	11-12	10-11
(Above	D	2-3	580-761	15-22	16-21	15-19
17"	D	3-1	331-366	8-9	9-10	8-9
Vacuum)	1	2-1	1122-1320	29-34	30-36	28-33
To Detent	D	1-2	1031-1240	27-32	28-34	26-31
(Torque	D	2-3	1690-1852	44-48	46-50	42-46
Demand)	D	3-2	1459-1614	38-42	40-44	36-40
Through	D	1-2	1516-1755	39-45	41-48	38-44
Detent	D	2-3	2621-2901	68-75	71-79	66-73
(W.O.T.)	D	3-2	2359-2516	61-65	64-68	59-63
	D	3-1 2-1	1110-1341	29-35	30-36	28-34

Axle Ratio	Tire Size	Use Column No.
3.55:1	P215/75R15SL	1
	P235/75R15XL	2
	P195/75RSL	3

1984 3.8L BRONCO II/RANGER 4x4—50 STATES/CANADA

Throttle	Range	Shift	OPS—R.P.M.	1	2	3	4
Closed	D	1-2	399-446	9-10	9-10	8-9	8-9
(Above	D	2-3	826-1120	18-25	19-25	17-23	17-23
17"	D	3-1	331-366	7-8	7-8	7-8	7-8
Vacuum)	1	2-1	1420-1695	32-38	32-38	29-35	30-36
To Detent	D	1-2	898-1435	20-32	20-33	18-30	19-30
(Torque	D	2-3	1660-2166	37-48	38-49	34-45	35-
Demand)	D	3-2	1595-1885	35-42	36-43	33-39	34-40
Through	D	1-2	1921-2250	43-50	44-51	39-46	40-47
Detent	D	2-3	3156-3564	70-79	72-81	65-73	66-75
(W.O.T.)	D	3-2	2817-3082	63-69	64-70	58-63	59-65
	D	3-1 2-1	1423-1727	32-38	32-39	29-36	30-36

Axle Ratio	Tire Size	Use Column No.
3.45:1	P195/75R15SL	1
	P205/75R15SL	2
3.73:1	P195/75R15SL	3
	P205/75R15SL	4

1984 2.8L BRONCO II/RANGER 4x4—ALTITUDE

Throttle	Range	Shift	OPS—R.P.M.	1	2
Closed (Above 17" Vacuum)	D	1-2	398-449	8-9	8-9
	D	2-3	823-1128	17-23	17-24
	D	3-1	331-366	7-8	7-8
	1	2-1	1420-1695	29-35	30-36
To Detent (Torque Demand)	D	1-2	848-1466	17-30	18-31
	D	2-3	1607-2207	33-45	34-46
	D	3-2	1559-1895	32-39	33-40
Through Detent (W.O.T.)	D	1-2	1912-2257	39-46	40-48
	D	2-3	3144-3577	65-74	66-75
	D	3-2	2805-3091	58-64	59-65
	D	3-1 2-1	1415-1733	29-36	30-36

Axle Ratio	Tire Size	Use Column No.
3.73:1	P195/75R15SL	1
	P205/75R15SL	2

GOVERNOR PRESSURE TEST

Testing

To perform a governor pressure test, use the following procedure as a guide.

1. Raise the vehicle and support safely so the drive wheels are clear of the floor. Install an oil pressure gauge.

2. Disconnect and plug the vacuum line to the modulator valve. Attach either a hand operated vacuum pump or an electrically operated vacuum pump hose to the nipple on the modulator valve.

3. With the engine operating, place the selector lever in the DRIVE position, with no load on the engine and apply 10 inches of vacuum to the modulator diaphragm.

4. Increase the engine speed slowly and watch the speedometer and the control pressure gauge.

5. The control pressure cutback should occur between 6-20 mph.

—— CAUTION ——

Do not exceed 60 mph speedometer speed and after each test, place the transmission in neutral and operate the engine at 1000 rpm to cool the transmission.

6. The governor is good if the cutback of the control pressure occurs within specifications. If the cutback does not occur within specifications, check the shift speeds to verify that it is the governor and not a stuck cutback valve. Service or replace the governor as required.

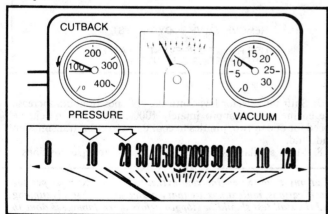

Checking shift points and governor light load tests (©Ford Motor Company)

STALL TEST

The stall test is used to check the maximum engine rpm (no more increase in engine rpm at wide open throttle) with the selector lever in the "D", "2", "1" and reverse positions and to determine if any slippage is occurring from the clutches, bands or torque converter. The engine operation is noted and a determination can be made as to its performance.

Performing the Stall Test

1. Check the engine oil level, start the engine and bring to normal operating temperature.

2. Check the transmission fluid level and correct as necessary. Attach a calibrated tachometer to the engine and a 0-400 psi oil pressure gauge to the transmission control pressure tap on the left side of the case.

3. Mark the specified maximum engine rpm on the tachometer cover plate with a grease pencil to immediately check if the stall speed is over or under specifications.

4. Apply the parking brake and block both front and rear wheels.

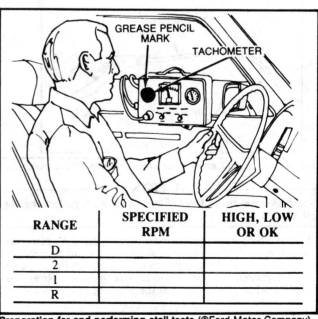

RANGE	SPECIFIED RPM	HIGH, LOW OR OK
D		
2		
1		
R		

Preparation for and performing stall tests (©Ford Motor Company)

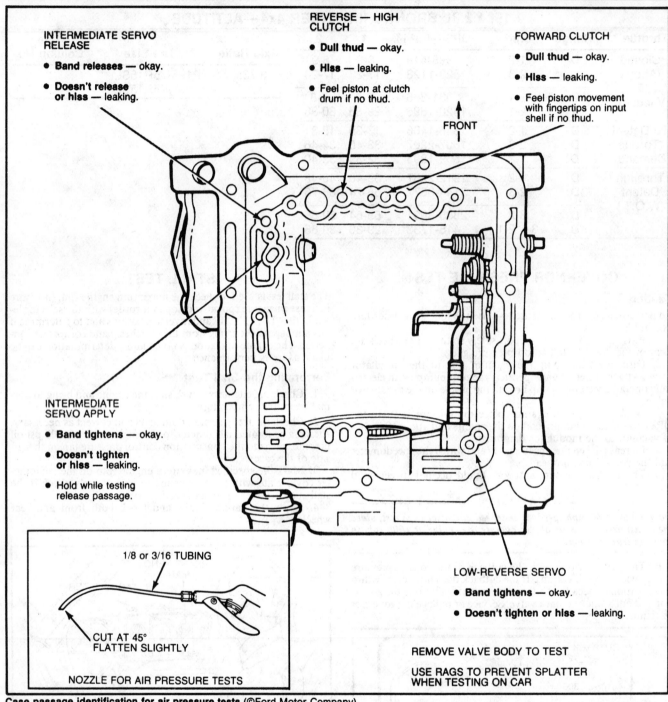

INTERMEDIATE SERVO RELEASE

- **Band releases** — okay.
- **Doesn't release or hiss** — leaking.

REVERSE — HIGH CLUTCH

- **Dull thud** — okay.
- **Hiss** — leaking.
- Feel piston at clutch drum if no thud.

FORWARD CLUTCH

- **Dull thud** — okay.
- **Hiss** — leaking.
- Feel piston movement with fingertips on input shell if no thud.

FRONT

INTERMEDIATE SERVO APPLY

- **Band tightens** — okay.
- **Doesn't tighten or hiss** — leaking.
- Hold while testing release passage.

1/8 or 3/16 TUBING

CUT AT 45° FLATTEN SLIGHTLY

NOZZLE FOR AIR PRESSURE TESTS

LOW-REVERSE SERVO

- **Band tightens** — okay.
- **Doesn't tighten or hiss** — leaking.

REMOVE VALVE BODY TO TEST

USE RAGS TO PREVENT SPLATTER WHEN TESTING ON CAR

Case passage identification for air pressure tests (©Ford Motor Company)

--- CAUTION ---

Do not allow anyone in front of the vehicle while performing the stall test.

5. While holding the brake pedal with the left foot, place the selector lever in "D" position and slowly depress the accelerator.

6. Read and record the engine rpm when the accelerator pedal is fully depressed and the engine rpm is stabilized. Read and record the oil pressure reading at the high engine rpm point.

--- CAUTION ---

The stall test must be made within five seconds.

7. Shift the selector lever into the "N" position and increase the engine rpm to approximately 1000-1200. Hold the engine speed for one to two minutes to cool the converter, transmission and fluid.

8. Make similar tests in the "2", "1" and reverse positions.

--- CAUTION ---

If at any time the engine rpm exceeds the maximum as per the specifications, indications are that a clutch unit or band is slipping and the stall test should be stopped before more damage is done to the internal parts.

9. Refer to the stall test results chart for further diagnostic information.

STALL SPEED SPECIFICATIONS—1982

Vehicle Application	Engine Disp.	Transmission Type	Converter Size	Stall Speed Min.	Max.
F-100	3.8L	C5	12″	1759	1961
F-100	4.2L	C5	12″	1869	2077

STALL SPEED SPECIFICATIONS—1983-84

Vehicle	Engine/Litre Displacement	Transmission Type	Converter Size (Inches)	ID	Stall Speed (RPM) Min.	Max.
Mustang/Capri/Fairmont/ Zephyr/LTD/Marquis	3.3L	C5	12″	FY	1527	1785
		C5	10¼″	GD	1503	1760
LTD/Marquis/Mustang/Capri/ Fairmont/Zephyr	3.8L	C5	12″	GB	1648	1911
Ranger 4x4, Bronco II	2.3L	C5	10¼″	BU①	2635③	3031③
F-100, F-150	3.8L	C5	12″	GB②	1737	2022

① 1984 GH ③ 1984 Min. stall speed—2539
② 1984 GE   Max. stall speed—2988

STALL SPEED TEST RESULTS CHART

Selector Positions	Stall Speed(s) High (Slip)	Stall Speeds Low
D only	Low (Planetary) One-Way Clutch	1. Does engine misfire or bog down under load?
D, 2 and 1	Forward Clutch	
All Driving Ranges	Perform Control Pressure Test	Check Engine for Tune-Up. If OK . . .
R Only	Reverse and High Clutch or Low-Reverse Band or Servo	2. Remove torque converter and bench test for reactor one-way clutch slip.

AIR PRESSURE TESTS

Air pressure testing is helpful in locating leak points during disassembly, and in verifying that the fluid circuits are not leaking during build-up. If the road test disclosed which clutch or servo isn't holding, that is the circuit to be air tested. Use air pressure regulated to about 25 psi and check for air escaping to detect leakage. If the pressures are found to be low in a clutch, servo or passageway, a verification can be accomplished by removing the valve body and performing an air pressure test. This test can serve two purposes:

1. To determine if a malfunction of a clutch or band is caused by fluid leakage in the system or is the result of a mechanical failure.
2. To test the transmission for internal fluid leakage during the rebuilding and before completing the assembly.

Air Pressure Test Procedure

1. Obtain an air nozzle and adjust for 25 psi.
2. Apply air pressure (25 psi) to the passages as listed in the accompanying chart.

AIR PRESSURE DIAGNOSIS CHART

Passage	Tests OK If	Leaking If
Reverse-and-high clutch	Dull thud or you can feel piston movement at the clutch drum	Hissing or no piston movement
Forward clutch	Dull thud or you can feel piston movement on the input shell	Hissing or no piston movement
Intermediate servo apply	Front band tightens	Hissing or no application
Intermediate servo release	Band releases while applying pressure to both passages 6 and 7	Hissing or no band release
Low-and-reverse servo apply	Rear band tightens	Hissing or no band apply

ON CAR SERVICES

ADJUSTMENTS

Vacuum Diaphragm

Adjustment

The vacuum diaphragm units used on the C-5 transmissions are non-adjustable. When a replacement unit is installed, the control line pressure must be checked. If the pressure is not to specifications, a longer or shorter throttle valve rod must be installed to bring the pressure within the specified limits. Five selective rods are available to obtain the proper pressure.

Length (inch)	Color code
1.5925-1.5875	Green
1.6075-1.6025	Blue
1.6225-1.6175	Orange
1.6375-1.6325	Black
1.6585-1.6535	Pink/White

NOTE: If the length of the rod is not known, it should be measured with a micrometer. To determine if a change in the length of the throttle valve rod is needed, use the following procedure.

1. Attach a tachometer to the engine.
2. Attach a hand vacuum pump to the vacuum diaphragm unit.

NOTE: An air bleed valve can also be used.

3. Attach an oil pressure gauge to the control pressure port on the side of the transmission case.
4. Apply the parking brake and the service brakes. Chock the wheels to avoid movement of the vehicle.
5. Start the engine and allow it to reach normal operating temperature.
6. Adjust the engine idle speed to specifications of 1000 rpm and apply 10 inches of vacuum to the vacuum diaphragm unit.
7. With the brakes applied, move the selector lever through all detent positions. Read and record the pressure readings in all selector positions.
8. Compare the pressure readings to the specified pressure reading on the line pressure charts.
9. If the pressure is within specifications, no rod change is required.
10. If the pressure is below specifications, use the next longer rod and retest.
11. If the pressure is above specifications, use the next shorter rod and retest.

Manual Linkage
COLUMN SHIFT

Adjustment

1. Place the selector lever in the DRIVE position. An eight (8) pound weight should be hung from the selector lever to be sure the lever is definitely against the DRIVE position stop during the linkage adjustment.
2. Loosen the adjusting nut located on the slotted rod end.
3. Position the transmission lever in the drive position which is the third detent position from the full counterclockwise position.
4. Be sure the slotted rod end has the flats aligned with the flats on the mounting stud.
5. Making sure that the selector lever has not moved from the DRIVE position, tighten the nut to 10-20 ft. lbs.
6. Move the selector lever through all detents to be certain that adjustment is correct.

CONSOLE OR FLOOR SHIFT

Adjustment
SOLID LINK TYPE

1. Position the transmission selector lever in the DRIVE position against the rearward DRIVE stop. Hold the selector lever in this position during the adjustment.
2. Raise the vehicle and loosen the manual lever shift rod retaining nut and move the transmission manual lever to the DRIVE position, third detent position from the back of the transmission.
3. With the transmission selector lever and the manual selector lever in the DRIVE position, tighten the attaching nut to 10-15 ft. lbs.
4. Check the operation of the shift lever in each transmission detent.

CABLE TYPE

1. Position the transmission selector lever in the DRIVE position against the rearward DRIVE Stop. Hold the selector lever in this position during the adjustment.
2. Raise the vehicle and loosen the manual shift lever cable retaining nut. Move the transmission manual lever to the DRIVE position, third detent position from the rear of the transmission.
3. With both the manual lever and the selector levers in the DRIVE positions, tighten the attaching nut to 10-15 ft. lbs.

TRANSMISSION VACUUM DIAPHRAGM ASSEMBLY SPECIFICATIONS

Diaphragm Type	Diaphragm Part No.	Identification	Throttle Valve Rod*		
			Part No. (7A380)	Length	Identification
S-HAD	D70P-7A377-AA	1 White Stripe	D3AP-JA	1.5925-1.5876	Green Daub
			D3AP-HA	1.6075-1.6025	Blue Daub
SAD	1983-E2DP-7A377-AA	No Identification	D3AP-KA	1.6225-1.6175	Orange Daub
	1982-E2DP-7A377-AA		D3AP-LA	1.6375-1.6325	Black Daub
S-SAD	D6AP-7A377-AA	1 Green Stripe	D3AP-MA	1.6585-1.6535	Pink/White Daub

*Selective Fit Rods
SAD—Single Area Diaphragm
S-SAD—Super Single Area Diaphragm
S-HAD—Super High Altitude Diaphragm

4. Check the operation of the selector lever in each detent position.

1983 RANGER

1. Position the selector lever in the DRIVE position and loosen the trunnion nut.

CAUTION

Be sure the selector lever is held against the rearward DRIVE detent stop during the linkage adjustment.

---

2. Position the transmission manual lever in the drive range by moving the bellcrank lever all the way rearward and then forward to the third detent position.

3. With the shift lever and the manual transmission lever in the DRIVE position, tighten the trunnion bolt to 13-23 ft. lbs. while holding a light forward pressure on the shift control tower arm.

NOTE: The forward pressure on the shifter arm will ensure correct positioning within the DRIVE detent.

4. After the adjustment, check for correct PARK engagement. The lever must move to the right when engaged in the PARK position. Operate the shift lever through all detents to assure proper transmission detent operation.

Shift Linkage Grommets

Urethane plastic grommets are used to connect the various rods and levers. It is most important to replace the old grommet with a new one whenever the rod is disconnected from a grommet. A special tool is needed to properly remove and replace the grommets from the levers. The proper procedure is as follows.

Replacement

1. Place the lower jaw of the tool between the lever and the

rod. Position the stop pin against the end of the rod and force the rod out of the grommet.

2. Remove the grommet from the lever by cutting off the large shoulder with a sharp knife.

3. Adjust the tool stop to ½ inch and coat the outside of the grommet with lubricant. Place a new grommet on the stop pin and force it into the lever hole. Turn the grommet several times to be sure the grommet is properly seated.

4. Squeeze the rod into the bushing until the stop washer seats against the grommet.

Neutral Start Switch

Adjustment

1. After the selector lever is properly adjusted, loosen the two switch attaching bolts.

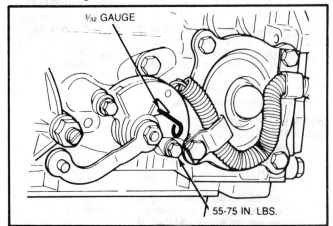

Neutral start switch adjustment (©Ford Motor Company)

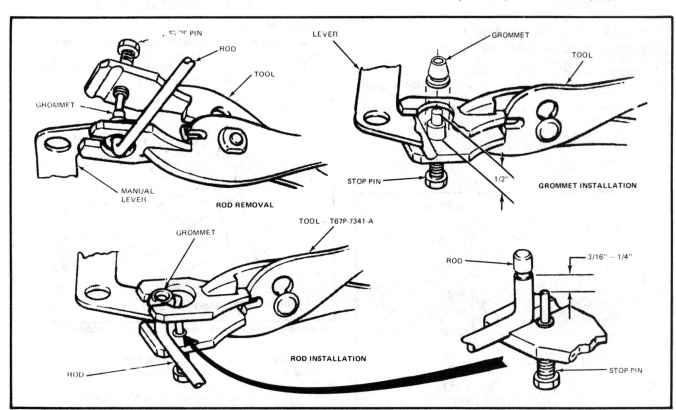

Removal and installation of shifting linkage grommets (©Ford Motor Company)

2. With the transmission selector lever in the NEUTRAL position, rotate the switch and insert a number 43 drill shank or a special gauge pin into the gauge pin holes of the switch. It is most important to install the drill or pin a full $31/64$ inch into the three holes of the switch components.

3. Tighten the switch attaching screws to 55-75 in. lbs. and remove the pin or drill from the switch.

4. Check the operation of the switch. The engine should only start in the PARK and NEUTRAL positions.

Kickdown Linkage

Adjustment

1982 GRANADA/COUGAR AND FAIRMONT/ZEPHYR, 1982-83 MUSTANG/CAPRI, 1983 FAIRMONT FUTURA/ZEPHYR AND LTD/MARQUIS—ALL WITH 2.3L ENGINE

1. With the carburetor linkage held at the wide open throttle (WOT) position, have the downshift rod held downward with a 6 pound weight holding the rod against the "Through Detent" stop.

2. Adjust the kickdown adjusting screw to obtain a 0.010-0.080 inch clearance between the screw and the throttle arm.

3. Return the system to the idle position and reinstall the kickdown rod retracting spring.

1983 MUSTANG/CAPRI AND THUNDERBIRD/XR-7—ALL WITH 3.8L ENGINE

1. With the carburetor linkage held at the wide open throttle (WOT) position, have the downshift rod held downward with a 4½ pound weight holding the rod against the "through detent" stop.

2. Adjust the kickdown adjusting screw head to obtain 0.010-0.080 inch between the screw head and the throttle arm.

3. Return the carburetor and downshift rod to the idle position and secure all pivot points. Reinstall retracting spring, if removed.

1982-83 F-100 AND F-150, 1983 RANGER

1. Rotate the throttle linkage to wide open throttle (WOT) position with the engine off.

2. Place a 6 pound weight in the downshift rod and insert a 0.060 inch spacer between the throttle lever and the adjusting screw.

3. Move the adjusting screw until contact is made between the screw and the spacer.

4. Remove the spacer and a gap of 0.010-0.080 inch should exist between the screw and the throttle rod.

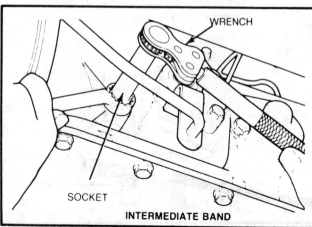

Adjustment of intermediate band (©Ford Motor Company)

5. Remove the weight from the downshift rod and return both the rod and the carburetor linkage to the idle position.

1982 THUNDERBIRD/XR-7 AND GRANADA/COUGAR, 1982 FAIRMONT/ZEPHYR AND MUSTANG/CAPRI, 1983 FAIRMONT FUTURA/ZEPHYR AND LTD/MARQUIS—ALL WITH 3.3L ENGINE

1. With the carburetor linkage held at the wide open throttle (WOT) position, hold the downshift rod downward with a 4¼ pound weight, against the "Through Detent" stop.

2. Adjust the kickdown adjusting screw to obtain a clearance of 0.010-0.080 inch between the screw head and the throttle arm. Lock the screw in position with the locknut.

3. Release the carburetor and downshift linkage to their free position.

4. Reinstall the downshift retracting spring on the 1982 models, if removed during the adjustment.

Band Adjustment

INTERMEDIATE BAND

Adjustment

1. Raise the vehicle and support safely. Clean dirt and foreign material from the band adjusting screw area.

2. Remove the locknut and discard. A new locknut should be used at each band adjustment procedure.

3. With the proper band adjusting tool, tighten the adjusting screw until either the tool clicks at a preset 10 ft. lbs. or indicates a torque of 10 ft. lbs. (13.5 N•m).

4. Back the adjusting screw off exactly 4¼ turns.

5. While holding the adjusting screw from turning, install a new locknut and tighten to 40 ft. lbs. (54 N•m).

LOW-REVERSE BAND

Adjustment

1. Raise the vehicle and support safely. Clean the dirt and foreign material from the band adjusting area.

2. Remove the locknut and discard. A new locknut should be used at each band adjustment procedure.

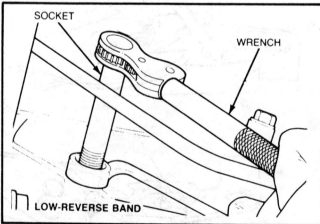

Adjustment of low/reverse band (©Ford Motor Company)

3. With the proper band adjusting tool, tighten the adjusting screw until either the tool clicks at a preset 10 ft. lbs. or indicates a torque of 10 ft. lbs. (13.5 N•m).

4. Back off the adjusting screw exactly 3 full turns.

5. While holding the adjusting screw from turning, install a new locknut and tighten to 40 ft. lbs. (54 N•m).

SERVICES

Fluid Changes

Ford Motor Company does not recommend the periodic changing of the automatic transmission fluid, because of normal maintenance and lubrication requirements performed on the automatic transmissions by mileage or time intervals. However, if the vehicle is used in continous service or driven under severe conditions, the automatic transmission fluid must be changed every 22,500 miles. No time limit is given, but as a rule of thumb, between 18 and 20 months would be an average in comparison to the mileage limit.

Severe conditions are described as extensive idling, frequent short trips of 10 miles or less, vehicle operation when the temperature remains below +10°F. for 60 days or more, sustained high speed operation during hot weather (+90°F.), towing a trailer for a long distance or driving in severe dusty conditions. If fleet vehicles and vehicles accumulating 2000 miles or more per month are not equipped with an auxiliary transmission oil cooler, the severe condition rating applies.

NOTE: Each vehicle operated under severe conditions should be treated individually.

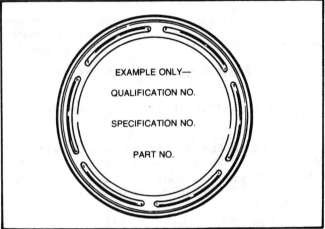

EXAMPLE ONLY—

QUALIFICATION NO.

SPECIFICATION NO.

PART NO.

Fluid container identification locations (©Ford Motor Company)

Type H automatic transmission fluid, meeting Ford Motor Company's specifications (ESP-M2C166-H), should be used in the C-5 automatic transmission. Failure to use the proper grade and type fluid could result in internal transmission damage.

When the transmission has to be removed for major repairs, the unit should be drained completely. The converter, cooler and cooler lines must be flushed to remove any particles or dirt that may have entered the components as a result of the malfunction or failure.

Oil Pan

Removal

1. Raise the vehicle and support safely. Place a drain pan under the transmission.
2. On pan filled transmissions, remove the filler tube from the oil pan to drain the transmission fluid. On case filled transmissions, loosen the oil pan bolts to drain the fluid from the pan.

NOTE: If the same fluid is to be used again, filter it through a 100 mesh screen. Reuse the fluid only if it is in good condition.

3. Remove the transmission oil pan and bolts.

— CAUTION —

Care must be exercised when working with automatic transmission fluid while it is hot. Resulting burns can cause severe personal injury.

---

4. Remove the filter screen attaching bolt and the filter screen.
5. Thoroughly clean and remove all gasket material from the oil pan and the mating surface of the transmission.

NOTE: Remove and discard the nylon shipping plug from the oil pan, if remaining. This plug was used to retain fluid in the transmission during manufacture and has no further use.

Installation

1. Install a new gasket on the oil pan and install the filter screen and bolt. Torque the filter screen bolt to 25-40 in. lbs.
2. Install the oil pan to the transmission mating surface. Install the oil pan retaining bolts and torque to 13-16 ft. lbs. Install the filler tube and tighten the fitting securely, if equipped.
3. If the converter has not been drained, install three quarts of type H fluid. If the transmission is dry, install five quarts of type H fluid.
4. Start the engine and check the fluid level. Add enough fluid to bring the level to the cold level on the dipstick.
5. Check the fluid level when the engine/transmission assembly reaches normal operating temperature. Correct the level as necessary.
6. Final check is to move the shifting lever through all detents and recheck the fluid level.

— CAUTION —

Do not overspeed the engine during the warm-up period.

DRAINING OF TORQUE CONVERTER

1. Remove the lower engine dust cover.
2. Rotate the torque converter until the drain plug is in view.
3. Remove the drain plug and allow the converter to drain.

NOTE: Due to the length of time needed for the draining of the converter, it is suggested this procedure be performed prior to other operations.

4. Install the drain plug and the lower engine dust cover.
5. Remove the cooler lines and flush lines and cooler.
6. Reinstall cooler lines and fill transmission as required to bring fluid to its proper level.

NOTE: When leakage is found at the oil cooler, the cooler must be replaced. When one or more of the cooler lines must be replaced, each must be fabricated from the same size steel line as the original.

Vacuum Diaphragm

Removal

1. Raise the vehicle and support safely.
2. Disconnect the vacuum (modulator) diaphragm hose.
3. Remove the unit retaining bracket and bolt. Pull the modulator assembly from the transmission case.

NOTE: Do not pry or bend the bracket during the removal procedure.

4. Remove the vacuum unit control rod from the transmission case.

Installation

1. Place the correct vacuum unit control rod in the transmission case passage.
2. Install the vacuum modulator unit into the transmission case. Secure the unit with the bracket and retaining bolt. Torque the bolt to 28-40 ft. lbs.

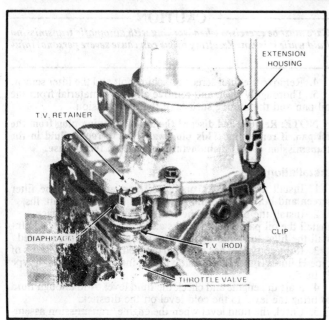

Removal or installation of diaphragm assembly and extension housing (©Ford Motor Company)

3. Install the vacuum hose to the modulator nipple, lower the vehicle and road test as required.

Control Valve Body

Removal

1. Raise the vehicle and support safely.
2. Drain the transmission oil pan as described in fluid change procedure.
3. Remove the transmission oil pan retaining bolts, oil pan and gasket.
4. Shift the transmission in the PARK position.
5. Remove the filter screen attaching bolt and the filter screen.
6. Remove the valve body to case attaching bolts. Hold the manual valve in the valve body and remove the valve body from the case.

─────────── CAUTION ───────────

Failure to hold the manual valve in the valve body during the removal could cause the manual valve to become bent or damaged.

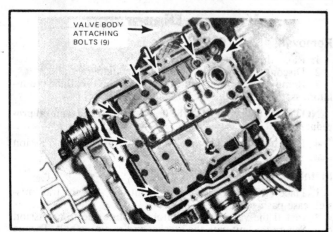

Location of valve body retaining bolts (©Ford Motor Company)

7. If the valve body is to be disassembled, refer to the appropriate outline in this section.

Installation

1. Thoroughly clean and remove all gasket material from the transmission pan mounting face and from the oil pan.

NOTE: Remove and discard the nylon plug from the oil pan. This was used to retain fluid in the transmission during manufacture and is no longer needed.

2. Position the transmission manual lever in the PARK detent. While holding the manual valve, position the valve body onto the transmission case. Be sure the inner downshift lever is between the downshift lever stop and the downshift valve and that the two lands on the end of the manual valve engage the actuating pin on the manual detent lever. Install the valve body retaining bolts (7) through the valve body and into the case. Do not tighten at this time.
3. Snug the valve body to case bolts evenly. Be sure all components are aligned. Torque the valve body to case bolts to 80-120 in. lbs.
4. Position the filter screen and install the bolt. Tighten the bolt to 25-40 in. lbs.
5. Install a new gasket on the oil pan and install in place on the transmission case. Install the retaining bolts and torque to 13-16 ft. lbs. If equipped with the pan filler tube, attach and securely tighten.
6. Lower the vehicle and fill the transmission with fluid, type H, to its proper level. Verify level at normal operating temperature. Check pan area for leakage.

Low-Reverse Servo

Removal

ALL MODELS, EXCEPT MUSTANG/CAPRI WITH 3.8L ENGINE

1. Raise the vehicle and support safely.
2. Loosen the low-reverse band adjusting screw locknut and tighten the band adjusting screw to 10 ft. lbs.

NOTE: The purpose of tightening the low-reverse adjusting screw is to insure the band strut will remain in place when the servo piston is removed.

3. Disengage the neutral start switch wiring harness from the routing clips. Note the position of the clips before removing.
4. Remove the servo cover retaining bolts. Remove the servo cover and seal from the case.
5. Remove the servo piston and return spring from the case. The piston seal cannot be replaced separately. The seal is bonded to the piston and both must be replaced as a unit.

Installation

1. Lubricate the piston/seal with transmission fluid and install in the case along with the return spring. Install a new seal on the cover and position the cover to the case with two longer bolts 180 degrees apart as installation bolts.
2. Install two cover retaining bolts, remove the two long bolts and install the remaining cover retaining bolts. Position the neutral start switch wiring harness routing clips on the proper cover bolts.
3. Torque the cover bolts to 13-20 ft. lbs.
4. Adjust the low-reverse band. Refer to the Band Adjusting procedure outline.

─────────── CAUTION ───────────

If the band cannot be adjusted properly, the low-reverse band struts have moved from their positions. It will be necessary to remove the oil pan, screen and valve body to properly install the struts and readjust the low-reverse band.

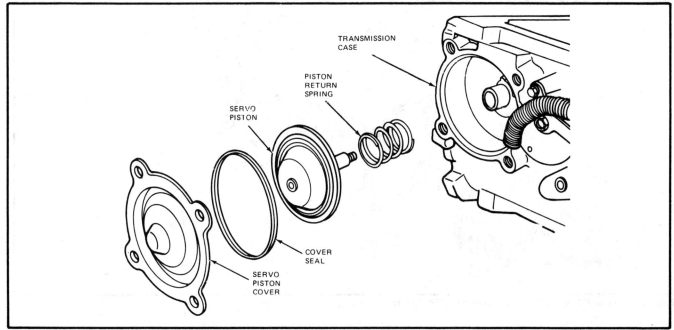

Exploded view of low/reverse servo assembly (©Ford Motor Company)

5. Lower the vehicle. If necessary, correct the fluid level.
6. Road test the vehicle as required.

Removal
MUSTANG/CAPRI WITH 3.8L ENGINE

1. Working from under the engine hood, remove the fan shroud attaching bolts and position the shroud back over the fan.
2. Raise the vehicle and support safely.
3. Support the transmission with and adjustable stand or a transmission jack.
4. Remove the number 3 crossmember to body bracket through bolts.
5. Carefully lower the transmission to obtain working clearance.
6. Loosen the low-reverse band adjusting screw locknut and adjust the band to 10 ft. lbs.

NOTE: The purpose of tightening the low-reverse band adjusting screw is to insure the band struts will remain in place when the servo piston is removed.

7. Disengage the neutral start switch wiring harness from the routing clips. Note the position of the routing clips on the servo cover bolts.
8. Remove the servo retaining bolts from the case and cover. Remove the servo cover and seal.
9. Remove the servo piston/seal and return spring. The piston seal is bonded to the piston and cannot be replace separately. Both must be replaced as a unit.

Installation

1. Install the piston return spring and piston assembly into the case bore. Install a new gasket on the cover.
2. Using two long bolts, 180 degrees apart to position the cover to the case, install two of the retaining bolts through the cover and into the case. Remove the two long bolts and install the remaining two bolts in their place. Position the wiring harness routing clips on their proper bolts and torque the cover bolts to 13-20 ft. lbs.
3. Install the neutral switch wiring in the routing clips.

4. Adjust the low-reverse band. Refer to the Band Adjusting procedure outline.

—————————— CAUTION ——————————

If the band cannot be adjusted properly, the low-reverse band struts have moved from their positions. It will be necessary to remove the oil pan, screen and valve body to properly install the struts and readjust the low-reverse band.

5. Raise the transmission back into position and install the crossmember through bolts. Remove the support from under the transmission.
6. Lower the vehicle, position the fan shroud and re-install the retaining bolts.
7. Inspect and correct as necessary, the transmission fluid level, check for leaks and road test as required.

Intermediate Servo

Removal

1. Working from underneath the engine hood, remove the fan shroud attaching bolts and position the shroud back over the fan.

NOTE: Not necessary on light trucks, steps 1, 2, 3, 4, 5, 6.

2. Raise and support the vehicle safely.
3. Support the transmission with an adjustable stand or a transmission jack.
4. Remove the number 3 crossmember to body bracket through bolts.
5. Carefully lower the transmission to obtain working clearance.
6. On vehicles equipped with the 3.8L engine, disconnect the transmission cooler lines using the correct special tool. Refer to "Cooling Line Removal and Installation" outline.
7. Remove the servo cover attaching bolts, along with the transmission I.D. tag. Note its position on cover.
8. Remove the servo cover and the piston assembly from the transmission case. Remove the piston return spring.

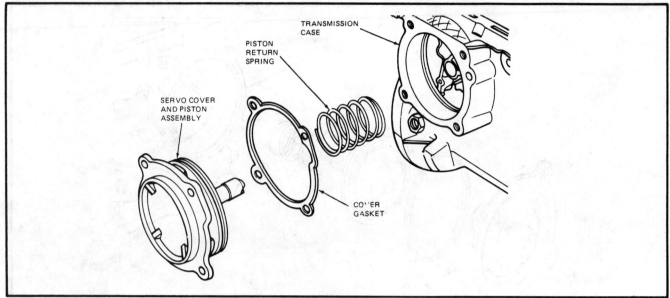

Exploded view of intermediate servo assembly (©Ford Motor Company)

9. Disassemble the servo as required and discard the cover gasket.

Installation

1. Position a new gasket on the servo cover. Align the notch in the gasket with a fluid passage in the case, during the installation.
2. Install the piston return spring, servo cover/piston assembly into the case. Using two long bolts 180 degrees apart, position the cover to the case. Install two retaining bolts and remove the two long bolts. Install the two remaining retaining bolts in the cover and torque to 16-22 ft. lbs. (12-20 ft. lbs. on light trucks).
3. Adjust the intermediate band as outlined in the Band Adjusting procedure outline.

NOTE: If the band cannot be adjusted properly, the band strut has dropped from it position. It is necessary to remove the oil pan, screen and valve body to reinstall the strut. Re-adjust the intermediate band.

Use the following procedures as applicable to the vehicle being serviced.
4. On models equipped with the 3.8L engine, connect the transmission cooler lines.
5. Raise the transmission and install the crossmember through bolts. Remove the transmission support.
6. Lower the vehicle and position the fan shroud. Install the shroud retaining bolts.
7. Inspect and correct as required, the transmission fluid level, check for leaks and road test, if necessary.

Extension Housing

Removal

1. Raise the vehicle and support safely.
2. Matchmark the drive shaft and remove from the vehicle.
3. Position an adjustable stand or transmission jack to support the transmission.
4. Remove the speedometer cable from the extension housing.
5. Remove the rear engine support to crossmember attaching nuts.
6. Raise the transmission and remove the rear support to body bracket through bolts. Remove the crossmember.

7. Loosen the extension housing attaching bolts and allow the unit to drain into a container. Remove vacuum tube clip with housing bolt.
8. Remove the extension housing attaching volts and remove the housing from the transmission. Discard the gasket.

Installation

1. Install a new gasket on the rear of the transmission case. Install the rear extension housing and install the retaining bolts. Install the vacuum tube clip with one bolt. Torque the bolts to 28-40 ft. lbs.
2. Position the crossmember and install the through bolts. Torque the nuts to 35-50 ft. lbs.
3. Lower the transmission and install the engine rear support to crossmember attaching nuts.
4. Remove the transmission support and install the speedometer cable in the extension housing.
5. Install the drive shaft in the same position following the matchmarks.
6. Lower the vehicle and fill the transmission with type H fluid to its proper level.
7. Check the extension housing for leakage, re-check the fluid level and road test, if required.

NOTE: The following operations can be accomplished with the rear extension housing on or off the transmission.

Seal

Removal and Installation

1. With the driveshaft removed, use a special puller tool to remove the seal assembly from the extension housing.
2. Using a seal installer tool, tap the seal into its seat in the extension housing. Lubricate the lip portion of the seal and install the driveshaft.

Bushing

Removal and Installation

1. Remove the extension housing seal with special puller tool. Using a special bushing remover tool, pull the bushing from the extension housing.

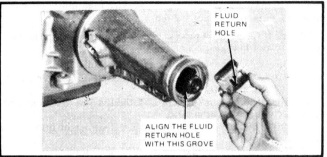

Installation of bushing in extension housing (©Ford Motor Company)

2. With a special bushing driver tool, tap the bushing into its seat in the extension housing. Install the seal in its seat in the housing by using the special seal driver tool.

NOTE: Be sure the fluid drain back hole in the bushing is aligned with the groove in the extension housing.

Governor

Removal

1. Remove the extension housing. Refer to the "Extension Housing Removal and Installation" procedure.
2. Remove the governor housing to governor distributor attaching bolts. Slide the governor away from the distributor body and off the shaft. Check governor screen.
3. The governor can be disassembled or replaced as required.

Installation

1. Slide the governor over the output shaft and position the governor on the governor distributor body. Install the retaining bolts and torque to 80-120 in. lbs. Be sure screen is in place.
2. Install the extension housing. Refer to the "Extension Housing Removal and Installation" procedure.

REMOVAL & INSTALLATION

C-5 AUTOMATIC TRANSMISSION

Removal
ALL CAR MODELS

1. Protect fender areas and disconnect the negative battery cable.
2. Remove the air cleaner assembly on vehicles equipped with the 3.8L engine.
3. After removing the attaching bolts, position the fan shroud back over the fan assembly.
4. Disconnect the thermactor air injection hose at the catalytic converter check valve on models equipped with the 3.8L engine and on the 1982 Mustang/Capri models equipped with the 4.2L engine.

NOTE: The check valve is located on the right side of the engine compartment, near the firewall.

5. Remove the two upper transmission to engine attaching bolts, accessible from the engine compartment, on vehicles equipped with the 3.8L engine.
6. Raise the vehicle and support safely.
7. Match-mark and remove the driveshaft.
8. While supporting the exhaust system, disconnect the muffler inlet pipe from the catalytic converter outlet pipe. Wire the exhaust system assembly to the vehicle's undercarriage.
9. Remove the exhaust pipe(s) from the manifold(s) and by pulling back on the converters, release the converter hangers from their mounting brackets. Lower the pipe assemblies and set aside.

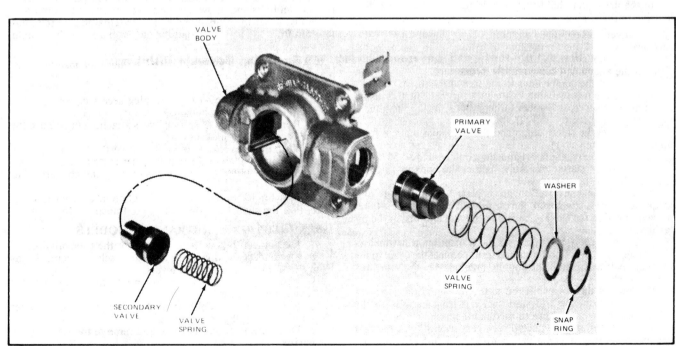

Exploded view of governor assembly (©Ford Motor Company)

10. Remove the speedometer driven gear from the extension housing.

11. Disconnect the neutral start switch wiring from the neutral start switch. Remove vacuum hose from modulator nipple.

12. Disconnect the downshift rod at the transmission manual lever. On floor mounted shift equipped vehicles, remove the shift cable routing bracket and disconnect the cable from the transmission manual lever.

13. Remove the converter housing dust shield and remove the converter to drive plate attaching nuts.

NOTE: The crankshaft must be turned to gain access to all the converter to drive plate attaching nuts.

14. Remove starter cable, remove starter attaching bolts and lower starter from the engine.

15. Loosen the attaching nuts from the rear support to the number three crossmember.

16. Position a transmission jack under the transmission and secure the transmission to the jack with a safety chain.

17. Remove the through bolts securing the number three crossmember to the body brackets.

18. Lower the transmission enough to gain working room and disconnect the oil cooler lines.

19. On vehicles equipped with the 3.8L engine and 1982 Cougar/Capri models, remove the four remaining transmission to engine attaching bolts. On all other models, remove the six remaining bolts.

20. Pull the transmission rearward to clear the converter studs from the drive plate and lower the transmission.

Installation

1. With the vehicle in the air and supported safely, raise the transmission on a transmission jack and into position to mate the converter studs with the drive plate holes and the transmission dowels on the rear of the engine to the transmission bell housing.

NOTE: It will be necessary to rotate the converter to align the converter studs and the converter drain plug to the holes in the drive plate. Be sure during the installation that the converter studs are in the drive plate holes before bolting the transmission bell housing to the engine.

2. On vehicles equipped with the 3.8L engine and 1982 Cougar/Capri models, install the four transmission to engine bolts. On all other models, install the six transmission to engine bolts. Tighten the attaching bolts to 40-50 ft. lbs.

3. Connect the cooler lines to the transmission.

4. Raise the transmission by the jack mechanism and install the number three crossmember through bolts. Install the attaching nuts and tighten to 20-30 ft. lbs.

5. Remove the safety chain and the transmission jack from under the vehicle.

6. Tighten the rear support attaching bolt nuts to 30-50 ft. lbs.

7. Install the starter assembly, tighten the bolts to 15-20 ft. lbs., and install the starter cable.

8. Install the converter to drive plate attaching nuts and torque to 20-30 ft. lbs. Install the dust shield and if previously removed, position the linkage bellcrank bracket and install the attaching bolts.

9. Connect the shift linkage to the transmission manual lever. If equipped with floor mounted shift, connect the cable to the manual lever and install the routing bracket with the attaching bolt.

10. Connect the downshift rod to the transmission lever.

11. Connect the neutral start switch wiring harness to the switch. Install vacuum hose to modulator nipple.

12. Install the speedometer driven gear assembly in the rear extension housing. Tighten the clamp bolt to 36-54 in. lbs.

13. Install the catalytic converters into their hanger brackets

and install the exhaust pipe(s) to the exhaust manifold. Install the attaching nuts, but do not tighten.

NOTE: If the exhaust pipe(s) were disconnected from the converters, install new gaskets or seals before connecting.

14. Disconnect the wire holding the exhaust system to the body undercarriage and connect the pipe to the converter outlet. Do not tighten the attaching nuts.

15. Align the exhaust system and tighten the manifold and converter outlet attaching nuts.

16. Install the driveshaft in the vehicle, aligning the previously made match-marks.

17. Check and if necessary, adjust the shift linkage.

18. Recheck undercarriage assembly and lower the vehicle.

19. On vehicles equipped with the 3.8L engine, install the two transmission to engine attaching bolts from the engine compartment.

20. On vehicles equipped with the 3.8L engine and 1982 Mustang/Capri models equipped with the 4.2L engine, connect the thermactor air injection hose to the converter check valve.

21. Position the fan shroud and install the retaining bolts.

22. Install the air cleaner on vehicles equipped with the 3.8L engine.

23. Connect the negative battery cable.

24. Add the correct type and amount of fluid to the transmission assembly as required.

25. Start the engine, being sure the starter will only operate in PARK or NEUTRAL positions.

26. Check and correct the transmission fluid level.

27. Raise the vehicle and inspect for leakage and correctness of assembly. Lower and road test as required.

Removal
LIGHT TRUCK AND RANGER MODELS

NOTE: Differences in the removal and installation procedures exist between models and model years. References are made to each throughout the removal and installation procedure outline.

1. Disconnect the negative battery cable. Raise the vehicle and support safely.

2. Place a drain pan under the transmission oil pan, loosen the oil pan bolts and allow the fluid to drain into the container. Carefully, lower the oil pan and allow the remainder of the fluid to drain from the pan. Reinstall the pan with a few bolts to hold the pan in place.

NOTE: Certain oil pans will have the filler tube installed. Remove to drain.

3. Remove the converter drain plug access cover from the bottom of the converter housing.

4. Remove the converter to flywheel attaching nuts. Turn the engine crankshaft to locate each nut.

5. Turn the converter to place the converter drain plug in the bottom position. Place a drain pan under the converter and remove the drain plug. Reinstall the plug after the converter has drained.

6. Matchmark the driveshaft and disconnect from the rear yoke. Pull the shaft from the rear of the transmission.

1982 LIGHT TRUCK AND RANGER MODELS

7. Disconnect the oil cooler lines from the transmission.

8. Disconnect the downshift and range selector control rods from the transmission manual levers.

9. Remove the speedometer driven gear assembly from the rear extension housing.

10. Remove the neutral start switch wires from the retainer clips and separate the connector assembly.

11. Disconnect the starter cable and remove the starter from the engine.

12. Remove the vacuum hose from the modulator nipple.

13. Position a transmission jack under the transmission and secure the unit with a safety chain.

14. Remove the two engine rear support crossmember-to-frame attaching bolts.

15. Remove the two engine rear support-to-extension housing attaching bolts.

1983 AND LATER LIGHT TRUCKS

16. Disconnect the starter cable and remove the starter from the engine.

17. Disconnect the neutral start switch wire at the connector.

18. Position a transmission jack under the transmission and remove the rear mount-to-crossmember insulator attaching nuts and the two crossmember-to-frame attaching bolts. Remove the right and left crossmember gussets.

19. Remove the two rear insulator-to-extension housing attaching bolts.

20. Disconnect the downshift and selector linkage from the transmission manual levers.

21. Remove the bellcrank bracket from the converter housing.

22. Raise the transmission assembly enough to gain clearance for removal of the crossmember. Remove the rear mount from the crossmember and remove the crossmember from the side supports.

23. Lower the transmission to gain access to the oil cooler lines and disconnect.

24. Disconnect the speedometer cable and remove the speedometer driven gear from the rear extension housing.

25. Remove the bolt holding the transmission filler tube to the engine block.

ALL MODELS

26. Be sure the safety chain is securing the transmission to the transmission jack. Remove the converter housing-to-engine bolts.

27. Pull the transmission to the rear while lowering the unit. Remove the transmission from under the vehicle.

Installation
ALL MODELS

1. Have the vehicle in the air and supported safely.

2. Position the transmission on the transmission jack and secure with the safety chain.

3. Move the transmission and jack assembly under the vehicle and position the transmission in-line with the engine block, having rotated the converter to align the attaching bolt holes in the flywheel with the attaching bolts of the converter. Push the assembly forward, mating the dowel pins on the engine with the holes in the converter housing.

4. Install the converter housing-to-engine attaching bolts. Torque to 40-50 ft. lbs.

———————— CAUTION ————————

Be sure drain plug is in place, tightened and positioned properly in the flywheel.

1983 AND LATER LIGHT TRUCKS

5. Install the bolt holding the filler tube to the engine block.

6. Install the speedometer driven gear assembly and speedometer cable to the rear extension housing.

7. Raise or lower the transmission to connect the oil cooler lines to the transmission case.

8. Raise the transmission assembly enough to install the crossmember to the side supports. Install the rear support to the crossmember.

9. Install the downshift and selector linkage to the transmission manual levers, while installing the bellcrank bracket to the converter housing.

10. Install the two rear insulator-to-extension housing attaching bolts.

11. Install the rear mount-to-crossmember insulator attaching nuts and install the left and right crossmember gussets. Remove the transmission jack.

12. Install the neutral start switch wiring connector and properly route the wiring through the retaining clips.

13. Install the starter assembly to the engine and install the starter cable.

1982 LIGHT TRUCK AND RANGER MODELS

14. Install the two engine rear support-to-extension housing attaching bolts. Install the crossmember to the frame and install the attaching bolts. Remove the transmission jack.

15. Install the vacuum hose to the nipple of the modulator.

16. Install the starter to the engine and install the starter cable.

17. Connect the neutral start wiring connector and route the wiring through the retainer clips as necessary.

18. Install the speedometer driven gear assembly and the speedometer cable to the rear extension housing.

19. Connect the downshift and range selector control rods to the transmission manual levers.

20. Connect the oil cooler lines at the transmission.

ALL MODELS

21. Install the drive shaft into the rear of the transmission and align the matchmarks at the yoke and secure.

22. Install the converter attaching nuts and tighten to 20-34 ft. lbs. Install the converter drain plug access cover.

23. Install the oil filler tube to the oil pan, if equipped. Otherwise, be sure a new gasket is on the oil pan and the retaining bolts are properly tightened.

24. Lower the vehicle, install the proper type and quantity of fluid into the transmission. Install the negative battery cable and start the engine. Recheck the fluid level and correct as required.

25. Raise the vehicle and check for leakage. Inspect the assembly for correct installation. Road test as required when lowered. Make any further adjustments.

BENCH OVERHAUL

C-5

AUTOMATIC TRANSMISSION

Before Disassembly

1. Clean the exterior of the transmission assembly before any attempt is made to disassemble the unit, to prevent the entrance of dirt or other foreign material from entering the transmission assembly or internal components during the disassembly and assembly phases.

NOTE: If steam cleaning is done to the exterior of the transmission assembly, immediate disassembly should be done to avoid rusting from condensation in the internal parts.

2. All screw, bolt and nut fasteners must be tightened to the torque indicated in the specification section, or as noted in the assembly outline.

3. When assembling the sub-assemblies, each component part should be lubricated with clean transmission fluid. Lubricate the sub-assemblies as they are installed in the transmission case.

4. Needle bearings, thrust washers and seals should be lightly coated with petroleum jelly during the assemble of the sub-assemblies and transmission.

5. During the assembly of the transmission and the sub-assemblies, always use new gaskets and seals.

6. Careful handling of the many components of the transmission is important to prevent the marring of the precision machined surfaces.

7. Whenever a seal is removed from a piston, shaft or servo, note the type of seal and the direction of the sealing lip. Look for modifications in seal or component application during the reassembly.

8. Keep the transmission service area clean and well organized. Provide a supply of lint-free shop clothes.

TRANSMISSION DISASSEMBLY

Converter

Removal

1. With the transmission secured on a work bench or in a transmission holding fixture, grasp the torque converter firmly and pull the assembly straight out of the transmission.

NOTE: The torque converter is a heavy unit and care must be exercised to be prepared to handle the weight.

Inspection
CHECKING CONVERTER END PLAY

1. Place the converter on a flat surface with the flywheel side down and the converter pump drive hub up.

2. Insert a special end play checking tool (Ford number T80L-7902-A or equivalent) into the drive hub opening of the converter pump, until the tool bottoms.

3. Tighten the threaded inner post of the tool, which will expand and lock the tool sleeve in the turbine spline.

4. Attach a dial indicator to the threaded post of the tool, with the indicator button resting on the converter pump housing. Set the dial indicator to zero.

5. Lift the tool and dial indicator assembly upward as far as possible and note the dial indicator reading. This reading is the total end play of the turbine and stator.

6. Replace the converter if the end play reading exceeds the specified limits of measurement.
End Play Specifications:
New or Rebuilt Converter—0.023 inch max.
Used Converter—0.050 inch max.

CHECKING CONVERTER ONE-WAY CLUTCH

1. The converter should be placed on a flat surface with the flywheel side down.

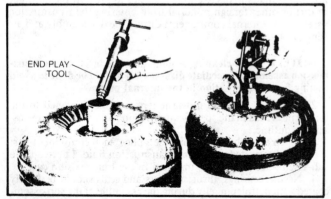

Checking converter end play (©Ford Motor Company)

Checking one-way clutch operation (©Ford Motor Company)

2. Insert a one-way clutch holding tool (Ford number T77L-7902-A or equivalent) in one of the grooves of the stator thrust washer, located directly under the converter pump drive hub.

3. Insert the one-way clutch torquing tool (Ford number T76L-7902-C or equivalent) into the converter pump drive hub and engage the one-way clutch inner race.

4. Attach a torque wrench to the one-way clutch torquing tool, and with the one-way clutch holding tool held stationary, turn the torque wrench counterclockwise. The converter should lock-up and hold a ten pound force.

5. Turn the torque wrench in a clockwise direction and the one-way clutch should rotate freely.

6. Repeat the operation in at least five different locations around the converter.

7. If the one-way clutch fails to lock-up, replace the converter assembly.

CHECKING STATOR TO IMPELLER INTERFERENCE

1. Position the oil pump assembly on a flat surface with the splined end of the stator shaft pointing up.

2. Mount the converter on the pump with the splines of the stator shaft engaged with the splines of the one-way clutch inner race. The converter hub should then engage the pump drive gear.

3. Hold the oil pump body stationary and rotate the converter counterclockwise. The converter should rotate freely without any signs of interference or scraping within the converter assembly.

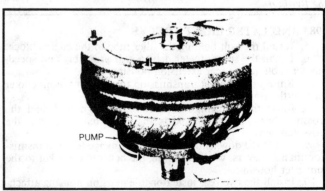

Checking stator to impeller interference (©Ford Motor Company)

4. If any indication of scraping or interference is noted, the converter should be replaced.

CHECKING THE STATOR-TO-TURBINE INTERFERENCE

1. Position the converter on the bench, flywheel side down, or in a vise using a holding fixture (Ford Number T83L-7902-3A or equivalent). When using the holding fixture, clamp it tightly in a vise. Place the converter on the holding fixture, aligning the pilot hub and one stud in the appropriate holes.

2. Install the holding wire into one of the grooves provided in the reactor thrust washer.

3. With the holding wire in position, spline the torque adapter tool in the converter. Make sure the shaft splines engage the splines in the turbine hub.

4. Install the pilot guide tool over the shaft and onto the impeller hub.

5. Hold the converter assembly and turn the torque converter turbine by rotating both clockwise and counterclockwise, using a torque wrench and a ¾ socket.

6. Replace the converter if there is a loud scraping noise or if the input shaft will not turn with 5 ft. lbs. of torque.

NOTE: If scraping or interference exists, the stator front thrust washer may be worn, allowing the stator to contact the turbine. If such is the case, the converter must be replaced.

VISUAL INSPECTION OF THE CONVERTER

Before installation of the converter, the crankshaft pilot should be inspected for nicks or burrs that could prevent the pilot from entering the crankshaft. Remove as necessary.

Inspect the converter front pump drive hub for nicks or burrs that could damage the pump oil seal during installation.

Checking stator to turbine interference (©Ford Motor Company)

Disassembly of Transmission

1. Mount the transmission in a suitable transmission holding fixture, separately or on a work bench area.

2. Remove the input shaft from the transmission assembly.

NOTE The input shaft may come out of the transmission when the converter is removed.

3. With the transmission inverted, remove the oil pan retaining bolts. Remove the pan and discard the gasket.

NOTE: Remove and discard the oil filler tube shipping plug found in the oil pan.

4. Remove the screw retaining the screen assembly. Lift the screen assembly from the valve body. Remove rubber grommet.

5. Remove the nine valve body retaining bolts and lift the valve body from the transmission case.

6. Remove the screen from the oil pump inlet bore, in the transmission case.

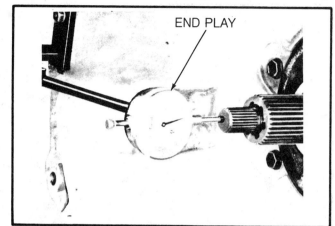

Checking end play of input shaft before disassembly (©Ford Motor Company)

7. Loosen the band adjusting locknuts on the adjusting screws. Remove the four band struts from inside the transmission.

8. In order to check the gear train end play bore disassembly, install the input shaft and force the gear train to the rear. Mount a dial indicator so that the indicator stem is touching the end of the input shaft and zero the indicator.

9. Using the rear brake drum for a fulcrum, not the aluminum planet carrier, pry the gear train forward. Read the indicated endplay on the dial indicator and record. The specified endplay is between 0.008 and 0.042 inch. If the endplay is incorrect, the thrust washer will have to be changed during the re-assembly. Remove the dial indicator.

10. Remove the input shaft and lay aside.

11. With the converter housing positioned so as not to cause a bind, remove the seven retaining bolts and remove the converter housing from the case assembly.

NOTE: On a car transmission, the converter housing bolts also hold the oil pump to the case. On truck transmissions, five bolts are used to retain the converter housing and seven bolts are used to retain the oil pump to the case.

12. Remove the oil pump assembly. If necessary, pry the gear assembly forward to loosen the pump assembly.

―――――――――――― CAUTION ――――――――――――

Pry on the rear drum, not on the aluminum planet carrier.

13. Remove the number one and number two thrust washers from the stator support. Remove and discard the gasket.

NOTE: Tag the thrust washers for assembly identification.

14. Reaching into the transmission case, align the intermedi-

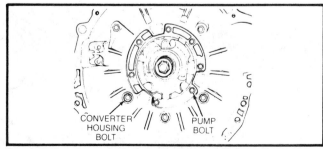

Converter housing and oil pump attachment on light truck transmission applications (©Ford Motor Company)

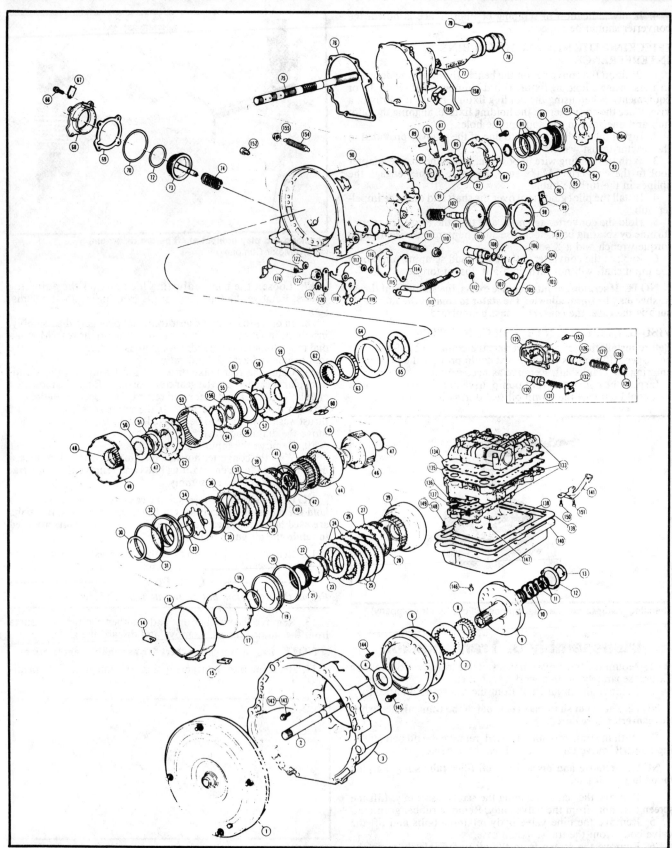

Exploded view of C-5 Automatic Transmission (©Ford Motor Company)

1. Converter Assembly
2. Shaft—Input
3. Housing Converter
4. Seal Assy.—Front Oil Pump
5. Body—Front Pump
6. Gasket—Front Oil Pump
7. Gear)—Frt. Oil Pump Driven
8. Gear—Frt. Oil Pump Drive
9. Stator Support—Frt. Oil Pump
10. Seal—Fwd. Cl. Cyl.
11. Seal—Rev. Cl. Cyl.
12. Washer—Frt. Pump Supt. Thrust (Sel.)—No. 1
13. Washer—Rev. Cl. Thrust—No. 2
14. Strut—Intermediate Brake
15. Strut—Intermediate Brake
16. Ban Assy.—Intermediate
17. Drum Assy.—Intermediate Brake
18. Seal—Rev. Cl. Inner
19. Piston Assy.—High Clutch
20. Seal—High Cl. Outer
21. Spring—High Cl. Piston
22. Retainer—Cl. Piston Spring
23. Ring
24. Plate—Cl. Ext. Spline—(Steel)
25. Plate Assy.—Cl. Int. Spline (Friction)
26. Plate—Clutch Pressure
27. Spring—Rev. Cl. Press. Plate Disc
28. Snap Ring (Selective)
29. Cylinder—Fwd. Clutch
30. O-Ring—Cl. Piston Oil
31. Seal—Fwd. Cl. Outer
32. Piston Assy.—Fwd. Clutch
33. Ring—Fwd. Cl. Pst. Spring Press
34. Spring—Fwd. Cl. Piston Disc
35. Ring)—Ret. Wave Int.
36. Plate—Fwd. Cl. Press
37. Plate—Cl. Ext. Spline (Steel)
38. Plate Assy. Cl. Int. Spline (Friction)
39. Plate—Clutch Press
40. Ring—Retaining Ext. (Sel.)
41. Washer—Fwd. Cl. Hub—Thrust—No. 3
42. Retainer
43. Hub & Bshg. Assy.—Fwd. Cl.
44. Gear—Output Shaft Ring
45. Washer—Rev. Planet Pinion—Thrust—No. 4
46. Planet Assy.—Fwd.
47. Retainer
48. Gear—Sun
49. Shell—Input
50. Washer—Input Shell Thrust—No. 5
51. Washer—Rev. Plt. Carrier—Thrust (Frt.)—No. 6
52. Planet Assy.—Reverse
53. Gear—Output Shaft Ring
54. Washer—Rev. Drum Thrust—No. 7
55. Hub—Output Shaft
56. Ring
57. Washer—Rev. Drum Thrust—No. 8
58. Drum Assy.—Rev. Brake
59. Band Assy.—Reverse
60. Strut—Rev. Brake
61. Strut—Rev. Band (Anchor)
62. Race—Overrun Clutch—Inner
63. Spring & Roller Assy.—OWC
64. Race—Overrun Clutch—Outer
65. Washer—Rev. Drum to Case—Thrust—No. 9.
66. Bolt
67. Tag—Service Identification
68. Cover—Interm. Band Servo
69. Gasket—Interm. Band Servo
70. Seal—Interm. Band Servo Cover—Large
71. Not Used
72. Seal—Interm. Band Servo Piston—Small
73. Piston Assy. Interm. Band Servo
74. Spring—Interm. Band Servo Piston
75. Shaft Assy.—Output
76. Gasket—Extension
77. Extension Assy.
78. Seal Assy.—Ext. Oil
79. Bolt
80. Body—Gov. Oil Collector
80A. Screen Assy.—Gov. Oil
81. Ring—Governor Seal
82. Ring (Not Used After Mid-1982 Production)
83. Bolt
84. Sleeve—Oil Distributor
85. Gear—Output Shaft Parking
86. Washer—Output Shaft Thrust—RR—No. 10
87. Spring—P.P. Return
88. Pawl Assy.—Parking
89. Pin
90. Case Assembly
91. Tube—Oil Distributor
92. Tube—Oil Distributor
93. Clip—Throt. Cntl. Valve Diaphragm
94. Diaphragm Assy.—T.V. Control
95. Rod—T.V. Control
96. Valve—Throttle Control
97. Bolt
98. Cover—Rev. Band Servo
99. Clip—Electrical Wiring Harness
100. Seal—Rev. Bnd. Servo Piston Cover
101. Piston Assy.—Rev. Band Servo
102. Spring—Rev. Servo Piston
103. Nut
104. Washer
105. Lever Assy.—Dwnshift Control—Outer
106. Switch Assy.—Neutral Start
107. Screw & Wshr. Assy.
108. Seal
109. Lever Assy.—Manual Control
110. Nut
111. Screw
112. Ring
113. Rod Assy.—Park Lever Actuating
114. Spacer—Parking Lever
115. Lever Assy.—Parking Actuating
116. Washer
117. Ring Retaining
118. Roller—Park Lever Actuating Rod
119. Lever Assy.—Man. Vlv. Detent—Inner
120. Link—P.P. Toggle Oper. Lever
121. Ring
122. Nut
123. Ring
124. Lever Assy.—Dwnshft. Detent—Inner
125. Body—Governor Valve
126. Valve—Governor Primary
127. Spring—Gov. Primary Valve
128. Washer
129. Retaining Ring—Internal
130. Valve—Gov. Secondary
131. Spring—Gov. Secondary Valve
132. Retainer—Gov. Sec. Valve Spring
133. Control Assy.—Main
134. Plate—Control Vlv. Body Sep.
135. Gasket—Control Vlv. Body Sep.
136. Plate—3-2 Timing Body Sep.
137. Gasket—3-2 Timing Body Sep.
138. Screen & Grommet Assy.—Oil Pan
139. Gasket—Oil Pan
140. Pan Assy.—Oil Pan
141. Spring Assy.—Main Vlv. Detent
142. Bolt
143. Bolt
144. Screw
145. Bolt
146. Bolt
147. Screw
148. Screw
149. Screw
150. Screw
151. Screw
152. Plug
153. Bolt
154. Screw
155. Nut
156. Ring
157. Body Assy. Governor
158. Tube—Vent
159. Clamp—Vent Tube

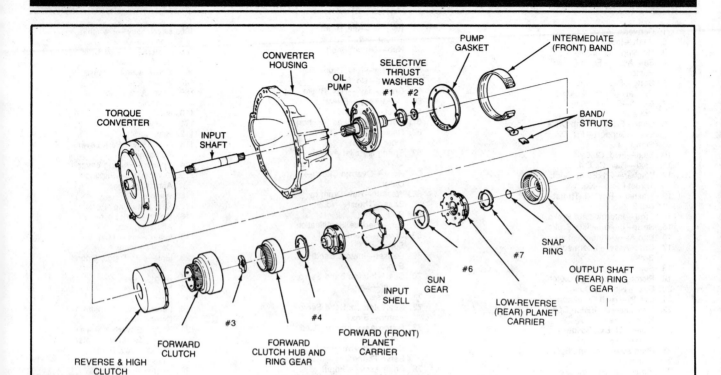

Exploded view of front internal components of transmission (©Ford Motor Company)

ate (front) band ends with the clearance hole in the case and remove the band.

15. Remove the clutch packs, front planetary and input shell as a unit from the transmission case. To prevent the assembly from rolling off bench, place the unit on the bench with the sun gear up.

16. Remove the reverse planet carier assembly. Remove the number six and number seven thrust washers from the carrier.

NOTE: Tag the thrust washers for assembly identification.

17. Align the low-reverse (rear) band ends with the clearance hole in the case and remove the band.

18. Place the transmission case face down on the bench or with the fixture.

19. Remove the extension housing bolts, the diaphragm retainer and tubing clip and the T.V. diaphragm with the diaphragm rod.

20. Using a magnet, remove the throttle valve from its bore in the case.

NOTE: If necessary, push the valve out from the inside of the case.

21. Lift the extension housing from the transmission case. Discard the gasket and remove the seal from the housing and discard.

NOTE: A spline seal may be found on the end of the output shaft. This is used for shipping only. Remove and discard.

22. Place the manual valve lever in the PARK position to lock the output shaft. Unbolt and remove the governor assembly.

23. Remove the governor screen and check for contamination.

24. Place the transmission in a horizontal position and remove the hub snapring retaining the reverse ring gear and hub assembly to the output shaft. Remove the reverse ring gear and hub assembly.

25. Remove the low-reverse drum. After removal, remove the number eight thrust washer from the low-reverse drum.

NOTE: Tag the thrust washer for assembly identification.

26. Place the transmission case face down. Lift the output shaft and governor collector body from the case (1982).

27. (Early 1982 models) Remove the snapring from the output shaft and separate the collector body from the output shaft. Remove the seals from the selector body.

NOTE: Mid 1982 and later output shafts do not use the retaining snapring holding the governor collector body to the shaft. The shaft can be lifted from the collector body separately and then the collector body can be removed.

28. Remove the four bolts holding the distributor sleeve and tubes. Remove the sleeve and tubes, being careful not to bend any of the tubes.

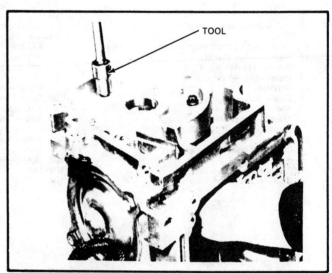

Removal of the one-way clutch retaining bolts with special socket (©Ford Motor Company)

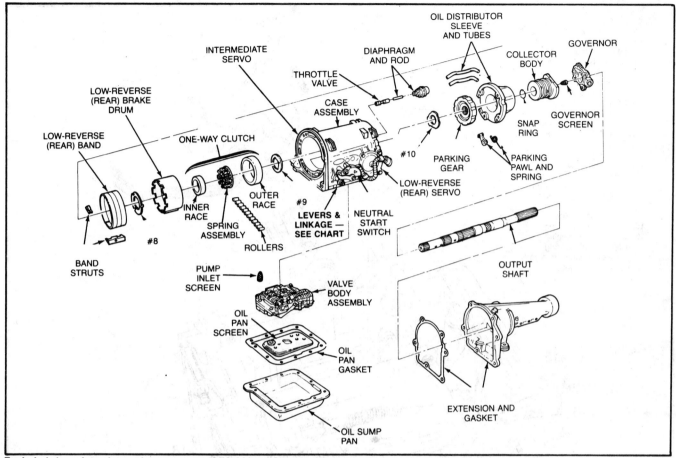

Exploded view of rear internal components (©Ford Motor Company)

29. Remove the parking gear and the number ten thrust washer.

NOTE: Tag the thrust washer for assembly identification.

30. Remove the spring, pawl and pivot pin.

31. With the transmission case still setting on its front area, reach into the rear of the transmission case and grasp the one-way clutch and keep it from falling, while removing the six special retaining bolts.

—————— CAUTION ——————
Use a special tool socket, Ford number T65P-7B456-B or its equivalent. A regular 5/16 inch socket may break due to the high torque necessary to remove the bolts.

32. Carefully, remove the one-way clutch assembly from the case. Remove the number nine thrust washer.

NOTE: Tag the thrust washer for assembly identification.

33. Remove the four bolts holding the intermediate servo and carefully remove the cover from the case. Remove the piston and spring from the cover. Remove and discard the gasket. Remove the piston seals.

34. Remove the low-reverse servo cover retaining bolts. Note the location of the wiring clips for installation purposes.

35. Remove the cover and discard the seal. Push the piston rod outward from inside the case. Remove the piston assembly and spring.

NOTE: The piston seal is bonded to the piston and cannot be removed.

36. Remove the cooler line fittings from the case. Remove the "O" rings and discard.

NOTE: This completes the usual disassembly of the C-5 Automatic Transmission case.

Case Internal Linkage

Removal

NOTE: The case internal linkage should only be disassembled if the components or case is damaged or the case must be replaced.

1. Position the downshift lever and note its relation to the case. Lubricate the outer lever nut with a penetrating oil and remove the nut, washer and the downshift lever.

2. Remove the "O" ring and inner downshift lever.

3. Remove the two retaining bolts and remove the neutral start and/or back-up lamp switch.

4. From inside the case, remove the hex nut from the inner manual lever and shaft. Push the manual lever out and remove the inner lever.

5. Noting the outer lever's position in relation to the case, remove the manual lever and shaft. Remove and discard the lever oil seal.

6. Remove the retainer and pull the manual lever link off the pin in the case.

7. At the rear of the case, remove the lower retainer and washer from the park pawl actuating lever. Remove the lever and link, attached to the actuating rod, through the rear of the case.

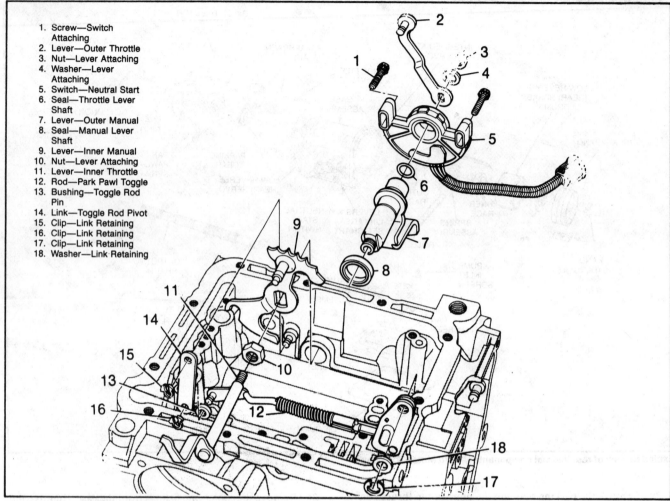

1. Screw—Switch Attaching
2. Lever—Outer Throttle
3. Nut—Lever Attaching
4. Washer—Lever Attaching
5. Switch—Neutral Start
6. Seal—Throttle Lever Shaft
7. Lever—Outer Manual
8. Seal—Manual Lever Shaft
9. Lever—Inner Manual
10. Nut—Lever Attaching
11. Lever—Inner Throttle
12. Rod—Park Pawl Toggle
13. Bushing—Toggle Rod Pin
14. Link—Toggle Rod Pivot
15. Clip—Link Retaining
16. Clip—Link Retaining
17. Clip—Link Retaining
18. Washer—Link Retaining

Manual and shift linkage components (©Ford Motor Company)

Inspections

1. No further disassembly is needed unless a part is damaged or bent. Replace as required.

Assembly

1. Lubricate the manual lever seal and install with the garter spring towards center of case. Install squarely into the seal bore with a seal installer tool.

2. Install the park actuating linkage. Install the retainer on the pin over the manual lever link.

3. Install the flat washer and the retainer over the park pawl actuating lever.

4. Install the outer manual lever and shaft in the original disassembled location.

5. Install the inner manual lever onto the flats of the outer manual lever shaft in its original position. Tighten the nut to 30-40 ft. lbs.

6. Install the neutral start and/or back-up lamp switch. Do not tighten the screws until the switch is adjusted.

7. Install the inner downshift lever assembly and shaft seal "O" ring. Install the outer downshift lever in its original position and install the nut and lockwasher. Tighten the nut to 12-16 ft. lbs.

UNIT DISASSEMBLY AND ASSEMBLY

Oil Pump

Disassembly

1. Remove the number one and two thrust washers from the stator support.

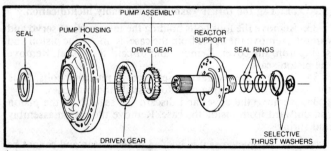

Exploded view of oil pump assembly (©Ford Motor Company)

2. Remove the teflon and cast iron rings from the stator support.

3. Remove the stator support to pump body attaching bolts and lift the stator support from the pump body.

4. Remove the pump gears from the pump body, noting their direction. The chamfer on the drive gear and the dot on the driven gear must face the pump body.

5. Remove the oil seal from the pump body, using a hammer and punch.

6. The bushing can be removed with the use of a bushing remover tool and a press.

Inspection

1. Inspect the mating surfaces of the pump body and cover for burrs.

2. Inspect the drive and driven gear bearing surfaces for scores and check the gear teeth for burrs.

3. Check all fluid passages for obstructions.

4. If any parts are scored deeply, worn or damaged, replace the pump assembly as a unit. Minor burrs and scores may be removed with crocus cloth.

Assembly

1. With the use of a bushing installer tool and press, install the bushing into the pump body. When the bushing is seated, stake the bushing to the pump body, using a chisel.

NOTE: Two notches are located in the bushing bore to be used as the stacking points.

2. Using a seal installer, install the oil seal in the front of the pump body.

—————— CAUTION ——————

The seal must be square in the bore. If the seal is not correct or the garter spring is out of place, replace the seal.

3. Lubricate the inside of the pump body with petroleum jelly and install the gears.

NOTE: The chamfered side of the drive gear and the dot on the driven gear must face the pump body.

4. Bolt the reactor support to the pump body with the five bolts. Torque the bolts to 12-20 ft. lbs.

5. Mount the pump on the converter hub and check the freeness of rotation of the pump gears.

6. Remove the pump from the converter and install the seals on the stator support. Place the iron seals in the lower grooves and the teflon seals in the upper grooves.

—————— CAUTION ——————

When installing the teflon rings, make sure the scarf ends overlap properly and when installing the cast iron rings, make sure the ends are securely interlocked.

7. Install the number one and number two thrust washers.

NOTE: These thrust washers are selective thicknesses and are used to limit the transmission end play to a specific tolerance. When installing, the thrust washers should be installed in pairs to obtain the desired end play of 0.008-0.042 inch (0.020-1.07 mm). Use petroleum jellly to lubricate.

8. If the end play is known, select the proper thrust washers from the chart and install.

9. If the end play is not known, install the original thrust washers. The end play will be checked and if necessary, corrected during the assembly of the transmission.

Identification of drive and driven oil pump gears
(©Ford Motor Company)

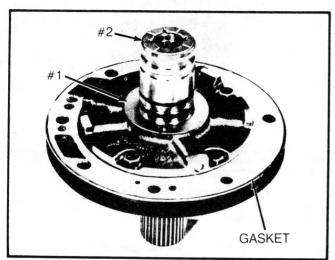

Location of numbers one and two thrust washers
(©Ford Motor Company)

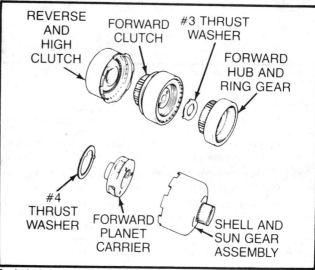

Exploded view of forward gear train and clutch assemblies
(©Ford Motor Company)

Forward (Front) Gear Train and Clutches

Disassembly

1. Lift the input shell and sun gear from the reverse-high clutch.
2. Remove the forward clutch hub and ring gear from the forward clutch. Separate and remove the front planet carrier and the number four thrust washer.

NOTE: The front planetary carrier is removed with the forward clutch hub.

3. Remove the number three thrust washer from the forward clutch. Separate the forward clutch from the direct (reverse-high) clutch.

NOTE: Assembly procedures for the Forward (Front) Gear Train and Clutches follow the disassembly and assembly of the sub-units.

Disassembly of Sub-Units

FORWARD CLUTCH

1. Remove the clutch pack retaining snapring and remove the clutch pack from forward clutch drum.
2. Using a pry tool, disengage the piston retaining ring from the clutch drum ring groove.
3. Remove the piston retaining ring, the Belleville piston return spring and the thrust ring from the forward clutch drum.
4. To remove the piston from the clutch drum, turn the piston clockwise or if necessary, blow the piston from the drum with air pressure.
5. Remove and discard the piston and the drum hub seals.

Inspection

1. Inspect the thrust surfaces, piston bore and clutch plate splines for scores and burrs. Replace the clutch cylinder if it is badly scored or damaged.

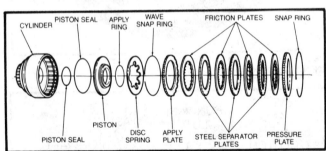

Exploded view of forward clutch assembly (©Ford Motor Company)

2. Check all fuid passages for obstructions. Inspect the clutch piston for scores and replace if required. Check the piston check ball for freedom of movement and proper seating.
3. Check the Belleville clutch release spring for distortion and cracks. Replace the spring if required.

NOTE: It is a good practice to replace the Belleville spring during an overhaul of the transmission unit.

4. Inspect the composition plates and steel clutch plates for worn or scored bearing surfaces. Inspect the clutch pressure place for worn or scored surface. Replace as required.
5. Check the steel plates for flatness and for freedom of movement on the clutch hub serrations. Replace as required.

NOTE: It is a good practice to replace the composition and steel clutch plates during an overhaul of the transmission unit.

6. Check the clutch hub thrust surfaces for scores and check the splines of the clutch hub and stator support for wear. Replace as required.
7. Inspect the bushing in the stator support for scores. Inspect the input shaft for damage or worn splines.

Assembly

1. Using petroleum jelly, lubricate and install a new seal on the clutch drum hub.
2. Using petroleum jelly, lubricate and install a new seal on the clutch piston. Note the direction of the sealing lip. The lip should face into the cylinder.

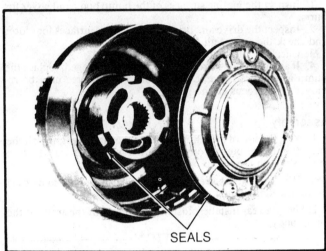

Installation of forward clutch piston and positioning of seals (©Ford Motor Company)

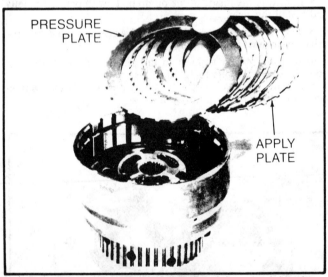

Installation of pressure plate and clutch pack into clutch housing (©Ford Motor Company)

3. Being sure both seals are lubricated, install the clutch piston into the clutch drum. To seat the piston properly into the drum, rotate the piston while pressing downward on it.

─────────── **CAUTION** ───────────

Be sure the seal lip is started into the clutch drum bore properly.

4. Install the thrust ring and Belleville piston return spring into the clutch drum.

NOTE: The dish of the Belleville spring must be down.

5. Install the wave snapring, working it firmly into the groove of the clutch drum. Be sure the ring is fully seated in the clutch drum groove.

6. Install the forward clutch pressure plate with the dished side of the plate facing the piston.

7. Install the clutch pack, starting with a friction plate and alternating with the steel plate until all plates are installed. The last plate to be installed is the rear pressure plate. The number of clutch plates vary with the transmission application.

NOTE: Before installing new clutch plates, soak in automatic transmission fluid at least 15 minutes.

8. Install the rear pressure plate and the clutch pack retaining ring. Be sure the retaining ring is properly seated in its groove on the forward clutch drum.

NOTE: The clutch pack retaining ring is a selective type snapring.

9. With the use of a feeler gauge, check the clearance between the pressure plate and the clutch pack retaining ring while holding the pressure plate downward. The clearance should be 0.025-0.050 inch (0.64-1.3 mm). If the clearance is not within the specified clearance, a selective snaprings must be used to obtain the correct clearance. The snaprings are available in the following thicknesses:
0.050-0.054 inch
0.064-0.068 inch
0.078-0.082 inch
0.092-0.096 inch
0.104-0.108 inch
Recheck the clutch pack clearance after installing the selective snapring.

Measuring clutch pack clearance on the forward clutch assembly (©Ford Motor Company)

10. Using air pressure, check the clutch assembly for operation. The clutch should be heard and felt, apply smoothly and have no leakage. As the air is stopped, the piston should return to the released position.

REVERSE-HIGH CLUTCH

Disassembly

1. Remove the clutch pack retaining snapring. Remove the clutch pack from the direct clutch drum, along with the Belleville disc spring and pressure plate.

2. Using a clutch spring compressor tool, compress the piston return spring and remove the spring retaining ring, using a set of external snapring pliers.

3. Remove the clutch piston from the clutch drum. If the piston is difficult to remove by turning it, air pressure can be used as required.

4. Remove the seal from the clutch piston and the seal from the clutch drum hub. Discard both.

Inspection

1. Inspect the drum band surface, bushings and thrust surfaces for scores. Badly scored parts must be replace. Minor scores can be removed with crocus cloth.

2. Inspect the clutch piston bore and the piston inner and outer bearing surfaces for scores. Check the air bleed valve for freeness, located in the clutch piston. Check the orifice to be sure it is not plugged.

3. Check all fluid passages for obstructions.

4. Inspect the clutch pressure plate for scores. Replace if necessary.

5. Inspect the composition and steel clutch plates. Check their fit on the splines of the clutch hub.

NOTE: It is a good practice to replace the composition and steel clutch plates during the overhaul of the transmission unit.

Assembly

1. Using petroleum jelly, lubricate and install a new seal on the clutch drum hub.

2. Using petroleum jelly, lubricate and install a new seal on the clutch piston, noting the direction of the seal lip.

NOTE: The seal lip should face into the clutch drum.

3. Being sure both seals are lubricated and the piston bore is lubricated with petroleum jelly, install the piston into the clutch drum bore with the use of a seal protector tool. Push the piston to the bottom of the bore with an even thumb pressure.

4. Position the piston return spring and the spring retainer on the clutch piston and compress the return spring, using the spring compressor tool. Install the spring retainer ring.

— CAUTION —
Be sure the retainer ring is in its groove properly before releasing the spring compressor tool.

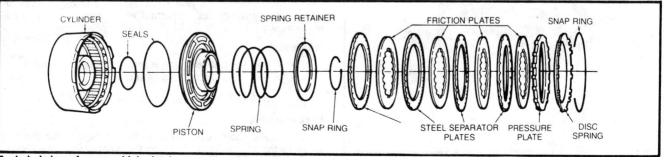

Exploded view of reverse-high clutch assembly (©Ford Motor Company)

5. To check the clutch pack clearance, install the clutch pack in the following order;

NOTE: The order of clutch pack installation for the check is not the correct installation sequence. This is done only for the check procedure. After checking the pack clearance, remove the plates and install in their proper sequence.

   a. Install a metallic plate.
   b. Alternately install the composition and metallic plates until two composition plates remain.
   c. Install the two remaining composition plates together.
   d. Install the disc spring.
   e. Install the pressure plate.

6. Install the clutch pack retaining ring. Using a feeler gauge, check the clearance between the pressure plate and the clutch pack retaining ring, while holding the pressure plate downward as the clearance is checked. The proper clearance is 0.025-0.050 inch (0.64-1.35 mm). If the clearance is not correct, selective snaprings are available in the following thicknesses:

   0.050-0.054 inch
   0.064-0.068 inch
   0.078-0.082 inch
   0.092-0.096 inch

7. Remove the clutch pack retaining ring and lift the clutch pack from the clutch drum. The clutch pack can now be installed into the clutch drum in the proper sequence, as follows;

   a. Install a metallic plate.
   b. Alternately intall the composition and metallic plates.
   c. Install the pressure plate.

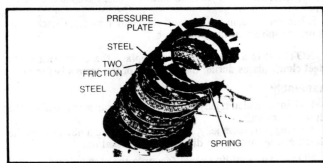

Temporary clutch pack assembly to make clearance check (©Ford Motor Company)

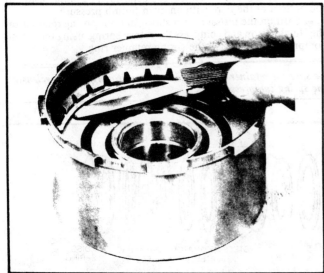

Making clutch pack clearance check (©Ford Motor Company)

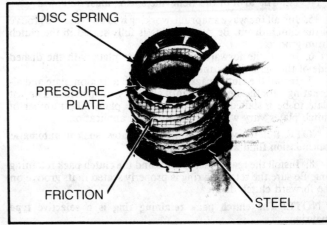

Correct order of clutch pack assembly after clutch pack clearance check (©Ford Motor Company)

   d. Install the disc spring with the splines facing the snapring.
   e. Install the clutch pack retaining ring into its groove in the clutch drum.

8. Using air pressure, check the operation of the clutch assembly. The clutch plates should be heard to engage and felt to apply smoothly and without leakage. When the air pressure is removed, the clutch pack should release smoothly and completely.

NOTE: During the air test, two holes must be blocked and the air pressure applied to the third.

FORWARD CLUTCH HUB AND RING GEAR

Disassembly

1. Remove the snapring retaining the hub to the ring gear.
2. Separate the hub from the ring gear.

Inspection

1. Inspect the splines, gear teeth and all mating surfaces for scores, pitting, chips and abnormal wear. Replace the components as required.

Assembly

1. Assemble the ring gear and forward clutch hub together.
2. Install the retaining snapring into its groove in the ring gear.

REVERSE RING GEAR AND HUB

Disassembly

1. Remove the snapring retaining the hub and flange to the ring gear.
2. Separate the hub and flange from the ring gear.

Inspection

1. Inspect the splines, gear teeth and all mating surfaces for scores, pitting, chips and abnormal wear. Replace the components as required.

Assembly

1. Assemble the hub and flange into the ring gear splines.
2. Install the snapring retainer into its groove in the ring gear.

INPUT SHELL AND SUN GEAR

Disassembly

1. With the use of external snapring pliers, remove the rear snapring from the sun gear. Remove the number five thrust washer.

2. Remove the sun gear from the input shell. If necessary, remove the snapring from the front of the sun gear.

Inspection

1. Inspect the splines, gear teeth and all mating surfaces for scores, pitting, chips and abnormal wear. Check the input shell for cracks and distortion. Replace the components as required.

Assembly

1. Install the front snapring on the sun gear. Install the sun gear into the input shell.

2. Install the number five thrust washer and the rear snapring retainer on the sun gear.

ONE-WAY CLUTCH

Disassembly

1. While pressing downward, rotate the inner race to separate it from the outer race spring retainer cage ring.

——————— CAUTION ———————

Note position of components for assembly identification. Do not lose rollers or springs during the disassembly.

Inspection

1. Inspect the rollers and mating surfaces on the inner and outer races for scoring, indentations and other abnormal wear indicators.

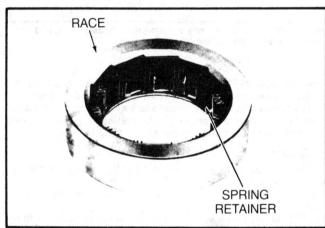

Position outer race with bolt holes down and install spring retainer (©Ford Motor Company)

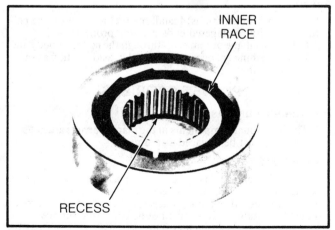

Install inner race with recessed spline down (©Ford Motor Company)

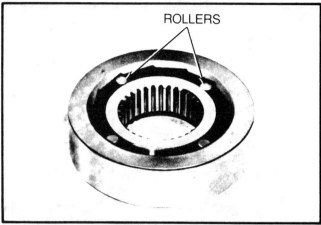

Installation of four rollers, 90° apart (©Ford Motor Company)

2. Inspect the spring and roller case for bent or damaged spring retainers.

Assembly

1. Position the one-way clutch outer race an the workbench with the bolt holes down.

2. Install the spring retainer with the cage ring down.

3. Install the inner race with the recessed end of the splines down.

4. Install four rollers spaced 90° apart between the inner and outer races. Install the remaining eight rollers and rotate the inner race to seat the components.

GOVERNOR

Disassembly

1. With the use of snapring pliers, remove the snapring from the governor bore containing the primary governor valve.

2. Remove the primary valve spring and the spring seat washer from the bore.

3. Remove the primary valve from the governor bore.

4. Remove the secondary valve spring retaining plate and pull the secondary valve and spring from the governor bore.

Installation of one-way clutch into transmission case (©Ford Motor Company)

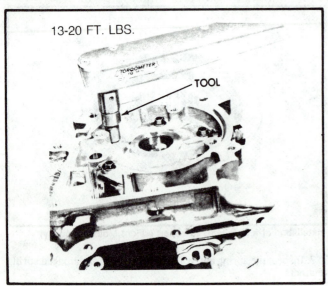

13-20 FT. LBS.

TOOL

Use of special socket and torque wrench to install bolts to the one-way clutch assembly (©Ford Motor Company)

Inspection

1. Inspect the governor valves and bores for scores, pitting or chips.
2. Inspect for free movement of the valves in their bores and fluid passages for cleanliness.
3. Inspect the springs for distortion.
4. Replace components as required.

Assembly

1. Install the secondary valve in the governor bore.
2. Position the secondary valve spring and depress, installing the spring retainer plate in its groove.

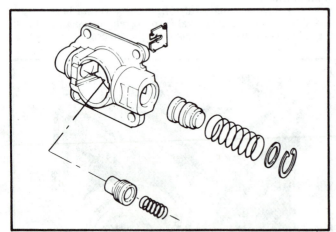

Exploded view of governor assembly (©Ford Motor Company)

─────── CAUTION ───────

Note the direction in which the plate is installed. The plate should be installed with the concave area facing the spring, thereby holding the spring in the correct position.

3. Install the primary valve in the governor bore, install the primary valve spring, the spring seat washer and the snapring.

OUTPUT SHAFT/GOVERNOR COLLECTOR BODY

Disassembly

1982

1. Remove the retaining snapring from the output shaft.
2. Separate the collector body from the output shaft.
3. Remove the three sealing rings from the collector body and discard.

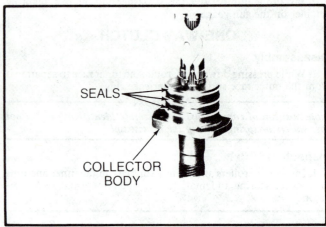

SEALS

COLLECTOR BODY

Removal of snapring on early transmission models. Snapring not used since mid-1982 (©Ford Motor Company)

NOTE: Mid-1982 and later C-5 transmissions do not use the retaining snapring on the output shaft. During the disassembly of the later transmissions, the output shaft and the governor collector body are removed separately. The sealing rings are removed and replaced in the same manner as the 1982 models.

Inspection

1. Inspect the mating surfaces of the collector body, seals and the output shaft.
2. Inspect the mating surface of the distributor sleeve at the point of sealing ring contact.
3. Should any major scores be evident, it is a good policy to replace the distributor sleeve and the collector body to prevent fluid pressure loss after reassembly.

Assembly

1. (All C-5 Transmissions) Install the sealing rings on the collector body with the tapered ends meeting properly.
2. (1982) Install the output shaft through the collector body and install the retaining snapring. Carefully lay aside until transmission unit assembly.

PLANET CARRIERS

Disassembly and Assembly

1. The individual components of the planet carriers are not serviceable and must be replaced as a unit.

Inspection

1. The pins and shafts in the planet assemblies should be checked for loose fits or poor engagement.
2. Check the shaft retaining pins should be checked for proper staking. The retaining pins must not be below the surface of the carrier more than 0.040 inch (1.0 mm).
3. Inspect the pinion gears for abnormal wear, chips and freeness of rotation.

Control Valve Body Assembly

Inspection

1. As each subsection of the control valve body assembly is disassembled, the internal components should be cleaned in solvent and blown dry with moisture free air pressure. The internal components should be inspected for any of the following conditions and immediately reassembled in the appropriate subsection to prevent the mixing of the internal components of one subsection to an other.

a. Inspect all valve and plug bores of scores. Check all fluid passages for obstructions.

b. Inspect the check valves for burrs and/or distortion

c. Inspect the plugs and valves for burrs or scores.

d. Inspect all springs for distortion or breakage.

e. Check all valves and plugs for freedom of movement in their respective bores. When dry, they should fall from their own weight in their bores.

f. Roll the manual valve on a flat surface to check for a bent condition.

g. Replace the valve body to screen gasket during the assembly.

TIMING VALVE BODY

Disassembly

1. Prepare a clean area on a work bench with room to spread the valve body components onto during the disassembly. All valves, springs and check balls must be kept in their correct order for ease in the assembly procedure.

2. Remove the ten screws from the timing valve body to lower valve body (Nine 5/16 inch heads and one 3/8 inch head bolts). Note the location of the short screw.

3. Carefully lift the timing valve body off the lower valve body. Remove the converter relief valve and spring.

4. Invert the timing valve body and remove the screw, separator plate and gasket. Note the position of the check ball and orifice "puck" and remove both.

5. Remove the retainers, the 1-2/3-2 shift timing valve, the 3-2 timing valve and the cut back valve from the valve body. Keep the springs with each valve.

NOTE: As each section of the valve body is disassembled, cleaned and inspected, each should be reassembled in their correct sequence and set aside until the complete assembly of the valve body is required. This procedure prevents mixing of section components and improper valve body operation after the overhaul is completed.

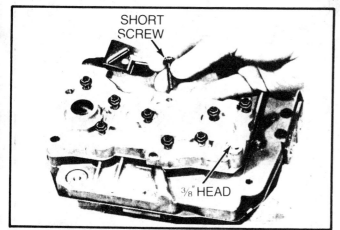

Removal of timing body attaching screws. Note location of short screw (©Ford Motor Company)

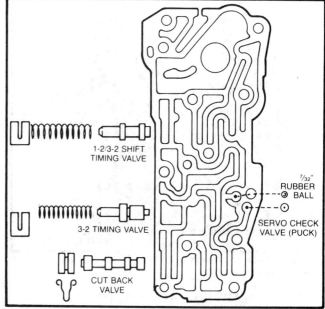

Exploded views of timing body component locations (©Ford Motor Company)

Assembly

1. Inspect the valve body and valves as indicated at the beginning of the valve body outline.

2. Install the valves, springs and retainers in their respective bores.

3. Install the check ball and check valve in the timing valve body.

NOTE: Depending upon vehicle application, there may be either a black or tan check valve (puck) used. During the reassembly, or when the puck replacement is required, be sure the correct colored puck is installed.

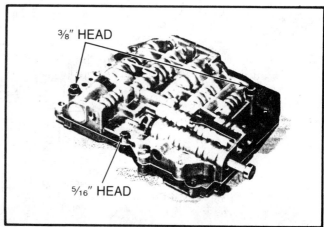

Remove three bolts, invert and remove the nine bolts to separate the upper and lower valve body (©Ford Motor Company)

4. Position the separator and gasket plate on the timing body and using either alignment pins or tapered punches, install the attaching screw and tighten to 25-40 in. lbs. (3-5 N•m).

5. Place the converter relief valve and the gasket with the timing valve body and lay aside until ready for complete valve body assembly.

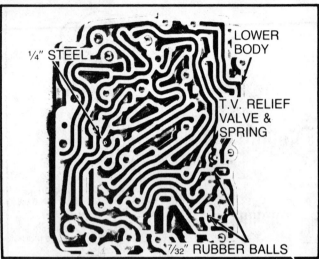

Lower valve body check balls and check valve locations
(©Ford Motor Company)

LOWER VALVE BODY

Disassembly

──────── CAUTION ────────

The lower valve body contains check balls which have to be held in position when the upper and lower valve body sections are separated. Follow this outline for the proper separation technique.

1. Remove the upper valve body to lower valve body attaching screws (three).

2. Turn the valve body assembly over and remove the nine lower body to upper body attaching screws. Remove the detent spring.

3. Carefully grip the separator plate and the lower valve body together and lift the body and plate away from the upper valve body. Turn the lower valve body and separator plate over with the separator plate facing up.

4. Remove the separator plate and the gasket. Discard the gasket.

5. Important: Identify the locations of the check balls in the lower valve body to aid in the reassembly. Remove the check balls and the throttle pressure limit valve and spring from the lower valve body.

6. **Important:** Identify the location of the check ball in the upper valve body to aid in the reassembly. Do not remove at this time. Set aside until the upper valve body is to be disassembled and assembled.

NOTE: A multi-channeled wooden block is a good holding tool for the disassembly of the valve bodies, holding the valves in a channel as they are removed from the valve body. This method keeps the valves in order of removal without fear of them rolling against each other and becoming mixed.

7. Clean, inspect and reinstall the throttle pressure limit valve and spring into its seat and position the black check balls in their seats in the lower valve body. Set the lower valve body aside until the complete assembly is done.

8. Noting the position of the checkball in the upper valve body, remove the ball and set aside.

9. A general removal procedure is given for the removal of the varied valves, retainers and springs. Lay the components of each bore out in order to prevent mixing them. After the complete removal of the components, the cleaning and inspection, of the

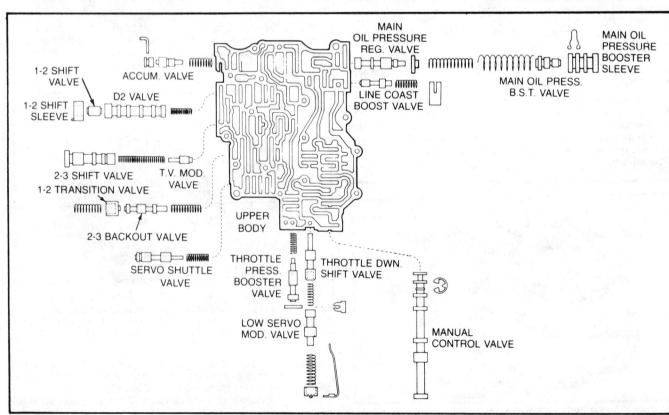

Exploded view of upper valve body components (©Ford Motor Company)

components and valve body, reassemble the valves, springs and retainers back into the valve body, in the reverse order of their removal.

NOTE: A wooden dowel pin or an aluminum fabricated tool is useful to relieve spring pressure when attempting to remove the retainers in the valve body bores.

CAUTION

Added or deleted internal components are sometimes found in a valve body bore that do not coincide with the components shown in an illustration. This can occur as a production change by the manufacturer for many various reasons. Consequently, components of a valve body bore must be kept in their proper order for correct reassembly.

a. Manual Control Valve—Remove the "E" clip retainer with an appropriate tool and remove the manual valve from its bore.

b. Low Servo Modulator Valve—Push inward on the end of the bore plug with a small probe and release the valve retainer. Lift the retainer from the valve body. Remove the bore plug, the spring, and the low servo modulator valve from the bore.

c. Throttle Downshift Valve—Using a long spring tool, such as the wooden dowel, lightly push inward and with the use of a magnet, remove the valve spring retainer from the opening of the valve body. Remove the throttle downshift valve and spring.

d. Throttle Pressure Boost Valve—Holding the valve body in a stationary position, and covering the bore, pull the bore plug lock pin from the valve body to release the bore plug. Remove the bore plug, the throttle pressure boost valve and the spring.

NOTE: During reassembly, tap lock pin in with small hammer.

e. Intermediate Servo Accumulator Valve—Push inward on the bore plug and remove the bore plug retainer with a magnet. Remove the bore plug, the intermediate servo accumulator valve and spring from the valve body bore.

f. Drive 2 Valve, 1-2 Shift Valve—Remove the three cover plate screws and remove the cover plate. Taking each bore separately, remove the sleeve, 1-2 shift valve, drive 2 valve and spring. From the second bore, remove the 2-3 shift valve, spring and the T.V. modulator valve.

NOTE: During the installation of the cover plate screws, torque to 25-40 in. lbs. (3-4.5 N•m).

g. Manual Low Control Valve/2-3 Backout Valve/Servo Shuttle Valve—Remove the two cover plate attaching screws and remove the cover. Taking each bore separately, remove the spring, manual control valve, backout valve and the spring. From the second bore, remove the servo shuttle valve and spring.

NOTE: During the installation of the cover plate screws, torque to 25-40 in. lbs. (3-4.5 N•m).

h. Main Pressure Boost Valve/Oil Pressure Regulator Valve—Pushing inward on the bore plug, remove the retaining clip from the plug groove. Remove the main pressure booster sleeve, the main pressure boost valve, the spring, inner spring and spring seat and the main oil pressure regulator valve.

NOTE: Some models do not use an inner spring.

1. Line Pressure Coasting Valve—Pushing inward with a spring tool, remove the spring retainer plate with a magnet. Remove the spring and line pressure coasting boost valve.

10. With the valves back in their respective bores, position the lower valve body with the channelled side up. Place the check balls and the pressure limiting valve in the lower valve body using petroleum jelly to retain in place.

NOTE: The steel check ball is larger than the others and must be placed in its proper location as was noted upon removal.

11. Place the gasket on the valve body and install locating pins (¼ inch drill bits or special pins). Install the separator plate and temporarily attach the plate to the valve body with the oil filter screen screw and tighten. Remove the alignment pins or drill bits.

12. Install the check ball in the upper valve body, using petroleum jelly to retain it.

13. Grasp the lower valve body and the separator plate. While holding the separator plate against the body, turn the assembly over and position the lower valve body on the upper valve body.

14. Install aligning pins and install two screws. Tighten the screws to 80-100 in. lbs. (9-13.5 N•m). Install the 5/16 inch head screw and tighten to 40-60 in. lbs. (4-5.7 N•m). Remove the oil filter screen screw and the alignment pins from the separator plate.

15. Position the detent spring and roller on the lower valve body and install the screw. Tighten to 40-60 in. lbs. (4.5-6.7 N•m). Use a drift to hold the assembly in place while the attaching screw is being tightened. Install the nine remaining screws and tighten to 40-60 in. lbs. (4.5-6.7 N•m).

NOTE: Do not forget to tighten the screw in the suction passage under the valve body.

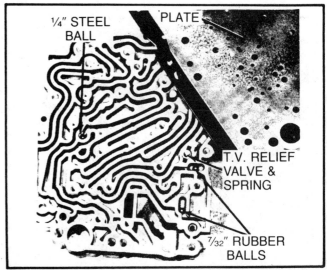

Lower valve body check balls and check valve locations before plate and gaskets are installed (©Ford Motor Company)

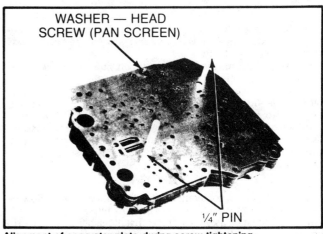

Alignment of separator plate during screw tightening (©Ford Motor Company)

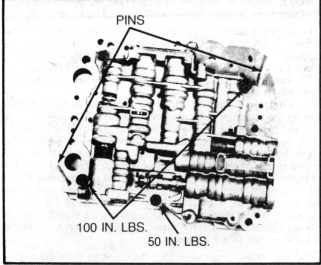

Alignment of torquing of bolts for lower and upper valve body assembly (©Ford Motor Company)

Installation of converter relief valve and spring (©Ford Motor Company)

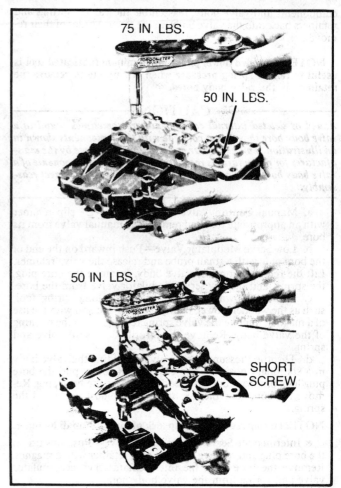

Torquing of timing valve body bolts to the upper/lower valve body assembly (©Ford Motor Company)

16. Install the gasket on the lower valve body. Install the converter relief valve in the lower body and position the timing body on the lower valve body.

17. Align the timing valve body with only one screw, tightened finger tight only. Visually check the alignment.

18. Install the remaining timing body attaching screws and tighten to 40-60 in. lbs. (4.5-6.7 N•m). Tighten the ¼ inch attaching screw to 52-72 in. lbs. (5.9-8.1 N•m).

19. Place the assembled valve body aside and cover the assembly to prevent dirt or other foreign objects from entering the passages, until ready for installation on the transmission case.

Bushings

Numerous bushings are available for installation in the following components. However, the proper bushing removing and installation tools must be used to properly install the bushings in a professional manner. If the bushing bore is damaged during the removal procedure or the new bushing is scored, cocked in the bore or crimped on the ends, assembly and operation of the transmission components will be affected.

    a. Sun Gear Bushing
    b. Case Bushing
    c. Pump Housing Bushing
    d. Forward Clutch Hub Bushing
    e. Low and Reverse Brake Drum Bushing

ASSEMBLY OF TRANSMISSION

1. Inspect the transmission case for re-use. Check the installation and operation of the shift linkage.

2. With the sealing portion of the threads wrapped with teflon tape, install the oil cooler pipe fittings into their respective passages on the transmission case. Torque to 18-24 ft. lbs. (24-31 N•m).

NOTE: Be sure to use new "O" rings in each fitting.

3. Position the transmission case on its side with the low-reverse servo bore up. Lubricate and install the spring and piston assembly. Install cover with a new cover seal. Lubricate the seal with petroleum jelly.

4. Install the retaining bolts in the cover with the wiring harness hangers in its proper position. Tighten the bolts to 12-20 ft.

lbs. (17-27 N•m). Position the wiring harness from the neutral start switch/back-up lamp switch into the wiring harness clips.

5. Install new seals on the intermediate piston and position a new gasket on the servo cover.

NOTE: A lubricant can be used to hold the gasket in place on the servo cover. Align the gasket with the fluid port in the case.

Installation of low/reverse servo assembly (©Ford Motor Company)

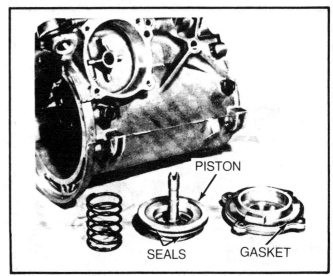

Installation of intermediate servo components
(©Ford Motor Company)

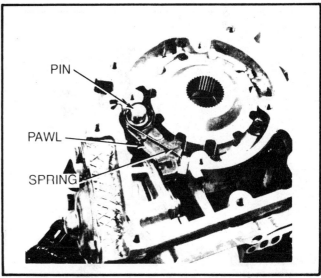

Installation of parking pawl and gear (©Ford Motor Company)

6. Install the piston into the cover and install the spring on the piston spring. Position the cover and piston in the case. Install the retaining bolts in the cover along with the transmission I.D. tag. Tighten the bolts to 16-22 ft. lbs. (22-30 N•m).

7. Lightly coat the number nine thrust washer with petroleum jelly and install in the transmission case at the rear.

8. Having previously assembled the one way clutch, position the transmission case front end down and install the one-way clutch over the number nine thrust washer. Install the one-way clutch retaining bolts and torque to 13-20 ft. lbs. (17-27 N•m), using the special socket available for this operation.

9. Check the parking pawl assembly for proper operation and install the number ten thrust washer and parking gear on the rear of the case. Spring load the parking pawl by looping the bend in the spring over the spring seat provided in the case. Check operation with manual lever.

10. Position the distributor sleeve on the case, making sure the oil tubes are fully seated in the case oil passages. During the installation of the distributor sleeve, be sure the parking pawl spring remains seated against the case. Torque the retaining bolts to 12-20 ft. lbs. (16-27 N•m).

11. Lubricate the oil seals and install the governor oil collector body into the distributor sleeve.

NOTE: On the 1982 model transmissions, the output shaft and the governor oil collector body must be assembled before the collector body is installed in the case. Both are then installed as a unit.

12. Position the transmission case on its top with the oil pan flange up. Lightly coat the number eight thrust washer with petroleum jelly and install in the low-reverse drum. Install the low-reverse drum into the transmission case and over the output shaft.

NOTE: Check the one-way clutch operation. The drum should turn free clockwise and lock up when turned counterclockwise.

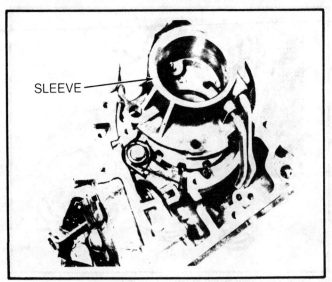

Installation of distributor sleeve on the transmission case
(©Ford Motor Company)

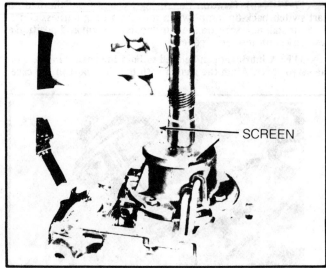

Installation of governor screen (©Ford Motor Company)

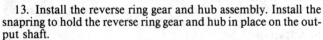

Installation of governor oil collector body and output shaft on early
1982 transmission models. Mid-1982 and later models can have the
individual components installed separately because snapring is not
used (©Ford Motor Company)

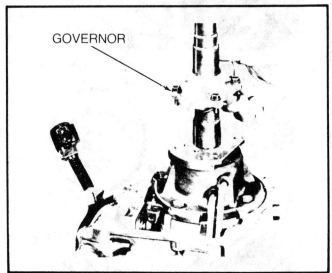

Installation of governor (©Ford Motor Company)

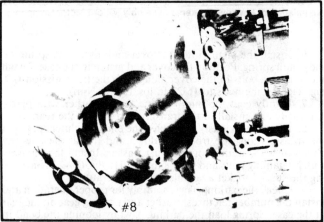

Installation of number eight thrust washer to the low/reverse drum
(©Ford Motor Company)

13. Install the reverse ring gear and hub assembly. Install the snapring to hold the reverse ring gear and hub in place on the output shaft.

NOTE: It may be necessary to push the output shaft forward to gain access to the snapring groove.

14. Install a new locknut on the rear band adjusting screw and start the screw into the case threads. Install the band strut stop into its passage in the case.

15. Align the band lugs with the relief clearance provided in the case and install the low-reverse band with the double lug of the band facing the adjusting screw.

16. Install the band struts and hand tighten the adjuster screw to hold the band in position.

17. Position the transmission case on its front flange. Install the governor screen into its passage in the governor distributor body. Position the governor assembly onto the governor distribu-

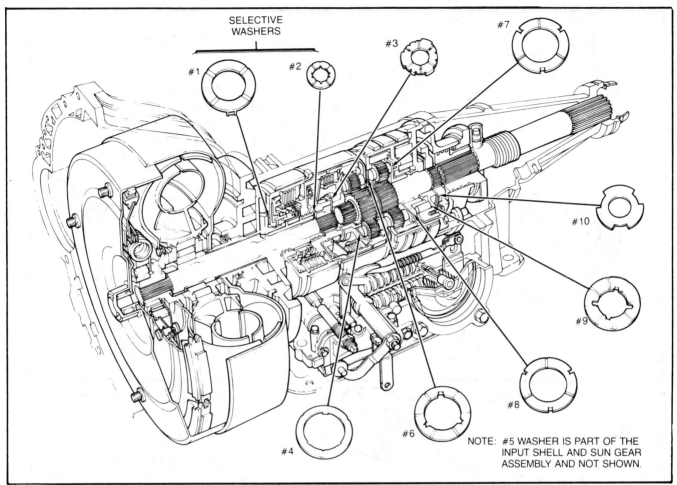

Location of thrust washers in the C-5 automatic transmission (©Ford Motor Company)

NOTE: #5 WASHER IS PART OF THE INPUT SHELL AND SUN GEAR ASSEMBLY AND NOT SHOWN.

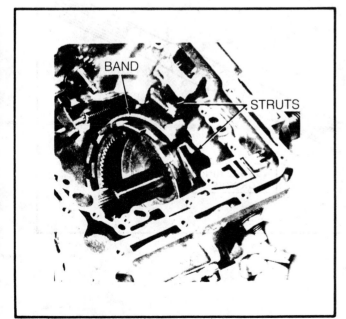

Installation of low/reverse band and struts
(©Ford Motor Company)

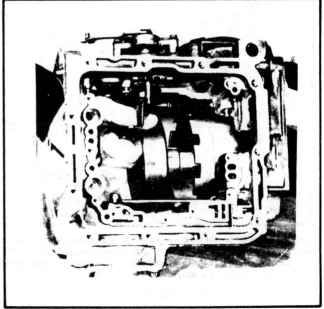

Installation of the hub and ring gear assembly
(©Ford Motor Company)

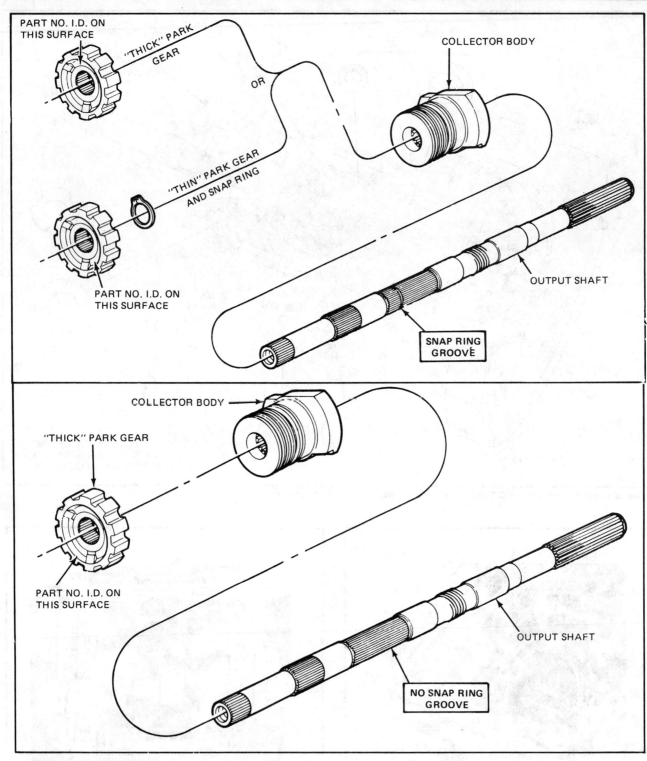

Installation of governor oil collector body and output shaft on early 1982 transmission models. Mid-1982 and later models can have the individual components installed separately because snapring is not used (©Ford Motor Company)

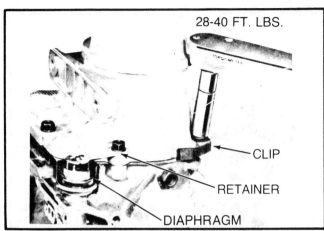

Installation of rear extension housing assembly
(©Ford Motor Company)

Installation of reverse planetary carrier (©Ford Motor Company)

tor body and install the attaching screws. Torque to 80-120 in. lbs. (9-13 N•m).

18. Position a new extension housing gasket on the case and position the extension housing without installing the attaching bolts.

19. Install the throttle valve and install the throttle valve rod and vacuum diaphragm.

20. Install the extension housing attaching bolts and torque to 28-40 ft. lbs. (38-54 N•m). Position the transmission identification tag and the modulator clamp in their proper locations.

21. Position the transmission with the oil pan flange up. Lightly lubricate the number six and seven thrust washers with petroleum jelly and position each on the reverse planetary gear assembly.

NOTE: The number seven thrust washer is placed on the gear side of the planetary gear assembly while the number six thrust washer is placed on the opposite side.

22. Install the planetary gear assembly into the transmission case, making sure the lugs are fully engaged in the low-reverse drum slots.

23. Install the forward clutch into the reverse-high clutch. Position the number three thrust washer on the forward clutch hub after lightly lubricating the washer with petroleum jelly.

24. Lightly lubricate the number four thrust washer and install on the front planetary assembly and install the front planetary

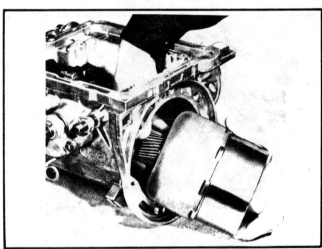

Installation of the front gear train (©Ford Motor Company)

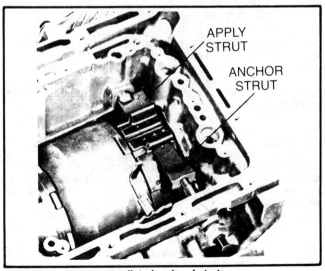

Installation of the intermediate band and struts
(©Ford Motor Company)

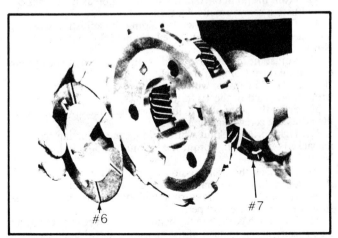

Installation of number six and seven thrust washers on the reverse planetary carrier (©Ford Motor Company)

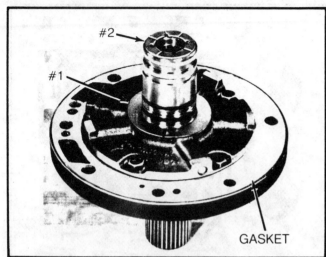

Preparation for installation of oil pump assembly
(©Ford Motor Company)

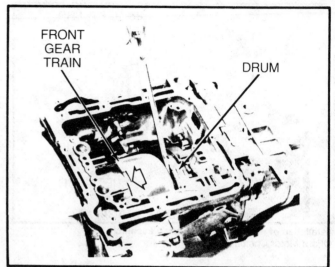

Gear train movement procedure for end play check
(©Ford Motor Company)

Checking input shaft end play (©Ford Motor Company)

unit in the forward clutch hub. Install the forward clutch hub and ring gear in the forward clutch.

25. Install the input shell and sun gear assembly onto the reverse-high clutch. Check the assembly for proper assembly by inserting and rotating the input shaft. The shaft should rotate in both directions. Remove the input shaft and lay aside.

26. Install the clutch packs, front planetary and input shell as an assembly into the transmission.

27. Align the intermediate band lugs with the relief clearance provided in the case and install the band.

28. Install a new locknut on the band adjusting screw and install the screw into its hole in the case. Install the intermediate band struts and tighten the adjusting screw finger tight to hold the band in place.

29. Install a new gasket on the front pump and position the pump assembly in the case and install two converter housing/pump attaching bolts. Snug the bolts.

30. Install the input shaft through the stator support. Position a dial indicator stylus against the end of the input shaft. Position a prying tool against a lug on the reverse-high clutch and push the gear train rearward.

31. Be sure the input shaft is fully seated and zero the indicator dial. Place the pry tool between the input shell and the reverse planetary gear assembly. Pry the input shell forward and read the dial indicator results.

32. If the end play is not within the specifications of 0.008 to 0.042 inch (0.20 to 1.07 mm), the number one and number two thrust washers must be changed. These thrust washers must be changed in pairs to obtain the specified clearances. After washer replacement, recheck end play. Check specification chart.

33. If the end play is within the specified limits, remove the dial indicator and the bolts holding the front pump assembly. Position the converter housing and install the attaching bolts. Tighten to 28-40 ft. lbs. (38-54 N•m).

NOTE: On the Truck models, the oil pump housing attaching bolts are installed and torqued to 28-38 ft. lbs. (38-51 N•m). Then position the converter housing on the transmission and bolt into place. Torque to 28-40 ft. lbs. (38-54 N•m).

34. Adjust the intermediate band by turning the adjusting screw out of the case several turns. Adjust the screw to 10 ft. lbs. (13.55 N•m) or with the use of an overrun or breakaway wrench, adjusted to the 10 ft. lbs. rating.

35. From the point of 10 ft. lbs. or wrench breakaway, back off the adjusting screw exactly 4¼ turns.

36. Tighten the locknut while holding the adjusting screw in its backed off position. Torque the locknut to 35-45 ft. lbs. (47-61 N•m).

37. Adjust the low-reverse band by turning the adjusting screw out several turns from its finger tight position. Adjust the screw to 10 ft. lbs. (13.55 N•m) or with the use of an overrun or breakaway wrench, adjusted to the 10 ft. lbs. rating.

38. From the point of 10 ft. lbs. or wrench breakaway, back off the adjusting screw exactly 3 turns.

39. Tighten the locknut while holding the adjusting screw in its backed off position. Torque the locknut to 35-45 ft. lbs. (47-61 N•m).

40. To check the transmission for proper assembly, install a slip yoke on the output shaft and turn the shaft in both directions. If the output shaft does not turn in both directions, the transmission is not assembled properly.

41. The clutch packs and the band servos can be checked by an air pressure check to each of the appropriate hydraulic passageways, using an air pressure of approximately 25 psi. The clutches should be heard and felt to apply smoothly and without leakage.

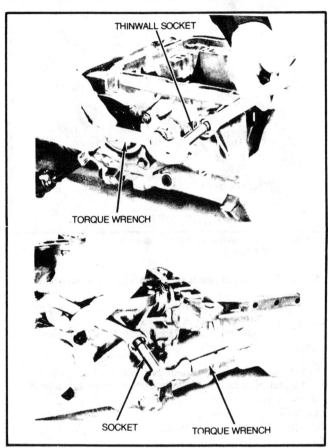

When the air pressure is released, the clutches should return to their released position.

42. When the air pressure is applied to the band servo passages, the band can be seen applying and releasing when the air pressure is removed.

43. Install the front oil pump inlet screen in the bore of the transmission.

44. Install the valve body carefully, making sure as the valve body is lowered into the case, the manual valve engages the manual lever and the downshift lever is positioned to engage the throttle downshift valve.

45. Install the valve body attaching bolts and torque to 80-120 in. lbs. (9-13 N•m).

NOTE: Place the two long bolts in their proper position at the left and right forward location on the timing valve body.

Adjustment of intermediate and low/reverse band
(©Ford Motor Company)

Installation of the valve body and filter screen
(©Ford Motor Company)

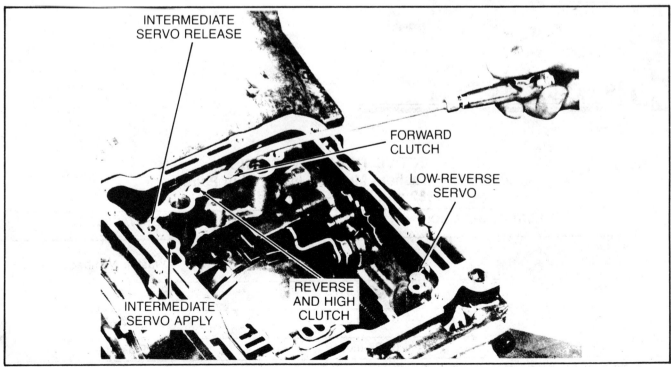

Checking the internal units with air pressure (©Ford Motor Company)

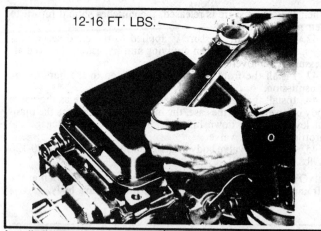

12-16 FT. LBS.

3/32 IN.

Installation of oil pan and torquing of the retaining bolts
(©Ford Motor Company)

Adjustment of the neutral start/back-up switch
(©Ford Motor Company)

46. Install the filter screen on the valve body and be sure the oil seal is in its proper position. Install the attaching screw and torque to 25-40 in. lbs. (3-4.5 N•m).

47. Position a new gasket on the oil pan and install the pan on the flange of the case.

48. Install the oil pan attaching bolts and torque to 12-16 ft. lbs. (16-22 N•m).

49. Install the input shaft and be sure it is fully seated.

50. Install the torque converter assembly, making sure the converter is fully seated, engaging the input shaft, stator support and the oil pump drive gear.

51. Place the transmission on its oil pan and place the manual lever in the neutral position and hold in place.

NOTE: Neutral position is two detents from the park lock position.

52. Insert a 3/32 inch (0.089 inch diameter minimum) drill or gauge pin through the hole in the switch.

53. Tighten the bolts to 55-75 in. lbs. and remove the drill or gauge.

NOTE: A continuity test can be performed before the transmission is installed in the vehicle to be certain the switch is properly adjusted.

54. The transmission can be placed on a transmission jack, safety chained and installed in the vehicle.

SPECIFICATIONS

SELECTIVE THRUST WASHERS
(Selective Washers Must Be Installed In Pairs)
1982 and Later—All Models

THRUST WASHER NO. 1		THRUST WASHER NO. 2
Color of Washer	Thickness	Washer Number
Red	0.053-0.0575	2
Green	0.070-0.0745	3
Neutral	0.087-0.0915	2 or 3 Plus Spacer ①

① This is a selective spacer used with washer 2 or 3. When used, install next to stator support.

CLUTCH PLATES
1982—Car Models

Model	Forward Clutch			Reverse Clutch		
	External Spline (Steel)	Internal Spline (Comp.)	Free Pack Clear (Inches)	External Spline (Steel)	Internal Spline (Comp.)	Free Pack Clear (Inches)
PEN Fairmont/Zephyr, Thunderbird/XR7, Granada/Cougar, Mustang Capri	4	5		3	3	
PEP Granada/Cougar,	4	5		3	3	.030-.055
PEM Mustang/Capri Fairmont/Zephyr	4	5		4	4	

CLUTCH PLATES
1983—Car Models

Model	Forward Clutch			Reverse Clutch		
	External Spline (Steel)	Internal Spline (Comp.)	Free Pack Clear (Inches)	External Spline (Steel)	Internal Spline (Comp.)	Free Pack Clear (Inches)
PEN Fairmont, Futura/ Zephyr, Thunderbird/XR7, LTD/Marquis, Mustang Capri	4	5	—	3	3	0.025-0.050
PEP LTD/Marquis	4	5	—	3	3	0.025-0.050

CLUTCH PLATES
1982—F100, F150 Trucks

Engine	Forward Clutch			Reverse Clutch		
	External Spline (Steel)	Internal Spline (Comp.)	Free Pack Clear (Inches)	External Spline (Steel)	Internal Spline (Comp.)	Free Pack Clear (Inches)
3.8L	4	5	0.025-0.050	3	3	0.050-0.071
4.2L	4	5	0.025-0.050	4	4	0.050-0.071

CLUTCH PLATES
1983—F100 Truck

Engine	Forward Clutch			Reverse and High Clutch		
	External Spline (Steel)	Internal Spline (Comp.)	Free Pack Clear (Inches)	External Spline (Steel)	Internal Spline (Comp.)	Free Pack Clear (Inches)
3.8L	4	5	0.025-0.050	3	3	0.025-0.050

CLUTCH PLATES
1983 and Later—Ranger

Engine	Forward Clutch			Reverse Clutch		
	External Spline (Steel)	Internal Spline (Comp.)	Free Pack Clear (Inches)	External Spline (Steel)	Internal Spline (Comp.)	Free Pack Clear (Inches)
2.3L	3	4	0.025-0.050	3	3	0.0250-0.050

VACUUM DIAPHRAGM ASSEMBLY SPECIFICATIONS
1982—F100, F150

Diaphragm Type	Diaphragm Part No.	Identification	Throttle Valve Rod*		
			Part No. (7A380)	Length	Identification
S-SAD	D6AP-7A377-AA	1 Green Stripe	D3AP-JA	1.5925-1.5876	Green Daub
			D3AP-HA	1.6075-1.6025	Blue Daub
			D3AP-KA	1.6225-1.6175	Orange Daub
			D3AP-LA	1.6375-1.6325	Black Daub
			D3AP-MA	1.6585-1.6535	Pink/White Daub

*Selective Fit Rods
S-SAD—Super Single Area Diaphragm

CHECKS AND ADJUSTMENTS
1982—Car Models

Operation	Specification
Transmission End Play	0.008-0.042 inch (Selective Thrust Washers Available)
Torque Converter End Play	New or Rebuilt 0.023 maximum. Used 0.050 maximum.
(Intermediate) Band Adjustment Front	Remove and discard lock nut. Install new nut. Adjust screw to 13.55 N•m (10 lb-ft), then backoff 4¼ turns. Hold screw and tighten lock nut to 54.33 N•m (40 lb-ft).
(Reverse) Band Adjustment Rear	Remove and discard lock nut. Adjust screw to 13.55 N•m (10 lb-ft), then back off 3 turns. Install new lock nut and tighten to 54.33 N•m (40 lb-ft).
Selective Snap Ring Thickness (Fwd. or Rev. Clutch)	0.050-0.054, 0.064-0.068, 0.078-0.082, 0.092-0.096, 0.104-0.108

CHECKS AND ADJUSTMENTS
1983—Car Models

Operation	Specification
Transmission End Play	0.008-0.042 inch (Selective Thrust Washers Available)
Torque Converter End Play	New or Rebuilt 0.023 maximum. Used 0.050 maximum.
(Intermediate) Band Adjustment Front	Remove and discard locknut. Install new nut. Adjust screw to 13.55 N•m (10 lb-ft), then backoff 4¼ turns. Hold screw and tighten locknut to 54.33 N•m (40 lb-ft).
(Reverse) Band Adjustment Rear	Remove and discard locknut. Adjust screw to 13.55 N•m (10 lb-ft), then back off 3 turns. Install new locknut and tighten to 54.33 N•m (40 lb-ft).
Selective Snap Ring Thickness (Fwd. or Rev. Clutch)	0.050-0.054, 0.064-0.068, 0.078-0.082, 0.092-0.096, (0.104-0.108—Fwd Clutch Only).

CHECKS AND ADJUSTMENTS
1982—F100, F150

Operation	Specification
Transmission End Play	0.008-0.042 inch (Selective Thrust Washers Available)
Torque Converter End Play	New or Rebuilt 0.023 maximum. Used 0.050 maximum.
(Intermediate) Band Adjustment Front	Remove and discard lock nut. Install new nut. Adjust screw to 13.55 N•m (10 lb-ft), then backoff 4¼ turns. Hold screw and tighten lock nut to 54.33 N•m (40 lb-ft).
(Reverse) Band Adjustment Rear	Remove and discard lock nut. Adjust screw to 13.55 N•m (10 lb-ft), then back off 3 turns. Install new lock nut and tighten to 54.33 N•m (40 lb-ft).
Selective Snap Ring Thickness (Fwd. or Rev. Clutch)	0.050-0.054, 0.064-0.068, 0.078-0.082, 0.092-0.096, 0.104-0.108

CHECKS AND ADJUSTMENTS
1983—F100

Operation	Specification
Transmission End Play	0.008-0.042 inch (Selective Thrust Washers Available)
Torque Converter End Play	New or Rebuilt 0.023 Maximum. Used 0.050 Maximum.
(Intermediate) Band Adjustment Front	Remove and discard locknut. Install new nut. Adjust screw to 13.55 N•m (10 ft-lbs), then back off 4¼ turns. Hold screw and tighten locknut to 54.33 N•m (40 ft-lbs).
(Reverse) Band Adjustment Rear	Remove and discard locknut. Adjust screw to 13.55 N•m (10 ft-lbs), then back off 3 turns. Install new locknut and tighten to 54.33 N•m (40 ft-lbs).
Selective Snap Ring Thickness (Fwd. or Rev. Clutch)	0.050-0.054, 0.064-0.068, 0.078-0.082, 0.092-0.096, 0.104-0.108

CHECKS AND ADJUSTMENTS
Ranger 1983 and Later

Operation	Specification
Transmission End Play	0.008-0.042 inch (Selective Thrust Washers Available)
Torque Converter End Play	New or Rebuilt 0.023 maximum. Used 0.050 maximum.
(Intermediate) Band Adjustment Front	Remove and discard locknut. Install new nut. Adjust screw to 13.55 N•m (10 ft-lbs), then back off 4¼ turns. Hold screw and tighten locknut to 54.33 N•m (40 ft-lbs).
(Reverse) Band Adjustment Rear	Remove and discard locknut. Adjust screw to 13.55 N•m (10 ft-lbs), then back off 3 turns. Install new locknut and tighten to 54.33 N•m (40 ft-lbs).
Selective Snap Ring Thickness (Fwd. or Rev. Clutch)	0.050-0.054, 0.064-0.068, 0.078-0.082, 0.092-0.096, 0.104-0.108

VACUUM DIAPHRAGM ASSEMBLY SPECIFICATIONS
1982 and Later—Car Models

Diaphragm Type	Diaphragm Part No.	Identification	Part No. (7A380)	Length	Identification
S-HAD	D70P-7A377-AA	1 White Stripe	D3AP-JA	1.5925-1.5876	Green Daub
			D3AP-HA	1.6075-1.6025	Blue Daub
SAD	EZDP-7A377-AA	No Identification	D3AP-KA	1.6225-1.6175	Orange Daub
			D3AP-LA	1.6375-1.6325	Black Daub
S-SAD	D6AP-7A377-AA	1 Green Stripe	D3AP-MA	1.6585-1.6535	Pink/White Daub

*Selective fit rods   SAD—Single Area Diaphragm   S-SAD—Super Single Area Diaphragm   S-HAD—Super High Altitude

VACUUM DIAPHRAGM ASSEMBLY SPECIFICATIONS
1983—F100

Diaphragm Type	Diaphragm Part No.	Identification	Part No. (7A380)	Length	Identification
S-SAD	D6AP-7A377-AA	1 Green Stripe	D3AP-JA	1.5925-1.5876	Green Daub
			D3AP-HA	1.6075-1.6025	Blue Daub
			D3AP-KA	1.6225-1.6175	Orange Daub
			D3AP-LA	1.6375-1.6325	Black Daub
			D3AP-MA	1.6585-1.6535	Pink/White Daub

*Selective fit rods
S-SAD—Super Single Area Diaphragm

VACUUM DIAPHRAGM ASSEMBLY SPECIFICATIONS
Ranger 1983 and Later

Diaphragm Type	Diaphragm Part No.	Identification	Part No. (7A380)	Length	Identification
S-SAD	D6AP-7A377-AA	1 Green Stripe	D3AP-JA	1.5925-1.5876	Green Daub
S-HAD	D70P-7A377-AA	No Color	D3AP-HA	1.6075-1.6025	Blue Daub
			D3AP-KA	1.6225-1.6175	Orange Daub
			D3AP-LA	1.6375-1.6325	Black Daub
			D3AP-MA	1.6585-1.6535	Pink/White Daub

*Selective fit rods
S-SAD—Super Single Area Diaphragm

TORQUE SPECIFICATIONS

Description	N•m	in. lbs.
End Plates To Valve Body	2.82-4.51	25-40
Separator Plate To Timing Valve Body	2.82-4.51	25-40
Lower Body To Upper Body (10-24)	4.51-6.77	40-60
Screen To Timing Valve Body	2.82-4.51	25-40
Governor To Governor Oil Collector Body	9.03-12.55	80-120
Pump Assembly To Case	2.25-3.95	20-38
Main Control To Case	9.03-13.55	80-120
Neutral Switch To Case	6.21-8.47	55-75

TORQUE SPECIFICATIONS

Description	N•m	in. lbs.
Upper Body To Lower Body (Long) (¼-20)	9.03-12.55	80-120
Upper Body To Lower Body (Short) (10-24)	4.51-6.77	40-60
3-2 Timing Valve Body To Upper Body (10-24)	4.51-6.77	40-60
3-2 Timing Valve Body To Lower Body (¼-20)	5.9-8.1	52-72
Detent Spring and Lower Body To Upper Body	4.51-6.77	40-60
Detent Spring and Main Control To Case	9.03-12.55	80-120
3-2 Timing Valve Body To Lower Body (10-24)	4.51-6.77	50-60
Speedometer Clamp Bolt	4-6	36-54

TORQUE SPECIFICATIONS

	N•m	ft. lbs.
Overrunning Clutch Race To Case	18-27	13-20
Push Connector To Transmission Case	24-31	18-23
Oil Pan To Case	16-22	12-16
Stator Support To Pump	17-27	12-20
Converter to Flywheel	27-46	20-34
Converter Housing Cover To Converter Housing	17-21	12-16
Converter Housing To Case	38-55	28-40
Engine Rear Cover Plate To Transmission	17-21	12-16
Rear Servo Cover To Case	17-27	12-20
Intermediate Servo Cover To Case	22-30	16-22
Oil Distributor Sleeve To Case	16-27	12-20

TORQUE SPECIFICATIONS

Description	N•m	ft. lbs.
Extension Housing To Case	38-54	28-40
Pump and Converter Housing To Case	38-51	28-38
Engine To Transmission (3.8L)	38-51	28-38
Transmission To Engine (3.3L, 4.2L)	55-67	40-50
Outer Throttle Lever To Shaft	17-21	12-16
Band Adjusting Screws To Case	13.5	10
Inner Manual Lever To Shaft	41-54	30-40
Pump Pressure Plug To Case	9-16	6-12
Intermediate Band and Reverse Band Adjusting Screw Locknut	47-61	35-45
Drain Plug To Converter Cover	20-24	15-18

 **SPECIAL TOOLS**

SPECIAL SERVICE TOOLS

Description	Tool Number	Description	Tool Number
Impact Slide Hammer	T50T-100-A	Extension Housing Seal Installer	T61L-7657-A
Puller Attachment	T58L-101-A	Extension Housing Bushing Installer	T77L-7697-B
Bench Mounted Holding Fixture	T57L-500-B	Extension Housing Bushing Remover	T77L-7697-A
Air Nozzle Assembly	Tool-7000-DE	Lip Seal Protector—Reverse Clutch	T82L-77404-A
Shift Linkage Insulator Tool	T67P-7341-A	Converter Clutch Holding Tool	T77L-7902-A
Output Shaft Retainer Pliers	T73P-77060-A	Converter Clutch Torquing Tool	T76L-7902-C
Clutch Spring Compressor	T65L-77515-A	End Play Checking Tool	T80L-7920-A
Pressure Gauge 0-400 P.S.I.	T57L-77820-A	Band Adjustment Torque Wrench Set	T71P-77370-A
Pump Seal Remover	Tool-1175AC	Shift Lever Seal Installer	Tool-77288
Pump Seal Installer	T63L-77837-A	Case Bolt Socket 5/16 Hex	T65P-7B-456-B
Extension Housing Seal Remover	T74P-77248-A	Transmission Bushing Set	T66L-7003-B

VALVE BODY SPRING TOOL FABRICATION

When assembling a valve body, the tool shown below will ease installation of those springs retained by a flat, slotted plate. The tool is not available from a manufacturer and must, therefore, be fabricated in the shop. The tool is cut from a ⅜-inch aluminum rod to the dimensions shown on the illustration. To simplify cutting and measuring, the illustration is actual size and can be used to check the accuracy of the fabricated tool.

To use the tool, position the spring in the valve body bore and compress it with the tool. While holding the spring compressed install the retainer plate over the end of the tool. The shape of the tool allows the plate slot to fit over the end of the tool while it is holding the spring compressed against the valve.

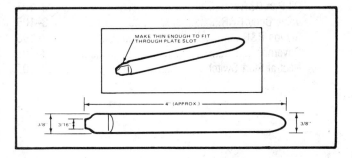

<h1>INDEX</h1>

FORD MOTOR COMPANY
ATX
Automatic Transaxle

APPLICATIONS

1981-83
Escort/Lynx, EXP/LN-7

1984
Tempo/Topaz, Escort/Lynx, EXP

GENERAL DESCRIPTION

The ATX automatic transaxle is a front wheel drive unit, housing both an automatic transmission and a differential in a single housing, bolted to the engine and mounted transversely in the vehicle. The ATX unit uses three friction clutch units, one band and a single one-way clutch. A compound planetary gear set is used to transmit the engine torque through the unit as the varied internal units are applied and released, providing three forward and one reverse gear ratios. Unique features of the transaxle are:

1. The valve body is mounted on top of the transaxle case.
2. The oil pump is mounted opposite the torque converter.
3. The parking gear is installed on the final drive unit.
4. The torque converter contains a planetary gear set, which is used to split the input power between mechanical and hydraulic drive, in most gears.
5. Two input shafts are used from the converter to the gear train.

3–75

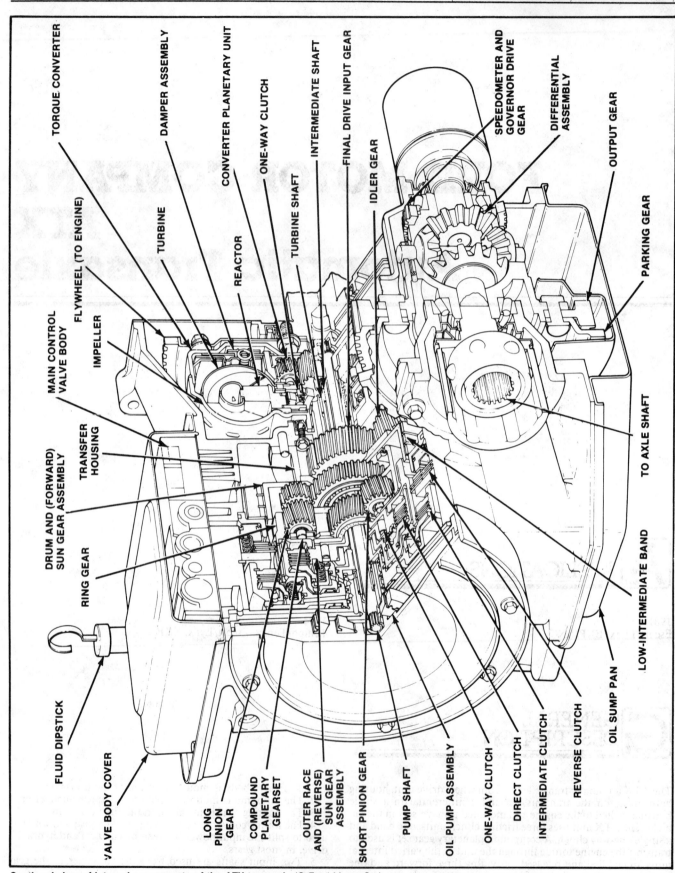

Sectional view of internal components of the ATX transaxle (© Ford Motor Co.)

TRANSAXLE AND CONVERTER IDENTIFICATION

Transaxle

The ATX automatic transaxle is identified by the letter "B" stamped on the vehicle certification label, mounted on the left driver's door body pillar post. The identification tag is located on one of the valve body pan retaining screws, above the oil pump location. The identification tag indicates the model code, the part number prefix and suffix, build code and transaxle serial number.

Converter

The ATX torque converter is identified by either a reference or part number stamped on the converter body and is matched to a specific engine. The torque converter is a welded unit and is not repairable. If internal problems exists, the torque converter must be replaced.

Transmission Fasteners

Metric bolts and nuts are used in the construction of the transaxle, along with the familiar inch system fasteners. The dimensions of both systems are very close and for this reason, replacement fasteners must have the same measurement and strength as those removed. Do not attempt to interchange metric fasteners for inch system fasteners. Mismatched or incorrect fasteners can result in damage to the transaxle unit through malfunctions, breakage or personal injury. Care should be exercised to replace the fasteners in the same locations as removed.

CAPACITIES

Year	Models	Quart	Liter
1981-82	All	10	9.46
1983-84	All	8.3	7.9

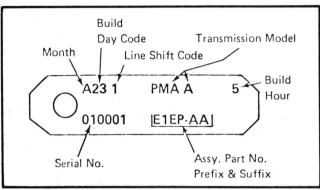

Typical ATX transaxle identification tag (© Ford Motor Co.)

Fluid Type Specifications

Only Motorcraft Dexron® II, Series D or CJ fluids, meeting Ford Motor Company's specifications ESP-M2C138-CJ, should be used in the ATX transaxle. Failure to use the proper fluid could result in internal transaxle damage.

Checking the ATX Transaxle Fluid Level

The ATX transaxle is designed to operate with the fluid level between the ADD and FULL mark on the dipstick, with the transaxle unit at normal operating temperature of 155 to 170 degrees Fahrenheit. If the fluid level is at or near the bottom indicator on the dip stick, either cold or hot, do not drive the vehicle until fluid has been added.

TRANSAXLE AT ROOM TEMPERATURE

70° to 95° F., Dipstick cool to the touch
1. With the vehicle on a level surface, engine idling, wheels blocked, foot brakes applied, move the selector lever through the gear positions to engage each gear and to fill the oil passages with fluid.

(ENGLISH) INCH SYSTEM Bolt, 1/2-13x1

G- Grade Marking
  (bolt strength)
L- Length, (inches)**
T- Thread Pitch
  (thread/inch)
D- Nominal Diameter
  (inches)

METRIC SYSTEM Bolt M12-1.75x25

--9.8

P- Property Class*
  (bolt strength)
L- Length (millimeters)**
T- Thread Pitch (thread width
  crest to crest mm)
D- Nominal Diameter
  (millimeters)

*The property class is an Arabic numeral distinguishable from the slash SAE English grade system.
**The length of all bolts is measured from the underside of the head to the end.
Examples of differences between inch and metric bolts (© Ford Motor Co.)

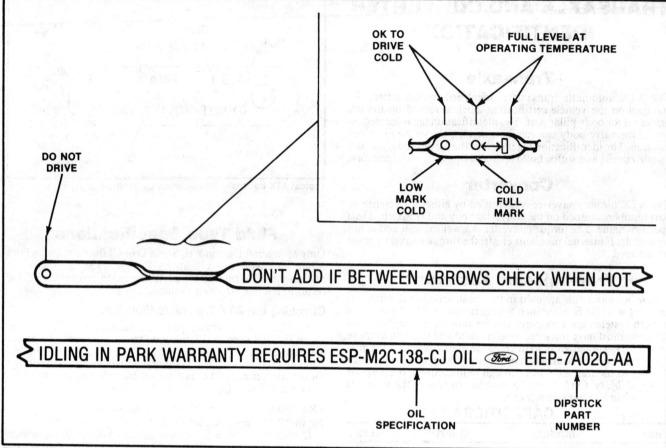

Fluid dipstick markings (© Ford Motor Co.)

2. Place the selector lever in the PARK position and apply the parking brakes, allowing the engine to idle.

3. Clean the dipstick area of dirt and remove the dipstick from the filler tube. Wipe the dipstick clean and re-insert it back into the filler tube and seat it firmly.

4. Remove the dipstick from the filler tube again and check the fluid level as indicated on the dipstick. The level should be between the cold low mark and the cold full mark on the dipstick indicator.

5. If necessary, add enough fluid to bring the level to its cold full mark. Re-install the dipstick and seat it firmly in the filler tube.

6. When the transaxle reaches normal operating temperature of 155° to 170° F., re-check the fluid level and correct as required to bring the fluid to its hot level mark.

TRANSAXLE AT NORMAL OPERATING TEMPERATURE

155° to 170° F., Dipstick Hot to the Touch

1. With the vehicle on a level surface, engine idling, wheels blocked, foot brake applied, move the selector lever through the gear positions to engage each gear and to fill the passageways with fluid.

2. Place the selector lever in the PARK position and apply the parking brake, allowing the engine to idle.

3. Clean the dipstick area of dirt and remove the dipstick from the filler tube. Wipe the dipstick clean, re-insert the dipstick into the filler tube and seat firmly.

4. Again remove the dipstick from the filler tube and check the fluid level as indicated on the dipstick. The level should be between the ADD and FULL marks. If necessary, add enough fluid to bring the fluid level to the full mark.

5. When the fluid level is correct, fully seat the dipstick in the filler tube.

NOTE: When the automatic transaxle fluid has become hotter than 155° to 170° F., the fluid should be allowed to cool before checking the level. Such causes of overheating are extended periods of high speed driving, trailer towing or stop and go traffic during periods of hot weather.

M MODIFICATIONS

FORD ATX TRANSAXLE MODIFICATIONS

Main Control Baffle Plate

1981 Escort/Lynx

Some ATX transaxles will be built with an oil baffle plate attached to the valve body. This baffle prevents fluid from spilling out of the breather vent in the oil pan. When servicing the transaxles equipped with this baffle plate, be sure to position the plate properly and do not discard.

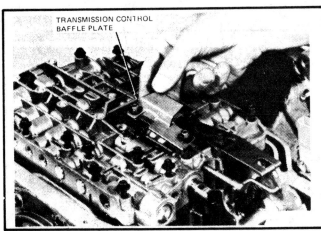

Transaxle control baffle plate location on valve body
(© Ford Motor Co.)

Valve Body To Case Attaching Bolt Torque Pattern

1981 Escort/Lynx

Proper torquing of the valve body attaching bolts will minimize the possibility of cross fluid leakage, sticking valves and erratic shifts. The bolt torque is 72-96 *inch pounds* (8-11 N•m).

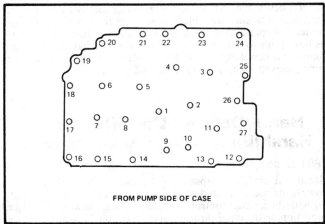

Valve body torque sequence (© Ford Motor Co.)

Cold Weather Cold Start/Shift

1981 Escort/Lynx

To improve cold starting and transaxle shifts on the ATX transaxle vehicles in sub-freezing weather, a new multi-viscosity automatic transaxle fluid should be used. The regular factory lubricant should be drained and the unit refilled with Motorcraft Ford Type MV automatic transaxle fluid, XT-3-MV, or equivalent, which meets Ford Specification number ESP-M2C164-A.

Drain and refill the transaxle as outlined in this transaxle section. In addition, have the vehicle in such a position as to be able to disconnect the cooler line return fitting, located at the pump end of the transaxle. With a routing hose into a container, start the engine and run for approximately 20-25 seconds, draining about three quarts of fluid from the transaxle, flushing the cooler and lines. Stop the engine and re-connect the fluid line and torque to 18-23 ft. lbs. Lower the vehicle and refill and transaxle with the multi-viscosity fluid to its correct level.

Shift Concerns

1981 Escort/Lynx

If concerns of shifting on new Escorts and Lynxs exists, this could be concerned to idle speeds and T.V. linkage. Verify and reset, if necessary, the idle speeds and T.V. linkage prior to performing any internal transaxle servicing.

Proper Seating of the Torque Converter

1981 Escort/Lynx

If the ATX torque converter is not properly engaged to the oil pump drive shaft, damage may occur to the oil pump. To determine whether the torque converter is fully seated, measure the distance between the converter stud face and the converter housing face. The dimension should read 1.205 to 1.126 inches (30.6 to 28.6 mm).

Powertrain Related Noise Transmitted Through Speedometer Cable and Engine Mounts

1981 Escort/Lynx

A transaxle gear noise, described as a buzzing, ticking or gear whine, can be transmitted into the passenger compartment via the speedometer cable conduit and/or by the engine mounts that can be grounded or bound up.

Prior to performing any service or repairs on the transaxle for gear noise, it must be determined that the noise is originating from the transaxle. To determine if the noise is coming from the transaxle, perform the following test;

1. Stop the vehicle.
2. Place the gear selector in neutral.
3. Increase the engine idle speed.
4. If the noise remains, the transaxle is not at fault. If the noise level changes or goes away, proceed with the following checks.

ENGINE MOUNTS

1. Verify the engine mounts are neutralized by loosening the bottom two retaining nuts on the left front engine mount. Loosen the bottom retaining nut on the left rear engine mount.
2. Position the engine mounts in the mounting brackets.
3. Maintain the correct engine mount alignment in the brackets with a prybar when tightening all the bottom retaining nuts to 55 ft. lbs.

SPEEDOMETER CABLE CONDUIT

1. Disconnect the speedometer cable at the transaxle and road test the vehicle. If the noise stops, the noise is being transmitted from the transaxle to the passenger compartment through the speedometer cable.
2. If noise is being trransmitted through the speedometer cable, the installation of a 14 inch piece of ⅜ inch diameter fuel hose over the speedometer cable will reduce the noise.
3. Spirally cut the hose and position it between the speedometer head and the dash panel grommet. Position the hose so that it overlaps the metal portion of the disconnect ferrule assembly by ¼ inch. The hose can be installed from the interior of the vehicle without disconnecting either end of the speedometer cable.

Transfer Gear Housing Removal

1981 Escort/Lynx

Idler gear damage can be done during the transfer gear housing removal, if the technician prys down upon the idler gear while removing the housing. To remove the transfer gear housing correctly, the technician must pry upward, never downward.

Throttle Valve (T.V.) Linkage Adjustment Procedure

1981 Escort/Lynx and 1982 EXP/LN7

When ever curb idle speeds are adjusted to the specifications indicated on the Emission Control Decal and the engine RPM increase is more than 100 rpm, or for any decrease in rpm, the transaxle throttle valve linkage must be checked and readjusted as required. The adjustments must be made at the T.V. control rod assembly sliding trunnion block.

The following procedure must be followed:

1. After the curb idle set to specification, turn the engine off and insure that the carburetor throttle lever is against the hot engine curb idle stop (the choke must be OFF).

NOTE: The linkage cannot be properly set if the choke is allowed to cool and the throttle lever allowed to be on the choke fast idle cam.

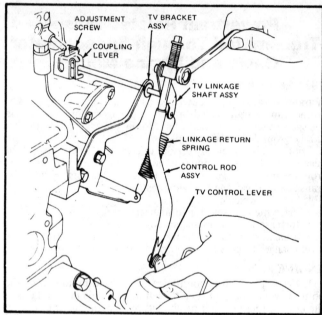

Adjustment procedure for T.V. linkage control (© Ford Motor Co.)

2. Set the coupling lever adjustment screw at its approximate midrange. Insure that the TV linkage shaft assembly is fully seated upward into the coupling lever.

3. Loosen the bolt on the sliding trunnion block on the TV control rod assembly one turn minimum.

4. Remove any corrosion from the control rod and free-up the trunnion block so that it slides freely on the control rod.

5. Rotate the transaxle TV control lever up using one finger and a light force, 2.2 Kg (approximately 5 pounds), to insure that the TV control lever is against its internal idle stop. Without relaxing the force on the TV control lever, tighten the bolt on the trunnion block to specification.

6. Verify that the carburetor throttle lever is still against the hot engine curb idle stop. If not, repeat Steps 1 through 6.

Revised Low-Intermediate Servo Rod and Band

1981 Escort/Lynx

The low-intermediate servo rod and band and seat has been changed to increase the ease of assembly. The new servo rod is round tipped while the old servo rod was flat tipped. The band

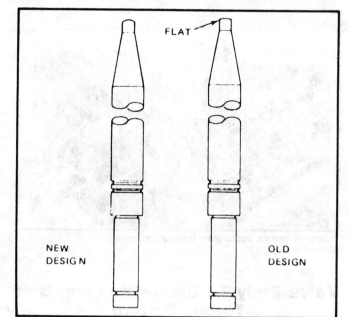

Difference in servo rod ends between old and new designs (© Ford Motor Co.)

seat has been changed to accept the new rounded servo rod. The old and new servo rods are not interchangcable and car must be exercised to be sure the replacement rod is the same as the original used. Replacement of the low-intermediate band on transaxles built prior to February 17, 1981, will necessitate the replacement of the new design servo rod. The new servo rods will use the same grooved identification as the old design.

Harsh/Delayed Upshifts and/or Harsh/Early Coasting downshifts

1981 Escort/Lynx

Harsh and/or delayed upshifts and harsh and/or early coasting downshifts maybe caused by the outer throttle valve lever binding with the manual control lever shaft because of an over-torque condition on the outer throttle valve lever attaching nut and lock washer. The torque specifications should be revised from 12-15 *foot* pounds to 88-115 *inch* pounds.

To check for possible binding, disconnect the outer throttle valve lever from the throttle control rod and check the throttle valve lever for a binding or hanging up condition. Also, check for clearance between the outer throttle valve lever and the manual control lever (0.010-0.015 inches minimum). If the throttle valve lever is binding or there is insufficient end clearance between the levers, it will be necessary to remove and discard the old lever. Install a new outer throttle lever (E1FZ-7A394-A) on the inner throttle lever shat and torque the attaching nut and lock washer to 88-115 *inch* pounds.

If the lever is not binding, the following other components should be checked for proper operation:

1. Check the throttle linkage system for freedom of movement (i.e., throttle control rod and carburetor throttle shaft).

2. Check the throttle kicker for proper operation (A/C and/or Power Steering equipped vehicles—RPM increases at idle when A/C is turned on or by turning steering wheel either direction).

3. Check engine idle RPM.

4. Check the throttle linkage adjustment.

Powertrain Noise Transmitted Through The Transaxle Shift Cable

1981 Escort/Lynx and 1982 EXP/LN7

Should a powertrain noise remain after insulating the speedometer cable and neutralizing the engine mounts, proceed to the transaxle shift cable stand-off (control) bracket mounting attachment bolts.

1. Remove the two attachment bolts for the stand-off (control) bracket and examine the insulators for correct positioning in the bracket or evidence of grounding or overtorquing. Replace if damaged. If the insulators are to be reused or new ones installed, torque the attaching bolts to 15-25 ft. lbs.

——————— CAUTION ———————

Do not over-torque, as the overtorquing will eliminate the effectiveness of the insulators.

Governor Gear Damage

1981 Escort/Lynx

ATX governor driven gear will exhibit two major patterns of identified damage and to eliminate the possibility of repeat service, perform one of the following two procedures:

1. When all teeth on the governor driven gear are in an hour glass or apple core shape, it may have been caused by a nick and/or burr on the speedometer drive gear located on the differential.

a. Remove the lower oil pan and inspect the speedometer drive gear for nicks and/or burrs by running a finger along each gear tooth. Since the drive gear is not a hardened gear, most nicks and/or burrs can be serviced b dressing with a file.

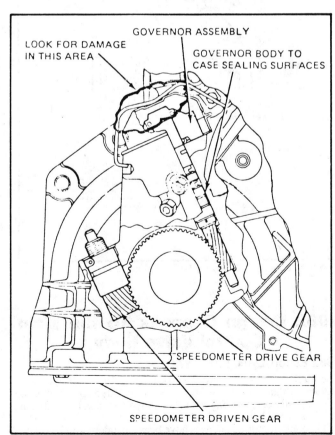

LOOK FOR DAMAGE IN THIS AREA

GOVERNOR ASSEMBLY

GOVERNOR BODY TO CASE SEALING SURFACES

SPEEDOMETER DRIVE GEAR

SPEEDOMETER DRIVEN GEAR

Governor gear damage appraisal (© Ford Motor Co.)

NOTE: Make sure that all fillings are cleaned from the transaxle and differential before assembling the lower oil pan to the case.

b. If the damaged speedometer drive gear can be serviced, then replace the governor assembly. If the damaged speedometer drive gear cannot be serviced, then replace both the governor assembly and the speedometer drive gear.

c. Because the speedometer drive gear is not a hardened gear, it can be easily damaged if struck or dropped. Care must be taken to not damage the drive gear during disassembly or assembly of the ATX transaxle.

d. Since the drive gear turns both the governor assembly and the speedometer driven gear assembly, the speedometer driven gear assembly should be inspected for damage and replaced if necessary.

2. When only one or two teeth on the governor are affected, it is the result of the governor assembly being momentarily stopped while the speedometer drive gear continues to turn. When this occurs, the technician must determine the reason for the damage.

a. Check for excessive metal and/or fiber debris in the transaxle fluid which could indicate bearing or clutch damage. Check the fluid level. Inspect the governor body to case sealing surfaces for gouging or distortions.

b. Check to see if the governor screen was improperly installed or omitted. The governor screen prevents miscellaneous debris from entering the governor hydraulic circuit.

c. Check the governor bore in the case for distortion in the areas of the no. 3 transaxle to engine mounting bolt hole. Some cases were found to have a shallow thread depth in the no. 3 bolt hole. When the attaching bolt is installed into a shallow hole, the bolt cuts into the case and causes the distortion in the case governor bore. When this happens, the case must be replaced, as well as the governor.

d. Check for damage to the governor cover and/or governor top, i.e., flyweight cage. Replace if necessary.

Service Oil Pump Insert

1981 Escort/Lynx and 1982 EXP/LN7

In the vent of ATX oil pump insert damage, the pump insert is serviced separately from the complete oil pump assembly (part number E1FZ-7F402-A).

Water In Transaxle Fluid

1981 Escort/Lynx and 1982 EXP/LN7

Reports of water being found in the transaxle fluid of the ATX has been recorded. Under normal operating conditions, water entry is restricted by the valve located on the top of the valve body cover. However, if water is forcefully directed at the valve, the resulting splash from the cover may allow be allowed to enter the valve. When water is being used to clean or cool the engine, care must be taken to prevent the vent from becoming sprayed with water.

——————— CAUTION ———————

If water is found in the transaxle fluid, a total drain, flush and refill of the transaxle and converter assembly is required.

Engine Speed-up During The 2-3 Upshift

1981 Escort/Lynx and 1982 EXP/LN7

A complaint of engine speed-up during the 2-3 upshift can be serviced as follows:

1. Remove the main control valve assembly from the transaxle in the prescribed manner.

2. Remove the separator plate from the main control valve body.

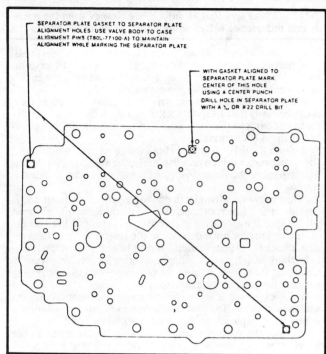

Reworking separator plate (© Ford Motor Co.)

3. Remove and discard the number two check ball.

4. Some main control valve body assemblies were manufactured without the number two check ball. In these instances, the separator plate must be reworked as shown in the accompanying illustration.

5. Assemble the reworked separator plate and new gasket to the main control valve body.

6. Install the main control valve body assemby in the transaxle and complete the assembly of the components.

Delay 1-2 upshift and/or 3-2 Coasting Downshift shudder (49 States Only)

1981 Escort/Lynx and 1982 EXP/LN7
(Transaxle Models PMA-A, A1, A2, K, and K1)

To correct delayed 1-2 upshifts and/or 3-2 coasting downshift shudder, a shift modification spring kit, part number E2FZ-7F415-A, is available to re-work the main control valve body assembly and the governor.

─────────── CAUTION ───────────

Before installing the shift modification spring kit, be sure the shifting problem is not being caused by binding/damaged/mis-adjusted transaxle throttle linkage and/or improper engine idle settings.

VALVE BODY RE-WORK

1. Remove the valve body from the transaxle.

2. Remove the separator plate and gasket from the main control valve body.

3. Note the position of the six check balls and the relief valve, then remove and set aside.

4. Remove the production 2-3 shift valve spring and install the service replacement spring (color-Black).

5. Remove the production 1-2 shift valve spring and the 1-2 shift T.V. modulator valve spring and install the service replacement 1-2 shift T.V. modulator valve spring (color-White). Discard the 1-2 shift valve spring.

6. Remove the production 2-1 scheduling valve spring and install the service replacement 2-1 scheduling valve spring (color-Purple).

7. Remove the production 3-2 shift control valve spring and install the service replacement 3-2 shift control valve spring (Size and color-Large Diameter, Purple).

8. Remove the production 2-3 shift T.V. modulator valve spring and install the service replacement 2-3 shift T.V. modulator valve spring (color-White).

9. Remove the production 1-2 shift valve accumulator valve spring and install the service replacement 1-2 shift accumulator valve spring (color-Dark Green).

10. Install the six check balls and relief valve in the main control valve body.

11. Assemble the separator plate and gasket to the main control valve body. Torque the bolts to 90 in. lbs.

12. Assemble the valve body to the transaxle in the prescribed manner.

GOVERNOR ASSEMBLY

1. Remove the governor assembly from the transaxle.

2. Remove and discard the two governor flyweight springs.

3. Install the service replacement governor spring (color-Brown) into either one of the governor spring positions. The other governor spring is not replaced.

4. Re-install the modified governor assembly into the transaxle assembly.

5. An authorized modification decal should be obtained and installed next to the Vehicle Emission Control Label.

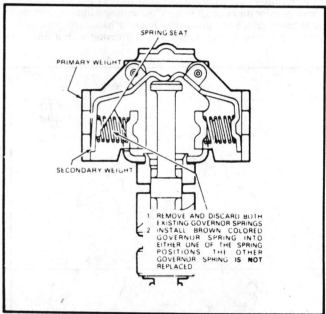

Reworking governor assembly (© Ford Motor Co.)

1982 Design Changes To the Main Control Valve Body

1982 Escort/Lynx, EXP/LN7

The 1982 ATX main control valve body differs from the 1981 models in that it does not have a 1-2 shift valve spring, the 1-2 T.V. modulator valve has a shorter stem and the valve body has one less check ball.

The manufacturing changes have been mistaken for improperly built main control valve body assemblies.

New Design ATX Flywheel

1982 EXP/LN7 Equipped w/1.6L H.O. Engines

The new designed flywheel can be identified by the following features:

1. Reduced converter to flywheel attaching nut access hole diameters.
2. Stamped "H.O." on the engine side of the flywheel and can be viewed when the lower engine dust cover is removed.
3. Four daubs of green paint on the transaxle side of the flywheel between the spokes.
4. Part number is E2FZ-6375-A (1.6L H.O. only)

--- CAUTION ---

The 1.6L H.O. engine cannot use the E1FZ-6375-C flywheel which is used on the standard 1.6L engines.

The new flywheel requires a 15mm-6 point-⅜ drive-thin walled socket with an O.D. of 20.55mm or less (Snap-on FSM 151 or equivalent). In some instances the flywheel-to-converter attaching nut access hole may be too small for the thin walled socket. If this occurs, find a socket with a smaller O.D., (20.5mm or less) or grind off some material on this present socket. A 15mm socket with an O.D. of 20.5mm will fit all cases. However, a new flywheel to converter fastener with a 13mm hex drive will be introduced into all ATX vehicle production and service in March 1982, that will permit the use of a standard wall socket-13mm 6 point.

Governor Drive Gear Service

1981 Escort/Lynx and 1982 EXP/LN7

A governor drive gear is now serviced separately, rather than the complete governor assembly replacement when the drive gear teeth have become damaged. The removal and installation procedures are outlined in the disassembly and assembly section.

New Design Plastic Speedometer Drive Gear

1981 Escort/Lynx and 1982 EXP/LN7

A new design plastic speedometer drive gear has replaced the present design steel gear for both production and service use.

--- CAUTION ---

The steel speedometer gear must not be used in an ATX transaxle that originally was manufactured with a plastic gear. Improper use of the steel speedometer gear may result in damage to the differential housing.

Misdiagnosis of Differential Seal Leakage

1981 Escort/Lynx and 1982 EXP/LN7

Field reports have indicated that fluid leaks from the transaxle input shaft seal, valve body cover gasket, engine main seal or oil pan gasket, will collect and give the appearance of being a differential seal leakage. In the event a transaxle exhibits a lubricant leak, clean off all traces of the fluid and allow the vehicle to stand, observing the location of the seepage. If no signs of a leak appear, run the engine in neutral and recheck the area again. if no leak occurs, drive the vehicle rechecking the area again. If seepage originates near the bottom of the bell housing, then the differential seals are not the cause of the leak.

If servicing the transaxle requires halfshaft removal, the halfshafts should be removed carefully and with the proper tools so that the differential seals are not damaged. The differential

seals will then not need to be replaced when the transaxle is reinstalled.

Torque Converter Cleaning Procedure
All Converter Assemblies Without Drain Plugs

1981 Escort/Lynx and 1982 EXP/LN7

This article releases a revised service procedure for torque converter cleaning to be used in addition to the Rotunda Model No. 60081-A torque converter cleaner for all torque converter assemblies without a drain plug.

Elimination of the converter drain plug requires a revised cleaning procedure. The lack of this plug increases the amount of residual flushing solvent retained in the converter after cleaning. The internal design of the converter does not allow for the drilling of a service drain hole. The following procedure is to be used after removal of the torque converter from the cleaning equipment.

1. Thoroughly drain remaining solvent through hub.
2. Add one quart of clean transmission fluid to the converter and hand agitate.
3. **THOROUGHLY** drain solution through converter hub.

Breakage of Gear Shift Lever

1981 Escort/Lynx and 1982 EXP/LN7

When a gear shift lever is replaced due to breakage failure, the shift cable should also be replaced. A new shift cable assembly has been released for production and service applications. The revised shift cable assembly can be identified by a white tie tag affixed to the cable housing assembly.

No 1-2 Upshift

1981-83 Escort/Lynx and 1981-83 EXP/LN7

Some vehicles may exhibit a no 1-2 upshift condition during acceleration. The 1-2 upshift can only be obtained by the release of the accelerator. This condition is also accompanied with an engine overrun (neutral feeling) during the partial throttle backout at speeds near 30 mph. The cause of this condition may be a stuck 1-2 T.V. modulator valve. Since the governor may cause a similar condition, it is important to verify the neutral interval condition during the release of the throttle at the 30 mph speed.

The stuck 1-2 T.V. modulator valve can only be verified by the valve body assembly removal. With the separator plate and gasket removed, check and verify the 1-2 T.V. modulator valve is either stuck or free. If the condition exists, clean up the valve body bore and install one of the following kits, depending upon the vehicle model year and valve body prefix. The kit contains one (1) 1-2 shift valve and one (1) 1-2 T.V. modulator valve. The applications for each kit are as follows:

1981-82 MODELS YEARS

Part Number E2FZ-7F417-A for models PMA-k, R
PMB-A with E2EP, E2GP
Part Number E3FZ-7F417-A for models PMA-K,R,Y
PMB-A1 with E3EP

A brinelling effect or bore deformation will be apparent on the bore surface where the 1-2 shift valve stops. It is this brinelling or deformation condition that causes the 1-2 T.V. modulator valve to stick at its most outward position.

Servicing Bearing Assembly For Transfer Housing

1981-83 Escort/Lynx and 1981-83 EXP/LN7

A revised service procedure and parts packaging method for the case and transfer housing assembly. The transfer housing bearing may be damaged when removed from the case and transfer housing assembly during the service. To avoid this condition in the future, the bearing will be included with the case and transfer housing assembly.

NOTE: The case and transfer housing are lined bored and matched during assembly. Therefore, neither are interchangeable with other cases or transfer housing.

Final Gear Set Individual Component Release

1981-83 Escort/Lynx and 1981-83 EXP/LN7

The final drive matched gear set has been revised to release the gears individually. The input gear, idler gear and bearing assembly, and the output gear may be serviced separately. Each gear as released, will service all current and past model applications. However, the final drive input gear will incorporate the use of a spacer for all 1.6L, 1.6L H.O. and 1.6L EFI units that contain the 25/32 inch long input gear bearing as opposed to the currently released 27/32 inch long bearing. It will be necessary to measure the bearing lengths to determine if the spacer is required for models built prior to March 31, 1983.

To install the spacer, turn the case 90° so the pump side is facing up and follow this procedure:

1. Install the thrust (needle) bearing number one (1) on the reactor support.
2. Install the cage needle bearing on the reactor support.
3. Install spacer on the reactor support.
4. Install the input gear over the spacer and cage bearing.

NOTE: For the 1983½ 2.3L applications, the spacer is not required.

Revised Procedure For Transaxle Halfshaft Removal

1981-83 Escort/Lynx and 1981-83 EXP/LN7

If extreme resistance is encountered using the current prybar method of removal of the halfshafts from the differential, use the following procedure to avoid damage or broken ATX transaxle cases and/or bent oil pans, resulting in fluid leaks.

In addition to the prescribed procedures for halfshaft removal, the following additions to the procedure are listed:

1. With the vehicle on a lift, remove the transaxle oil pan and discard the old gasket.
2. Insert a large bladed prybar between the differential pinion shaft and the inboard C.V. joint stub shaft.
3. Give a sharp tap to the handle of the prybar to dislodge the circlip from the sidegear, thus freeing the halfshaft from the differential.

Prior to the installation of the halfshaft into the differential, install a new circlip on the inboard stub shaft. In addition, a new transaxle oil pan gasket and transaxle fluid must be installed.

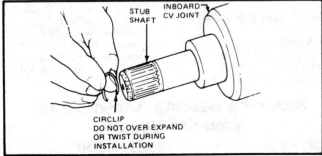

Installation of new circlip on stub axle shaft (© Ford Motor Co.)

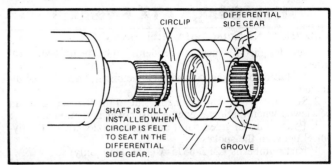

Circlip seat in differential gear groove (© Ford Motor Co.)

Torque the oil pan bolts to 15-19 ft. lbs. and fill the transaxle to its specified level.

NOTE: Use the dipstick to insure proper transaxle fluid level to avoid over or under filling of the transaxle.

Cooler Line Disconnect Tool Usage Push Connect Fittings— Transaxle End Only

To service (transaxle cooler lines, a Tool, Ford Number T82L-9500-AH or equivalent, is required. The illustration shows the tool end and its proper orientation for disassembly of tube from fitting. The purpose of the tool is to spread the "duck bill" retainer to disengage the tube bead. The following steps are necessary for use of the tool:

To facilitate use of the tool, clean the road dirt from the fitting before inserting the tool into the fitting. Also, it is important to avoid any contamination of the fitting and transaxle, dirt in the fitting could cause an O-ring leak.

1. Slide the tool over the tube.
2. Align the opening of the tool with one of the two tabs on the fitting "duck bill" retainer.
3. Firmly insert tool into fitting until it seats against the tube bead (a definite click should be heard).
4. With a thumb held against the tool, firmly pull back on the tube until it disengages from the fitting.

——————— CAUTION ———————

Do not attempt to separate the cooler line from the fitting by prying with another tool. This will break the plastic insert in fitting and bend the cooler lines at the junction to the fitting.

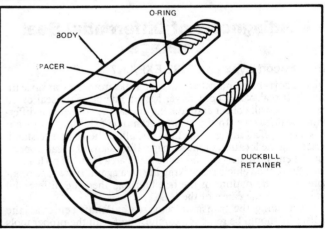

New type cooler line fitting (© Ford Motor Co.)

Before assembly of the lines in the fitting, visually inspect the plastic retainer in the fitting for a broken tab. If a tab is broken, the fitting must be replaced. Also visually inspect the cooler lines to make sure they are not bent at the junction of the fitting.

Tube assembly is accomplished by inserting the tube into the fitting until the retainer engages the tube head (a definite click should be heard). Pull back on the tube to ensure full engagement.

TROUBLE DIAGNOSIS

A logical and orderly diagnosis outline and charts are provided with clutch and band applications, shift speed and governor pressures, main control pressures and oil flow circuits to assist the repairman in diagnosing the problems, causes and the extent of repairs needed to bring the automatic transaxle back to its acceptable level of operation.

Preliminary checks and adjustments should be made to the manual valve linkage, accelerator and downshift linkages.

Transaxle oil level should be checked, both visually and by smell, to determine whether the fluid level is correct and to observe any foreign material in the fluid, if present. Smelling the fluid will indicate if any of the bands or clutches have been burned through excessive slippage or overheating of the trans.

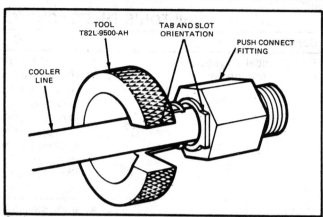
Cooler line disconnect tool usage (© Ford Motor Co.)

It is most important to locate the defect and its cause, and to properly repair them to avoid having the same problem reoccur.

In order to more fully understand the ATX automatic transaxle and to diagnose possible defects more easily, the clutch and band application chart and a general description of the hydraulic control system is given.

CLUTCH AND BAND APPLICATION CHART
ATX Automatic Transaxle

Range and Gear		Band	Direct Clutch	Intermediate Clutch	Reverse Clutch	One-Way Clutch
Park		—	—	—	—	Applied
Reverse		—	Applied	—	Applied	Applied
Neutral		—	—	—	—	Applied
D	1st	Applied	—	—	—	Applied
	2nd	Applied	—	Applied	—	—
	3rd	—	Applied	Applied	—	—
2	1st	Applied	—	—	—	Applied
	2nd	Applied	—	Applied	—	—
1	1st	Applied	Applied	—	—	Applied

CHILTON'S THREE "C's" TRANSAXLE DIAGNOSIS CHARTS

Condition	Cause	Correction
Slow initial engagement	a) Improper fluid level b) Damaged or improperly adjusted manual linkage c) Incorrect T.V. linkage adjustment d) Contaminated fluid e) Improper clutch and band application, or oil control pressure f) Dirty valve body	a) Add fluid as required b&c) Service or adjust linkage d) Change fluid and filter e) Perform control pressure test f) Clean, repair, or replace valve body

CHILTON'S THREE "C's" TRANSAXLE DIAGNOSIS CHARTS

Condition	Cause	Correction
Rough initial engagement in either forward or reverse	a) Improper fluid level b) High engine idle c) Auto. choke on (warm temp) d) Looseness in halfshafts, CV joints, or engine mounts e) Improper clutch or band application, or oil control pressure f) Incorrect T.V. linkage adjustment g) Sticky or dirty valve body	a) Perform fluid check b) Adjust idle to specs. c) Disengage choke d) Service as required e) Perform control pressure test f) Service or adjust linkage g) Clean, repair, or replace valve body
No drive in any gear	a) Improper fluid level b) Damaged or improperly adjusted manual linkage c) Improper clutch or band application, or oil control pressure d) Internal leakage e) Valve body loose f) Damaged or worn clutches or band g) Sticking or dirty valve body	a) Perform fluid check b) Service or adjust linkage c) Perform control pressure test d) Check and repair as required e) Tighten to specs. f) Perform air pressure test g) Clean, repair, or replace valve body
No forward drive— reverse OK	a) Improper fluid level b) Damaged or improperly adjusted manual linkage c) Improper one-way clutch, or band application, or oil pressure control system d) Damaged or worn band, servo or clutches e) Valve body loose f) Dirty or sticking valve body	a) Perform fluid level check b) Service or adjust linkage c) Perform control pressure test d) Perform air pressure test e) Tighten to specs. f) Clean, service or replace valve body
No drive, slips, or chatters in reverse— forward OK	a) Improper fluid level b) Damaged or improperly adjusted manual linkage c) Looseness in half shafts, CV joints, or engine mounts d) Improper oil pressure control e) Damaged or worn reverse clutch f) Valve body loose g) Dirty or sticking valve body	a) Perform fluid level check b) Service or adjust linkage c) Service as required d) Perform control pressure test e) Perform control pressure test f) Tighten to specs. g) Clean, service, or replace valve body
Car will not start in neutral or park	a) Neutral start switch improperly adjusted b) Neutral start wire disconnected/damaged c) Manual linkage improperly adjusted	a) Service or adjust neutral start switch b) Replace/repair c) Service or adjust linkage
No drive, slips or chatters in first gear in D	a) Damaged or worn one-way clutch b) Improper fluid level c) Damaged or worn band d) Incorrect T.V. linkage adjustment	a) Service or replace one-way clutch b) Perform fluid level check c) Service or replace band assembly d) Service or adjust linkage

CHILTON'S THREE "C's" TRANSAXLE DIAGNOSIS CHARTS

Condition	Cause	Correction
No drive, slips, or chatters in second gear	a) Improper fluid level b) Incorrect T.V. linkage adjustment c) Intermediate friction clutch d) Improper clutch application e) Internal leakage f) Dirty or sticking valve body g) Polished, glazed band or drum	a) Perform fluid level check b) Service or adjust linkage c) Service clutch d) Perform control pressure test e) Perform air pressure test f) Clean, service, or replace valve body g) Replace or service as required
Starts up in 2nd or 3rd	a) Improper fluid level b) Damaged or improperly adjusted manual linkage c) Improper band and/or clutch application, or oil pressure control system d) Damaged or worn governor e) Valve body loose f) Dirty or sticking valve body g) Cross leaks between valve body and case mating surface	a) Perform fluid level check b) Service or adjust linkage c) Perform control pressure test d) Perform governor check Replace or service governor and e) Tighten to specifications f) Clean, service, or replace valve body g) Replace valve body and/or case as required
Shifts points incorrect	a) Improper fluid level b) T.V. linkage out of adjustment c) Improper clutch or band application, or oil pressure system d) Damaged or worn governor e) Dirty or sticking valve body	a) Perform fluid level check b) Service or adjust linkage c) Perform shift test and control pressure test d) Service or replace governor and clean screen e) Clean, service, or replace valve body
No upshifts at any speed in D	a) Improper fluid level b) T.V. linkage out of adjustment c) Improper band or clutch application, or oil pressure control system d) Damaged or worn governor e) Dirty or sticking valve body	a) Perform fluid level check b) Service or adjust linkage c) Perform control pressure test d) Service or replace governor and clean screen e) Clean, service or replace valve body
Shifts 1-3 in D	a) Improper fluid level b) Intermediate friction clutch c) Improper clutch application, or oil pressure control system d) Dirty or sticking valve body	a) Perform fluid level check b) Service c) Perform control pressure test d) Clean, service or replace valve body

CHILTON'S THREE "C's" TRANSAXLE DIAGNOSIS CHARTS

Condition	Cause	Correction
Engine over-speeds on 2-3 shift	a) Improper fluid level b) Improper band or clutch application, or oil pressure control system c) Damaged or worn direct clutch and/or servo d) Dirty or sticking valve body	a) Perform fluid level check b) Perform control pressure test c) Perform air pressure test and service as required d) Clean, service or replace valve body
Mushy 1-2 shift	a) Improper fluid level b) Incorrect engine performance c) Improper T.V. linkage adjustment d) Improper intermediate clutch application, or oil pressure control system e) Damaged intermediate clutch f) Dirty or sticking valve body	a) Perform fluid level check b) Tune adjust engine idle ad required c) Service or adjust d) Perform control pressure test e) Perform air pressure test and service as required f) Clean, service or replace valve body
Rough 1-2 shift	a) Improper fluid level b) Improper T.V. linkage adjustment c) Incorrect engine idle or performance d) Improper intermediate clutch application or oil pressure control system e) Dirty or sticking valve body	a) Perform fluid level check b) Service or adjust linkage c) Tune and adjust engine idle d) Perform control pressure test e) Clean, service or replace valve body
Rough 2-3 shift	a) Improper fluid level b) Incorrect engine performance c) Improper band release or direct clutch application, or oil control pressure system d) Damaged or worn servo release and direct clutch piston check ball e) Improper T.V. linkage adjustment f) Dirty or sticking valve body	a) Perform fluid level check b) Tune and adjust engine idle c) Perform control pressure test d) Air pressure test the servo apply and release and the direct clutch piston check ball. Service as required e) Service or adjust linkage f) Clean, service, or replace valve body
Rough 3-2 shift at closed throttle in D	a) Improper fluid level b) Incorrect engine idle or performance c) Improper T.V. linkage adjustment d) Improper band or clutch application, or oil pressure control system	a) Perform fluid level check b) Tune and adjust engine idle c) Service or adjust linkage d) Perform control pressure test

CHILTON'S THREE "C's" TRANSAXLE DIAGNOSIS CHARTS

Condition	Cause	Correction
Rough 3-2 shift at closed throttle in D	e) Improper governor operation	e) Perform governor test. Service as required
	f) Dirty or sticking valve body	f) Clean, service or replace valve body
No forced downshifts	a) Improper fluid level	a) Perform fluid level check
	b) Improper clutch or band application, or oil pressure control system	b) Perform control pressure test
	c) Damaged internal kickdown linkage	c) Service internal kickdown linkage
	d) T.V. linkage out of adjustment	d) Service or adjust T.V. linkage
	e) Dirty or sticking valve body	e) Clean, service or replace valve body
	f) Dirty or sticking governor	f) Clean or replace governor
Runaway engine on 3-2 or 3-1 downshift	a) Improper fluid level	a) Perform fluid level check
	b) T.V. linkage out of adjustment	b) Service of adjust T.V. linkage
	c) Band out of adjustment	c) Check and adjust servo rod travel
	d) Improper band or clutch application, or oil pressure control system	d) Perform control pressure test
	e) Damaged or worn servo	e) Air pressure test check the servo. Service servo and/or seals
	f) Polished, glazed band or drum	f) Service or replace as required
	g) Dirty or sticking valve body	g) Clean, service or replace valve body
No engine braking in manual first gear	a) Improper fluid level	a) Perform fluid level check
	b) T.V. linkage out of adjustment	b&c) Service or adjust linkage
	c) Damaged or improperly adjusted manual linkage	
	d) Band or clutch out of adjustment	d) Check direct clutch and service as required and check servo rod travel
	e) Oil pressure control system	e) Perform control pressure test
	f) Polished, glazed band or drum	f) Service or replace as required
	g) Dirty or sticking valve body	g) Clean, service or replace valve body
No engine braking in manual second gear	a) Improper fluid level	a) Perform fluid level check
	b) T.V. linkage out of adjustment	b&c) Service or adjust linkage
	c) Damaged or improperly adjusted manual linkage	
	d) Improper band or clutch application, or oil pressure control system	d) Perform control pressure test
	e) Servo leaking	e) Perform air pressure test of servo for leakage and service as required

CHILTON'S THREE "C's" TRANSAXLE DIAGNOSIS CHARTS

Condition	Cause	Correction
No engine braking in manual second gear	f) Polished, glazed band or drum	f) Service or replace as required
Transaxle noisy— valve resonance Note; Gauges may aggravate any hydraulic resonance. Remove gauge and check for resonance level	a) Improper fluid level b) T.V. linkage out of adjustment c) Improper band or clutch application, or oil pressure control system d) Cooler lines grounding e) Dirty or sticking valve body f) Internal leakage or pump cavitation	a) Perform fluid level check b) Service or adjust T.V. linkage c) Perform control pressure test d) Free cooler lines e) Clean, service or replace valve body f) Service as required
Transaxle overheats	a) Excessive tow loads b) Improper fluid level c) Incorrect engine idle or performance d) Improper clutch or band application, or oil pressure control system e) Restriction in cooler or lines f) Seized converter one-way clutch g) Dirty or sticking valve body	a) Check Owner's Manual for tow restriction b) Perform fluid level check c) Tune or adjust engine idle d) Perform control pressure test e) Service restriction f) Replace converter g) Clean, service or replace valve body
Transaxle fluid leaks	a) Improper fluid level b) Leakage at gaskets, seals, etc.	a) Perform fluid level check b) Remove all traces of lube on exposed surfaces of transaxle. Check for free-breathing. Operate transaxle at normal temperatures and inspect for leakage. Service as required.

THE HYDRAULIC CONTROL SYSTEM

The hydraulic Control System of the ATX automatic transaxle is used to supply fluid for the torque converter operation, to direct fluid under pressure to apply the servo bands and clutches, to lubricate the transaxle parts and to remove heat generated by the internal components of the transaxle and the torque converter.

The Main Components

The main components are listed, along with a description of their function in the valve body and transaxle.

1. **SUMP**—The transmission oil pan contains a supply of hydraulic fluid for the system.
2. **SCREEN**—Protects the pump inlet from dirt and other foreign material that may cling to the fluid.
3. **OIL PUMP**—Pumps hydraulic fluid to the system when the engine is running.
4. **MAIN OIL PRESSURE BOOSTER VALVE**—Increases or decreases main control pressure in relation to throttle opening.

Also provides different main control pressures depending on range or gear ratio.

5. **MAIN OIL PRESSURE REGULATOR VALVE**—Regulates main (line) control pressure in the system.
6. **CONVERTER RELIEF VALVE**—Prevents excess pressure build-up in the torque converter.
7. **COOLER**—Removes heat generated in the torque converter and transmission.

NOTE: Cooler return fluid is used for lubrication before it returns to the sump.

8. **MANUAL VALVE**—Moves with the shift selector and directs control pressure to various passages to apply clutches and servos, and to provide automatic functions of the hydraulic system.
9. **THROTTLE PLUNGER**—Varies spring force on throttle valve with throttle opening. Also operates "kickdown" system at wide-open throttle.
10. **THROTTLE CONTROL VALVE**—Regulates throttle pressure as an engine load signal to the hydraulic system.
11. **T.V. LIMIT VALVE**—Regulates maximum T.V. (throttle) pressure in the throttle control circuit.
12. **1-2 ACCUMULATOR VALVE**—And . . .

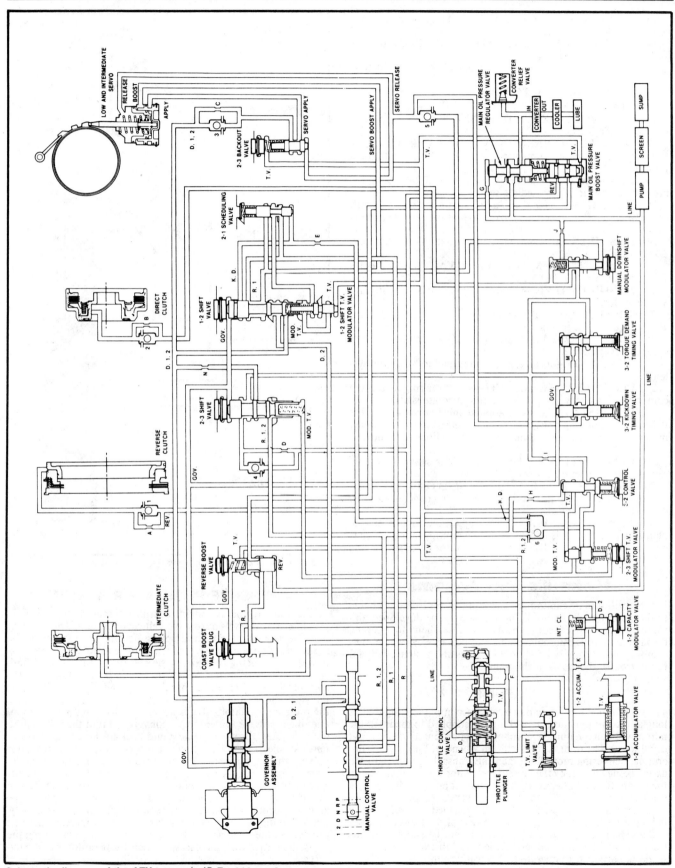

Schematic diagram of the ATX transaxle (© Ford Motor Co.)

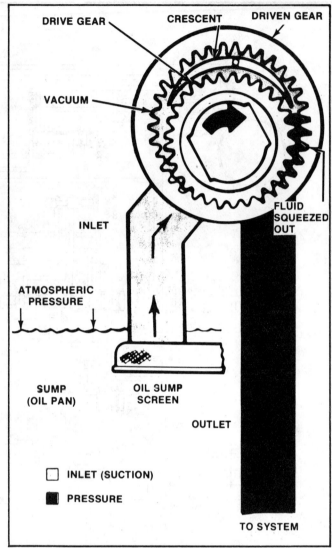

13. **1-2 CAPACITY MODULATOR VALVE**—Operate together to smooth the 1-2 upshift.

14. **2-3 SHIFT T.V. MODULATOR VALVE**—Modulates T.V. pressure acting on the 2-3 shift valve.

15. **3-2 CONTROL VALVE**—Regulates 3-2 downshift timing.

16. **3-2 KICKDONW TIMING VALVE**—Operates to smooth the 3-2 downshift during kickdown (full throttle).

17. **3-2 TORQUE DEMAND TIMING VALVE**—Operates to smooth the 3-2 downshift at part throttle.

18. **MANUAL DOWNSHIFT MODULATOR VALVE**—Provides modulated line pressure to the direct clutch in Manual Low (1).

19. **2-3 BACKOUT VALVE**—Controls feed rate of apply pressure to low and intermediate servo.

20. **2-1 SCHEDULING VALVE**—Determines 2-1 downshift speed when the shift selector is moved to Manual Low (1) from D range.

21. **1-2 SHIFT VALVE**—Controls automatic 1-2 upshift and 2-1 downshift.

22. **1-2 SHIFT T.V. MODULATOR VALVE**—Modulates T.V. pressure on 1-2 shift valve.

23. **2-3 SHIFT VALVE**—Controls automatic 2-3 upshift and 3-2 downshift.

24. **REVERSE BOOST VALVE**—Provides increased main control pressure in reverse gear.

25. **COAST BOOST VALVE PLUG**—No hydraulic function except to plug a hole.

26. **GOVERNOR**—Road speed input signal to hydraulic system.

27. **INTERMEDIATE CLUTCH**—Locks the ring gear of the compound planetary gear set to the intermediate shaft.

28. **REVERSE CLUTCH**—is applied only in reverse gear. It's purpose is to hold the ring gear stationary, therefore, it is installed in the transaxle case rather than in a rotating cylinder like the other clutches.

29. **DIRECT CLUTCH**—Locks the turbine shaft to the low-reverse sun gear with no free-wheeling.

30. **LOW AND INTERMEDIATE SERVO**—When under fluid pressure, forces a piston rod to tighten the band around the drum and lock it stationary.

31. **TORQUE CONVERTER**—Couples the engine to the planetary gear train. Also provides torque multiplication which is equivalent to additional gear reduction during certain drving conditions.

Operation of oil pump assembly (© Ford Motor Co.)

Hydraulic Sub-Systems

Before the transaxle can transfer the input power from the engine to the drive wheels, hydraulic pressures must be developed and routed to the varied components to cause them to operate, through numerous internal systems and passages. A basic understanding of the components, fluid routings and systems, will aid in the trouble diagnosis of the transaxle.

OIL PUMP

The oil pump is a positive displacement pump, meaning that as long as the pump is turning and fluid is supplied to the inlet, the pump will deliver fluid in a volume proportionate to the input drive speed. The pump is in operation whenever the engine is operating, delivering more fluid than the transaxle needs, with the excess being bled off by the pressure regulator valve and routed to the sump. It should be remembered, the oil pump is driven by a shaft which is splined into the converter cover and through a drive gear insert. The gears, in turn, are installed in a body, which is bolted to the pump support at the rear of the transaxle case.

Should the oil pump fail, fluid would not be supplied to the transaxle to keep the converter filled, to lubricate the internal working parts and to operate the hydraulic controls.

PRESSURE REGULATOR SYSTEM

The pressure regulator system controls the main line pressure at pre-determined levels during the vehicle operation. The main oil pressure regulator valve and spring determines the psi of the main line control pressure. The main line control pressure is regulated by balancing pressure at the end of the inner valve land, against the valve spring. When the pump begins fluid delivery and fills the passages and transaxle components, the spring holds the valve closed and there is no regulation. As the pressure rises, the pressure regulator valve is moved against the spring tension, opening a passage to the torque converter. Fluid then flows into the converter, the cooler system and back to the lubrication system. As the pressure continues its rise, the pressure regulator valve is moved further against the spring tension, and at a predetermined psi level and spring tension rate, the valve is moved further to open a passage, allowing excess pressurized fluid to return to the sump. The valve then opens and closes in a vibrating type action, dependent upon the fluid requirements of the transaxle. A main line ol pressure booster valve is used to increase the line pressure to meet the needs of higher pressure, required when the transaxle torque load increases, to operate the clutches and band servo.

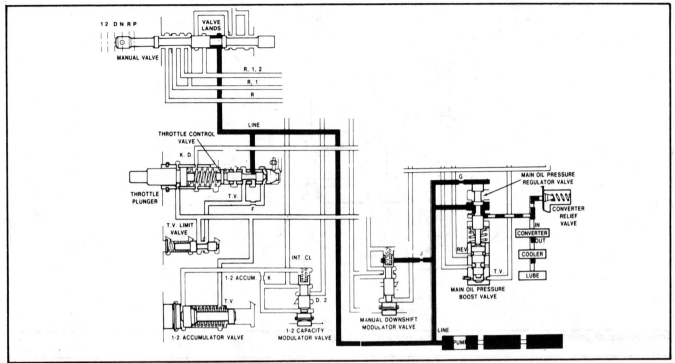

Main line control pressure routing (© Ford Motor Co.)

THE MANUAL CONTROL VALVE

Main line control pressure is always present at the manual control valve. Other than required passages, such as to the converter fill, the lubricating system and to certain valve assemblies, the manual valve must be moved to allow the pressurized fluid to flow to the desired components or to charge certain passages in order to engage the transaxle components in their applicable gear ratios.

GOVERNOR ASSEMBLY

The governor assembly reacts to vehicle road speed and provides a pressure signal to the control valves. This pressure signal causes automatic upshifts to occur as the road speed increases and permits downshifts as the road speed decreases. The governor has three hydraulic passages, an exhaust governor pressure out and line pressure in, controlled by springs and weights, with the weight position determined by centrifugal force as the governor assembly rotates.

THROTTLE VALVE

The ATX transaxle uses a manually controlled throttle valve to prvide the T.V. or throttle pressure signal that is proportional to the throttle opening of the carburetor. This throttle pressure signal is needed for the hydraulic control system to know what the engine load is, so the shifting from one ratio to another can be done at its proper time.

The T.V. system operates during closed throttle, light throttle and wide open throttle operations. The kickdown is actuated with the throttle in the wide open position.

NOTE: In case of the linkage becoming disconnected or other linkage failure, the system will go to full T.V. limit pressure to protect the clutches and band from slippage and burn-out.

T.V. pressure has several functions in the system, to delay upshifts and to boost line pressure when higher engine loads exist. Shift T.V. modulator valves convert T.V. pressure to lesser modulated pressure, necessary to match the requirements for balancing pressure and spring force against governor pressure. At closed throttle and at light throttle, there may not be enough T.V. pressure to cause the modulator valves to modulate. In either case, the only delay forces will be from the line pressure on the differential areas plus, of course, spring force. Thus, upshifts can occur at minimum road speeds with the throttle closed or nearly closed. Two kinds of forced downshifts can be obtained, *torque demand* and *wide open throttle kickdown*. With our understanding that shift points depend upon the balance between governor pressure and downshift delay pressure, and should we change that balance to overcome governor presssure, downshifts will occur.

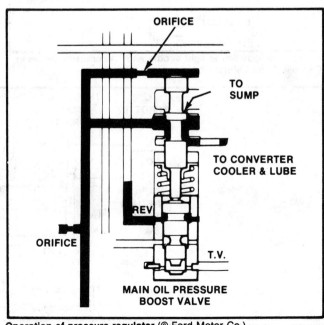

Operation of pressure regulator (© Ford Motor Co.)

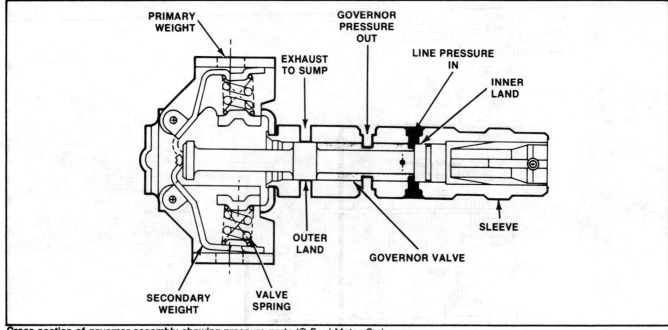

Cross section of governor assembly showing pressure ports (© Ford Motor Co.)

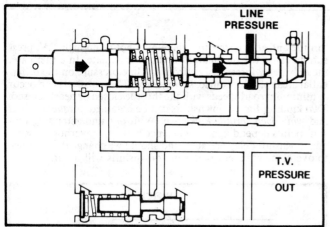

Throttle valve position at light throttle and T.V. pressure between 10-85 psi (© Ford Motor Co.)

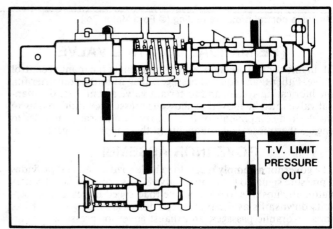

Throttle valve position at T.V. limit of 85 psi and up (© Ford Motor Co.)

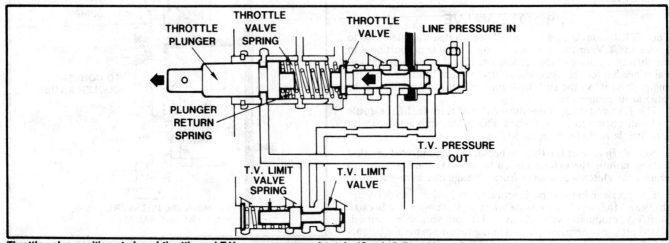

Throttle valve position at closed throttle and T.V. pressure approximately 10 psi (© Ford Motor Co.)

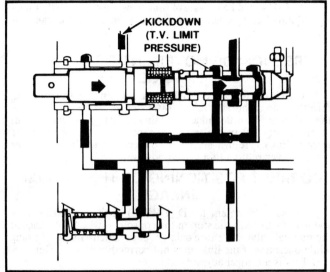

KICKDOWN
(T.V. LIMIT
PRESSURE)

Throttle valve position at wide open throttle (W.O.T.) kickdown (© Ford Motor Co.)

DIAGNOSIS TESTS

General Diagnosis Sequence

A general diagnosis sequence should be followed to determine in what area of the transaxle a malfunction exists. The following sequence is suggested by Ford Motor Company to diagnose and test the operation of the ATX transaxle.

1. Inspect fluid level and condition.

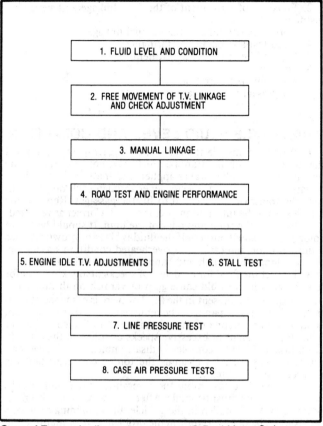

1. FLUID LEVEL AND CONDITION

2. FREE MOVEMENT OF T.V. LINKAGE AND CHECK ADJUSTMENT

3. MANUAL LINKAGE

4. ROAD TEST AND ENGINE PERFORMANCE

5. ENGINE IDLE T.V. ADJUSTMENTS   6. STALL TEST

7. LINE PRESSURE TEST

8. CASE AIR PRESSURE TESTS

General Transaxle diagnosis sequence (© Ford Motor Co.)

HYDRAULIC SUB-SYSTEM FUNCTION IN RELATION TO GEARING

Sub-System	Function	Gear(s)
Fluid Supply	Pump oil to operate, lubricate, cool	All
Main Oil Pressure Regulator	Match line pressure to system pressure needs; charge converter-cooler-lube	All
Governor	Road speed signal to control valves	1,2,3
Throttle (T.V.)	Engine load signal to control valves	All
Modulated T.V.	Control automatic upshift road speeds	1,2
Kickdown	Force wide-open throttle downshifts	2,3
Shift Valve	Control automatic upshifts and downshifts	1,2,3
1-2 Lockout	Prevent 1-2 shift valve from "upshift" movement	R and Manual 1
2-3 Lockout	Prevent 2-3 shift valve from "upshift" movement	R and Manual 1,2
1-2 Accumulator and Intermediate Clutch Apply	Cushion 1-2 upshift	1 (Ranges D and 2)
Servo Apply	Cause servo to apply the band	1 and 2, R
Servo Boost	Increase apply force of servo.	R and Manual 1
Servo Release	Cause servo to release the band	R and 3
Direct Clutch	Cause direct clutch to apply	3 and R, Low
Reverse Clutch	Cause reverse clutch to apply	R
Reverse Pressure Boost	Cause maximum line pressure in reverse	R
3-2 Downshift	Control exhaust of direct clutch and servo apply	3

2. Freedom of movement of the T.V. linkage and verify adjustment.
3. Correct positioning of the manual linkage.
4. Road test and engine performance.
5. Engine idle and T.V. adjustment.
6. Stall test.
7. Main line pressure test.
8. Air pressure test of case.

TRANSAXLE FLUID LEVEL AND CONDITION

1. With the transaxle at normal operating temperature, inspect the fluid level with the transaxle in PARK position, the engine operating at curb idle, brakes applied and the vehicle on a level surface. Move the shift selector lever through each range to engage the transaxle gearing and to fill the passages. Remove the dipstick from the tube and inspect the level. Correct as required.

2. Observe the color and odor of the fluid. It should be red in color and not smell burned. If the fluid is black or brown and has a burned odor, indications of overheated condition having occurred, along with clutch and band failure are evident.

3. Inspect the fluid for evidence of specks of any kind and for anti-freeze, which would cause gum or varnish on th dipstick.

4. If specks are present in the fluid or there is evidence of anti-freeze, the transaxle pan must be removed for further inspection. If fluid contamination or transaxle failure is confirmed by further evidence of coolant or excessive specks or solids in the oil pan, the transaxle must be completely disassembled and cleaned and serviced. This includes cleaning the torque converter and the transaxle cooling sytem.

5. During the disassembly and assembly, all overhaul checks and adjustments must be made. After the transaxle has been assembled and re-installed in the vehicle, the remaining diagnosis tests should be made to confirm the problem and/or malfunction has been corrected.

NOTE: It would be a waste of time to perform any further tests should anti-freeze or excessive specks be found in the transaxle fluid.

FREEDOM OF MOVEMENT OF THE T.V. LINKAGE

1. Check for wide open carburetor and linkage travel at full throttle. The carburetor full throttle stop must be contacted by the carburetor throttle linkage and there must be a slight amount of movement left in the transaxle throttle linkage. Be sure the throttle linkage return spring is connected and the carburetor throttle lever returns to a closed position.

CORRECT POSITIONING OF THE MANUAL LINKAGE

1. Be sure the detent for D (DRIVE) in the transaxle corresponds exactly with the stop in the console, Hydraulic leakage at the manual valve can cause delay in engagement and/or slipping while operating if the linkage is not correctly adjusted. Remember: This is a critical adjustment.

ROAD TEST AND ENGINE PERFORMANCE

When making the road test, note the quality of the shift and its engagement. If soft or mushy shifts occur, or if initial engagement or shift engagement is too harsh, an inoperative or improperly adjusted T.V. linkage could be the cause. Correct the condition and road test again. During the road test, check the shift points with the engine at normal operating temperature to avoid fast idle operation of the engine, which would affect the shift timing. Perform the road test over a predetermined course, keeping traffic safety in mind. Evaluate engine performance during the road test.

Check the minimum throttle upshifts in Drive. The transaxle should start in first gear, shift to second and then shift to third, within the shift points of the Service Specifications.

ROAD TEST

Range		Check for	Condition (OK or Not OK)
1		Engagement	
		No 1-2 Upshift	
		Engine Braking in 1st Gear	
		Slipping	
2		Engagement and Shift Feel	
		Automatic 1-2 Upshift	
		Automatic 2-1 Downshift	
		Slipping	
D		Engagement and Shift Feel	
	Look Up Specs. For Shift Points	Minimum Throttle 1-2	
		Minimum Throttle 2-3	
		Minimum Throttle 3-2	
		Minimum Throttle 2-1	
		W.O.T. 1-2	
		W.O.T. 2-3	
		W.O.T. 3-2	
		W.O.T. 2-1	
		Slipping	
R		Engagement	
		Back-up Without Slip	

With transaxle in third gear, depress the accelerator pedal to the floor. The transaxle should shift from third to second or third to first, depending on vehicle speed.

Check the closed throttle downshifts from third to first by coasting down from about 30 mph (48 km/h), in third gear. The shifts should occur within the limits of the Service Specifications.

When the selector lever is at 2 (Second), the transaxle will operate in first and in second gears.

With the transaxle in third gear and road speed over approximately 30 mph (48 km/h), the transaxle shift to second gear when the selector lever is moved from Drive to 2 (Second), to 1 (First). The transaxle will shift into 1 (First) when road speeds are less than 30 mph (48 km/h).

When the selector lever is moved from drive (D) to second (2), the transaxle will shift into Second (2), regardless of vehicle speed.

This check will determine if the governor pressure and shift control valves are functioning properly. During the shift point check operation, if the transaxle does not shift within specifications, or certain gear ratios cannot be obtained, refer to the Diagnosis Guides to resolve the problem.

A shift test can be performed in the shop to check shift valve operation, governor circuits, shift delay pressures, throttle boost and downshift valve action.

--- CAUTION ---

Never exceed 60 mph (97 km/h) speedometer speed.

1. Raise the vehicle, place the transaxle in Drive and make a minimum throttle 1-2, 2-3 shift test. At this point of shift you will see the speedometer needle make a momentary surge and feel the driveline bump. If the shift points are within specification, the 1-2 and 2-3 shift valves and governor are OK. If the shift points are not within specification, perform a Governor Check to isolate the problem.

Governor Check

Accelerate vehicle to 30-40mph (48-64 km/h) then back off throttle completely. If the governor is functioning properly, the transaxle will shift to third gear.

ENGINE IDLE AND T.V. ADJUSTMENT

A condition of too high or too low engine idle could exist and be found during the road test. Should idle adjustment be necessary

by more than 50 rpm, the T.V. linkage must be re-checked for proper adjustment. Refer to the adjustment section for T.V. linkage adjustment procedures.

STALL TEST

The stall test can be made in Drive 2, 1 (First) or reverse at full throttle, to check engine performance converter operation and the holding abilities of the direct clutch, reverse clutch, the low-intermediate band and the gear train one-way clutch.

1. To perform the stall test, start the engine and allow it to come to normal operating temperature. Apply both the parking and service brakes during the test.

--- CAUTION ---

Do not allow any one to stand either in front of or behind the vehicle during the stall test. Personal injury could result.

2. Install a tachometer to the engine. Place the selector lever in the desired detent and depress the accelerator to the wide open position, noting the total rpm achieved.

--- CAUTION ---

Do not hold the throttle open for more than five (5) seconds at a time during the test.

3. After the test, move the selector lever to the Neutral position and increase the engine speed to 1000 rpm and hold approximately 15-30 seconds to cool the converter before making a second or third test.

4. During the test, if the engine rpm exceeds the maximum limits, release the accelerator immediately because clutch or band slippage is indicated and repairs to the transaxle should be done.

MAIN LINE PRESSURE TEST

1. Place the selector lever in the Park position and apply the parking brake.

2. Attach a 300 psi pressure gauge to the line port on the transaxle case with enough flexible hose to make the gauge accessible while operating the engine.

3. Start and operate the engine until normal operating temperature is reached.

4. With the selector lever in the desired position, refer to the pressure specifications chart and compare pressure reading with manufacturer's specifications.

STALL TEST

Range	Specified Stall Speed	Actual (High or Low)
D or 2		
1		
R		

Ranges(s)	Stall Speed(s) High (Slip)	Stall Speeds Low
D,2	Turbine Shaft One-Way Clutch	
D,2,1	Low-Intermediate Band or Servo	1. Check Engine for Tune-up. If OK . . .
R	Reverse Clutch	2. Remove Torque Converter and Bench Test for Reactor One-Way Clutch Slip
All Driving Ranges	1. Check T.V. Adjustment 2. Perform Control Pressure Test	

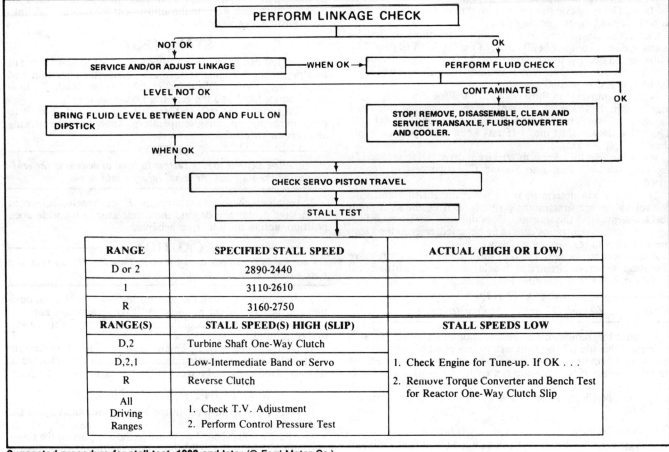

PERFORM LINKAGE CHECK

NOT OK → SERVICE AND/OR ADJUST LINKAGE — WHEN OK → PERFORM FLUID CHECK ← OK

LEVEL NOT OK → BRING FLUID LEVEL BETWEEN ADD AND FULL ON DIPSTICK

CONTAMINATED → STOP! REMOVE, DISASSEMBLE, CLEAN AND SERVICE TRANSAXLE, FLUSH CONVERTER AND COOLER. — OK

WHEN OK → CHECK SERVO PISTON TRAVEL → STALL TEST

RANGE	SPECIFIED STALL SPEED	ACTUAL (HIGH OR LOW)
D or 2	2890-2440	
1	3110-2610	
R	3160-2750	

RANGE(S)	STALL SPEED(S) HIGH (SLIP)	STALL SPEEDS LOW
D,2	Turbine Shaft One-Way Clutch	
D,2,1	Low-Intermediate Band or Servo	1. Check Engine for Tune-up. If OK . . .
R	Reverse Clutch	2. Remove Torque Converter and Bench Test for Reactor One-Way Clutch Slip
All Driving Ranges	1. Check T.V. Adjustment 2. Perform Control Pressure Test	

Suggested procedure for stall test, 1983 and later (© Ford Motor Co.)

NOTE: Wide open throttle (W.O.T.) readings are to be made at full stall. However, be sure to run the engine at fast idle in neutral for cooling between tests. To avoid operating at stall, have a helper manually force the T.V. linkage to the kickdown position, which will simulate W.O.T.

Keep in mind that clutch and servo leakage may or may not show up on the control pressure test. This is because (1) the pump has a high output volume and the leak may not be severe enough to cause a pressure drop; and (2) orifices between the pump and pressure chamber may maintain pressure at the source, even with a leak down-stream. Pressure loss caused by a less-than-major leak is more likely to show up at idle than at W.O.T. where the pump is delivering full volume.

Conversely, manipulating the T.V. linkage to simulate W.O.T., but actually testing at idle, the leak is more likely to cause a pressure loss in the W.O.T. position.

To further isolate leakage in a clutch or servo circuit, it is necessary to remove the oil pan (to drain the fluid) and the valve body; and to perform case air pressure tests.

5. To determine control pressure variation causes, refer to the Control Pressure Diagnosis chart.

AIR PRESSURE TEST OF CASE

A NO DRIVE condition can exist, even with the correct transaxle fluid pressure, because of inoperative clutches or band. Erratic shifts could be caused by a stuck governor valve. The inoperative units can be located through a series of checks by substituting air pressure for the fluid pressure to determine the location of the malfunction.

A NO DRIVE condition in Drive and 2 may be caused by an inoperative band or one-way clutch. When there is no drive in 1, the difficulty could be caused by improper functioning of the direct clutch or band and the one-way clutch. Failure to drive in re-

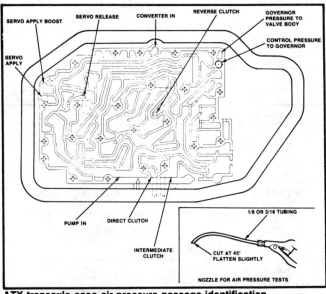

ATX transaxle case air pressure passage identification (© Ford Motor Co.)

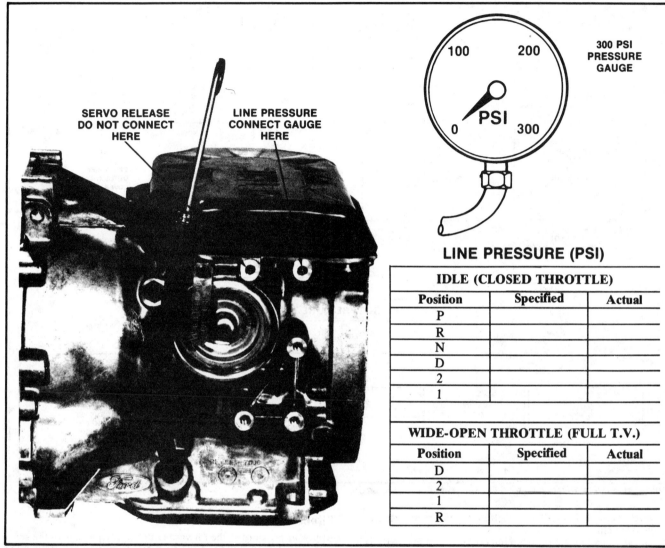

LINE PRESSURE (PSI)

IDLE (CLOSED THROTTLE)		
Position	Specified	Actual
P		
R		
N		
D		
2		
1		

WIDE-OPEN THROTTLE (FULL T.V.)		
Position	Specified	Actual
D		
2		
1		
R		

Line pressure test ports, gauge and test results shown on typical chart (© Ford Motor Co.)

LINE PRESSURE TEST RESULTS
1981-82

Control Pressure Condition	Possible Cause(s)
Low in P	Valve body
Low in R	Direct clutch, reverse clutch, valve body
Low in N	Valve body
Low in D	Servo, valve body
Low in 2	Servo, valve body
Low in 1	Servo, direct clutch, valve body
Low at idle in all ranges	Low fluid level, restricted inlet screen, loose valve body bolts, pump leakage, case leakage, valve body, excessively low engine idle, fluid too hot.
High at idle in all ranges	T.V. linkage, valve body
Okay at idle but low at W.O.T.	Internal leakage, pump leakage, restricted inlet screen, T.V. linkage, valve body (T.V. or T.V. limit valve sticking)

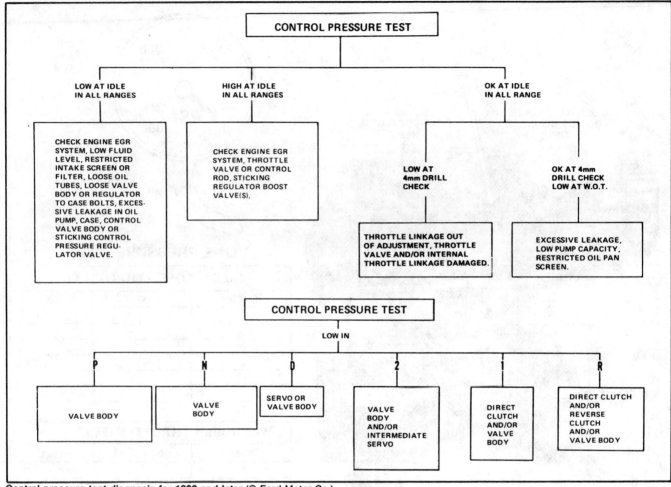

Control pressure test diagnosis for 1983 and later (© Ford Motor Co.)

verse range could be caused by a malfunction of the reverse clutch or one-way clutch.

When you have a slip problem but do not know whether it is in the valve body or in the hydraulic system beyond the valve body, the air pressure tests can be very valuable.

To properly air test the Automatic Transaxle a main control to case gasket and the following special service tools or equivalent will be required.

- Adapter Plate (Ford number, T82P-7006-B)
- Adapter Plate Attaching Screws (Ford number, T82L-7006-C)
- Air Nozzle TOOL (Ford number, 7000-DE)
- Air Nozzle Rubber Tip TOOL (Ford number, 7000-DD)

With the main control body removed, position the adapter plate and gasket on the transmission. Install the adapter plate attaching screws and tighten the screws to 80-100 in. lbs. (9-11 N•m) torque. Note that each passage is identified on the plate. Using the air nozzle equipped with the rubber tip, apply air pressure to each passage in the following order:

Band Apply Servo

Apply air pressure to the servo apply passage in the service tool plate. The band should apply, however, because of the cushioning effect of the servo release spring the application of the band may not be felt or heard. The servo should hold the air pressure without leakage and a dull thud should be heard when air pressure is removed allowing the servo piston to retrun to the release position.

Direct Clutch

Apply air pressure to the forward clutch apply passage in the service tool plate. A dull thud can be heard or movement of the piston can be felt on the case as the clutch piston is applied. If the clutch seal(s) are leaking a hissing sound will be heard.

Intermediate Clutch

Apply air pressure to the intermediate clutch apply passage in the service tool plate. A dull thud can be heard or movement of the piston can be felt on the case as the clutch piston is applied. If the clutch seal(s) are leaking a hissing sound will be heard.

Reverse Clutch

Apply air pressure to the reverse clutch apply passage in the service tool plate. A dull thud can be heard or movement of the piston can be felt on the case as the clutch piston is applied. If the clutch seal(s) are leaking a hissing sound will be heard.

Converter In

This passage can only be checked for blockage. If the passage holds air pressure remove the service tool plate and check for an obstruction or damage.

Control Pressure to Governor

Remove the governor cover and while applying air pressure to the passage in the service to plate watch for movement of the governor valve.

Governor to Control Pressure

This passage can only be checked for blockage. If the passage holds air pressure remove the service tool plate and check for an obstruction or damage.

Pump In (Bench Test)

With the transmission removed from the vehicle and the converter removed, the rotating pump gears should be heard when air pressure is applied to this passage. This check is normally performed during the assembly of an overhauled transmission.

TRANSAXLE FLUID COOLER

Flow Check

The linkage, fluid and control pressure must be within specifications before performing this flow check.

Remove the transaxle dipstick from the filler tube. Place a funnel in the transaxle filler tube. Raise the vehicle, remove the cooler return line from its fitting in the case. Attach a hose to the cooler return line and fasten the free end of the hose in the funnel installed in the fillr tube. Start the engine and set idle speed at 1000 rpm with the transaxle in Neutral.

Observe the fluid flow at the funnel. When the flow is "solid" (air bleeding has been completed), the flow should be liberal. If there is not a liberal flow at 1000 rpm in Neutral, low pump capacity, main circuit system leakage, or cooler system restriction is indicated.

To separate transaxle trouble from cooler system trouble, observe the flow at the transaxle case converter-out fitting.

Transaxle Fluid Leakage

The transaxle assemblies present other than the usual fluid leakages found on the rear drive transmission assemblies. Again, as a rule of thumb, start at the top and work downward when attempting to correct transaxle fluid leaks. Some of the possible leakage areas are listed as follows:

1. Speedometer cable connection at the transaxle.
2. Oil pan gasket, pan flange or bolts.
3. Filler tube connection at transaxle case.
4. Fluid lines and fittings.
5. Transaxle fluid cooler.
6. Throttle control lever and/or manual lever shaft seals.
7. Converter drain plug, converter hub seal or defective welds.
8. Engine or power steering oil leaks, dropping on the transaxle, causing the appearance of transaxle fluid leakage.
9. Differential Bearing retainer "O" rings, gaskets and governor cover.
10. Oil pump gasket area.
11. Servo cover seal.

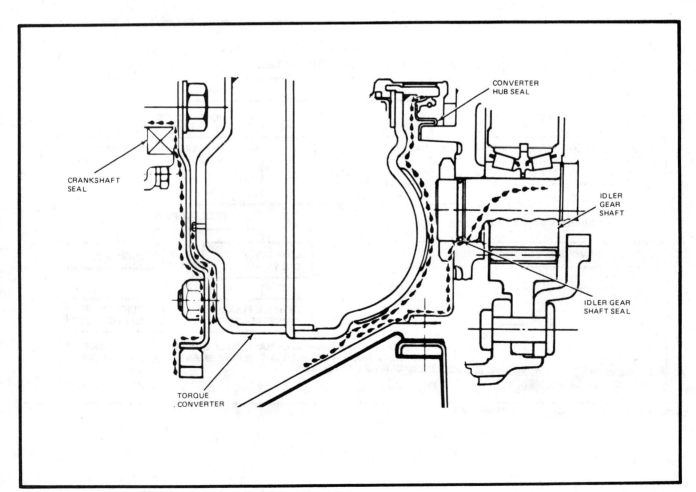

Possible fluid leakage around converter assembly (© Ford Motor Co.)

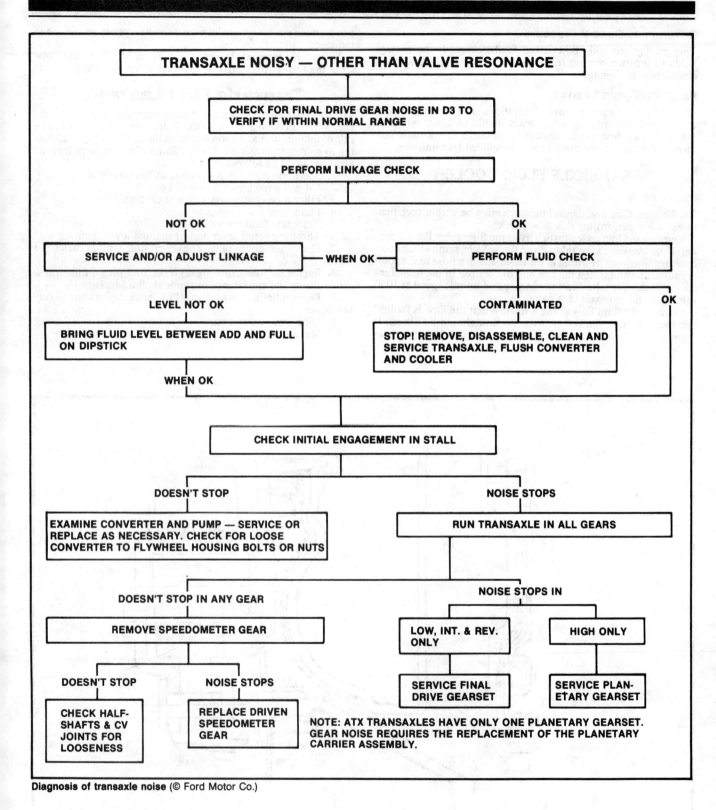

TRANSAXLE NOISY — OTHER THAN VALVE RESONANCE

CHECK FOR FINAL DRIVE GEAR NOISE IN D3 TO VERIFY IF WITHIN NORMAL RANGE

PERFORM LINKAGE CHECK

NOT OK — SERVICE AND/OR ADJUST LINKAGE — WHEN OK —

OK — PERFORM FLUID CHECK

LEVEL NOT OK — BRING FLUID LEVEL BETWEEN ADD AND FULL ON DIPSTICK — WHEN OK —

CONTAMINATED — STOP! REMOVE, DISASSEMBLE, CLEAN AND SERVICE TRANSAXLE, FLUSH CONVERTER AND COOLER

OK

CHECK INITIAL ENGAGEMENT IN STALL

DOESN'T STOP — EXAMINE CONVERTER AND PUMP — SERVICE OR REPLACE AS NECESSARY. CHECK FOR LOOSE CONVERTER TO FLYWHEEL HOUSING BOLTS OR NUTS

NOISE STOPS — RUN TRANSAXLE IN ALL GEARS

DOESN'T STOP IN ANY GEAR — REMOVE SPEEDOMETER GEAR

DOESN'T STOP — CHECK HALF-SHAFTS & CV JOINTS FOR LOOSENESS

NOISE STOPS — REPLACE DRIVEN SPEEDOMETER GEAR

NOISE STOPS IN

LOW, INT. & REV. ONLY — SERVICE FINAL DRIVE GEARSET

HIGH ONLY — SERVICE PLANETARY GEARSET

NOTE: ATX TRANSAXLES HAVE ONLY ONE PLANETARY GEARSET. GEAR NOISE REQUIRES THE REPLACEMENT OF THE PLANETARY CARRIER ASSEMBLY.

Diagnosis of transaxle noise (© Ford Motor Co.)

 ON CAR SERVICES

ADJUSTMENTS
T.V. Linkage Adjustment

Two methods of T.V. linkage adjustment can be done. One method is by the manual adjustment of the linkage and the second method is the linkage adjustment using line pressure.

INFORMATION COMMON TO ALL THROTTLE LINKAGE SYSTEMS ON THE ATX

The control rod is adjusted to proper length during initial assembly. The external TV control lever actuates the internal TV control mechanism which regulates the TV control pressure. The external TV control lever motion is controlled by stops internal to the transaxle at idle and beyond wide open throttle (WOT). The linkage return spring must overcome the transaxle TV lever load (due to spring loading to WOT).

Adjustment of fine adjustment screw for T.V. linkage
(© Ford Motor Co.)

The TV control linkage is set to its proper length during initial assembly using the sliding trunnion block on the TV control rod assembly. Any required adjustment of the TV control linkage can normally be accomplished using this sliding trunnion block. Whenever it is necessary to set the TV linkage using a line pressure gauge, use the adjustment screw at the coupling lever at the carburetor, throttle body or bell-crank, depending on application.

When the linkage is within its adjustment range at a nominal setting, the TV control lever on the transaxle will just contact its internal idle stop position (lever up as far as it will travel when the carburetor throttle body is at its hot engine curb idle position with the A/C Off if so equipped).

At Wide-Open Throttle (WOT), the TV control lever on the transmission will not be at its wide-open stop. The wide-open throttle position must not be used as the reference point in adjusting linkage.

Linkage Adjustment
1.6L CARBURETED ENGINES

The TV control linkage must be adjusted at the TV control rod assembly sliding trunnion block using the following procedure

except in the case where a line pressure gauge is used for linkage adjustment.

1. Set the engine curb idle speed to specification. Refer to the Engine/Emissions Diagnosis manual.
2. After the curb idle check, turn the engine off and insure that the carburetor throttle lever is against the hot engine curb idle stop (the choke must be Off).

NOTE: The linkage cannot be properly set if the throttle lever allowed to be on the choke fast idle cam.

3. Set the coupling lever adjustment screw at its approximate midrange.

─────── CAUTION ───────
The following steps involve working in proximity to the EGR system. Allow the EGR system to cool before proceeding.

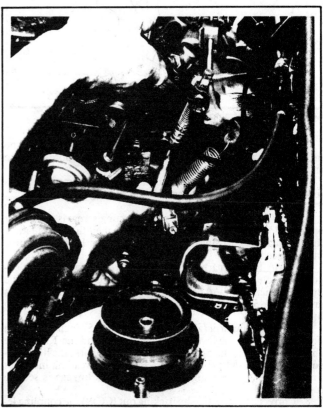

Adjusting linkage arm (© Ford Motor Co.)

4. Loosen the bolt on the sliding trunnion block on the TV control rod assembly one turn minimum.
   Remove any corrosion from the control rod and free-up the trunnion block so that it slides freely on the control rod.
5. Rotate the transaxle TV control lever up using one finger and a light force, (approximately 5 pounds to insure that the TV control lever is against its internal idle stop. Without relaxing the force on the TV control lever, tighten the bolt on the trunnion block to specification.
6. Verify that the carburetor throttle lever is still against the hot engine curb idle stop. If not, repeat Steps 2 through 6.

Linkage Adjustment Using Line Pressure

The following procedure may be used to check and/or adjust the TV control linkage using a fine pressure gauge.
1. Place the shift selector lever in the Park position.
2. Apply the emergency brake.

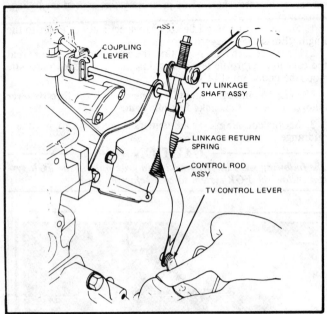

Adjusting T.V. control rod (© Ford Motor Co.)

3. Attach a 0-300 psi (0-2500 KPa) pressure gauge to the line press port on the transaxle with sufficient flexible hose to make gauge accessible while operating engine.

4. Operate engine until normal operating temperature is reached and throttle lever is against the hot engine curb idle stop (with A/C Off if so equipped).

5. Verify that the coupling lever adjusting screw is in contact with the TV linkage shaft assembly. If not, then the linkage must first be readjusted using the procedure under Linkage Adjustment procedure.

6. Verify that the carburetor throttle lever is against its hot engine curb idle stop. With engine operating at idle and in Park, line pressure must be 52-66 psi (357-455 KPa). If the line pressue is greater than 66 psi (455 KPa), the TV control linkage is set too long.

7. Place a 4mm drill (a 5/32 inch drill or 0.157 inch gauge pin) between the coupling lever adjustment screw and the TV linkage shaft. With the engine operating at idle and in park, the line pressure must be 72-88 psi (496-606 KPa). A low reading indicates linkage is set short. A high reading indicates linkage is set too long.

8. Correct a long setting by backing out (CCW) the coupling lever adjustment screw; turn in (CW) the adjustment screw for a short rod condition. This adjusting screw will change line pressure by approximately 2 psi per turn. If insufficient adjusting capacity is available, the TV control rod length must be reset using the Linkage Adjustment procedure.

Shift Trouble Diagnosis Related to Throttle Linkage Adjustment

If there is a complaint of poor transaxle shift quality, the following diagnostic procedure should be followed:

A. **Symptoms:** Excessively early and/or soft upshifts with or without slip-bump feel. No forced downshift (kickdown) function at appropriate speeds.

Cause: TV control linkage is set too short.

Remedy: Adjust linkage using Linkage Adjustment procedure.

B. **Symptoms:** Extremely delayed and harsh upshifts and harsh idle engagement.

Cause: TV control linkage is set too long.

Remedy: Adjust linkage using Linkage Adjustment procedure.

C. **Symptoms:** Harsh idle engagement after engine warm up. Shift clunk when throttle is backed off after full or heavy throttle acceleration. Harsh coasting downshifts (automatic 3-2, 2-1 shifts in D range). Delayed upshifts at light acceleration.

Cause: Interference due to hoses, wires, etc. prevents return of TV control rod or TV linkage shaft.

Remedy: Correct interference area. Check or reset linkage using the Linkage Adjustment procedure.

Cause: Excess friction due to binding of grommets prevents return of TV control linkage.

Remedy: Check for bent or twisted rods or levers causing misalignment of grommets. Repair or replace defective components (replace grommets if damaged). Reset TV control linkage using the Linkage Adjustment procedure.

D. **Symptoms:** Erratic/delayed upshifts, possibly no kickdown, harsh engagements.

Cause: Clamping bolt on trunnion at upper end of TV control rod is loose.

Remedy: Reset TV control linkage using the Linkage Adjustment procedure.

E. **Symptoms:** No upshifts and harsh engagements.

Cause: TV control rod disconnected. (Transaxle is at maximum TV pressure.)

Remedy: Reconnect TV control rod. Replace grommet(s) if rod disconnect was due to defective grommet(s).

Cause: Linkage return spring broken or disconnected.

Remedy: Reconnect or replace spring.

Shift Linkage Adjustment

1. Move the selector lever into the "D" position and against the gate stop.

NOTE: Be sure to hold the lever against the stop during the adjustment.

2. Raise the vehicle and support safely. Loosen the nut retaining the control cable to the manual lever of the transaxle.

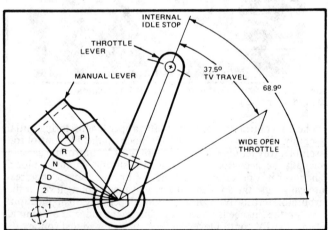

Throttle and manual linkage lever positions at the transaxle (© Ford Motor Co.)

3. Move the Transaxle lever rearward to the last detent, then forward *two* steps to the "D" detent.

4. Tighten the nut to 10-15 ft. lbs. torque.

5 Lower the vehicle and verify the adjustment.

Band Adjustment

The band servo piston rod is measured during the assembly of the transaxle and selected from a graduated group of rods to a specific length. No other adjustments are possible.

SERVICES

Oil Pan

Removal and Installation
(including Fluid Drain and Refill)

1. Raise the vehicle and support safely. Place a drain pan under the oil pan and loosen the pan attaching bolts.
2. Allow the fluid to drain from the oil pan, to the level of the pan flange. Remove the pan attaching bolts in such a manner as to allow the oil pan to drop slowly, draining more of the fluid from the oil pan.
3. Remove the oil pan and drain the remaining fluid into the container. Discard the old gasket and clean the oil pan.
4. Clean or replace the oil filter screen. Install a new gasket on the oil pan and install the pan on the transaxle. Tighten the oil pan attaching bolt to 15-19 ft. lbs.
5. Lower the vehicle and fill the transaxle to the correct level with the specified fluid. Re-check the level as required, using the room temperature checking procedure.

— CAUTION —

Using fluid other than specified, could result in transaxle malfunction and/or failure.

NOTE: The A/T fluid should be changed every 30,000 miles (48,000 Km) if the vehicle accumulates 5,000 miles (8,045 Km) or more per month or is used in continuous stop and go service.

Converter Draining
(When equipped with drain plug)

1. With the vehicle raised and supported safely, remove the converter drain access plate.
2. Rotate the engine in the normal direction of rotation until the converter drain plug is accessible.
3. Place a drain pan under the converter and remove the plug.
4. After the torque converter has been drained, install the drain plug and torque to 8-12 ft. lbs. (10-16 N•m).

Valve Body

Removal

1. Position the vehicle, open the hood and set the parking brake.
2. Remove the battery and battery tray. Remove the ignition coil.
3. Remove the dipstick from the transaxle. Remove the supply hoses and the vacuum lines from the managed air valve. Remove the valve from the valve body cover.
4. Disconnect the neutral start switch connector and the fuel evaporator hose at the frame rail.
5. Disconnect the fan motor and water temperature sending unit wiring.
6. Remove the valve body cover retaining bolts and remove the valve body cover. Discard the gasket.
7. Remove the valve body attaching bolts and remove the valve body, with gasket, from the transaxle.

NOTE: Disassembly and assembly of the valve body can be found in the transaxle disassembly and assembly section.

Installation

1. Install two valve body alignment pins in the transaxle.
2. Install the valve body to case gasket.
3. Temporarily remove one of the alignment pins while positioning the valve body, to allow the attachment of the manual valve. After the manual linkage is attached to the manual valve, re-install the alignment pin.

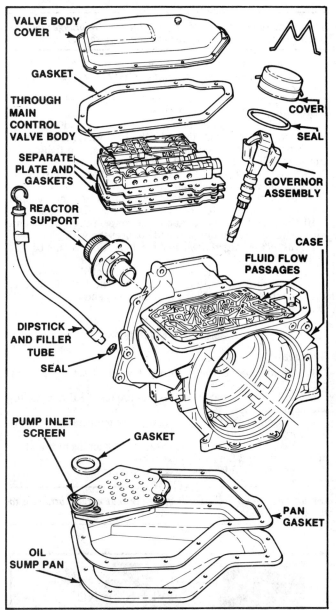

Arrangement of the hydraulic components (© Ford Motor Co.)

Valve body position on the transaxle case (© Ford Motor Co.)

─────────── CAUTION ───────────

Be sure the roller on the end of the throttle valve plunger has engaged the cam on the end of the throttle lever shaft.

───────────────────────────────

4. Connect the throttle valve control spring.

5. Install the valve body retaining bolts (27), the detent spring and the oil pressure regulator exhaust plate.

6. Remove the valve body alignment pins. Tighten the attachment bolts to 6-8 ft. lbs. (8-11 N•m).

7. Install a new gasket on the transaxle case and install the valve body cover. Install the valve body cover retaining bolts (10), along with the transaxle identification tag, and tighten the attaching bolts to 7-9 ft. lbs. (9-12 N•m).

8. Attach the neutral start switch connector and install the managed air valve on the main control cover.

9. Connect the managed air valve supply hoses and vacuum lines. Connect the fuel evaporator hoses at the frame rail.

10. Connect the fan motor and water temperature sending unit wiring. Install the ignition coil and connect the wiring.

11. Install the battery tray and the battery. Connect the battery cables.

12. Start the engine and cycle the fluid through the transaxle by moving the shift lever through all detents. Check the fluid level and correct as required. Check for fluid leaks and road test vehicle, if necessary.

Servo Cover

Removal

1. Position the vehicle, open the hood and set the parking brake.

2. Disconnect the battery cables and unplug the FM capacitor wiring, if equipped.

3. Disconnect the fan motor and water temperature sending unit wire connectors.

4. Disconnect the fan shroud to radiator attaching nuts and remove the fan and fan shroud assembly.

5. Remove the filler tube to case attaching bolt.

─────────── CAUTION ───────────

Do not lose the service identification tag that may be attached to the filler tube bolt.

───────────────────────────────

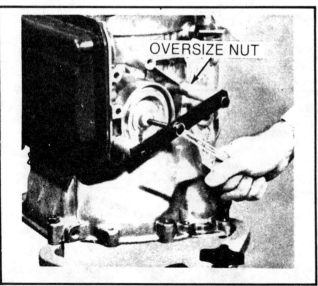

Removal or installation of the servo cover using special tool bar (© Ford Motor Co.)

6. Remove the filler tube and dipstick. Have container under transaxle to catch small amount of transaxle fluid that will drain from dipstick opening.

7. Remove the lower left mount to case attaching bolt from the left front (number 1) engine mount.

8. Remove the servo cover and snap ring with the use of the servo cover removing and installation tool, Ford number T81P-70027-A or its equivalent.

NOTE: Some fluid will leak from the case when the servo cover is removed.

Installation

1. Install new seals on the servo cover and install in place, using the servo installation and removal tool, Ford Number T81P-70027-A or its equivalent.

2. Install the mount to case attaching bolt in the left front engine mount.

3. With a new seal, install the filler tube and install the retaining bolt with the identification tag attached.

4. Install the fan and fan shroud assembly. Install the retaining nuts.

5. Connect the fan motor and temperature sending unit wiring connectors and connect the FM capacitor wiring, if equipped.

6. Connect the battery cables and start the engine. Cycle the transaxle fluid. Check the fluid level and correct as required.

7. Check the assembly for fluid leaks and road test the vehicle as required.

Governor

Removal

1. Position the vehicle and raise the hood. Set the parking brake.

2. Disconnect the battery cables and remove the two managed air valve supply rear hoses and all vacuum lines from the managed air valve.

3. Remove the attaching screw for the managed air valve supply hose band from the intermediate shift control bracket.

4. Remove the air cleaner.

5. With a long pry bar, remove the governor cover retaining clip and remove the governor cover. The governor can then be removed for service.

Installation

1. With the governor installed, install the governor cover with a new "O" ring seal in its bore of the transaxle housing.

2. With a long pry bar, position the governor cover retaining clip in its place on the cover.

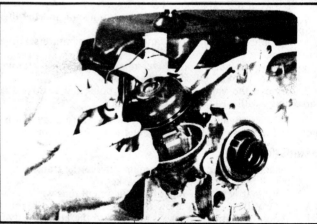

Removal or installation of governor cover (© Ford Motor Co.)

3. Install the managed air valve supply hose band to the intermediate shift control bracket attaching screw.

4. Connect the managed air valve supply hoses (2) to the rear of the valve and all vacuum lines to the valve.

5. Connect the battery cables and start the engine.

6. Cycle the transaxle fluid and check the level. Correct as required.

7. Check the transaxle for fluid leakage and road test the vehicle as required.

Neutral Start Switch

Removal

1. Position the vehicle and open the hood. Set the parking brake.

2. Disconnect the battery cables. Remove the managed air valve supply hoses from the rear of the valve and the vacuum lines from the valve. Remove the supply hose band retaining screw from the intermediate shift control bracket.

3. Remove the air cleaner assembly.

4. Disconnect the neutral start switch wire connector.

5. Remove the neutral start switch retaining bolts and remove the switch from the manual shaft.

Installation

1. Position the neutral start switch over the manual shaft and loosely install the two retaining bolts and washer through the switch and into the case.

2. Set the neutral start switch, using a number 43 drill (0.089 inch) and tighten the attaching bolts to 7-9 ft. lbs. (9-12 N•m).

3. Connect the neutral switch wiring connector.

4. Install the air cleaner assembly.

5. Install the supply hose band retaining screw into the intermediate shift control bracket.

6. Connect the two managed air valve supply hoses and all the vacuum lines to the managed air valve.

7. Connect the battery cables and start the engine. Cycle the fluid through the transaxle and check the operation of the neutral start switch. The engine should only start with the transaxle in either Park or Neutral shift positions.

REMOVAL & INSTALLATION

Transaxle Removal

ALL EXCEPT 1984 MODELS EQUIPPED WITH 2.3L H.S.C. ENGINE

NOTE: The 1984 ATX transaxle and the 2.3L H.S.C. engine must be removed and installed as a unit. Should any attempt be made to remove either component separately, damage to the ATX transaxle or to the lower engine compartment metal structure may result.

1. Position the vehicle on a lifting device, open the hood and protect the fender surfaces. Disconnect the negative battery cable.

2. Remove the bolts retaining the managed air valve, to the valve body cover.

3. Disconnect the wiring harness from the neutral start switch.

4. Disconnect the throttle valve linkage and the manual valve lever cable at their respective levers.

5. Remove the two transaxle to engine upper bolts, located below and on either side of the distributor.

6. Raise the vehicle on the lifting device. Remove the nut from the control arm to steering knuckle attaching bolt, at the ball joint, both right and left sides.

Neutral start switch assembly (© Ford Motor Co.)

7. With the use of a punch and hammer, drive the bolt out of the knuckle, both right and left sides.

NOTE: The bolt and nut from the right and left sides must be discarded and new ones used during the assembly.

8. With the use of a pry bar, disengage the control arm from the steering knuckle, on both the right and left sides.

─────────── CAUTION ───────────
Do not use a hammer on the knuckle to remove the ball joints.

NOTE: The plastic shield installed behind the rotor contains a molded pocket into which the lower control arm ball joint fits.

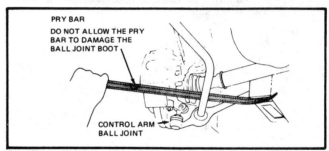

Correct position of pry bar (© Ford Motor Co.)

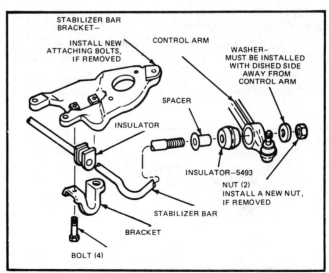

Stabilizer bar to control arm attachment (© Ford Motor Co.)

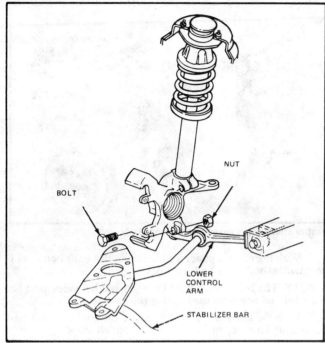

Steering knuckle pinch bolt and nut (© Ford Motor Co.)

When disengaging the control arm from the knuckle, clearance for the ball joint can be provided by bending the shield back towards the rotor. Failure to provide clearance for the ball joint can result in damage to the shield.

9. Remove the bolts attaching the stabilizer bar to the frame rail to both sides. Discard the bolts.

10. Remove the stabilizer bar to control arm attaching nut and washer from both sides. Discard the nuts. Pull the stabilizer bar out of the control arms.

11. Remove the bolt attaching the brake hose routing clip to the suspension strut bracket, on both sides.

12. Remove the steering gear tie rod to steering knuckle attaching nut and disengage the tie rod from the steering knuckle on both sides.

13. Pry the halfshaft from the right side of the transaxle. Position the halfshaft on the transaxle housing.

NOTE: Due to the configuration of the ATX transaxle case, the right halfshaft assembly must be removed first.

14. Insert the differential tool, Ford number T81P-4026-A or its equivalent into the right side halfshaft bore and drive the left halfshaft from the transaxle differential side gear.

15. Pull the left halfshaft from the transaxle and support the end of the shaft by wiring the shaft to the underbody of the vehicle.

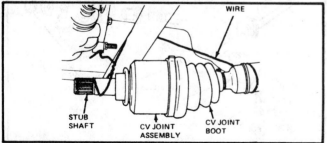

Support shaft and CV joint assembly with wire (© Ford Motor Co.)

CAUTION

Never allow the halfshaft to hang unsupported as damage to the outboard CV joint may result.

---

16. Install seal plugs into the bores of the left and right halfshafts.

17. Remove the starter support brackets and disconnect the starter cable. Remove the starter attaching bolts and remove the starter.

18. Remove the transaxle support bracket and the outer dust cover from the torque converter housing.

19. Remove the torque converter to flywheel retaining nuts. Matchmark torque converter to flywheel. Turn the crankshaft pulley bolt to bring the retaining nuts into an accessible position.

20. Remove the nuts attaching the left front (number 1) insulator mount to the body bracket.

21. Remove the bracket to body attaching bolts and remove the bracket.

22. Remove the left rear (number 4) insulator mount bracket attaching nut.

23. Disconnect the transaxle cooler lines and remove the bolts attaching the manual lever bracket to the transaxle case.

24. Position a transmission jack or other lifting device under the transaxle and remove the four remaining transaxle to engine attaching bolts.

25. Separate the transaxle from the engine enough for the torque converter studs to clear the flywheel and lower the transaxle approximately 2 to 3 inches.

26. Disconnect the speedometer cable and continue lowering the transaxle from the vehicle.

CAUTION

When moving the transaxle away from the engine, if the number one insulator mount contacts the body before the converter studs clear the flywheel, remove the insulator mount.

Installation

1. With the transaxle on a lifting device or transmission jack, position the assembly under the vehicle and slowly raise the unit into position to mate with the engine.

2. With the unit almost in position, attach the speedometer cable to the transaxle assembly. Rotate the converter or the flywheel until the matchmarks made during the removal, are in alignment. Raise the transaxle and position to the engine. Install the four lower transaxle to engine retaining bolts. Torque to 40-50 ft. lbs. (54-68 N•m).

3. Install the oil cooler lines and bolt the manual lever bracket to the transaxle case.

4. Install the left rear (number 4) insulator mount bracket attaching nut.

5. Install the left front insulator mount bracket attaching bolts and install the nuts attaching the left front (number 1) insulator mount to body bracket.

6. Install the torque converter to flywheel retaining nuts and torque to 17-29 ft. lbs. (23-39 N•m).

7. Install the transaxle support bracket and the dust cover to the torque converter housing.

8. Install the starter and retaining bolts. Attach the starter cable and install the starter support brackets.

9. Remove the left halfshaft bore seal plug, install a new circlip on the CV joint stub shaft and insert the stub shaft into the differential. Carefully align the splines of the stub axle with those of the differential gear. Push the CV joint until the circlip is felt to seat in the differential side gear.

CAUTION

Use care during the stub axle installation, not to damage the differential oil seal.

---

10. Using the same method, install the right side stub shaft into the differential side gear splines. Be sure the new circlip seats in the groove of the differential side gear.

11. Install the steering gear tie rod to steering knuckle attaching nut, after installing the tie rod stud in its bore on the steering knuckle, on both the right and left sides. Lock in place as required.

12. Install the brake hose routing clip retaining bolts on the left and right side suspension strut brackets.

13. Install the stabilizer bar into the control arms and install new nuts and washers.

14. Install new bolts and attach the stabilizer bar to the frame rails on both sides.

15. Install the lower control arm ball joints into the steering knuckle assemblies on the right and left sides. Using new bolts, install them into the steering knuckle, locking the ball joint stud to the steering knuckle. Torque to 37-44 ft. lbs. (50-60 N•m) by tightening the nut. Do not tighten the bolt.

16. Lower the vehicle and install the two upper engine to transaxle bolts, torquing them to 40-50 ft. lbs. (54-68 N•m).

17. Connect the two linkages, the throttle and manual controls, to their respective levers.

18. Connect the neutral start switch wire connector and install the bolts retaining the managed air valve to the valve body cover.

19. Connect the negative battery cable and verify the transaxle fluid level.

20. Start the engine and cycle the fluid through the transaxle by moving the manual valve control lever.

21. Re-check the assembly, the fluid level and road test as required.

Removal

2.3L HIGH SWIRL COMBUSTION (H.S.C.) ENGINE AND TRANSAXLE ASSEMBLY

1. Mark the position of the hood and remove the hood from the vehicle.

2. Disconnect the negative battery cable and remove the air cleaner.

3. Position a drain pan under the lower radiator hose and remove the lower hose. Allow the coolant to drain into the pan.

CAUTION

Do not drain the cooling system at this point if the coolant is at normal operation temperature. Personal injury can result, due to excessive heat of the coolant.

---

4. Remove the upper radiator hose from the engine.

5. Disconnect the oil cooler lines at the rubber hoses below the radiator.

6. Remove the coil assembly from the cylinder head.

7. Disconnect the coolant fan electrical connector, remove the radiator shroud and cooling fan as an assembly. Remove the radiator.

8. If equipped with air conditioning, discharge the system and remove the pressure and suction lines from the air conditioning compressor.

CAUTION

Refrigerant R-12 is contained in the air conditioning system under high pressure. Extreme care must be used when discharging the system.

---

9. Identify and disconnect all electrical and vacuum lines as necessary.

10. Disconnect the accelerator linkage, the fuel supply and return hoses on the engine and the thermactor pump discharge hose at the pump. Disconnect T.V. linkage at transaxle.

11. If equipped with power steering, disconnect the pressure and return lines at the power steering pump. Remove the power steering lines bracket at the cylinder head.

12. Install an engine holding or support tool device to the engine lifting eye. Raise the vehicle on a hoist or other lifting device.

13. Remove the starter cable from the starter.

14. Remove the hose from the catalytic converter.

15. Remove the bolt attaching the exhaust pipe bracket to the oil pan.

16. Remove the exhaust pipes to exhaust manifold retaining nuts. Pull the exhaust system from the rubber insulating grommets.

17. Remove the speedometer cable from the transaxle.

18. Position a coolant drain pan under the heater hoses and remove the heater hose from the water pump inlet tube. Remove the remaining heater hoses from the steel tube on the intake manifold.

19. Remove the water pump inlet tube clamp attaching bolt at the engine block and remove the two clamp attaching bolts at the underside of the oil pan. Remove the inlet tube.

20. Remove the bolts retaining the control arms to the body. Remove the stabilizer bar brackets retaining bolts and remove the brackets.

21. Remove the bolt retaining the brake hose routing clip to the suspension strut.

22. From the right and left sides, remove the nut from the ball joint to steering knuckle attaching bolt. Drive the bolt out of the steering knuckle with a punch and hammer.

CAUTION

Discard the bolt and nut. DO NOT re-use the bolt or nut.

---

23. Separate the ball joint from the steering knuckle by using a pry bar. Position the end of the pry bar outside of the bushing pocket to avoid damage to the bushing or ball joint boot.

NOTE: The lower control arm ball joint fits into a pocket formed in the plastic disc brake shield. This shield must be bent back, away from the ball joint while prying the ball joint out of the steering knuckle.

24. Due to the configuration of the ATX transaxle housing, the right side halfshaft must be removed first. Position the pry bar between the case and the shaft and pry outward.

CAUTION

Use extreme care to avoid damaging the differential oil seal or the CV joint boot.

---

25. Support the end of the shaft by suspending it from a convenient underbody component with a length of wire.

CAUTION

Do not allow the halfshaft to hand unsupported; damage to the outboard CV joint may occur.

---

26. Install a driver, Ford number T81P-4026-A or its equivalent, in the right halfshaft bore of the transaxle, and tap the left halfshaft from its circlip retaining groove in the differential side gear splines. Support the left halfshaft in the same manner as the right halfshaft. Install plugs in the left and right halfshaft bores.

27. Disconnect the manual shift cable clip from the lever on the transaxle. Remove the manual shift linkage bracket bolts from the transaxle and remove the bracket.

28. Remove the left hand rear (number 4) insulator mount bracket from the body bracket by removing the two nuts.

29. Remove the left hand front (number 1) insulator to transaxle mounting bolts.

30. Lower the vehicle and attach the lifting equipment to the two existing eyes on the engine. Remove the engine holding or support tool.

NOTE: Do not allow the front wheels to touch the floor.

31. Remove the right hand (number 3A) insulator intermediate bracket to engine bracket bolts, intermediate bracket to insu-

lator attaching nuts and the nut on the bottom of the double ended stud which attaches the intermediate bracket to the engine bracket. Remove the bracket.

32. Carefully lower the engine/transaxle assembly to the floor. Raise the vehicle from over the assembly. Separate the engine from the transaxle and do the necessary repair work to the transaxle assembly.

Installation

1. Raise the vehicle on a hoist or other lifting device.
2. Position the assembled engine/transaxle assembly directly under the engine compartment.
3. Slowly and carefully, lower the vehicle over the engine/transaxle assembly.

NOTE: Do not allow the front wheels to touch the floor.

4. With lifting equipment in place and attached to the two lifting eyes on the engine, raise the engine/transaxle assembly up through the engine compartment and position it to be bolted fast.
5. Install the right hand (number 3A) insulator intermediate attaching nuts and intermediate bracket to the engine bracket bolts. Install the nut on the bottom of the double ended stud that attaches intermediate bracket to the engine bracket. Tighten to 75-100 ft. lbs. (100-135 N•m).
6. Install an engine support fixture to an engine lifting eye to support the engine/transaxle assembly. Remove the lifting equipment.
7. Raise the vehicle and position a lifting device under the engine. Raise the engine and transaxle assembly into its operating position.
8. Install the insulator to bracket nut and tighten to 75-100 ft. lbs. (100-135 N•m).
9. Tighten the left hand rear (number 4) insulator bracket to body bracket nuts to 75-100 ft. lbs. (100-135 N•m).
10. Install the starter cable to the starter.
11. Install the lower radiator hose and install the retaining bracket and bolt. Tighten to specifications.
12. Install the manual shift linkage bracket bolts to the transaxle. Install the cable clip to the lever on the transaxle.
13. Connect the lower radiator hose to the radiator. Install the thermactor pump discharge hose at the pump.
14. Install the speedometer cable to the transaxle.
15. Position the exhaust system up and into the insulating grommets, located at the rear of the vehicle.
16. Install the exhaust pipe to the exhaust manifold bolts and tighten to specifications.
17. Connect the gulp valve hose to the catalytic converter.
18. Position the stabilizer bar and the control arms assemblies in position and install the attaching bolts. Tighten all fasteners to specifications.
19. Install new circlips in the stub axle inboard spline grooves on both the left and right halfshafts. Carefully align the splines of the stub axle with the splines of the differential side gears and with some force, push the halfshafts into the differential unit until the circlips can be felt to seat in their grooves in the differential side gears.
20. Connect the control arm ball joint stud into its bore in the steering knuckle and install new bolts and nuts.

CAUTION ---

Do not use the old bolts and nuts. Their torque holding capacity is no longer of any value.

---

21. Tighten the new bolt and nut to 37-44 ft. lbs. (50-60 N•m).

NOTE: Tighten the nut only and hold the bolt from turning.

22. Position the brake hose routing clip on the suspension components and install their retaining bolts.
23. Lower the vehicle and remove the engine support tool.

24. Connect the vacuum and electrical lines that were disconnected during the removal procedure.
25. Install the disconnected air conditioning components.
26. Connect the fuel supply and return lines to the engine and connect the accelerator cable.
27. Install the power steering pressure and return lines. Install the brackets.
28. Connect the T.V. linage at the transaxle.
29. Install the radiator shroud and the coolant fan assembly. Tighten the bolts to specifications.
30. Install the coil and connect the coolant fan electrical connector.
31. Install the upper radiator hose to the engine and connect the transaxle cooler lines to the rubber hoses under the radiator. Fill the radiator and engine with coolant.
32. Install the negative battery cable and the air cleaner assembly.
33. Install the hood in its original position.
34. Check all fluid levels and correct as required.
35. Start the engine and check for leakage. Correct any fluid levels as required.
36. Charge the air conditioning system and road test the vehicle as necessary.

GENERAL SERVICE PRECAUTIONS

When servicing this unit, it is recommended that as each part is disassembled, it is cleaned in solvent and dried with compressed air. All oil passages should be blown out and checked for obstructions. Disassembly and reassembly of this unit and it parts must be done on a clean work bench. As is the case when repairing any hydraulically operated unit, cleanliness is of the utmost importance. Keep bench, tools, parts, and hands clean at all times. Also, before installing bolts into aluminum parts, *always dip the threads into clean transmission oil.* Anti-seize compound can also be used to prevent bolts from galling the aluminum and seizing. Always use a torque wrench to keep from stripping the threads. Take care with the seals when installing them, especially the smaller O-rings. The slightest damage can cause leaks. Aluminum parts are very susceptible to damage so great care should be exercised when handling them. The internal snap rings should be expanded and the external snap rings compressed if they are to be re-used. This will help insure proper seating when installed. Be sure to replace any O-ring, gasket, or seal that is removed, although often the Teflon seal rings, when used, will not need to be removed unless damaged. Lubricate all parts with Dexron® II when assembling.

Transaxle Disassembly

1. Mount the transaxle in a fixture, if available to the repairman. The fixture should be free to rotate as required.

NOTE: The following disassembly procedures will be outlined with the assumption that the transaxle is mounted to a fixture.

2. With the use of universal converter handles, lift the converter out of the transaxle assembly. Remove the handles and set aside.
3. Pull the oil pump drive shaft from the input shaft.
4. Rotate the assembly 180° and remove the differential end seal and allow the remaining transaxle fluid to drain into a container. Remove the dipstick from the case.

5. Remove the oil pan retaining bolts (13), remove the pan and discard the gasket.

6. Remove the oil filter retaining bolts (3), remove the filter and discard the gasket.

7. Remove the retaining bolts (6) from the differential retainer housing. Using small pry bars, wiggle the retainer to break it loose.

—————— CAUTION ——————

Do not damage the shims.

Lifting converter from transaxle (© Ford Motor Co.)

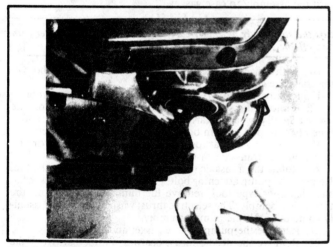

Installation of plastic plug in the differential opening
(© Ford Motor Co.)

8. Lift the differential assembly towards the retainer to free the retainer without damaging the shims.

9. Lift the differential retainer with the inner seal and the outer "O" ring, from the axle assembly.

10. Remove the selective shim and copper taper shim.

—————— CAUTION ——————

Note the position of the notch in the taper shim for correct installation during the assembly.

11. Remove the differential assembly by lifting up and out. Lay aside for later disassembly, if necessary.

12. Remove the valve body cover retaining bolts (10), remove the cover and gasket and discard the gasket.

13. Disconnect the throttle lever return spring and loosen the valve body attaching bolts (27).

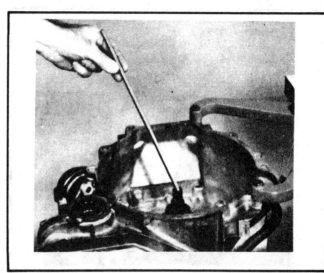

Removal of oil pump shaft (© Ford Motor Co.)

Removal of oil filter and gasket (© Ford Motor Co.)

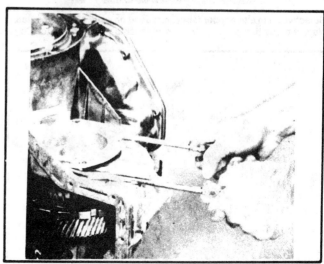

Breaking differential retainer loose after removal of retaining bolts
(© Ford Motor Co.)

3–111

Removal of differential retainer with two seals, one inner and one outer "O" ring (© Ford Motor Co.)

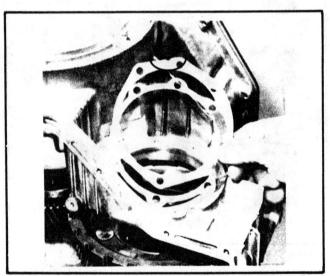

Selective shim and copper taper shim used on early ATX transaxles. Note the position of the notch for re-assembly (© Ford Motor Co.)

Removing baffle plate (© Ford Motor Co.)

Removal of the differential assembly (© Ford Motor Co.)

14. Remove the main oil pressure regulator and transaxle control baffle plate with the 7 special bolts.

NOTE: Because of their length, keep the 7 special bolts separate from the remaining valve body bolts.

15. Remove the detent spring and roller assembly, remove the remaining bolts and lift the valve body from the transaxle. Discard the assembly to case gasket.

─────── **CAUTION** ───────

Lift the valve body assembly carefully from the T.V. plunger cam and off the "Z" link on the rooster comb.

16. With a piece of wire, lift the governor filter from its bore in the transaxle case.

17. Remove the governor cover and remove the governor.

18. Remove the speedometer driven gear retaining pin. The use of a small pry bar to start the pin out and the use of side cutter type pliers to pull the pin out of the case, is recommended.

19. With the use of a wooden hammer handle, tap the driven gear from the transaxle case from the inside out.

20. Rotate the transaxle to place the pump assmbly on the top. Remove the pump attaching bolts (7) and with the use of a special slide hammer type tool, remove the pump assembly from the transaxle assembly. The selective thrust washer number 12 usually come out with the pump assembly.

21. Remove the pump to case gasket and discard.

22. Remove the needle thrust bearing, number 11, from the top of the intermediate clutch. Lift the intermediate clutch assembly from the case assembly.

23. Remove the number 10 thrust bearing from the top of the direct clutch. Lift out the ring gear assembly along with the direct clutch assembly.

─────── **CAUTION** ───────

The ring gear assembly and the direct clutch parts are loose and can separate during handling.

NOTE: During the disassembly, the ring gear and direct clutch can be removed separately.

24. Remove the number 7 thrust washer from the surface of the planetary assembly.

25. Remove the reverse clutch snap ring and remove the reverse clutch pack.

26. Remove the planetary assembly and the number 5 thrust washer.

Disconnecting inner T.V. control spring (© Ford Motor Co.)

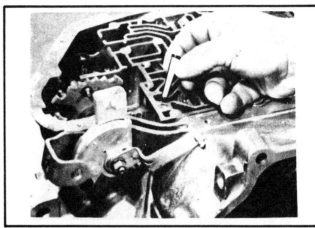

Location of governor filter screen (© Ford Motor Co.)

Removal of the detent spring (© Ford Motor Co.)

Removing speedometer gear retaining pin (© Ford Motor Co.)

Removal of pump from case with slide hammer type tools
(© Ford Motor Co.)

Lifting pump assembly from the case, exposing thrust washer number 11 (© Ford Motor Co.)

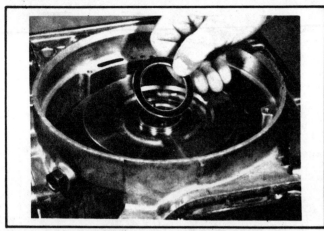

Removal of thrust washer number 11 (© Ford Motor Co.)

Removal of intermediate clutch assembly (© Ford Motor Co.)

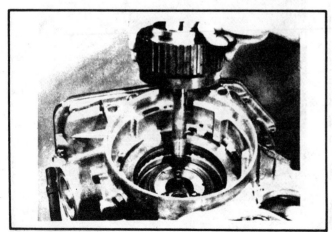

Removal of ring gear and direct clutch assemblies
(© Ford Motor Co.)

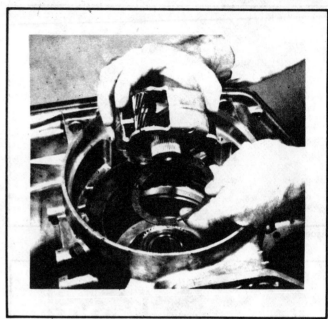

Removal of the planetary assembly and number 5 thrust washer
(© Ford Motor Co.)

Removal of the reverse clutch pack (© Ford Motor Co.)

27. Remove the reverse clutch return springs and holder assembly. Remove the reverse clutch piston.

28. Pry the reverse clutch cylinder upward to loosen, and then remove from the case, along with the piston and seals.

29. Install a servo cover remover/installer tool, Ford number T81P-70027-A or its equivalent and remove the servo retaining snap ring. Back the tool off and allow the spring pressure to neutralize. Remove the cover, servo piston and tool.

30. Remove the low-intermediate band and the sun gear with the drum assembly.

NOTE: Lift the sun gear assembly and rotate clockwise to remove.

31. Remove the number 4 thrust washer from the transfer housing.
32. Remove the transfer housing attaching bolts (55), pry the transfer housing from the idler gear shaft and remove the transfer housing.

CAUTION

When prying against the transfer housing, lift upward against the housing only. Prying downward can result in damage to the transfer gear teeth.

33. Remove the number 3 thrust needle bearing from the input gear and remove the gear. Remove the caged needle bearing, number 2, from the hub of the case.
34. Lift out the number 1 thrust needle bearing.
35. Rotate the case as necessary. Install an allen wrench into the idler gear shaft, wedging the handle between the band anchor strut and the case. With a socket and breaker bar, remove the idler shaft locknut from the converter side of the case. Tap on the end of the shaft and remove the idler gear assembly from the case.

NOTE: Do not disassemble the gear assembly.

Remove the transfer housing (© Ford Motor Co.)

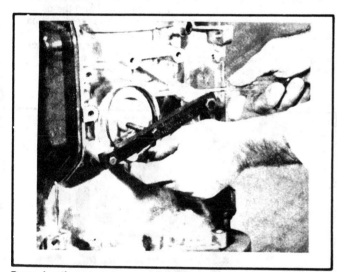

Removing the servo cover retaining snapring (© Ford Motor Co.)

Removal of the drum and sun gear assembly (© Ford Motor Co.)

Removal of the reverse clutch return spring and damper assembly (© Ford Motor Co.)

Remove the low and intermediate band (© Ford Motor Co.)

Removal of the idler gear assembly (© Ford Motor Co.)

36. Inspect the reactor shaft for damage. Do not remove unless the shaft is considered unserviceable. Special puller tools are required, should this operation be needed. Proceed as follows:

a. Remove the reactor support retaining bolts (5).

b. With the special tool mounted in place, force the reactor support out of the case.

c. To install a new reactor support, place a guide pin in the new support and seat into the case with the special tools used as a press. Install the reactor support bolts and remove the guide pin. Torque the bolts to 6-8 ft. lbs. (8-11 N•m).

Sub Assemblies

MANUAL AND THROTTLE LINKAGE

Disassembly

NOTE: The manual and throttle linkage are normally not disassembled unless damaged linkage or leaking seals are encountered.

1. Remove the throttle valve outer lever retaining nut while holding the lever stationary.

—————— CAUTION ——————

Should the throttle lever be allowed to rotate, damage to the inner lever can result.

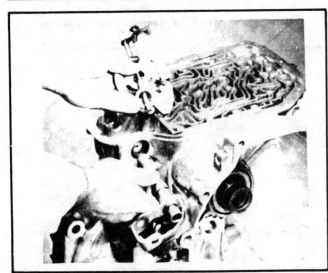

Manual and throttle linkage components (© Ford Motor Co.)

2. Remove the neutral start switch retaining screws and the neutral start switch.

3. Remove the manual lever retaining pin from the lever and shaft.

4. Remove the parking pawl ratcheting spring. Loosen and remove the inner manual lever (detent) and parking pawl actuating lever to manual shaft retaining nut.

5. Remove the manual lever and shaft assembly. With the manual lever out, the throttle valve lever and components can be removed.

6. Remove the parking pawl return spring. With the shaft out of the case, remove the manual lever shaft seal from the case. Remove the throttle valve lever shaft seal from the manual lever.

Assembly

1. Install the manual lever shaft seal in the case using a special seal installer, Ford Number T81P-70337-A or its equivalent.

2. Install the throttle lever shaft seal using an appropriate sized socket.

3. Install the parking pawl return spring.

4. Install the inner manual lever detent and parking pawl actuator attaching nut, the inner manual lever detent and the parking pawl actuator in the case, on the throttle shaft and insert the manual lever and shaft assembly into the case, over the throttle lever.

5. Position the parking pawl actuator and inner manual lever detent on the manual lever shaft and install the attaching nut. Tighten the nut securely.

6. Install the parking pawl ratcheting spring.

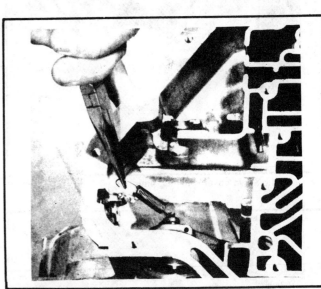

Installation of paking pawl spring (© Ford Motor Co.)

7. Install the manual lever retaining pin through the manual lever and into the shaft.

8. Install the neutral start switch and retaining screws. Do not tighten the retaining screws.

9. Install the outer throttle valve lever and adjust the neutral start switch, refering to the On-Car Adjustment section.

PUMP

Disassemby

1. Remove the number 12 thrust washer from the clutch support.

2. Remove the sealing rings from the clutch support.

3. Remove the pump to case seal from the outer circle of the pump. Matchmark the clutch support to the pump body.

Oil pump drive gear, driven gear and pump drive insert
(© Ford Motor Co.)

Assembling driven gear to pump body with chamfered teeth and center punch mark to the pump body (© Ford Motor Co.)

Seal locations on oil pump (© Ford Motor Co.)

4. Remove the clutch support to pump body bolts (5) and separate the clutch support from the pump body.

5. Remove the insert from the pump drive gear and remove the pump driven gear. Remove the pump drive gear.

Inspection

1. Inspect the mating surfaces of the pump body and cover for burrs.

2. Inspect the drive and driven gear bearing surface for scores and check the gear teeth for burrs.

3. Check the fluid passages for obstructions.

4. If any of the parts are found to be worn or damaged, the pump should be replaced as a unit. Minor scores and burrs can be removed with crocus cloth.

Assembly

1. Install the pump drive and driven gears. Install the pump drive insert.

2. Position the clutch support on the pump body in its proper position. Use guide pins to align.

3. Install the clutch support to pump body retaining bolts (5) and torque to 6-8 ft. lbs. (8-11 N•m).

4. Install a new pump body to case seal in its groove on the pump body.

5. Install the seal rings on the clutch support. Overlap the scarf cuts properly.

6. Install the number 12 thrust washer on the clutch support.

INTERMEDIATE CLUTCH

Disassembly

1. Remove the intermediate shaft retaining snapring.

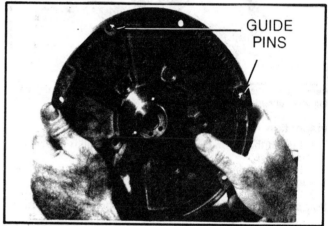

Use of guide pins to properly locate pump components during bolt tightening (© Ford Motor Co.)

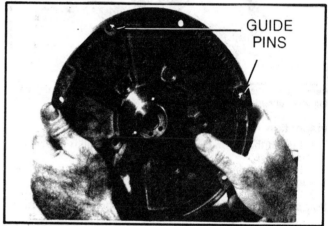

Removal of intermediate clutch pack (© Ford Motor Co.)

2. Remove the intermediate shaft from the intermediate clutch cylinder. Inspect the shaft stop ring and replace if damaged.

3. Remove the clutch pack retaining ring, remove the pressure plate and clutch pack. Remove the seal rings from the clutch cylinder hub.

4. With the use of a clutch spring compressor tool, Ford number T65L-77515-A or its equivalent, disengage the piston retaining snap ring. Slowly release the spring pressure and remove the tool from the clutch cylinder.

5. Remove the snapring and piston return spring retainer and remove the piston from the cylinder.

6. Remove the piston seal from the clutch cylinder and the seal from the clutch piston.

Removal or installation of the intermediate clutch return spring, using special spring removal tool (© Ford Motor Co.)

Inspection

1. Inspect the piston check ball. Be sure the ball is present and free in its cage.

2. Inspect the clutch piston bore and the piston inner and outer bearing surfaces for burrs or scores.

3. Check the clutch pressure plate for scores on the clutch plate bearing surfaces. Check the clutch release springs for distortion.

Assembly

1. Lubricate and install the seal, with the lip facing up, on the clutch piston.

2. Lubricate and install the seal, with the lip facing down, on the clutch cylinder hub.

3. Apply a light film of petroleum jelly to the piston seals, clutch cylinder seal area and the clutch piston inner seal area. Install the clutch piston by pushing downward and rotating it into its bore in the clutch cylinder.

Installation of intermediate clutch return spring assembly into clutch cylinder (© Ford Motor Co.)

4. Position the piston return springs, retainer and piston retaining ring on the clutch cylinder.

5. Install the clutch spring compressor tool, depress the piston and return springs and install the piston retaining ring. Remove the clutch spring compressor tool carefully, being sure the retaining ring is in its groove.

6. Install the seal rings on the clutch cylinder hub. Be sure the scarf cut seals overlap at the bevel edge.

7. Soak the fiber clutch plates in transmission fluid and install the clutch plates in the following order:
    a. one steel plate next to piston
    b. one fiber plate
    c. one steel plate
    d. one fiber plate
    e. pressure plate
    f. selective snapring

8. Two methods of measuring clutch pack clearance of the intermediate clutch can be used.

Feeler Gauge Method—Measure the free play clearance between the pressure plate and the first fiber plate in two places, 180° apart, with a feeler gauge blade. If the feeler gauge reading is 0.020-0.035 inch (0.51-0.88 mm), the clearance is within specifications. If the reading is under 0.020 inch (0.51 mm), install a smaller selective snapring and re-measure. If the reading is over 0.035 inch (0.88 mm), install a thicker selective snapring and repeat the measurement procedure.

Measuring the clutch pack with the feeler gauge method (© Ford Motor Co.)

Install intermediate clutch shaft and retain with snapring (© Ford Motor Co.)

Dial Indicator Method—Position a dial indicator on the clutch cylinder hub with the stem touching the top of the pressure plate. Push downward on the clutch pack with at least 10 pounds of pressure. Release the pressure and zero the dial indicator. Lift the pressure plate with the thumbs of both hands and note the dial indicator reading. Take two readings, 180° apart and use the average of the two readings. The clearance should be 0.030-0.044 inch (0.75-1.22 mm) for three plates. If the clearance is not within limits, selective snaprings are available. Install the correct snapring and re-check the clearance.

9. The selective snaprings are available in the following thicknesses:

    0.049-0.053 inch (1.245-1.346 mm)
    0.059-0.063 inch (1.499-1.600 mm)
    0.070-0.074 inch (1.778-1.880 mm)

10. If removed, install the stop ring on the intermediate shaft and install the shaft into the clutch cylinder hub.

11. Install the intermediate shaft retaining ring.

Exploded view of direct and one-way clutch assembly (© Ford Motor Co.)

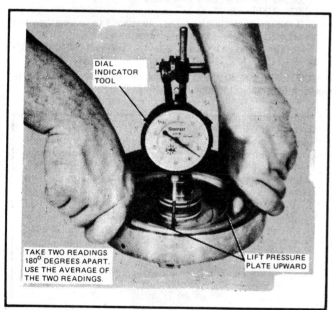
Measurement of the clutch pack with dial indicator method (© Ford Motor Co.)

Use of special spring compression tool to remove and install the spring assembly and its retaining ring (© Ford Motor Co.)

DIRECT CLUTCH

Disassembly

1. Remove the ring gear off the one way clutch and direct clutch assembly, if not already off.

2. Lift the one way clutch outer race and the sun gear from the direct clutch.

3. Remove the number 8 thrust washer and the one way clutch.

4. Remove the clutch pack retaining snapring. Remove the pressure plate and clutch pack. Remove the number 9 brass thrust washer.

5. With the use of a clutch spring compressor tool, Ford number T81P-70235-A or its equivalent, remove the piston retaining snapring from the clutch cylinder.

6. Remove the piston return snapring retainer and remove the clutch piston.

NOTE: Snapring pliers can be used to aid in the piston removal.

7. Remove the piston seal and the clutch cylinder hub seal. Discard the seals.

Installation of the spring assembly and the retaining snapring in the clutch housing (© Ford Motor Co.)

Inspection

1. Inspect the piston check ball. Be sure the ball is present and free in its cage.

2. Inspect the clutch piston bore and the piston inner and outer bearing surfaces for burrs or scores.

3. Check the clutch pressure plate for scores on the clutch plate bearing surfaces. Check the clutch return springs for distortion.

Assembly

1. Lubricate and install the clutch cylinder hub seal with the lip facing downward.

2. Lubricate the install the seal on the piston with the seal lip facing upward.

3. Apply a light film of petroleum jelly to the piston and hub seals, the sealing areas and install the clutch piston into its bore in the clutch cylinder.

4. Position the piston return spring retainer and the piston retaining snapring in position on the clutch cylinder.

5. With the use of the clutch spring compressor tool, compress the clutch return springs and install the piston retaining snapring. Install the number 9 thrust washer.

6. Soak the fiber clutch plates in transaxle fluid. Install the clutch pack, starting with a steel plate, alterating the fiber and steel plates for a total of four fiber and four steel plates. Install the pressure plate and the selective retaining snapring.

7. Install the one way clutch and the number 8 thrust washer.

NOTE: When properly installed, the thrust washer tabs will be against the shoulder of the inner race.

8. Two methods of measuring the clutch pack clearance of the direct clutch can be used.

Feeler Gauge Method—Measure the free play clearance between the pressure plate and the first fiber plate, in two places, 180° apart, with a feeler gauge blade.

NOTE: If necessary, the one way clutch and the number 8 thrust washer can be removed until the measurements are completed.

If the feeler gauge reading is 0.031-0.047 inch (0.78-1.20 mm), the clearance is within specifications. If the reading is under 0.031

inch (0.78 mm), install a smaller selective snapring and remeasure. If the reading is over 0.047 inch (1.20 mm), install a thicker selective snapring and repeat the measurement procedure.

Dial Indicator Method—Position a dial indicator on the clutch cylinder hub with the stem touching the top of the pressure plate. Push downward on the clutch pack with at least a 10 pound pressure. Release the pressure and zero the dial indicator. Lift the pressure plate with the thumbs of both hands and note the dial indicator reading. Take two readings and average the results of the two readings, take 180° apart. The clearance, with four fiber friction plates, should be 0.040-0.056 inch (1.01-1.43 mm). If the clearance is not within specifications, selective snaprings are available. Install the correct snapring and re-check the clearance.

9. The selective snaprings are available in the following thicknesses:

0.050-0.054 inch (1.26-1.36 mm)
0.062-0.066 inch (1.58-1.68 mm)
0.075-0.079 inch (1.90-2.00 mm)

10. If the one way clutch and the number 8 thrust washer was removed, install then into the clutch cylinder.

11. Install the sun gear/clutch race assembly into the clutch cylinder.

NOTE: The one way clutch allows the sun gear/clutch race to rotate in one direction only.

Assembling number 9 thrust washer and cage, roller and spring assembly with the smooth surface down (© Ford Motor Co.)

Installation of thrust washer number 8 and one-way clutch outer race and sun gear (© Ford Motor Co.)

Measurement of the direct clutch pack with the feeler gauge method (© Ford Motor Co.)

REVERSE CLUTCH

Disassembly

1. Remove the seals from the clutch cylinder and the clutch piston.

Inspection

1. Check the clutch cylinder and the clutch piston for scores or burrs.

Assembly

1. Install the new seal on the clutch piston with the lip facing upward.
2. Install the inner seal on the clutch cylinder with the lip facing down.
3. Install the outer clutch cylinder seal. The seal is square cut, making the direction of the seal unimportant.

BAND APPLY SERVO

Disassembly

1. Remove the piston spring and remove the servo piston from the cover.
2. Remove the piston rod retaining clip. Remove the rod, cushion spring and spring retainer washer from the piston.
3. Remove the seals from the servo cover. Remove the seals from the servo piston.

Inspection

1. Inspect the servo bore for cracks, burrs or scores.
2. Check the fluid passages for obstructions.
3. Inspect the servo spring and servo band struts for distortion or damage.
4. Check the servo piston for burrs or scores.

Assembly

1. Position the spring retainer washer and cushion spring on the servo rod. Install the spring and rod assembly in the servo piston.
2. Compress cushion spring and install the retaining clip.
3. Install the seals on the servo piston.
4. Install the seals on the servo cover.
5. Lubricate the piston seals with petroleum jelly and install the piston into the cover.
6. Install the piston return spring on the piston rod.

VALVE BODY

Disassembly

There is no definite disassembly procedure for the valve body components. When removing the valves and springs from the valve body, a channeled type holder should be be used to hold the valves and springs in their correct order, until their re-assembly. A general disassembly and assembly outline is suggested, but the step by step procedure will depend upon the discretion of the repairman.

— CAUTION —

During the disassembly or assembly of the valve body, do not turn the throttle valve adjusting screw. This adjustment is set during the manufacture of the valve body and must not be altered.

1. Remove the separator plate attaching screws, the separator plate and gasket.
2. Note the location of the check balls and the relief valve. Remove and set aside.

NOTE: 1981-82 transaxle models use six check balls in the valve body, while the 1983 and later models use only five check balls.

Installation of seals on the reverse clutch cylinder (© Ford Motor Co.)

Servo cover and piston seal locations (© Ford Motor Co.)

3. Remove the valve plug retainer, the valve plug, reverse boost valve and spring.
4. Remove the valve plug retainer, the valve plug, the 2-3 shift valve and valve spring.
5. Remove the valve plug retainer, the valve plug, the 1-2 shift valve, the 1-2 T. V. modulator valve and spring.
6. With the aid of a fabricated valve spring depressing tool, remove the valve spring retainer, the 2-1 scheduling valve and valve spring.
7. Remove the valve plug retainer, the valve plug, the 2-3 back-out valve and valve spring.
8. Remove the valve sleeve retainer, the main pressure boost sleeve, the main oil regulator valve, spring and spring retainer.
9. Remove the plug retainer, the valve plug, the manual low downshift valve and spring.
10. With the use of the fabricated valve spring depressing tool, remove the spring retainer, the 3-2 torque demand timing control valve and spring.
11. Depress the valve spring and remove the 3-2 kickdown timing valve and spring.
12. Depress the valve spring and remove the 3-2 control valve and spring.
13. Remove the plug retainer, the valve plug, the 2-3 shift T.V. modulator valve and spring.
14. Remove the plug retainer and remove the valve plug, the 1-2 capacity modulator valve and spring.

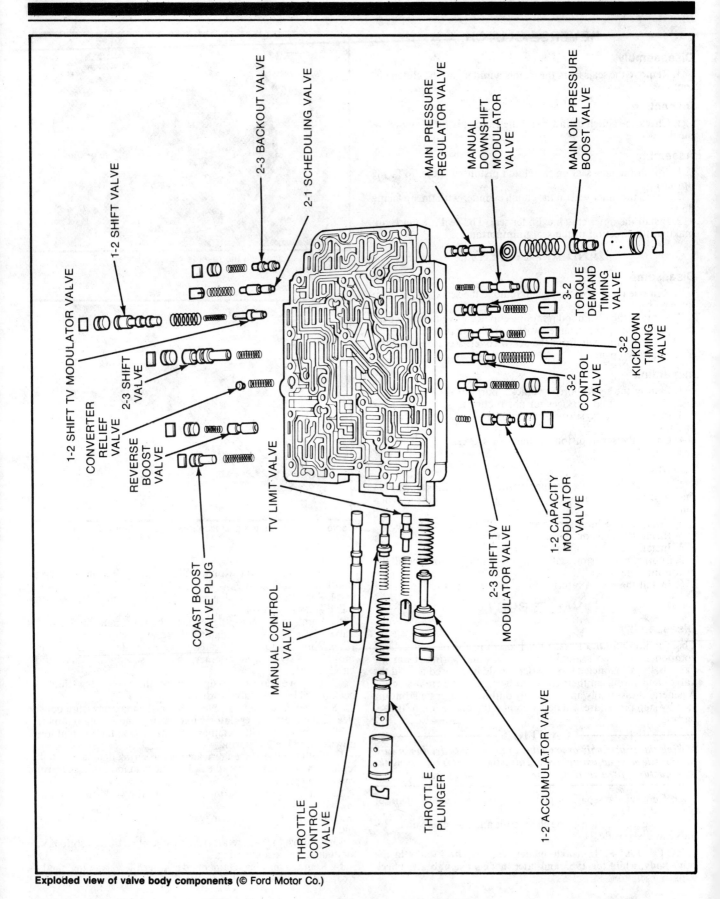

Exploded view of valve body components (© Ford Motor Co.)

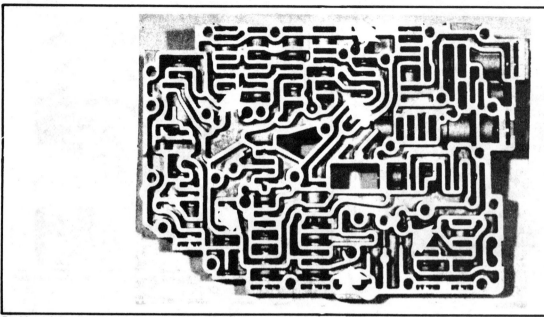

Position of the six check balls in the 1981-82 ATX transaxle models (© Ford Motor Co.)

15. Remove the valve plug retainer, the valve plug, the 1-2 accumulator valve and spring.
16. Depress the valve spring and remove the valve spring retainer, the T.V. limit valve and spring.
17. Remove the valve sleeve retainer, the throttle valve plunger sleeve, throttle pressure valve, throttle plunger return spring, throttle pressure valve and spring. Remove the washer and throttle pressure adjusting sleeve.

--- CAUTION ---

Do not turn the throttle valve adjusting screw. This adjustment is set during the manufacture and must not be altered.

18. Remove the manual control valve.

Inspection

1. Inspect all valves and plug bores for scores. Check all fluid passages for obstructions. Inspect the check valves for freedom of movement. Inspect all mating surfaces for burrs or distortion. If needed, polish the valves and plugs with crocus cloth to remove minor burrs or scores.
2. Inspect all springs for distortion.
3. Check all valves and plugs in their bores for freedom of movement. Valves and plugs, when dry, must fall from their own weight in their respective bores.
4. Roll the manual valve on a flat surface to check for a bent condition.

Assembly

The assembly of the valve body should follow, in principle, the disassembly procedure. However, the assembly will depend upon the discretion of the repairman. When the valves have all been installed, the completion of the assembly can be done as follows:
1. Install the check balls and relief valve in their proper seats in the valve body.
2. Install two alignment pins and the separator plate gasket.
3. Install the separator plate and attaching bolts. Torque the bolts to 80-90 *inch pounds* and remove the alignment pins.

TRANSFER HOUSING BEARING
Removal

1. Position the transfer housing in the case and install the attaching bolts to secure the housing.

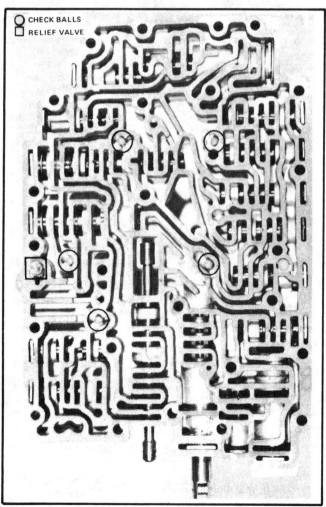

Position of the five check balls in the 1983 and later ATX transaxle models (© Ford Motor Co.)

Temporarily install the transfer housing in the case to remove the bearing assembly (© Ford Motor Co.)

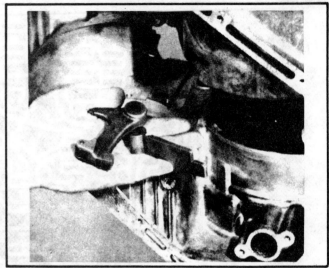

With the pin removed, the park pawl and band anchor strut can be removed (© Ford Motor Co.)

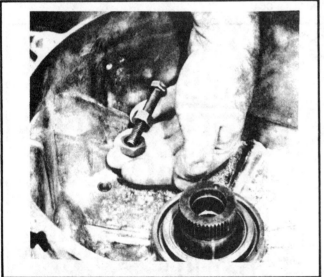

After threading the cup, install bolt and nut, along with an oversize nut as a spacer, to remove the cup from the case (© Ford Motor Co.)

2. Using a slide hammer type tool, remove the bearing from the transfer housing.

Installation

1. Remove the transfer housing from the case and support the housing from underneath.

2. Install the new bearing in the transfer housing with the use of a bearing installer tool.

NOTE: The transfer housing and the ATX transaxle case are matched parts. If one is damaged, both must be replaced.

PARKING PAWL AND BAND STRUT

Removal

1. Working in the converter housing, thread a ⅜ inch national coarse, bottoming type tap into the welsh plug, holding the parking pawl and band anchor pin in position.

2. Position a nut over the plug to act as a spacer.

NOTE: The hole in the nut has to be larger than the diameter of the plug.

3. With the spacer nut in position, install a nut on a ⅜ inch—16 national coarse bolt and thread the bolt into the newly made threads of the plug.

4. Thread the nut on the bolt downward, against the spacer nut, causing the plug to move upward and out of the case.

5. With the use of a magnet, remove the parking pawl and band anchor pin from the case. Remove the parking pawl and band anchor from the case.

Installation

1. Position the parking pawl and band anchor in the case, with the band strut positioned closest to the pump housing end of the transaxle.

2. Install the parking pawl and band anchor pin in the transaxle case.

3. Position a new welsh plug in the case bore and seat with an appropriate tool.

NOTE: The threaded welsh plug can be re-used if necessary, if not damaged.

GOVERNOR DRIVEN GEAR

Removal

1. Support the governor assembly on a vise and remove the roll pin from the driven gear with a ³/₃₂ inch drift.

2. Clamp the plastic driven gear in a vise, grip the governor body firmly, twist and pull the body at the same time, separating the driven gear from the governor drive shaft.

Installation

1. Align the driven gear to the governor drive shaft. Press the gear on the shaft as far as possible with hand pressure.

2. Be sure the governor gear is properly aligned with the governor drive shaft and tap the driven gear on the shaft with a soft faced hammer. The driven gear is in its proper position when the molded shoulder is seated against the governor body.

3. With the use of a drill press, if possible, drill a ⅛ inch hole through the plastic gear, in line with the hole in the governor drive shaft.

4. Supporting the governor assembly to protect the plastic driven gear, install a new roll pin through the gear and shaft.

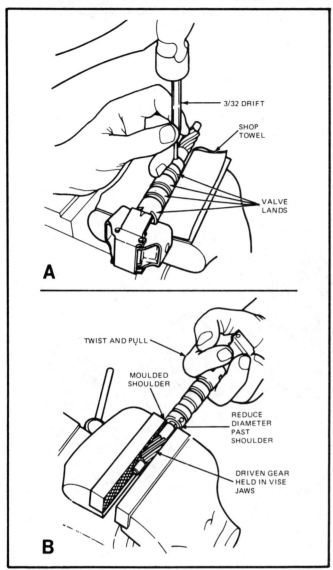

A—Removing the pin from the governor driven gear. B—Twisting the governor shaft from the driven gear (© Ford Motor Co.)

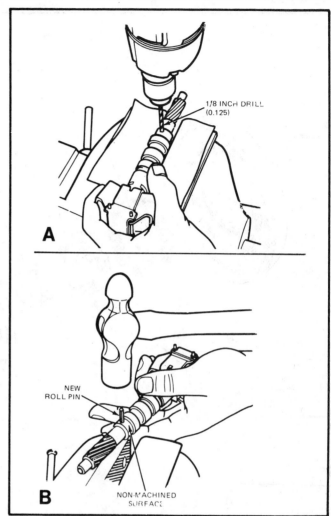

A—Drilling hole through the new governor driven gear. B—Installation of new roll pin (© Ford Motor Co.)

CAUTION

Do Not re-use the old roll pin.

CONVERTER ASSEMBLY

Reactor One-Way Clutch Check

NOTE: Special Tools are needed for this check.

1. Align the slot in the thrust washer with the slot in the holding lug.

NOTE: To align the slots, use tool, Ford number T81P-7902-B or its equivalent to turn the reactor.

2. Position the holding wire, Ford number T81P-7902-A or its equivalent, in the holding lug.

3. While holding the wire in position in the lug, install the one-way clutch torquing tool, Ford number T81P-7902-B or its equivalent, in the reactor spline.

4. Continue holding the wire and turn the torquing tool counterclockwise with a torque wrench. If the torquing tool begins to

Insert holding wire with tab and slot aligned in the converter, in order to lock the reactor (© Ford Motor Co.)

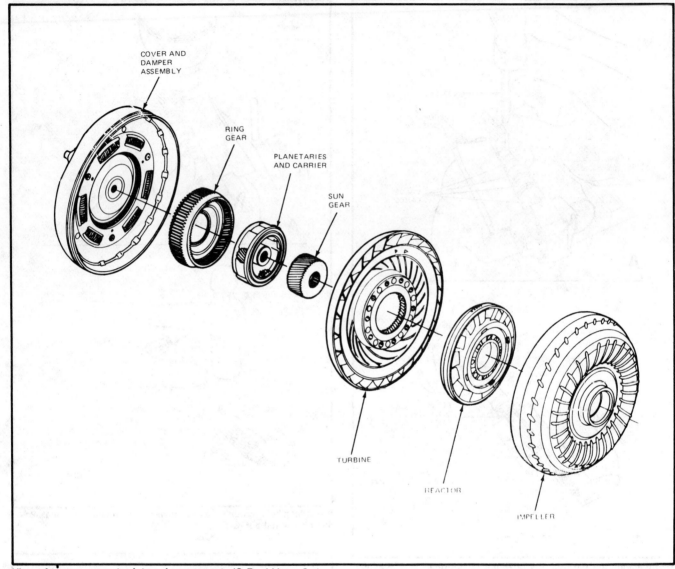

View of torque converter internal components (© Ford Motor Co.)

Install one way clutch torquing tool into the converter hub (© Ford Motor Co.)

With special end play checking tool in place, mount the dial indicator. Zero it and obtain the end play reading (© Ford Motor Co.)

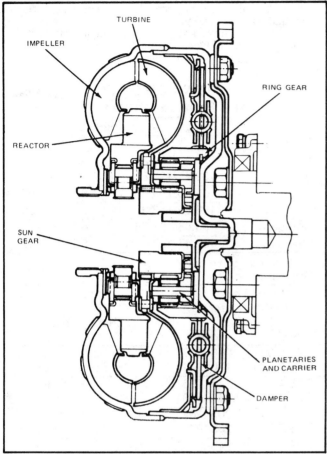

IMPELLER

TURBINE

REACTOR

RING GEAR

SUN GEAR

PLANETARIES AND CARRIER

DAMPER

Sectional view of torque converter (© Ford Motor Co.)

turn before the torque wrench reaches 10 ft. lbs. (13.5 N•m), replace the converter.

5. Remove the special tools.

Checking End Play of Converter

1. Insert the end play checking tool, T81P-7902-D or its equivalent, into the converter hub. Tighten the nut enough to permit lifting the converter by the handles of the tool.

2. Mount a dial indicator on the special mounting block and zero the dial indicator, with the stylus on the outer portion of the converter.

3. Lift the converter and observe the dial indicator reading. The reading should not exceed 0.023 inch (0.58 mm). If the reading is exceeded, replace the converter.

──────── CAUTION ────────

In cases where the end play is zero, the internal rotational friction must not exceed 5 ft. lbs. (6.7 N•m). Check by using torquing tool T81P-7902-B or its equivalent, but without using the holding wire.

TRANSAXLE ASSEMBLY

1. Place the transaxle case in the holding tool and position the case with the converter housing upright.

2. Install a new "O" ring on the idler gear shaft and install the shaft into the case. Position the idler gear on the shaft from the inside of the case.

Holding the idler gear shaft with an Allen wrench (© Ford Motor Co.)

3. Install an Allen wrench in the idler gear shaft and allow it to turn until the wrench catches on the band anchor strut. Install thread locking sealant to the attaching nut and tighten to 110-130 ft. lbs. (149-176 N•m) for the 1981-82 models, and 80-100 ft. lbs. (108-136 N•m) for 1983 and later models. Remove the Allen wrench.

4. Turn the transaxle case over and lock into position. Install the number 1 thrust bearing on the support assembly.

5. Install the input gear caged bearing on the support assembly. Install the input gear. Install the number 3 needle thrust bearing on the input gear.

6. Position the transfer housing in the case, making sure it is firmly seated on the alignment dowels. Install the transfer housing bolts and tighten to 15-19 ft. lbs. (20-26 N•m) for 1981-82 models and 18-23 ft. lbs. (24-32 N•m) for 1983 and later models.

──────── CAUTION ────────

Before installing the transfer housing, be sure the band strut is rotated to its operating position.

NOTE: The transaxle case and the housing are matched parts. If one is damaged, both must be replaced.

7. Install the number 4 thrust washer on the transfer housing.

8. Install the sun gear and the drum assembly. Install the intermediate band, making sure the band lug engages the strut.

Installing the drum and sun gear into position, within the band (© Ford Motor Co.)

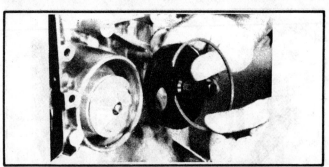

Install servo piston selection disc tool in place and retain with snap ring (© Ford Motor Co.)

Taking measurement with dial indicator and torque wrench (© Ford Motor Co.)

9. The servo travel check must be made at this time, if any of the following components have been changed during the overhaul.

NOTE: It is a good practice to do the servo travel check when ever the unit is disassembled and assembled.

    a. Transaxle case
    b. Band assembly
    c. Drum and sun gear assembly
    d. Servo piston
    e. Servo piston rod
    f. Band anchor strut

Special tools are required to perform the servo travel check. Ford Tool numbers are as follows;
    a. Return spring—T81P-70027-A
    b. Servo piston selector tool—T81P-70023-A
    c. Dial indicator—TOOL-4201-C

These tools or their equivalents must be used to determine the length of the servo piston rod to be used.

Procedure

    a. If necessary, clean and assemble the servo piston. Do not install the piston seals. This check is performed without the seals on the piston.
    b. Install the return spring, Ford number T81P-70027-A or its equivalent, on the servo rod and position the rod in the servo bore of the case.
    c. Install the servo piston selector tool, Ford number T81P-70023-A or its equivalent, in the servo bore and secure it with the servo cover snapring.
    d. Tighten the gauge disc screw to 10 ft. lbs. (13.5 N•m).
    e. Mount the dial indicator, Ford number TOOL-4201-C or its equivalent, to the transaxle case and position the indicator stylus through the hole in the gauge disc. Be sure the stylus has contacted the servo piston and zero the dial indicator.

    f. Back off the gauge disc screw until the piston movement stops and read the dial indicator total piston movement. The amount of piston travel as shown on the dial indicator will determine the servo rod length to be installed.
    g. Select a piston rod from the accompanying chart.

PISTON ROD SIZES
With Paint Identification

I.D. Color①	Rod Length②	
	MM	Inch
Yellow	160.52-160.22	6.319/6.307
Green	159.91-159.61	6.295/6.283
Red	159.30-159.00	6.271/6.259
Black	158.69-158.39	6.247/6.235
Orange	158.08-157.78	6.223/6.211
Blue	157.47-157.17	6.199/6.187

① Daub of paint on tip or rod.
② From far end of snap ring groove to end of rod.

For This Dial Indicator Reading		Install A New Piston Rod That Is . . .
MM	Inch	
2.45-2.98	.096-.117	5th Size Shorter
3.05-3.58	.120-.141	4th Size Shorter
3.65-4.18	.144-.165	3rd Size Shorter
4.25-4.78	.167-.188	2nd Size Shorter
5.12-5.14	.202-.202	1 Size Shorter
5.15-6.28	.203-.247	NO CHANGE
6.29-6.31	.248-.248	1 Size Longer
6.65-7.18	.262-.283	2nd Size Longer
7.25-7.78	.285-.306	3rd Size Longer
7.85-8.38	.309-.330	4th Size Longer
8.45-8.98	.333-.353	5th Size Longer

NOTE: For readings not on Table, select nearest one and repeat measurement with the new rod.

PISTON ROD SIZES
With Groove Identification

I.D	Rod Length①	
	mm	Inch
0 Groove	160.22-160.52	6.313-6.324
1 Grooves	159.61-159.90	6.289-6.300
2 Grooves	159.00-159.30	6.265-6.276
3 Grooves	158.39-158.69	6.240-6.252
4 Grooves	157.78-158.08	6.216-6.189
5 Grooves	157.17-157.47	6.197-6.209

① From far end of snap ring groove to end of rod.

If the dial indicator reads:

Less than 5.15mm (used) Less than 5.15mm (new) (.203 in.) The piston rod is too long. A shorter rod (more grooves) will have to be installed.	More than 2.04mm (used) More than 6.28mm (new) (.247 in.) The piston rod is too short. A longer rod (less grooves) will have to be installed.	5.15-2.04mm (used) 5.15-6.28mm (new) (.203-.247 in.) The piston rod is the correct length and no change is required.

10. Install the piston seals on the piston and install it into its bore in the case. Using the servo piston remover/installing tool, Ford Number T81P-70027-A or its equivalent, compress the piston spring far enough to allow installation of the retaining ring. Remove the tool.

NOTE: Before removing the compressing tool, be sure the piston rod has engaged to band lug.

11. Position the reverse clutch cylinder in the case, tapping it in place with a hammer handle.

Measuring reverse clutch pack clearance (© Ford Motor Co.)

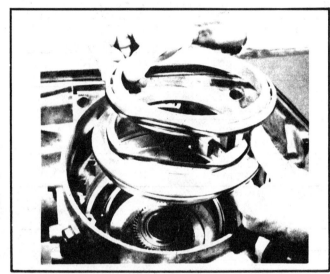

Preparing seals on the reverse clutch piston and cylinder (© Ford Motor Co.)

12. Install the reverse clutch piston in the clutch cylinder, using the seal protector, Ford Number T81P-70402-A or its equivalent. Apply even pressure when installing the piston in the drum.

13. Install the number 5 thrust washer on the planetary assembly, holding it in place with petroleum jelly. Install the planetary on the sun gear in the case.

14. Install the reverse clutch return spring and holder assembly.

15. To check the reverse clutch pack clearance, use the following procedure.

a. This check has to be made without the piston return spring and holder assembly in position and with the clutch pack installed. After the clutch pack clearance is made, the clutch pack is removed and the spring and retainer holder assembly installed.

b. Install the clutch pack wave spring, the clutch pack, starting with a steel plate, fiber plate, steel plate, fiber plate, pressure plate and the selective sized snapring.

c. Using a feeler gauge, measure the clearance between the snapring and the pressure plate in two places, 180° apart while holding a downward pressure of 10 pounds (40 N) on the clutch pack.

d. If the average reading is 0.018-0.039 inch (0.46-1.00 mm) for the 1981-82 models, and 0.030-0.053 inch (0.76-1.35 mm) for the 1983 and later models, the clutch pack clearance is within specifications.

e. If the reading is below 0.018 inch (0.46 mm) on the 1981-82 models, and below 0.030 (0.76 mm) for the 1983 and later models, install a thinner selective snapring and re-check the clearance.

f. If the measurement is more than 0.039 inch (1.00 mm) for the 1981-82 models, and more than 0.053 inch (1.35 mm) on the 1983 and later models, install a thicker selective snapring and re-check the clutch pack clearance.

g. Remove the retaining snapring and clutch pack assembly. Install the reverse clutch return spring and holder assembly, the clutch pack assembly and the retaining snapring.

16. Install the number 7 thrust needle bearing on the planetary gear assembly.

17. Install the intermediate clutch hub and ring gear assembly.

18. Install the direct clutch assembly and install the number 10 thrust needle bearing on the direct clutch housing.

19. Install the intermediate clutch with a rotating movement side to side, to engage the clutch plates with the hub.

REVERSE CLUTCH
Selective Snap Rings

Thickness	Part No.
1.89-1.94 mm (.074-.076 inch)	N800654
2.33-2.43 mm (.092-.096 inch)	N800655
2.77-2.87 mm (.109-.113 inch)	N800656
3.21-3.31 mm (.126-.130 inch)	N800657
For This Clearance Reading	**Install a New Ring That Is . . .**
0.12-0.46 mm (.005-.018 inch)	One Size THINNER
1.02-1.32 mm (.040-.052 inch)	One Size THICKER
1.45-1.75 mm (.057-.069 inch)	2nd Size THICKER
1.88-2.26 mm (.074-.089 inch)	3rd Size THICKER

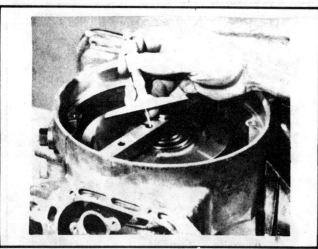

Placement of measuring bar and depth micrometer for end play check (© Ford Motor Co.)

Measurement of end play in two places, necessary to obtain an average reading (© Ford Motor Co.)

20. To check the clutch for proper engagement, use the number 11 thrust bearing. Position the thrust bearing on one of the machined tabs and push it up against the case. If the thrust bearing is flush with or slightly below the machined pump housing surface, the clutch is fully engaged.

21. Install the number 11 thrust needle bearing. Install alignment pins, Ford number T80L-77100-A or their equivalent, into the transaxle case oil pump bolt holes, 180° apart. Install the pump housing gasket on the case.

22. Check the transaxle end play with the appropriate special tools, Ford numbers T81P-77389-A (end play alignment cup), T80L-77003-A (gauge bar) or their equivalents.

NOTE: The end play setting is critical. It must be properly checked to assure success in the rebuilding process.

a. Position the assembled tools in the intermediate clutch center bore. Be sure the gauge bar rests on the pump to case gasket.

b. Position a depth micrometer (0-1 inch) on the gauge bar and through the inner hole in the bar, to measure against the surface of the thrust washer bearing.

c. Seat the micrometer and measure in two places, 180° apart, and average the readings. Locate the average reading in the accompanying chart and select the number 12 thrust washer from the selection.

NOTE: If needed, refer to the accompanying chart for thicknesses of the number 12 thrust washer.

d. Remove the tools from the transaxle assembly.

END PLAY ADJUSTMENT MEASUREMENTS

For This Reading	Use This Washer Part ID
.779-.796 inch (19.78-20.22 mm)	AA
.789-.804 inch (20.04-20.42 mm)	BA
.797-.812 inch (20.24-20.62 mm)	CA
.807-.825 inch (20.50-20.95 mm)	CA

WASHER THICKNESS

Inch	MM	ID
.055-.057	1.40-1.45	AA
.063-.065	1.60-1.65	BA
.071-.073	1.80-1.85	CA
.081-.083	2.05-2.10	DA

23. Install the correct number 12 thrust washer on the pump body.

NOTE: The use of petroleum jelly will hold the thrust washer to the pump body during its installation.

24. Position the pump body in the case, using the guide pins, and tap into position. Start the retaining bolts and remove the two guide pins. Torque the bolts to 7-9 ft. lbs. (9-12 N•m).

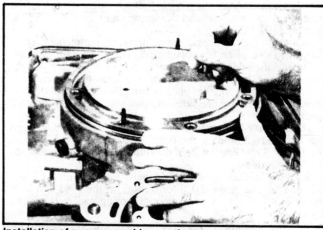

Installation of pump assembly over the two guide pins (© Ford Motor Co.)

NOTE: The washers on the pump retaining bolts provide the bolt seal and must not be substituted. Failure to use the sealing washers may result in a transaxle fluid leak.

25. Install the differential assembly by sliding the unit in and then down.

26. Differential Bearing End Play Check—The differential bearing end play is set during the manufacture and does not need to be checked or adjusted unless one or more of the following components have been replaced: transaxle case, differential case, differential bearings and differential bearing retainer.

Tapered roller differential bearings were used on early transaxle models and later changed to ball bearings. Both types could be encountered when overhauling numerous ATX transaxles. The end play adjustment is basically the same for both type bearings, but with different measurement specifications.

a. Remove the differential seal and bearing retainer "O" ring from the differential bearing retainer.

b. Position the differential assembly in place and install the differential retainer in the transaxle case.

c. Install the shim selector spacer tool, Ford number T81P-4451-A or its equivalent, in units equipped with tapered bearings and shim selector spacer tool, Ford number T83P-4451-BA or its equivalent, in units equipped with ball bearings. Place the shim selector spacer tool in the center of the differential seal bore of the differential retainer.

NOTE: The measurement of the T83P-4451-BA spacer or its equivalent must be 0.053 inch (1.35 mm) thick.

d. Position the gauge bar, Ford number T81P-4451-A or its equivalent, accross the differential bearing retainer and snugly hand tighten two attaching bolts to hold the gauge bar in place.

e. Tighten the center screw on the gauge bar and rotate the differential assembly several times to seat the bearings. Torque the center screw to 10 *inch pounds,* rotate the differential assembly and recheck the center screw torque.

f. With the use of feeler gauge blades, measure the clearance between the bearing retainer and the case.

NOTE: Be sure there are no burrs on the case mounting surfaces to hinder the blade measurement.

g. Obtain measurements from three positions around the bearing retainer to case gap. Take an average of the three clearance readings.

Example:

1st Reading	0.075 inch (1.91 mm)
2nd Reading	0.074 inch (1.88 mm)
3rd Reading	0.076 inch (1.91 mm)
Average of the three readings	0.075 inch (1.91 mm)

h. Tapered roller bearing type units—To determine the shim thickness needed, subtract the standard "Interference Factor" of 0.048 inch (1.22 mm), which includes 0.018 inch (0.46 mm) preload, plus the 0.030 inch (0.76 mm) tapered shim.

Example:

Average Reading	0.075 inch (1.91 mm)
Interference factor	0.048 inch (1.22 mm)
Shim size needed	0.027 inch (0.69 mm)

Select the proper shim from the accompanying chart. For odd numbered shim requirements, use the next *larger* shim.

i. Ball bearing type units—To determine the shim thickness needed, subtract the average reading from the thickness of the shim spacer tool, which is constant.

Example:

Constant reading	0.053 inch (1.35 mm)
Average reading	0.025 inch (0.63 mm)
Shim required	0.028 inch (0.72 mm)

Select the proper shim from the accompanying chart. For odd numbered shim requirements, use the next *smaller* shim.

Use of special tools to measure differential bearing end play
(© Ford Motor Co.)

DIFFERENTIAL BEARING END PLAY SHIMS

Part No.	Inch	MM
E1FZ-4067-A	0.012	0.30
B	0.014	0.35
C	0.016	0.40
D	0.018	0.45
E	0.020	0.50
F	0.022	0.55
G	0.024	0.60
H	0.026	0.65
J	0.028	0.70
K	0.030	0.75
L	0.032	0.80
M	0.033	0.85
N	0.035	0.90
P	0.037	0.95
R	0.039	1.00
S	0.041	1.05
T	0.043	1.10
U	0.045	1.15
V	0.047	1.20
W	0.049	1.25
X	0.051	1.30

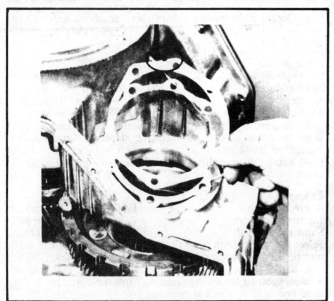

Position notch in proper position when using the copper colored tapered shim (© Ford Motor Co.)

Installation of plastic plug in the differential opening (© Ford Motor Co.)

j. Remove the special tools and differential bearing retainer from the transaxle. Install the oil seal and the "O" ring on the bearing retainer for installation.

27. Position the differential shims, as required by model and install a new seal on the differential retainer and position the retainer in the case. It will be necessary to tap the retainer into the case to properly seat it.

28. Apply a hardening sealer to the threads of the retaining bolts and install. Torque to 15-19 ft. lbs. (20-26 N•m).

29. Install a new seal on the filter and install to the case. Install the retaining bolts and torque to 7-9 ft. lbs. (9-12 N•m).

30. Install the oil pan with a new gasket, torquing the pan bolts to 15-19 ft. lbs. (20-26 N•m). Rotate the transaxle housing assembly and install new differential end seals and the converter hub seal, using seal installer tools. Install the plastic plug in the differential opening.

Installation of governor assembly (© Ford Motor Co.)

31. Install a new seal on the speedometer gear retainer and install it in the case. Tap into position and install the retaining pin.

NOTE: Be sure the flat of the speedometer gear retainer is properly positioned to the case before installing the pin.

32. Install the governor in its bore of the case. Install a new governor cover seal and install the cover. Be sure the cover is properly seated and install the retaining wire clip.

33. Rotate the gear box as required and install the governor filter in the valve body channel of the case.

34. Using alignment pins, install the valve body gasket. Install the valve body. Connect the throttle valve control spring to the separator late and install the Z-link in the manual valve while positioning the valve body on the case. Be sure the roller on the end of the throttle valve plunger has engaged the cam on the end of the throttle lever shaft.

NOTE: One alignment pin has to be removed to allow the Z-link to be installed in the manual valve. When installed, replace the alignment pin.

35. Install the detent spring and roller assembly to the valve body. Install the main oil pressure regulator and transmission control baffle plates.

NOTE: The main oil pressure regulator plate uses longer bolts.

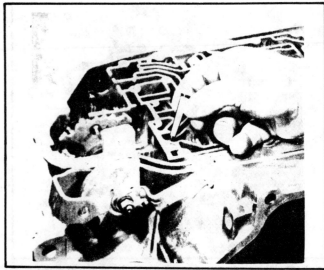

Installation of governor filter (© Ford Motor Co.)

36. Tighten the valve body attaching bolts in a specific sequence and to specifications of 72-96 *inch pounds* (8-11 N•m).

37. Install the throttle lever return spring to the spring anchor on the throttle lever.

NOTE: Be sure of installation of the spring as this spring applies T.V. if the T.V. linkage should disconnect or break.

38. Position the manual lever in the Neutral position and hold it there. Insert a $^3/_{32}$ inch drill or pin through the aligning hole in the switch. Move the neutral switch until the drill or pin seats in the case. Torque the two retaining bolts to 7-9 ft. lbs. (9-12 N•m).

39. Using guide pins, install a new valve body cover gasket on the transaxle case. Install the valve body cover and install the retaining bolts. Torque the bolts to 7-9 ft. lbs. (9-12 N•m).

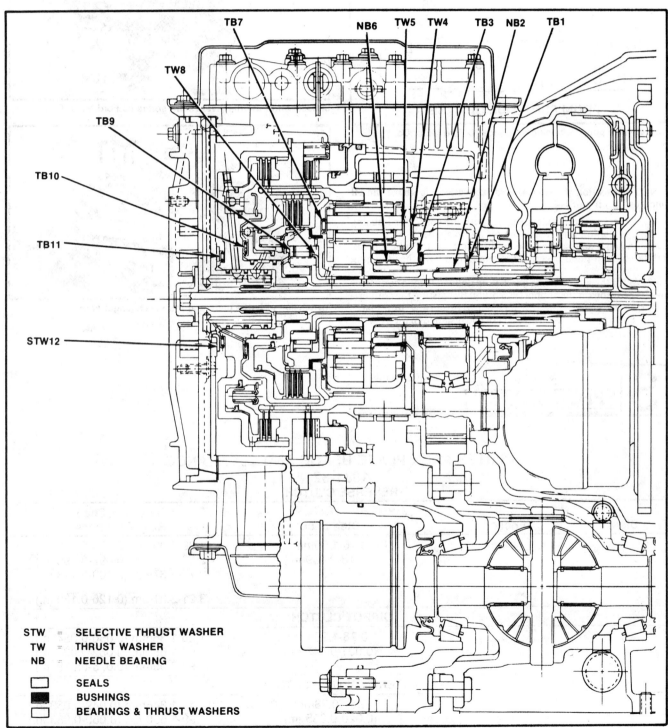

STW	=	SELECTIVE THRUST WASHER
TW	=	THRUST WASHER
NB	=	NEEDLE BEARING

☐ SEALS
■ BUSHINGS
☐ BEARINGS & THRUST WASHERS

Location of thrust washers, bushings and needle bearings in the ATX transaxle (© Ford Motor Co.)

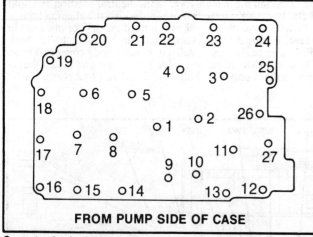

FROM PUMP SIDE OF CASE

Correct valve body tightening sequence (© Ford Motor Co.)

Installation of throttle lever return spring (© Ford Motor Co.)

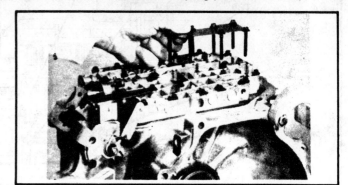

Installation of baffle plate and the seven longer screws
(© Ford Motor Co.)

NOTE: Position the transaxle identification tag above the oil pump on an oil pan bolt.

40. Rotate the transaxle assembly with the converter housing in the up position and install the oil pump shaft.

41. Install a new seal on the fluid level indicator tube and install the tube in the transaxle case, retaining it with the attaching bolt.

42. With the aid of lifting handles, install the torque converter into its position on the support assembly. By twisting the converter during the installation, it will be felt to drop twice into place.

43. The transaxle assembly can now be installed in the vehicle.

S SPECIFICATIONS

CLUTCH PACK PLATE USAGE AND CLEARANCE
1981-82
REVERSE CLUTCH

Steel	Friction	Clearance	Selective Snap Ring Thickness
2	2	0.46-1.00mm (0.018-0.039 in.)	1.89-1.99mm (0.074-0.078 in.)
			2.33-2.43mm (0.092-0.096 in.)
			2.77-2.87mm (0.109-0.113 in.)
			3.21-3.31mm (0.126-0.130 in.)
DIRECT CLUTCH			
3	3	0.78-1.20mm (0.031-0.047 in.)	1.26-1.36mm (0.050-0.054 in.)
			1.58-1.68mm (0.062-0.066 in.)
			1.90-2.00mm (0.075-0.079 in.)
INTERMEDIATE CLUTCH			
2	2	0.51-0.88mm (0.020-0.035 in.)	1.245-1.346mm (0.049-0.053 in.)
			1.499-1.600mm (0.060-0.063 in.)
			1.778-1.880mm (0.070-0.074 in.)

CLUTCH PACK PLATE USAGE AND CLEARANCE
1983

Steel	Friction	Clearance	Selective Snap Ring Thickness
REVERSE CLUTCH			
2 or 3	2 or 3	0.76-1.40mm (0.030-0.055 in.)	1.89-1.99mm (0.074-0.078 in.) 2.33-2.43mm (0.092-0.096 in.) 2.77-2.87mm (0.109-0.113 in.) 3.21-3.31mm (0.126-0.130 in.)
DIRECT CLUTCH			
3	3	0.78-1.20mm (0.031-0.047 in.)	1.26-1.36mm (0.050-0.054 in.) 1.58-1.68mm (0.062-0.066 in.) 1.90-2.00mm (0.075-0.079 in.)
4	4	1.01-1.43mm (0040-0.056 in.)	
INTERMEDIATE CLUTCH			
2	2	0.56-0.88mm (0.022-0.035 in.)	1.24-1.34mm (0.049-0.053 in.) 1.51-1.61mm (0.060-0.064 in.) 1.78-1.88mm (0.071-0.075 in.)
3	3	0.75-1.12mm (0.030-0.044 in.)	

1984

Steel	Friction	Clearance	Selective Snap Ring Thickness
REVERSE CLUTCH			
3*	3*	0.76-1.40mm (0.030-0.055 in.)	1.89-1.99mm (0.074-0.078 in.) 2.33-2.43mm (0.092-0.096 in.) 2.77-2.87mm (0.109-0.113 in.) 3.21-3.31mm (0.126-0.130 in.)

*With Cushion Spring

Steel	Friction	Clearance	Selective Snap Ring Thickness
DIRECT CLUTCH			
4	4	1.01-1.43mm (0.040-0.056 in.)	1.26-1.36mm (0.050-0.054 in.) 1.58-1.68mm (0.062-0.066 in.) 1.90-2.00mm (0.075-0.079 in.)
INTERMEDIATE CLUTCH			
3	3	0.75-11.12mm (0.030-0.044 in.)	1.24-1.34mm (0.049-0.053 in.) 1.51-1.61mm (0.060-0.064 in.) 1.78-1.88mm (0.071-0.075 in.)

SERVO PISTON TRAVEL

Year	Acceptable Travel[1]	Available Rod Lengths[2]	Identification
1981-82	5.5-6.28mm (0.203-0.247)	157.4-157.7mm (6.197-6.209 in.)	0 Groove
		156.8-157.1mm (6.173-6.185 in.)	1 Groove
		156.2-156.5mm (6.161-6.173 in.)	2 Groove
		155.6-155.9mm (6.125-6.138 in.)	3 Groove
		155.0-155.3mm (6.102-6.144 in.)	4 Groove
		154.4-154.7mm (6.079-6.091 in.)	5 Groove

SERVO PISTON TRAVEL

	Acceptable Travel①	Available Rod Lengths②	Identification
1983-84	5.15-7.04mm (0.203-0.277 inch)	160.22-160.52mm (6.313-6.324 inch)	0 Groove
		159.61-159.90mm (6.289-6.300 inch)	1 Groove
		159.00-159.30mm (6.265-6.276 inch)	2 Groove
		158.39-158.69mm (6.240-6.252 inch)	3 Groove
		157.78-158.08mm (6.216-6.189 inch)	4 Groove
		157.17-157.47mm (6.197-6.209 inch)	5 Groove

① Rod Stroke—not piston stroke
② Measured from far end of snap ring groove to end of rod

TRANSAXLE END PLAY
1981-84

Measured Depth	Thrust Washer Required	Identification Code
2.00-1.77mm (0.079-0.070 inch)	1.40-1.45mm (0.055-0.057 inch)	AA
2.20-2.00mm (0.087-0.079 inch)	1.60-1.65mm (0.063-0.065 inch)	BA
2.41-2.20mm (0.095-0.087 inch)	1.80-1.85mm (0.071-0.073 inch)	CA
1.77-1.46mm (0.070-0.057 inch)	1.15-1.20mm (0.045-0.047 inch)	EA

SHIFT POINTS
1983

		PMA-K3		PMA-P		PMA-R1	
		Base Engine		EFI Engine		HO Engine	
Drive Range		km/h	mph	km/h	mph	km/h	mph
Idle:	1-2	16-30	10-19	16-30	10-19	16-30	10-19
	2-3	24-50	15-31	24-50	15-31	24-50	15-31
	3-2	22-40	14-25	22-40	14-25	22-40	14-25
	2-1	16-24	10-15	16-24	10-15	16-24	10-15
Part Throttle:	1-2	18-27	11-17	18-27	11-17	18-27	11-17
	2-3	43-72	27-45	38-67	24-42	43-72	27-45
	3-2	38-67	24-42	32-61	20-38	38-67	24-42
WOT:	1-2	46-72	29-45	54-82	34-51	54-82	34-51
	2-3	94-118	59-74	98-125	61-78	98-125	61-78
	3-2	83-109	52-68	88-117	55-73	88-117	55-73
	2-1	35-61	22-38	45-70	28-44	45-70	28-44
Manual Low:	2-1	32-61	20-38	32-61	20-38	32-61	20-38

Axle Ratio: 3.3:1
Tire Size: P165/70R13, P165/80R13

SHIFT POINTS
1981-82

Drive Range		MPH
Idle:	1-2	10-17
	2-3	16-31
	3-2	14-22
	2-1	9-14
Part Throttle:	1-2	17-32
	2-3	29-46
	3-2	24-42
	2-1	13-20
WOT:	1-2	29-46
	2-3	58-75
	3-2	51-67
	2-1	19-35
Manual Low:	2-1	21-38

SHIFT POINTS
1984

		PMA-N	
		2.3L HSC Engine	
Drive Range		km/h	mph
Idle:	1-2	17-29	11-18
	2-3	25-50	16-31
	3-2	26-40	16-25
	2-1	15-23	9-14
Part Throttle:	1-2	19-44	12-27
	2-3	42-70	26-43
	3-2	36-45	22-41
WOT:	1-2	38-66	23-41
	2-3	84-112	52-69
	3-2	76-102	49-64
	2-1	24-51	15-32
Manual Low:	2-1	32-61	20-38

Axle Ratio: 3.3:1
Tire Size: P165/70R13, P165/80R13

LINE PRESSURE
1981-82

Range	Pressure (At Idle)		Pressure (WOT Stall)	
	kPa	PSI	kPa	PSI
D-2-1	296-400	43-58	724-875	105-127
R	483-724	70-105	1585-1965	230-285
P-N	296-400	43-58	NA	

NOTE: Governor Pressure is at zero (vehicle stationary). Transaxle is at operating temperature.

LINE PRESSURE
1983

	Pressure (At Idle)		Pressure (WOT Stall)			
	All Engines		Base		HO and EFI	
Range	kPa	PSI	kPa	PSI	kPa	PSI
D-2-1	377-455	54-66	669-779	97-113	731-841	106-122
R	531-765	77-111	1434-1793	208-260	1600-1958	232-284
P-N	377-455	54-66	669-779	97-113	731-841	106-122

NOTE: Governor Pressure is at zero (vehicle stationary). Transaxle is at operating temperature.

LINE PRESSURE
1984

Transaxle Model	Range	Pressure (At Idle)		Pressure (WOT Stall)	
		kPa	PSI	kPa	PSI
PMA-N	D-2-1-P-N	338-420	49-61	655-765	95-111
	R	455-689	66-100	1413-1772	205-257
PMA-U, V	D-2-1-P-N	—	54-66	—	106-122
PMB-C, D	R	—	77-111	—	232-284

NOTE: Governor Pressure is at zero (vehicle stationary). Transaxle is at operating temperature.

TORQUE CHART

Description	N•m	ft. lbs.
Reactor Support to Case	8-11	6-8
Separator Plate to Valve Body	8-11	6-8
Filler Tube Bracket to Case	9-12	7-9
Filter to Case	9-12	7-9
Valve Body Cover to Case	9-12	7-9
Pump Support to Pump Body	8-11	6-8
Natural Safety Switch to Case	9-12	7-9
Pump Assembly to Case	9-12	7-9
Valve Body to Case	8-11	72-96 (in. lbs.)
Oil Pan to Case	20-26	15-19
Lower Ball Joint to Steering Knuckle	50-60	37-44
Transfer Housing to Case ①	24-32	18-23
Differential Retainer to Case (with Sealant)	20-26	15-19
Pressure Test Port Plugs to Case ②	5-11	4-8
Cooler Tube Fitting to Case	24-31	18-23
Outer Throttle Lever to Shaft Nut ③	10-13	7.5-9.5
Inner Manual Lever to Shaft Nut	43-65	32-48
Idler Shaft Attaching ④	108-136	80-100
Converter Drain Plug	10-16	8-12
Valve Body Retaining Bolts	8-11	6-8
T.V. Adjuster Locknut	2.7-4.1	24-36 (in. lbs.)

① 1981-82—20-26 N•m, 15-19 ft. lbs.
② 1981-82—9-15 N•m, 7-11 ft. lbs.
③ 1981-82—16-20 N•m, 12-15 ft. lbs.
④ 1981-82—149-176 N•m, 110-130 ft. lbs.

PISTON ROD SIZES With Paint Identification

I.D. Color①	Rod Length②	
	MM	Inch
Yellow	160.52-160.22	6.319/6.307
Green	159.91-159.61	6.295/6.283
Red	159.30-159.00	6.271/6.259
Black	158.69-158.39	6.247/6.235
Orange	158.08-157.78	6.223/6.211
Blue	157.47-157.17	6.199/6.187

① Daub of paint on tip or rod.
② From far end of snap ring groove to end of rod.

For This Dial Indicator Reading		Install A New Piston Rod That Is . . .
MM	Inch	
2.45-2.98	.096-.117	5th Size Shorter
3.05-3.58	.120-.141	4th Size Shorter
3.65-4.18	.144-.165	3rd Size Shorter
4.25-4.78	.167-.188	2nd Size Shorter
5.12-5.14	.202-.202	1 Size Shorter
5.15-6.28	.203-.247	NO CHANGE
6.29-6.31	.248-.248	1 Size Longer
6.65-7.18	.262-.283	2nd Size Longer
7.25-7.78	.285-.306	3rd Size Longer
7.85-8.38	.309-.330	4th Size Longer
8.45-8.98	.333-.353	5th Size Longer

NOTE: For readings not on Table, select nearest one and repeat measurement with the new rod.

STALL SPEED SPECIFICATIONS

Year and Vehicle	Engine/Litre Displacement	Transmission Type	Converter Size (Inches)	ID	Stall Speed (RPM) Min.	Max.	K— Factor
1981-82							
Escort/Lynx	—	ATX	9.25	—	1850	2550	—
1983							
Escort/Lynx,	1.6L	ATX	9.25	B01	2440	2890	300
EXP/LN7	EFI 1.6L	ATX	9.25	D12	2750	3160	320
	HO 1.6L	ATX	9.25	D12	2610	3110	320
1984							
Tempo/Topaz	2.3L HSC	ATX	9.25	E05	2272	2664	320
Escort/Lynx, EXP	1.6L (EFI)	ATX	9.25	E04	2627	3095	—
	1.6L (HO)	ATX	9.25	E04	7655	3147	—

PISTON ROD SIZES
With Groove Identification

I.D	Rod Length[1]	
	mm	Inch
0 Groove	160.22-160.52	6.313-6.324
1 Grooves	159.61-159.90	6.289-6.300
2 Grooves	159.00-159.30	6.265-6.276
3 Grooves	158.39-158.69	6.240-6.252
4 Grooves	157.78-158.08	6.216-6.189
5 Grooves	157.17-157.47	6.197-6.209

[1] From far end of snap ring groove to end of rod.

If the dial indicator reads:

Less than 5.15mm (used) Less than 5.15mm (new) (.203 in.) The piston rod is too long. A shorter rod (more grooves) will have to be installed.	More than 2.04mm (used) More than 6.28mm (new) (.247 in.) The piston rod is too short. A longer rod (less grooves) will have to be installed.	5.15-2.04mm (used) 5.15-6.28mm (new) (.203-.247 in.) The piston rod is the correct length and no change is required.

DIFFERENTIAL BEARING END PLAY SHIMS

Part No.	Inch	MM
E1FZ-4067-A	0.012	0.30
B	0.014	0.35
C	0.016	0.40
D	0.018	0.45
E	0.020	0.50
F	0.022	0.55
G	0.024	0.60
H	0.026	0.65
J	0.028	0.70
K	0.030	0.75
L	0.032	0.80
M	0.033	0.85
N	0.035	0.90
P	0.037	0.95
R	0.039	1.00
S	0.041	1.05
T	0.043	1.10
U	0.045	1.15
V	0.047	1.20
W	0.049	1.25
X	0.051	1.30

REVERSE CLUTCH
Selective Snap Rings

Thickness	Part No.
1.89-1.94 mm (.074-.076 inch)	N800654
2.33-2.43 mm (.092-.096 inch)	N800655
2.77-2.87 mm (.109-.113 inch)	N800656
3.21-3.31 mm (.126-.130 inch)	N800657

For This Clearance Reading	Install a New Ring That Is . . .
0.12-0.46 mm (.005-.018 inch)	One Size THINNER
1.02-1.32 mm (.040-.052 inch)	One Size THICKER
1.45-1.75 mm (.057-.069 inch)	2nd Size THICKER
1.88-2.26 mm (.074-.089 inch)	3rd Size THICKER

END PLAY ADJUSTMENT MEASUREMENTS

For This Reading	Use This Washer Part ID
.779-.796 inch (19.78-20.22 mm)	AA
.789-.804 inch (20.04-20.42 mm)	BA
.797-.812 inch (20.24-20.62 mm)	CA
.807-.825 inch (20.50-20.95 mm)	CA

WASHER THICKNESS

Inch	MM	ID
.055-.057	1.40-1.45	AA
.063-.065	1.60-1.65	BA
.071-.073	1.80-1.85	CA
.081-.083	2.05-2.10	DA

1984 STALL SPEED SPECIFICATION
ATX Transaxle

Vehicle Application	Engine Disp.	Converter Size	Converter ID	Stall Speed Min.	Stall Speed Max.
Escort/Lynx/EXP/LN7	1.6L (EFI)	9¼"	E04	2627	3095
Tempo/Topaz	2.3L	9¼"	E05	2272	2664
Escort/Lynx/EXP/LN7	1.6L (HO)	9¼"	E04	2655	3147

1984 LINE PRESSURE SPECIFICATIONS
ATX Transaxle

Transmission Model	Range	Line Pressure at Idle	WOT Stall
PMA-U, V	D, 2, 1, P, N	54-66	106-122
PMB-C, D	R	77-111	232-284
PMA-N	D, 2, 1, P, N	49-61	95-111
	R	66-100	205-257

 SPECIAL TOOLS

TOOL NO.	TOOL NAME	APPLICATION
T81P-77000-AB	Plastic Storage Case	New special tool storage
T57L-500-B	Bench Mounting Fixture	Hold and turn transmission
T81P-7902-C	Converter Handles	Lift Converter
T81P-78103-A	Remover Adapter	Remove Pump
T81P-78103-B	Adapter Bolts	Remove Pump
T50T-100-A	Slide Hammer (large)	Various
—	12 mm Allen	Idler Gear
—	32 mm, 12-point Socket	Idler Gear
—	Breaker Bar	Idler Gear
TOOL-1175-AC	Seal Remover	Various
T81P-70027-A	Servo Installer Tool	Servo snap ring
T71P-19703-C	O-Ring Pick	Various
—	30 mm End Wrench	Park mechanism

TOOL NO.	TOOL NAME	APPLICATION
T80L-77100-A	Guide Pins (2)	Various
T65L-77515-A	Spring Compressor	Intermediate & Direct
T81P-70222-A	Compressor Adapter	Inter. Clutch snap ring
TOOL-7000-DE	Air Nozzle	Various
D81L-4201-A	Feeler Gauge, Metric	Various
T80L-77515-A	Adapter Extension	Direct Clutch snap ring
T81P-70235-A	Spring Compressor	Direct Clutch snap ring
T73P-77060-A	Snap Ring Plier	Various
T81P-7902-A	Holding Wire	Converter clutch
T81P-7902-B	1-Way Clutch Torquing Tool	Converter
D81L-600-B	Ft. Lb. Torque Wrench	Various

SPECIAL TOOLS

TOOL NO.	TOOL NAME	APPLICATION	TOOL NO.	TOOL NAME	APPLICATION
T81P-7902-D	End Play Checking Tool	Converter	T81P-79363-A3	Collar	Reactor Support
TOOL-4201-C	Dial Indicator	Converter; Servo	T81P-70363-A4	Adapter	Reactor Support
T77F-1102-A	Bearing Puller	Remove bearing	T81P-70363-A5	Screw	Reactor Support
T81P-77380-A	Housing Bearing Replacer	Transfer Bearing Housing	T81P-70363-A6	Guide Pins	Reactor Support
T77F-4220-B1	Differential Bearing Remover	Differential Bearing	T81P-70023-A	Selection Tool	Servo Rod Travel
T57L-4220-A	Bearing Cone Remover	Differential Bearing	T81P-70402-A	Seal Protector	Reverse Clutch Piston
T81P-4220-A	Step Plate	Differential Bearings	T80L-77003-A	Gauge Bar	End Play
T81P-4221-A	Bearing Installer	Differential Bearings	T81P-77389-A	Align. Cup	End Play
T81P-4451-A	Shim Selection Tool	Differential Preload	D80P-4201-A	0-1" Depth Microm.	End Play
D81L-600-D	In. Lb. Torque Wrench	Various	T81P-1177-A	Differential Seal Replacer	Differential Seals
—	3/8" Natl. Coarse Bottom Tap	Park Pawl, Band Strut	T81P-1177-B	Plastic Plug	Differential Seal
—	Bar Magnet	Pin removal	T81P-70401-A	Converter Hub Seal Replacer	Converter Hub Seal
T81P-70363-A	Spacer	Reactor Support	T81P-70337-A	TV Seal Replacer	Control Lever
T81P-70363-A1	Receiver	Reactor Support	—	3/32" Drill Bit	Neutral Adj.
T81P-70363-A2	Sleeve	Reactor Support	T81P-4026-A	Differential Rotator	Differential
			T80L-77030-B	Servo Piston Remover	Air Check Puck

INDEX

GENERAL MOTORS
LOCK-UP TORQUE CONVERTERS

GENERAL MOTORS TRANSMISSION IDENTIFICATION
"Automatic"

Parts Book Code	Service Identity	Manufacturer	Function	Physical Identification
M34	125	Hydra-matic	FWD	I.D. Plate on Rear Section of Case
MD9	125C	Hydra-matic	FWD-TCC	I.D. Plate on Rear Section of Case
MD2	180C	Strasbourg	RWD-TCC	I.D. Plate on Case
M29	200	Hydra-matic	RWD	I.D. Plate on Rear Section of Case
MV9	200C	Hydra-matic	RWD-TCC	I.D. Plate on Rear Section of Case
MW9	200-4R	Hydra-matic (3-Rivers)	RWD-OD-TCC	I.D. Plate on Rear Section of Case
M-31	250C	Chevrolet	RWD-TCC	I.D. Stamping on Gov. Cover
M-57	325-4L	Hydra-matic	RWD-OD-TCC	I.D. Plate on Trans. Case
M-33	350	Buick	RWD	I.D. Stamping on Gov. Cover
M-38	350	Chevrolet & Canada	RWD	I.D. Stamping on Right Side of Pan
M-40	400	Hydra-matic	RWD	I.D. Plate on Case
MV4	350C	Chevrolet	RWD-TCC	I.D. Stamping on Right Side of Pan
MX2	350C	Buick	RWD-TCC	I.D. Stamping on Gov. Cover or 1-Z Acc. Cover
MX3	350C	Buick	RWD-TCC	I.D. Stamping on Gov. Cover or 1-Z Acc. Cover
MX5	350C	Buick	TCC	Forward Clutch Apply Switch
MD-8	700-R4	Chevrolet	RWD-OD-TCC	Stamped on Case Boss
ME-9	440-T4	Hydra-matic	FWD-OD-TCC	Stamped or Inked on Case

FWD—Front Wheel Drive
OD—Overdrive
TCC—Torque Converter Clutch
RWD—Rear Wheel Drive
I.D.—Identification

General Information

The computer command control is a system that controls emissions by close regulation of the air-fuel ratio and by the use of a three-way catalytic converter which lowers the level of oxides of nitrogen, hydrocarbons and carbon monoxide.

The essential components are an exhaust gas oxygen sensor (OS), an electronic control module (ECM), an electronically controlled air-fuel ratio carburetor and a three-way catalytic converter (ORC).

To maintain good idle and driveability under all conditions, input signals are used to modify the computer output signal. These input signals are supplied by the engine temperature sensor, the vacuum control switch(es), the throttle position switch (TPS), the distributor (engine speed), the manifold absolute pressure sensor (MAP) and the barometer pressure sensor (BARO).

Why should this system affect the Automatic Transmission repairman? With the use of the Torque Converter Clutch (TCC) to provide a direct mechanical link-up between the engine and the drive wheels, the means of applying its engagement and disengagement modes must be controlled, to provide the optimum advantage in operation and fuel economy. This control of the Torque Converter Clutch operation was included in the overall control of the engine and its components by the Electronic Control Module.

To aid the repairman in understanding the system operation, as it applies to transmission/transaxle converter clutch application, a brief outline is given. Should the need arise to diagnose and repair the system, refer to the appropriate Chilton's Professional Automotive Repair Manual.

The Computer Command Control System was used on certain vehicles sold in California, beginning in 1981 and is used on all 1982 and later General Motors vehicles except diesel engines and certain throttle body injection type carbureted engines, although the Electronic Control System is similar in appearance and operation.

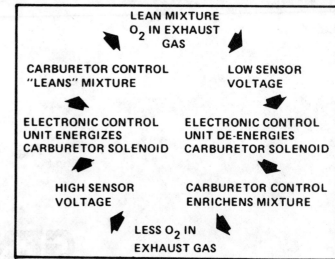

Computer Command Control (CCC) basic cycle of operation (©General Motors Corp.)

System Operation

There are two primary modes of operation for the Computer Command Control System:
1. Open Loop.
2. Closed Loop.

Open Loop Mode of Operation

In general terms, each system will be in the open loop mode of operation (or a variation of) whenever the engine operating conditions do not conform with the programmed criteria for closed loop operation, such as when the engine is first started through its reaching normal operating temperature mode.

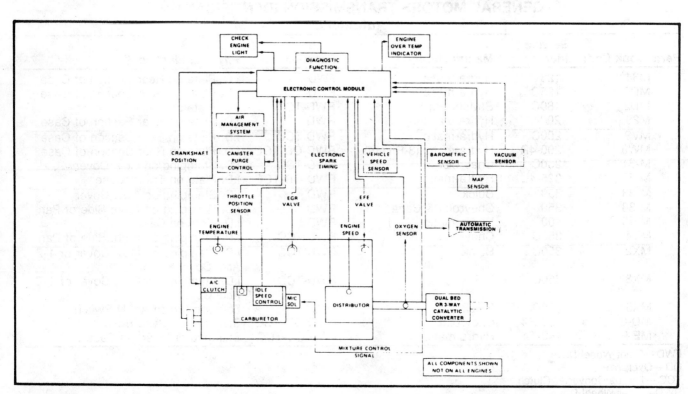

Computer Command Control system (©General Motors Corp.)

During open loop operation the air/fuel mixture is maintained at a programmed ratio that is dependent on the type of engine operation involved. The oxygen sensor data is not accepted by either system during this mode of operation. The following conditions involve open loop operation.
1. Engine Start-UP
2. Coolant or Air Temperature Too Low
3. Oxygen Sensor Temperature Too Low
4. Idle
5. Wide Open Throttle (WOT)
6. Battery Voltage Too Low

Closed Loop Mode of Operation

When all input data conforms with the programmed criteria for closed loop operation, the oxygen content output voltage from the oxygen sensor is accepted by the microprocessor. This results in an air/fuel mixture that will be optimum for the current engine operating condition and also will correct any pre-existing too lean or too rich mixture condition.

NOTE: A high oxygen content in the exhaust gas indicates a lean air/fuel mixture. A low oxygen content indicates a rich air/fuel mixture. The optimum air/fuel mixture ratio is 14.7:1.

System Components & Operation

ELECTRONIC CONTROL MODULE (ECM)

The electronic control module (ECM) monitors the voltage output of the oxygen sensor, along with information from other input signals, to generate a control signal to the carburetor solenoid. The control signal is continually cycling the solenoid between "ON" (lean command) and "OFF" (rich command). When the solenoid is on (energized), the solenoid pulls down a metering rod which reduces fuel flow. When the solenoid is off (deenergized), the spring-loaded metering rod returns to the up position and fuel flow increases. The amount of time on relative to time off is a function of the input voltage from the oxygen sensor.

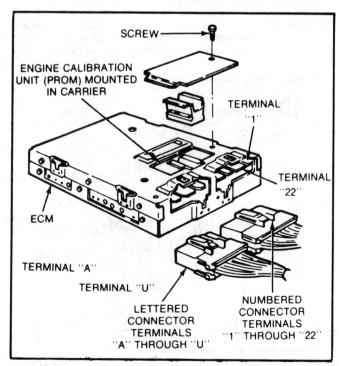

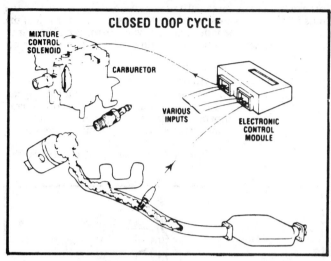

On 3.8 liter V-6 engines, the ECM also controls the electronic spark timing system (EST). On 5.7 liter V-8 engines, the ECM also controls the electronic module retard (EMR) system. The EMR module has the capability of retarding the engine timing 10 degrees during certain engine operations to reduce the exhaust emissions.

During other engine operations, the module functions the same as a standard HEI module. The terminal "R" on the module is connected to the ECM and the retard is accomplished by an internal ground. The timing is retarded 10 degrees only when the engine coolant temperature is between 19°C (66°F) and 64°C (147°F), with the throttle opening position below 45% and the engine speed above 400 rpm.

NOTE: The ECMs are not the same, with PROMs being separately programmed for a specific vehicle/engine combination. Do not attempt to interchange.

Tachometer Signal to Computer

The computer monitors the engine crankshaft position signal in order to determine engine rpm. This signal is generated as a pulse from the HEI distributor.

The tachometer signal comes from the tach terminal of the distributor. A tachometer signal filter is located between the distributor and the computer to reduce radio noise.

A tachometer cannot be connected in the line between the tach filter and the computer, or the computer may not receive a tach signal. The presence of a tach signal from the distributor can be determined by connecting a tachometer to the distributor tach terminal.

ENGINE COOLANT SENSOR

The coolant temperature sensor in the engine block sends the ECM information on engine temperature which can be used to vary the air-fuel ratio as the engine coolant temperature varies with time during a cold start. It also accomplishes various switching functions at different temperatures (EGR, EFF, etc.), provides a switch point for hot temperature light indication and varies spark advance.

The coolant temperature sensor has a connector which lets the ground return lead surround the signal lead. This design provides an interference shield to prevent high voltage in the area (such as spark plug leads) from affecting the sensor signal to the computer.

NOTE: The ground return wire goes to the computer which internally grounds the wire.

EXHAUST OXYGEN SENSOR

The oxygen sensor located in the exhaust manifold compares the oxygen content in the exhaust stream to the oxygen content in the outside air. This shows that there is a passage from the top of the oxygen sensor to the inner chamber which permits outside air to enter. When servicing the sensor, do not plug or restrict the air passage.

A rich exhaust stream is low in oxygen content and will cause the oxygen sensor to send a rich signal, approximatley one volt, to the computer. A lean exhaust stream will result in a lean signal, less than half a volt, from the oxygen sensor to the computer.

As the sensor temperature increases during engine warm-up, the sensor voltage also increases. Because the minimum voltage required to operate this circuit is half a volt, the computer will not use the oxygen sensor signal until the sensor has reached 600°F.

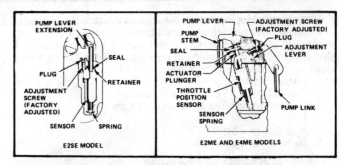

Throttle positioner sensor (TPS) (©General Motors Corp.)

MIXTURE CONTROL (M/C)

The mixture control solenoid actuates two spring-loaded rods, controlling fuel flow to the idle and main metering circuits of the carburetor. Energizing the solenoid lowers the metering rod into the main metering jet. This makes the air-fuel mixture in the Dualjet and Quadrajet carburetors leaner. The Varajet carburetor has a solenoid operated fuel control valve.

The mixture control solenoid changes the air-fuel ratio by allowing more or less fuel to flow through the carburetor. When no electrical signal is applied to the solenoid, maximum fuel flow to the idle and main metering circuits. When an electrical signal is applied to the solenoid, the mixture is leaned. (Leaning means reducing the amount of fuel mixed with the air.)

COMPUTER COMMAND CONTROL SYSTEM CARBURETORS

Three types of Rochester carburetors are used for system applications. The Varajet is a two barrel, staged opening carburetor. The Quadrajet is a four barrel staged opening carburetor. The Dualjet is a two barrel non-staged carburetor, essentially the primary side of a Quadrajet.

The metering rods and an idle bleed valve are connected to a 12 volt mixture control solenoid. The model E2SE carburetor, used with the computer command control system, is a controlled air-fuel ratio carburetor of a two barrel, two stage down-draft design with the primary bore smaller in size than the secondary bore. Air-fuel ratio control is accomplished with a solenoid controlled on/off fuel valve which supplements the preset flow of fuel

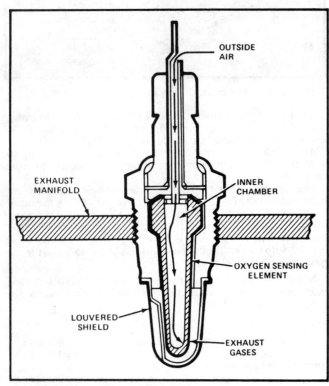

Typical oxygen sensor (©General Motors Corp.)

THROTTLE POSITION SENSOR (TPS)

This sensor is located in the carburetor body and is actuated by the accelerator pump lever. The stem of the sensor projects up through the air horn, contacting the underside of the lever. As the throttle valves are opened, the pump lever presses down proportionately on the sensor, thus indicating throttle position.

The throttle position sensor changes the voltage in circuit E (reference voltage) to G (voltage input to the computer) as the sensor shaft moves up or down. This is similar to the operation of the gas tank gauge sending unit, except that the throttle position sensor permits the computer to read throttle position.

BAROMETRIC PRESSURE SENSOR

The barometric pressure sensor provides a voltage to the computer to allow ambient pressure compensation of the controlled functions. This unit senses ambient barometric pressure and provides information to the computer on atmospheric pressure changes due to weather and/or altitude.

The computer uses this information to adjust the air-fuel ratio. The sensor is mounted under the instrument panel near the right-hand A/C outlet and is electronically connected to the computer. The atmospheric opening is covered by a foam filter.

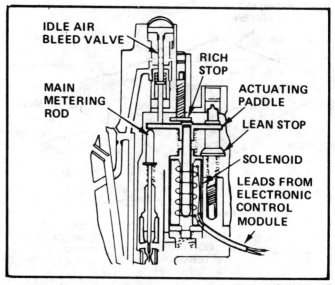

Typical mixture control solenoid used with E2ME and E4ME carburetors (©General Motors Corp.)

which supplies the idle and main metering systems. The solenoid on/off cycle is controlled by a 12 volt signal from the computer. The solenoid also controls the amount of air bled into the idle system. The air bleed valve and fuel control valve work together so that the fuel valve is closed when the air bleed valve is open, resulting in a leaner air-fuel mixture. Enrichment occurs when the fuel valve is open and air bleed valve closed.

The Quadrajet-Dualjet arrangement is such that the level of metering is dependent on the positioning of rods in the orifices. The Varajet system is different in that it features a non-moving-part main system for lean mixtures and a supplemental system to provide for rich mixture.

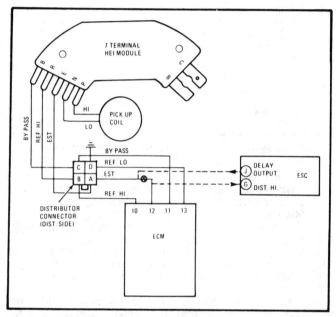

Electronic Spark Control schematic (©General Motors Corp.)

AIR FLOW CONTROL SYSTEMS

Two types of air systems are used on computer command control engines, the pulse air injection reactor (PAIR) and the belt-driven air pump (AIR). Both types are controlled by the computer through solenoid valves. The PAIR system uses an on/off solenoid which is open during cold operation and wide open throttle (WOT). Air is injected into the exhaust ports when the solenoid valves are open.

AIR MANAGEMENT SYSTEM

The computer controlled solenoid can divert air during any desired operating mode. The valves diverting and switching the air flow are the air diverter valve and the air select valve. With the air divert valve, a rapid increase of engine manifold vacuum diverts air to the air cleaner and high air system pressure is diverted to the air cleaner. The air select valve switches air between the catalytic converter and exhaust ports.

DISTRIBUTOR HEI MODULE

The computer will control the module above 200 rpm by applying a voltage to the by-pass line and signaling terminal E.

Current loss at terminals R or B will cause the distributor (HEI) module to take over. Loss of terminal E electronic spark timing will cause the engine to stop (assuming by-pass voltage is present).

If the engine is equipped with electronic spark control, the computer electronic spark timing line would go to the electronic spark control distributor high. The electronic spark control delay output would go to the HEI electronic spark timing input.

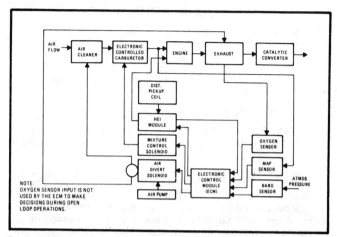

Typical CCC system functional block schematic, non turbo equipped (©General Motors Corp.)

ELECTRONIC SPARK TIMING (EST)

Electronic spark timing is a computer controlled system that has all the engine spark timing information stored in memory. At various engine operating conditions as determined by rpm and manifold pressure, the system determines (from a table) the proper spark advance. It then produces the firing signal at the desired crankshaft position. Other parameters, such as coolant temperature and barometric pressure, can be sensed and this information used to modify, as appropriate, the spark advance number from the table. The system provides a much more flexible and accurate spark timing control than the conventional centrifugal and vacuum advance mechanisms in the distributor.

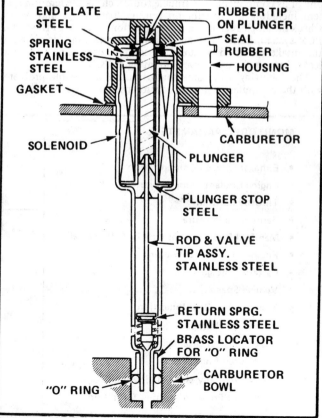

Typical mixture control solenoid used with E2SE carburetor (©General Motors Corp.)

VACUUM SENSORS

The vacuum sensors measure changes in manifold pressure and provide this information (in the form of an electrical signal) to the ECM. The pressure changes reflect need for adjustments in the air/fuel mixture, spark timing (EST) and other controlled operations to maintain good vehicle performance under various driving conditions.

VEHICLE SPEED SENSOR—VSS

The VSS is mounted behind the speedometer in the instrument cluster. It provides a series of pulses to the ECM which are used to determine vehicle speed.

IDLE SPEED CONTROL SYSTEM—ISC

An Idle Speed Control—ISC is used on some engines to control idle speed. ISC maintains low idle speed while preventing stalls due to engine load changes. A motor assembly mounted on the carburetor moves the throttle lever to open or close the throttle blades. The ECM monitors engine load to determine proper idle speed. To prevent stalling, the ECM monitors the air conditioning compressor switch, transmission, park/neutral switch and the ISC throttle switch. With this information the ECM will control the ISC motor and vary the engine idle as necessary.

Trouble Diagnosis

BUILT-IN DIAGNOSTIC SYSTEM

The computer command control system should be considered as a possible trouble source of engine performance, fuel economy and exhaust emissions complaints only after diagnostic checks, which apply to engines without the computer command control system, have been completed.

Before suspecting the computer command control system or any of its components as a trouble source, check the ignition system including the distributor, timing, spark plugs and wires. Check the air cleaner, evaporative emissions system, EFE system, PCV system, EGR valve and engine compression. Also inspect the intake manifold, vacuum hoses and hose connections for leaks. Inspect the carburetor mounting bolts.

The following symptoms could indicate a possible problem with the computer command control system.

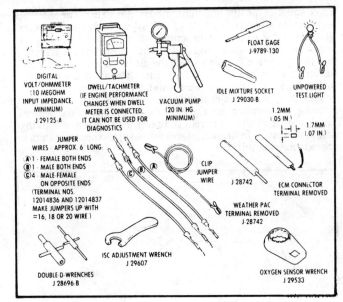

Typical CCC system diagnostic tools (©General Motors Corp.)

1. Detonation
2. Stalls or rough idle, cold
3. Stalls or rough idle, hot
4. Missing
5. Hesitation
6. Surges
7. Sluggish performance
8. Poor gasoline mileage
9. Hard starting, cold
10. Hard starting, hot
11. Objectionable exhaust odor
12. Cuts out

A built-in diagnostic system catches problems which are most likely to occur. The self-diagnostic system lights a "CHECK ENGINE" light on the instrument panel when there is a problem in the system.

Since the self-diagnostics do not include all possible faults, the absence of a code docs not mean there is no problem with the system. To determine this, a system performance check is necessary. It is made when the "CHECK ENGINE" light does not indicate a problem but the computer command system is suspected because no other reason can be found for a complaint. By grounding a

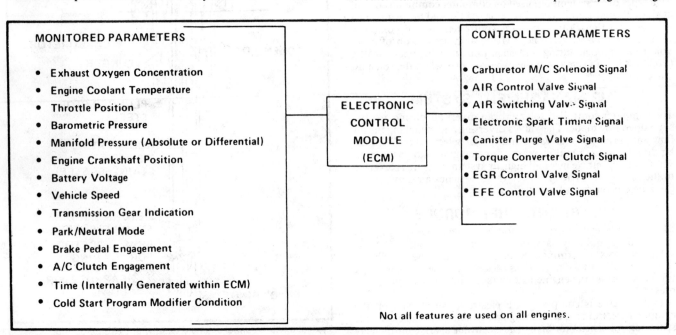

CCC system monitored and controlled parameters (©General Motors Corp.)

MONITORED PARAMETERS

- Exhaust Oxygen Concentration
- Engine Coolant Temperature
- Throttle Position
- Barometric Pressure
- Manifold Pressure (Absolute or Differential)
- Engine Crankshaft Position
- Battery Voltage
- Vehicle Speed
- Transmission Gear Indication
- Park/Neutral Mode
- Brake Pedal Engagement
- A/C Clutch Engagement
- Time (Internally Generated within ECM)
- Cold Start Program Modifier Condition

ELECTRONIC CONTROL MODULE (ECM)

CONTROLLED PARAMETERS

- Carburetor M/C Solenoid Signal
- AIR Control Valve Signal
- AIR Switching Valve Signal
- Electronic Spark Timing Signal
- Canister Purge Valve Signal
- Torque Converter Clutch Signal
- EGR Control Valve Signal
- EFE Control Valve Signal

Not all features are used on all engines.

"TROUBLE CODE" test lead under the instrument panel, the "CHECK ENGINE" light will flash a numerical code if the diagnostic system has detected a fault.

As a bulb and system check, the light will come on when the ignition is turned on with the engine stopped. The "CHECK ENGINE" light will remain on for a few seconds after the engine is started. If the "TROUBLE CODE" test lead is grounded with the ignition switch on and the engine stopped, the light will flash a code "12" which indicates the diagnostic system is working. This consists of one flash followed by a pause, and then two more flashes. After a long pause, the code will be repeated two more

times. The cycle will then repeat itself until the engine is started or the ignition is turned off.

If the "TROUBLE CODE" test lead is grounded with the engine running and a fault has been detected by the system, the trouble code will flash three times. If more than one fault has been detected, its code will be flashed three times after the first code set. The series will then repeat itself.

A trouble code indicates a problem with a given circuit. For example, code 14 indicates a problem in the coolant sensor circuit. This includes the coolant sensor, harness and electronic control module (ECM).

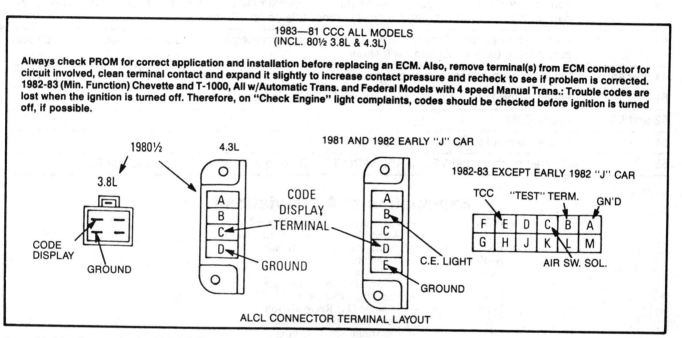

1983—81 CCC ALL MODELS
(INCL. 80½ 3.8L & 4.3L)

Always check PROM for correct application and installation before replacing an ECM. Also, remove terminal(s) from ECM connector for circuit involved, clean terminal contact and expand it slightly to increase contact pressure and recheck to see if problem is corrected. 1982-83 (Min. Function) Chevette and T-1000, All w/Automatic Trans. and Federal Models with 4 speed Manual Trans.: Trouble codes are lost when the ignition is turned off. Therefore, on "Check Engine" light complaints, codes should be checked before ignition is turned off, if possible.

ALCL CONNECTOR TERMINAL LAYOUT

Assembly Line Communication Link (ALCL) connector terminal arrangement, typical (©General Motors Corp.)

TROUBLE CODE IDENTIFICATION CHART CCC
All Except Chevette

Code #	Problem Indicated
12	No tachometer or reference signal to computer. This code will only be present while a fault exists, and will not be stored if the problem is intermittent.
13	Oxygen sensor circuit. The engine has to operate for about five minutes (eighteen minutes 3.8L V6) at part throttle before this code will show.
14	Shorted coolant sensor circuit. The engine has to run two minutes before this code will show.
15	Open coolant sensor circuit. The engine has to operate for about five minutes at part throttle before this code will show.
21	Shorted wide open throttle switch and/or open closed-throttle switch circuit (when used). Throttle position sensor circuit. (Must run 25 seconds below 800 rpm to set code.)
23	Open or grounded carburetor solenoid circuit.
24	Trouble in the Vehicle Speed Sensor (VSS) circuit—The vehicle must operate up to 5 minutes at road speed for this code to set.
32	Barometric pressure sensor (BARO) output low.
34	Manifold absolute pressure (MAP) sensor output high. (Engine must operate up to 5 minutes below 800 rpm to set code).

TROUBLE CODE IDENTIFICATION CHART CCC
All Except Chevette

Code #	Problem Indicated
35	Idle speed control (ISC) circuit shorted (operate engine over ½ throttle for over two seconds to set code.
42	Electronic Spark Timing (EST) bypass circuit grounded.
43	Throttle position sensor out of adjustment.
44	Lean oxygen sensor. Engine must be run for approximately five minutes in closed loop mode and part throttle at roadload (vehicle drive) before this code will show.
45	Rich oxygen sensor. Engine must be run for approximately five minutes in closed loop mode and part throttle before this code will show.
44 and 45	Faulty oxygen sensor or open sensor circuit.
51	Faulty calibration unit (PROM) or improper PROM installation.
52 and 53	Faulty ECM.
54	Faulty carburetor solenoid and/or computer.
55	Shorted or grounded VSS, MAP, BARO TPS. If not grounded or shorted replace ECM.

Explanation of Abbreviations

CCC	— Computer Command Control
ALCL	— Assembly Line Communication Link
BAT+	— Battery Positive Terminal
BARO.	— Barometric
Conv.	— Converter
ECM	— Electronic Control Module
EFE	— Early Fuel Evaporation
EGR	— Exhaust Gas Recirculation
ESC	— Electronic Spark Control
EST	— Electronic Spark Timing
HEI	— High Energy Ignition
ISC	— Idle Speed Control
MAP	— Manifold Absolute Pressure
M/C	— Mixture Control
OEM	— Original Equipment Manufacture
PCV	— Positive Crankcase Ventilation
P/N	— Park, Neutral
Port.	— Exhaust Ports
PROM	— Programmable Read Only Memory (engine calibration unit)
TCC	— Torque Converter Clutch
"Test" lead or terminal	— Lead or ALCL connector terminal which is grounded to obtain a trouble code
TPS	— Throttle Position Sensor (on carburetor)
Vac.	— Vacuum
VIN	Vehicle Identification Number
VSS	Vehicle Speed Sensor (signals load speed)
WOT	Wide Open Throttle

GENERAL MOTORS TORQUE CONVERTER CLUTCH DIAGNOSIS

Conventional Torque Converter System

Three basic components make up the conventional torque converter, a pump bolted to the engine's flywheel, a turbine connected to the automatic transmission's input shaft and a stator unit mounted between the pump and the turbine. When the engine is operating, the pump vanes directs hydraulic fluid to the vanes of the turbine, forcing the turbine to rotate. As the turbine rotates, the turbine vanes directs the fluid to the stator vanes, which in turn adds torque multiplication to the automatic transmission input shaft by redirecting the fluid back to the pump vanes. When the transmission is in first or second gear, the pump rotates faster than the turbine, 1.5 to 2 times for every rotation of the turbine. This difference in speed between the pump and the turbine, along with the action of the stator, multiplies the torque of the engine. As the transmission reaches third or top gear, the pump and the turbine are spinning at nearly the same speed, cancelling out the action of the stator since torque multiplication is not needed. The pump can now rotate approximately 1.1 times for every rotation of the turbine.

Torque Converter Clutch System

To eliminate the difference in speed between the converter pump and turbine in the top gear application, a mechanical type clutch was added between the turbine and the torque converter pump cover on the flywheel side. Additional components were added to complete the mechanical link-up between the engine and the transmission' planetary gear sets and their holding components. This mechanical link-up eliminates the difference in speed between the pump and turbine, resulting in greater fuel economy for the vehicle.

PRINCIPLES OF OPERATION

The additional components added to the converter unit are; the pressure plate, spring and dampener assembly and friction material. Transmission oil pressure is routed, at specific times, to apply or disengage the converter clutch unit. The friction material is bonded to the outer inch around the cover side of the plate and around the center at the spline area. The center friction material is not bonded in a full circle, since this friction area must supply support against the converter cover during the clutch apply so that the plate does not flex. It must also allow oil flow to the entire surface of the plate during the release phase. If this area was bonded in a complete circle, the release pressure area would not release the clutch. The turbine thrust spacer is located in the hub area of the pressure plate and contacts the turbine and the converter cover. The pressure plate moves in relation to the thrust spacer.

The thrust spacer has "O" ring seals that are important to the

shift feel of the clutch. The outer "O" ring seals the pressure plate hub. The inner "O" ring seal the inner area of the thrust spacer.

A firm clutch apply depends upon the releasing of all pressure on the engine side of the pressure plate. Damaged or missing "O" ring seals in the turbine thrust spacer can cause a pressure in the release side and cause the clutch to slip or intermittently apply and release. This condition can cause a vibration similar to a wheel balance problem should it happen during the clutch apply speeds.

It should be noted there are two types of pressure plates used. Both types have spring type torsional dampening components. The pressure plate used with the diesel engine has additional valves on the pressure plate. These valves are used to equalize the oil pressure on both sides of the pressure plate during a disengagement phase, because the diesel engine is compressing a full charge of air on every stroke, even during deceleration. This rapid slowing of the diesel engine causes a reverse rotation of the plate on the hub and opens two valves which allows the oil to move to the front of the plate, effecting a quicker clutch release. This results in a smoother clutch release and less rpm drop. Other converter clutch units do not use the poppet valve in the pressure plate assembly.

Diesel engine converters are identified by code markings and by having either three or six nuts, each of which are welded halfway around the nut (180 degrees), while the gasoline engine converters have three nuts and are welded in one spot.

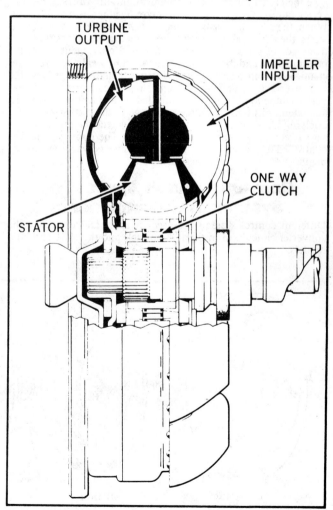

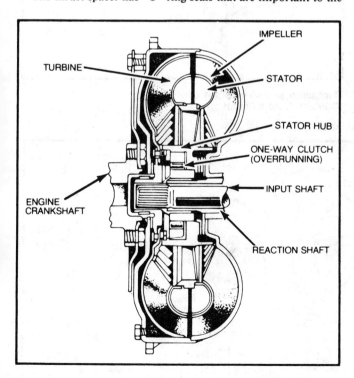

Cross section of typical torque converter assembly
(©General Motors Corp.)

Cut-away view of impeller input (pump), stator and turbine output
(©General Motors Corp.)

To aid in reducing torsional shock during converter clutch apply, a damper assembly is incorporated in the converter clutch pressure plate. The spring loaded damper is splined to the converter turbine assembly. The converter clutch pressure plate is attached to the pivoting mechanism of the damper assembly. This pivoting action allows the pressure plate to rotate independent of the damper assembly, up to approximately 45°. The rate of independent rotation is controlled by the pivoting mechanism acting on the springs in the damper assembly.

Converter Clutch Engagement and Disengagement

The converter clutch should engage when the engine is warm enough to handle the slight extra load and when the vehicle speed is high enough to allow operation to be smooth and the vehicle free of engine pulses. The converter clutch should release when the emissions would be affected as in a coast condition, torque multiplication is needed in the converter and when coming to a stop and a mechanical connection from engine to the rear wheels would be detrimental.

Clutch Operation Hydraulically

When a Torque converter clutch equipped vehicle is in the top speed (may apply in 2nd, 3rd or 4th speeds, depending upon the transmission/transaxle model) an apply valve directs converter feed oil pressure to the turbine side of the pressure plate. The fluid pressure forces the pressure plate against the housing cover to engage the clutch for a direct mechanical link-up, causing the converter pump and the turbine to move together as a single unit. When the vehicle coasts or the transmission/transaxle downshifts, the apply valve redirects the converter feed pressure to the front side of the pressure plate. This fluid movement forces the pressure plate and the housing cover apart to break the direct mechanical link-up. The converter clutch is then disengaged and the converter clutch acts as any conventional torque converter. The converter clutch apply valve is controlled by a solenoid, and when activated, opens an oil pressure passage to the converter clutch, applying it. When the solenoid is deactivated, the pressure passage is closed and the clutch disengages. The apply solenoid is controlled by a series of electrical, electronic and vacuum components.

Converter Clutch Application Systems And Components

Different control systems are used to apply and disengage the converter clutch unit in the 1980 models and the 1981 and later models.

Basically, the torque converter clutch system has developed in two stages. The first stage, which started in the 1980 model year, used two electrical and vacuum switches to control the operation of the TCC. The second stage, introduced at the beginning of the 1981 model year, continues for 1982. It uses four sensors and the ECM of the Computer Command Control system to control the operation of the torque converter clutch.

The 1980 system typically uses a brake switch and low vacuum switch in the feed circuit to the TCC solenoid in the transmission.

As long as the brakes aren't applied and engine vacuum is above 5 in. Hg, the TCC solenoid receives a voltage feed. What it doesn't have is a ground, which it needs to operate. The ground for the system is controlled by the governor switch. When governor line pressure is high enough, which is relative to vehicle road speed, the contacts in the switch will close. This provides the TCC solenoid with a ground which, in turn, applies the TCC.

The system used on the 1980 diesel engines used this same system with one addition: a high vacuum switch was installed in the feed circuit to the TCC solenoid. This switch opens the feed circuit to the TCC solenoid during periods of high engine vacuum, as in a coastdown situation. This basic system is used up to the present time on diesel engines, except the high vacuum switch function is replaced by the poppet valves, in the torque converter.

Since 1981, the ECM on most Computer Command Control

CLUTCH APPLIED

TURBINE — PUMP
HOUSING COVER
OUTPUT →
HYDRAULIC PRESSURE
PRESSURE PLATE

Mechanical lock-up of torque converter clutch
(©General Motors Corp.)

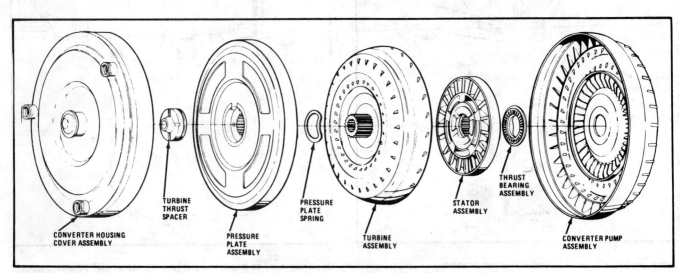

Exploded view of torque converter clutch assembly (©General Motors Corp.)

CONVERTER HOUSING COVER ASSEMBLY

TURBINE THRUST SPACER

PRESSURE PLATE ASSEMBLY

PRESSURE PLATE SPRING

TURBINE ASSEMBLY

STATOR ASSEMBLY

THRUST BEARING ASSEMBLY

CONVERTER PUMP ASSEMBLY

equipped vehicles has controlled the ground for the TCC solenoid, replacing the governor switch of the earlier system. Typically, in ECM controlled systems, a TCC brake switch is installed in the feed circuit to the TCC solenoid.

Provided the brakes aren't applied, voltage is fed to the TCC solenoid. But, again, the TCC solenoid needs a ground for operation. So, when the ECM has the proper input from four sensors (the coolant temperature sensor, throttle position sensor, vacuum sensor and vehicle speed sensor), it provides the TCC solenoid with a ground. In turn, that allows operation of the TCC. To prevent application of the TCC in lower gears, this system may also use hydraulically actuated electrical switches installed in the valve body.

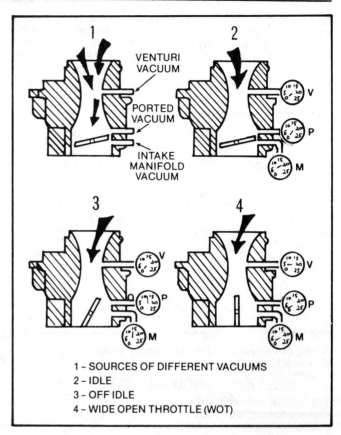

1 - SOURCES OF DIFFERENT VACUUMS
2 - IDLE
3 - OFF IDLE
4 - WIDE OPEN THROTTLE (WOT)

Sources of engine related vacuum (©General Motors Corp.)

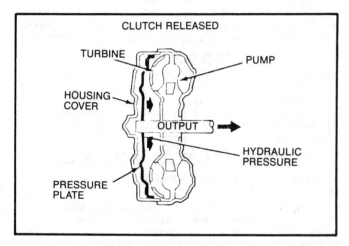

Torque converter clutch release, reverting converter to a conventional torque converter operation (©General Motors Corp.)

Control Components

To control the application and release of the converter clutch during various driving conditions: vacuum, electrical and electronic controls are used. Most of the controls are external to the transmission/transaxle assemblies. It should be noted that all controlling components may not be on each vehicle equipped with the converter clutch.

PORTED VACUUM

Ported vacuum is used on the 1980 TCC system and it's simply a vacuum port positioned above the throttle plates. It provides the vacuum source for the engine vacuum switch. When the throttle plate is closed during deceleration, the vacuum signal to the engine vacuum switch drops to zero. This opens the contacts in the vacuum switch, releasing the torque converter clutch.

VACUUM DELAY VALVE

The vacuum delay valve is used on some of the 1980 TCC systems with gasoline engines. Its purpose is to delay the response of the vacuum switch to slight, momentary changes in engine vacuum. This helps avoid unnecessary disengagement and engagement of the torque converter clutch.

The vacuum delay valve is positioned in the vacuum line between the ported vacuum source on the carburetor and the vacuum switch.

LOW VACUUM SWITCH

The low vacuum switch is used in 1980 TCC systems. This normally open switch consists of a steel housing with a diaphragm and port on one end and a set of electrical contacts and blade terminals on the other. A hose from a ported vacuum source on the carburetor or throttle body is connected to the port on the vacuum switch. When ported vacuum is high, the diaphragm is pulled

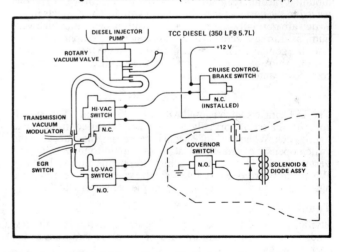

Typical converter clutch controls with diesel engine usage (©General Motors Corp.)

so that the electrical contacts touch each other. This provides the circuit for a voltage feed to the TCC solenoid. However, when ported vacuum is low, as during closed throttle deceleration, the diaphragm is released. This caused the contacts to separate, breaking the circuit to the TCC solenoid. Now the torque converter clutch is released.

Testing

The low vacuum switch can be tested with an ohmmeter and a vacuum pump. Connect the test leads from the ohmmeter to the blade terminals on the switch. Without any vacuum applied to the port on the switch, resistance will be infinite. Now, connect the hose from the vacuum pump to the vacuum port. Pump the vacuum pump until the gage reads between approximately 3 and 9 in. Hg. on gasoline engines and 5 to 6 in. Hg. on diesel engines. Now the resistance between the terminals will be zero. If the switch doesn't pass either of these tests, replace it.

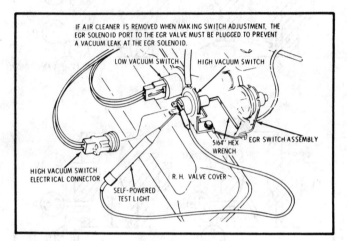

Testing and adjustment of the high vacuum switch (©General Motors Corp.)

HIGH VACUUM SWITCH

The high vacuum switch is used only with diesel engines. It consists of a housing with a diaphragm and port on one end and a set of electrical contacts and blade terminals on the other.

A hose from the vacuum pump (via the injection pump-mounted vacuum regulator valve and low vacuum switch) is connected to the port on the high vacuum switch. The contacts in the switch are normally closed. However, during periods of high vacuum, as during a zero throttle coastdown, the diaphragm separates the contacts. This breaks the circuit to the torque converter clutch, thereby avoiding feeling the engine pulses during coastdown and reducing emissions.

TCC BRAKE SWITCH

The TCC brake switch is used with all TCC systems. This switch consists of a set of normally closed contacts housed in the brake light switch. When the brake pedal is depressed, the plunger in the switch opens the contacts, breaking the circuit to the torque converter clutch. This avoids stalling the engine during braking. When the brake pedal is released, the contacts close and once again complete the circuit to the torque converter clutch.

Testing

Use an ohmmeter to test the TCC brake switch. Place the test leads from the ohmmeter on each of the blade terminals at the rear of the switch.

With the plunger fully extended, resistance should be infinite. With the plunger pushed in, resistance should be zero. If the switch fails either one of those tests, it should be replaced.

Adjustments

The adjustment of the switch is important. If the switch is screwed in too far, the contacts that complete the circuit to the torque converter clutch will not open when the brakes are applied. This can cause the engine to stall at idle. The switch should be adjusted so that the plunger just contacts the brake lever when the brake pedal is in the released position.

GOVERNOR SWITCH

The governor switch is used on the 1980 TCC system and some later diesel engine applications. This normally open switch consists of a housing, diaphragm, port, electrical contacts and terminals. The governor switch is positioned in the pressure line from the governor. As vehicle speed increases, so does the oil pressure from the governor. When the oil pressure reaches a predetermined level (which usually coincides with approximately 40 MPH) the normally open contacts in the governor switch close. This completes the circuit through the switch and gives the TCC solenoid a ground. In turn, this allows the application of the torque converter clutch.

THIRD GEAR SWITCH

The third gear switch consists of a switch body, blade terminals, diaphragm and a set of normally open contacts

When the transmission is in third gear, transmission oil acts on the diaphragm forcing the contacts to close. This completes the electrical circuit to the TCC solenoid, allowing TCC operation.

TCC SOLENOID

The TCC solenoid is used with all TCC systems. This assembly consists of an electrical solenoid, check ball, seat, and O-ring.

When the TCC solenoid is energized, the check ball moves, redirecting transmission oil to the converter clutch apply valve. The apply valve routes transmission oil to apply the torque converter clutch.

ELECTRONIC CONTROL MODULE (ECM)

NOTE: The ECM is a part of the Computer Command Control system.

The operation of the electronic control module discussed here is only relative to torque converter clutch operation.

The ECM has controlled the operation of the torque converter clutch since the beginning of the 1981 model year in all passenger cars equipped with gasoline engines.

The electronic control module provides a ground for the TCC solenoid in the transmission when four conditions are met:

1. Engine coolant temperature is above a predetermined value.
2. The throttle position sensor and vacuum sensor indicate the engine isn't under a heavy load.
3. The throttle isn't closed.
4. Vehicle speed is above a predetermined value.

COOLANT TEMPERATURE SENSOR

The coolant temperature sensor is a part of the Computer Command Control system and is positioned in the engine coolant stream. This sensor works on the principle of varying resistance.

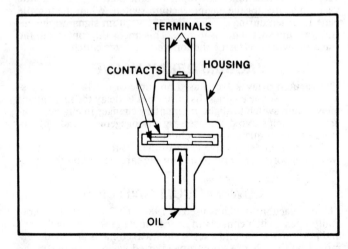

Typical governor switch—Note contact appearance (©General Motors Corp.)

The ECM delivers a five-volt electrical signal to the coolant temperature sensor. Depending on the resistance offered by the coolant temperature sensor, an electrical signal is sensed by the ECM. The difference in voltage between the five-volt output signal from the ECM and the voltage at the coolant temperature sensor represents the coolant temperature to the ECM. The voltage at the coolant temperature sensor drops as coolant temperature increases.

A faulty sensor that indicates engine coolant is always below a predetermined temperature (usualy between 130°-150°F) will prevent the engagement of the TCC. If the sensor is improperly indicating that coolant temperature is always above the predetermined temperature, the TCC will engage before the engine is warmed up and can cause drivability problems.

When the coolant is cold, the sensor has a resistance of approximately 100,000 ohms. When the coolant is at normal operating temperature, the sensor has a resistance of less than 1,000 ohms.

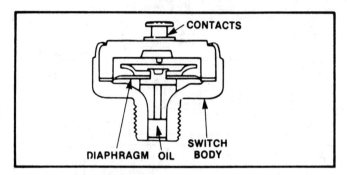

Typical top gear switch—Note contact appearance
(©General Motors Corp.)

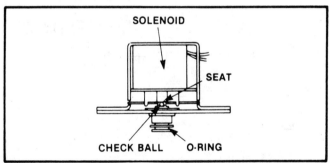

Typical torque converter clutch solenoid (©General Motors Corp.)

Testing

A faulty coolant temperature sensor is indicated by a Code 14 or 15 and can be tested by accessing the stored trouble codes in the ECM.

THROTTLE POSITION SENSOR (TPS)

The throttle position sensor (TPS) is a part of the Computer Command Control system. The TPS is a variable resistor that converts the degree of throttle plate opening to an electrical signal that the ECM uses.

The resistance between terminals A and B of the TPS ranges from about 10,000-15,000 ohms when the throttle is closed, to approximately 1,000 ohms when the throttle is wide open. The resulting voltages returned to the ECM vary from under one volt at closed throttle to nearly five volts at wide open throttle.

The degree of throttle opening, together with the amount of engine vacuum sensed by the vacuum sensor, provides the ECM with information directly relating to engine load. If the engine load is high enough, the ECM will break the circuit to the TCC. It does this by denying a ground for the TCC circuit.

The TPS is located in the carburetor on carbureted engines. In electronic fuel injection systems, the TPS is mounted on the throttle body. On EFI systems that use dual throttle bodies, the TPS is mounted on the outboard side of the left throttle body.

Testing

When testing for a malfunctioning TPS, keep in mind that it's part of the Computer Command Control system. If the TPS is malfunctioning, either a Code 21 or 22 will be stored in the memory of the ECM. To check for this, you'll have to access the memory of the ECM.

VACUUM SENSOR (VS)

The vacuum sensor (also known as a MAP sensor) is part of the Computer Command Control system. Its purpose is to sense changes in engine vacuum and supply a voltage to the ECM relative to the strength of the vacuum signal.

Testing

A vacuum sensor that's faulty is indicated by a Code 33 or 34 which can be determined by accessing the trouble code(s) stored in the ECM.

VEHICLE SPEED SENSOR (VSS)

The vehicle speed sensor is part of the Computer Command Control system and consists of a reflective blade, light emitting diode, photo cell and buffer/amplifier.

The reflective blade of the VSS is part of the speedometer cable/head assembly. When the vehicle starts moving, the speedometer cable starts rotating. Since the reflective blade is attached to the cable, it also begins to rotate. As the blade enters the light beam provided by the L.E.D., light is reflected back at the photo cell. The photo cell "sees" the light which causes a low power electrical signal to be sent to the buffer/amplifier. The buffer/amplifier conditions and amplifies the electrical signal and then sends it to the ECM. The number of electrical signals sent to the ECM is a fixed period of time indicates road speed.

The vehicle speed sensor is located in the speedometer head behind the instrument panel. To service the VSS, the instrument panel cluster must be removed.

Testing

Since the VSS is part of the Computer Command Control system, if it's faulty a Code 24 will be stored in the memory of the ECM. To verify a faulty VSS, access the memory of the ECM.

EGR BLEED SOLENOID

The EGR bleed solenoid is a normally closed switch that's used in some Computer Command and non-Computer Command Control applications. It controls an air bleed in the EGR control vacuum passage. This solenoid is used because less EGR is needed at TCC engagement speeds and helps avoid feeling engine pulsing during TCC engagement.

When the TCC solenoid is energized, so is the EGR bleed solenoid. With the EGR bleed solenoid energized, the air bleed in the EGR control vacuum passage is opened. This adds air to the EGR signal vacuum, reducing or eliminating EGR and allowing smoother engine operation.

THERMAL VACUUM VALVE

The thermal vacuum valve is used in the non-Computer Command Control system and consists of a body, two vacuum ports and a thermally activated valve. The end of the valve is immersed in engine coolant. When coolant temperature is below approximately 130°F, the valve blocks the vacuum from the ported vacuum source on the carburetor. When coolant temperature reaches 130°F, the valve moves, connecting carburetor ported vacuum to the vacuum switch. In turn, this allows operation of the TCC. The thermal vacuum valve prevents the application of the torque converter clutch at engine coolant temperatures below 130°F.

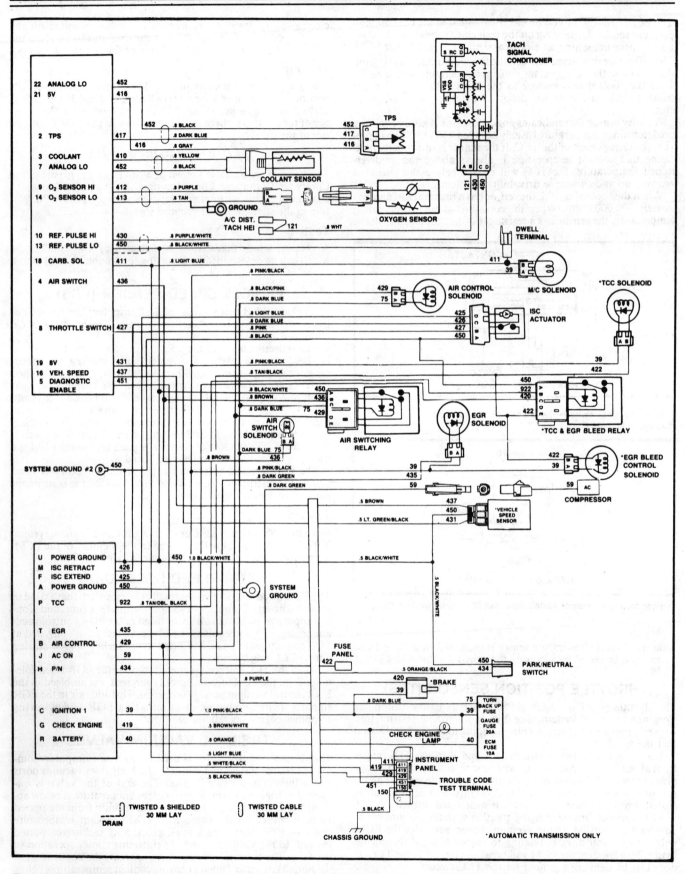

Typical wiring schematic for torque converter equipped vehicles (©General Motors Corp.)

VACUUM REGULATOR VALVE

The vacuum regulator valve replaced the rotary switch on 1982 diesel engines. This valve uses two sets of contacts wired in series. The first set of contacts are normally open when the engine is at or near idle speed. This prevents a voltage feed from reaching the TCC solenoid, which, in turn, doesn't activate the TCC. When the throttle is opened partway, the first set of contacts closes. This completes the feed circuit to the TCC solenoid and the TCC applies. The second set of contacts is normally closed. They open near and at full throttle, breaking the feed circuit to the TCC solenoid.

Adjustments

The adjustment of the vacuum regulator valve is critical to proper TCC operation. This procedure relies on using a carburetor angle gage and vacuum gage to properly test and adjust the VRV.

FOURTH GEAR SWITCH

The fourth gear switch operates in the same manner as the third gear switch, except that the contacts close when the transmission is in fourth gear. It looks exactly like the third gear switch.

4-3 PULSE SWITCH

The normally closed 4-3 pulse switch is a pressure-activated electrical switch that opens the circuit to the TCC momentarily on a 4-3 downshift. This allows the engine to receive torque multiplication from the torque converter and reduce downshift harshness.

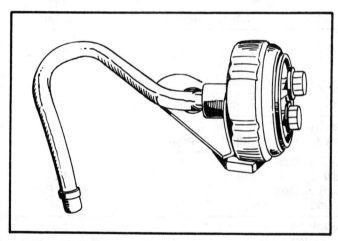

4-3 pulse switch (©General Motors Corp.)

TCC RELAY

The TCC relay is used on some Computer Command Control systems. This normally closed relay allows control of two solenoids without overloading the ECM. The two solenoids are used for the TCC and EGR bleed.

ROTARY SWITCH

The rotary switch is used on 1981 diesel engines with a TCC transmission. Two types of rotary switches are used. The first type contains one set of electrical contacts that are normally closed. When the accelerator is near or at full throttle, the contact points open, breaking the feed circuit to the TCC solenoid. In turn, this releases the TCC. The second type of rotary switch uses two sets of contacts wired in series. The first set of contacts are normally open when the engine is at or near idle speed. This prevents a voltage feed from reaching the TCC solenoid, which, in turn, doesn't activate the TCC. When the throttle is opened partway, the first set of contacts close. This completes the feed circuit

to the TCC solenoid and the TCC applies. The second set of contacts are normally closed. They open near and at full throttle, breaking the feed circuit to the TCC solenoid. The rotary switch was replaced by the vacuum regulator valve on 1982 diesel engines with a TCC transmission.

Testing

The rotary switch can be tested using an ohmmeter. Disconnect the electrical connectors and connect the test leads from the ohmmeter to the TCC terminals leading to the switch. The switch that uses one set of contacts should have no resistance until the throttle is at or near wide open. When the throttle is at this point, resistance should be infinite. The switch that uses two sets of contacts should show infinite resistance with the throttle at or near the closed position. When the throttle is opened halfway, resistance should be zero. And when the throttle is near or at wide open, resistance should be infinite once again. Replace the rotary switch if it doesn't pass these tests.

TCC COLD OVERRIDE SWITCH

The TCC cold override switch is in the TCC solenoid feed circuit on 1982 5.7L V-8 diesels. This normally open switch prevents TCC engine engagement when engine temperature is below 67° C (152° F).

Diagnosis of the Converter Clutch Systems

The diagnosing of the converter clutch system problems can be separated into four major catagories.
1. External electrical controls, less the CCC system.
2. Internal electrical controls.
3. Grounding circuit for the ECM, when equipped with the CCC system.
4. Internal hydraulic system.

PRELIMINARY INSPECTION AND ADJUSTMENT PROCEDURES

Before performing any diagnostic procedures on the converter clutch system, inspect and adjust the following components, as required.
1. Inspect all electrical and vacuum connections, wires and hoses.
2. Be certain the engine is properly turned. Can be diagnosed as a poorly operating transmission/transaxle.
3. Inspect the fluid level and correct as required. Inspect the fluid condition.
4. Inspect the manual linkage and throttle cable (TV) adjustments.
5. Road test the vehicle, with owner, if possible, to verify complaint.

CONVERTER CLUTCH DRIVING CHARACTERISTICS

The converter clutch units have a specific operational characteristics that should not be diagnosed as a transmission/transaxle malfunction. One characteristic, called vehicle "chuggle" is best described as a varying fore and aft motion of the vehicle while driving at a steady speed with the converter clutch engaged. This condition has been misdiagnosed as engine surge, engine miss and tire and wheel balance problems. Another driving characteristic is called a "bump," which occurs when the driver releases the accelerator quickly. The bump is the result from the reversal of the engine torque during deceleration. Both "chuggle" and "bump" characteristics are normal of a vehicle equipped with a converter clutch and repairs are not necessary.

GOVERNOR SWITCHES

If problem is encountered when coverter clutch engages, causing

CONVERTER CLUTCH ELECTRICAL AND HYDRAULIC CONTROLS
1980 Models

Engines	Transmission & Codes	Oil to Solenoid T.C.C. Valve	Switches		Vacuum Controls		Test Lead
			Not In Trans.	In Trans.		Source	
Gas	200-C 250-C 350-C	Direct Clutch	1. Brake N.C. 2. Lo. Vac. N.O.	Gov. Press. SW-N.O.	1. TVS-EGR 2. TVS-EFE 3. Relay Valve 4. Delay Valve	EGR-Ported	Near Fuse Panel ①
Diesel	200-C 350-C	Direct Clutch	1. Brake-N.C. 2. Lo. Vac.-N.O. 3. Hi Vac-N.C.	Gov. Press. SW-N.O.	Rotary Valve At Inj. Pump	Vacuum Pump	Near Fuse Panel ①

① Test lead in feed side of solenoid
② Test lead in ground side of circuit
N.O. = Normally open
N.C. = Normally closed

CONVERTER CLUTCH ELECTRICAL AND HYDRAULIC CONTROLS
1981 Models

Transmissions & Codes	Oil to Solenoid T.C.C. Valve	Switches		Sensors			Test Lead
		Not In Trans.	In Trans.	V.S.S.	Coolant	T.P.S.-Vacuum	
200-C 250-C 350-C MV-4	Direct Clutch	Brake N.C.	None	E.C.M. Input	E.C.M. Input	E.C.M. Input	In Fuse Panel ②
350-C MX-2	Forward Clutch	Brake N.C.	3rd Gear Switch Normally Open	E.C.M. Input	E.C.M. Input	E.C.M. Input	②
350-C MX-3	Forward Clutch	Brake N.C.	2nd Gear Switch Normally Open	E.C.M. Input	E.C.M. Input	E.C.M. Input	②
350-C Lt. Truck MV-4	Direct Clutch	1. Brake N.C. 2. Lo Vac-N.O. 3. 3rd Gear On Trans (EGR)	Gov. Press. Switch-Normally Open Speed & EGR Solenoid Ground				Near Fuse Panel ②
350-C Diesel	Forward Clutch	Brake & Rotary Valve	3rd Gear Switch-N.O. Gov. Press. Switch-N.O. (Speed & EGR Solenoid)				Near Fuse Panel ①
200-C Diesel	Direct Clutch	Brake & Rotary Valve	Gov. Pressure-Veh. Speed & E.C.R. Solenoid Ground-Normally Open				Near Fuse Panel ①
200-4R MW-9	2nd Clutch & T.C.C. Shift Valve	Brake N.C.	1-4th Gear N.C. 2-4-3 Pulse N.C.	E.C.M. Input	E.C.M. Input	E.C.M. Input	In Fuse Panel ②
125-C MD-9	Line	Brake N.C.	3rd Gear Switch Normally Open	E.C.M. Input	E.C.M. Input	E.C.M. Input	In Fuse Panel ②

① Test lead in feed side of solenoid

② Test lead in ground side of circuit
N.O. = Normally open
N.C. = Normally closed

CONVERTER CLUTCH ELECTRICAL AND HYDRAULIC CONTROLS
1982 Without CCC Systems

Transmission and Code	Oil to Solenoid & TCC Valve	Switches Not in Trans.	Switches In Trans.	Apply in Gear	Sensor Device Speed	Sensor Device Temp	Sensor Device Load	Test Lead
350C-MX2 5.7L Diesel	Forward Clutch	Ign-Brake VRV at Inj. Pump	3rd Gear-N.O. Gov.-N.O.	3	Gov. Sw. 40 mph	None	Sw. at Inj. Pump (Heavy Load)	In ALDL①
200-4R-MW9 5.7L Diesel	2nd Clutch Oil Thru TCC Shift Valve	Ign-Brake VRV at Inj. Pump (Open Near W.O.T.)	Gov.-N.O. (Used for EGR Bleed Only)	2-3-4	TCC Shift Valve (Hyd.)	None	Sw. at Inj. Pump (Heavy Load Open)	In ALDL①
325-4L-M57 5.7L Diesel	2nd Clutch Oil Thru TCC Shift Valve	Ign-Brake VRV at Inj. Pump (Open Near W.O.T.)	Gov.-N.O. (Used for EGR Bleed Only)	2-3-4	TCC Shift Valve (Hyd.)	None	Sw. at Inj. Pump (Heavy Load Open)	In ALDL①
700-R4-MD8 6.2L Diesel	2nd Clutch Oil Thru TCC Shift Valve	Ign-Brake TPS at Inj. Pump (Open at Light Throttle)	4-3 Pulse N.C.	2-3-4	TCC Shift Valve (Hyd.)	None	TPS at Inj. Pump (Light Load Open)	Wire Near① Fuse Panel 6" Lt. Gn.)
700-R4-MD8 6.2L Diesel	2nd Clutch Oil Thru TCC Shift Valve	Ign-Brake TPS at Inj. Pump (Open at Light Throttle) Relay-4 W.D. N.C. 4 W.D. Sw. at Transfer Case	4-3 Pulse-N.C. 4th Gear-N.O. (Bypasses Relay to Provide TCC in 4th Only When in 4 W.D.	2-3-4 (ex. 4 W.D.) 4th on 4 W.D.	TCC Shift Valve (Hyd.)	None	TPS at Inj. Pump (Light Load Open) Stays Applied at Heavy Throttle	Wire Near① Fuse Panel 6" Lt. Gn.)
700-R4-MD8 Gas-	2nd Clutch Oil Thru TCC Shift Valve	Ign-Brake Low VAC	4-3 Pulse-N.C. 4th Gear-N.O. (Bypass Lo VAC in 4th)	2-3-4	TCC Shift Valve (Hyd.)	Thermal Vacuum Valve to Lo VAC Sw.	Lo VAC Sw. and Ported VAC Sig. (Stays on in 4th)	Wire Near① Fuse Panel 6" Lt. Gn.)
		3rd Clutch-N.O. TCC Signal-N.O.	EGR Bleed Control					

①Test lead in feed side of solenoid.
N.O.=Normally open.
N.C.=Normally closed.
*Assembly line diagnostic link mounted under dash.

G.M. LOCK-UP TORQUE CONVERTERS

CONVERTER CLUTCH ELECTRICAL AND HYDRAULIC CONTROLS
1982 With CCC Fuel Systems—EMC Control

Transmissions and Codes	Oil to Solenoid & TCC Valve	Switches		Apply Possible Gear	ECM Input Sensors			Test Lead
		Not in Trans.	In Trans.		Speed	Temp.	Load	
125-C (MD9)	Line	Ign-Brake	3rd Gear-N.O.	3rd	Vehicle Speed Sensor	Coolant Sensor	TPS and Vacuum	Note ②③
200C-MV9 250C-M31	Direct Clutch	Ign-Brake	None	3rd	Vehicle Speed Sensor	Coolant Sensor	TPS and Vacuum	ALDL*
350C-MV4	Direct Clutch	Ign-Brake	None	3rd	Vehicle Speed Sensor	Coolant Sensor	TPS and Vacuum	ALDL*
350C-MX2	Forward Clutch	Ign-Brake	3rd Gear-N.O.	3rd	Vehicle Speed Sensor	Coolant Sensor	TPS and Vacuum	ALDL*
350C-MX3	Forward Clutch	Ign-Brake	2nd Gear-N.O.	2nd 3rd	Vehicle Speed Sensor	Coolant Sensor	TPS and Vacuum	ALDL*
350C-MX5	Forward Clutch	Ign-Brake	2nd Gear-N.C. 3rd Gear-N.C.	1st 2nd 3rd	Vehicle Speed Sensor	Coolant Sensor	TPS and Vacuum	ALDL*
200-4R MW9	2nd Clutch Thru TCC Shift Valve	Ign-Brake	4-3 Pulse-N.C. 4th Clutch-N.C.	2-3-4	Vehicle Speed Sensor	Coolant Sensor	TPS and Vacuum	ALDL*
325-4L M57	2nd Clutch Thru TCC Shift Valve	Ign-Brake	4-3 Pulse-N.C. 4th Clutch-N.C.	2-3-4	Vehicle Speed Sensor	Coolant Sensor	TPS and Vacuum	ALDL*
700-R4 MD8	2nd Clutch Thru TCC Shift Valve	Ign-Brake	4-3 Pulse N.C. 4th Clutch-N.C.	2-3-4	Vehicle Speed Sensor	Coolant Sensor	TPS and Vacuum	ALDL*

① Test Lead in feed side of solenoid.
② Test Lead in ground side of solenoid.
③ J-Car in fuse panel—others in ALCL.

N.O.=Normally open.
N.C.=Normally closed.
*Assembly line diagnostic link mounted under dash.

TORQUE CONVERTER CLUTCH

Number Code	Part Number	Color Code	On/Off Psi	Transmission
33	8641197	Green	33-30	350C
34	8633398	Green	34-31	200C
36	8630740	Tan	36-33	350C
38	8633363	White	38-35	350C
40	8641265	Violet	40-36	350C
42	8630819	Yellow	42-38	350C
44	8633361	Pink	44-40	350C
46	8633362	Silver	46-42	200C
48	8633359	Orange	48-44	200C
50	8633397	White	50-46	200C
52	8633360	Yellow	52-50	200C
54	8633364	Lt. Brown	54-50	200C

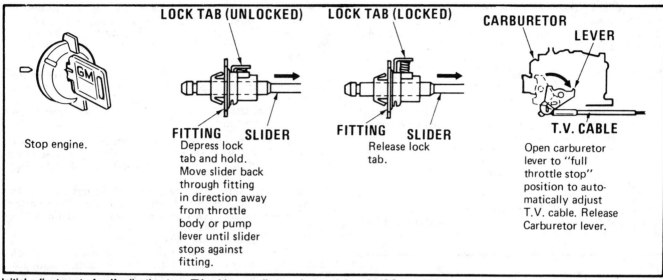

Initial adjustment of self-adjusting type TV cable, gasoline engine models only (©General Motors Corp.)

a torsional vibration or chuggle, an improved coverter clutch engagement can be attained by changing the Governor switch, which is located on the valve body of the 350C-250C transmissions, or on the case next to the valve body on the 200C Transmissions.

Use this procedure: First road test the vehicle and see what road speed the clutch engages (check speedometer). Then remove governor switch and check number that is stamped on switch. Add eight to the number on switch and select the switch closest to the sum of those numbers. Example: 36 is on switch you remove, add 8, then replace with a number 44 (8633361). Re-road test.

Throttle Valve (T.V.) Cable Adjustment

GASOLINE ENGINE WITH MANUAL CABLE

1. Engine must be stopped.
2. Unlock the T.V. cable snap lock by pushing upon the snap lock button.
3. Rotate the carburetor lever by hand to the wide open throttle position.
4. Lock the T.V. cable by pressing down on the cable snap button.

GASOLINE ENGINES WITH SELF-ADJUSTING CABLE

1. Engine must be stopped.
2. Depress the T.V. cable lock tab. With the lock tab depressed, move the cable slider back through the cable fitting in a direction away from the carburetor throttle body or accelerator pump lever.
3. Release the lock tab to lock the cable slider in its position.

4. Rotate the carburetor lever by hand to the wide open throttle position to automatically adjust the T.V. cable.

DIESEL ENGINES WITH MANUAL T.V. CABLE

1. The engine must be stopped.
2. Remove the cruise control rod, if equipped.
3. Unlock the T.V. cable snap lock by pushing up on the snap lock button.
4. Remove the T.V. cable from the bell crank.
5. Remove the throttle rod from the bell crank by moving it away from its attaching pin on the bell crank.
6. Rotate the bell crank to the wide open throttle stop position and hold it in this position.

NOTE: If the bell crank cannot be placed in the wide open throttle stop position when the accelerator pedal is depressed, all wide open throttle stop adjustments must be made with the accelerator in the completely depressed position, instead of rotating the bell crank by hand.

7. Push the throttle rod and pump lever to the wide open throttle stop position. Adjust the throttle rod to meet the bell crank pin at the wide open throttle stop position.

NOTE: Do not connect the throttle rod to the bell crank at this time.

8. Release the bell crank and reconnect the T.V. cable to the bell crank.
9. Rotate the bell crank to the wide open throttle stop position and hold in place.

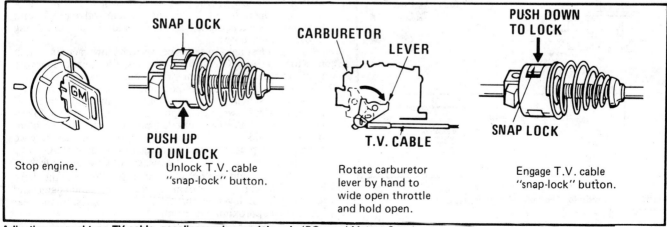

Adjusting manual type TV cable, gasoline engine models only (©General Motors Corp.)

10. Lock the T.V. cable snap lock by pushing down on the cable snap lock button. Release the bell crank.

11. Reconnect the throttle rod and the cruise control rod (if equipped) to the bell crank assembly.

Diagnosis of Control Components— 1980

Should the results of the road test indicate a problem in the converter clutch system, the following procedure can be followed.

1. Operate the engine at idle.

2. Locate and remove the electrical connection on the low vacuum switch. If the vehicle is equipped with a diesel engine, perform this operation on the high vacuum switch.

3. Ground the negative lead of a test light.

4. Locate the transmission side of the low vacuum switch connection by probing both female terminals inside the connector. The terminal that doesn't light is the transmission side of the circuit.

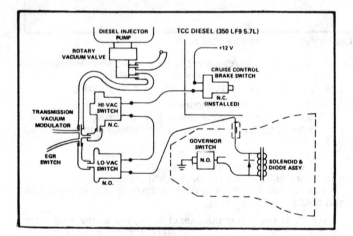

Electrical schematic for gasoline engines with vacuum delay valve
(©General Motors Corp.)

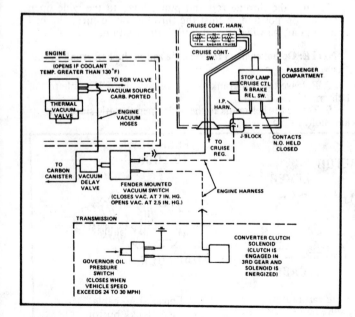

Electrical schematic for 1980 diesel engine, typical
(©General Motors Corp.)

NOTE: Certain vehicle models have the low vacuum switch and connector with three terminals rather than two. The center terminal of the switch and connector is used by the factory for specific tests, with the center terminal being on the transmission side of the circuit.

5. Reconnect the low vacuum switch connector to the switch. On the diesel engines, connect the connector the high vacuum switch.

6. Insert the test light probe into the low vacuum switch connector on the transmission side of the circuit.

NOTE: When testing a three terminal connector and low vacuum switch, insert the probe into the connector at the two green wires. For diesel engines, probe the transmission side of the high vacuum switch.

a. The test lamp should be out, but if the test lamp glows, check the vacuum hoses and their routings from the vehicle's Emission Control label. Check the engine speed for proper idling.

b. If the light does not glow, continue the diagnosis.

7. In order to increase the engine speed, activate the fast idle solenoid by turning on the air conditioning system, if equipped. The test light should remain off. On diesel equipped vehicles, disconnect the pink and green wire connector from the engine coolant switch to activate the fast idle solenoid. The test light at the high vacuum switch should remain off.

a. Should the test light glow, adjust the fast idle speed to proper specifications. On diesel engines, it is necessary to adjust the fast idle speed and the high vacuum switch (procedures follow this testing outline).

NOTE: Should the test light continue to glow at the transmission/transaxle side of the electrical circuit, on the low vacuum or high vacuum switches after a fast idle speed or high vacuum switch adjustment, check the operation of the low vacuum switch and double check the vacuum hose routings against the vehicle's Emission Control label.

b. If the test light remains off with the engine at idle, with the fast idle solenoid activated, continue the diagnosis.

8. With the test light still in place on either the low or high vacuum switch, slowly raise the engine speed. The test light should glow when the engine speed is increased and go out when the engine speed is decreased back to idle. With the engine idle speed above idle and the light glowing, depress the brake pedal and the light should go out.

a. If the test lamp indicates the vacuum switches are operating properly, continue the diagnosis.

b. Should the test light glow above engine idle speed, but not go out when the brake pedal is depressed, check the brake switch for proper operation.

c. If the test light does not glow at any engine speed, check for a blown fuse in the fuse panel, broken or loose wires and connectors in the converter clutch control circuit or check the operation of the low and/or high vacuum switches. Also, check for proper operation of the thermal vacuum valve (TVV) and the relay valve, if equipped, and recheck for the proper routing of the vacuum hoses.

9. Insert a test light in series between the low vacuum switch connector terminals. On the diesel models, also connect a jumper wire between the high vacuum switch connector terminals. Turn the ignition switch to the ON position.

a. If the test light glows, remove the transmission oil pan and check the wire between the apply valve solenoid and the governor switch. If the wire is grounded, pinched or cut, the problem has been found. However, should this not be the case, replace the governor pressure switch and again test the circuit.

b. If the test light does not glow, continue with the test.

10. If the test light did not glow, as in test 9b, raise the drive wheels and blocking the remaining wheels, start the engine and with the selector lever in the DRIVE position, spin the rear wheels up to approximately 50 mph.

11. The test light between the low vacuum switch electrical connector terminals should begin to glow between approximately 35 to 50 mph and should stop glowing at speed below 30 mph.

    a. If the test light does not glow as stated, check the wire between the low vacuum switch and the transmission electrical connector for an open circuit. If this is found, repair or replace the wire as required.

    b. If the circuit is without defect, remove the transmission oil pan and check for a loose, disconnected or cut wire. Check for a defective solenoid. If the solenoid is defective, replace it. If the solenoid is good, replace the governor pressure switch.

    c. If the test light between the low vacuum switch connector terminals begins to glow between 35 and 50 mph and shuts off at speeds below 30 mph, the electrical circuits and components are operating properly. Remove the test lamp and jumper wire. Re-connect the wire connectors as required.

12. If a vacuum or electrical malfunction has not been found, to correct the problem with the converter clutch unit, an internal hydraulic/mechanical controls check will have to be done.

NOTE: Refer to diagnostic pages for 1980 procedures.

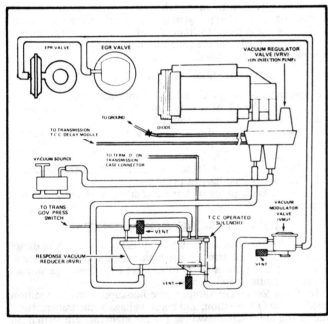

Typical diesel engine EGR system. Differences exist between V-6 and V-8 engines and from year to year (©General Motors Corp.)

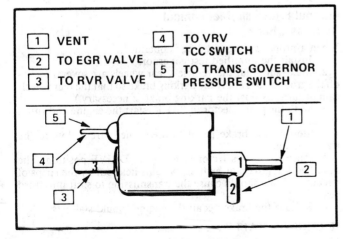

1 VENT	4	TO VRV TCC SWITCH
2 TO EGR VALVE		
3 TO RVR VALVE	5	TO TRANS. GOVERNOR PRESSURE SWITCH

TCC solenoid assembly (©General Motors Corp.)

Engine Idle and Fast Idle Adjustments
GASOLINE AND DIESEL ENGINE—1980
Refer to the appropriate Chilton Automotive Service manual and/or to the vehicle's Emission Control Information label for the proper engine idle and fast idle adjusting procedures for the particular vehicle being serviced. Adjust to specifications as listed on the Emission Control Information label.

High Vacuum Switch Adjustment
5.7L DIESEL ENGINES
1. Using a self-powered test light, connect the test light lead to either terminal at the high vacuum switch. Connect the test probe to the other switch terminal.
2. With the engine at high idle speed, energize the fast idle solenoid by disconnecting the pink and green wire connector from the coolant temperature switch, normally located on the intake manifold at the rear of the engine.
3. Remove the seal cap on the rear of the high vacuum switch.

— CAUTION —

If the air cleaner assembly is removed when making the switch adjustment, the EGR solenoid port to the EGR valve must be plugged to prevent a vacuum leak at the EGR solenoid.

4. Check the test lamp. If the light glows, this indicates the high vacuum switch has a closing circuit. If the test light is off, turn the switch adjustment screw clockwise until the test light begins to glow. An Allen type hex wrench is needed, sized to 5/16 inch.
5. Adjust the vacuum switch by slowly turning the adjustment screw counterclockwise until the light goes off.
6. With the wrench and screw at just the light off position, turn the wrench clockwise an additionmal 1/8 to 3/16 inch to properly adjust the high vacuum switch.

High Vacuum Switch Check
5.7 L DIESEL ENGINE
1. Disconnect the vacuum hose and electrical connector from the high vacuum switch (if they're not already disconnected).
2. Attach one lead of a test light to either of the two high vacuum switch terminals.
3. Ground the remaining high vacuum switch terminal.
4. Attach the remaining test light lead to the **hot** side of the battery.
5. Attach a hand vacuum pump with a gauge to the high vacuum switch vacuum port.
6. Work the hand vacuum pump.

    a. The test light should remain glowing until the vacuum reading on the hand pump gauge reaches 11.5 to 13 inches and then goes out.

    b. Slowly let air back into the hand pump. The test light should go on when vacuum drops below 11.5 to 13 inches.

7. If the test light does not turn on and off at the specified vacuum values, the high vacuum switch is inoperative and must be replaced.

Low Vacuum Switch Check
1. Disconnect the vacuum hose and electrical connector from the low vacuum switch (if not already disconnected).
2. Attach one lead of a test light to either of the two low vacuum switch terminals.
3. Ground the remaining low vacuum switch terminal.
4. Attach the remaining test light lead to the hot side of the low vacuum switch connector.
5. Attach a hand vacuum pump with a gauge to the vacuum port of the low vacuum switch.
6. Turn the ignition switch to the ON position.
7. Work the hand vacuum pump.

a. The test light should remain off until the hand vacuum pump gauge reads between:
- 5.5 and 6.5 inches for the 3.8 Liter engine
- 6.5 and 7.5 inches for the 4.9 Liter engine
- 7.5 and 8.5 inches for the 5.7 Liter engine
- 5.0 and 6.0 inches for the 5.7 Liter diesel engine

b. The test light should begin glowing between the vacuum values given above.

8. Slowly let air into the vacuum pump.

a. The test light should remain on until the hand vacuum pump gauge drops to between:
- .3 and 1.3 inches for the 3.8 Liter engine
- 1.2 and 2.2 inches for the 4.9 Liter engine
- 1.5 and 2.5 inches for the 5.7 Liter engine
- 3.5 and 4.5 inches for the 5.7 Liter diesel engine

b. The test light should go out between the values given above.

9. If the low vacuum switch does not turn the test light on and off at the vacuum values given above, the switch must be replaced.

NOTE: The high vacuum limit (the point at which the test light begins glowing) and the low vacuum limit (the point at which the test light turns off) must have at least 4 inches of difference for all gasoline engines. On diesel engines, the low vacuum switch will turn the test light on and off at the same vacuum level.

Brake Switch Check

NOTE: Make sure that the brake switch is properly adjusted before performing the following check.

1. Disconnect the electrical connector from the rear of the brake switch. (The exposed terminals are for converter clutch release and cruise control, if so equipped.)
2. Turn the ignition switch to the ON position.
3. Check for current at the connector. (Current should flow from only one terminal; otherwise, the switch is defective and must be replaced.)
4. Ground one of the terminals of the brake release switch with a jumper wire.
5. Connect one test light lead to the remaining brake release switch terminal.
6. Attach the remaining test light lead to the brake switch connector wire. The test light should now glow.
7. Depress brake pedal.

a. If the test light goes out, the brake switch is okay.

b. If the test light is off before brake application, or if the test light does not go off during brake application, the brake switch is inoperative and must be replaced.

Thermal Vacuum Valve (TVV) Check

1. Disconnect the vacuum hose at the thermal vacuum valve EFE port.
2. Attach a vacuum gauge to the EFE port.

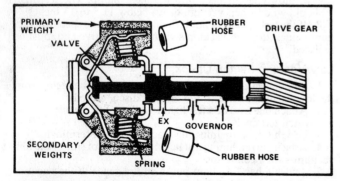

Rubber hose location on governor assembly to test TCC operation, typical (©General Motors Corp.)

3. Start the engine and check the vacuum gauge reading.

a. With a cold engine, the vacuum gauge should read a minimum of 10 inches vacuum.

b. With a warm engine (approximately 5 minutes running at fast idle), the vacuum reading should be zero.

4. If these readings are **not** obtained, the TVV switch is inoperative and must be replaced.

NOTE: An engine is cold when engine coolant temperature is below 150 degrees Fahrenheit for 4.9 Liter engines, and below 120 degrees Fahrenheit for 3.8 and 5.7 Liter engines. Conversely, an engine is considered warm when engine coolant temperature is above 150 degrees Fahrenheit for 4.9 Liter engines, and above 120 degrees Fahrenheit for 3.8 and 5.7 Liter engines.

Diagnosis of Internal Hydraulic/ Mechanical Controls—1980 Models

GENERAL INFORMATION

An internal hydraulic/mechanical controls check may be performed with the transmission in the vehicle; however, this should only be done after all external diagnostic checks have been made. In addition, the internal hydraulic/mechanical controls check can only be performed after the governor has been removed from the transmission and modified.

GOVERNOR MODIFICATIONS

For governors from 250 C and 350 C transmissions:
1. Obtain a length of ½ inch outer diameter vacuum hose.
2. Cut two ¾ inch lengths off of the hose.
3. Insert one piece of hose under each of the two governor weights.

For governors from 200 C transmissions:
1. Obtain a length of 5/16 inch outer diameter vacuum hose.
2. Cut two 3/8 inch lengths off of the hose.
3. Insert each piece of hose under each of the governor weights.

JUMPER WIRE CONNECTIONS

Remove the low vacuum switch electrical connector. (On diesel engine models, also remove the high vacuum switch electrical connector.) Use jumper wires to connect both female terminals on each vacuum switch connector.

To check for proper jumper wire hookups, turn the ignition switch to the ON position, and check voltage at the transmission side of the low vacuum switch, (or the high vacuum switch on diesel engine models)—a twelve volt reading should be obtained. If not, doublecheck the jumper wire connections and check the voltage again.

Internal Hydraulic/Mechanical Controls Check

When a proper voltage reading is obtained:
1. Install the modified test governor.
2. Make sure that the vehicle's rear wheels are several inches off the ground, and apply the parking brake so that the rear wheels cannot turn. (Adjust the parking brake if necessary.)
3. With the gear selector in PARK, start the engine and run it at idle.
4. Step on the brake pedal to break the current flow to the transmission.
5. Place the gear selector lever in the DRIVE position. The modified test governor, with its weights held out by the strips of vacuum tubing, should cause the transmission to shift into third gear.
6. Release the brake pedal; the engine should stall.

IF ENGINE STALLS . . .

If the engine stalls, the transmission converter clutch's internal hydraulic and mechanical controls are working properly. The diagnosis is complete.

Remove the modified test governor from the transmission and take out the piece of vacuum tubing under each of the governor weights.

It is also extremely important to make sure that the governor weight springs are seated in their proper locations before reinstall the governor.

Also, be sure to remove the jumper wire at the low vacuum switch electrical connector and reconnect the connector to the switch. (On diesel engine models, also do the same at the high vacuum switch.)

IF THE ENGINE DOES NOT STALL . . .

If the engine does **not** stall during the internal controls check, check for loose solenoid mounting bolts.

1. On 250 C and 350 C transmissions, also check for a missing check ball in the solenoid.
2. For 200 C transmissions, check for a missing check ball or "O" ring.

CONVERTER CLUTCH SOLENOID LOCATIONS

THM 180C	On valve body
THM 200C	On oil pump
THM 200-4R	Through case and into oil pump
THM 250C	Valve body
THM 350C	Valve body
THM 700-R4	Through case and into oil pump
THM 125C	Valve body
THM 325-4L	On accumulator housing
THM 440	Valve body

Solenoid Checks

WITH REMOVABLE CHECK BALL ASSEMBLY

1. Remove the solenoid from the transmission/transaxle assembly.
2. Press on the bottom of the solenoid to remove the plastic check ball assembly.
3. Disassemble the check ball assembly and check for scores, nicks or scratches on the check ball and ball seat. If necessary, replace the solenoid assembly.
4. When reassembling the check ball assembly, check for the ball seat being offset and the check ball positioned directly under the ball seat.
5. Reinstall the check ball assembly into the solenoid and reinstall the solenoid into the transmission/transaxle assembly.

WITH NON-REMOVABLE CHECK BALL ASSEMBLY

1. Remove the solenoid from the transmission/transaxle assembly.
2. Hold the solenoid up to a light and visually inspect the check ball and the "O" ring.
3. If either the ball or the "O" ring seems scratched, nicked or scored, replace the solenoid.
4. Replace the solenoid into the transmission/transaxle.

Other Possible Problems

Other possible problems that can effect the converter clutch system, as well as other major assemblies in the transmission/transaxle units include:

1. An apply valve which is sticking, binding or damaged.
2. A turbine shaft "O" ring which is missing or damaged.
3. Direct clutch oil passages which are blocked, restricted or interconnected.

Having completed all diagnostic and servicing procedures, be sure to road test the vehicle a second time to make sure the problem is resolved.

Diagnosis of Control Components— 1981 and Later With and Without Electronic Control Module (ECM)

GENERAL INFORMATION

Tachometers and Ohmmeters are used to test the varied circuits in the 1981 and later converter clutch systems. Because of the many transmission/transaxle applications, it is important to identify the unit and its controlling components, in order to properly test and diagnose the converter clutch operation.

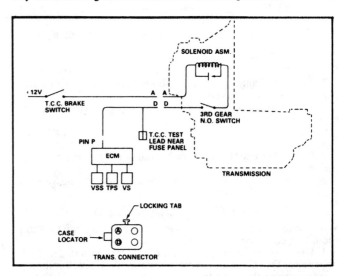

1981 and later TCC wiring schematic when equipped with Computer Command Control, typical (©General Motors Corp.)

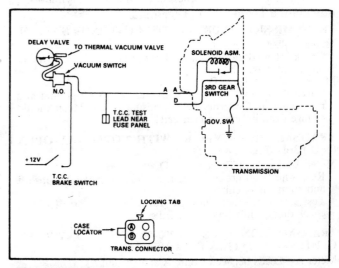

1981 and later TCC wiring schematic except with Computer Command Control, typical (©General Motors Corp.)

TACHOMETER DIAGNOSIS

The tachometer is used to verify the converter clutch operation during a road test with the vehicle weight and resistance acting upon the converter. An on-hoist test is inconclusive because the rpm changes in light load conditions are hard to detect.

The tachometer to be used should be one of the following;

1. Equipped with a primary pick-up lead.
2. Equipped with a secondary pick-up lead.
3. Balancer mag-tach type.

The tachometer that is used should be mounted in the passenger compartment, with the leads routed and taped to prevent damage. Place the tachometer in a position that it is easily readable. Observe all driving safety regulations.

Testing With the Tachometer
1. Verify the tachometer is operating properly.
2. Road test the vehicle, using the drive position.
3. Allow the transmission/transaxle to shift into the top gear.
4. Hold the road speed steady and read the tachometer.
5. While holding the speed steady, depress the brake pedal just enough to open the converter clutch circuit and read the tachometer.
6. Shift to the next lowest manual range and repeat steps 4 and 5.
7. Shift back into the Drive position and monitor all shift points, as required.

Results of Test

1. The rpm change is step 4 to 5 should increase. This indicates slippage when the converter clutch is disengaged.
2. The rpm change will be greater if the top gear is an overdrive gear.
3. The rpm change will be greater with greater engine load, such as on an incline.
4. The tachometer must be read at the same mph indicated on the speedometer.
5. If no rpm change is indicated, converter clutch engagement is not occurring.

OHMMETER DIAGNOSIS

When testing with an ohmmeter, the following information should be observed;
1. Use only an ohmmeter with a needle movement. The high impedance DVOM does not operate because the small miliamp current passing through the solenoid coil, reacts as though the resistance was zero.
2. Disconnect the wire harness at the transmission/transaxle.

Resistance Testing With Ohmmeter

TRANSMISSION TYPE A—WITH GOVERNOR SWITCH
Meter A—Leads Term A—D.
   Check solenoid diode—no need to run engine. Reverse meter leads and make 2nd check.
Meter B—Leads Term D to ground.
   Check Governor Switch requires rear wheels off ground and run vehicle to Governor Switch closing speed—Meter should change from infinity to near zero.

TRANSMISSION TYPE B—WITH NORMALLY OPEN (N.O.) 3rd GEAR SWITCH
Meter A—Leads to Term A and D.
   Rear wheels off ground, engine running to make 2-3 upshift and complete circuit.
   Test solenoid, diode and 3rd gear switch. Reverse leads to check diode while holding in 3rd gear.

TRANSMISSION TYPE C—200-4R WITH CONVERTER CLUTCH OPERATION IN FOURTH GEAR
Meter A—Leads between Conn. A and ground.
   Engine stopped, check continuity of solenoid and diode. Reverse leads and check diode.
Meter B—Leads Term D to ground.
   Check Governor Switch requires rear wheels off ground and run vehicle to Governor Switch closing speed—Meter should change from infinity to near zero.

TRANSMISSION TYPE D—
WITH NORMALLY CLOSED (N.C.) 4-3 PULSE SWITCH
WITH NORMALLY OPEN (N.O.) 4TH GEAR SWITCH
WITH NORMALLY OPEN (N.O.) 3RD GEAR SWITCH
AND NORMALLY
OPEN (N.O.) CONVERTER CLUTCH SIGNAL SWITCH

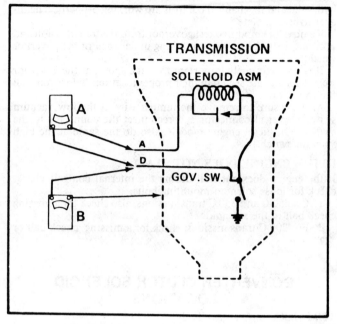

Transmission type A (©General Motors Corp.)

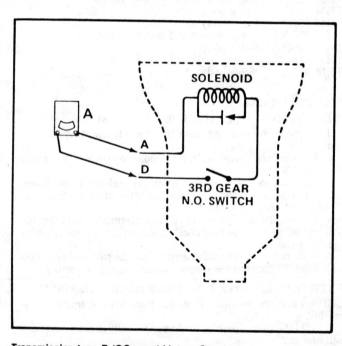

Transmission type B (©General Motors Corp.)

Meter A—Leads between Conn. A and ground.
   Step 1—Engine stopped, check continuity of solenoid, diode and 4-3 sw. Reverse leads and check diode.
   Step 2—Run engine with rear wheels off ground and make downshift to check operation of 4-3 sw.
Meter B—Leads between Conn. and A & B.
   Check 4th clutch switch requires rear wheels off ground. Run engine to make 3-4 upshift—Meter should change from infinite to zero.
Meter C—Leads Term D to ground.
   Requires rear wheels off ground. Run engine and vehicle to close 3rd clutch and TCC signal switches. Meter change from infinite to near zero.

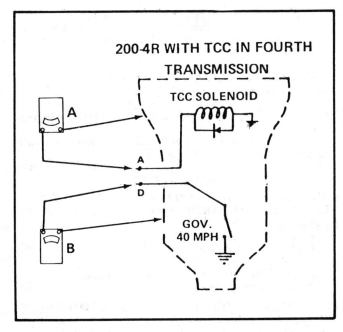

Transmission type C (©General Motors Corp.)

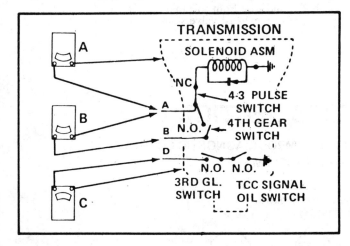

Transmission type D (©General Motors Corp.)

TRANSAXLE TYPE A—
WITH NORMALLY CLOSED (N.C.) 4-3 DOWNSHIFT
SWITCH
WITH NORMALLY CLOSED (N.C.) 4TH CLUTCH
SWITCH
WITH THREE WIRE CONNECTOR
Meter A—Leads between Conn. A and B.
  Step 1—Engine stopped, check continuity of solenoid, diode
  and 4-3 sw. Reverse leads and check diode.
  Step 2—Run engine with rear wheels off ground and make
  downshift to check operation of 4-3 sw.
Meter B—Check 4th clutch switch requires rear wheels off
  ground. Run engine to make 3-4 upshift—Meter should change
  from zero to infinite.

TRANSAXLE TYPE B—
WITH NORMALLY CLOSED (N.C.) 4-3 DOWNSHIFT
SWITCH
WITH NORMALLY OPEN (N.O.) GOVERNOR SWITCH
WITH TWO WIRE CONNECTOR
Meter A—Leads between Conn. A and ground.
  Step 1—Engine stopped, check continuity of solenoid, diode
  and 4-3 switch. Reverse leads and check diode.
  Step 2—Rear wheels off ground, run engine and make down-
  shift to check operation of 4-3 pulse switch.
Meter B—Leads Term D to ground.
  Requires rear wheels off ground. Run engine and vehicle to
  Governor Switch closing speed—Meter change from infinite to
  near zero.

Solenoid Diode Check

The transmission/transaxle solenoids should not be bench tested
by touching the leads to an automobile battery. The internal diode will be destroyed by touching the negative terminal (black,
marked with a −) to the positive battery terminal and the positive terminal (red, marked with a +) to the negative battery terminal. The diodes used in the solenoids must only be checked using a meter reading or scale type ohmmeters, set on the X1 scale.
Electronic or digital type meters canot be used because of false indications being obtained.

Testing Procedure

  1. Set the ohmmeter on the X1 scale and zero the needle.
  2. Connect the positive meter and solenoid leads together and
the negative meter and solenoid leads together.
  3. If the meter reading is 20 to 40 ohms (depending upon solenoid temperature), the diode and/or coil is not shorted.
  4. If the meter reading is 0 ohms, the diode or coil is shorted.
  5. If the meter reading indicates an open circuit reading, the
coil is open.

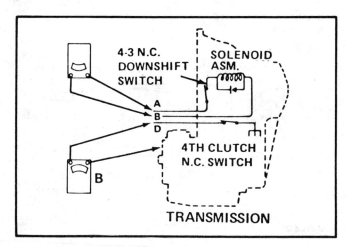

Transaxle type A (©General Motors Corp.)

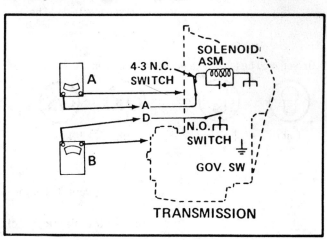

Transaxle type B (©General Motors Corp.)

6. When the ohmmeter leads are reversed, the solenoid is operative if the reading is 2 to 15 ohms less than those in step 3.

7. If the meter reading is zero, the diode is open.

Converter Clutch Diagnose When Equipped with C.C.C., E.C.M. or Governor Switch Control

Preliminary Checks

1. Check transmission/transaxle oil level and correct as required.

2. Check manual linkage and throttle cable adjustments as required.

3. Road test vehicle to verify complaint at normal operating temperature.

NOTE: If the engine performance indicates the need for an engine tune-up, this should be done before further testing or diagnosing of the unit is made. Poor engine performance can result in rough shifting or other malfunctions.

Road Test Results

CONVERTER CLUTCH APPLIES ERRATICALLY

(This may be described as a shudder, jerking, jumping or rocking sensation).

Possible causes:

a. Vacuum hose leakage.

b. Vacuum switch malfunction.

c. Release oil exhaust orifice at pump blocked or restricted.

d. Turbine shaft "O" ring damaged or missing.

e. Converter malfunction, such as pressure plate warped, etc.

f. "O" ring at solenoid damaged or missing.

g. Solenoid bolts loose, 200C, 200-4R, 125C

CONVERTER CLUTCH APPLIED IN ALL RANGES

(Engine will stall when put in gear).

Possible causes:

a. Converter clutch valve in pump stuck in the apply position. 200C, 200-4R.

b. Converter clutch valve in the auxiliary valve body stuck in the apply position. 250C, 350C Units.

c. Converter clutch control valve or converter clutch regulator valve stuck in the applied position. 125C Unit.

CONVERTER CLUTCH APPLIES AT VERY LOW OR VERY HIGH 3RD GEAR SPEEDS

Possible causes:

a. Governor switch malfunction.

b. Governor Malfunction.

c. High Main line pressure.

d. Converter clutch valve sticking or binding.

e. Solenoid malfunction.

CONVERTER CLUTCH APPLIED AT ALL TIMES IN 3RD GEAR

Possible causes:

a. Governor pressure switch shorted to ground.

b. Ground wire from solenoid shorted to case.

c. Solenoid exhaust valve stuck closed.

HOW TO USE THIS SECTION

THIS SECTION CONTAINS A SERIES OF PICTURE/SYMBOL DIAGNOSTIC CHARTS THAT WILL HELP YOU QUICKLY FIND THE CAUSE OF A PROBLEM.

The charts use symbols, like those used on highway signs. For example, you may be told to:

Connect Disconnect Check Repair Replace Or Adjust Stop

Or, you may be told that a particular condition is:

Light No Limit OK Not OK

In addition, the charts use pictures like these:

Ignition On Dwell Meter Voltmeter Digital Voltmeter

Plus a few words to tell you what to do.

To find the diagnostic chart you need:

• Perform Diagnostic Circuit Check

TORQUE CONVERTER CLUTCH DIAGNOSIS
DIAGNOSIS CIRCUIT CHECK
1981-83 CCC System, All Models
(incl. 1980½ 3.8L and 4.3L Engines)

Step/Sequence   **Result**

Always check PROM for correct application and installation before replacing an ECM. Also, remove terminal(s) from ECM connector for circuit involved, clean terminal contact and expand it slightly to increase contact pressure and recheck to see if problem is corrected.

1982-83 (Min. Function) Chevette and T-1000, All w/Automatic Trans. and Federal Models with 4 speed Manual Trans.:
Trouble codes are lost when the ignition is turned off. Therefore, on "Check Engine" light complaints, codes should be checked before ignition is turned off, if possible.

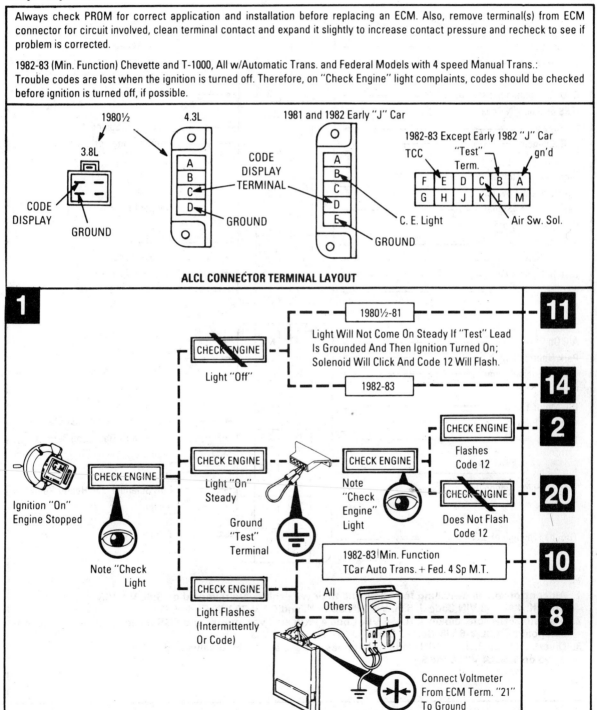

ALCL CONNECTOR TERMINAL LAYOUT

ECM TERMINAL IDENTIFICATION
1981 CCC System Production
(Note: Not all terminals used on all engine applications)

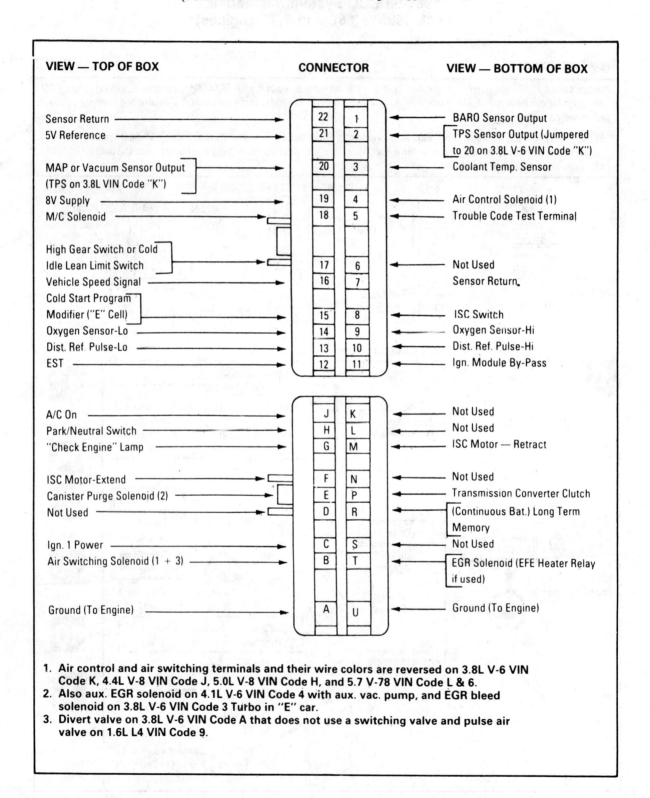

VIEW — TOP OF BOX	CONNECTOR	VIEW — BOTTOM OF BOX

1. **Air control and air switching terminals and their wire colors are reversed on 3.8L V-6 VIN Code K, 4.4L V-8 VIN Code J, 5.0L V-8 VIN Code H, and 5.7 V-78 VIN Code L & 6.**
2. **Also aux. EGR solenoid on 4.1L V-6 VIN Code 4 with aux. vac. pump, and EGR bleed solenoid on 3.8L V-6 VIN Code 3 Turbo in "E" car.**
3. **Divert valve on 3.8L V-6 VIN Code A that does not use a switching valve and pulse air valve on 1.6L L4 VIN Code 9.**

ECM TERMINAL IDENTIFICATION
1982 CCC System
(Note: Not all terminals used on all engine applications)

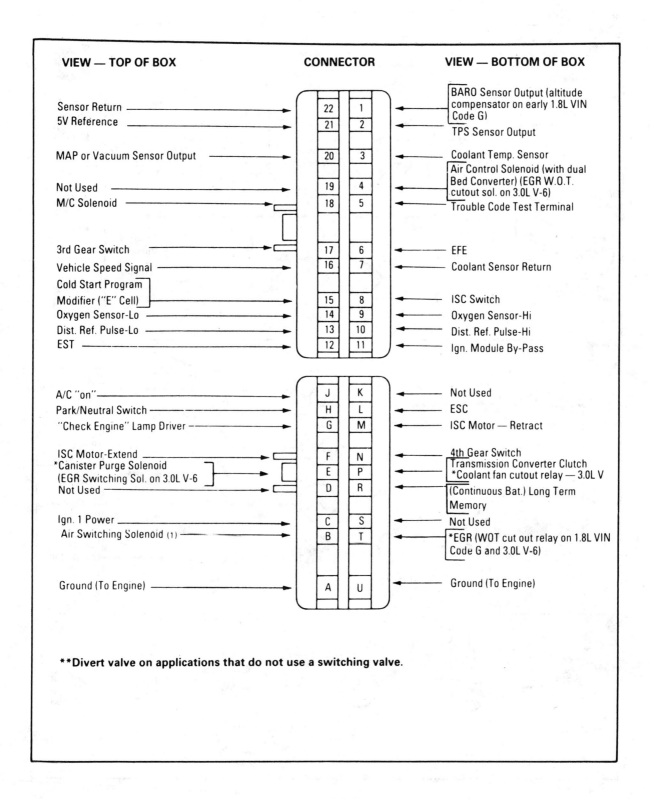

VIEW — TOP OF BOX

Terminal	Signal
22	Sensor Return
21	5V Reference
20	MAP or Vacuum Sensor Output
19	Not Used
18	M/C Solenoid
17	3rd Gear Switch
16	Vehicle Speed Signal
15	Cold Start Program Modifier ("E" Cell)
14	Oxygen Sensor-Lo
13	Dist. Ref. Pulse-Lo
12	EST

Terminal	Signal
J	A/C "on"
H	Park/Neutral Switch
G	"Check Engine" Lamp Driver
F	ISC Motor-Extend
E	*Canister Purge Solenoid (EGR Switching Sol. on 3.0L V-6)
D	Not Used
C	Ign. 1 Power
B	Air Switching Solenoid (1)
A	Ground (To Engine)

VIEW — BOTTOM OF BOX

Terminal	Signal
1	BARO Sensor Output (altitude compensator on early 1.8L VIN Code G)
2	TPS Sensor Output
3	Coolant Temp. Sensor
4	Air Control Solenoid (with dual Bed Converter) (EGR W.O.T. cutout sol. on 3.0L V-6)
5	Trouble Code Test Terminal
6	EFE
7	Coolant Sensor Return
8	ISC Switch
9	Oxygen Sensor-Hi
10	Dist. Ref. Pulse-Hi
11	Ign. Module By-Pass

Terminal	Signal
K	Not Used
L	ESC
M	ISC Motor — Retract
N	4th Gear Switch
P	Transmission Converter Clutch *Coolant fan cutout relay — 3.0L V
R	(Continuous Bat.) Long Term Memory
S	Not Used
T	*EGR (WOT cut out relay on 1.8L VIN Code G and 3.0L V-6)
U	Ground (To Engine)

**Divert valve on applications that do not use a switching valve.

TROUBLE CODE 24—VEHICLE SPEED SENSOR
Carbureted Engines

| Step/Sequence | Note: Be Sure Speedometer Works Before Proceeding. | Result |

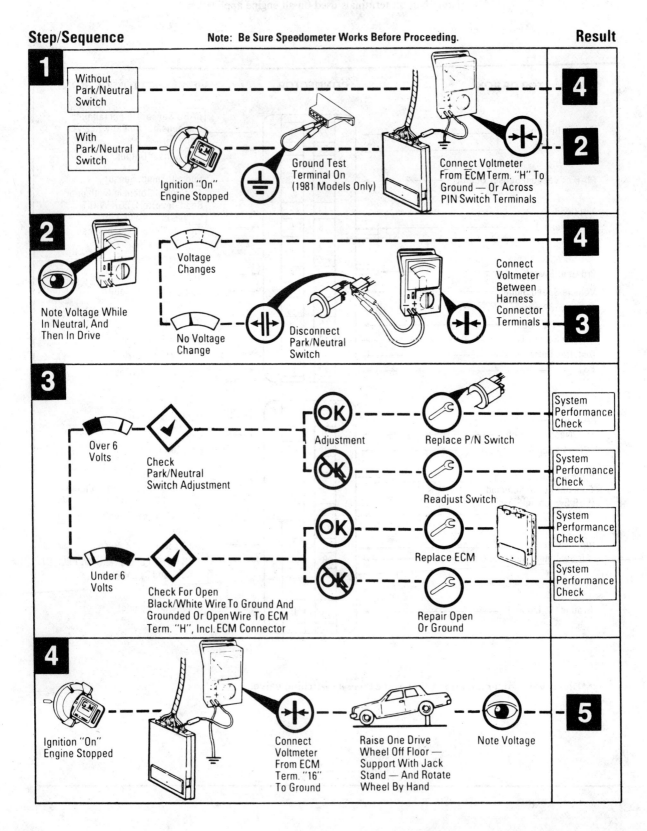

TROUBLE CODE 24—VEHICLE SPEED SENSOR (Cont.)
Carbureted Engines

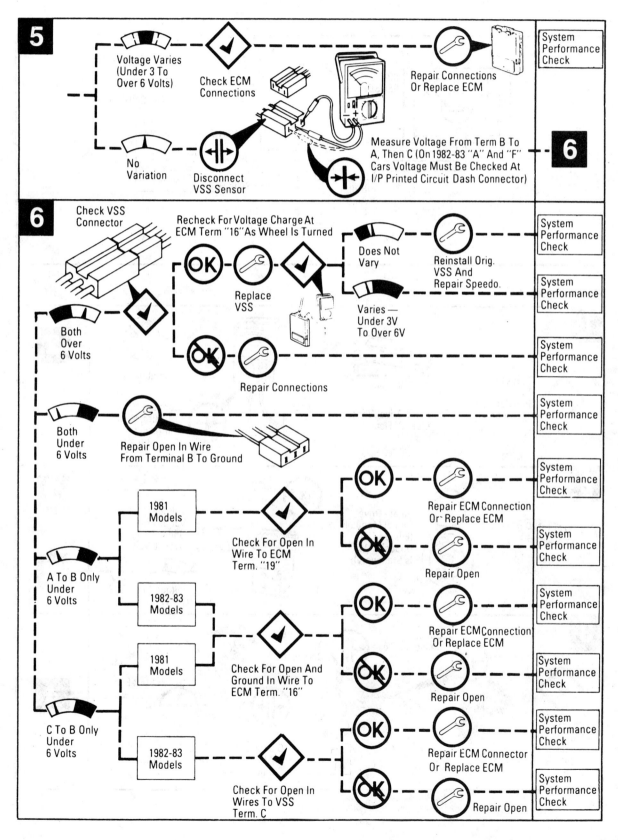

TROUBLE CODE 24—VEHICLE SPEED SENSOR
Throttle Bore Injected Engines

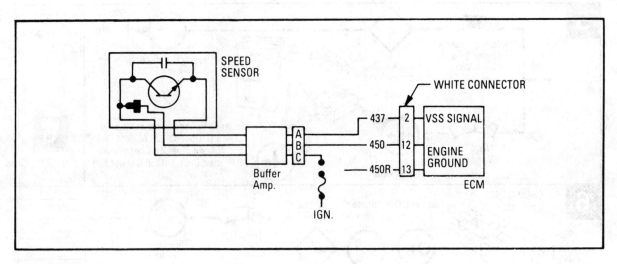

Step/Sequence   **Result**

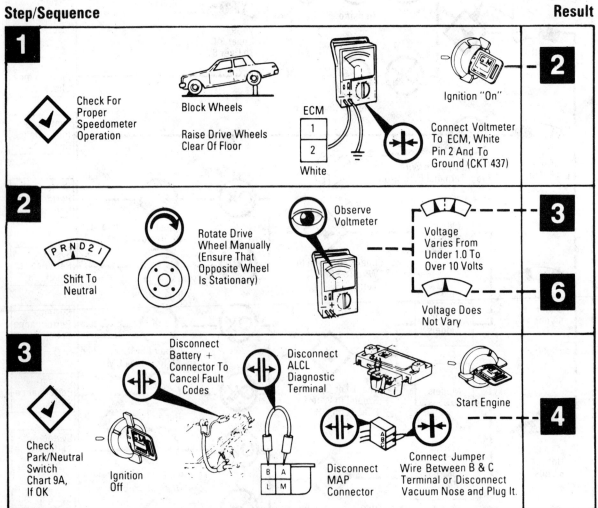

TROUBLE CODE 24—VEHICLE SPEED SENSOR (Cont.)
Throttle Bore Injected Engines

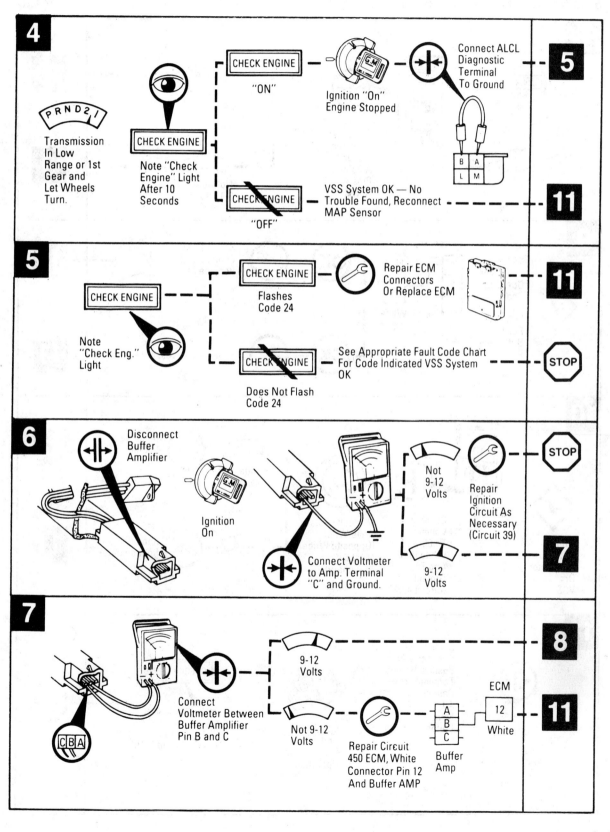

TROUBLE CODE 24—VEHICLE SPEED SENSOR (Cont.)
Throttle Bore Injected Engines

Step/Sequence   **Result**

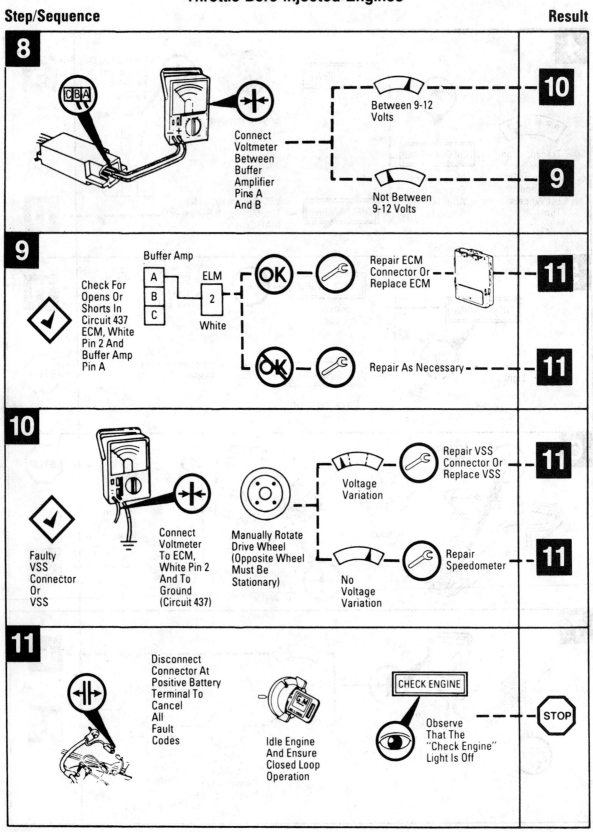

TORQUE CONVERTER CLUTCH (TCC) ELECTRICAL DIAGNOSIS
Fuel Injected Engines

Note: Mechanical checks, such as linkage, oil level, etc., should be performed prior to using this diagnostic procedure.

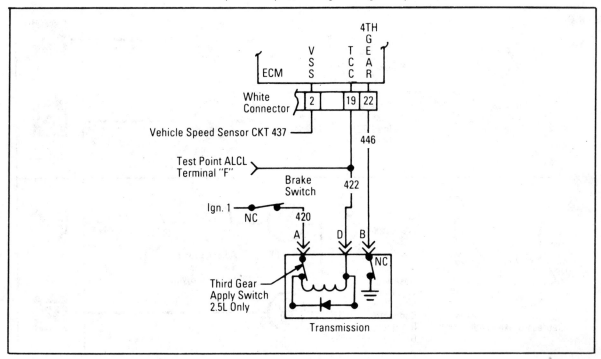

Step/Sequence   **Result**

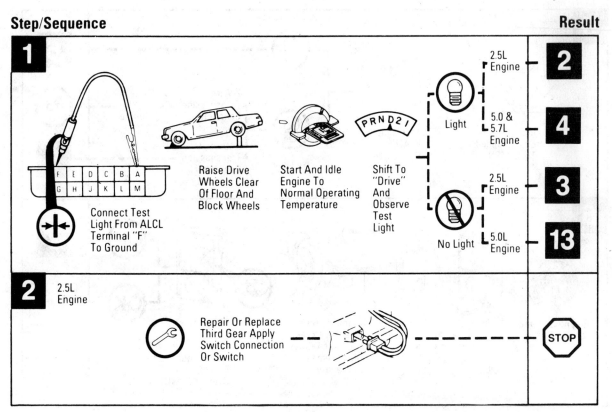

TORQUE CONVERTER CLUTCH (TCC) ELECTRICAL DIAGNOSIS (Cont.)
Fuel Injected Engines

Step/Sequence   **Result**

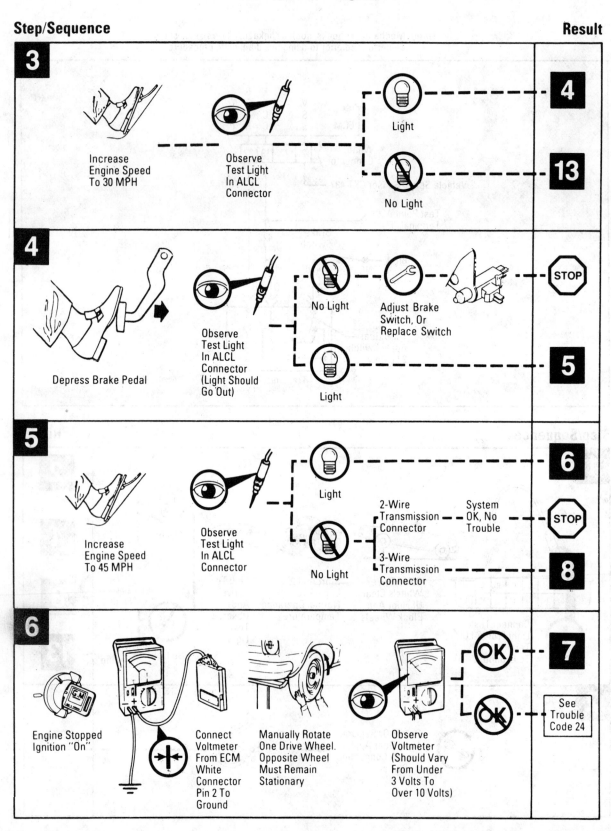

3 — Increase Engine Speed To 30 MPH → Observe Test Light In ALCL Connector → Light → **4** / No Light → **13**

4 — Depress Brake Pedal → Observe Test Light In ALCL Connector (Light Should Go Out) → No Light → Adjust Brake Switch, Or Replace Switch → **STOP** / Light → **5**

5 — Increase Engine Speed To 45 MPH → Observe Test Light In ALCL Connector → Light → **6** / No Light → 2-Wire Transmission Connector → System OK, No Trouble → **STOP** / 3-Wire Transmission Connector → **8**

6 — Engine Stopped Ignition "On" → Connect Voltmeter From ECM White Connector Pin 2 To Ground → Manually Rotate One Drive Wheel. Opposite Wheel Must Remain Stationary → Observe Voltmeter (Should Vary From Under 3 Volts To Over 10 Volts) → OK → **7** / Not OK → See Trouble Code 24

TORQUE CONVERTER CLUTCH (TCC) ELECTRICAL DIAGNOSIS (Cont.)
Fuel Injected Engines

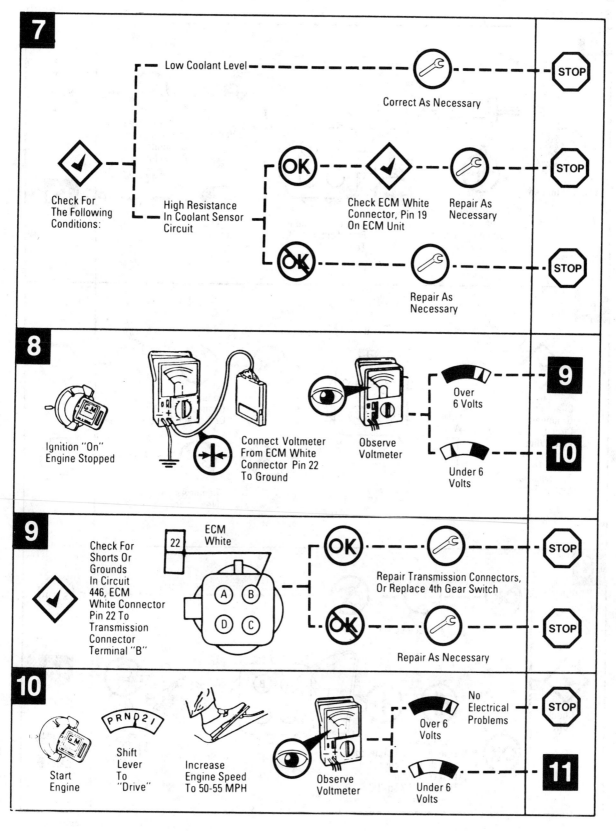

7

Low Coolant Level — Correct As Necessary — STOP

Check For The Following Conditions:

High Resistance In Coolant Sensor Circuit

OK — Check ECM White Connector, Pin 19 On ECM Unit — Repair As Necessary — STOP

OK — Repair As Necessary — STOP

8

Ignition "On" Engine Stopped

Connect Voltmeter From ECM White Connector Pin 22 To Ground

Observe Voltmeter

Over 6 Volts — **9**

Under 6 Volts — **10**

9

Check For Shorts Or Grounds In Circuit 446, ECM White Connector Pin 22 To Transmission Connector Terminal "B"

22 ECM White

A B
D C

OK — Repair Transmission Connectors, Or Replace 4th Gear Switch — STOP

OK — Repair As Necessary — STOP

10

Start Engine

PRND21 Shift Lever To "Drive"

Increase Engine Speed To 50-55 MPH

Observe Voltmeter

Over 6 Volts — No Electrical Problems — STOP

Under 6 Volts — **11**

TORQUE CONVERTER CLUTCH (TCC) ELECTRICAL DIAGNOSIS (Cont.)
Fuel Injected Engines

Step/Sequence   **Result**

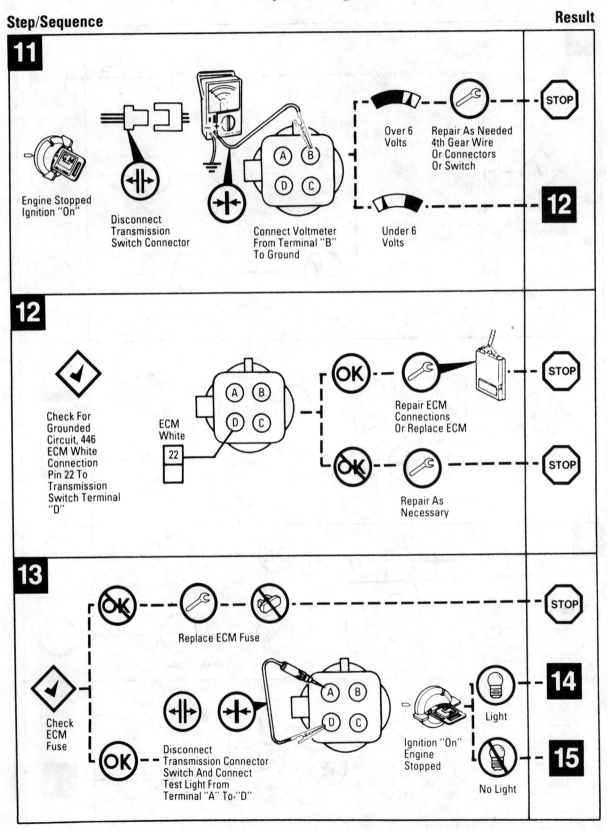

11

Engine Stopped Ignition "On"

Disconnect Transmission Switch Connector

Connect Voltmeter From Terminal "B" To Ground

Over 6 Volts — Repair As Needed 4th Gear Wire Or Connectors Or Switch → STOP

Under 6 Volts → **12**

12

Check For Grounded Circuit, 446 ECM White Connection Pin 22 To Transmission Switch Terminal "D"

ECM White 22

OK — Repair ECM Connections Or Replace ECM → STOP

NOT OK — Repair As Necessary → STOP

13

Check ECM Fuse

NOT OK — Replace ECM Fuse → STOP

OK — Disconnect Transmission Connector Switch And Connect Test Light From Terminal "A" To "D"

Ignition "On" Engine Stopped

Light → **14**

No Light → **15**

TORQUE CONVERTER CLUTCH (TCC) ELECTRICAL DIAGNOSIS (Cont.)
Fuel Injected Engines

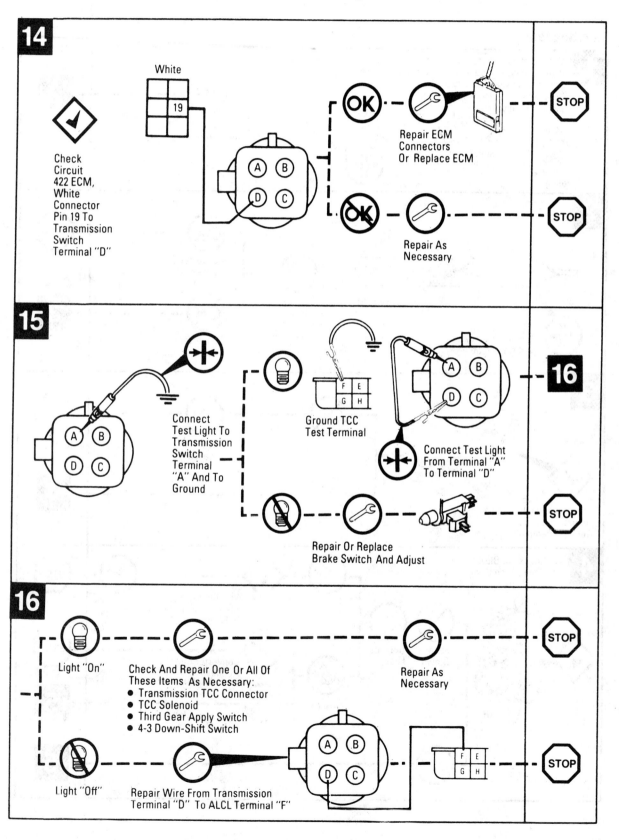

14

White

19

✓ Check Circuit 422 ECM, White Connector Pin 19 To Transmission Switch Terminal "D"

A B
D C

OK — Repair ECM Connectors Or Replace ECM — STOP

OK — Repair As Necessary — STOP

15

Connect Test Light To Transmission Switch Terminal "A" And To Ground

Ground TCC Test Terminal

Connect Test Light From Terminal "A" To Terminal "D"

16

Repair Or Replace Brake Switch And Adjust — STOP

16

Light "On" — Check And Repair One Or All Of These Items As Necessary:
● Transmission TCC Connector
● TCC Solenoid
● Third Gear Apply Switch
● 4-3 Down-Shift Switch

Repair As Necessary — STOP

Light "Off" — Repair Wire From Transmission Terminal "D" To ALCL Terminal "F"

STOP

TORQUE CONVERTER CLUTCH DOES NOT APPLY
1980 Gasoline and Diesel Engine Applications

Step/Sequence   **Result**

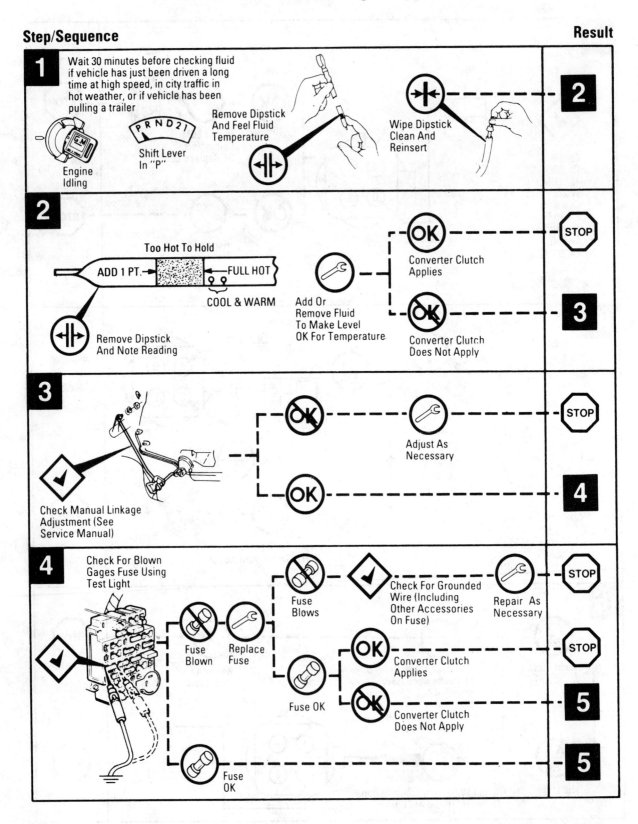

TORQUE CONVERTER CLUTCH DOES NOT APPLY (Cont.)
1980 Gasoline and Diesel Engine Applications

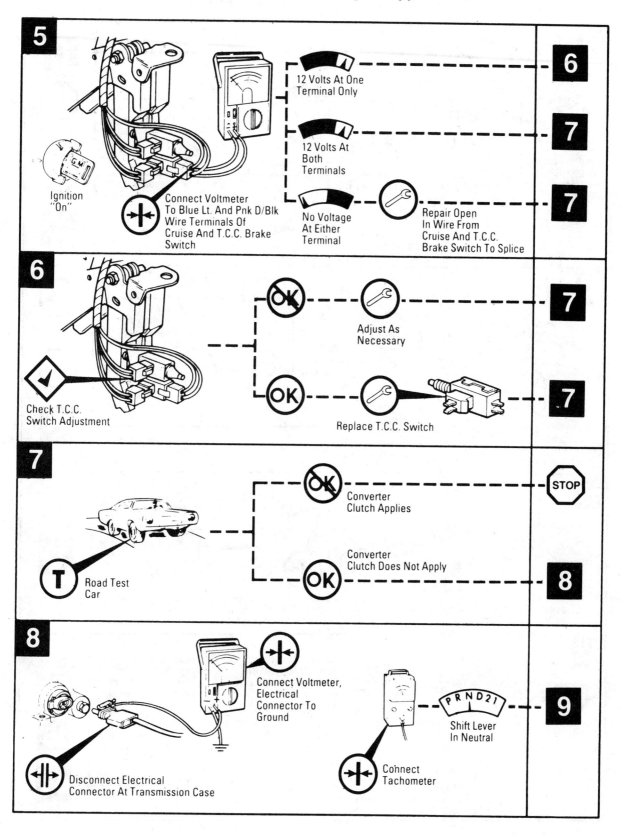

5 — Ignition "On"
Connect Voltmeter To Blue Lt. And Pnk D/Blk Wire Terminals Of Cruise And T.C.C. Brake Switch

- 12 Volts At One Terminal Only → **6**
- 12 Volts At Both Terminals → **7**
- No Voltage At Either Terminal → Repair Open In Wire From Cruise And T.C.C. Brake Switch To Splice → **7**

6 — Check T.C.C. Switch Adjustment

- OK → Adjust As Necessary → **7**
- OK → Replace T.C.C. Switch → **7**

7 — Road Test Car

- Converter Clutch Applies → STOP
- Converter Clutch Does Not Apply → **8**

8 — Disconnect Electrical Connector At Transmission Case
Connect Voltmeter, Electrical Connector To Ground
Connect Tachometer
Shift Lever In Neutral (P R N D 2 1) → **9**

TORQUE CONVERTER CLUTCH DOES NOT APPLY (Cont.)
1980 Gasoline and Diesel Engine Applications

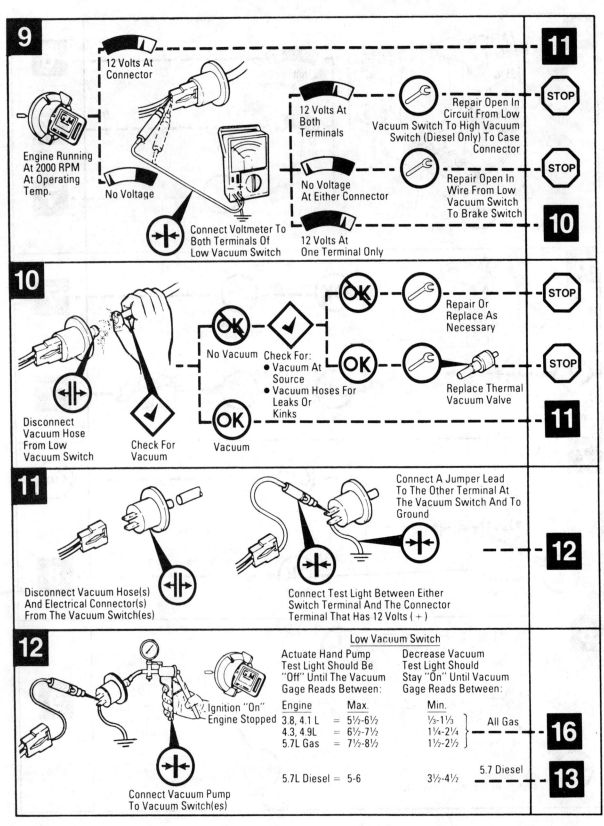

9

12 Volts At Connector

Engine Running At 2000 RPM At Operating Temp.

No Voltage

Connect Voltmeter To Both Terminals Of Low Vacuum Switch

12 Volts At Both Terminals

No Voltage At Either Connector

12 Volts At One Terminal Only

11

Repair Open In Circuit From Low Vacuum Switch To High Vacuum Switch (Diesel Only) To Case Connector **STOP**

Repair Open In Wire From Low Vacuum Switch To Brake Switch **STOP**

10

10

Disconnect Vacuum Hose From Low Vacuum Switch

Check For Vacuum

No Vacuum

Check For:
● Vacuum At Source
● Vacuum Hoses For Leaks Or Kinks

Vacuum

Repair Or Replace As Necessary **STOP**

Replace Thermal Vacuum Valve **STOP**

11

11

Disconnect Vacuum Hose(s) And Electrical Connector(s) From The Vacuum Switch(es)

Connect Test Light Between Either Switch Terminal And The Connector Terminal That Has 12 Volts (+)

Connect A Jumper Lead To The Other Terminal At The Vacuum Switch And To Ground

12

12

Ignition "On" Engine Stopped

Connect Vacuum Pump To Vacuum Switch(es)

Low Vacuum Switch

	Actuate Hand Pump Test Light Should Be "Off" Until The Vacuum Gage Reads Between:	Decrease Vacuum Test Light Should Stay "On" Until Vacuum Gage Reads Between:	
Engine	Max.	Min.	
3.8, 4.1 L =	5½-6½	$\frac{1}{3}$-1$\frac{1}{3}$	All Gas
4.3, 4.9L =	6½-7½	1¼-2¼	
5.7L Gas =	7½-8½	1½-2½	
			16
5.7L Diesel =	5-6	3½-4½	5.7 Diesel **13**

TORQUE CONVERTER CLUTCH DOES NOT APPLY (Cont.)
1980 Gasoline and Diesel Engine Applications

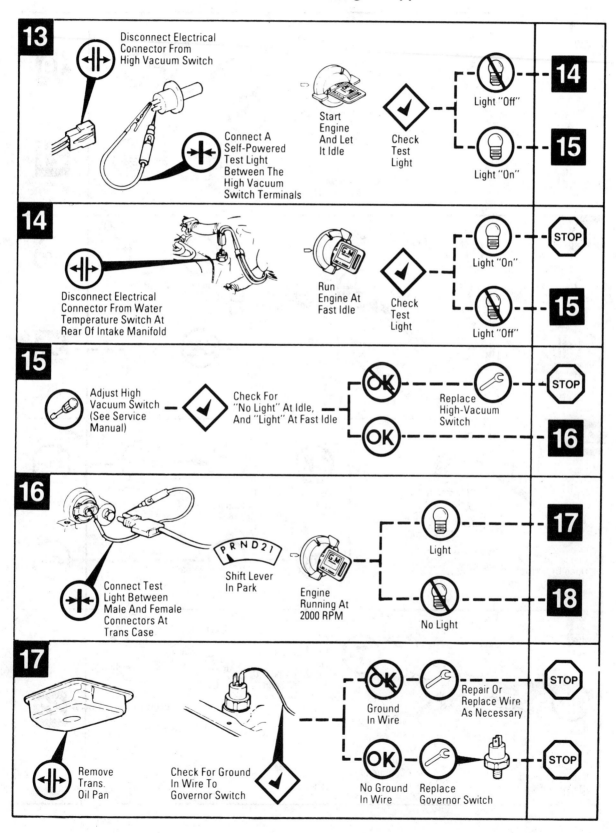

TORQUE CONVERTER CLUTCH DOES NOT APPLY (Cont.)
1980 Gasoline and Diesel Engine Applications

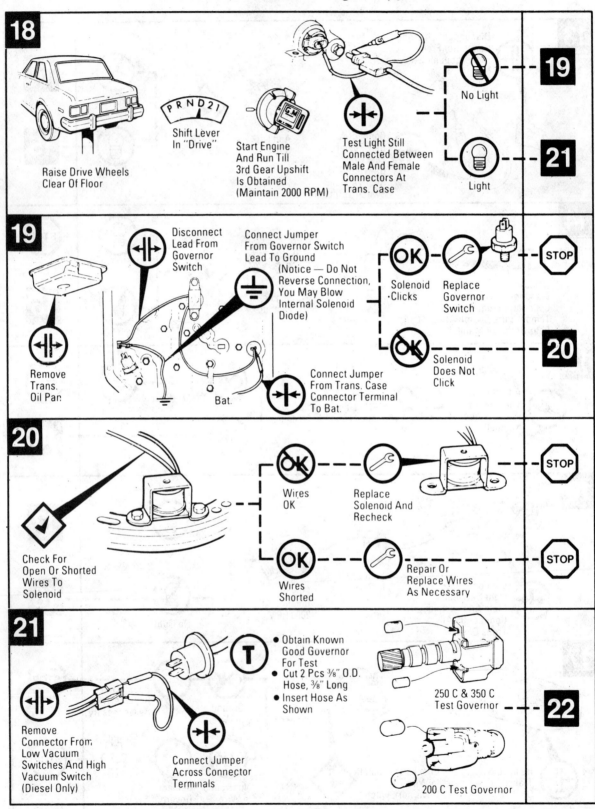

18

Raise Drive Wheels Clear Of Floor

PRND21 — Shift Lever In "Drive"

Start Engine And Run Till 3rd Gear Upshift Is Obtained (Maintain 2000 RPM)

Test Light Still Connected Between Male And Female Connectors At Trans. Case

No Light — **19**

Light — **21**

19

Remove Trans. Oil Pan

Disconnect Lead From Governor Switch

Connect Jumper From Governor Switch Lead To Ground (Notice — Do Not Reverse Connection, You May Blow Internal Solenoid Diode)

Connect Jumper From Trans. Case Connector Terminal To Bat.

Bat.

OK Solenoid Clicks — Replace Governor Switch — STOP

OK Solenoid Does Not Click — **20**

20

Check For Open Or Shorted Wires To Solenoid

OK Wires OK — Replace Solenoid And Recheck — STOP

OK Wires Shorted — Repair Or Replace Wires As Necessary — STOP

21

Remove Connector From Low Vacuum Switches And High Vacuum Switch (Diesel Only)

Connect Jumper Across Connector Terminals

T
• Obtain Known Good Governor For Test
• Cut 2 Pcs ⅜" O.D. Hose, ⅜" Long
• Insert Hose As Shown

250 C & 350 C Test Governor

200 C Test Governor

22

TORQUE CONVERTER CLUTCH DOES NOT APPLY (Cont.)
1980 Gasoline and Diesel Engine Applications

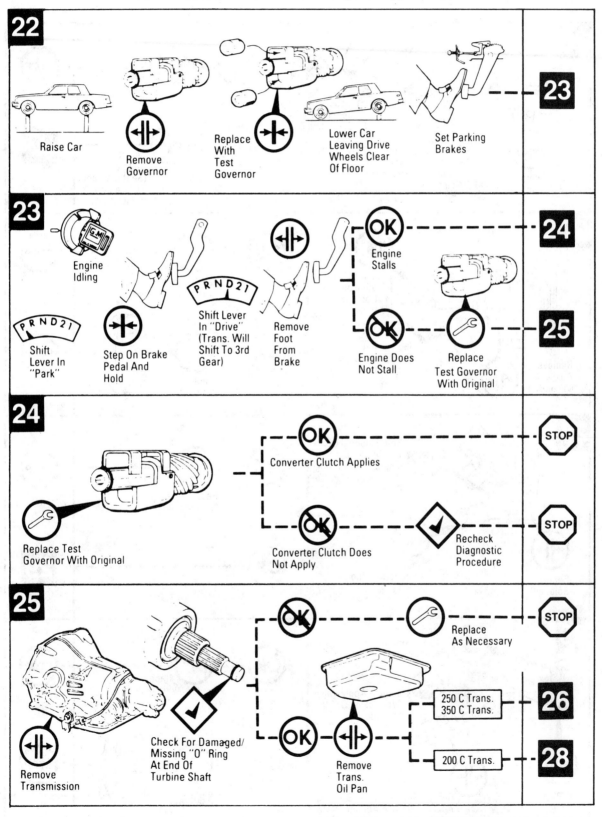

22

Raise Car — Remove Governor — Replace With Test Governor — Lower Car Leaving Drive Wheels Clear Of Floor — Set Parking Brakes → **23**

23

Engine Idling — PRND21 Shift Lever In "Park" — Step On Brake Pedal And Hold — PRND21 Shift Lever In "Drive" (Trans. Will Shift To 3rd Gear) — Remove Foot From Brake — Engine Stalls **OK** → **24** — Engine Does Not Stall — Replace Test Governor With Original → **25**

24

Replace Test Governor With Original — Converter Clutch Applies **OK** → **STOP** — Converter Clutch Does Not Apply — Recheck Diagnostic Procedure → **STOP**

25

Remove Transmission — Check For Damaged/Missing "O" Ring At End Of Turbine Shaft — Replace As Necessary → **STOP** — Remove Trans. Oil Pan — 250 C Trans. 350 C Trans. → **26** — 200 C Trans. → **28**

TORQUE CONVERTER CLUTCH DOES NOT APPLY (Cont.)
1980 Gasoline and Diesel Engine Applications

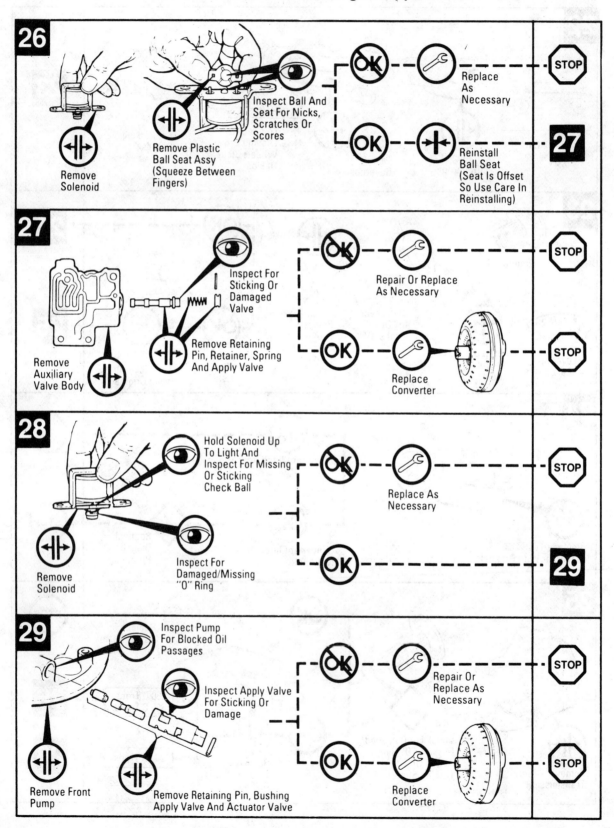

26

Remove Solenoid

Remove Plastic Ball Seat Assy (Squeeze Between Fingers)

Inspect Ball And Seat For Nicks, Scratches Or Scores

OK — Replace As Necessary — STOP

OK — Reinstall Ball Seat (Seat Is Offset So Use Care In Reinstalling) — **27**

27

Remove Auxiliary Valve Body

Inspect For Sticking Or Damaged Valve

Remove Retaining Pin, Retainer, Spring And Apply Valve

OK — Repair Or Replace As Necessary — STOP

OK — Replace Converter — STOP

28

Remove Solenoid

Hold Solenoid Up To Light And Inspect For Missing Or Sticking Check Ball

Inspect For Damaged/Missing "O" Ring

OK — Replace As Necessary — STOP

OK — **29**

29

Remove Front Pump

Inspect Pump For Blocked Oil Passages

Inspect Apply Valve For Sticking Or Damage

Remove Retaining Pin, Bushing Apply Valve And Actuator Valve

OK — Repair Or Replace As Necessary — STOP

OK — Replace Converter — STOP

TORQUE CONVERTER CLUTCH DOES NOT APPLY
1981 and later TCC Diagnosis without EGR Bleed Solenoid

Step/Sequence   **Result**

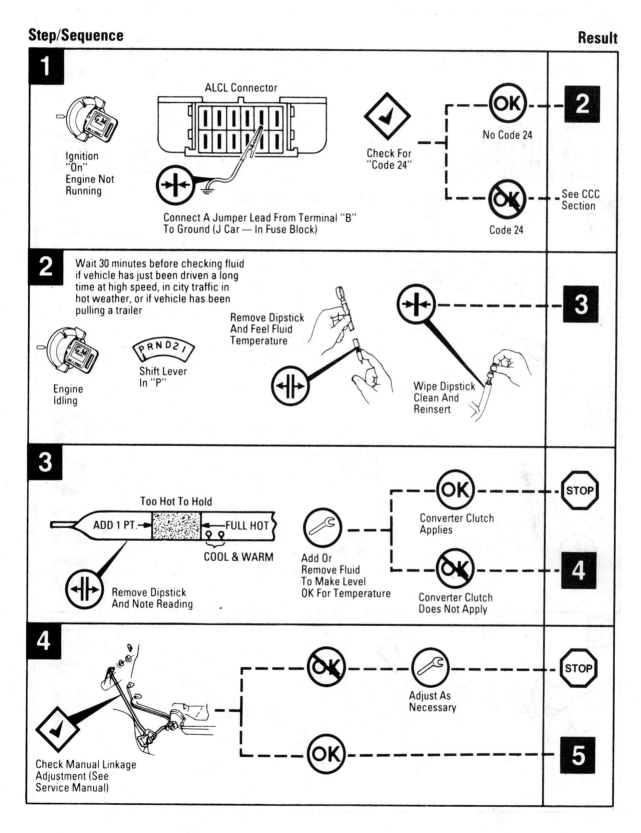

TORQUE CONVERTER CLUTCH DOES NOT APPLY (Cont.)
1981 and Later TCC Diagnosis Without EGR Bleed Solenoid

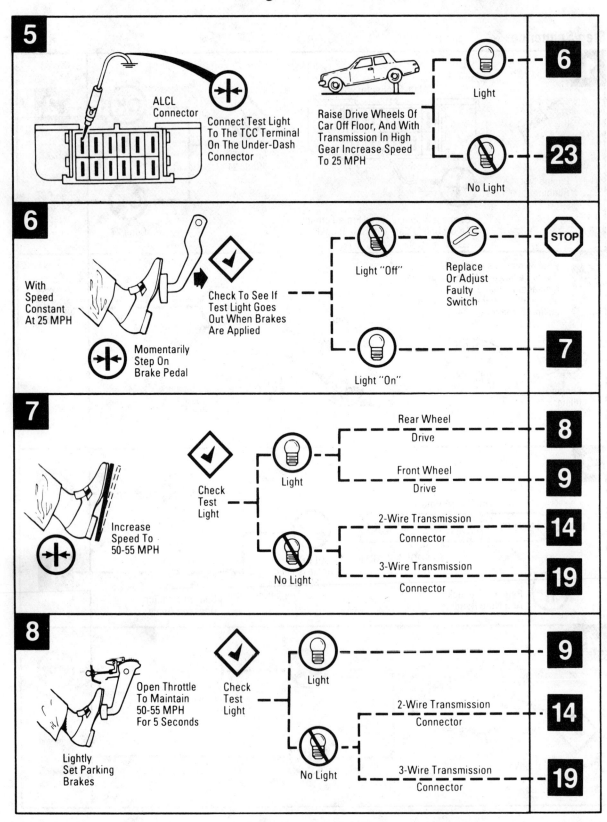

5 — ALCL Connector — Connect Test Light To The TCC Terminal On The Under-Dash Connector — Raise Drive Wheels Of Car Off Floor, And With Transmission In High Gear Increase Speed To 25 MPH — Light → **6** — No Light → **23**

6 — With Speed Constant At 25 MPH — Momentarily Step On Brake Pedal — Check To See If Test Light Goes Out When Brakes Are Applied — Light "Off" → Replace Or Adjust Faulty Switch → **STOP** — Light "On" → **7**

7 — Increase Speed To 50-55 MPH — Check Test Light — Light → Rear Wheel Drive → **8** — Front Wheel Drive → **9** — No Light → 2-Wire Transmission Connector → **14** — 3-Wire Transmission Connector → **19**

8 — Open Throttle To Maintain 50-55 MPH For 5 Seconds — Lightly Set Parking Brakes — Check Test Light — Light → **9** — No Light → 2-Wire Transmission Connector → **14** — 3-Wire Transmission Connector → **19**

TORQUE CONVERTER CLUTCH DOES NOT APPLY (Cont.)
1981 and Later TCC Diagnosis Without EGR Bleed Solenoid

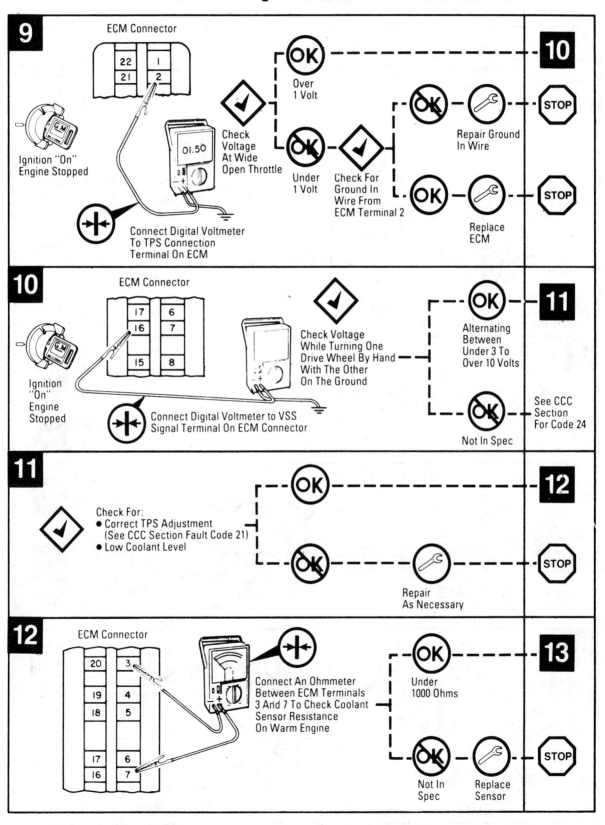

TORQUE CONVERTER CLUTCH DOES NOT APPLY (Cont.)
1981 and Later TCC Diagnosis Without EGR Bleed Solenoid

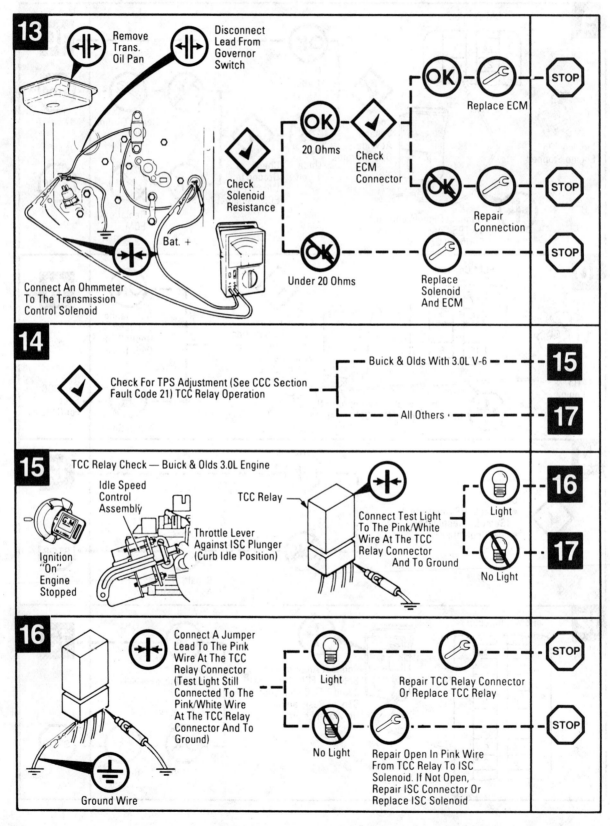

TORQUE CONVERTER CLUTCH DOES NOT APPLY (Cont.)
1981 and Later TCC Diagnosis Without EGR Bleed Solenoid

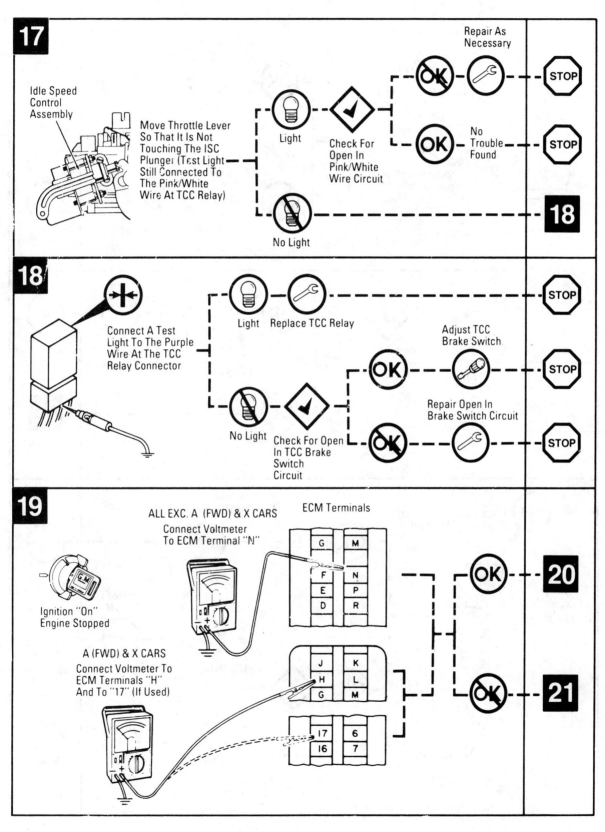

17

Idle Speed Control Assembly

Move Throttle Lever So That It Is Not Touching The ISC Plunger (Test Light Still Connected To The Pink/White Wire At TCC Relay)

Light — Check For Open In Pink/White Wire Circuit — OK̶ — Repair As Necessary — STOP

OK — No Trouble Found — STOP

No Light — **18**

18

Connect A Test Light To The Purple Wire At The TCC Relay Connector

Light — Replace TCC Relay — STOP

No Light — Check For Open In TCC Brake Switch Circuit — OK — Adjust TCC Brake Switch — STOP

OK̶ — Repair Open In Brake Switch Circuit — STOP

19

ALL EXC. A (FWD) & X CARS
Connect Voltmeter To ECM Terminal "N"

ECM Terminals

G	M
F	N
E	P
D	R

OK — **20**

A (FWD) & X CARS
Connect Voltmeter To ECM Terminals "H" And To "17" (If Used)

Ignition "On" Engine Stopped

J	K
H	L
G	M

| 17 | 6 |
| 16 | 7 |

OK̶ — **21**

TORQUE CONVERTER CLUTCH DOES NOT APPLY (Cont.)
1981 and Later TCC Diagnosis Without EGR Bleed Solenoid

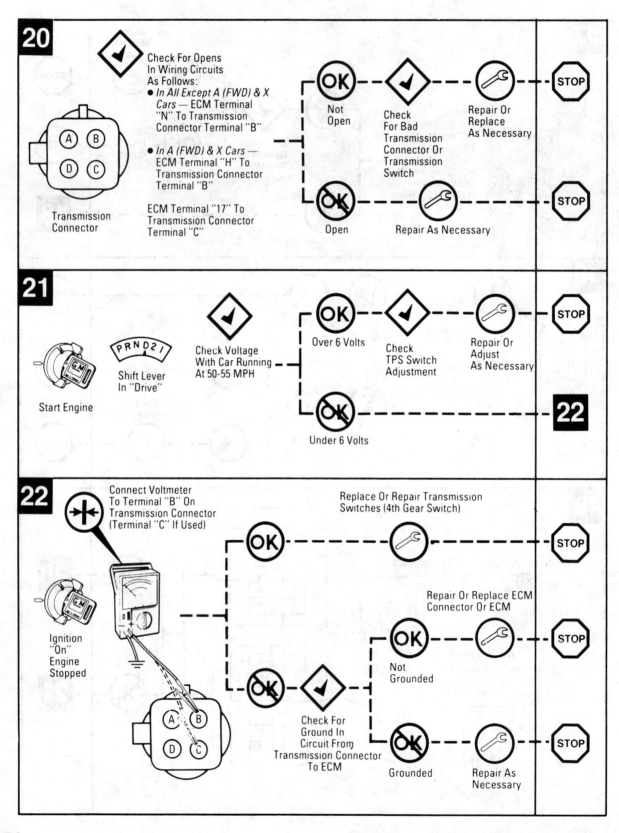

20

Check For Opens In Wiring Circuits As Follows:
- *In All Except A (FWD) & X Cars* — ECM Terminal "N" To Transmission Connector Terminal "B"
- *In A (FWD) & X Cars* — ECM Terminal "H" To Transmission Connector Terminal "B"

ECM Terminal "17" To Transmission Connector Terminal "C"

Transmission Connector

OK Not Open

Check For Bad Transmission Connector Or Transmission Switch

Repair Or Replace As Necessary

STOP

OK Open

Repair As Necessary

STOP

21

Start Engine

PRND21 Shift Lever In "Drive"

Check Voltage With Car Running At 50-55 MPH

OK Over 6 Volts

Check TPS Switch Adjustment

Repair Or Adjust As Necessary

STOP

OK Under 6 Volts

22

22

Connect Voltmeter To Terminal "B" On Transmission Connector (Terminal "C" If Used)

Ignition "On" Engine Stopped

Replace Or Repair Transmission Switches (4th Gear Switch)

OK

STOP

Repair Or Replace ECM Connector Or ECM

OK Not Grounded

STOP

OK

Check For Ground In Circuit From Transmission Connector To ECM

OK Grounded

Repair As Necessary

STOP

TORQUE CONVERTER CLUTCH DOES NOT APPLY (Cont.)
1981 and Later TCC Diagnosis Without EGR Bleed Solenoid

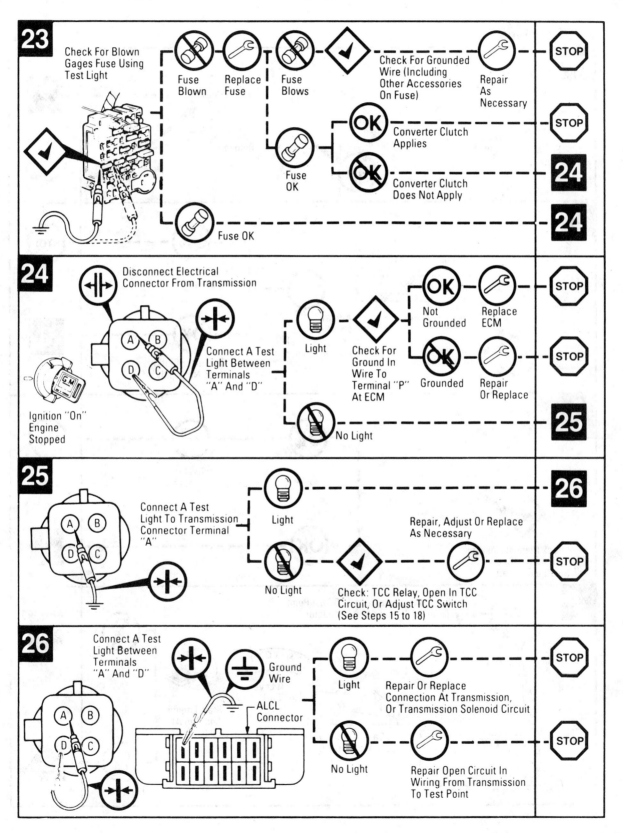

TORQUE CONVERTER CLUTCH DOES NOT APPLY
1981 and Later TCC w/5.7L Diesel Engine

Step/Sequence   **Result**

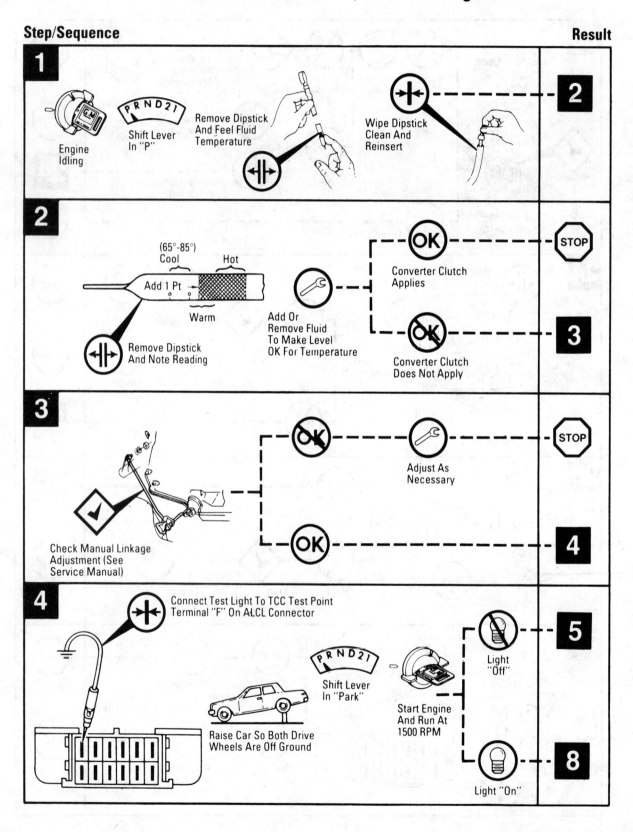

TORQUE CONVERTER CLUTCH DOES NOT APPLY (Cont.)
1981 and Later TCC w/5.7L Diesel Engine

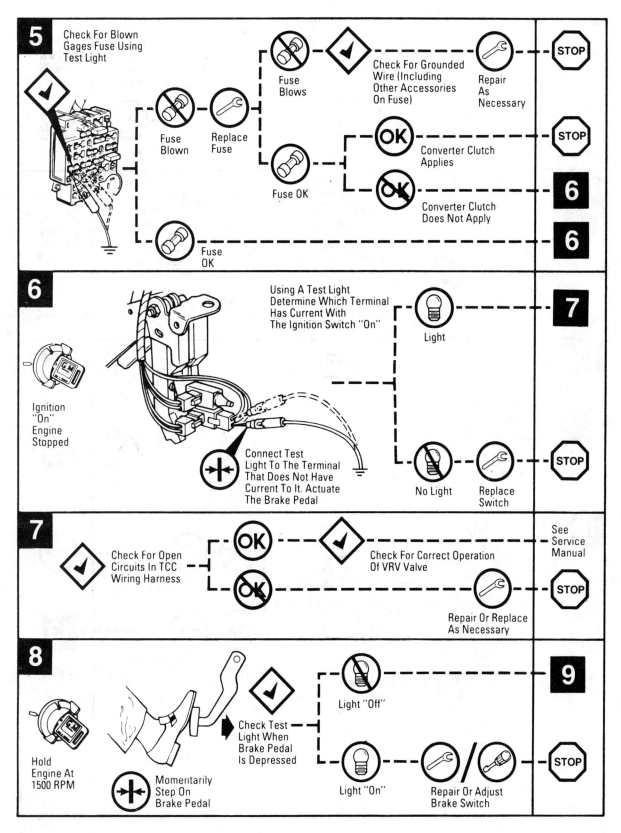

5 Check For Blown Gages Fuse Using Test Light

Fuse Blown — Replace Fuse

Fuse Blows → Check For Grounded Wire (Including Other Accessories On Fuse) → Repair As Necessary → STOP

Fuse OK → OK Converter Clutch Applies → STOP

Fuse OK → NOT OK Converter Clutch Does Not Apply → **6**

Fuse OK → **6**

6 Ignition "On" Engine Stopped

Using A Test Light Determine Which Terminal Has Current With The Ignition Switch "On"

Connect Test Light To The Terminal That Does Not Have Current To It. Actuate The Brake Pedal

Light → **7**

No Light → Replace Switch → STOP

7 Check For Open Circuits In TCC Wiring Harness

OK → Check For Correct Operation Of VRV Valve → See Service Manual

NOT OK → Repair Or Replace As Necessary → STOP

8 Hold Engine At 1500 RPM

Momentarily Step On Brake Pedal

Check Test Light When Brake Pedal Is Depressed

Light "Off" → **9**

Light "On" → Repair Or Adjust Brake Switch → STOP

TORQUE CONVERTER CLUTCH DOES NOT APPLY (Cont.)
1981 and Later TCC w/5.7L Diesel Engine

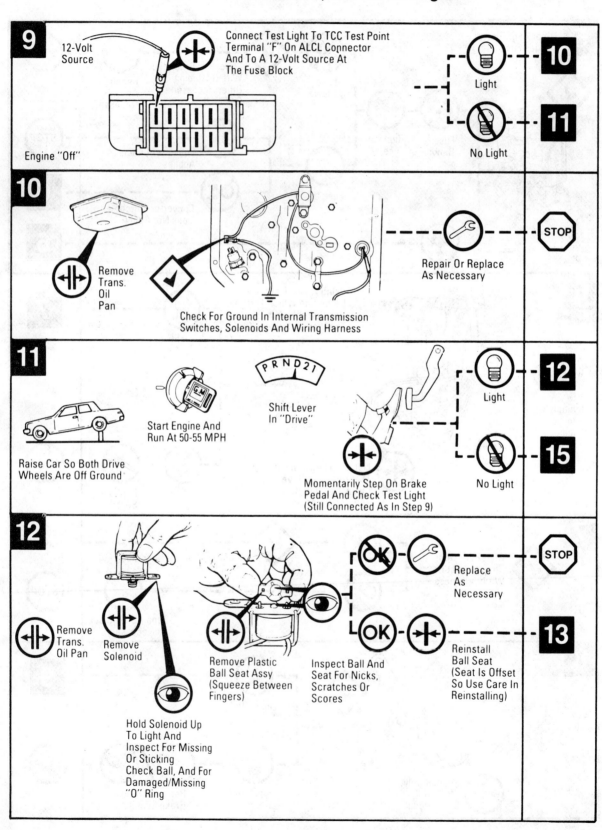

9 12-Volt Source — Connect Test Light To TCC Test Point Terminal "F" On ALCL Connector And To A 12-Volt Source At The Fuse Block — Engine "Off"

10 Light

11 No Light

10 Remove Trans. Oil Pan — Check For Ground In Internal Transmission Switches, Solenoids And Wiring Harness — Repair Or Replace As Necessary — STOP

11 Raise Car So Both Drive Wheels Are Off Ground — Start Engine And Run At 50-55 MPH — PRND21 — Shift Lever In "Drive" — Momentarily Step On Brake Pedal And Check Test Light (Still Connected As In Step 9)

12 Light

15 No Light

12 Remove Trans. Oil Pan — Remove Solenoid — Hold Solenoid Up To Light And Inspect For Missing Or Sticking Check Ball, And For Damaged/Missing "O" Ring — Remove Plastic Ball Seat Assy (Squeeze Between Fingers) — Inspect Ball And Seat For Nicks, Scratches Or Scores — OK / Replace As Necessary — STOP

OK — Reinstall Ball Seat (Seat Is Offset So Use Care In Reinstalling) — **13**

TORQUE CONVERTER CLUTCH DOES NOT APPLY (Cont.)
1981 and Later TCC w/5.7L Diesel Engine

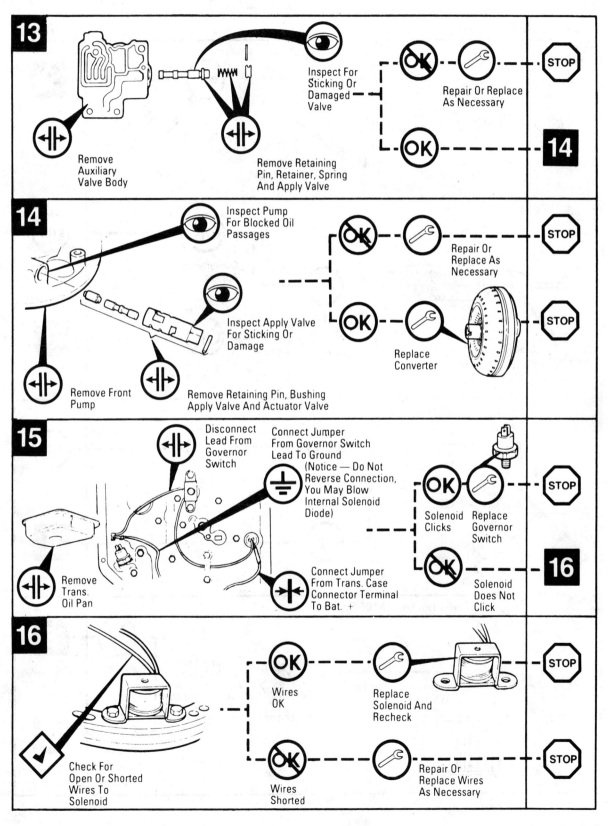

13

Remove Auxiliary Valve Body

Remove Retaining Pin, Retainer, Spring And Apply Valve

Inspect For Sticking Or Damaged Valve

Repair Or Replace As Necessary

STOP

OK → 14

14

Inspect Pump For Blocked Oil Passages

Inspect Apply Valve For Sticking Or Damage

Remove Front Pump

Remove Retaining Pin, Bushing Apply Valve And Actuator Valve

Repair Or Replace As Necessary

STOP

Replace Converter

STOP

15

Disconnect Lead From Governor Switch

Connect Jumper From Governor Switch Lead To Ground (Notice — Do Not Reverse Connection, You May Blow Internal Solenoid Diode)

Remove Trans. Oil Pan

Connect Jumper From Trans. Case Connector Terminal To Bat. +

Solenoid Clicks — Replace Governor Switch

STOP

Solenoid Does Not Click → 16

16

Check For Open Or Shorted Wires To Solenoid

Wires OK → Replace Solenoid And Recheck → STOP

Wires Shorted → Repair Or Replace Wires As Necessary → STOP

TORQUE CONVERTER CLUTCH DOES NOT APPLY
1982 and Later TCC w/4.3L Engine

Step/Sequence   **Result**

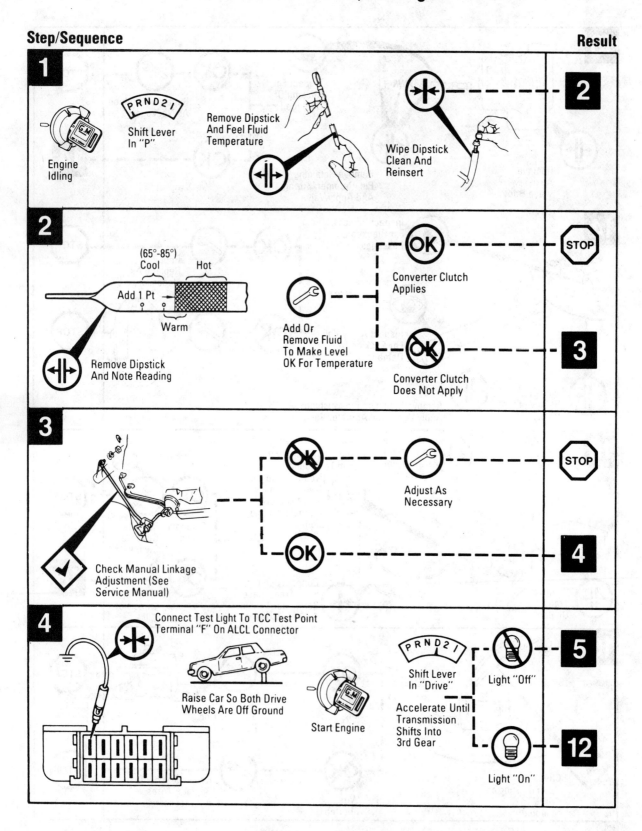

1
Engine Idling · Shift Lever In "P" · Remove Dipstick And Feel Fluid Temperature · Wipe Dipstick Clean And Reinsert → **2**

2
Remove Dipstick And Note Reading — (65°-85°) Cool / Warm / Hot / Add 1 Pt · Add Or Remove Fluid To Make Level OK For Temperature
- OK Converter Clutch Applies → **STOP**
- Converter Clutch Does Not Apply → **3**

3
Check Manual Linkage Adjustment (See Service Manual)
- Adjust As Necessary → **STOP**
- OK → **4**

4
Connect Test Light To TCC Test Point Terminal "F" On ALCL Connector · Raise Car So Both Drive Wheels Are Off Ground · Start Engine · Shift Lever In "Drive" · Accelerate Until Transmission Shifts Into 3rd Gear
- Light "Off" → **5**
- Light "On" → **12**

TORQUE CONVERTER CLUTCH DOES NOT APPLY (Cont.)
1982 and Later TCC w/4.3L Engine

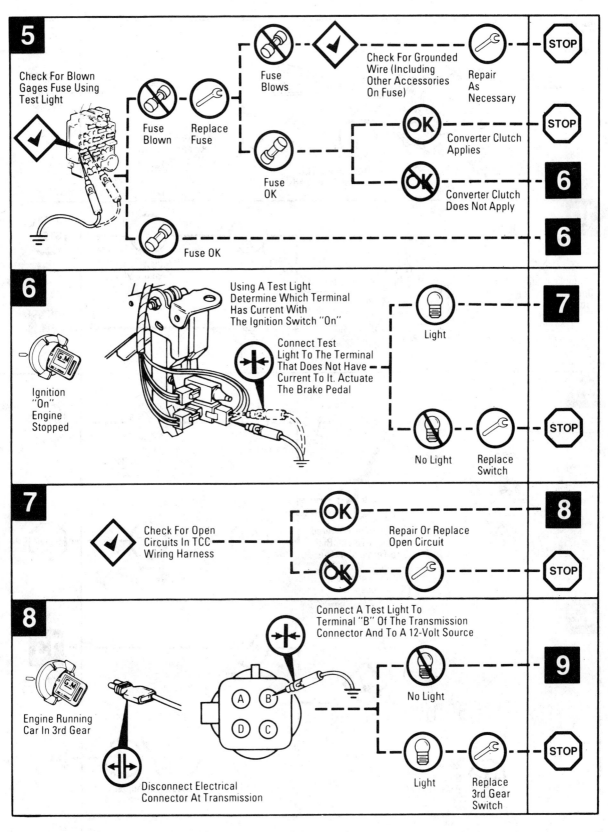

TORQUE CONVERTER CLUTCH DOES NOT APPLY (Cont.)
1982 and Later TCC w/4.3L Engine

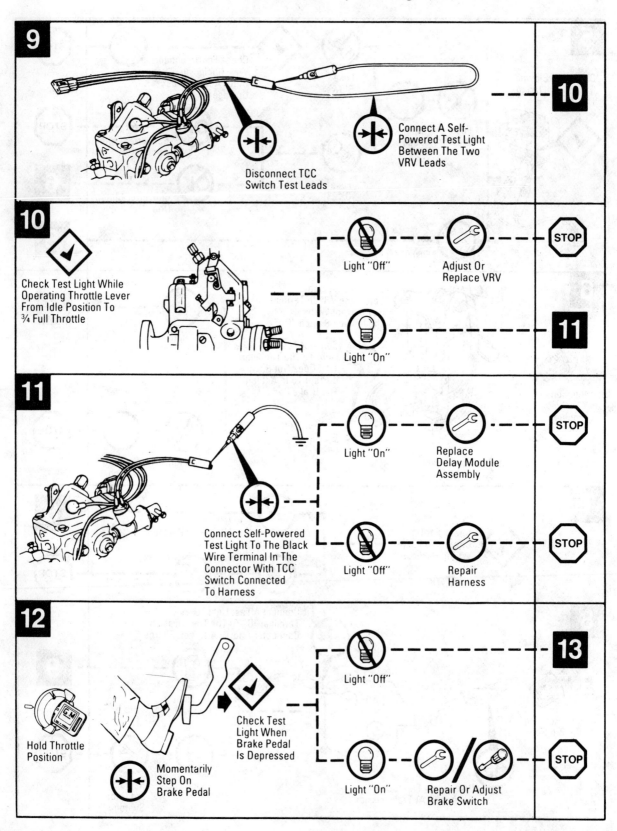

9

Disconnect TCC Switch Test Leads

Connect A Self-Powered Test Light Between The Two VRV Leads

10

10

Check Test Light While Operating Throttle Lever From Idle Position To ¾ Full Throttle

Light "Off" — Adjust Or Replace VRV — **STOP**

Light "On" — **11**

11

Connect Self-Powered Test Light To The Black Wire Terminal In The Connector With TCC Switch Connected To Harness

Light "On" — Replace Delay Module Assembly — **STOP**

Light "Off" — Repair Harness — **STOP**

12

Hold Throttle Position

Momentarily Step On Brake Pedal

Check Test Light When Brake Pedal Is Depressed

Light "Off" — **13**

Light "On" — Repair Or Adjust Brake Switch — **STOP**

TORQUE CONVERTER CLUTCH DOES NOT APPLY (Cont.)
1982 and Later TCC w/4.3L Engine

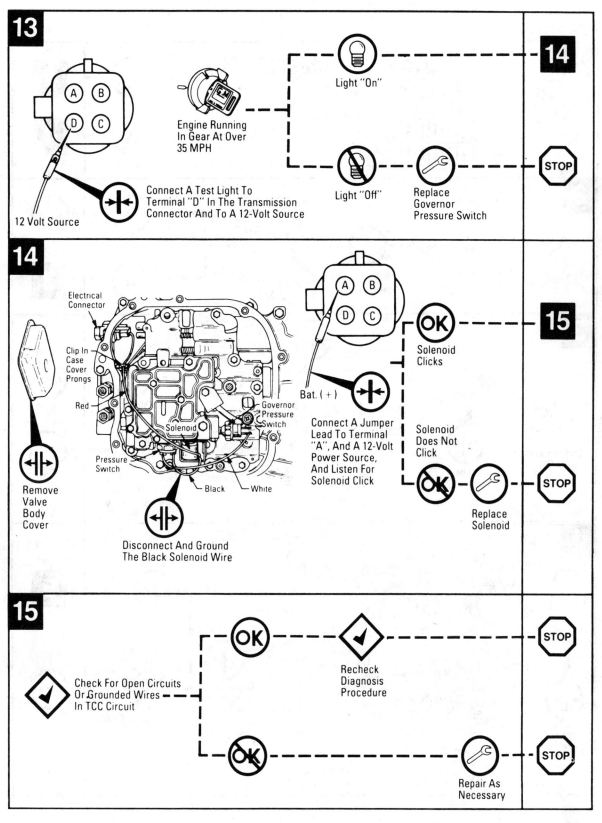

13

A B
D C

12 Volt Source

Connect A Test Light To Terminal "D" In The Transmission Connector And To A 12-Volt Source

Engine Running In Gear At Over 35 MPH

Light "On"

Light "Off"

Replace Governor Pressure Switch

14

STOP

14

Electrical Connector

Clip In Case Cover Prongs

Red

Solenoid

Pressure Switch

Governor Pressure Switch

Black   White

Remove Valve Body Cover

Disconnect And Ground The Black Solenoid Wire

A B
D C

Bat. (+)

Connect A Jumper Lead To Terminal "A", And A 12-Volt Power Source, And Listen For Solenoid Click

Solenoid Clicks

Solenoid Does Not Click

Replace Solenoid

15

STOP

15

Check For Open Circuits Or Grounded Wires In TCC Circuit

OK — Recheck Diagnosis Procedure — STOP

OK — Repair As Necessary — STOP

TORQUE CONVERTER CLUTCH OPERATION IN SECOND GEAR AS WELL AS IN THIRD GEAR
1982 and Later TCC w/4.3L Diesel Engine

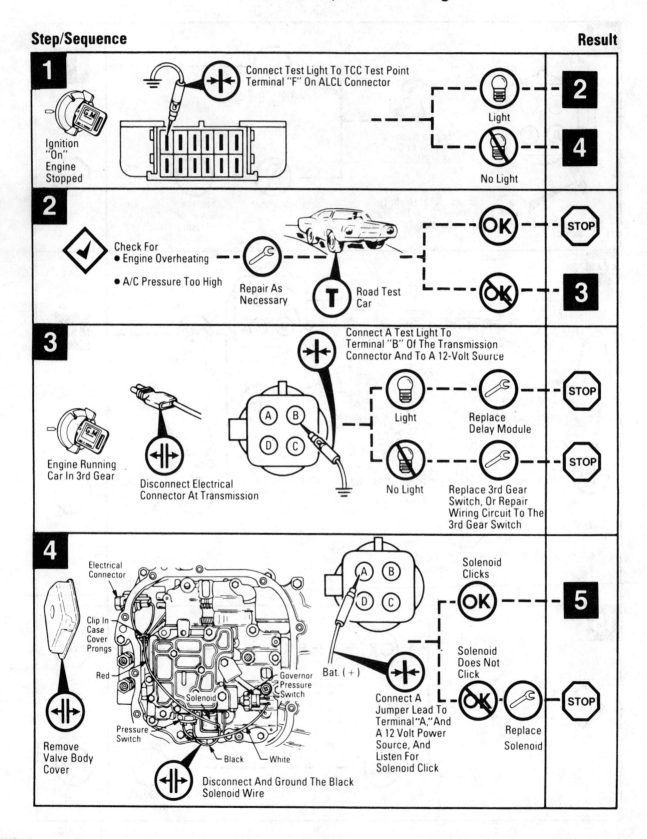

TORQUE CONVERTER CLUTCH OPERATION IN SECOND GEAR AS WELL AS IN THIRD GEAR
(Cont.)
1982 and Later TCC w/4.3L Diesel Engine

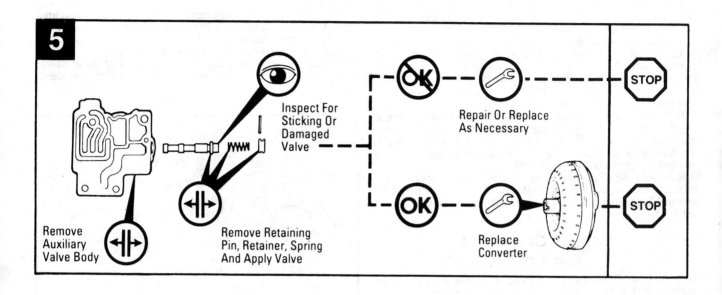

Verification Of Converter Clutch Valve Operation—1980

1. Using a tee fitting, install an oil pressure gauge to the cooler out line at the radiator.

2. Remove the vacuum line from the vacuum switch.

3. Raise the wheels and run the vehicle in DRIVE range until the 3rd gear upshift is obtained. Maintain 2000 rpm and note the oil pressure reading.

4. While maintaining 2000 rpm, re-connect the vacuum hose to the vacuum switch and observe the oil pressure reading. The pressure should drop 5 to 15 psi. This indicates the converter clutch valve is operating and if the converter clutch is inoperative, a missing "O" ring at the end of the turbine shaft or a defective converter could be the cause.

5. If the pressure does not drop, check the solenoid for a missing check ball or "O" ring, loose solenoid bolts, defective solenoid ball seat or ball. The converter clutch valve could be sticking, binding or damaged, or the direct clutch oil passages could be blocked, restricted or interconnected.

6. Should the oil pressure drop to zero, the cooler feed orifice in the converter clutch valve bushing is blocked, restricted or missing. Transmission failure could result due to lack of lubricating fluid.

Torque Converter Evaluation

Before a torque converter is replaced, verify that all of the other external and internal component checks have been made. A torque converter should only be replaced if one of the following conditions exists:

1. Front oil pump or body are badly scored. When these components become scored, cast-iron grindings enter the torque converter and oil circuit. This scoring is usually the result of the drive gear in the pump wearing into the crescent, or down into the pocket, or the outer gear wearing against the pocket. A cracked flexplate can also cause the drive lugs on the pump drive gear to become badly damaged.

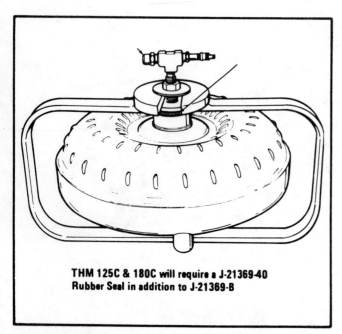

THM 125C & 180C will require a J-21369-40 Rubber Seal in addition to J-21369-B

Air pressure checking tools installed on torque converter
(©General Motors Corp.)

2. An internal converter failure, such as the converter clutch not engaging. However, all the external checks and internal checks must be made before condemning the converter. Some other types of internal converter failure are: the stator overrun clutch not locking or a thrust bearing that's failed. These types of failures are usually associated with aluminized oil present in the converter.

3. An external leak is apparent, such as at the hub weld area. However, a converter that's been in service for some time and didn't leak, probably never will.

4. End play in the converter exceeds .050 inch. This measurement can't be estimated, but must be made with Kent Moore tools J-21371, J-21371-8, J-25020, J-29060 or their equivalent.

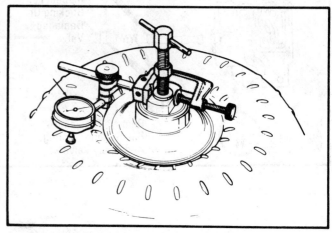

Checking end play of torque converter components
(©General Motors Corp.)

5. A scored or otherwise damaged hub. A damaged hub can cause a front seal failure or front pump housing failure.

6. A broken, damaged or poorly fitting converter pilot. This can cause the converter to not fit into the crankshaft bore properly or to be misaligned with the crankshaft centerline.

7. The converter has an imbalance problem which cannot be corrected. Most converter imbalance problems are minor and can be corrected by following the converter vibration procedure. Also, check for missing balance weights on the converter and flexplate. Replace the converter if the balance weights are missing. If these checks and test don't solve the problem, try balancing the flexplate.

A torque converter should NOT be replaced if it displays any of the following conditions:

1. The transmission oil has an odor or is discolored but there is no evidence of metallic particles. This is not an indication of converter or front pump damage. However, if the oil is discolored with engine coolant, the converter with a converter clutch must be replaced, as well as the friction materials and seals in the transmission.

2. Fretting wear on the hub where the oil pump drive gear locates. A small amount of wear is normal and does not require the replacement of the converter or front pump.

3. A defective oil cooler which allowed engine coolant to mix with the transmission oil. Conventional torque converters should be drilled, drained and repaired.

NOTE: A TCC torque converter must be replaced if engine coolant is found in the transmission oil.

4. Damaged threads in the torque converter attaching lugs. Repair the threads using a reputable thread repair kit.

Diagnosis of Converter Clutch Internal Control Components On Specific Transmission/Transaxle Assemblies

1980 THM 200C, 250C, 350C CHEVROLET BUILT TRANSMISSION—ALL GAS ENGINES WITHOUT DELAY VALVE

1. Direct clutch oil is fed to the TCC apply valve via the TCC solenoid.
2. The ignition switch, brake switch and low vacuum switch control the feed voltage to the TCC solenoid.
3. The governor switch acts as a speed sensor for TCC operation and provides the ground for the TCC solenoid.
2. The ignition switch, brake switch and low vacuum switch control the feed voltage to the TCC solenoid.
3. The governor switch acts as a speed sensor for TCC operation and provides the ground for the TCC solenoid.
4. The delay valve slows the response of the low vacuum switch.

1980 THM 200C, 350C with 5.7L DIESEL

1. Direct clutch oil is fed to the TCC apply valve via the TCC solenoid.
2. The ignition switch, brake switch, low vacuum switch and high vacuum switch control the feed voltage to the TCC solenoid.
3. The governor switch acts as a speed sensor for TCC operation and provides a ground for the TCC solenoid.

1981 THM 200C 5.7L DIESEL WITH EGR BLEED. TCC APPLYS IN 3RD SPEED ONLY

1. The direct clutch oil is fed to the TCC apply valve via the TCC solenoid.
2. The ignition switch, brake switch and rotary vacuum switch control the feed voltage to the TCC and EGR bleed solenoids.
3. The governor switch acts as a speed sensor for TCC operation and controls the ground for the EGR bleed solenoid.

Diagnosis:
A. Test light between test lead and ground.
1. Test light lights when all of the following are met:
   a. Ignition on.
   b. Brake switch closed.
   c. Rotary switch at specifications.

1980 THM 200C, 250C, 350C CHEVROLET BUILT TRANSMISSION—ALL GAS ENGINES WITH DELAY VALVE

1. Direct clutch oil is fed to the TCC apply valve via the TCC solenoid.
2. Test light will not go out when TCC engages.
3. Test light does not monitor EGR solenoid.
4. Test light at transmission side of EGR solenoid can test governor switch without pan removal. Disconnect EGR solenoid &

connect test light. Light should be on until governor switch closes & TCC engages.

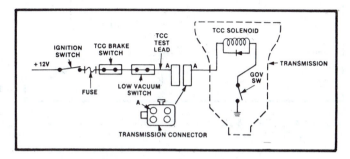

1980 THM 200C, 250C, 350C Chevrolet built, with gasoline engine less vacuum delay valve (©General Motors Corp.)

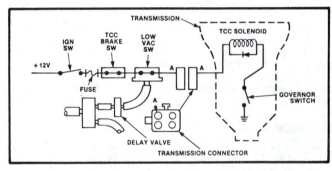

1980 THM 200C, 250C, 350C Chevrolet built, with gasoline engine and vacuum delay valve (©General Motors Corp.)

B. Jumper Wire Diagnosis—

NOTE: Do not use jumper wire between test lead and ground.

1. Connect jumper wire between EGR bleed solenoid transmission side and ground.
2. Ignition on—
   a. Listen for EGR solenoid click.
   b. Disconnect EGR solenoid connector, maintain test ground and listen for TCC solenoid click.
1. No click—TCC solenoid problem.
2. Clicks—verify TCC engagement in 3rd gear by driving.
   a. Engages—governor switch problem.
   b. No engage—TCC solenoid seal or ball seat.

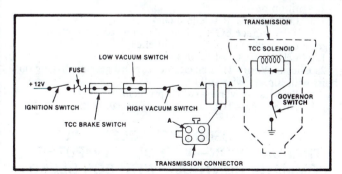

1980 THM 200C, 350C with 5.7L diesel engine (©General Motors Corp.)

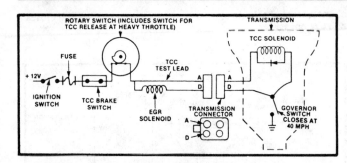

1981 THM 200C with 5.7L diesel engine and EGR bleed (©General Motors Corp.)

THM 350C-5.7L DIESEL WITH EGR BLEED. TCC APPLIES IN 3RD SPEED ONLY

1. The forward clutch oil is fed to the TCC apply valve via the TCC solenoid.
2. The third gear switch is necessary to restrict TCC operations to third gear.
3. The governor switch acts as a speed sensor for TCC operation and provides a ground for the EGR bleed solenoid.
4. The ignition switch, brake switch and rotary switch control the feed voltage to the TCC and EGR bleed solenoids.

Diagnosis:

A. Test light between test lead and ground.
1. Test light lights when all of the following are met:
  a. Ignition on.
  b. Brake switch closed.
  c. Rotary switch at specifications.
2. Test light will not go out when TCC engages.
3. Test light does not monitor EGR solenoid.
4. Test light at transmission side of EGR solenoid and ground can be used to test operation or failure of 3rd gear switch and governor switch but cannot isolate which is at fault if test shows failure.
  a. Disconnect EGR solenoid connector & connect test light. Light should be on until 3rd gear switch & governor switch close and TCC engages.
B. Jumper wire diagnosis—

NOTE: Do not use jumper wire between test and 2nd ground.

1. Connect jumper wire between EGR bleed solenoid transmission side and ground.
2. Ignition on—
  a. Listen for EGR solenoid click.
  b. Disconnect EGR solenoid connector, maintain test ground & listen for TCC solenoid click.
1. No click—TCC solenoid problem.
2. Clicks—Verify TCC engagement in 3rd gear by driving on hoist with jumper wire still on.
  a. Engages—governor switch problem.
  b. No engage—TCC solenoid seal or ball seat.

1981 THM 350C CHEVROLET-BUILT TRANSMISSION—WITHOUT COMPUTER COMMAND CONTROL—WITH EGR BLEED

1. Direct clutch oil is fed to the TCC apply valve via the TCC solenoid.

2. The governor switch acts as a speed sensor for TCC operation and provides the ground for the TCC and EGR bleed solenoids.
3. The ignition switch, brake switch and low vacuum switch control the feed voltage to both the TCC and EGR bleed solenoids.
4. The high gear switch only controls the EGR solenoid. The high gear switch is mounted externally on the transmission.

Diagnosis:

A. Test light between test lead and ground.

NOTE—Test lead is in TCC solenoid feed circuit.

1. Test light will light when all the following are met:
  a. Ignition on.
  b. Brake switch closed.
  c. Vacuum switch closed (vac. over spec.)
2. Test light will not go out when TCC engages.
3. Test light does not monitor EGR bleed solenoid.
4. Test light between EGR solenoid high gear switch terminal and ground will test EGR solenoid circuit. This means we can test governor switch without removing the oil pan if TCC does not work:
  a. Install test light at EGR solenoid.
  b. If needed, bypass high gear switch with a wire.
  c. Operate vehicle to 45-50 MPH—light going off means governor switch is OK since light is in ground side of EGR solenoid circuit.
B. Jumper wire diagnosis.

NOTE: Do not use jumper between test lead and ground.

1. Jumper wire between transmission terminal D and ground.
  a. Ignition on-listen for solenoid click.
  1. No click—TCC solenoid problem.
  2. Click—drive on hoist to verify TCC engage in 3rd gear.
  a. No engage—TCC solenoid seal or ball seat.
  b. Engages—disconnect ground wire & verify.
  a. No engage—governor switch problem.
  b. Engage—no problem found.

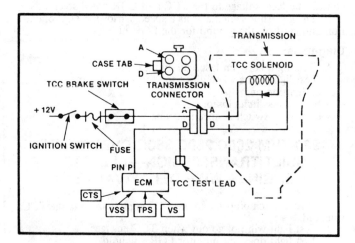

1981-82 THM 200C, 250C, 350C Chevrolet built with CCC system, without EGR bleed (©General Motors Corp.)

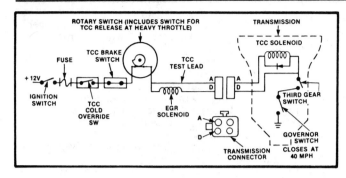

THM 350C with 5.7L diesel engine and EGR bleed. TCC applies in third speed only (©General Motors Corp.)

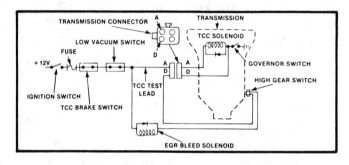

1981 THM 350C Chevrolet built without CCC system and with EGR bleed (©General Motors Corp.)

1981-82 THM 200C, 250C, 350C CHEVROLET-BUILT WITH COMPUTER COMMAND CONTROL—WITHOUT EGR BLEED

1. Direct clutch oil is fed to the TCC apply valve via the TCC solenoid.
2. No internal electrical switches are used.
3. The ignition and brake switches control the feed voltage to the TCC solenoid.
4. The ECM controls the TCC solenoid ground.

Diagnosis:

A. Test light between test lead and ground.

1. Ignition switch on—test light goes on (if not, check at brake switch).
2. Tap brake pedal—test light goes off and on.
3. Start engine—light stays on at all times until the following are met:

   a. Coolant temperature up to specifications.
   b. Vehicle speed to specifications.
   c. Throttle position/vacuum switch signal to specifications.
Note that solenoid can be energized even before third gear is achieved—TCC then comes on with 2-3 shift.

4. When vehicle TCC is on, test light is off.

   a. Goes on with change in throttle position, then back off with slight delay.
   b. Tap brake pedal—light stays off—ECM keeps ground.
   c. Downshift causes light to go on via TPS sensor and vacuum.

B. Jumper wire between test lead and ground.

NOTE: Use jumper wire if test light stays on at all times in A-3 above. Jumper wire bypasses ECM & causes full time solenoid on.

1. Ignition on engine off—Touch wire between test lead and ground and listen for solenoid click.
2. Road test or drive wheels on rack to verify TCC coming on as trans. shifts to 3rd gear. If TCC comes on, refer to CCC charts for ECM & input diagnosis.

THM 200C OR CHEVROLET BUILT 250C, 350C WITH E.G.R BLEED

1. Direct clutch oil to solenoid and apply valve.
2. Ignition switch controls solenoid feed (both).
3. TCC relay controls ground for both solenoids.
4. Brake switch controls TCC relay coil feed.
5. ECM controls TCC relay coil ground.

Diagnosis:

A. Test light between test lead and ground, EGR bleed solenoid disconnected.

1. Ignition on—test light on—no brake switch effect.
2. Start engine-drive position—light stays on until:

   a. Coolant temperature to specifications.
   b. Vehicle speed to specifications.
   c. TPS-VAC signal to specifications.
Note that TCC relay maybe energized before 3rd gear is reached in transmission.

3. When TCC is on, TCC relay is energized and light is off.

   a. Goes on with throttle changes and out with slight delay.
   b. Tap brake pedal—light goes on and off—no delay. (tests relay operation).

B. Jumper wire between test light & ground.

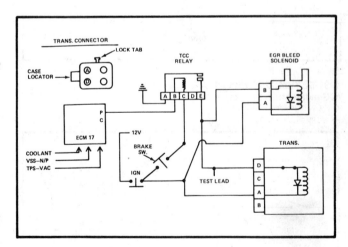

THM 200C, 250C, 350C with EGR bleed (©General Motors Corp.)

NOTE: Use only if test light stays on any time ignition switch is on. Jumper wire by-passes TCC relay points (EGR solenoid disconnected).

1. Turn ignition on—listen for TCC solenoid click.

   a. No click—problems in transmission.
   b. Does click—road test or test on hoist to verify TCC engagement.

1. No engagement—check TCC solenoid seal & ball seat.
2. Engages—check TCC relay operation by moving jumper wire to TCC relay terminal B.
   a. TCC relay should click & TCC solenoid should click.
   b. No click—ground TCC relay terminal E—TCC solenoid should click.

1981-82 THM 125C WITHOUT COMPUTER COMMAND CONTROL OR EGR BLEED

1. Line oil is fed to the TCC apply valve via the TCC solenoid.
2. The ignition switch, brake switch and low vacuum switch control the feed voltage to the TCC solenoid.
3. The governor switch acts as a speed sensor for TCC operation and provides the ground for the TCC solenoid.

Test Light Diagnosis—
Test light between test lead and ground.

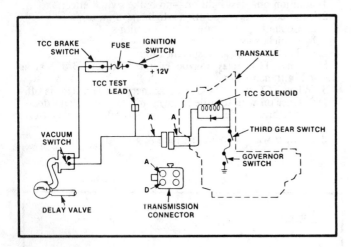

1981-82 THM 125C without CCC or EGR bleed
(©General Motors Corp.)

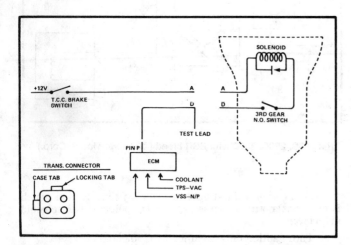

1981-82 THM 350C, Model MX-2, Buick built
(©General Motors Corp.)

NOTE: Test lead is in TCC solenoid feed circuit.

1. Test light will light when all the following are met:
   a. Ignition on.
   b. Brake switch closed.
   c. Vacuum switch closed.
2. Test light will not go out when TCC engages.
3. Test light at terminal D in transaxle connector.
   a. Test one conditions met cause light to go on.
   b. Light goes out as 3rd gear switch and governor switch close and TCC engages.
   c. The vehicle drive wheels off the ground to drive the vehicle and allow TCC engagement test.
4. Jumper wire—Do not use jumper wire on this system.
   a. Do not use wire at test lead—electrical damage may result.
   b. Do not use wire at transmission terminal D since TCC may engage any time.

1981-82 THM 200C OR CHEVROLET-BUILT 250C-350C WITH COMPUTER COMMAND CONTROL AND EGR BLEED

1. Direct clutch oil is fed to the TCC apply valve via the TCC solenoid.
2. The ignition switch controls the feed voltage to both the TCC and EGR bleed solenoids.
3. The brake switch controls the TCC relay coil feed.
4. The TCC relay controls the ground for both the TCC and EGR bleed solenoids.
5. The ECM controls the ground of the TCC relay coil.

1981-82 THM 350C WITH MX-2 VALVE BODY—BUICK BUILT

1. Forward clutch oil to solenoid and apply valve.
2. Requires 3rd clutch oil switch.
3. ECM controls solenoid ground.
4. Brake Switch— Ignition Switch controls solenoid feed.

Diagnosis:

A. Test light between test lead and ground.
1. Test light does not go on until transmission reaches 3rd gear to complete circuit to test lead (18-20 mph) (if not-test at brake switch).
2. Test light will then come on until the following are met:
   a. Coolant up to specifications.
   b. TPS-VAC to specifications.
   c. VSS to speed.

Note that all ECM requirements may be met before 3rd gear is reached—light may not come on.
3. When TCC is on, test light is off. While in 3rd gear.
   a. Tap brake pedal causes light to stay out— ECM keeps ground.

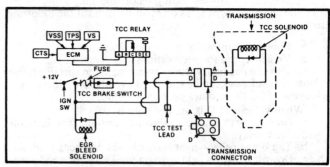

1981-82 THM 200C or Chevrolet built 250C, 350C with CCC and EGR bleed (©General Motors Corp.)

b. Throttle position change causes light to go on and off with slight delay.

c. Downshift causes light to stay out because internal switch goes open.

B. Jumper wire between test lead and ground.

NOTE: Use jumper wire if light stays on at all times in A-2 above. Jumper wire by-passes ECM and causes TCC engage in 3rd at all loads.

1981-82 THM (MX-3)—BUICK BUILT

1. Forward clutch oil to solenoid and apply valve.
2. Requires 3rd or 2nd clutch oil switch.
3. ECM controls solenoid ground.
4. Brake Switch—Ignition Switch controls solenoid feed.

Diagnosis:

A. Test light between test lead and ground.

1. Test light does not go on until transmission reaches 2nd gear to complete circuit to test lead (18-20 mph) (if not-test at brake switch).

2. Test light will then come on until the following are met:
   a. Coolant up to specifications.
   b. TPS-VAC to specifications.
   c. VSS to speed.
Note that all ECM requirements may be met before 3rd gear is reached—light may not come on.

3. When TCC is on, test light is off. While in 2nd gear.
   a. Tap brake pedal causes light to stay out—ECM keeps ground.
   b. Throttle position change causes light to go on and off with slight delay.
   c. Downshift causes light to stay out because internal switch goes open.

B. Jumper wire between test lead and ground.

NOTE: Use jumper wire if light stays on at all time in A-2 above. Jumper wire by-passes ECM and causes TCC engage in 3rd at all loads.

1981-82 THM 125C WITH COMPUTER COMMAND CONTROL—WITHOUT EGR BLEED

1. Line oil is fed to the TCC apply valve via the TCC solenoid.
2. The third gear switch is required to prevent TCC application in lower gears.
3. The ignition and brake switches control the feed voltage to the TCC solenoid.
4. The ECM controls the solenoid ground.

Diagnosis:

A. Test light between test lead and ground.

1. Test light does not go on until transmission reaches third gear to complete circuit to test lead (18-20 mph).

2. Test light will then come on until the following are met:
   a. Coolant up to specifications.
   b. TPS-VAC to specifications.
   c. VSS to speed.
Note that all ECM may be met before 3rd gear is reached—light may not come on. .

3. When TCC is on, test light is off. While in 3rd gear:
   a. Tap brake pedal causes light to stay out—ECM keeps ground.
   b. Throttle position change causes light to go on and off with slight delay.
   c. Downshift causes light to stay out because internal switch goes open.

B. Jumper wire between test lead and ground.

NOTE: Use jumper wire if light stays on at all times in A-2 above. Jumper wire by-passes ECM & causes TCC engage in 3rd at all leads.

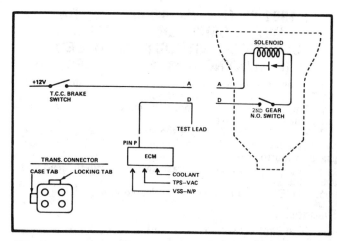

1981-82 THM 350C, Model MX-3, Buick built (©General Motors Corp.)

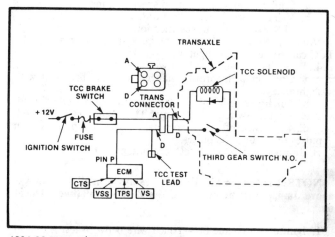

1981-82 THM 125C with CCC, less EGR bleed (©General Motors Corp.)

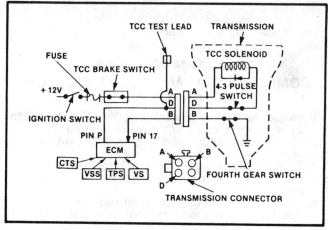

1981-82 THM 200-4R, 700-R4 with CCC system, less EGR bleed. Includes Corvette with EFI (©General Motors Corp.)

a. Vehicle on hoist or on road test—accelerate to 25 mph & listen for solenoid click as transmission shifts 3rd gear.

b. Feel for engine pulsations on coast until trans. shifts out of 3rd.

1981-82 THM 200-4R AND 700-R4 WITH COMPUTER COMMAND CONTROL—WITHOUT EGR BLEED (INCLUDING CORVETTE W/EFI)

1. Second gear clutch oil is fed through the TCC shift valve to the TCC apply valve via the TCC solenoid.
2. The 4-3 pulse switch is normally closed. It opens momentarily during a 4-3 downshift.
3. When the ECM recognizes the fourth gear switch is activated, it provides more TPS movement before the TCC is disengaged in fourth gear.
4. The ignition switch and brake switch control the feed voltage to the TCC solenoid.
5. The ECM controls the TCC solenoid ground.

Diagnosis:

A. Test light between test lead and ground.
1. Ignition on-brake off—test light comes on.
2. Tap brake pedal—test light goes off and on.
3. Start engine and drive—light stays on until the following:
   a. Coolant temperature up to specifications.
   b. Vehicle speed to specifications.
   c. TPS-VAC signal to specifications.
Note that the solenoid can be energized in second gear and the TCC can come on. The light will go out.
4. When TCC is on, test light will be out.
   a. Goes on with throttle changes and then back out with slight delay.
   b. Tap brake pedal—light stays out since only feed affected.
   c. 4th gear switch signals larger TPS change TCC affected.
   d. 4-3 downshift may leave light off as 4-3 pulse switch pulses open for TCC pulses.
B. Jumper wire between test lead and ground.

NOTE: Use jumper wire if test light stays on at all times in A-3 above. Jumper wire by-passes ECM and causes full time solenoid on.

1. Ignition on engine off—Touch wire between test lead and ground and listen for solenoid click.
2. Road test or drive wheels—drive to verify that TCC does engage after 1-2 shift and holds engaged in 3rd and 4th at all loads.
   a. If not—check solenoid seals and ball seat.
   b. If yes—refer to CCC diagnosis for ECM and input sensors.

1981-82 THM 200-4R WITH COMPUTER COMMAND CONTROL AND EGR BLEED

1. Second gear clutch oil is fed through the TCC shift valve to the TCC apply valve via the TCC solenoid.
2. The 4-3 pulse switch is normally closed. It opens momentarily during a 4-3 downshift.
3. When the ECM recognizes the fourth gear switch is activated, it provides more TPS movement before the TCC is engaged in fourth gear.
4. The ignition switch controls the solenoid feed.
5. The TCC relay controls the ground for both the TCC and EGR bleed solenoids.
6. The brake switch controls the TCC relay coil feed.
7. The ECM controls the ground of the TCC relay coil.

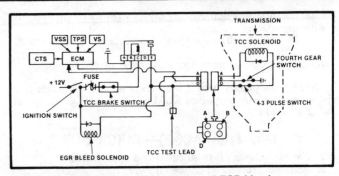

1981-82 THM 200-4R with CCC system and EGR bleed
(©General Motors Corp.)

Diagnosis:

A. Test light between test lead and ground.
1. Ignition on—test light on—no brake pedal effect.
2. Start engine and drive—light stays on until:
   a. Coolant temperature to specifications.
   b. Vehicle speed to specifications.
   c. TPS-VAC signal to specifications.
Note that the relay and both solenoids can be energized in second gear and TCC can come on. Test light goes off as relay closes.
3. When TCC is on, relay is energized and test light is off.
   a. Goes on with throttle changes and goes out with slight delay.
   b. Tap brake pedal and light goes on and off—no delay.
   c. 4-3 downshift leaves light out as relay may stay on, but full throttle downshift may use TPS to affect ECM and light goes on.
   d. 4th gear switch causes wide throttle movement before TCC is affected in 4th gear.
B. Jumper wire between test lead and ground.

NOTE: Use jumper wire if test light stays on at all times in A-3 above. Jumper wire by-passes ECM and causes full time solenoid on.

1. Turn ignition on—listen for TCC solenoid click.
   a. No click—problem in transmission.
   b. Does click—road test or test on hoist to verify TCC engage.
   1. No engage—check TCC solenoid seal and ball seat.
   2. Engages—check TCC relay operation by moving jumper wire to TCC relay terminal B.
      a. TCC relay should click and TCC solenoid should click.
      b. No click—ground TCC relay terminal E—TCC solenoid should click.

1982 THM 700-R4 CORVETTE CCC OR EFI

1. 2nd clutch oil through TCC shift valve to solenoid and apply valve.
2. 4-3 pulse switch is normally closed. Opens only momentarily on 4-3 downshifts.
3. 4th gear switch is input to ECM (N.C.) provides wider TPS movement before TCC is affected while in 4th
4. Ignition switch and brake switch control feed to solenoid.
5. ECM controls solenoid ground.

Diagnosis:

A. Test light between test lead and ground.
1. Ignition on-brake off—test light comes on.
2. Tap brake pedal—test light goes off and on.
3. Start engine and drive—light stays on until the following.
   a. Coolant temperature up to specifications.
   b. Vehicle speed to specifications.
   c. TPS-VAC signal to specifications.

Note that the solenoid can be energized in second gear and the TCC can come on. The light will go out.
4. When TCC is on, test light will be out.
   a. Goes on with throttle changes and then back on with slight delay.
   b. Tap brake pedal—light stays out since only feed affected.
   c. 4th gear switch signals larger TPS change before TCC affected.
   d. 4-3 downshift may leave light off as 4-3 pulse switch pulses open for TCC pulses.

B. Jumper wire between test lead and grounds.

NOTE: Use jumper wire if light stays on at all times in A-3 above. Jumper wire by-passes ECM & causes full time solenoid on.

1. Ignition on engine off—touch wire between test lead and ground and listen for solenoid click.
2. Road test or drive wheels—drive to verify that TCC does engage after 1-2 shift and holds engaged in 3rd and 4th at all loads.
   a. If not—check solenoid seals and ball seat.
   b. If yes—refer to CCC diagnosis for ECM and input sensors.

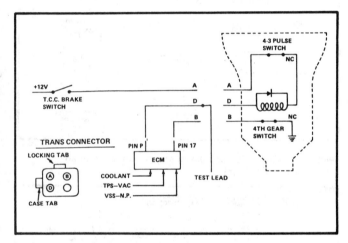

1982 THM 700-R4—Corvette equipped with CCC or EFI (©General Motors Corp.)

1982 THM 325-4L—5.7L DIESEL

1. 2nd clutch oil through TCC shift valve to solenoid—apply valve area.
2. Solenoid has constant ground—and will be energized when key is turned "on" unless throttle switch is open.
3. TCC will engage as soon as signal oil from TCC shift valve is available.
4. Load control at heavy throttle switch on injector pump.

Diagnosis—

A—test light at test lead monitors feed circuit.
B—DO NOT use jumper wire at test lead.

1982 THM 350-C (MX-5)—BUICK BUILT

1. Forward clutch oil to solenoid and apply valve.
2. ECM controls solenoid ground.
3. Brake switch—Ignition switch controls solenoid feed.
4. 2nd and 3rd clutch oil switches used to match engagement load with gear used.

Diagnosis—

A. Test light between test lead and ground.

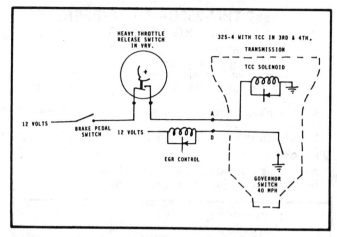

1982 THM 325-4L with 5.7L Diesel engine (©General Motors Corp.)

1. Ignition switch on—test light goes on (if not check at brake switch).
2. Tap brake pedal—test light goes off and on.
3. Start engine—light stays on at all time until the following are met:
   a. Coolant temperature up to specifications.
   b. Vehicle speed to specifications.
   c. Throttle position/vacuum switch signal to specifications.
Note that solenoid can be energized even before 2nd gear is achieved—TCC then comes on in first.
4. When vehicle TCC is on, test light is off.
   a. Goes on with change in throttle position, then back off with slight delay.
   b. Tap brake pedal—light stays off—ECM keeps ground.
   c. Downshift causes light to go on via TPS sensor and vacuum.
B. Jumper wire between test lead and ground.

NOTE: Use jumper wire if test light stays on at all times in A-3 above. Jumper wire bypasses ECM and causes full time solenoid on.

1. Ignition on engine off—Touch wire between test lead and ground and listen for solenoid click.

NOTE: The jumper wire at test lead will apply TCC as soon as shift lever is moved to any forward gear. An idling engine will stall.

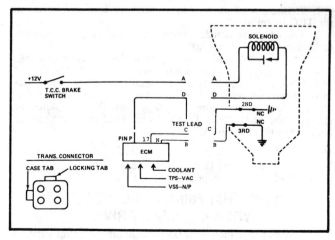

1982 THM 350C, Model MX-5, Buick built (©General Motors Corp.)

1982 THM-350C—5.7L DIESEL WITH EGR BLEED

Forward clutch oil to solenoid and apply valve.
1. Requires 3rd gear switch to keep apply to 3rd.
2. Governor switch for vehicle speed information.
3. Release switch at injection pump for Heavy Throttle Release.
4. EGR bleed controlled through 3rd clutch and governor switches.
5. TCC remains engaged on coast until 3-2 downshift.

Diagnosis—

A—Test light at test lead monitors feed circuit.
B—DO NOT use jumper wire at test lead.

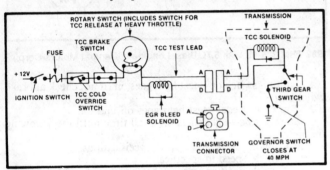

1982 THM 350C with 5.7L diesel engine with EGR bleed (©General Motors Corp.)

1982 THM 200-4R 5.7L DIESEL WITH EGR BLEED

1. Second gear clutch oil is fed through the TCC shift valve to the TCC apply valve via the TCC solenoid.
2. The TCC solenoid has a constant ground. It is energized when the ignition switch is turned "ON," unless the throttle switch in the rotary switch is open.
3. The TCC will engage as soon as the signal oil from the TCC apply valve is available.

Diagnosis—

A—Test light at test lead monitors feed circuit.
B—DO NOT use jumper wire at test lead.

1982 THM 700-R4 6.2L DIESEL 2-WHEEL DRIVE

1. Second gear clutch oil is fed through the TCC shift valve via the TCC solenoid.
2. The 4-3 pulse switch is normally closed. It opens momentarily during a 4-3 downshift.
3. The TCC solenoid is directly grounded at case mounting.
4. The rotary switch contains two sets of contacts. The first set of contacts is open at light throttle, while the second set of contacts are normally closed. The first set of contacts closes anytime the throttle is opened more than 10 percent. This completes the circuit to the TCC solenoid. The second set of contacts opens at heavy throttle to release the TCC.

Diagnosis—

A—Test light at test lead monitors feed circuit.
B—DO NOT use jumper wire at test lead.

1982 THM 700-R4 6.2L DIESEL WITH 4-WHEEL DRIVE

1. Second gear clutch oil is fed through the TCC shift valve to the TCC apply valve via the TCC solenoid.

2. The 4-3 pulse switch is normally closed. It opens momentarily during a 4-3 downshift.
3. The TCC solenoid is directly grounded to the transmission case.
4. The rotary switch contains two sets of contacts.
5. When in 2-wheel drive, the feed to the TCC solenoid is through the ignition switch, brake switch, rotary switch and 4-wheel drive relay.
6. When in 4-wheel drive, the contacts in the 4-wheel drive relay open and the fourth gear switch must close to feed the TCC solenoid.
7. A switch in the transfer case closes when the vehicle is in 4-wheel drive. In turn, this energizes the 4-wheel drive relay, opening the contacts. This opens the circuit and provides the ground for the 4-wheel drive indicator light. The transfer case switch also prevents TCC operation in second and third gears when in 4-wheel drive.

Diagnosis—

A—test light at test lead monitors feed circuit
B—DO NOT use jumper wire at test lead

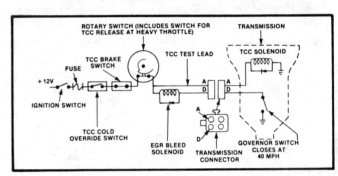

1982 THM 200-4R with 5.7L diesel engine with EGR bleed (©General Motors Corp.)

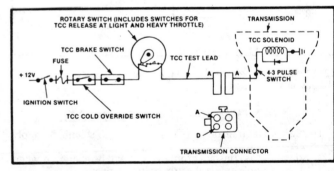

1982 THM 700-R4 with 6.2L diesel engine, Two wheel drive (©General Motors Corp.)

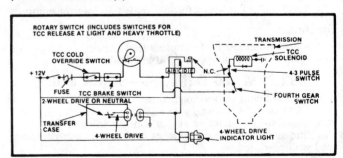

1982 THM 700-R4 with 6.2L diesel engine, Four wheel drive (©General Motors Corp.)

1982 THM 700-R4 LIGHT TRUCK WITH GAS ENGINE 2- OR 4-WHEEL DRIVE

1. Second gear clutch oil is fed through the TCC shift valve to the TCC apply valve via the TCC solenoid.

2. The 4-3 pulse switch is normally closed. It opens momentarily during a 4-3 downshift.

3. The fourth gear switch is used to bypass the vacuum switch when the transmission is in fourth gear. This keeps the TCC engaged at all throttle openings in fourth gear.

4. The vacuum switch opens when vacuum drops to approximately 1-3 in. Hg near wide open throttle. The switch is also open at idle.

5. The EGR bleed solenoid is energized when the transmission is in third or fourth gear and both the third clutch and TCC signal switches close.

Diagnosis—

A—test light at test lead monitors feed circuit
B—DO NOT use jumper wire at test lead

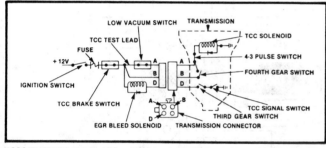

1982 THM 700-R4, light truck with gasoline engine, two or four wheel drive (©General Motors Corp.)

1982 THM 350C 5.7L DIESEL WITHOUT EGR BLEED

1. Direct clutch oil is fed to the TCC apply valve via the TCC solenoid.

2. The third gear switch is necessary to restrict TCC operation to third gear.

3. The governor switch acts as a speed sensor for TCC operation and provides the ground for the third gear switch and TCC solenoid.

4. A release switch is included in the rotary switch on the injection pump to release the TCC during heavy throttle applications.

5. The TCC solenoid remains energized during coastdown until the governor switch opens. The TCC clutch is released by the poppet valves inside the converter.

Diagnosis—

A—test light at test lead monitors feed circuit
B—DO NOT use jumper wire at test lead

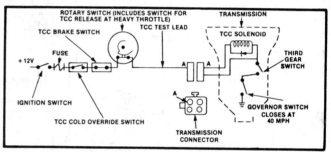

1982 THM 350C with 5.7L diesel engine with EGR bleed (©General Motors Corp.)

TORQUE CONVERTER CLUTCH OIL FLOW

1983 And Later Wiring Information Refer To Individual Automatic Transmission Sections.

GENERAL INFORMATION

In all converter clutches, fluid pressure delivered to the stator side of the pressure plate applies the clutch, while fluid pressure applied to the engine side of the pressure plate releases the clutch.

REAR DRIVE VEHICLES

In the automatic transmissions installed in the rear drive vehicles, the release fluid is delivered through a drilled passage in the turbine shaft. The release fluid is fed between the stator shaft bushing and the front oil delivery ring on the turbine shaft. The bushing acts as the front seal as fluid is directed to the drilled turbine shaft and to the engine side of the pressure plate.

Apply fluid is delivered between the converter hub and the stator shaft. Apply fluid is delivered between the converter hub and the stator shaft. Apply fluid moves the pressure plate against the converter cover to apply the clutch. Fluid is sealed at the oil pump drive gear, the pressure plate friction area and at the turbine thrust spacer "O" rings in both areas.

FRONT DRIVE VEHICLES

In the Automatic transaxles installed in the front drive vehicles, release fluid is delivered through a hollow turbine shaft around the pump shaft to the engine side of the pressure plate. Apply fluid is delivered through the turbine shaft by way of a sleeve pressed in the shaft. The sleeve allows apply fluid to be routed

around the sprocket area in the link cavity without losing pressure. There are two teflon rings at the valve body end of the turbine shaft sleeve and one at the converter end of the sleeve on the turbine shaft. Missing and/or damaged seals could cause improper clutch apply.

Hydraulic Converter Clutch Controls

A converter clutch apply (control) valve is used to control the direction of oil flow throughout the torque converter and therefore the apply or release of the converter clutch.

All apply valves are moved to the apply position by oil pressure through an orifice to an area between the valve and a bleed (TCC) Solenoid. The restriction is smaller than the bleed in the solenoid so no pressure is available to move the valve when the solenoid is not energized electrically.

The 200-C apply valve train (2-piece) is located in the oil pump. Line oil on the small end holds the valve in the release position. Direct clutch oil (in 3rd gear) acts on the large end of the valve train when the solenoid is energized.

The 250-C—350-C apply valve is a one piece valve located in an auxiliary valve body on the front of the separator plate. The valve is held in the release position by spring pressure. 2-3 clutch oil, acting on the ring land difference, moves the valve to the apply position. In the Chevrolet built units (M-31 & M-38). Buick built units (MX2, MX3, MX5) use forward clutch oil to move the

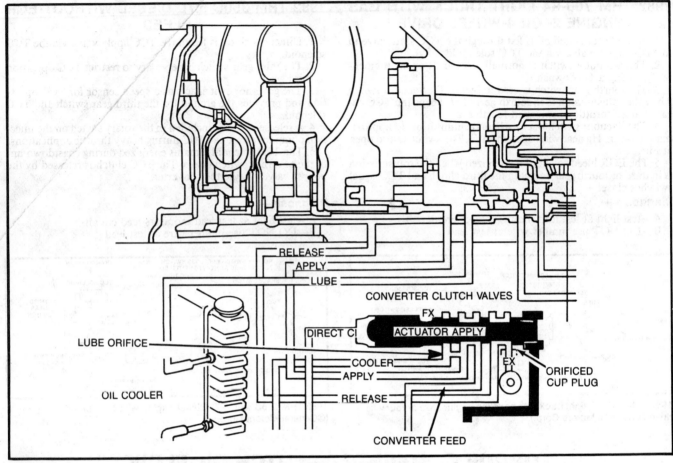

THM 200C oil flow schematic to apply Torque Converter Clutch (©General Motors Corp.)

valve to the apply position. This allows the clutch to be applied in a lower gear than 3rd.

The overdrive transmissions use a one piece valve spring loaded to the release position. Oil to move it to the apply position is controlled by a TCC shift valve that operates like any shift valve. Governor pressure must rise to move the shift valve before oil called signal can move the apply valve. The 1-2 shift must also be made to supply 2nd clutch oil to the TCC shift valve. The screen in the signal oil passages helps to keep the apply valve and the solenoid bleed clean.

The 125-C apply valve is a one piece valve held released by line oil pressure. Line oil controlled by the solenoid is used to move the valve to the apply position. A TCC regulator valve is used to regulate line oil to the converter during clutch apply. This limits converter oil pressure much like the converter feed orifice does in the other transmissions and in release position in the 125-C.

In all cases, oil returning from the converter during the release operation is routed to the oil cooler, then to the lube passages.

SOURCE OF CONVERTER CLUTCH OIL SUPPLY

Oil to the solenoid-apply valve area to move the valve to the apply position is different on different transmissions.
1. 200-C, 250-C and Chev. built 350-C—direct clutch oil.
2. 350-C Buick built—forward clutch oil.
3. 125-C—line oil
4. Overdrive Trans.—2nd. clutch oil through a TCC shift valve.

ACTION OF RELEASE OIL DURING CONVERTER CLUTCH APPLY

In all cases oil from the converter release passage during the apply:
1. Passes through an orifice to slow the oil allowing a timing-dampening action during the apply. The orifice may be
    a. A scallop in a check ball seat (in turbine shaft)
    b. A hole near the converter valve in transfer plate, valve body, or pump assy.
2. Is exhausted without going through cooler and lube passages. Lube oil is passed to the cooler and lube passages through another orifice.

The direction of oil flow to the converter is reversed when the apply valve is moved. The valve is held in the release position by spring pressure or by line oil pressure.

BLOCKED SOLENOID VALVE CONDITIONS

With the various pressure sources of fluid to the apply valve, a stuck closed solenoid valve can cause the following conditions:
1. Direct clutch oil operated will have TCC apply whenever 3rd gear is achieved even at an 18 mph light throttle shift. This can cause rough engine and stalling. It could also cause vehicle jerk on coast down before a 3-2 or 3-1 downshift.
2. The 350-C with forward clutch oil operated TCC apply, will cause the engine to stall as the shift lever is moved to any forward gear.
3. The overdrive transmissions would have TCC shift valve shifts to direct signal oil to the apply valve. We could also feel the coast jerkiness.

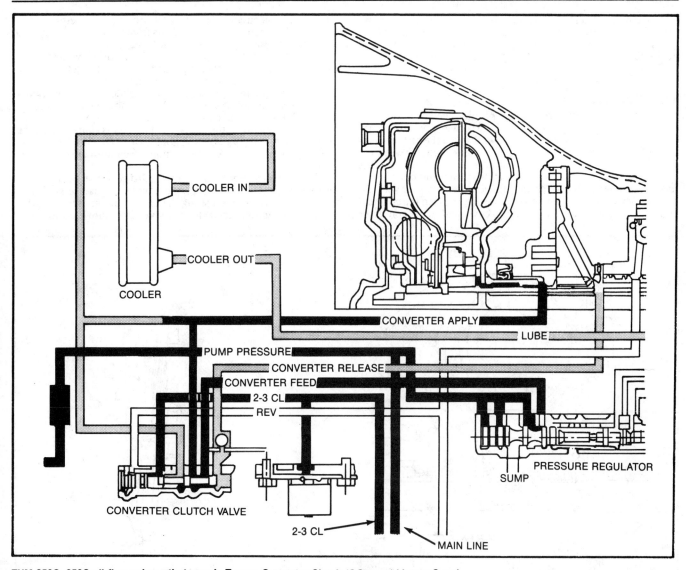

THM 250C, 350C oil flow schematic to apply Torque Converter Clutch (©General Motors Corp.)

4. The 125-C uses line oil to the apply valve and we could expect engine stall in a shift to any gear (forward or reverse) when the driveline connection is made to the drive wheels.

NOTE: If the apply valve is stuck in the apply position, the transmission will react in the same manner as the 125-C in step four. If the apply valve is stuck in the released position, the converter clutch will not apply at any time.

OIL FLOW—CONVERTER CLUTCH

THM 200-C

In Park, Neutral, Reverse, and First gear, the converter clutch apply valve, located in the pump, is held in the release position by line pressure. That is, the apply valve in this position takes converter feed oil, coming from the pressure regulator valve, and sends it into the release passage, through the turbine shaft, and into the cavity between the converter clutch pressure plate and the converter cover. This moves the clutch pressure plate away from the converter cover, releasing the converter clutch.

After the oil releases the converter clutch, it flows from behind the pressure plate and through the converter. From the converter, the oil flows backwards through the apply passage to the apply

valve. The apply valve then sends oil returning from the converter to the transmission cooler in the radiator. Oil, returning from the cooler, is then directed to the transmission lubrication system.

The converter clutch apply valve is controlled by a solenoid operated exhaust valve that is located in the direct clutch oil passage leading to the converter clutch actuator valve. When the vehicle is in drive range, third gear, direct clutch oil which applies the direct clutch also passes through an orifice to the solenoid exhaust valve and to the converter clutch actuator valve.

When vehicle speed in drive range, 3rd gear, reaches a pre-determined speed, (35-45 mph), the governor pressure switch or ECM switch closes, completing the ground circuit and energizes the solenoid closing the exhaust valve. This allows direct clutch oil to move the converter clutch valve or apply valve to the apply position.

With the converter clutch valve in the apply position, converter feed oil is redirected from the release passage to the apply passage. Converter feed oil then flows from the apply passage to the converter by flowing between the converter hub and stator shaft. The converter is now being charged with oil from the apply side of the converter clutch pressure plate.

As the pressure plate begins to move to its applied position, release oil on the front side of the pressure plate is redirected back into the turbine shaft down the release passage and is exhausted at the converter clutch apply valve orifice to dampen apply.

Oil to cooler and lube system is provided through an orifice in the apply line to cooler passage.

THM 250-C and 350-C

With the apply valve in the release position, converter feed oil flows through the open apply valve to the converter clutch release passage in the pump cover. It then flows through the turbine shaft to the front or release side of the converter clutch, between the converter clutch pressure plate and the converter cover. This moves the clutch pressure plate away from the converter cover, releasing the converter clutch and charging the converter with oil. The oil then leaves the converter by flowing between the converter hub and stator shaft into the pump cover and to the converter

clutch apply oil circuit. The apply oil circuit is now being used in a reverse direction. The oil then flows from the apply passage into the cooler passage and to the lubrication system.

The converter clutch apply valve is controlled by a solenoid operated exhaust valve that is located in the direct clutch oil passage leading to the converter clutch actuator valve. When the vehicle is in drive range, third gear, direct clutch oil which applies the direct clutch also passes through an orifice to the solenoid exhaust valve and to the converter clutch acutator valve.

In the Buick built 350-C, forward clutch oil is supplied past the orifice to the apply valve—solenoid area.

This oil is exhausted by the solenoid bleed until the solenoid is energized.

When vehicle speed in drive range, 3rd gear, reaches a pre-determined speed, (35-45 mph), the governor pressure switch or ECM switch closes, completing the ground circuit and energizes the solenoid, closing the exhaust valve. This allows oil to move

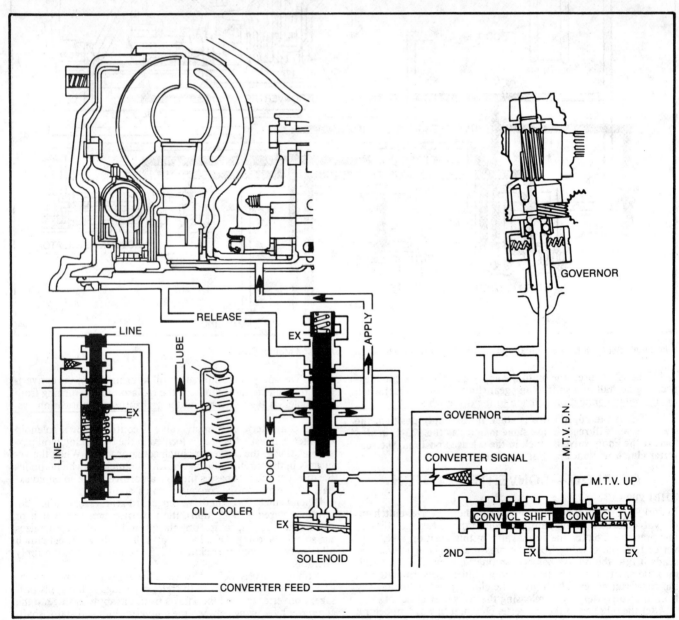

THM 200-4R, 325-4L, 700-R4 oil flow schematic to apply Torque Converter Clutch (©General Motors Corp.)

the converter clutch valve or apply valve to the apply position.

With the converter clutch valve in the apply position, converter feed oil is redirected from the release passage to the apply passage. Converter feed oil then flows from the apply pasage to the converter hub and stator shaft. The converter is now being charged with oil from the apply side of the converter clutch pressure plate.

As the pressure plate begins to move to its applied position, release oil on the front side of the pressure plate is redirected back into the turbine shaft down the release passage and is exhausted at the converter clutch apply valve. Check ball orifice to dampen appy.

Oil to cooler and lube system is provided through an orifice in the apply line to cooler passage.

OVERDRIVE UNITS

In Park, Neutral, Reverse, and First gear, the converter clutch apply valve located in the pump, is held in the release position by the converter clutch apply valve spring. The apply valve in this position takes converter feed oil, coming from the pressure regulator valve, and sends it into the release passage, through the turbine shaft, and into the release passage, through the turbine shaft, and into the cavity between the converter clutch pressure plate and the converter cover. This moves the clutch pressure plate away from the converter cover, releasing the converter clutch.

After the oil releases the converter clutch, it flows from behind the pressure plate and through the converter. From the converter, the oil flows backwards throughout the apply passage to the apply valve. The apply valve then sends oil returning from the converter, to the transmission cooler in the radiator. Oil, returning from the cooler, is then directed to the transmission lubrication system.

The apply valve is controlled by the converter clutch shift valve and will stay in the release position until it receives converter clutch signal oil from the converter clutch shift valve in the valve body.

It should be noted that the converter clutch apply valve can shift to apply the converter only when the converter clutch solenoid is energized electrically. When the solenoid is not energized, the converter clutch signal oil is exhausted and the converter will stay in the release position. The converter clutch solenoid would be off under conditions such as: high engine vacuum (idle), low engine vacuum (full throttle), braking, or cold engine operation.

In the hydraulic system as described thus far, the 1-2. 2-3. 3-4. and converter clutch shifts will always take place at the same vehicle speeds: that is, whenever the governor pressure overcomes the force of the springs on the shift valves. When accelerating under a heavy load or for maximum performance, it is desirable to have the shifts occur at higher vehicle speeds.

As the pressure plate begins to move to its applied position, release oil on the front side of the pressure plate is redirected back into the turbine shaft past the scallop in the check ball seat, then down the release passage and is exhausted at the converter clutch apply valve to dampen the apply.

Oil to cooler and lube system is provided through an orifice in the apply line to cooler passage.

THM 125-C

The apply or release of the converter clutch is determined by the direction that the converter feed oil is routed to the converter. The converter feed oil from the pressure regulator valve flows to the converter clutch control valve. The position of the converter clutch control valve controls which direction converter feed oil flows to the converter.

The converter clutch control valve is held in the release position in park, neutral, reverse, drive range 1st gear and 2nd gear, by line pressure acting on the end of the converter clutch apply valve. With the converter clutch control valve in the release position, converter feed oil flows into the converter clutch release passages. It then flows between the pump drive shaft and turbine shaft to the front or release side of the converter clutch, between the converter clutch pressure plate and the converter cover. This

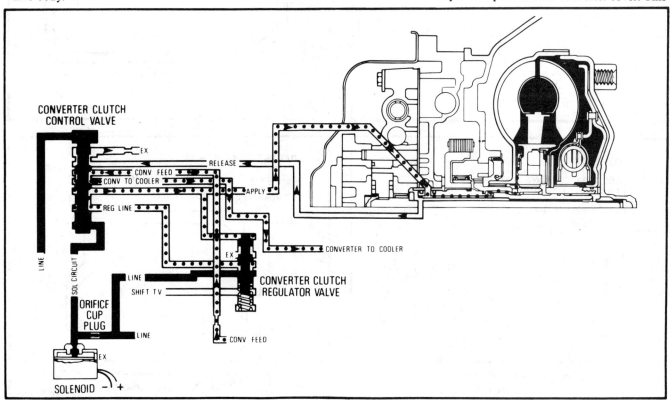

THM 125C oil flow schematic to apply Torque Converter Clutch (©General Motors Corp.)

moves the converter clutch pressure plate away from the converter cover, releasing the converter clutch and charging the converter with oil. The oil then leaves the converter by flowing through the turbine shaft into the converter clutch apply oil circuit. The apply oil circuit is now being used in a reverse direction. The oil then flows from the apply passage into the cooler passage and to the lubrication system.

To prevent the converter clutch from applying in drive range 3rd gear, at car speeds below converter clutch engagement speeds, the C3 or governor pressure switch (depending on system) will break the circuit to the solenoid exhaust valve. This de-energizes the solenoid and opens the exhaust valve to the exhaust the solenoid circuit oil at the converter clutch control valve. Line pressure then holds the converter clutch control valve in the release position.

When car speed in drive range 3rd gear, reaches converter clutch engagement speed, the C3 or governor pressure switch (de-pending on system) will activate the solenoid, closing the exhaust valve. This allows solenoid circuit to move the converter clutch control valve against line pressure. With the converter clutch control valve in the apply position, regulated line oil, from the converter clutch regulator valve, is allowed to pass into the converter apply passage. It then flows through the turbine shaft to the apply side of the converter clutch. The regulated line oil from the converter clutch regulator valve, controls the apply feel of the pressure plate.

As the pressure plate begins to move to its applied position, release oil on the front side of the pressure plate is redirected back between the turbine shaft and pump drive shaft and exhausted at the converter clutch control valve through an orifice, to time the clutch apply. When the converter clutch control valve moved to the apply position, orificed converter feed oil entered the converter to cooler passage to provide oil to the lubrication system.

TRANSMISSION/TRANSAXLE OIL PAN IDENTIFICATION

Parts Book Code	Service Identity	Parts Book Code	Service Identity	Parts Book Code	Service Identity
MD9	125C	M-31	250C	MX3	350C
MD2	180C	M-57	325-4L	MX5	350C
MV9	200C	MV4	350C	MD-8	700-R4
MW9	200-4R	MX2	350C	ME-9	400-T4

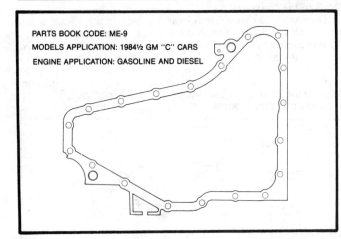

PARTS BOOK CODE: ME-9

MODELS APPLICATION: 1984½ GM "C" CARS

ENGINE APPLICATION: GASOLINE AND DIESEL

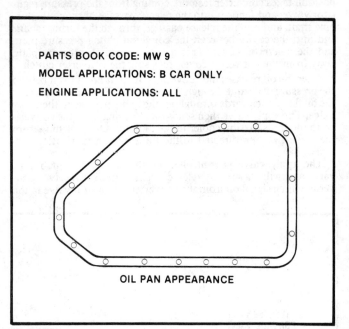

PARTS BOOK CODE: MW 9

MODEL APPLICATIONS: B CAR ONLY

ENGINE APPLICATIONS: ALL

OIL PAN APPEARANCE

PARTS BOOK CODE: MD 8

MODEL APPLICATIONS: B WAGON, Y CAR AND L.D. TRUCKS

ENGINE APPLICATIONS: GASOLINE AND DIESEL

OIL PAN APPEARANCE

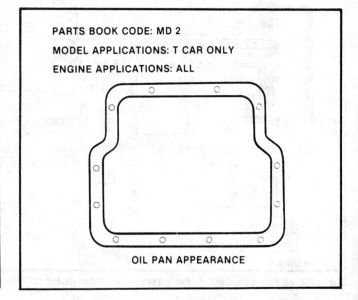

PARTS BOOK CODE: MD 2

MODEL APPLICATIONS: T CAR ONLY

ENGINE APPLICATIONS: ALL

OIL PAN APPEARANCE

ENGINE TRANSMISSION MARRIAGE I. D. CODE

(I. D. Plate or Ink Stamp on Bell)

Engine Mfg.	Hydramatic Built Transmission	Buick Built 350-350C Transmission	Chevrolet & Canadian Built 350-350C Transmission
Chevrolet	C-	J-	W-T-X M- 1980-83 350C Comp. V- 350C
Buick	B-	K-	—
Oldsmobile	O-	L-	—
Pontiac	P-	M- THRU 1979	—
Cadillac	A-	—	—
Jaguar	Z-	—	—
Non ECM Export 1981 Chevrolet 5.0	E-	—	—
Non ECM Export 1982-83 Chevrolet 5.0	H-	W-	—

PARTS BOOK CODE: MV 4

MODEL APPLICATIONS: '81 A, B, G, Y, L.D. TRUCKS

ENGINE APPLICATIONS: GASOLINE AND DIESEL

ID LOCATION: STAMPED ON RIGHT SIDE OF OIL PAN

OIL PAN APPEARANCE

PARTS BOOK CODE: MV 9

MODEL APPLICATIONS: '81 A, B, F, G, H, AND T CARS AND S TRUCK

ENGINE APPLICATIONS: GASOLINE AND DIESEL

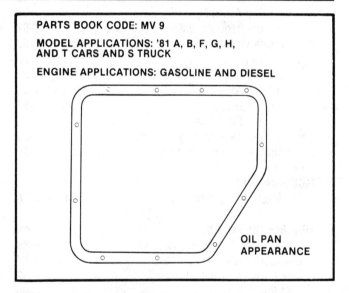

OIL PAN APPEARANCE

PARTS BOOK CODE: M 31

MODEL APPLICATIONS: '81 A, B AND G CARS

ENGINE APPLICATIONS: ALL

ID LOCATION: STAMPED ON GOV COVER

OIL PAN APPEARANCE

PARTS BOOK CODE: MD 9

MODEL APPLICATIONS: X, J AND '82 A CARS

ENGINE APPLICATIONS: ALL

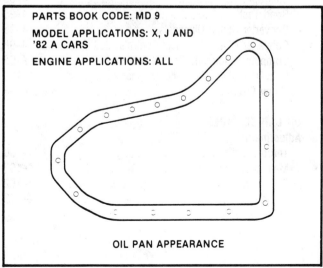

OIL PAN APPEARANCE

INDEX

GENERAL MOTORS
TURBO HYDRA-MATIC 125C
Automatic Transaxle

 **APPLICATIONS**

GENERAL MOTORS THM-125C AUTOMATIC TRANSAXLE APPLICATION CHART

Year	Make	Model
1982 and Later	Buick	Skylark (X body)
	Buick	Century (A body)
	Buick	Skyhawk (J body)
1982 and Later	Chevrolet	Citation (X body)
	Chevrolet	Celebrity (A body)
	Chevrolet	Cavalier (J body)
1982 and Later	Oldsmobile	Omega (X body)
	Oldsmobile	Ciera (A body)
	Oldsmobile	Firenza (J body)
1982 and Later	Pontiac	Phoenix (X body)
	Pontiac	6000 (A body)
	Pontiac	2000 (J body)

GENERAL MOTORS 125C AUTOMATIC TRANSAXLE APPLICATION CHART

Year	Make	Model
1982 and Later	Cadillac	Cimarron (J body)

 **GENERAL DESCRIPTION**

The General Motors THM 125C automatic transaxle is a fully automatic transaxle consisting of a compound planetary gear set and dual sprocket, drive link assembly and a four element hydraulic torque converter. Also contained in the transaxle assembly are the differential and final drive gear set.

The friction elements required to obtain the desired function of the planetary gear sets are provided for by three multiple disc clutches, a roller clutch and a band.

The hydraulic system is pressurized by a vane type pump which provides the working pressure required to operate the automatic controls and the friction elements.

Contained in the four element torque converter assembly are a pump, a pressure plate splined to the turbine, the turbine and a stator assembly. The pressure plate provides a mechanical direct drive coupling of the engine to the planetary gear, when applied.

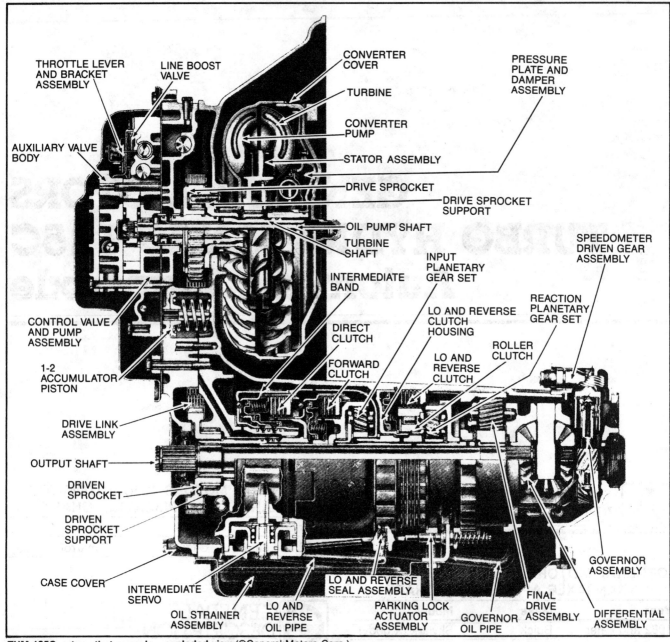

THROTTLE LEVER AND BRACKET ASSEMBLY

LINE BOOST VALVE

CONVERTER COVER

PRESSURE PLATE AND DAMPER ASSEMBLY

TURBINE

CONVERTER PUMP

AUXILIARY VALVE BODY

STATOR ASSEMBLY

DRIVE SPROCKET

DRIVE SPROCKET SUPPORT

OIL PUMP SHAFT

TURBINE SHAFT

SPEEDOMETER DRIVEN GEAR ASSEMBLY

INPUT PLANETARY GEAR SET

INTERMEDIATE BAND

REACTION PLANETARY GEAR SET

LO AND REVERSE CLUTCH HOUSING

CONTROL VALVE AND PUMP ASSEMBLY

DIRECT CLUTCH

ROLLER CLUTCH

LO AND REVERSE CLUTCH

FORWARD CLUTCH

1-2 ACCUMULATOR PISTON

DRIVE LINK ASSEMBLY

OUTPUT SHAFT

DRIVEN SPROCKET

DRIVEN SPROCKET SUPPORT

GOVERNOR ASSEMBLY

CASE COVER

INTERMEDIATE SERVO

LO AND REVERSE SEAL ASSEMBLY

OIL STRAINER ASSEMBLY

LO AND REVERSE OIL PIPE

PARKING LOCK ACTUATOR ASSEMBLY

GOVERNOR OIL PIPE

FINAL DRIVE ASSEMBLY

DIFFERENTIAL ASSEMBLY

THM 125C automatic transaxle—exploded view (©General Motors Corp.)

TRANSAXLE AND CONVERTER IDENTIFICATION

Transaxle

The model identification code is located on top of the transaxle, near the manual control lever shaft. The serial number is stamped on the oil pan flange pad, to the right of the oil dipstick.

Torque Converter

The torque converter is a welded unit and cannot be disassembled for repairs. Should this unit need to be replaced, both new and rebuilt units are available.

Metric Fasteners

The THM 125C automatic transaxle uses metric fasteners. Metric fastener dimensions are very close to the dimensions of the inch system fasteners, and for that reason, replacement fasteners must have the same measurement and strength as those removed. Do not attempt to interchange metric fasteners for inch fasteners. Mismatched or incorrect fasteners can result in damage to the transaxle unit through malfunctions or breakage, or even personal injury. Care should be taken to reuse the fasteners in the same location as removed.

Fluid Capacities

The THM 125C automatic transaxle has a fluid capacity of 9

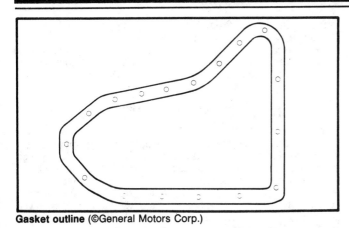

quarts including the torque converter. If the torque converter has not been drained the fluid capacity is 4 quarts. Be sure to use Dextron® II automatic transmission fluid in the THM 125C transaxle. Always bring the transaxle up to operating temperature and recheck the fluid level. Never overfill the unit.

Check fluid Levels.

1. Position the selector lever in "P."
2. Apply brakes and start engine.
3. Move the selector lever through each of the selector ranges.
4. Check the oil level on the dipstick. Add fluid as required.
5. The fluid level at room temperature should be ½ inch above the full mark or between the two dimples on the transaxle dipstick.
6. At normal operating temperature the fluid level should be between the add and full marks on the transaxle dipstick.

NOTE: Due to the shape of the filler tube, oil level readings may be misleading. Look carefully for a full oil ring on both sides of the dipstick. If there is any doubt, recheck the level. Remember that in this unit the cold level will be higher than the hot level. Do not overfill this unit or foaming, loss of fluid and possible overheating may occur.

MODIFICATIONS

Solenoid And Clip Assembly

After August 1981, all THM 125C automatic transaxles are being built with a new design solenoid and clip assembly. This new design solenoid and clip assembly includes a plastic tubular clip to prevent the wire from moving off its location on the valve body during assembly of the unit. The new design solenoid and clip assembly and the old design assembly, which uses a metal retainer, can be interchanged. However, due to the difference of the wire length, the plastic clip, which is available for service separately, cannot be used with the old design solenoid assembly. When installing a new design solenoid assembly be sure to remove and discard the metal clip from the old design solenoid assembly.

Torque Converter Change

Starting in 1982, all THM 125C automatic transaxles will be built with a new design 245 millimeter torque converter. This new torque converter will replace the present 254 millimeter torque converter.

Turbine Shaft Design Change

Starting in 1982, all THM 125C automatic transaxles will be built using a new design turbine shaft. The spline of the new shaft will be 5.1 millimeters longer than the old style turbine shaft, and will incorporate an O-ring seal groove.

Redesigned Oil Pump Drive Shaft

Beginning in January 1982, a new style oil pump drive shaft will be used in all THM 125C automatic transaxles. The new shaft is 23.7 millimeters longer and has 15 teeth on the oil pump spline rather than the old shaft which has 20 teeth on the oil pump spline and is 23.7 millimeters shorter in length.

New Design Torque Converter Housing Oil Seal

Due to the usage of a new turbine shaft and oil pump drive shaft a new design torque converter housing oil seal is being used in all units produced after December 1981. The new oil seal can be identified by the part number 8637420 stamped on the front face of the seal. The old design seal will have either the part number 8631158 stamped on it or no part identifcation at all.

Oil Weir Usage

During the month of September 1981 all THM 125C automatic transaxles were assembled with a new part called an oil weir. This part was used by General Motors for one month as a production trial run. The oil weir is located in the rear case oil pan area. It's function is to revise the lubrication flow around the differential

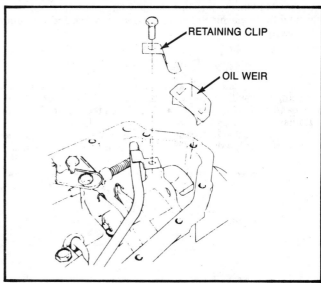

assembly. The oil weir part number is 8637836 and is held in position by a retaining clip, part number 8637837. Both of these parts are available for service. Automatic transaxles built without an oil weir do not require the addition of the part during service.

Intermediate Band and Direct Clutch Assembly

In late August 1981 some 1982 automobiles built with the THM 125C automatic transaxle used a new wide design intermediate band assembly, part number 8637623 and a new type direct clutch housing and drum assembly, part number 8637976. This new design direct clutch housing has a wider surface finish area on the drum outside diameter in order to accommodate the new wider intermediate band assembly.

The narrow design direct clutch housing assembly, part number 8631928 can only be used with the narrow design intermediate band assembly, part number 8631030. Do not use the narrow design direct clutch housing assembly with the wider design intermediate band assembly or interference will result.

The narrow design intermediate band is 1.49 inch in width and the new design intermediate band is 1.74 inch in width.

The following is a list of automatic transaxle models incorporating this design change; BE, BL, PL, PK, PI, CT, CL, CV, CS, LE, OP, HM, HW, HS.

Pressure Regulator Valve Retaining Pin

When diagnosing the 125C automatic transaxle for no drive or harsh shifts (high line pressure), check the control valve assembly for a worn or missing pressure regulator valve retaining pin.

If, after checking the pressure regulator valve retaining pin, it is found to be either missing or worn, the following repair must be performed.

1. Position the control valve and oil pump assembly with the machined portion face down. Be certain that the machined face is protected in order to prevent damage to its surface.

2. Using a ⅜ drift punch and hammer, close the pressure regulator valve retaining pin hole. Close the pin hole only enough to hold the new retaining pin in place after assembly.

3. Reassemble the pressure regulator and reverse boost valve train assembly.

4. Retain the valve train with a new steel retaining pin, part number 112496. Be sure that the new pin is inserted from and flush with the machined face of the control valve and oil pump assembly.

New Style Band Plug

Beginning mid-June 1981, all THM 125C automatic transaxles are being built with a new design band anchor plug. This new design band anchor plug has a tab which holds the part in place. All automatic transaxles built prior to mid June 1981 have the old design style band anchor plug. A staking operation was required to hold the old plug in place. This staking operation is not re-

quired with the new design band anchor plug as due to the tab extension it is held in place by the reverse oil pipe.

When replacing the new design band anchor plug, the reverse oil pipe must be removed first. When repairing automatic transaxles prior to mid June 1981 use the new style band anchor plug, part number 8637640.

Torque Converter "Shudder"

Some 1982 THM 125C automatic transaxles may experience a "shudder" feel immediately following the engagement of the torque converter clutch. If this occurs, check for the following:

1. Inspect the turbine shaft seal and O-ring for damage. The O-ring is located at the long end of the turbine shaft (spline end) with one large diameter yellow teflon seal located below the torque converter feed lube holes. Two other teflon seals both yellow with green speckles are located at the short end of the turbine shaft. Replace the damaged seals as required.

2. Inspect the pump shaft seal for damage. Replace as required.

3. Check the torque converter clutch control regulator valve in the auxiliary control valve assembly for freeness. Replace as required.

In the Spring of 1982 a new governor pressure switch went into production for diesel equipped vehicles. In order to correct the "shudder" the new governor pressure switch raises the torque converter clutch apply speed. In servicing automatic transaxles equipped with a diesel engine, refer to the following chart for the proper governor pressure switch usage.

GOVERNOR PRESSURE SWITCH APPLICATION CHART THM-125C

Automatic Transaxle Model	Governor Pressure Switch Number
OP, HU, HY, H6, HR	8643369
HW	8637296
HS	8643368
CD, HI, HC	8643367

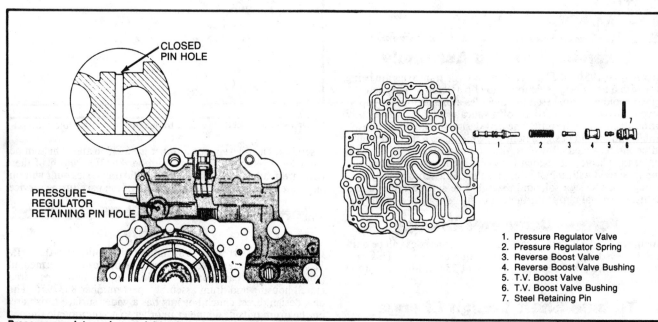

1. Pressure Regulator Valve
2. Pressure Regulator Spring
3. Reverse Boost Valve
4. Reverse Boost Valve Bushing
5. T.V. Boost Valve
6. T.V. Boost Valve Bushing
7. Steel Retaining Pin

Pressure regulator valve retaining pin installation (©General Motors Corp.)

No Torque Converter Clutch Release

Some 1982 THM 125C automatic transaxles may experience a no-torque converter clutch release condition. If this is the case, check the auxiliary control valve assembly for the proper location of the solenoid orifice cup plug. After checking the solenoid orifice cup plug, the following conditions could occur:

1. The solenoid orifice cup plug could be missing or damaged with a hole in it. This will cause the no-torque converter clutch release condition. Repair or replace as required.

2. The solenoid orifice cup plug could be installed too far into the bore past the third land in the auxiliary control valve assembly. Correct as required.

Service Case Packages With Missing Plugs

Some 1982 THM 125C automatic transaxle service case packages, part number 8631916 may have been assembled without three cup plugs and one pipe plug. If you encounter a service case package without these four parts, see the list below for the correct part and number.
Parking pawl shaft cup plug—part number 8631016.
Third oil cup plug—part number 8611710.
Case servo orifice cup plug—part number 8628864.
Governor pressure pipe plug—part number 0444612 or part number 044613.

Service Package For Intermediate Band/Direct Clutch Housing Assembly

A new intermediate band and direct clutch housing service package has been assembled and released to repair all THM 125C automatic transaxles. This new service package, part number 8643941 consists of a direct clutch housing, drum assembly, and an intermediate band assembly. These parts must be used together as a complete unit. These items are no longer available individually and can only be obtained as a set.

New Design Oil Pump Shaft Seal

At the start of production, all 1982 "A" and "X" body vehicles equipped with the THM 125C automatic transaxle are being built with a new design oil pump shaft seal. This new seal will assure adequate torque converter clutch apply oil pressure. Also included in this seal design change are the 1982 "J" body vehicles built after July 31, 1981.

When replacing the control valve oil pump assembly, be sure that the oil pump shaft is completely seated in the control valve oil pump assembly. The new design oil pump shaft seal may be used on all 125C automatic transaxles.

Design Change Low—Reverse Clutch Assembly

Beginning around the middle of July 1981, some 1982 vehicles produced with the THM 125C automatic transaxle were assembled with a new design low and reverse clutch assembly. This design change produces a more desirable neutral to reverse shift. The new design assembly consists of a modified low and reverse piston which eliminates the need for an apply ring. A smaller low and reverse clutch housing feed orifice is also used, as is a waved steel clutch plate, located next to the low and reverse piston. The new waved steel clutch plate eliminates the one flat steel clutch plate.

Burnt Band and Direct Clutch Assembly Condition

Some THM 125C automatic transaxle equipped vehicles may experience a burnt band and direct clutch condition. A possible cause of the burnt band and direct drive condition might be the third accumulator check valve not seating properly. This condition allows the intermediate band to drag while the direct clutch is applied, causing excessive friction. If the third accumulator is found to be defective, order service package part number 8643964, which contains a new dual land third accumulator check valve and conical spring. Refer to the following procedure to replace the accumulator assembly.

1. Remove the intermediate servo cover and gasket.
2. Remove the third accumulator check valve and spring. Inspect the third accumulator valve bore for wear and damage to the valve seat and also for the presence of the valve seat.
3. Plug both the feed and exhaust holes in the bore using petroleum jelly.
4. Replace the third accumulator check valve with the new dual land check valve. Center the valve to be sure that it is seated properly.

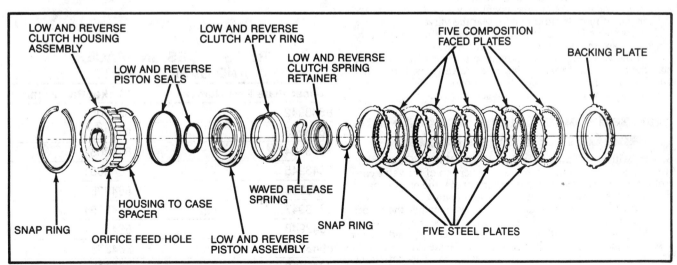

Old design Low and Reverse clutch housing and piston assembly (©General Motors Corp.)

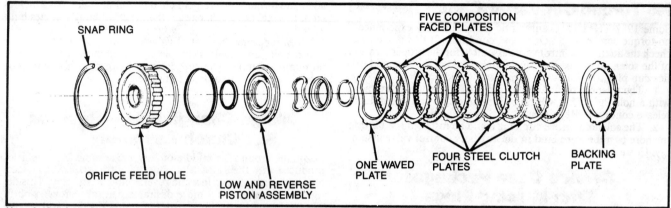

New design Low and Reverse clutch housing and piston assembly (©General Motors Corp.)

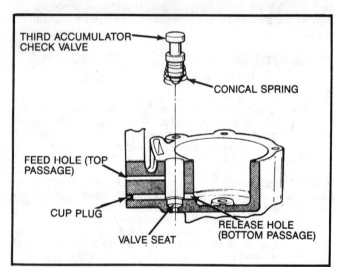

Third accumulator check valve replacement
(©General Motors Corp.)

OLD DESIGN SPACER AND GASKET CHART	
Spacer Plate Part Number	Gasket Part Number
8637119	8637096
8637121	8637096
8637122	8637096
8637123	8637096
8637124	8637096
8637125	8637096
8637127	8637096
8637128	8637096
8637129	8637096
8637130	8637096
8637815	8637096
8637816	8637096
8637817	8637096
8637818	8637096

5. Leak test the valve seat by pouring solvent into the accumulator check valve bore. Check for a leak on the inside of the case. A small amount of leakage is acceptable.

6. If the valve leaks tap the assembly with a brass drift and rubber mallet in order to try and reseat the valve.

7. Repeat the leak test procedure. It may be necessary to replace the case should the valve continue to leak.

8. If the valve does not leak, remove the check valve and install the new conical valve spring onto the valve, with the small end first. Install the valve into the case bore.

9. Using a new gasket install the servo cover.

Revised Valve Control Spacer Plate and Gasket Service Package

Starting with the month of March 1982, all THM 125C automatic transaxles are being built with a revised version of the valve control spacer plate and gasket. It is important that the proper spacer plate and gasket be used when servicing the unit.

The revised design spacer plate package will include the proper gasket and can be identified with a yellow stripe. Be sure to use the old design gasket with the old design spacer plate and the revised design gasket with the revised spacer plate, as these parts are not interchangeable. Use the following chart to determine proper usage.

REVISED DESIGN SPACER AND GASKET CHART	
Spacer Plate Part Number	Gasket Part Number
8643942	8643051
8643943	8643051
8643944	8643051
8643945	8643051
8643946	8643051
8643947	8643051
8643948	8643051
8643949	8643051
8643950	8643051

TROUBLE DIAGNOSIS

CLUTCH AND BAND APPLICATION CHART
THM 125C

Range	Gear	Direct Clutch	Intermediate Band	Forward Clutch	Roller Clutch	Low-Reverse Clutch
Park—Neut.	—	—	—	—	—	—
Drive	First	—	—	Applied	Holding	—
	Second	—	Applied	Applied	—	—
	Third	Applied	—	Applied	—	—
Int.	First	—	—	Applied	Holding	—
	Second	—	Applied	Applied	—	—
Low	First	—	—	Applied	Holding	Applied
	Second	—	Applied	Applied	—	—
Rev.	—	Applied	—	—	—	Applied

CHILTON'S THREE "C"'s" TRANSMISSION DIAGNOSIS CHART

Condition	Cause	Correction
No drive in Drive range	a) Low fluid level b) Manual linkage c) Clogged oil stainer d) Pressure regulator stuck e) Faulty oil pump f) Manual valve disconnected g) Faulty forward clutch h) Faulty roller clutch	a) Add as required b) Adjust as required c) Change screen and fluid d) Clean valve and bore e) Overhaul pump f) Repair link g) Overhaul h) Overhaul or replace
Oil pressure too high or too low	a) Throttle valve cable broken or not adjusted b) Throttle valve bracket bent or broken c) Line boost valve binding d) Throttle valve boost valve binding e) Reverse boost valve binding f) Pressure regulator binding g) Pressure relief valve spring damaged h) Manual valve disconnected i) Faulty oil pump	a) Adjust or replace as required b) Repair or replace c) Clean valve body d) Clean valve body e) Clean valve body f) Clean valve and bore g) Replace h) Repair link i) Overhaul pump
No drive or slipping in reverse	a) Throttle valve cable broken or not adjusted b) Manual linkage c) Throttle valve binding d) Reverse boost valve binding e) Malfunction in Low/Reverse clutch assembly	a) Adjust or replace as required b) Adjust as required c) Clean valve body d) Clean valve body e) Overhaul

CHILTON'S THREE "C"s TRANSMISSION DIAGNOSIS CHART

Condition	Cause	Correction
No drive or slipping in reverse	f) Reverse oil pipe plugged or out of place	f) Remove, clean, replace seal
	g) Check ball #4 out of place	g) Restore to proper location
	h) Malfunction in direct clutch assembly	h) Overhaul
	i) Spacer plate clogged	i) Clean valve body
No engine braking in L2	a) Faulty intermediate servo	a) Replace seal or assembly
	b) Faulty intermediate band	b) Replace band
No engine braking in L1 Note: No reverse would be a complaint with this condition	a) Malfunction in Low/Reverse clutch assembly	a) Overhaul
No part throttle or detent downshifts	a) Throttle valve cable broken or not adjusted	a) Adjust or replace as required
	b) Throttle valve binding	b) Clean valve body
	c) Shift T.V. valve binding	c) Clean valve body
	d) Spacer plate clogged	d) Clean valve body
	e) 2-3 T.V. bushing passages clogged	e) Clean valve body
	f) Throttle plunger bushing passages clogged	f) Clean valve body
Low or high shift points	a) Throttle valve cable broken or not adjusted	a) Adjust or replace as required
	b) Throttle valve binding	b) Clean valve body
	c) Shift T.V. valve binding	c) Clean valve body
	d) Line boost valve binding	d) Clean valve body
	e) 1-2 or 2-3 Throttle valve binding	e) Clean valve body
	f) Spacer plate or gasket out of position	f) Replace gaskets, align plate
	g) Throttle valve bracket bent or broken	g) Repair or replace
Note: Look for external leak	h) Governor seal or cover worn or damaged	h) Replace affected parts
First speed only—no 1-2 shift	a) Malfunction in governor assembly or feed circuits	a) Service or replace governor, clean passages
	b) 1-2 Shift train binding	b) Clean valve body
	c) Faulty intermediate servo	c) Replace seal or assembly
First and second speed only—no 2-3 shift	a) Malfunction in pump or control valve assembly	a) Overhaul pump, clean valve body as required
	b) 2-3 Valve train binding	b) Clean valve body
	c) Case and cover leakage	c) Check for missing 3rd oil cup plug, direct clutch accumulator check valve or servo bleed cup plug
	d) Oil seals leaking on driven sprocket support	d) Replace seals, check for clogged passages
	e) Malfunction in direct clutch assembly	e) Overhaul
	f) Faulty intermediate servo	f) Replace seal or assembly
	g) Governor seal or cover work or damaged	g) Replace affected parts
Third speed only	a) 2-3 Shift valve binding in upshift position	a) Clean valve body
	b) Malfunction in governor assembly or feed circuits	b) Service or replace governor, clean passages

CHILTON'S THREE "C"s TRANSMISSION DIAGNOSIS CHART

Condition	Cause	Correction
Drive in Neutral	a) Manual linkage b) Forward clutch will not release c) Cross leakage in case	a) Adjust as required b) Overhaul c) Clean valve body mating surfaces, replace gaskets
Won't hold in Park	a) Manual linkage b) Broken internal linkage c) Loose parts on internal linkage, detent roller	a) Adjust as required b) Replace affected parts c) Adjust or replace as required
Slips on 1-2 shift	a) Low fluid level b) Throttle valve cable broken or not adjusted c) Spacer or gasket out of position d) 1-2 Accumulator valve binding e) Faulty 1-2 accumulator assembly f) Faulty intermediate servo g) T.V. valve or shift T.V. valve binding h) Faulty intermediate band	a) Add as required b) Adjust or replace as required c) Replace gaskets, align plate d) Clean valve body e) Replace seal or piston as required f) Replace seal or assembly g) Clean valve body h) Replace band
Slips on 2-3 shift	a) Low fluid level b) Throttle valve cable broken or not adjusted c) Throttle valve binding d) Spacer plate or gasket out of position e) Faulty intermediate servo f) Malfunction in direct clutch assembly	a) Add as required b) Adjust or replace as required c) Clean valve body d) Replace gaskets, align plate e) Replace seal or assembly f) Overhaul
Rough 1-2 shift	a) Throttle valve cable broken or not adjusted b) Throttle valve, T.V. plunger or shift T.V. valve binding c) 1-2 accumulator valve binding d) Faulty 1-2 accumulator assembly e) Faulty intermediate servo	a) Adjust or replace as required b) Clean valve body c) Clean valve body d) Replace seal or piston as e) Replace seal or assembly
Rough 2-3 shift	a) Throttle valve cable broken or not adjusted b) Throttle valve, T.V. plunger or shift T.V. valve binding	a) Adjust or replace as required b) Clean valve body
Transaxle noisy	a) Pump noise due to low fluid level, cavication b) Pump noise due to damage c) Gear noise	a) Add as required b) Overhaul pump c) Check for grounding to body, worn roller bearings

Note: If noisy in 3rd gear or on turns only, check differential and final drive unit

HYDRAULIC CONTROL SYSTEM

Major Components

FLUID RESERVOIRS

The hydraulic control system requires a supply of transmission fluid. Due to the low profile of the transaxle, a reservoir other than the oil pan is required to maintain a specific fluid level during hot and cold transaxle operation, so that the oil pump can provide pressured fluid to the hydraulic control system. This added reservoir, located in the lower section of the valve body cover, is controlled by a thermostatic element, operating by opening and closing according to the temperature of the fluid that is trapped in the reservoir portion of the valve body cover. As the temperature of the fluid in the valve body reservoir increases, the volume of the fluid increases. As the temperature decreases, the thermostatic element opens and allows the fluid to drain into the lower sump reservoir or oil pan.

OIL PUMP ASSEMBLY

The oil pump is a variable capacity vane-type pump, driven by the engine. The pump is located within the control valve assembly. A slide is incorporated in the pump that automatically regulates the pump output according to the needs of the transaxle. Maximum pump output is attained when the priming spring has been fully extended and has the slide held against the side of the pump body. As the slide moves towards the center, the pump output is reduced until the minimum pump output is reached.

PRESSURE REGULATOR VALVE

Movement of the pump slide is accomplished by directing fluid from the pressure regulator to the pump side opposite the priming spring. When the engine is stopped, the slide is held in the maximum output position by the priming spring. As the engine is started and the pump rotor is operated, its fluid output is directed to the pressure regulator valve. When the fluid output is below the desired pressure, the regulator valve is held in its bore by the pressure regulator spring. With the pressure regulator valve held in this position, the pump slide is held by the priming spring for maximum pressure output. As the pump output and the control pressure increases, the pressure regulator valve is moved against the pressure regulator spring. This allows fluid to be directed to the slide and causes the slide to move against the priming spring, decreasing the pump output. Fluid is also directed from the pressure regulator valve to fill the converter. When filled, the fluid is directed to the transaxle fluid cooler, located in the radiator. The fluid returning from the cooler is then directed to the lubrication system.

THROTTLE VALVE

The requirements of the transaxle for the apply of the band and clutches will vary with the engine torque and throttle opening. Under heavy throttle operation, the control pressure (approximately 70 psi) is not sufficient to hold the band or clutches in the applied mode without slipping. To provide a higher control (line) pressure when the throttle is opened, a throttle valve, relating to the throttle opening and engine torque, is provided in the valve body. As the accelerator pedal is depressed and the throttle plates are opened, the mechanical linkage (T.V. cable) relays the movement to the throttle plunger and increases the force on the T.V. spring and throttle valve, increasing T.V. pressure.

NOTE: T.V. pressure can be regulated from 0 to 105 psi, approximately.

SHIFT T.V. VALVE

A shift T.V. valve is used to control the T.V. pressure to a maximum of 90 psi for shift control, by exhausting the excess pressure.

T.V. BOOST VALVE

The T.V. boost valve is used to boost line pressure from 70 to 140 psi as the throttle opening directs.

LINE BOOST VALVE

A feature has been included in the T.V. system that will prevent the transaxle from being operated with low or minimum line pressure in the event that the T.V. cable becomes broken or disconnected. This feature is the line boost valve, which is located in the control valve and oil pump assembly at the T.V. regulating exhaust port. The line boost valve is held off its seat by the throttle lever and bracket assembly, allowing the T.V. pressure to regulate normally when the T.V. cable is properly adjusted. Should the T.V. cable become broken, disconnected or not adjusted properly, the line boost valve will close the T.V. exhaust port and keep the T.V. and line pressure at full line pressure psi.

GOVERNOR ASSEMBLY

The governor is driven by the differential and final drive carrier and, dependent upon vehicle speed, signals the valve body to make the shifts by increasing the governor pressure to oppose the main control and throttle pressure until the pressure forces are equal. This allows the valve springs to move the shifting valve and causes the transaxle to shift into another gear ratio. The governor pressure is developed from drive pressure, which is metered through two orifices and directed through the governor shaft to primary and secondary check balls, seated opposite of each other, and tends to exhaust through the check ball seats. The governor weights are so arranged that the primary weight, assisted by the primary spring, acts on one check ball, while the secondary weight, assisted by the secondary spring, acts on the other check ball. As the governor turns, the weights are moved outward by centrifugal force. This force is relayed to the check balls and seats them to control the governor pressure being exhausted. As the speed of the governor increases, so does the force relayed to the check balls. The heavier or primary weight and spring are more sensitive to changes in differential and final drive carrier speeds at lower rpm than the secondary weight. At greater vehicle speeds, as centrifugal force increases on the primary weight, the primary ball check is held tighter to its seat and cannot exhaust any fluid. From this point on, the secondary weight and spring are used to apply force to the secondary check ball and regulates the exhausting of the governor oil pressure.

1-2 AND 2-3 SHIFT VALVES

The 1-2 and 2-3 shift valves are used to change the gear ratios at predetermined speeds, fluid pressures and engine torque demands.

1-2 ACCUMULATOR VALVE

Controlling of the intermediate band apply pressure is accomplished by the 1-2 accumulator valve which provides a variable accumulator pressure to cushion the band apply in relation to throttle opening. At light throttle operation, the engine develops a small amount of torque, and as a result, the band requires less apply force to hold the direct clutch housing. At heavy throttle, the engine develops a large amount of torque which requires a greater apply pressure to lock the band on the direct clutch housing. If the band locks too slowly, it will slip excessively and burn due to the heat created by the slippage.

2-3 SHIFT ACCUMULATION

The 2-3 shift accumulation system operates in the same manner as the 1-2 accumulator valve operation, only to soften the direct clutch application during the 2-3 shift, by acting upon the release side of the intermediate servo piston.

TRANSAXLE DOWNSHIFTING

Transaxle downshifting is accomplished at part throttle and full throttle openings. Throttle, governor and control pressures are used in relation to each other, to provide the downshifting and movement of the necessary valves. An examination of the flow circuits for downshifting will show the operation of the various valves during the downshifting procedure. The transaxle will automatically downshift during the coast down with the throttle closed.

Diagnosis Tests

Automatic transaxle malfunctions may be caused by three major operating conditions. These conditions are improper transaxle adjustments, poor engine performance and hydraulic or mechanical malfunctions.

The suggested sequence for transaxle diagnosis is:
1. Check and adjust transaxle fluid level as required.
2. Check and adjust T.V. cable as required.
3. Check, correct and adjust manual linkage as necessary.
4. Check vehicle engine performance.
5. Install an oil pressure gauge and tachometer and check the transaxle control pressure.
6. Road test in all transaxle shift selector ranges and note any changes in operation and oil pressure.
7. Attempt to isolate the unit that is involved in the malfunction.
8. If the road test indicates that the engine is in need of a tune

up, it should be performed before any corrective action is taken to repair the transaxle.

Control Pressure Test

To test the control pressure on the THM 125C automatic transaxle, connect the oil pressure gauge to the transaxle control pressure port. This port is located on top of the transaxle above the valve body oil pan. Install a tachometer to the engine and verify that the manual control linkage and the throttle control linkage are correct. Be sure that both the engine and the automatic transaxle are at operating temperature. Check the transaxle fluid level and adjust as required, before testing.

Minimum T. V. Line Pressure Test

1. Adjust the T.V. cable to specifications.
2. Be sure that the brakes are applied.
3. Install the oil pressure gauge to the transaxle.
4. Take and record the line pressure readings in the ranges and at the engine RPM as indicated in the Line Pressure Specification Chart.
5. Total testing time must not exceed two minutes for all combinations.

Maximum T. V. Line Pressure Test

1. Tie or hold the T.V. cable to the full extent of its travel.
2. Be sure that the brakes are applied.
3. Install the oil pressure gauge to the transaxle.
4. Take and record the line pressure readings in the ranges and at the engine RPM as indicated in the Line Pressure Specification Chart.
5. Total testing time must not exceed two minutes for all combinations.

LINE PRESSURE SPECIFICATIONS
1982 Model THM-125C Automatic Transaxle

Range	Model	Normal Oil Pressure At Maximum T.V.	Normal Oil Pressure At Minimum T.V.
Park at 1000 RPM	CI, HY, HC, CJ, EF, EA, EB, EQ, EC, EK	No T.V. pressure in Park. Line pressure is equal to Park at minimum T.V.	50-70
	PO, PL, PZ, HX, PK, PI, EL		60-80
	CL, CE, CD, CV, HM, HS, HW, OP, BL, BF		70-90
Reverse at 2000 RPM	EF, EA, EB, EQ, EC	234-254	95-115
	CI, CJ, HC, HY, EK	199-219	95-115
	PO, PL, PZ, PI, PK, HX, EL	216-236	112-132
	CL, CE, CD, CV, HM, HS, HW, OP, BL, BF	267-287	128-148
Neutral at 1000 RPM	EF, EA, EB, EQ, EC	130-150	50-70
	CI, CJ, HC, HY, EK	110-130	50-70
	PO, PL, PZ, PI, PK, HX, EL	120-140	60-80
	CL, CE, CD, CV, HM, HS, HW, OP, BL, BF	150-170	70-90
Drive at 1000 RPM	EF, EA, EB, EQ, EC	130-150	50-70
	CI, CJ, HC, HY, EK	110-130	50-70
	PO, PL, PZ, PI, PK, HX, EL	120-140	60-80
	CL, CE, CD, CV, HM, HS, HW, OP, BL, BF	150-170	70-90
Inter. at 1000 RPM	CI, HY, HC, CJ, EF, EA, EB, EQ, EC, EK	92-113	93-113
	PO, PL, PZ, HX, PK, PI, EL	110-130	110-130
	CL, CE, CD, CV, HM, HS, HW, OP, BL, BF	125-145	125-145
Low at 1000 RPM	CI, HY, HC, CJ, EF, EA, EB, EQ, EC, EK	No T.V. pressure in Low. Line pressure is equal to interrange at minimum T.V.	93-113
	PO, PL, PZ, HX, PK, PI, EL		110-130
	CL, CE, CD, CV, HM, HS, HW, OP, BL, BF		125-145

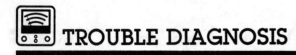

LINE PRESSURE SPECIFICATIONS
1983 Model THM-125C Automatic Transaxle

Range	Model	Normal Oil Pressure At Maximum T.V.	Normal Oil Pressure At Minimum T.V.
Park at 1000 RPM	EM, EN, EI, EF, EK, EB, EQ	No T.V. pressure in Park. Line pressure is equal to Park at minimum T.V.	58-62
	CA, CB, CF, HC, HY, PG, PW, EL, HW		67-75
	CE, CL, CT, HM, HS, CK, OP, HD, HV, BF, BL		75-85
	EP		75-85
Reverse at 1000 RPM	EM, EN	200-220	100-107
	EI	220-240	100-107
	EF, EK	235-255	100-107
	EB, EQ	240-280	110-117
	CA, CB, CF, HC, HY, PG, PW, EL, HW	217-240	118-130
	CE, CL, CT, HM, HS, CK, OP, HD, HV	240-295	130-150
	BF, BL	240-285	140-160
	EP	230-260	133-147
Neutral at 1000 RPM	EM, EN	115-125	58-62
	EI	125-135	58-62
	EF, EK, EB, EQ	135-145	58-62
	CA, CB, CF, EL, HC, HY, HW, PG, PW	123-140	67-75
	CE, CK, CL, CT, HD, HM, HS, HV, OP	150-170	75-85
	BF, BL	130-147	75-85
	EP	132-147	75-85
Drive at 1000 RPM	EM, EN	115-125	58-62
	EI	125-135	58-62
	EF, EK, EB, EQ	135-145	58-62
	CA, CB, CF, EL, HC, HY, HW, PG, PW	123-140	67-75
	CE, CK, CL, CT, HD, HM, HS, HV, OP	150-170	75-85
	BF, BL	130-147	75-85
	EP	132-147	75-85
Inter. at 1000 RPM	EM, EN, EI, EF, EK	No T.V. pressure in Inter. Line pressure is equal to Interrange at minimum T.V.	105-110
	EB, EQ		125-130
	CA, CB, CF, EL, HC, HY, HW, PG, PW		115-132
	CE, CK, CL, CT, HD, HM, HS, HV, OP		130-150
	BF, BL		160-183
	EP		135-150
Low at 1000 RPM	EM, EN, EI, EF, EK	No T.V. pressure in Low. Line pressure is equal to Low Range at minimum T.V.	105-110
	EB, EQ		125-130
	CA, CB, CF, EL, HC, HY, HW, HW, PG, PW		115-132
	CE, CK, CL, CT, HD, HM, HS, HV, OP		130-150
	BF, BL		160-183
	EP		135-150

Air Pressure Test

The positioning of the THM 125C transaxle in the vehicle and the valve body location will cause the air pressure tests to be very difficult. It is advisable to make any air pressure tests during the disassembly and assembly of the transaxle to ascertain if a unit is operating.

Stall Speed Test

General Motors Corporation does not recommend performing a stall test because of the excessive heat that is generated within the transaxle by the converter during the tests.

Recommendations are to perform the control pressure test and road test to determine and localize any transaxle malfunctions.

Road Test

Drive Range

Position selector lever in Drive range and accelerate the vehicle. A 1-2 and 2-3 shift should occur at throttle openings. (The shift points will vary with the throttle openings). Check part throttle 3-2 downshift at 30 MPH by quickly opening throttle approximately three-fourths. The transmission should downshift at 50 MPH, by depressing the accelerator fully.

Intermediate Range

Position the selector lever in Intermediate range and accelerate the vehicle. A 1-2 shift should occur at all throttle openings. (No 2-3 shift can be obtained in this range.) The 1-2 shift point will vary with throttle opening. Check detent 2-1 downshift at 20 MPH the transaxle should downshift 2-1. The 1-2 shift in Intermediate range is somewhat firmer than in Drive range. This is normal.

Low Range

Position the selector level in the Low range and accelerate the vehicle. No upshift should occur in this range.

Intermediate Range (Overrun Braking)

Position the selector lever in Drive range, and with the vehicle speed at approximately 50 MPH, with closed or zero throttle, move the selector lever to Intermediate range. The transmission should downshift to 2nd. An increase in engine rpm and an engine braking effect should be noticed.

Low Range (Overrun Braking)

At 40 MPH, with throttle closed, move the selector lever to Low. A 2-1 downshift should occur in the speed range of approximately 40 to 25 MPH, depending on valve body calibration. The 2-1 downshift at closed throttle will be accompanied by increased engine rpm and an engine braking effect should be noticed. Stop vehicle.

Reverse Range

Position the selector lever in Reverse position and check for reverse operation.

Torque Converter Stator Operation Diagnosis

The torque converter stator assembly and its related roller clutch can possibly have one of two different type malfunctions.
1. The stator assembly freewheels in both directions
2. The stator assembly remains locked up at all times

Malfunction Type One

If the stator roller clutch becomes ineffective, the stator assembly freewheels at all times in both directions. With this condition, the vehicle will tend to have poor acceleration from a standstill. At speeds above 30-35 MPH, the vehicle may act normal. If poor acceleration problems are noted, it should first be determined that the exhaust system is not blocked, the engine is in good tune and the transmission is in 1st gear when starting out.

If the engine will freely accelerate to high rpm in Neutral, it can be assumed that the engine and exhaust system are normal. Driving the vehicle in Reverse and checking for poor performance will help determine if the stator is freewheeling at all times.

Malfunction Type Two

If the stator assembly remains locked up at all times, the engine rpm and vehicle speed will tend to be limited or restricted at high speeds. The vehicle performance when accelerating from a standstill will be normal. Engine over-heating may be noted. Visual examination of the converter may reveal a blue color from the overheating that will result.

CONVERTER CLUTCH OPERATION AND DIAGNOSIS

Converter Clutch Operation

The "apply" or "release" of the torque converter clutch assembly is determined by the direction that the feed oil is distributed to the torque converter. When transmission oil is routed to the cov-

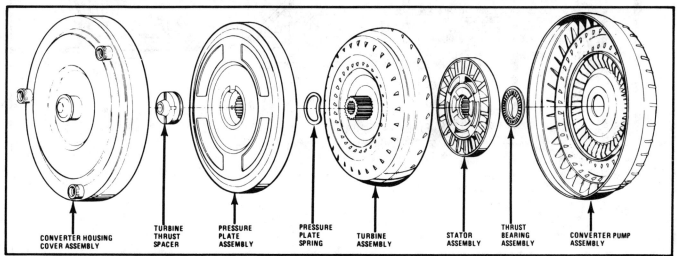

CONVERTER HOUSING COVER ASSEMBLY   TURBINE THRUST SPACER   PRESSURE PLATE ASSEMBLY   PRESSURE PLATE SPRING   TURBINE ASSEMBLY   STATOR ASSEMBLY   THRUST BEARING ASSEMBLY   CONVERTER PUMP ASSEMBLY

Torque converter clutch assembly—exlploded view (©General Motors Corp.)

er assembly and the clutch plate the torque converter clutch is in the released position. When transmission oil is routed to the turbine side of the clutch plate the torque converter clutch is in the applied position.

To assist in the reduction of torsional shock during torque converter clutch "apply" a damper assembly has been installed in the torque converter clutch pressure plate. The spring loaded damper assembly is splined to the torque converter turbine assembly. The torque converter clutch pressure plate is attached to the pivoting mechanism, which is incorporated within the damper assembly. The pivoting action allows the clutch pressure plate to rotate independently of the damper assembly up to about forty-five degrees. The rate of this independent rotation is controlled by a pivoting mechanism acting on the springs that are assembled in the damper assembly. The spring cushioning effect of the damper assembly aids in the reduction of the engagement feel of the torque converter clutch "apply" function. To further aid in the "apply" and "release" function of the torque converter clutch during various driving situations other types of controls have been incorporated in the electrical system.

Torque Converter Clutch (T.C.C.)— Released Position

The release of the torque converter clutch is determined by the direction that the torque converter feed oil is routed to the torque converter. The torque converter feed oil from the pressure regulator flows to the torque converter clutch control valve. The position of the torque converter clutch control valve controls the direction that the torque converter feed oil flows to the torque converter.

The torque converter clutch control valve is held in the release position in "Park," "Neutral," "Reverse," "Drive," "First" and "Second," by line pressure acting on the end of the torque con-

verter clutch apply valve. With the torque converter clutch control valve in the released state, the converter feed oil flows into the torque converter clutch release passages. The oil then flows between the pump driveshaft and the turbine shaft to the front or release side of the torque converter clutch. The oil flow then continues between the torque converter clutch pressure plate and the torque converter cover. This operation moves the converter clutch pressure plate away from the converter cover, releasing the converter clutch and charging the converter with oil. The oil then leaves the converter by flowing through the turbine shaft into the converter clutch apply oil circuit. The apply oil circuit is now being used in a reverse direction. The oil then flows from the apply passage into the cooler passage and to the lubrication system.

Torque Converter Clutch (T.C.C.)— Applied Position

At vehicle speeds below torque converter clutch engagement speed, the C3 or governor pressure switch, depending on the system, will break the circuit to the solenoid exhaust valve. This operation is done in order to prevent the torque converter clutch from applying in the drive- third gear mode. This de-energizes the solenoid and opens the exhaust valve to exhaust the solenoid circuit oil at the converter clutch control valve. Line pressure then holds the converter clutch control valve in the release position.

When the vehicle speed in the drive range- third gear reaches torque converter clutch engagement speed, the C3 or governor pressure switch, depending on the system, will activate the solenoid, closing the exhaust valve. This allows solenoid circuit oil to move the converter clutch control valve against line pressure. With the converter clutch control valve in the apply position, regulated line oil, from the converter clutch regulator valve, is allowed to pass into the converter apply passage. It then flows through the turbine shaft to the apply side of the converter clutch.

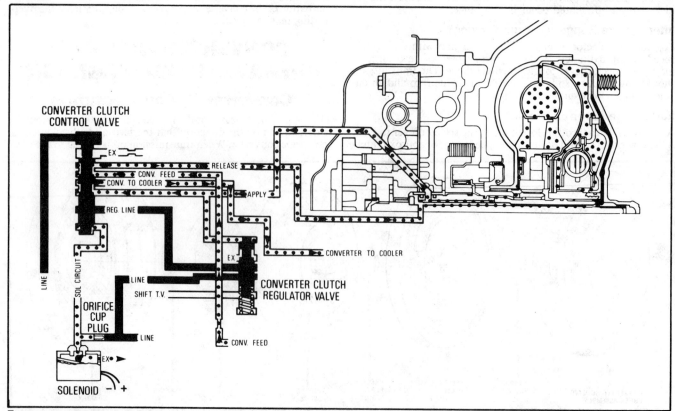

Torque converter clutch (T.C.C.)—released position (©General Motors Corp.)

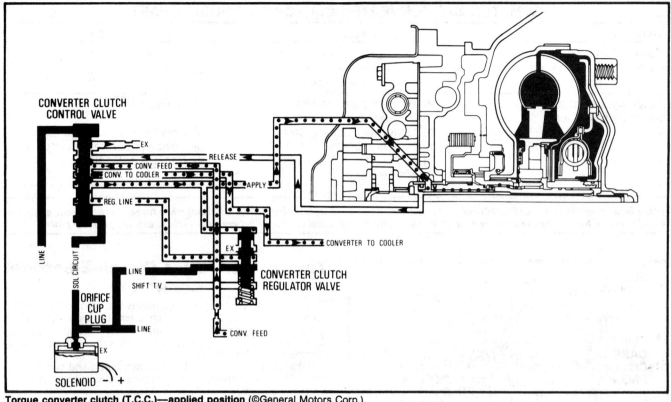

Torque converter clutch (T.C.C.)—applied position (©General Motors Corp.)

The regulated line oil, from the converter clutch regulator valve, controls the apply feel of pressure plate. As the pressure plate begins to move to its applied position, release oil in the front side of the pressure plate is redirected back between the turbine shaft and pump drive shaft and exhausted at the converter clutch control valve through an orifice, to time the clutch apply. When the converter clutch control valve moves to the apply position, orificed converter feed oil enters the torque converter to cooler passage in order to provide oil to the lubrication system.

Troubleshooting the Torque Converter Clutch

Before diagnosing the torque converter clutch system as being at fault in the case of rough shifting or other malfunctions, make sure that the engine is in at least a reasonable state of tune. Also, the following points should be checked:

1. Check the transmission fluid level and correct as necessary.
2. Check the manual linkage adjustment and correct as necessary.
3. Road test the vehicle to verify the complaint. Make sure that the vehicle is at normal operating temperature.
4. If the problem has been traced to the torque converter clutch system, refer to the G.M. Torque Converter Clutch Diagnosis Chart.

NOTE: When diagnosing a torque converter clutch problem on the 1983 Oldsmobile Ciera equipped with a 3.0 liter engine, disconnect the engine cooling fan relay.

G. M. TORQUE CONVERTER CLUTCH DIAGNOSIS CHART

Condition	Cause	Correction
Clutch applied in all ranges (engine stalls when put in gear)	a) Converter clutch valve stuck in apply position	a) R&R oil pump and clean valve—R&R auxiliary valve body and clean valve
Clutch does not apply: applies erratically or at wrong speeds	a) Electrical malfunction in most instances	a) Follow troubleshooting procedure to determine if problem is internal or external to isolate defect.
Clutch applies erratically; shudder and jerking felt	a) Vacuum hose leak b) Vacuum switch faulty c) Governor pressure malfunction	a) Repair hose as needed b) Replace switch c) Replace switch

G. M. TORQUE CONVERTER CLUTCH DIAGNOSIS CHART

Condition	Cause	Correction
Clutch applies erratically; shudder and jerking felt	d) Solenoid loose or damaged e) Converter malfunction; clutch plate warped	d) Service or replace e) Replace converter
Clutch applies at a very low or high 3rd gear	a) Governor switch shorted to ground b) Governor malfunction c) High line pressure d) Solenoid inoperative or shorted to case	a) Replace switch b) Service or replace governor c) Sevice pressure regulator d) Replace solenoid

─────────── CAUTION ───────────

When inspecting the stator and turbine of the torque converter clutch unit, a slight drag is normal when turned in the direction of free-wheel rotation because of the pressure exerted by the waved spring washer, located between the turbine and the pressure plate.

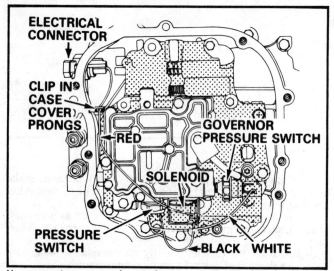

Non computer command control system component location inside the auxiliary valve body (©General Motors Corp.)

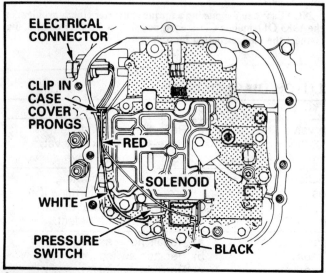

Computer command control system component location inside the auxiliary valve body (©General Motors Corp.)

Torque Converter Clutch-Electrical Controls

Two types of electrical control systems are used to control the "apply" function of the torque converter clutch assembly. Both systems use the third clutch pressure switch and a solenoid. The difference in the two systems is that the vehicle speed sensing controls are not the same.

Vehicles equipped with the computer command control system (C3) use this system to energize the solenoid when certain vehicle speed has been reached.

Vehicles that are not equipped with the computer command system (C3) use a governor pressure switch to energize the solenoid when certain vehicle speed has been reached.

Computer Command Control System

Vehicles equipped with the computer command control system utilize the following components to accomplish the "apply" function of the torque converter clutch assembly.

1. Vacuum sensor—which sends engine vacuum information to the electronic control module.

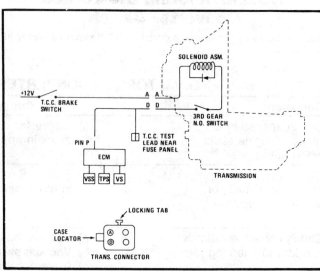

Computer command control system wiring diagram (©General Motors Corp.)

2. Throttle position sensor—which sends throttle position information to the electronic control module.

3. Vehicle speed sensor—which sends vehicle speed information to the electronic control module.

4. Electronic control module—which energizes and grounds the transaxle electrical system.

5. Brake release switch—which avoids stalling the engine when braking. Any time that the brakes are applied the torque converter clutch is released.

6. Third gear switch—which prevents operation until third gear speed is obtained.

Non Computer Command Control System—Diesel Engines

Vehicles not equipped with the computer command control system utilize the following components to accomplish the "apply" function of the torque converter clutch assembly.

1. Engine coolant fan temperature switch—which is a two position switch that closes when engine coolant temperature is about 246 degrees F. This causes the cooling fan to operate at high speed and bypass the delay feature in the torque converter clutch delay module. The torque converter clutch will operate when the governor pressure switch is closed.

2. Air condition high pressure switch—which closes when the air conditioner high pressure side reaches 370 psi. It performs the same functions as the engine coolant fan temperature switch.

3. Vacuum regulator valve—which opens at about ¾ or more throttle and disengages the torque converter clutch on heavy acceleration.

4. Governor pressure switch—which completes the ground circuit for the torque converter clutch and EGR solenoids when the vehicle reaches about 35 mph.

5. Third gear switch—which signals the torque converter clutch assembly when the transaxle is operating in third gear. This switch also prevents the torque converter clutch from operating in first or second gear unless the engine coolant fan temperature switch is closed, or the air condition high pressure switch is closed.

6. Torque converter control module—which delays the torque converter control "apply" function to keep the torque converter clutch and the direct clutch from applying at the same time. This

would cause an objectionable thump. The delay feature is overridden when either the air condition high pressure switch or the dual temp high switch is closed. When either switch is closed, the torque converter clutch assembly will "apply" as soon as the governor pressure switch closes.

7. Brake release switch—which avoids stalling the engine when braking. Any time that the brakes are applied, the torque converter clutch is released.

Non Computer Command Control System—Gas Engines

Vehicles not equipped with the computer command control system utilize the following components to accomplish the "apply" function of the torque converter clutch assembly.

1. Ported vacuum—which opens the vacuum switch to release the clutch during a closed throttle coast down.

2. Vacuum delay valve—which slows the vacuum switch response to vacuum changes.

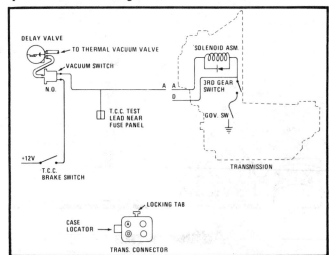

Non computer command control system wiring diagram—gas engines (©General Motors Corp.)

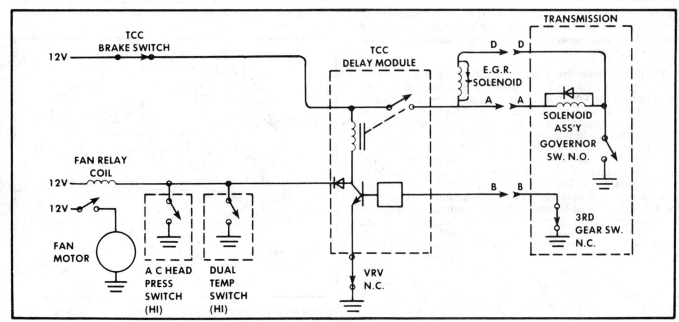

Non computer command control system wiring diagram—diesel engines (©General Motors Corp.)

3. Engine vacuum switch—which releases the torque converter clutch when the engine vacuum drops to approximately 1.5-3.0 inches during moderate acceleration, prior to a part throttle or detent downshift.

4. Thermal vacuum valve—which prevents the torque converter clutch from applying until the engine coolant temperature reaches 130 degrees F.

5. Brake release switch—which avoids stalling the engine when braking. Any time that the brakes are applied, the torque converter clutch is released.

Torque Converter Clutch Electrical Diagnosis

NOTE: When diagnosing a torque converter clutch problem on the 1983 Oldsmobile Ciera equipped with a 3.0 liter engine, be sure to disconnect the engine cooling fan relay.

Before making any electrical checks on the THM 125C automatic transaxle, the following points should be checked in order.
1. Check and adjust transaxle fluid level.
2. Check and adjust manual linkage.
3. Road test the vehicle and verify the complaint.
4. If the problem has been traced to the electrical function of the torque converter clutch assembly, refer to the proper Transaxle Converter Clutch Electrical Diagnosis Chart.

NOTE: When using the torque converter clutch electrical diagnosis chart for vehicles equipped with the computer command control system be sure to check for the presence of code 24 before performing the torque converter electrical diagnosis checks outlined in the chart. Refer to the proper Chilton Manual for the correct procedure to determine the presence of code 24.

 ON CAR SERVICES

Adjustments
THROTTLE LINKAGE

Adjustment

The T.V. cable controls line pressure, shift points, part throttle downshifts and detent downshifts. The T.V. cable operates the throttle lever and bracket assembly, which is located within the valve body assembly.

The throttle lever and bracket assembly have two functions. The primary function of this assembly is to transfer throttle movement to the T.V. plunger in the control valve pump assembly as related by the T.V. cable and linkage. This causes T.V. pressure and line pressure to increase according to throttle opening; it also controls part throttle and detent downshifts. The proper adjustment of the T.V. cable is based on the T.V. plunger being fully depressed (flush with the T.V. bushing) at wide open throttle. The secondary function of this assembly involves the line boost lever and line boost valve. The function of this system is to prevent the transmission from operating at low (idle position) pressures, if the T.V. cable should become broken or disconnected. If the cable is connected (not broken or stretched), the line boost lever will not move from its normal spring-loaded "up" position which holds the line boost valve off its seat. The line boost lever will drop down to allow the line boost valve to seat only if the cable is broken, disconnected or extremely out of adjustment. With the valve body cover removed, it should be possible to pull down on the line boost lever and, when released, the lever spring should return the lever to the normal "up" position. If the throttle lever

and bracket assembly binds or sticks so the line boost lever cannot lift the line boost valve off its seat, high line pressures and delayed upshifts will result.

The cable should be checked for freeness by pulling out on the upper end of the cable. The cable should travel a short distance with slight spring resistance. This light resistance is caused by the

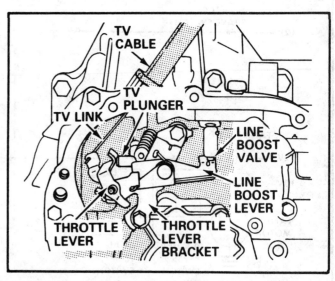

Throttle lever and bracket assembly (©General Motors Corp.)

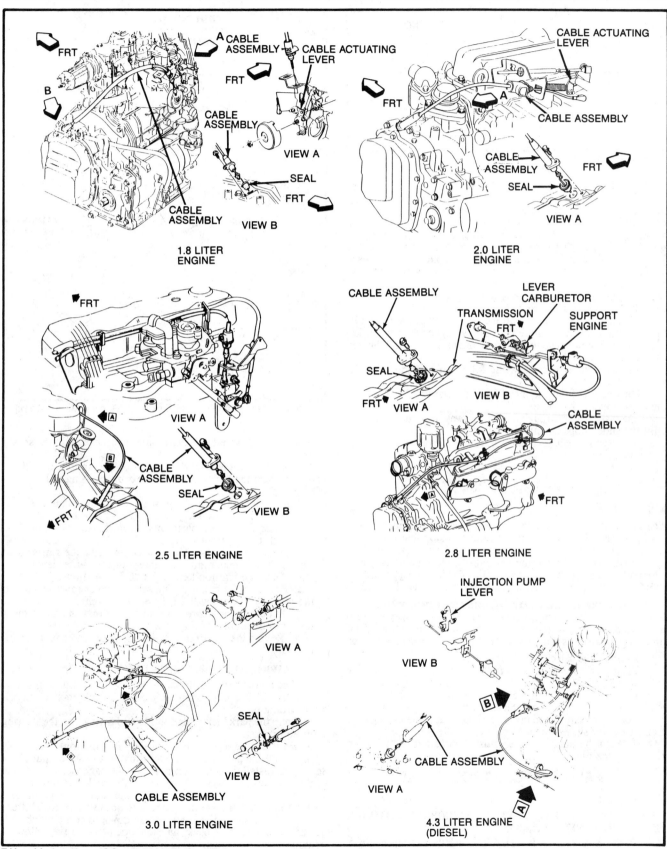

1.8 LITER ENGINE

2.0 LITER ENGINE

2.5 LITER ENGINE

2.8 LITER ENGINE

3.0 LITER ENGINE

4.3 LITER ENGINE (DIESEL)

T.V. cable locations (©General Motors Corp.)

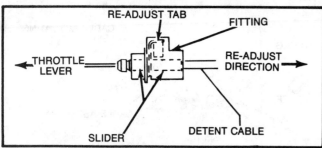

T.V. cable adjuster (©General Motors Corp.)

small coiled return spring on the T.V. lever and bracket that returns the lever to zero T.V., or closed throttle position. Pulling the cable farther out moves the lever to contact the T.V. plunger which compresses the T.V. spring which has more resistance. By releasing the upper end of the T.V. cable, it should return to the zero T.V. position. This test checks the cable in its housing, the T.V. lever and bracket, and the throttle valve plunger in its bushing for freeness. To check on the adjustment and verify that it is correct, use the following procedure:

1. Install line pressure gauge. Adjust engine speed to 1000 rpm with the selector in Park, and check line pressure.
2. Check line pressure in Neutral at 1000 rpm. Pressure should be in the same as or no more than 10 psi higher than in Park.
3. Adjust engine speed to 1400 rpm and make sure that there is an increase in line pressure.

If readjustment is necessary, the following procedure is suggested:

1. Depress and hold the metal lock tab that will be found on the cable adjuster by the idler lever.
2. Move the slider through the fitting away from the idler lever until the slider stops against the fitting.
3. Release the metal lock tab. As a double check, repeat the adjustment.

MANUAL LINKAGE

Adjustment

The THM 125C automatic transaxle manual linkage must be adjusted so that the shift selector indicator and stops correspond to the automatic transaxle detent positions. If the manual linkage is not adjusted correctly, an internal transaxle leak could develop. This could cause a clutch or band to slip. The following procedure should be used when adjusting the transaxle manual linkage.

NOTE: If the manual linkage adjustment is made with the selector lever in "P", the parking pawl should freely engage the reaction internal gear in order to keep the vehicle from rolling. If not properly adjusted, transaxle, vehicle or personal injury may occur.

1. Position the selector lever in "N".
2. Position the transaxle lever in "N". Obtain "N" position by turning the transaxle lever clockwise from "P" through "R" into "N". Or, turn the lever counterclockwise from "L" through "S" ("1" through "2" on some vehicles) and "D" to "N".
3. Loosely assemble the pin part of the shift cable through the transaxle lever slotted hole.
4. Tighten the nut. The lever must be held out of "P" when tightening the nut.

PARK LOCK CONTROL CABLE

Adjustment

CENTURY, SKYLARK, CITATION, CELEBRITY, OMEGA, CIERA, PHOENIX AND 6000

1. Position the selector lever in "P". Lock the steering column.
2. Position the transaxle lever into the "P" position.

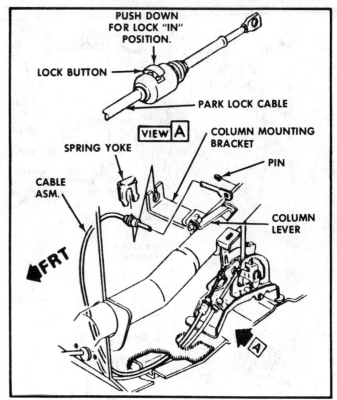

Park/lock cable routing—A and X series (©General Motors Corp.)

3. Install the cable-to-shifter mounting bracket with the spring yoke.
4. Install the cable-to-park lock pin on the shifter. Install the lock pin. Push the lock button on the cable housing in to set the cable length.
5. Check the cable operation in the following manner:
  a. Turn the ignition key to the "Lock" position.
  b. Press the detent release button in the shifter handle.
  c. Pull the shifter lever rearward.
  d. The shifter lock-hook must engage into the shifter base slot, within 2° maximum movement of the shifter lever.
  e. Turn the ignition key to the "Off" position.
  f. Repeat steps b and c. The selector lever must be able to move rearward to the "L" ("1" on some vehicles) position.
  g. Repeat Steps a through d to assure that the adjustment nut has not slipped during check.
  h. Return the key to the "Lock" position and check key removal.

SKYHAWK, CAVALIER, FIRENZA, CIMARRON AND 2000

1. The ignition lock cylinder on the steering column must be in the "Lock" position. Snap the park lock cable-to-steering column sliding pin.
2. Snap the park lock cable column end fitting into the steering column bracket.
3. Position the selector lever in the "P" position.
4. Install the park lock cable terminal to the shifter park lock lever pin. Install the retainer pin. Install the park lock cable to the shifter mounting bracket by pushing the lock button housing against the adjusting spring, dropping the cable through the slot in the mounting bracket and seating the housing to the shifter.
5. Push the lock adjustment button down to complete the adjustment procedure. Check the cable operation in the following manner:
  a. With the ignition in the "Lock" position, try to push the

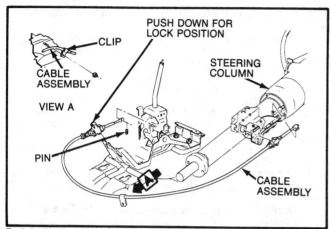

Park/lock cable routing— J series (©General Motors Corp.)

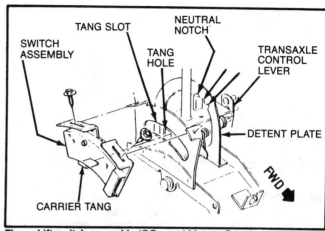

Floor shift switch assembly (©General Motors Corp.)

button on the shift handle. Button travel must not be enough to let the shifter move out of "P".

b. Turn the ignition key to the "Run" position. The shifter should select the gears when the handle turns rearward.

c. With the shifter in neutral, the ignition lock cylinder should not return to "Lock".

d. Return the shifter to "P" then return the ignition key to the "Lock" position.

NEUTRAL SAFETY AND REVERSE LIGHT SWITCH

Adjustment
FLOOR SHIFT MODELS

1. Remove center console and all related components in order to service the switch assembly.

2. Position the automatic transaxle control shifter assembly in "N".

3. Loosen the switch attaching screws.

4. Turn the switch on the shifter assembly to align the service adjustment hole with the carrier tang hole.

5. Install a .90 inch diameter gauge pin to a depth of about .060 inch.

6. Tighten the switch adjusting screws.

7. Remove the gauge pin. Check for correct adjustment.

COLUMN SHIFT MODELS

1. Locate the neutral safety switch at the bottom of the steering column.

2. With the switch installed move the housing all the way down toward the low gear position.

3. Adjust the switch by positioning the selector lever in "P". The main housing and the housing back should ratchet, providing proper switch adjustment.

SERVICES

FLUID CHANGES

The conditions under which the vehicle is operated is the main consideration in determining how often the transaxle fluid should be changed. Different driving conditions result in different transaxle fluid temperatures. These temperatures affect change intervals.

If the vehicle is driven under severe service conditions, change the fluid and filter every 15,000 miles. If the vehicle is not used under severe service conditions, change the fluid and replace the filter every 100,000 miles.

Do not overfill the transaxle. It only takes one pint of fluid to change the level from add to full on the transaxle dipstick. Overfilling the unit can cause damage to the internal components of the automatic transaxle.

OIL PAN

Removal and Installation

Some THM 125C automatic transaxles may be built using RTV silicone sealant in place of the usual oil pan gasket. RTV sealant is an effective substitute for this application, but its usage depends on the type of oil pan being used in the transaxle assembly. If RTV sealant is used on an oil pan, the flange surface must be either flat or have depressed stiffening ribs. Do not use RTV sealant on transaxle oil pans which have raised stiffening ribs.

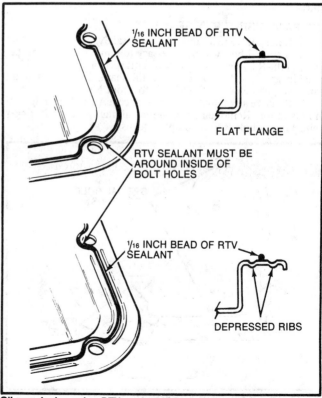

Oil pan design using RTV sealant (©General Motors Corp.)

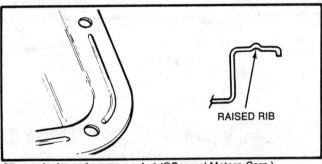

Oil pan design using pan gasket (©General Motors Corp.)

OIL PANS USING REGULAR PAN GASKET

1. Raise the vehicle and support it safely.
2. With a drain pan under the transaxle assembly, remove the oil pan attaching bolts from the front and side of the oil pan.
3. Loosen the rear pan bolts approximately four turns.
4. Carefully pry the transaxle oil pan loose with a suitable tool allowing the fluid to drain into the drain pan.
5. Remove the remaining bolts from the assembly. Remove both the oil pan and gasket from the vehicle.
6. Remove the transaxle filter and O-ring seal.
7. Install a new filter and O-ring seal locating the filter against the dipstick stop.
8. Install a new transaxle pan gasket on the oil pan. Install the oil pan to the transaxle assembly and tighten the bolts to 12 ft. lbs.
9. Lower the vehicle. Add about four quarts of the proper grade and type automatic transmission fluid.
10. With the selector lever in "P" apply the parking brake and block the wheels. Start the engine and let it idle. Do not race the engine.
11. Move the selector lever through all the ranges. With the selector lever in "P" check for proper fluid level and correct as required.

OIL PANS USING RTV SEALANT

1. Raise the vehicle and support it safely.
2. A special bolt will be necessary to remove the oil pan because of the RTV assembly process. This special bolt can be fabricated by grinding down a section of the shank diameter of an oil pan bolt, just below the bolt head, to about 3/16 inch.
3. Remove all of the oil pan bolts except A and B as indicated in illustration. Remove bolt A and install the special bolt, finger tight. Loosen bolt B four complete turns.

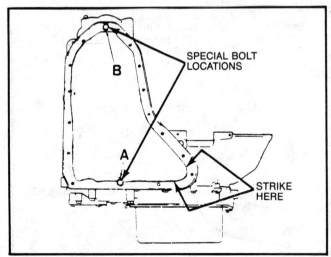

Oil pan removal procedure (©General Motors Corp.)

4. Do not try to pry the oil pan loose from the transaxle case, as damage may occur.
5. Using a rubber mallet strike the oil pan corner.
6. Remove the special bolt. Let the fluid drain. Remove the oil pan from the transaxle assembly.
7. Remove the transaxle filter and O-ring.
8. Install a new filter and O-ring seal locating the filter against the dipstick stop.
9. When installing the transaxle oil pan be sure to remove all the old RTV sealant from both the oil pan and the transaxle case flange.
10. Be sure that both of these surfaces are dry and free of any film. If not, leakage may result.
11. Install a 1/16 inch bead of RTV sealant on the oil pan flange. The bead of RTV sealant must also be applied around the bolt holes.
12. Install the oil pan to the transaxle assembly. Torque the attaching bolts to 12 ft. lbs.
13. Lower the vehicle. Add about four quarts of the proper grade and type automatic transmission fluid.
14. With the selector lever in "P" apply the parking brake and block the wheels. Start the engine and let it idle. Do not race the engine.
15. Move the selector lever through all of the ranges. With the selector lever in "P", check for proper fluid level. Correct as required.

LOW AND REVERSE PIPES

Removal and Installation

1. Raise the vehicle and support it safely.
2. Remove the transaxle fluid pan. Remove the fluid filter and O-ring seal.
3. Remove the reverse oil pipe, seal back-up ring and the O-ring.
4. Remove the low and reverse cup plug assembly.
5. To install, position the low and reverse cup plug assembly in its proper place.
6. Position the reverse oil pipe, seal back-up ring and O-ring in its proper place.
7. Install a new fluid filter and O-ring seal. Install the transaxle oil pan.
8. Lower the vehicle and fill with the proper grade and type automatic transmission fluid. Road test and correct as required.

PARKING PAWL SHAFT

Removal and Installation

1. Raise the vehicle and support it safely.
2. Remove the transaxle fluid pan. Remove the fluid filter and the O-ring.
3. Remove the dipstick stop, rod retainer and parking lock bracket.
4. Remove the clip, rod, pin and spring.
5. To install, position the clip, rod, pin and spring assembly to its proper mounting spot within the transaxle assembly.
6. Install the dipstick stop, rod retainer and parking lock bracket.
7. Install a new fluid filter and O-ring seal. Install the transaxle fluid pan.
8. Lower the vehicle. Fill the transaxle with the proper grade and type automatic transmission fluid. Road test and correct as required.

INTERMEDIATE SERVO ASSEMBLY

Removal and Installation

1. Raise the vehicle and support it safely.

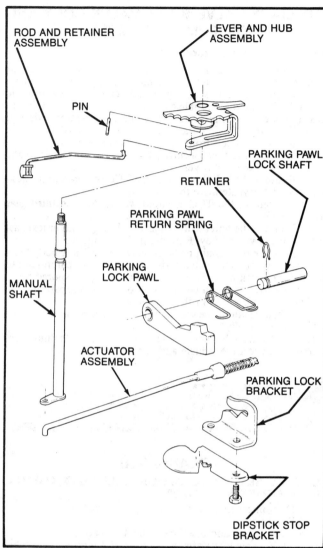

ROD AND RETAINER ASSEMBLY

LEVER AND HUB ASSEMBLY

PIN

PARKING PAWL LOCK SHAFT

RETAINER

PARKING PAWL RETURN SPRING

PARKING LOCK PAWL

MANUAL SHAFT

ACTUATOR ASSEMBLY

PARKING LOCK BRACKET

DIPSTICK STOP BRACKET

Parking pawl shaft assembly (©General Motors Corp.)

2. Remove the transaxle fluid pan. Remove the fluid filter and O-ring seal.

3. Remove the reverse oil pipe retaining brackets.

4. Remove the intermediate servo cover and gasket.

5. Remove the intermediate servo assembly from the automatic transaxle.

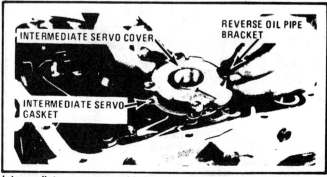

INTERMEDIATE SERVO COVER

REVERSE OIL PIPE BRACKET

INTERMEDIATE SERVO GASKET

Intermediate servo assembly location (©General Motors Corp.)

6. Position the intermediate servo assembly to the transaxle. Intsall the servo cover.

7. Install the reverse oil pipe retaining brackets.

8. Install a new fluid filter and O-ring seal. Install the transaxle fluid pan.

9. Lower the vehicle from the hoist. Fill the transaxle with the proper grade and type automatic transmission fluid. Road test and correct as required.

ACCUMULATOR CHECK VALVE

Removal and Installation

1. Raise the vehicle and support it safely.

2. Remove the transaxle fluid pan. Remove the fluid filter and O-ring seal.

3. Remove the reverse oil pipe retaining brackets. Remove the intermediate servo cover and gasket.

4. Remove the intermediate servo assembly from the transaxle assembly.

5. Remove the third accumulator check valve and spring from its mounting place inside the transaxle.

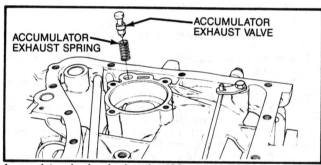

ACCUMULATOR EXHAUST VALVE

ACCUMULATOR EXHAUST SPRING

Accumulator check valve location (©General Motors Corp.)

6. Install the third accumulator valve and spring in its proper place.

7. Install the intermediate servo assembly in the automatic transaxle.

8. Install the intermediate servo cover and gasket. Install the reverse oil pipe brackets.

9. Install a new fluid filter and O-ring seal. Install the transaxle fluid pan.

10. Lower the vehicle . Fill the automatic transaxle with the proper grade and type automatic transmission fluid. Road test and correct as required.

SPEEDOMETER DRIVE GEAR

Removal and Installation

1. Raise the vehicle and support it safely.

2. Disconnect the speedometer cable from the transaxle assembly.

3. Remove the speedometer driven gear and the sleeve assembly from its mounting place.

4. Remove the transaxle governor cover and O-ring.

5. Remove the speedometer drive gear assembly from the automatic transaxle.

6. To install, position the speedometer drive gear assembly into the transaxle. Install the transaxle governor cover using a new O-ring.

7. Install the speedometer driven gear and sleeve assembly. Connect the speedometer cable to the transaxle case.

8. Lower the vehicle . Check the fluid level and correct as necessary. Road test the vehicle.

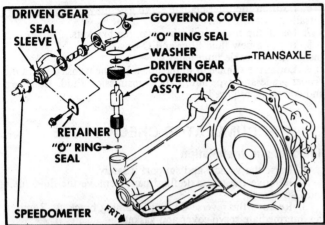

Speedometer drive and governor assembly—exploded view (©General Motors Corp.)

GOVERNOR ASSEMBLY

Removal and Installation

1. Raise the vehicle and support it safely.
2. Disconnect the speedometer cable from the transaxle assembly.
3. Remove the speedometer driven gear and sleeve assembly from the transaxle.
4. Remove the transaxle governor cover and O-ring. Carefully remove the governor assembly from its mounting in the transaxle.
5. To install, position the governor assembly in the transaxle. Install the governor cover using a new O-ring.
6. Install the speedometer driven gear and sleeve assembly. Connect the speedometer cable to the transaxle case.
7. Lower the vehicle. Adjust the transaxle fluid level as required. Road test the vehicle.

T.V. CABLE

Removal and Installation

1. Remove the air cleaner from the engine.
2. Disconnect the T.V. cable from the throttle lever.
3. Remove the bolt securing the T.V. cable to the transaxle assembly. Pull up on the cable cover at the transaxle until the cable is seen.
4. Disconnect the T.V. cable from the transaxle rod.
5. To install, reverse the removal procedure.
6. Check and adjust as required.

AUXILIARY VALVE BODY, VALVE BODY AND OIL PUMP ASSEMBLY

Removal and Installation

1. Disconnect the negative battery cable. Remove the air cleaner and disconnect the T.V. cable from its mounting place.
2. Remove the bolt securing the T.V. cable to the automatic transaxle. Pull up on the cable cover at the transaxle until the cable can be seen. Disconnect the T.V. cable from the transaxle rod.
3. Raise the vehicle on a hoist and support it safely.
4. Remove the left front tire and wheel assembly.
5. Remove the bolts securing the valve body cover to the transaxle. Remove the cover.
6. Remove the bolt that secures the TCC solenoid to the auxiliary valve body. Remove the solenoid.
7. Disconnect the TCC solenoid wires from the third gear pressure switch.
8. Remove the bolt securing the T.V. linkage and bracket assembly to the valve body. Remove the T.V. linkage.
9. Remove the remaining bolts securing the valve body to its mounting. Remove the valve body assembly, being careful not to lose the six check balls. Do not remove the green colored bolt.
10. Remove the green colored bolt only when separating the auxiliary valve body from the valve body.
11. When installing, be sure to hold the six check balls in place with petroleum jelly.
12. Adjust the T.V. cable. Add automatic transaxle fluid as required. Road test the vehicle and correct as required.

 REMOVAL & INSTALLATION

TRANSAXLE

CENTURY, SKYLARK, CITATION, CELEBRITY, OMEGA, CIERA, PHOENIX AND 6000

1. Disconnect the negative battery cable at the transaxle.
2. Remove:
   a. Air cleaner and disconnect the T.V. cable.
   b. T.V. cable lower attaching bolt and disconnect the cable from the transaxle.
   c. Strut shock bracket bolts from the transaxle.
   d. Oil cooler lines from the strut bracket.
   e. Transaxle-to-engine bolts, leaving the bolt near the starter installed loosely.

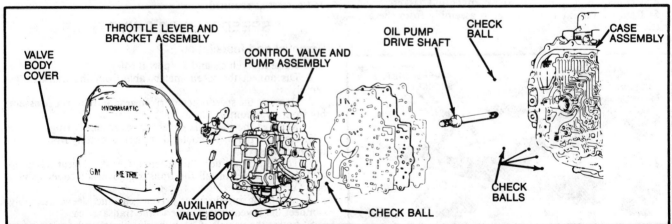

Auxiliary valve body, valve body and oil pump assembly (©General Motors Corp.)

f. Shift linkage retaining clip and washer at the transaxle.

g. Shift linkage bracket bolts.

3. Disconnect the speedometer drive cable at the upper and lower couplings (at the transducer if equipped with cruise control).

4. Disconnect the oil cooler lines at the transaxle.

5. Install the engine support fixture, locating it at the center of the cowl for four cylinder engines and on the strut towers for six cylinder engines.

NOTE: When installing a lift chain onto the aluminum cylinder head of the 4.3 liter V6 diesel engine, be sure that the bolt is tight or damage to the cylinder head may occur.

6. Rotate the steering wheel to position the steering gear stub shaft bolt in the upward position. Remove the bolt.

7. Raise the vehicle and place a jack under the engine to act as a support during removal and installation.

8. Remove the left front tire and wheel assembly.

9. Remove the power steering line brackets, remove the mounting bolts for the steering rack assembly, and support the assembly.

10. Disconnect the driveline vibration absorber, if so equipped.

11. Disconnect the left side lower ball joint at the steering knuckle.

12. Remove the front stabilizer bar reinforcements and bushings from the right and left cradle side members.

13. Using a drill with a ½ in. bit, drill through the spot weld located between the rear holes of the left side front stabilizer bar mounting.

14. Disconnect the engine and transaxle mounts from the cradle.

15. Remove the sidemember-to-cross-member bolts.

16. Remove the bolts from the left side body mounts.

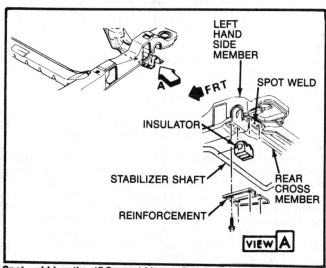

Spot weld location (©General Motors Corp.)

17. Remove the left side and front crossmember assembly. It may be necessary to carefully pry the crossmember loose.

18. Install axle shaft boot protectors, and using the appropriate special tools, pull the axle shaft cones out and away from the transaxle.

19. Pull the left axle shaft out of the transaxle.

20. Rotate the strut assembly so that the axle shaft is out of the way.

21. Remove:

a. Starter and converter shields.

b. Flywheel-to-converter bolts.

c. Two transaxle extension bolts from the engine-to-transaxle bracket.

d. Rear transaxle mount bracket assembly. It may be necessary to raise the transaxle assembly.

22. Securely attach a transaxle jack to the transaxle assembly.

23. Remove the two braces to the right end of the transaxle bolts.

24. Remove the remaining transaxle-to-engine bolt.

25. Remove the transaxle by moving it towards the driver's side, away from the engine.

26. Installation is performed in the reverse of the previous steps. Note the following:

a. When raising the transaxle into place, guide the right side axle shaft into the transaxle.

b. Check and adjust all front end alignment settings after the installation is complete.

c. Adjust the transmission detent cable.

SKYHAWK, CAVALIER, FIRENZA, CIMARRON, AND 2000

1. Disconnect the negative battery cable where it attaches to the transaxle.

2. Insert a ¼ x 2 in. bolt into the hole in the right front motor mount to prevent any mislocation during the transaxle removal.

3. Remove the air cleaner. Disconnect the T.V. cable.

4. Unscrew the bolt securing the T.V. cable to the transaxle. Pull up on the cable cover at the transaxle until the cable can be seen. Disconnect the cable from the transaxle rod.

5. Remove the wiring harness retaining bolt at the top of the transaxle.

6. Remove the hose from the air management valve and then pull the wiring harness up and out of the way.

7. Install an engine support bar. Raise the engine high enough to take the pressure off the motor mounts.

NOTE: The engine support bar must be located in the center of the cowl and the bolts must be tightened before attempting to support the engine.

8. Remove the transaxle mount and bracket assembly. It may be necessary to raise the engine slightly to aid in removal.

9. Disconnect the shift control linkage from the transaxle.

10. Remove the top transaxle-to-engine mounting bolts. Loosen, but do not remove, the transaxle-to-engine bolt nearest to the starter.

11. Unlock the steering column. Raise and support the front of the car. Remove the front wheels.

12. Pull out the cotter pin and loosen the castellated ball joint nut until the ball joint separates from the control arm. Repeat on the other side of the car.

13. Disconnect the stabilizer bar from the left lower control arm.

14. Remove the six bolts that secure the left front suspension support assembly.

15. Connect an axle shaft removal tool to a slide hammer.

16. Position the tool behind the axle shaft cones and then pull the cones out and away from the transaxle. Remove the axle shafts and plug the transaxle bores to reduce fluid leakage.

17. Remove the nut that secures the transaxle control cable bracket to the transaxle, then remove the engine-to-transaxle stud.

18. Disconnect the speedometer cable at the transaxle.

19. Disconnect the transaxle strut (stabilizer) at the transaxle.

20. Remove the four retaining screws and remove the torque converter shield.

21. Remove the three bolts securing the torque converter to the flex plate.

22. Disconnect and plug the oil cooler lines at the transaxle. Remove the starter.

23. Remove the screws that hold the brake and fuel line brack-

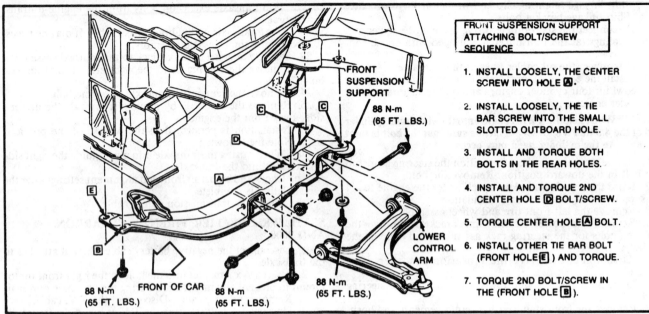

FRONT SUSPENSION SUPPORT ATTACHING BOLT/SCREW SEQUENCE

1. INSTALL LOOSELY, THE CENTER SCREW INTO HOLE A.
2. INSTALL LOOSELY, THE TIE BAR SCREW INTO THE SMALL SLOTTED OUTBOARD HOLE.
3. INSTALL AND TORQUE BOTH BOLTS IN THE REAR HOLES.
4. INSTALL AND TORQUE 2ND CENTER HOLE D BOLT/SCREW.
5. TORQUE CENTER HOLE A BOLT.
6. INSTALL OTHER TIE BAR BOLT (FRONT HOLE E) AND TORQUE.
7. TORQUE 2ND BOLT/SCREW IN THE (FRONT HOLE B).

FRONT SUSPENSION SUPPORT

88 N-m (65 FT. LBS.)

LOWER CONTROL ARM

88 N-m (65 FT. LBS.) FRONT OF CAR 88 N-m (65 FT. LBS.) 88 N-m (65 FT. LBS.)

Front suspension support installation sequence—Skyhawk, Cavalier, Firenza, Cimarron and 2000 (©General Motors Corp.)

ets to the left side of the underbody. This will allow the lines to be moved slightly for clearance during transaxle removal.

24. Remove the bolt that was loosened in Step 10.

25. Remove the transaxle to the left.

Installation is in the reverse order of removal. Please note the following:

a. Reinstall both axle shafts AFTER the transaxle is in position.

b. When installing the front suspension support assembly you must follow the tightening sequence shown in the illustration.

c. Check alignment when installation is complete.

BENCH OVERHAUL

Before Disassembly

Cleanliness is an important factor in the overhaul of the 125C automatic transaxle. Before opening up this unit, the entire outside of the transaxle assembly should be cleaned, preferably with a high pressure washer such as a car wash spray unit. Dirt entering the transaxle internal parts will negate all the time and effort spent on the overhaul. During inspection and reassembly all parts should be thoroughly cleaned with solvent then dried with compressed air. Wiping cloths and rags should not be used to dry parts since lint will find its way in to the valve body passages.

Wheel bearing grease, long used to hold thrust washers and lube parts, should not be used. Lube seals with Dexron® II and use ordinary unmedicated petroleum jelly to hold the thrust washers and to ease the assembly of seals, since it will not leave a harmful residue as grease often will. Do *not* use solvent on neoprene seals, friction plates if they are to be reused, or thrust washers. Be wary of nylon parts if the transaxle failure was due to failure of the cooling system. Nylon parts exposed to water or antifreeze solutions can swell and distort and must be replaced.

Before installing bolts into aluminum parts, always dip the threads into clean transmission oil. Anti-seize compound can also be used to prevent bolts from galling the aluminum and seiz-

ing. Always use a torque wrench to keep from stripping the threads. Take care with the seals when installing them, especially the smaller "O"-rings. The internal snap rings should be expanded and the external rings should be compressed, if they are to be reused. This will help insure proper seating when installed.

Torque Converter Inspection

TORQUE CONVERTER

Removal

1. Make certain that the transaxle is held securely.

2. The converter pulls out of the transaxle. Be careful since the converter contains a large amount of oil. There is no drain plug on the converter so the converter should be drained through the hub.

3. Place the transaxle in a holding fixture if possible. Turn the transaxle so that the right-hand axle opening is down so that the unit can drain.

4. If the oil in the converter is discolored but does not contain metal bits or particles, the converter is not damaged and need not be replaced. Remember that color is no longer a good indicator of transmission fluid condition. In the past, dark color was associated with overheated transmission fluid. It is not a positive sign of transmission failure with the newer fluids like Dexron® II.

5. If the oil in the converter contains metal particles, the converter is damaged internally and must be replaced. The oil may have an "aluminum paint" appearance.

6. If the cause of oil contamination was burned clutch plates or overheated oil, the converter is contaminated and should be replaced.

TORQUE CONVERTER INSPECTION

After the converter is removed from the transaxle, the stator roller clutch can be checked by inserting a finger into the splined inner race of the roller clutch and trying to turn the race in both directions. The inner race should turn freely in the clockwise direction, but not turn in the counterclockwise direction. The inner race may tend to turn in the counterclockwise direction, but with great difficulty, this is to be considered normal. Do not use such items as the driven sprocket support or the shafts to turn the race, as the results may be misleading. Inspect the outer hub lip and the inner bushing for burrs or jagged edges to avoid injury to your fingers when testing the torque converter.

Transaxle Disassembly

EXTERNAL COMPONENTS

1. Remove the speedometer driven gear and the sleeve assembly from the transaxle case.

2. Remove the governor cover O-ring. Discard the O-ring.

3. Remove the governor cover to speedometer drive gear thrust bearing assembly. Remove the speedometer drive gear assembly.

4. Remove the transaxle oil pan. If RTV sealant was used do not pry the oil pan loose from the transaxle case as damage to the pan flange or case will occur. Use a rubber mallet and strike the edge of the oil pan to shear the RTV seal.

5. Remove and discard the fluid filter and O-ring seal.

6. Remove the reverse oil pipe retaining brackets. Remove the intermediate servo cover and gasket. Discard the gasket.

7. Remove the intermediate servo assembly from its mounting in the transaxle.

8. Check for the proper intermediate band apply pin by installing special tool J-28535 or an equivalent intermediate band apply pin gauge. Hold the tool in place using the two intermediate servo cover screws. Remove the band apply pin from the intermediate servo assembly. Install special tool J-28535-4 or equivalent on the band apply pin. Position this assembly into the apply pin gauge. To check, compress the band apply 100 inch pounds torque. If any part of the white line appears in the window, the pin is the correct length. If the white line connot be seen, change the band apply pin and recheck. Replace parts as required.

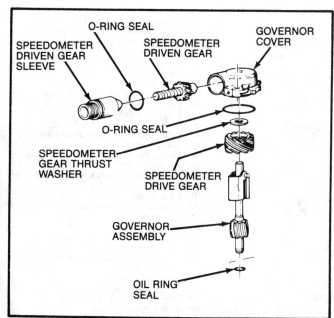

Speedometer drive and governor assembly—exploded view (©General Motors Corp.)

9. Remove the third accumulator check valve and spring assembly.

10. Remove the reverse oil pipe, seal back up ring and O-ring seal.

11. Remove the low and reverse cup plug assembly by grinding about ¾ inch from the end of a number four easy out. Using the modified tool, remove the seal assembly.

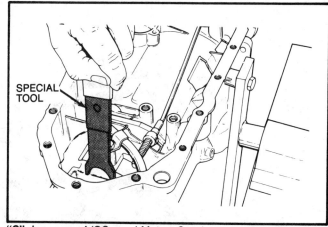

"C" ring removal (©General Motors Corp.)

12. Remove the dipstick stop and the parking lock bracket.

13. If equipped, remove the oil weir clip and remove the oil weir assembly.

14. Position the final drive unit so that the "C" ring is visible. Position the "C" ring so that the open side is facing the access window. Using special tool J-28583 or equivalent push the "C" ring partially off of the output shaft.

15. Carefully turn the output shaft so that the "C" ring is up. Remove the "C" ring by pulling it out with a needle nose pliers.

16. Remove the output shaft from the transaxle assembly.

VALVE BODY, CASE COVER AND SPROCKET LINK ASSEMBLY

1. Remove the valve body cover. If RTV sealant was used, do not pry the transaxle oil pan loose from the transaxle case as damage to the transaxle case or oil pan flange will occur.

2. Remove the two bolts retaining the throttle lever and bracket assembly.

3. Remove the throttle lever and bracket assembly with the T.V. cable link.

4. Remove the auxiliary valve body bolts except for the green colored bolt. Do not remove the green colored bolt unless it is necessary to separate the valve body from the auxiliary valve body.

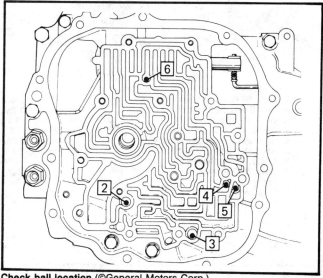

Check ball location (©General Motors Corp.)

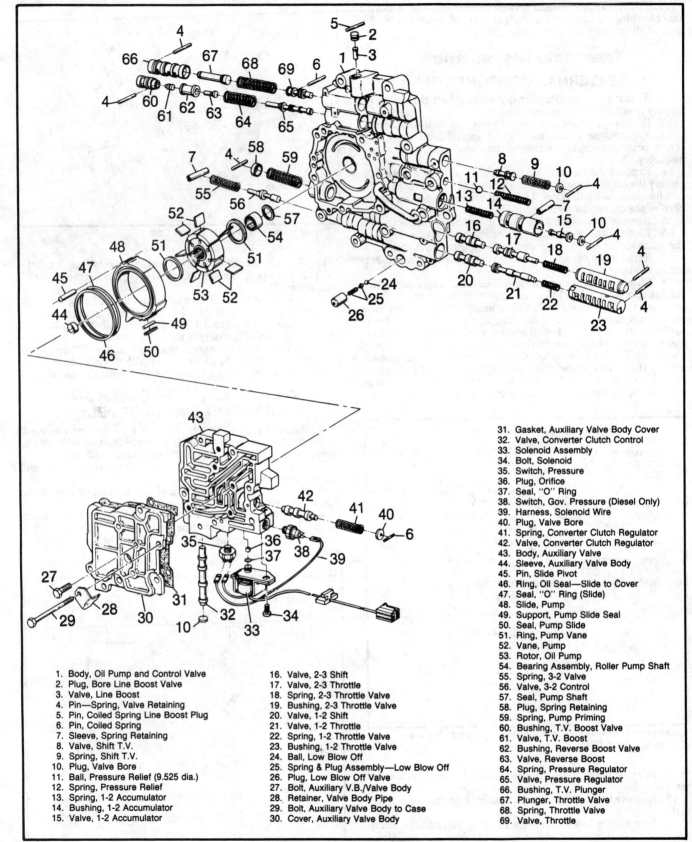

Valve body assembly—exploded view (©General Motors Corp.)

1. Body, Oil Pump and Control Valve
2. Plug, Bore Line Boost Valve
3. Valve, Line Boost
4. Pin—Spring, Valve Retaining
5. Pin, Coiled Spring Line Boost Plug
6. Pin, Coiled Spring
7. Sleeve, Spring Retaining
8. Valve, Shift T.V.
9. Spring, Shift T.V.
10. Plug, Valve Bore
11. Ball, Pressure Relief (9.525 dia.)
12. Spring, Pressure Relief
13. Spring, 1-2 Accumulator
14. Bushing, 1-2 Accumulator
15. Valve, 1-2 Accumulator

16. Valve, 2-3 Shift
17. Valve, 2-3 Throttle
18. Spring, 2-3 Throttle Valve
19. Bushing, 2-3 Throttle Valve
20. Valve, 1-2 Shift
21. Valve, 1-2 Throttle
22. Spring, 1-2 Throttle Valve
23. Bushing, 1-2 Throttle Valve
24. Ball, Low Blow Off
25. Spring & Plug Assembly—Low Blow Off
26. Plug, Low Blow Off Valve
27. Bolt, Auxiliary V.B./Valve Body
28. Retainer, Valve Body Pipe
29. Bolt, Auxiliary Valve Body to Case
30. Cover, Auxiliary Valve Body

31. Gasket, Auxiliary Valve Body Cover
32. Valve, Converter Clutch Control
33. Solenoid Assembly
34. Bolt, Solenoid
35. Switch, Pressure
36. Plug, Orifice
37. Seal, "O" Ring
38. Switch, Gov. Pressure (Diesel Only)
39. Harness, Solenoid Wire
40. Plug, Valve Bore
41. Spring, Converter Clutch Regulator
42. Valve, Converter Clutch Regulator
43. Body, Auxiliary Valve
44. Sleeve, Auxiliary Valve Body
45. Pin, Slide Pivot
46. Ring, Oil Seal—Slide to Cover
47. Seal, "O" Ring (Slide)
48. Slide, Pump
49. Support, Pump Slide Seal
50. Seal, Pump Slide
51. Ring, Pump Vane
52. Vane, Pump
53. Rotor, Oil Pump
54. Bearing Assembly, Roller Pump Shaft
55. Spring, 3-2 Valve
56. Valve, 3-2 Control
57. Seal, Pump Shaft
58. Plug, Spring Retaining
59. Spring, Pump Priming
60. Bushing, T.V. Boost Valve
61. Valve, T.V. Boost
62. Bushing, Reverse Boost Valve
63. Valve, Reverse Boost
64. Spring, Pressure Regulator
65. Valve, Pressure Regulator
66. Bushing, T.V. Plunger
67. Plunger, Throttle Valve
68. Spring, Throttle Valve
69. Valve, Throttle

5. Remove the remaining control valve body retaining bolts.

6. Remove the control valve and oil pump assembly. Place it on a clean work bench with the machined surface up.

7. Remove the number one check ball from the spacer plate. Remove the oil pump shaft. Remove the spacer plate and the gaskets.

8. Remove the five check balls which are located in the transaxle case cover.

9. Check the input shaft to case cover end play for the proper selective snap ring thickness.

10. To do this, rotate the transaxle assembly so that the right axle end is up. Install tool J-26958-10 or equivalent into the right hand axle end. Install an output shaft loading tool and bracket to the right hand axle end. Adjust the loading tool by turning the handle in until the knob bottoms. This is the correct load.

11. Rotate the transaxle assembly so that the transaxle case cover is up. Install an an input shaft lifter tool into the input shaft bore. Secure by turning the handle. Install the dial indicator and the dial indicator post on the transaxle assembly. Place the dial indicator extension on the end play tool. Press down on the lifting tool and zero in the dial indicator.

12. Pull up on the end play tool and take the end play reading. The end play should be 0.10-0.84 mm (.004"-.033"). The selective snap ring controlling the end play is located on the input shaft. If the end play is not in the proper range, select the proper snap ring from the input shaft selective snap ring chart. Measure the thickness for positive identification.

INPUT SHAFT SELECTIVE SNAP RING CHART

Thickness (inches)	Identification
.071-.076	White
.078-.084	Blue
.088-.092	Brown
.095-.099	Yellow
.103-.107	Green

13. Disconnect the manual valve rod from the manual valve.

14. Remove the remaining transaxle case cover bolts.

15. Install two bolts about two inches long into the transaxle case cover dowel pin holes. These bolts will self tap, bottom out on the dowel pins and separate the transaxle case cover from the transaxle case. Do not pry the case cover from the case, as damage to the mating surfaces will result.

16. Remove the transaxle case cover. Place the case cover on the work bench with the 1-2 accumulator side up. The 1-2 accumulator pin may fall out.

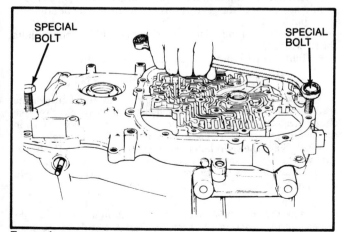

Transaxle case removal using special bolts (©General Motors Corp.)

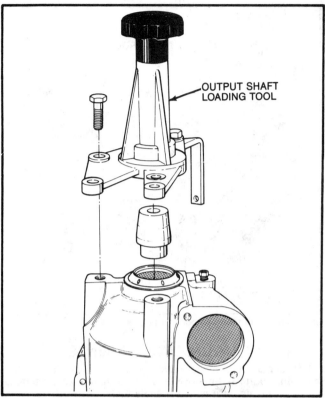

Output shaft loading tool installation (©General Motors Corp.)

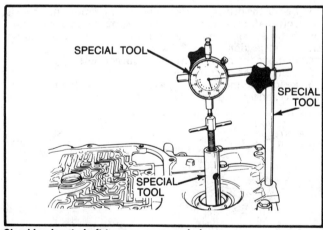

Checking input shaft to case cover end play
(©General Motors Corp.)

NOTE: To remove the gasket, spray gasket remover on the transaxle case and case cover gasket surface. Use a plastic edged gasket scraper to avoid damaging the machined surfaces.

17. Remove the 1-2 accumulator spring and the center case to case cover gasket.

18. Remove the case cover to drive sprocket thrust washer and the driven sprocket thrust bearing assembly. The case cover to drive sprocket thrust washer may be on the case cover.

19. Remove the turbine shaft "O" ring.

20. Remove the drive sprocket, driven sprocket and link assembly.

21. Remove the drive and driven sprocket to support thrust washers. These thrust washers may have come off with the sprockets.

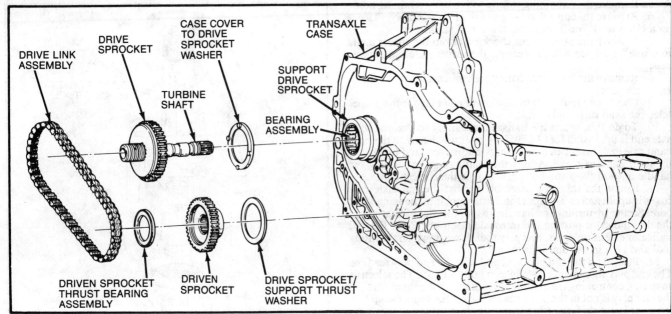

Drive link and related components (©General Motors Corp.)

INPUT UNIT COMPONENTS

1. Using a small punch, remove the detent lever to manual shaft pin. Locate and remove the manual shaft to case retaining pin by prying up on it with a suitable tool.

2. Remove the detent lever, manual shaft, and parking lock rod.

3. Locate the driven sprocket support and remove it. There is a support-to-direct clutch housing thrust washer under the support that may come out with the support. Remove it, and set aside for inspection.

4. Remove the intermediate band anchor hole plug.

5. Remove the intermediate band.

6. Reach into the case and grasp the input shaft and remove the direct and forward clutch assemblies. Separate the assemblies.

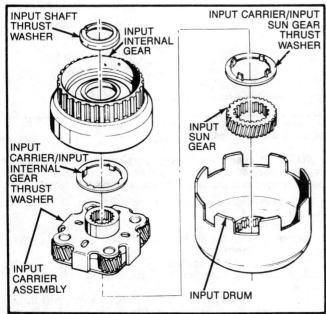

Input unit—exploded view (©General Motors Corp.)

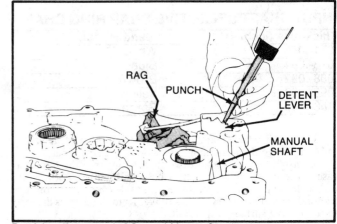

Retaining pin removal (©General Motors Corp.)

7. Remove the input internal gear and the thrust washer that should be on top of the gear. If it is not there, it may be stuck on the input shaft.

8. Remove the input planetary carrier assembly along with its thrust washers.

9. Remove the input sun gear and the input drum.

REACTION UNIT COMPONENTS

NOTE: There are selective fit snap rings located between the reaction sun gear and the input drum and also between the reverse clutch housing and the low roller clutch. Special tools are used in the factory procedure which involves a preload on the right hand axle end. Since the right hand axle was taken out when the transaxle was removed from the vehicle, the preload goes against an adapter plug that fits into the transaxle case and takes the place of the right hand axle for the end play check.

A dial indicator with a long extension will be needed. With a preload on the right-hand axle end, the dial indicator is brought to bear on the reaction sub gear, between the ends of the snap ring. Press down on the sun gear to make sure that it is seated, and

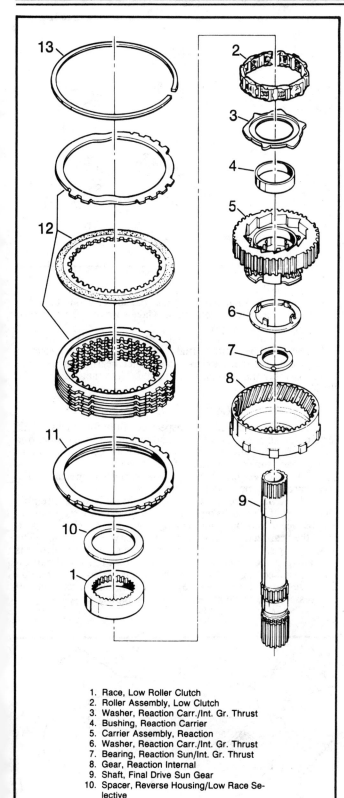

1. Race, Low Roller Clutch
2. Roller Assembly, Low Clutch
3. Washer, Reaction Carr./Int. Gr. Thrust
4. Bushing, Reaction Carrier
5. Carrier Assembly, Reaction
6. Washer, Reaction Carr./Int. Gr. Thrust
7. Bearing, Reaction Sun/Int. Gr. Thrust
8. Gear, Reaction Internal
9. Shaft, Final Drive Sun Gear
10. Spacer, Reverse Housing/Low Race Selective
11. Plate, Low & Reverse Clutch Backing
12. Plate, Low & Reverse Clutch
13. Snap Ring

Reaction Components—exploded view (©General Motors Corp.)

zero the indicator. Rotate the selective fit snap ring under the indicator and read the reaction sun gear to input drum play. It should be from 0.013 to −0.005″. If more or less snap ring thickness is required, select one that will bring the clearance into specification. They are available from the factory in sizes from 0.089-0.093″ up to 0.136-0.140″ in increments of 0.004″.

REACTION GEAR TO INPUT DRUM SELECTIVE SNAP RING CHART

Thickness (Inches)	Identification
.089-.093	Pink
.096-.100	Brown
.103-.107	Light Blue
.109-.113	White
.116-.120	Yellow
.123-.127	Light Green
.129-,133	Orange
.136-.140	No Color

In a similar manner, the reverse clutch housing to the Low roller clutch selective fit snap ring should be checked. A preload on the output shaft (right hand axle end) is again used and the dial indicator is set up as before, and zeroed. Place a suitable tool through the opening in the case next to the parking pawl and lift the reaction internal gear to check the low and reverse selective end play. The end play should be 0.003-0.046″. If more or less selective washer clearance is required, select one that will bring the clearance into specification. They are available from the factory in sizes from 0.039-0.043″ to 0.122-0.126″ in increments of 0.004″.

REVERSE CLUTCH HOUSING TO LOW RACE SELECTIVE WASHER CHART

Thickness (Inches)	Identification
.039-.043	One
.056-.060	Two
.072-.076	Three
.089-.093	Four
.105-.109	Five
.122-.126	Six

To continue with the disassembly of the reaction unit parts, proceed as follows:
1. Remove the reaction sun gear.
2. Locate and remove the low and reverse clutch housing-to-case snap ring, which is 0.092″ thick.
3. Remove the low and reverse clutch housing.
4. Remove the low and reverse clutch housing-to-case spacer ring which is 0.042″ thick.
5. Reach into the case and grasp the final drive sun gear shaft and pull out the reaction gear set.
6. Separate the parts by removing the roller clutch and reaction carrier assembly from off the final drive sun gear shaft.
7. Remove the four-tanged thrust washer from the end of the reaction carrier (or the inside of the internal gear if it has stuck there).
8. Remove the low and reverse clutch plates, taking note of their number and order.

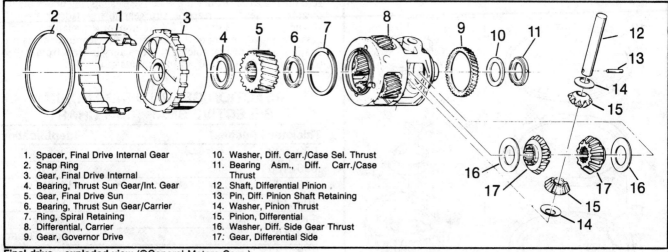

1. Spacer, Final Drive Internal Gear
2. Snap Ring
3. Gear, Final Drive Internal
4. Bearing, Thrust Sun Gear/Int. Gear
5. Gear, Final Drive Sun
6. Bearing, Thrust Sun Gear/Carrier
7. Ring, Spiral Retaining
8. Differential, Carrier
9. Gear, Governor Drive
10. Washer, Diff. Carr./Case Sel. Thrust
11. Bearing Asm., Diff. Carr./Case Thrust
12. Shaft, Differential Pinion
13. Pin, Diff. Pinion Shaft Retaining
14. Washer, Pinion Thrust
15. Pinion, Differential
16. Washer, Diff. Side Gear Thrust
17. Gear, Differential Side

Final drive—exploded view (©General Motors Corp.)

9. Remove the internal gear to sun gear thrust bearing from the reaction gear, and remove the internal gear from the final drive sun gear.

FINAL DRIVE COMPONENTS

The end play of the final drive to case should be checked before further disassembly. Again mount the dial indicator to the transaxle by first rotating the transaxle in the holding fixture so that the right-hand axle opening is facing up. No preload is needed for this check. Position the indicator to bear on the axle shaft and push down on the shaft before setting the indicator to zero. Using a suitable tool, position it in the governor bore and lift on the governor drive gear to read the final drive to case end play. The end play should be from 0.005-0.032″.

The selective fit washer that controls this end play is located between the differential carrier and the differential carrier case thrust bearing assembly. If the washer needs to be changed to bring the end play reading within the specifications, they are available in sizes from 0.055-0.59″ to 0.091-0.095″ in increments of 0.004″. After checking the end play, proceed as follows:

FINAL DRIVE TO CASE SELECTIVE WASHER CHART

Thickness (Inches)	Identification
.055-.059	Zero
.059-.062	One
.062-.066	Two
.066-.070	Three
.070-.074	Four
.074-.078	Five
.078-.082	Six
.082-.086	Seven
.086-.091	Eight
.091-.095	Nine

1. Rotate the transaxle so that the case cover side up as before and locate and remove the final drive internal gear spacer snap ring, which is 0.092″ thick.

2. Remove the final drive internal gear spacer. Do not bend or deform the spacer when removing. Make sure that the governor has been removed at this time.

3. Reach into the case and pull out the final drive unit.

4. There will be a differential-to-case thrust washer (selective fit) and a differential carrier-to-case roller bearing on the final drive assembly that should be removed. Note that they may be stuck in the case.

TRANSAXLE CASE

1. Clean the case well and inspect carefully for cracks. Make certain that all passages are clean and that all bores and snap ring grooves are clean and free from damage. Check for stripped bolt holes. Check the case bushings for damage.

2. A new converter seal can be installed at this time, unless the drive sprocket support is to be removed.

3. Check the drive sprocket support roller bearing assembly for damage. If it requires replacement, a slide hammer type puller will be needed to pull the bearing from the sprocket support. Once the bearing is out, inspect the bore for wear or damage. The new bearing should be driven in with care so that the bearing is not damaged and is in straight. Be sure that the bearing identification is installed "up." Also, if the bearing is replaced, the race on the drive sprocket must also be checked carefully. If replacement is necessary, the drive sprocket must be replaced.

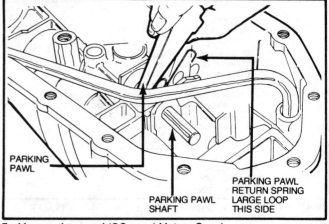

PARKING PAWL

PARKING PAWL SHAFT

PARKING PAWL RETURN SPRING LARGE LOOP THIS SIDE

Parking pawl removal (©General Motors Corp.)

4. If it proves necessary to remove the drive sprocket, rotate the transaxle so that the right-hand axle side is up. The converter seal must be removed. When removing the retaining bolts place one hand under the support. To reinstall the support, use a hardening type compound on the bolt threads and torque to 18 footpounds. Renew the converter seal.

5. If the parking pawl and related parts are to be removed, begin by turning the transaxle to the oil pan side up. Use a punch to remove the cup plug. Remove the parking pawl shaft retainer, then the shaft, pawl and return spring. Check the pawl carefully for cracks.

6. If it is necessary to remove the governor oil pipe, note that the pipe is held firmly in place with Loctite® and may require a great amount of force to remove. A large pry bar may be used to remove the pipe from its mounting in the transaxle case. Use Loctite® or equivalent on both ends of the pipe when reinstalling.

7. If the manual shaft oil seal is to be replaced, pry out the old seal and drive in the new seal at this time. Be sure that the seal is installed with the lip up.

8. Check the axle seal for damage. Replace as required.

Component Disassembly and Assembly

DIFFERENTIAL AND FINAL DRIVE

1. Remove the final drive internal gear from the differential carrier and also the thrust bearing from between the internal gear and the sun gear.

2. Remove the final drive sun gear and also the thrust bearing from between the sun gear and differential carrier.

3. Examine the governor drive gear. It is not necessary to remove it unless it is to be replaced. If removal is necessary, a gear puller will be required and a soft hammer will be needed to drive the gear back on.

4. Inspect the differential side gears and pinions for excessive wear or damage. If the differential side gears are to be replaced, a lock pin that is installed through the pinion shaft will have to be removed.

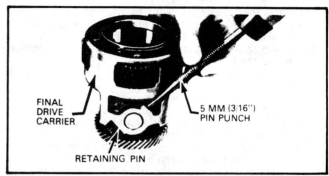

Differential shaft retaining pin removal (©General Motors Corp.)

NOTE: This pin can only be removed one way. A 3/16" pin punch can be used to drive out the pin. Drive from the carrier side to the governor gear side. When reinstalling the pin, turn the carrier assembly over and drive the pin in from the governor gear side.

5. With the lock pin removed, the pinion shaft can be removed, along with the pinion gears and thrust washers. Do this by rotating the gears until they are in the access "window" in the carrier where they can be removed. If any pinion does not have a thrust washer with it, then it has fallen off and needs to be removed from the carrier.

6. Remove the differential side gears. The gears should slide out. Be sure that the thrust washers are also removed with the differential side gears.

7. Carefully examine the pinion gears and the differential side gears along with their thrust washers. Look for excessive wear or scoring. If it is determined that all parts are useable, begin assembly by coating the thrust washers with petroleum jelly and install the differential gears and their thrust washers into the case. Coat the pinion gear thrust washers in the same way and install them on their gears and then carefully install the pinion gears through each "window."

8. Align the pinion gears by sliding the pinion shaft through the gears. With the gears in line, remove the pinion shaft. Rotate the pinions to seat them in place and insall the pinion shaft.

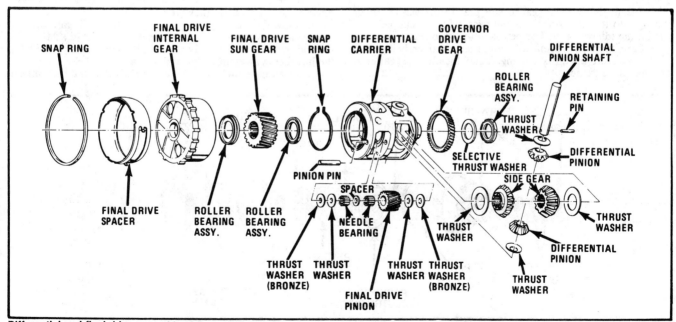

Differential and final drive assembly (©General Motors Corp.)

9. Install the "one way" lock pin by inserting it from the governor gear side and driving it into place with a ⅜" drift pin.

10. Final drive pinion end play should be checked at this time. The end play should be from .009 to .025 inch. If pinions must be removed, do the following.

a. Locate and remove the snap ring which retains the pinion pins.

b. Remove the pinion pins.

c. Remove the final drive pinions. Caution should be used since each pinion and thrust washer assembly contains 36 needle bearings. Try to hold the pinion and thrust washer assembly from the ends to prevent dropping the needle bearings. There are four of these pinion assemblies.

d. The pinion assemblies should be cleaned and the rollers carefully examined. Check the pinion for wear or damaged teeth. Make sure that the spacer between the two rows of roller bearings and the thrust washers is not scored or damaged.

11. Reassemble the final drive pinions and their related pieces by installing a thrust washer on a pinion pin. Retain it with petroleum jelly.

12. Apply a thin coat of petroleum jelly on the pin and carefully lay 18 needle rollers on the pinion shaft.

13. Install the needle spacer on the shaft again, with petroleum jelly to hold it in place. Carefully lay 18 more rollers on the shaft using the petroleum jelly as necessary to retain these parts. Add the other thrust washer.

14. Carefully insert the pinion shaft and its rollers into the pinion. Install a bronze thrust washer on both ends of the pinion, on top of the steel thrust washers.

15. Carefully pull out the pinion shaft.

16. Install the pinion assembly into the carrier and, hold the pinion from the end, so that all the pieces will stay together.

17. As each pinion assembly is installed, carefully install the pinion shaft, stepped end last. Align the step so that the step is on the outside.

18. Repeat the installation procedure for the remaining pinions and then install the snap ring.

19. Inspect the final drive sun gear to carrier case thrust bearing. If it is reusable, install it with the outer race against the carrier.

20. Inspect the sun gear, especially around the splines and gear teeth for excessive wear or damage. It is installed with the step side up.

21. Install the internal gear to sun gear thrust washer roller bearing into the final drive internal gear, with the cupped race side on the internal gear. Use petroleum jelly to retain it.

22. Carefully turn the final drive internal gear over, being careful not to disturb the thrust bearing. Install the internal gear onto the final drive carrier.

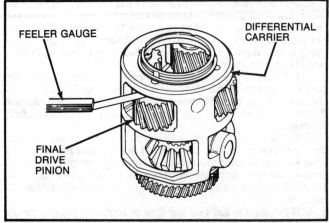

Differential pinion end play check (©General Motors Corp.)

23. Inspect the differential selective thrust washer for scored, rough or pitted surfaces. Also check the differential roller bearing thrust washer for damage. If these parts are serviceable, install first the selective washer, and then the thrust bearing assembly, with the inner race against the selective thrust washer.

FORWARD CLUTCH

1. Position the forward clutch assembly so that it is situated with the input shaft facing down. Remove the clutch pack snap ring.

2. Remove the backing plate, the clutch plates, and then compress the spring pack to remove the snap ring. Remove the retainer and spring assembly.

3. Remove the piston and discard the seals.

4. Clean all parts and check for damage or excessive wear. Make sure that the check ball is free and not sticking in its capsule. Check the piston for burrs or cracks and the housing for damaged snap ring grooves.

5. Inspect the input shaft splines and journals for damage. Check the shaft sleeve. The sleeve must not turn. The slot in the sleeve must be aligned with the hole in the input shaft.

6. Inspect the Teflon seals on the shaft for damage. Do not remove these seals unless they are damaged.

7. Allow the clutch plates to soak for at least 20 minutes before using them in assembly. Soak the plates in Dexron® II. Lube the seals with petroleum jelly and install them on the piston with the lips facing away from the apply ring side.

8. Install the clutch apply ring on the piston, and then install

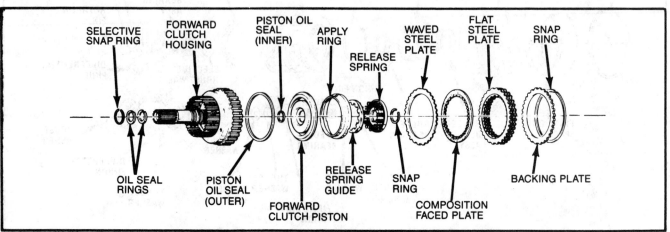

Forward clutch assembly—exploded view (©General Motors Corp.)

the piston into the forward clutch housing. Use care and plenty of lubricant to protect the seals.

9. Install the spring guide and then the spring and spring retainer. Compress the assembly and install the snap ring.

10. Install the clutch plates starting with a waved steel plate, then a friction plate, and alternate with steel plates until the clutch pack is installed. Install the backing plate with the flat side "up."

11. Install the snap ring and make certain that the friction plates turn freely.

12. Inspect the input shaft to driven sprocket snap ring for wear. Do not remove this snap ring unless replacement is necessary, as this is a selective fit snap ring.

DIRECT CLUTCH

1. Remove the large outer snap ring from the direct clutch housing. Remove the backing plate and clutch plates.

2. Remove the snap ring that holds the apply ring and the release spring assembly. Remove the two components.

3. Remove the direct clutch piston. Discard the seals. Remove the center seal from the housing.

4. Clean all parts and check for damage or excessive wear. Make sure that the check ball is free and not sticking in its capsule. Check the piston for burrs or cracks and the housing for damaged snap rings.

5. Allow the clutch plates to soak for at least 20 minutes before using them in assembly. Soak the plates in Dexron® II. Lube the seals with petroleum jelly and install them on the piston with the lips facing away from the clutch apply ring side.

6. Install the center seal on the housing with the lip facing "up."

7. Coat the seals with petroleum jelly and slide the piston into the housing. Be very careful since the sharp snap ring grooves could damage the seals.

8. An installation tool can be made from wire to ease the installation process. Once the piston is in place, rotate the piston to help seat the seals.

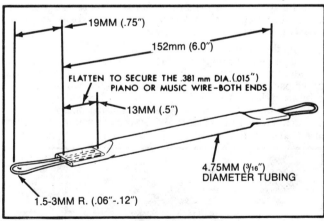

Piston installation tool (©General Motors Corp.)

9. Install the apply ring, retainer and springs and the snap ring.

10. Install the clutch plates, starting with the steel plate and alternating with the friction plates until they are all in place.

11. Install the backing plate making sure that the flat side is "up." Install the snap ring.

CONTROL VALVE AND PUMP ASSEMBLY

NOTE: As each part of the valve train is removed, place the pieces in order and in a position that is relative to the position on

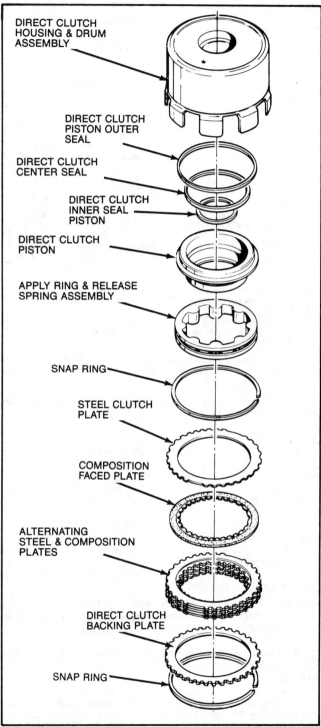

Front direct clutch—exploded view (©General Motors Corp.)

the valve body to lessen chances for error in assembly. None of the valves, springs or bushings are interchangeable.

1. Remove the roll pins by pushing through from the rough casting side of the control valve body, except, of course, the blind hole roll pins.

2. Lay the valve body on a clean work bench with the machined side up and the line boost valve at the top. The line boost valve should be checked for proper operation before removing it.

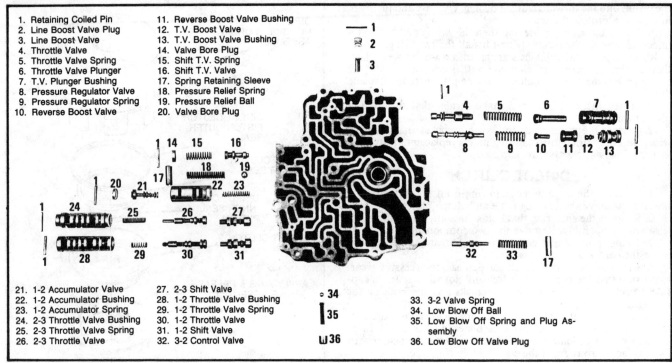

1. Retaining Coiled Pin
2. Line Boost Valve Plug
3. Line Boost Valve
4. Throttle Valve
5. Throttle Valve Spring
6. Throttle Valve Plunger
7. T.V. Plunger Bushing
8. Pressure Regulator Valve
9. Pressure Regulator Spring
10. Reverse Boost Valve
11. Reverse Boost Valve Bushing
12. T.V. Boost Valve
13. T.V. Boost Valve Bushing
14. Valve Bore Plug
15. Shift T.V. Spring
16. Shift T.V. Valve
17. Spring Retaining Sleeve
18. Pressure Relief Spring
19. Pressure Relief Ball
20. Valve Bore Plug

21. 1-2 Accumulator Valve
22. 1-2 Accumulator Bushing
23. 1-2 Accumulator Spring
24. 2-3 Throttle Valve Bushing
25. 2-3 Throttle Valve Spring
26. 2-3 Throttle Valve
27. 2-3 Shift Valve
28. 1-2 Throttle Valve Bushing
29. 1-2 Throttle Valve Spring
30. 1-2 Throttle Valve
31. 1-2 Shift Valve
32. 3-2 Control Valve
33. 3-2 Valve Spring
34. Low Blow Off Ball
35. Low Blow Off Spring and Plug Assembly
36. Low Blow Off Valve Plug

Control valve body and related components (©General Motors Corp.)

If it is necessary to remove the boost valve, grind the end of a #49 drill to a taper and lightly tap the drill into the roll pin. Pull the drill and roll pin out. Push the line boost valve out of the top of the valve body.

3. The throttle valve should be checked for proper operation before removing it, by pushing the valve against the spring. If it is necessary to remove the throttle valve, first remove the roll pin holding the T.V. plunger bushing and pull out the plunger and bushing. Remove the throttle valve spring. Remove the throttle valve.

4. Remove the roll pin and the throttle valve boost bushing and valve assembly. In the same valve train will be found the reverse boost valve and bushing, as well as the pressure regulator valve and spring.

5. Move to the other side of the valve body and from the top bore, remove the roll pin.

NOTE: This roll pin is under pressure. When removing this pin, use a rag to prevent the bore plug and spring from flying out and becoming lost. This is the shift T.V. spring and valve.

6. Move to the next bore down and remove the spring retaining sleeve with snap ring pliers. Be careful since the spring is under load.

7. Remove the pressure relief spring and check ball.

8. Move to the next bore down and remove the roll pin and the bore plug. Remove the 1-2 accumulator bushing, valve and spring.

9. At the next bore down, remove the roll pin and 2-3 throttle valve. Behind the 2-3 throttle valve will be the 2-3 shift valve.

10. At the next bore down, remove the roll pin and 1-2 throttle bushing, valve and spring. Behind the valve will be the 1-2 shift valve which is also removed.

11. At the opposite side of the valve body there is a spring retaining sleeve that holds the 3-2 control valve and spring. This sleeve is under load. Cover the open end of the bore when removing the sleeve to prevent loss of the spring. Remove the 3-2 control valve and the spring.

12. Use a ¼" punch to remove the low blowoff valve plug. Remove the valve spring, spring plug and ball.

NOTE: The low blowoff assembly must be removed and replaced if the control valve pump body is washed in solvent. These are some of the few parts that are serviced separately in the valve body. The technician should, therefore, use great caution when servicing the valve body and the pump so as not to lose or damage any of the components that may not be available except in the complete pump/control valve assembly.

13. If the pump assembly is to be serviced, remove the roll pin from the pump priming spring bore. Use caution because this pin is under spring pressure. Cover the end of the bore to prevent loss of the parts inside. Remove the priming spring cup plug and the priming spring.

14. Remove the pump cover screw and the pump cover.

15. Remove the auxiliary valve body cover screw, valve body, gasket and cover.

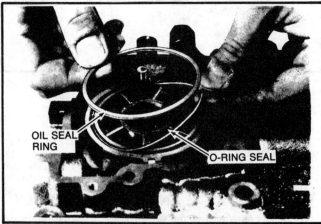

OIL SEAL RING

O-RING SEAL

Oil seal and O-ring installation (©General Motors Corp.)

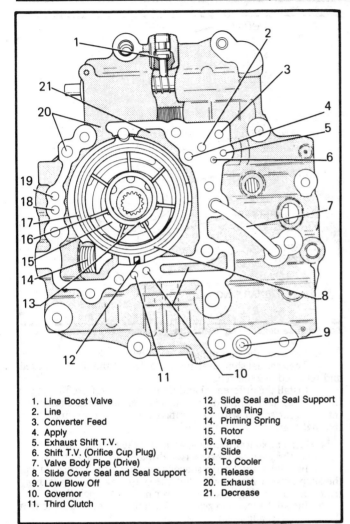

1. Line Boost Valve
2. Line
3. Converter Feed
4. Apply
5. Exhaust Shift T.V.
6. Shift T.V. (Orifice Cup Plug)
7. Valve Body Pipe (Drive)
8. Slide Cover Seal and Seal Support
9. Low Blow Off
10. Governor
11. Third Clutch
12. Slide Seal and Seal Support
13. Vane Ring
14. Priming Spring
15. Rotor
16. Vane
17. Slide
18. To Cooler
19. Release
20. Exhaust
21. Decrease

Variable capacity vane oil pump (©General Motors Corp.)

16. Remove the pump slide, rotor, the seven vanes and the two vane seal rings.

17. Clean all pump components as foreign matter in the pump can cause damage and be circulated to other transaxle components. Pump seals should not be washed in solvent.

18. If the pump shaft bearing is to be replaced, drive it out toward the pump pocket. The new bearing would then be driven in from the pump pocket side.

19. Install until the bearing cup is .017-.005 inch below the pump pocket face. Inspect all parts for wear or damage.

20. Turn the valve assembly so that the pump pocket side is up.

21. Install the pump slide and the pump slide seal support into the pocket. Install the pump slide seal. Retain with petroleum jelly. Make sure that the seal is located properly.

22. Position the pump slide with the pump slide pivot hole. Install the pump slide pivot pin.

23. Install a vane ring in the pocket and then the rotor. Install the seven vanes into the pump and make sure that each vane is seated flush with the rotor. Install the top vane ring.

24. Lube and install the O-ring into the pump slide. Over the O-ring, install the oil seal ring.

25. Check the auxiliary valve body sleeve for damage. Install the auxiliary valve body, gasket and cover.

26. Align the pump rotor step with the auxiliary valve body

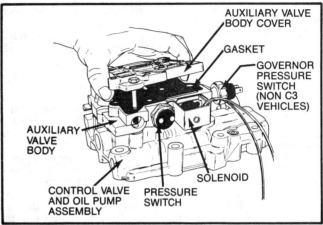

Auxiliary valve body cover alignment (©General Motors Corp.)

sleeve. Install and torque the auxiliary valve body screw to 9 foot pounds.

27. Install the pump primary spring and cup plug, flat side out. Compress the cup plug and install the roll pin.

28. Position the control valve and pump assembly so that the machined side is up and the pump pocket is on the right-hand side. All parts must be clean before assembly.

29. Starting at the lower right-hand side, install the 3-2 control valve, large end first, the spring and the spring retainer. Make sure that the retainer is level with or below the machined surface.

30. Move to the opposite side to the lower left-hand bore, and install the 1-2 shift valve large end first. Place the 1-2 T.V. spring in the 1-2 T.V. bushing and install the 1-2 T.V. valve and bushing assembly into the bore, making sure that the pin slot is aligned with the pin hole in the valve body. Install the roll pin.

31. Still on the left-hand side, move to the next bore up and install the 2-3 shift valve, large end first. Install the 2-3 T.V. spring into the bushing and install the 2-3 T.V. valve into the bushing with the larger end toward the valve body and insert the assembly into the bore. Make sure that the pin slot is aligned with the pin hole in the valve body. Install the roll pin.

32. Move to the next bore up and install the 1-2 accumulator spring. Install the 1-2 accumulator valve into the bushing, small end first. Install the bore plug with the flat side first. Install the assembly into the bore. Make sure that the pin slot is aligned with the pin hole in the valve body. Push in on the plug and install the roll pin.

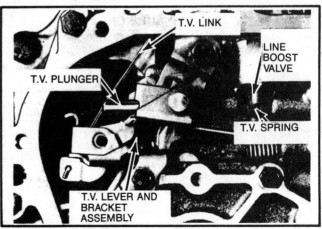

Throttle lever and bracket assembly installation
(©General Motors Corp.)

4-119

33. Move to the next bore up and install the pressure relief ball, the spring and the retainer.

34. Move to the next bore up and install the shift T.V. valve, small end first. Install the spring and bore plug with the flat side first and install the roll pin.

35. Move to the opposite side of the valve body and at the second bore from the top, install the pressure regulator with the small end to the outside, and follow with the spring. Install the reverse boost valve into the bushing and see that the small end is toward the outside of the valve body. Insert the boost valve and bushing assembly into the bore with the open end first. Now install the T.V. boost valve, small (stem) end first into the T.V. boost bushing and then install the assembly into the valve body bore with the open end first. Make sure that the pin slot is aligned with the pin hole in the valve body. Install the roll pin.

36. Move to the top bore on the right-hand side of the valve body and install the throttle valve with the stem end to the outside, and follow with the spring. Insert the T.V. plunger into the T.V. bushing, small end first. Install the T.V. plunger assembly into the bore with the open end first. Make sure that the pin slot that is in the T.V. plunger bushing is aligned with the pin hole in the valve body. Install the roll pin for the bushing, as well as the roll pin that belongs in the throttle valve pin hole.

37. At the top of the valve body, install the line boost valve with the stem end first into the bore. Install the bore plug with the ring end first, and insert the roll pin. Make certain that the line boost valve moves freely.

38. Install the low blowoff ball in the lower right-hand corner of the rough casting side of the valve body. Assemble the blowoff spring and the spring plug. Install the spring and plug assembly with the plug end first, and then install the outer plug with the cupped end first, using a plastic hammer until it seats in its bore.

Transaxle Assembly

1. If the parking pawl or its shaft has been removed, inspect the pawl for cracks before installing it. Place the return spring on the pawl and place into the case. Slide the shaft through the case into the pawl and check for free movement of the pawl. Install the parking pawl shaft retainer clip. With a ⅜″ rod, install the shaft cup plug.

2. If the governor pipe had been removed, coat both ends with Loctite® or its equivalent and install. Install the governor pipe retainer strap and its bolt.

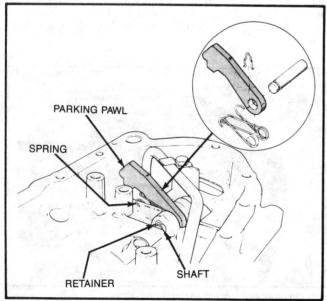

Parking pawl and related components (©General Motors Corp.)

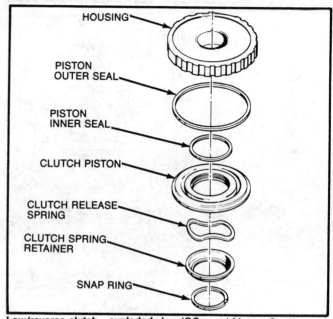

Low/reverse clutch—exploded view (©General Motors Corp.)

3. Reinstall the cup plug if so equipped.

4. The manual shaft seal and the axle seal should be checked and replaced as required.

6. Install the differential and final drive unit. Carefully lower the final drive unit into the transaxle case. Install the final drive internal gear spacer with the cupped side against the final drive internal gear.

NOTE: The spacer has an opening for the parking pawl. Make sure that the parking pawl passes through the spacer freely.

7. Install the final drive spacer snap ring into the case. Install the snap ring in such a way that the snap ring opening is away from the spacer ring gap. The snap ring is .092 inch thick.

8. Install the reaction gear set. Turn the transaxle case so that the case cover side is up.

9. Inspect the final drive sun gear shaft and journals for damage or wear. If the sun gear shaft and its reaction internal gear are usable, install the internal gear on the sun gear shaft.

10. Install the internal gear to sun gear roller thrust bearing so that the inner race goes against the internal gear.

11. The low roller clutch should be disassembled if necessary to inspect the rollers and also to check the planetary pinions for end play. To do this follow these steps:

a. Remove the Reverse clutch-to-low race selective washer. Remove the low roller clutch race and then the clutch cage assembly.

b. Remove the thrust washer that will be found in the reaction carrier. Inspect for excessive wear.

c. Check the planetary pinion end play with feeler gauges. End play should be 0.009″-0.027″.

d. If any of the clutch rollers have come out of the cage, install them by compressing the spring and inserting the roller from the outside. Install the clutch cage assembly into the reaction carrier.

e. Install the roller clutch inner race with the splined side up. It may have to be rotated clockwise to allow it to drop in.

f. Install the four-tanged internal gear thrust washer, using petroleum jelly to hold it in place.

g. Assemble the reaction carrier and clutch assembly onto the internal gear.

12. With the low roller clutch assembled the next item to be installed is the low/reverse clutch housing-to-roller clutch selective washer.

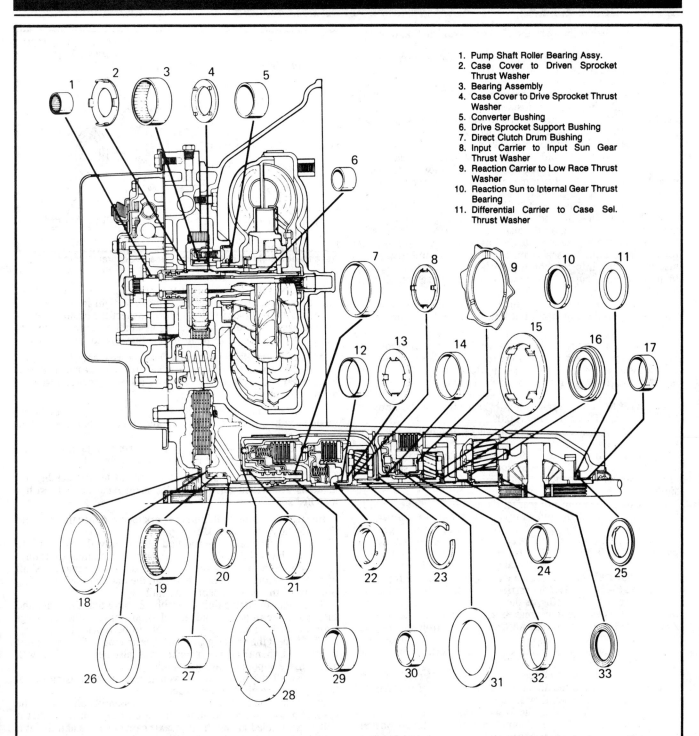

1. Pump Shaft Roller Bearing Assy.
2. Case Cover to Driven Sprocket Thrust Washer
3. Bearing Assembly
4. Case Cover to Drive Sprocket Thrust Washer
5. Converter Bushing
6. Drive Sprocket Support Bushing
7. Direct Clutch Drum Bushing
8. Input Carrier to Input Sun Gear Thrust Washer
9. Reaction Carrier to Low Race Thrust Washer
10. Reaction Sun to Internal Gear Thrust Bearing
11. Differential Carrier to Case Sel. Thrust Washer

12. Input Internal Gear Bushing
13. Input Carrier to Input Int. Gear Thrust Washer
14. Low and Reverse Clutch Housing Bushing
15. Reaction Carrier to Int. Gear Thrust Washer
16. Sun Gear to Internal Gear Thrust Bearing
17. Case Bushing

18. Driven Sprocket Thrust Bearing Assembly
19. Bearing Assembly
20. Selective Snap Ring
21. Direct Clutch Bushing
22. Input Shaft Thrust Washer
23. Selective Snap Ring
24. Final Drive Internal Gear Bushing
25. Differential Carrier to Case Thrust Brg. Assy.

26. Driven Sprocket Support Thrust Washer
27. Input Shaft Bushing
28. Thrust Washer
29. Driven Sprocket Support Bushing
30. Reaction Sun Gear Bushing
31. Reverse Housing to Low Race Selective Washer
32. Reaction Carrier Bushing
33. Sun Gear to Carrier Thrust Bearing

Thrust washer and bushing locations (©General Motors Corp.)

13. Grasp the assembly by the final drive sun shaft and install into the case.

14. The components making up the low and reverse clutch are assembled next. Inspect the backing plate for cracks and if it is usable, install into the case with the stepped side down.

15. New clutch plates should have been soaking in Dexron® II for at least 20 minutes before installation. Install the plates starting with a friction (composition) plate and following with a steel plate until the clutch pack has been installed.

16. Install the low and reverse spacer ring (0.0042″ thick).

17. The low/reverse clutch piston seals should be replaced. Begin by compressing the spring retainer and removing the snap ring. Then disassemble the spring retainer and remove the waved spring and piston. Discard the seals. Remove the apply ring.

18. Inspect all parts especially the housing for a plugged feed hole. Check the bushing and the piston for damage. Assemble by installing the apply ring on the piston and then by installing new seals. The seals are installed with the lips facing away from the apply ring side. Lube the seals with petroleum jelly and install the piston carefully into the housing. Install the waved spring and then the retainer, cupped side down. Compress the assembly and install the snap ring.

19. Lower the low/reverse clutch assembly into the case. Be sure to line up the clutch oil feed hole with the feed hole in the case.

NOTE: If it is difficult to get the low/reverse clutch housing past the snap ring groove, install the reaction sun gear to use as a tool. Rotate the sun gear while pushing down on the clutch housing until the housing drops below the snap ring groove.

20. With the low/reverse clutch in place, install the housing case snap ring. This snap ring is 0.092″ thick.

21. With the reaction sun gear in place, install the selective snap ring.

22. Inspect the input drum for damage. Note that there is an angle on the roll pins. This is normal and no attempt should be made to straighten them. Install the input drum on the reaction sun gear.

23. Inspect the input sun gear for damage or excessive wear. If it is usable, install the input sun gear into the input drum.

24. Inspect the input planetary carrier for wear or damage. The end play of the pinions should be checked with a feeler gauge. End play should be 0.009″-0.027″. Also check that the pinion pins are tight and that the pinion gears rotate freely. Check the two thrust washers that are used with the planetary gear. One is the input internal gear-to-input planetary carrier (it has four tangs and is the larger of the two). The other is the input planetary carrier-to-input sun gear (also has four tangs, but the smaller of the two).

25. Install both the thrust washers using petroleum jelly to hold them in place.

26. Install the input planetary carrier into the case. Install the input internal gear.

27. Place the forward clutch assembly on the bench with the input shaft up. Install the direct clutch assembly over it and onto the forward clutch. When they are properly seated, it will be 1⁷/₃₂ inches from the tang end of the forward clutch drum.

28. Install the input shaft-to-input internal gear thrust washer with the rounded side against the input shaft, and the stepped side facing out. Petroleum jelly will hold it in place.

29. Install the direct and forward clutch assemblies into the case. It may prove helpful to rotate the unit when installing rather than pushing down on the unit. To make sure that the assembly is properly installed, measure with a steel rule from the case face to the direct clutch housing. It should be approximately 1¹¹/₁₆″ (42mm).

30. Install the intermediate band. Make sure that the eye end is located in the case and that the lug end is aligned with the apply pin bore, otherwise the band will not operate.

31. Install the band anchor hole plug.

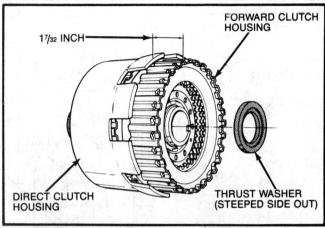

Assembled height of clutch assemblies (©General Motors Corp.)

32. Inspect the driven sprocket support for wear and pay close attention to the support sleeve. It must be tight in its bore and aligned with the holes in the support. Inspect the bushing for excessive wear as well as the bearing assemblies. Inspect the seal rings but do not remove them unless they are to be replaced.

NOTE: If the driven sprocket bearing assembly is to be replaced, the race on the driven sprocket must also be checked. If it is necessary to replace the race, the driven sprocket assembly must be replaced.

33. Install the direct clutch drum-to-driven sprocket thrust washer with plenty of lubricant and then install the driven sprocket support.

34. Inspect the manual shaft parts for wear. Slight raised edges on the detent lever can be removed with a fine file.

NOTE: The manual shaft and the detent lever assembly are made as a matched set. Any replacement must be done with a matched set.

Install the manual shaft and the park lock rod into the case and through the driven sprocket support.

35. If the manual valve rod had been removed, install it into the detent lever. Place the detent lever on the manual shaft hubside away from the sprocket support and push the manual shaft into place. With a small punch, install the manual shaft retaining pin. Install the manual shaft case nail.

36. Install the drive link assembly. Be sure that the gears and their parts are in good condition. Do not remove the seal rings from the turbine shaft unless they are to be replaced.

37. Inspect the driven gear thrust bearing race. If the race must be replaced, then the drive sprocket must also be replaced, as well as the drive support bearing assembly.

38. Install the drive and driven thrust washers using petroleum jelly to retain them in place.

39. Install the sprockets into the link assembly and locate the colored guide link which should have numbers on it, so that it will be assembled facing the transaxle case cover. Install the unit into the transaxle case.

40. Install the roller bearing thrust washer on the driven sprocket with the outer race against the sprocket.

41. Inspect the case cover for cracks or any damage. Check the bolt holes for damaged threads and see that the vent assembly is not damaged. If the vent must be replaced, use Loctite® or its equivalent on the replacement vent and tap into place with a plastic hammer. Coolant line connectors need not be removed unless they are to be replaced. Coat the threads of any replacement connector with thread sealer.

42. Install the manual detent spring and roller assembly and torque the retaining screw to 8 foot-pounds.

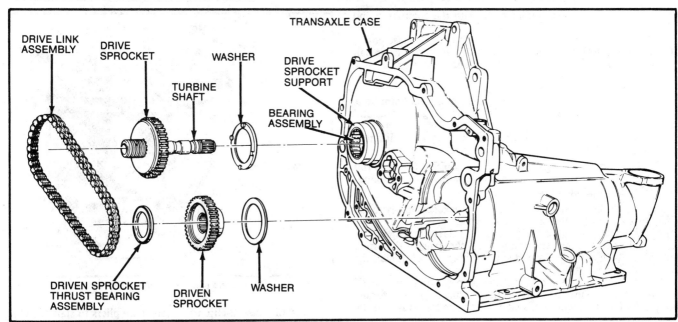

Drive link assembly and related components (©General Motors Corp.)

43. Inspect the axle oil seal and replace if necessary.

44. Inspect the case cover sleeve. Make sure that the hole in the sleeve aligns with the case cover passages that intersect the case cover pump shaft bore.

45. Inspect the manual valve. If it needs to be replaced, the cup plug will have to be removed so that the valve can be slid out. When installing a manual valve, be sure that the small end goes in first. Coat the cup plug with sealant before installing and tap it in until just below the surface.

NOTE: For a brief period, a production trial run was made using a steel manual valve instead of aluminum. The steel valve has an extra locating land on the shaft. If service replacement is required of the manual valve, the present design aluminum valve should be used.

46. Renew the seal in the 1-2 accumulator piston. Otherwise the seal should not be removed. Install the accumulator piston and make sure that it is installed with the flat side down.

47. Inspect the thermo-element valve and plate. If the valve has to be replaced, be sure that the roll pin and washers are installed to their specific heights. Carefully measure the pins, their height should be .21 inch.

48. To remove the valve use a pair of long nosed pliers and pry off the washer from the spring pin. Remove the element. Inspect the element and replace as required.

49. Install the original thermo-element and plate. The roll pin washers are to be installed to a specification height of exactly 5.24mm (0.21").

50. Install the tanged thrust washer on the case cover, and retain with petroleum jelly. This thrust washer protects the driven sprocket. If the two dowel pins were removed from the case cover, they should now be installed.

51. Install the 1-2 accumulator spring into the case. Install the center gasket on the case, using petroleum jelly to retain it.

52. Using a new gasket, install the case-to-case cover. All screws are torqued to 18 foot-pounds.

53. There are two screws that are installed from the torque converter side and they should be installed, with new washers, and torqued to 18 foot-pounds.

54. Install the manual valve rod into the manual valve and install the clip. The rod may have to be pulled up in place with long nose pliers.

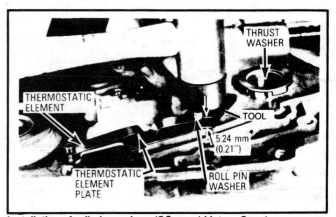

Installation of roll pin washers (©General Motors Corp.)

Transaxle case cover bolt locations (©General Motors Corp.)

4–123

55. Install the 1-2 accumulator pin with the chamfered end first.

56. Install the five check balls that belong in the case cover. Lay a new control valve pump body gasket in place, and install the spacer plate. Install the sixth check ball into the control valve and pump assembly.

57. Install the oil pump drive shaft.

58. It is helpful to have two guide pins made from 6mm bolts to help guide the valve body into place. Use the guide pins and install the valve body.

59. Install the valve body retaining bolts. They are of different lengths and different torque specifications.

NOTE: There is a capped roll pin in the center of the two roll pins by the element plate. If this pin is removed, it must be installed to a height of .24 inch.

60. Install the throttle valve lever and bracket assembly to the T.V. link and install the unit onto the case.

61. Inspect the valve body cover for damage, especially around the gasket flange and bolt holes. If the flange is distorted, straighten with a block of wood and a rubber mallet. Install the pan, using a new gasket.

62. Rotate the transaxle so that the oil pan side is up. Install the output shaft. Turn the shaft carefully so that the groove for the C-ring on the shaft will be visible through the "window" in the

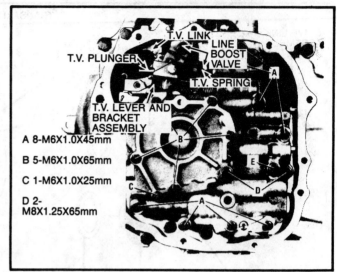

A 8-M6X1.0X45mm

B 5-M6X1.0X65mm

C 1-M6X1.0X25mm

D 2-M8X1.25X65mm

Valve body bolt locations (©General Motors Corp.)

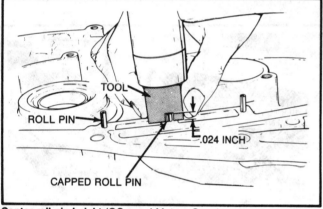

Center roll pin height (©General Motors Corp.)

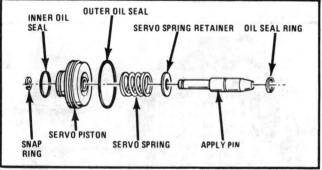

Intermediate servo—exploded view (©General Motors Corp.)

case. Install a new C-clip, using long nose pliers to start the clip and then a long tool to push the clip home. Make sure that it is seated in the groove.

63. Install the parking lock bracket and dipstick stop. Make sure that the parking actuator rod is positioned over the pawl and out of the park range. Torque the two retaining screws to 18 foot-pounds. Check the assembly for proper operation.

64. Install a new low/reverse seal assembly, using a piece of ⅜ inch rod as a driver.

65. Renew the O-ring seal in the reverse oil pipe and install the pipe with the plain end first, then the end with the seal.

66. Install the intermediate servo assembly. The seals should not be removed unless they are to be replaced. Make sure that the seals are seated in their grooves.

67. Install the servo cover with a new gasket, and install the three cover screws. Position the reverse oil pipe bracket and torque the screws to 8 foot-pounds.

68. Install a new oil strainer and O-ring seal. The screen should be positioned against the dipstick stop.

69. Install the pan using a new gasket or RTV sealant depending on which was used upon assembly.

70. Install the governor assembly. Install the speedometer drive gear and thrust washer on the governor. Renew the O-ring seal in the governor cover, lube with petroleum jelly and install the cover and retaining bolts.

NOTE: Make certain that the governor shaft is centered in the cover and not misaligned before tightening the bolts.

71. Install the speedometer driven gear assembly and retainer. Torque the screw to 75 inch-pounds.

72. Place the transaxle onto the jack and secure. Install the torque converter and see that it is fully seated and that all shafts are engaged.

73. Install the transaxle into the vehicle. Fill with the proper grade and type automatic transmission fluid. Adjust linkage as required. Roadtest the vehicle.

SPECIFICATIONS

1982 THM 125C AUTOMATIC CLUTCH PLATE AND APPLY RING CHART
Cimarron, Cavalier and 2000

Direct Clutch			Forward Clutch				Low & Reverse Clutch		
Flat Steel Plate	Comp. Faced Plate	Apply Ring	Waved Plate	Flat Steel Plate	Comp. Faced Plate	Apply Ring	Flat Steel	Comp. Faced Plate	Apply Ring
No. 4	No. 4	I.D.①7	No. 1	No.3	No. 4	I.D.①O	No. 5	No.5	I.D.①O or A
Thickness 2.3mm (0.09")		Width 19mm (0.75")	Thickness 1.6mm (0.06")	Thickness 1.9mm (0.08")		Width 12mm (0.47")			Width 15.4mm (0.61")

① Measure the width of the clutch apply ring for positive identification.

NOTE: The direct and forward clutch flat steel clutch plates and the forward clutch waved steel plate should be identified by their thickness. The direction of the forward production installed composition faced plates must not be interchanged. For service, the direct and forward clutch use the same composition faced plates.

1982 THM 125C AUTOMATIC TRANSAXLE CLUTCH PLATE AND APPLY RING CHART
Firenza and Skyhawk

Direct Clutch				Forward Clutch			Low & Reverse Clutch		
Flat Steel Plate	Comp. Faced Plate	Apply Ring	Waved Plate	Flat Steel Plate	Comp. Faced Plate	Backing Plate	Waved Plate	Flat Steel Plate	Comp. Faced Plate
No. 4	No. 4	I.D.①7	No. 1	No. 3	No. 4	I.D.②	No.1	No.4	No. 5
Thickness 2.3mm (0.09")		Width 19mm (0.75")	Thickness 1.6mm (0.06")	Thickness 1.9mm (0.08")		Width ②			

① Measure the width of the clutch apply ring for positive identification.
② Backing plates are selective. Code 1 is .24-.23
Code 2 is .21-.20
Code 3 is .19-.18

NOTE: The direct and forward clutch flat steel clutch plates and the forward clutch waved steel plate should be identified by their thickness. The direction of the forward production installed composition faced plates must not be interchanged. For service, the direct and forward clutch use the same composition faced plates.

1982 THM 125C AUTOMATIC TRANSAXLE CLUTCH PLATE AND APPLY RING CHART
Century, Skylark, Citation, Celebrity, Omega, Ciera, Phoenix and 6000

Direct Clutch			Forward Clutch				Low & Reverse Clutch			
Flat Steel Plate	Comp. Faced Plate	Apply Ring	Waved Plate	Flat Steel Plate	Comp. Faced Plate	Backing Plate	Waved Plate	Flat Steel Plate	Comp. Faced Plate	Apply Ring
No. 4	No. 4	I.D.①7	No. 1	No. 3	No. 4	I.D.②	No.1	No. 4	No. 5	I.D.①O or A
Thickness 2.3mm (0.09")		Width 19mm (0.75")	Thickness 1.6mm (0.06")	Thickness 1.9mm (0.08")		Width ②				Width 15.4mm (0.61")

① Measure the width of the clutch apply ring for positive identification.
② Backing plates are selective. Code 1 is .24-.23
Code 2 is .21-.20
Code 3 is .19-.18

NOTE: The direct and forward clutch flat steel clutch plates and the forward clutch waved steel plate should be identified by their thickness. The direction of the forward production installed composition faced plates must not be interchanged. For service, the direct and forward clutch use the same composition faced plates.

1983 and Later THM 125C AUTOMATIC TRANSAXLE CLUTCH PLATE AND APPLY RING
All Models

Clutch	Flat Steel Plate No.	Flat Steel Plate Thickness	Comp. Faced Plate No.	Waved Plate No.	Waved Plate Thickness	Backing Plate I.D.①	Backing Plate Width
Direct EF, EK, EP	3	2.3mm (0.09")	3	—	—	—	9.25mm (0.36")
All Others	4	2.3mm (0.09")	4	—	—	—	4.92mm (0.19")
Forward All	3	1.9mm (0.08")	4	1	1.6mm (0.06")	—	②
Low and Reverse All	4	—	5	1	1.94mm (0.08")	—	—

① Measure the width of the clutch apply ring for positive identification.
② Backing plates are selective. Code 1 is .24-.23
    Code 2 is .21-.20
    Code 3 is .19-.18

NOTE: The direct and forward clutch flat steel clutch plates and the forward clutch waved steel plate should be identified by their thickness. The direction of the forward production installed composition faced plates must not be interchanged. For service, the direct and forward clutch use the same composition faced plates.

TORQUE SPECIFICATION CHART

Description	Quantity	Torque Specification
Valve body to Case Cover	2	8 ft. lbs. (11 N•m)
Pump Cover to Case Cover	1	18 ft. lbs. (24 N•m)
Pump Cover to Valve Body	4	8 ft. lbs. (11 N•m)
Pump Cover to Valve Body	3	8 ft. lbs. (11 N•m)
Solenoid to Valve Body	1	8 ft. lbs. (11 N•m)
Valve Body to Case Cover	9	8 ft. lbs. (11 N•m)
Valve Body to Case	1	18 ft. lbs. (24 N•m)
Valve Body to Driven Sprocket Support	1	18 ft. lbs. (24 N•m)
Case Cover to Case	4	18 ft. lbs. (24 N•m)
Case Cover to Case	4	18 ft. lbs. (24 N•m)
Case Cover to Case	1	18 ft. lbs. (24 N•m)
Case Cover to Case	7	18 ft. lbs. (24 N•m)
Case Cover to Case	2	18 ft. lbs. (24 N•m)
Case to Drive Sprocket Support	4	18 ft. lbs. (24 N•m)
Oil Pan and Valve Body Cover	27	12 ft. lbs. (16 N•m)
Manual Detent Spring Assembly to Case	1	8 ft. lbs. (11 N•m)
Cooler Connector	2	23 ft. lbs. (38 N•m)
Line Pressure Take-Off	1	8 ft. lbs. (11 N•m)
Intermediate Servo Cover	4	8 ft. lbs. (11 N•m)
Parking Lock Bracket to Case	2	18 ft. lbs. (24 N•m)
Pipe Retainer to Case	2	18 ft. lbs. (24 N•m)
Governor Cover to Case	2	8 ft, lbs. (11 N•m)
Speedometer Driven Gear to Governor Cover	1	75 in. lbs. (9 N•m)
T.V. Cable to Case	1	75 in. lbs. (9 N•m)
Pressure Switch	2	8 ft. lbs. (11 N•m)

SPECIAL TOOLS

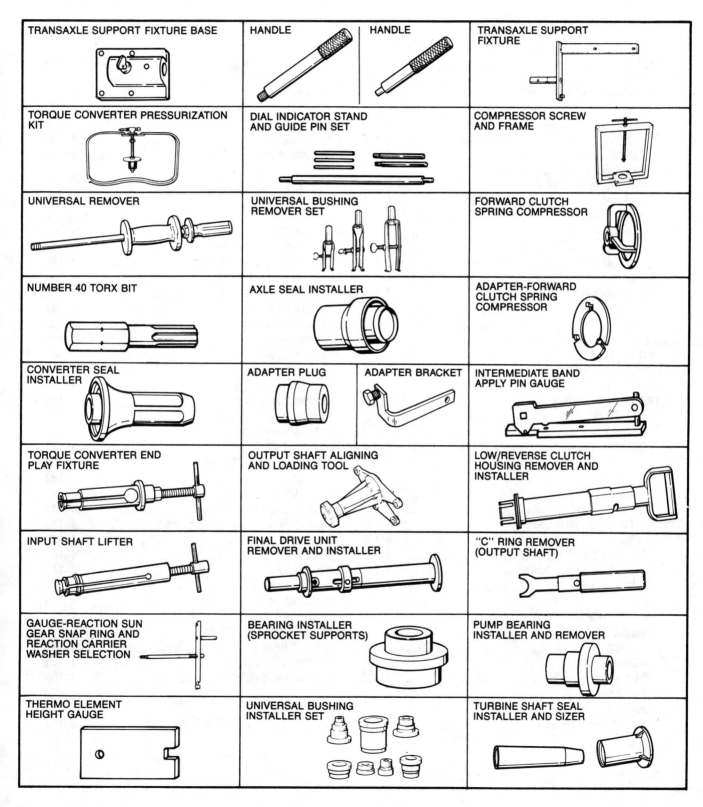

TRANSAXLE SUPPORT FIXTURE BASE	HANDLE	HANDLE	TRANSAXLE SUPPORT FIXTURE
TORQUE CONVERTER PRESSURIZATION KIT	DIAL INDICATOR STAND AND GUIDE PIN SET		COMPRESSOR SCREW AND FRAME
UNIVERSAL REMOVER	UNIVERSAL BUSHING REMOVER SET		FORWARD CLUTCH SPRING COMPRESSOR
NUMBER 40 TORX BIT	AXLE SEAL INSTALLER		ADAPTER-FORWARD CLUTCH SPRING COMPRESSOR
CONVERTER SEAL INSTALLER	ADAPTER PLUG	ADAPTER BRACKET	INTERMEDIATE BAND APPLY PIN GAUGE
TORQUE CONVERTER END PLAY FIXTURE	OUTPUT SHAFT ALIGNING AND LOADING TOOL		LOW/REVERSE CLUTCH HOUSING REMOVER AND INSTALLER
INPUT SHAFT LIFTER	FINAL DRIVE UNIT REMOVER AND INSTALLER		"C" RING REMOVER (OUTPUT SHAFT)
GAUGE-REACTION SUN GEAR SNAP RING AND REACTION CARRIER WASHER SELECTION	BEARING INSTALLER (SPROCKET SUPPORTS)		PUMP BEARING INSTALLER AND REMOVER
THERMO ELEMENT HEIGHT GAUGE	UNIVERSAL BUSHING INSTALLER SET		TURBINE SHAFT SEAL INSTALLER AND SIZER

INDEX

GENERAL MOTORS TURBO HYDRA-MATIC 180C

APPLICATIONS

GENERAL MOTORS 180C AUTOMATIC TRANSMISSION APPLICATION CHART

Year	Make	Model
1982	Chevrolet	Chevette
	Pontiac	T-1000
1983-84	Chevrolet	Chevette
	Pontiac	1000

GENERAL DESCRIPTION

Transmission and Torque Converter Identification

The THM 180C automatic transmission is a fully automatic unit consisting mainly of a four element hydraulic torque converter and a compound planetary gear set. Three multiple disc clutches, a roller clutch, and a band provide the friction elements required to obtain the desired function of the compound planetary gear set. The THM 180C automatic transmission also utilizes a hydraulic system pressurized by a gear type pump which provides the working pressure required to operate the friction elements and the automatic controls.

Transmission Identification

The THM 180C automatic transmission can be identified by the Vin Location Tag, which is positioned on the right side of the automatic transmission case. This tag incorporates the model and year of the unit, the serial number of the transmission and the transmission part number. There is also an I.D. Location Tag, which is positioned on the left side of the assembly, just after the torque converter housing.

Torque Converter Identification

The torque converter is a welded unit and cannot be disassembled. No specific data is given for quick identification by the repairman. Should the torque converter require repair, order a replacement unit from the information given on the transmission identification tag.

Metric Fasteners

Metric tools will be required to service this unit. Due to the large number of alloy parts used, torque specifications should be strictly observed. Before installing capscrews into aluminum parts, always dip the screws into oil to prevent the screws from galling the aluminum threads and to prevent seizing. Aluminum castings and valve body parts are very susceptible to nicks and burrs.

The metric fastener dimensions are very close to the dimensions of the familiar inch system fasteners, and for this reason, replacement fasteners must have the same measurement and strength as those removed.

Do not attempt to inter-change metric fasteners for inch system fasteners. Mismatched or incorrect fasteners can result in damage to the transmission unit through malfunctions, breakage or possible personal injury.

Care should be taken to re-use the fasteners in the same locations as removed.

Fluid Capacities

To refill the THM 180C automatic transmission after complete overhaul, add five quarts of Dexron® II automatic transmission

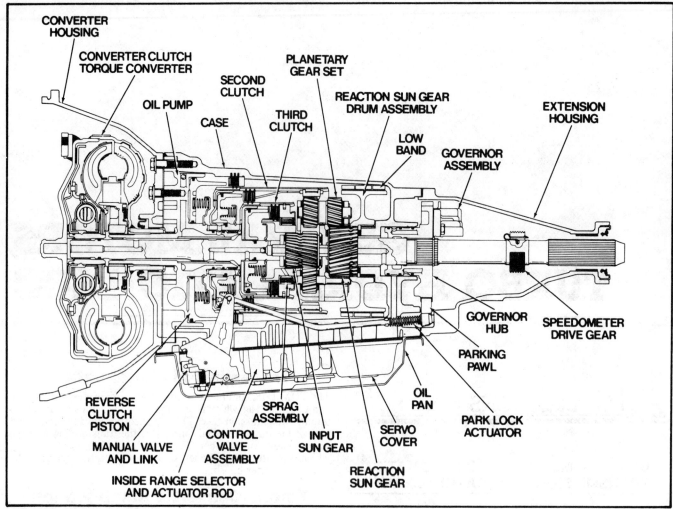

THM 180C automatic transmission— (©General Motors Corp.)

fluid. To refill the 180C automatic transmission after the pan has been removed and replaced, add three quarts of Dexron® II automatic transmission fluid. Recheck and correct level after starting engine.

Checking Fluid Level

TRANSMISSION AT OPERATING TEMPERATURE (HOT)

The THM 180C automatic transmission is designed to operate at the "FULL HOT" mark on the dipstick, at normal operating temperature. Normal operating temperature is obtained after at

least fifteen miles of highway type driving. To determine the proper fluid level, proceed as follows.

1. Make sure vehicle is parked level.
2. Apply parking brake and move selector lever to Park.
3. Start engine but do not race. Allow to idle.
4. Move selector through each range, back to Park, then check level. The fluid should read "FULL HOT".
5. Correct the fluid level as required.

TRANSMISSION AT ROOM TEMPERATURE (COLD)

Automatic transmissions are frequently overfilled because the fluid level is checked when the fluid is cold and the dipstick indicates fluid should be added. However, the low reading is normal since the level will rise as the fluid temperature increases. To determine the proper fluid level, proceed as follows.

1. Make sure that the vehicle is parked level.
2. Apply the parking brake. Move the selector lever into the park position.
3. With the engine running, remove the dipstick, wipe it clean and reinsert it until the cap seats.
4. Remove the dipstick and note the reading. It should be between the two dimples below the "ADD" mark.
5. Adjust the fluid level as required.

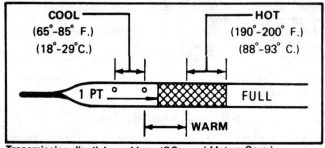

Transmission dipstick markings (©General Motors Corp.)

Fluid Drain Intervals

The transmission operating temperature resulting from the type of driving conditions under which the car is used is the main consideration in establishing the proper frequency of transmission fluid changes.

Change the transmission fluid and service the screen every 15,000 miles (24 000 km) if the car is usually driven under one or more of the following conditions.
1. In heavy city traffic.
2. Where the outside temperature regularly reaches 90°F. (32°C).
3. In very hilly or mountainous areas.
4. Commercial use, such as taxi, police car, or delivery service.

If you do not use your car under any of these conditions, change the fluid and service screen every 100,000 miles (160 000 km).

NOTE: DO NOT OVERFILL. It takes only one pint to raise lever from "ADD" to "FULL" with a hot transmission.

MODIFICATIONS

Information regarding any modifications to the THM 180C automatic transmission is not available at the time of publication.

TROUBLE DIAGNOSIS

CLUTCH AND BAND APPLICATION CHART

Range	Reverse Clutch	Second Clutch	Third Clutch	Low Band	Sprag
Park/Neutral	Released	Released	Released	Released	Locked
Drive range—first gear	Released	Released	Released	Applied	Locked
Drive range—second gear	Released	Applied	Released	Applied	Overrunning
Drive range—third gear	Released	Applied	Applied	Released	Locked
L₁ range	Released	Released	Applied	Applied	Locked
L₂ range	Released	Applied	Released	Applied	Overrunning
Reverse range	Applied	Released	Applied	Released	Locked

CHILTON'S THREE "C's" TRANSMISSION DIAGNOSIS CHART
THM 180C

Condition	Cause	Correction
Won't move in any range	a) Low fluid level b) Clogged filter screen c) Manual valve disconnected d) Input shaft broken e) Pressure regulator valve stuck open f) Faulty oil pump	a) Add as required b) Replace screen c) Repair link d) Replace affected parts e) Clean valve and bore f) Repair or replace
Must "jiggle" selector to move	a) Manual linkage adjustment b) Selector shaft retaining pin broken c) Manual valve link loose d) Selector shaft nut loose	a) Adjust as required b) Replace pin c) Repair or replace link d) Tighten or replace
Sudden start after RPM increase	a) Servo piston binding b) Low fluid level c) Faulty oil pump d) Check ball in valve body out of place	a) Repair or replace servo b) Add as required c) Repair or replace d) Remove valve body and check for proper locations of check balls

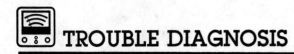

CHILTON'S THREE "C's" TRANSMISSION DIAGNOSIS CHART
THM 180C

Condition	Cause	Correction
Heavy jerking when starting out	a) Low oil pressure	a) Check fluid level, clogged oil screen. Replace if necessary
	b) Pressure regulator valve stuck	b) Clean valve and bore
	c) Check ball in valve body out of place	c) Remove valve body and check for proper location of check balls
Won't move in "D" or "L2". OK in "L1" and "R"	a) Input sprag failure	a) Replace affected parts
Won't move in "D", "L2" or "L1". OK in "R"	a) Faulty band	a) Replace band
	b) Servo piston binding	b) Repair or replace servo
	c) Parking pawl did not disengage	c) Repair faulty parking lock parts
Won't move in "R". OK in other ranges	a) Reverse clutch failure	a) Overhaul
Car moves in "N"	a) Manual linkage adjustment	a) Adjust as required
	b) Planetary gear set broken	b) Replace planetary and other parts if affected
	c) Faulty band	c) Check for proper position and adjustment
No 1-2 upshift (Stays in 1st gear)	a) Governor valves stuck	a) Service or replace governor
	b) 1-2 Shift valve stuck	b) Clean valve body
	c) Faulty oil pump hub seals	c) Replace hub seals
	d) Governor screen clogged	d) Remove and clean screen
No 2-3 upshift (Stays in 2nd gear)	a) 2-3 Shift valve stuck	a) Clean valve body
	b) Governor circuit leaking	b) Check governor seals
Upshifts at full throttle only	a) Faulty vacuum modulator	a) Replace modulator
	b) Vacuum leak	b) Check engine and accessories for vacuum leak
	c) Detent cable broken	c) Replace cable
	d) Detent valve stuck	d) Clean valve and bore
Upshifts only at part throttle	a) Detent pressure regulator valve stuck	a) Clean valve body
	b) Detent cable broken or out of adjustment	b) Repair or replace as necessary
Transmission keeps slipping to a lower gear	a) Manual linkage broken	a) Replace affected parts
	b) Pressure loss at governor	b) Replace governor seals
Slips during 1-2 upshift	a) Low oil pressure	a) Check fluid level, clogged oil screen. Replace if necessary
	b) Second clutch seals leak	b) Overhaul
	c) Faulty oil pump hub seals	c) Replace hub seals
Slips during 2-3 upshift	a) Low oil pressure	a) Check fluid level, clogged oil screen. Replace if necesary
	b) Faulty band	b) Adjust or replace
	c) Third clutch leaking	c) Replace seals
	d) Input shaft bushing worn	d) Replace bushing
	e) Check ball in valve body out of place	e) Remove valve body and check for proper locations of check balls

CHILTON'S THREE "C's" TRANSMISSION DIAGNOSIS CHART
THM 180C

Condition	Cause	Correction
Abrupt 1-2 upshift	a) High oil pressure	a) Check for stuck pressure regulator-Clean bore
	b) 1-2 Accumulator stuck	b) Clean valve and bore
	c) Second clutch spring broken	c) Overhaul clutch-Replace affected parts
Abrupt 2-3 upshift	a) High oil pressure	a) Check for stuck pressure regulator-Clean bore
	b) Faulty band	b) Adjust or replace
No engine braking in "L1"	a) Manual linkage adjustment	a) Adjust as required
	b) Manual low control valve stuck	b) Clean valve body
No engine braking in "L2"	a) Manual linkage adjustment	a) Adjust as required
Screeching noise when starting	a) Converter failure	a) Replace converter

HYDRAULIC CONTROL SYSTEM

The THM 180C automatic transmission uses a gear type pump to draw fluid through the filter screen in order to supply fluid to the different hydraulic units and valves. The pump drive gear is keyed to the converter pump hub. Since the converter is attached to the engine by means of a flex-plate, whenever the engine is turning, the oil pump in the automatic transmission is turning.

The hydraulic control system can be divided into four major types of elements.

1. Pressure regulating valves. These would include the main pressure regulating valve, the modulator valve, the detent pressure regulator valve, the 1-2 accumulator valve and the governor.

2. Selector valves (both manually and hydraulically controlled). These would include the manual valve, detent valve, 1-2 shift valve, 2-3 shift valve, 3-2 downshift control valve, the manual low and reverse control valve and the boost control valve.

3. Timing valves. These would include the low speed downshift timing valve, the high speed downshift timing valve and the second clutch orifice valve.

4. The accumulators. These would be the 1-2 accumulator and the low servo piston.

Main Pressure Regulator Valve

Transmission fluid under pressure from the oil pump is delivered to the line port of the main pressure regulator valve. This port is connected through a damping orifice to the regulator port at the end of the pressure regulator valve. As the pressure in this port increases, it moves the valve against the spring force until the second spool of the valve just opens to the line port. This allows the pump pressure to be bypassed into the pump suction passage. In this way, the valve will regulate at a fixed minimum pressure as determined by the spring force, and all excess pump delivery oil will be bypassed back into the pump suction passage. By moving from its "bottomed" position to the pressure regulating position, the valve also opens line pressure to the converter feed passage. This oil is then directed through the converter, through the oil cooler, to the gear box lubrication system, and then back to the oil sump.

It is desirable to have a variable line pressure for the clutches and band that will increase with engine torque. This is done by introducing a "modulator" pressure on the end of the boost valve. The force of the boost valve acts against the end of the regulator valve and increases the line pressure above the base pressure as established by the spring force. By allowing this line pressure to come into the stepped area between the spools of the boost valve, an additional pressure increase, over and above that previously described, is obtained. The regulated line pressure is then fed to the manual valve, the modulator valve and the detent pressure regulator valve.

Vacuum Modulator And Modulator Valve

Line pressure is brought to the second port of the modulator valve. This pressure passes between the spools of the valve and into the modulator port. The modulator port is connected to the regulating port at the end of the valve through a damping orifice. As the pressure in the regulating port increases, it moves the valve outward against the spring force of the modulator assembly until the end spool of the valve just closes the line port. If there is excess pressure building up in the regulator port, the valve will continue to move until the second spool just opens to the exhaust port. In this way, the valve tends to regulate between the line and exhaust ports.

Since the modulator spring force is a constant, thereby causing the modulator valve to regulate at a fixed pressure, as the car speed increases, the pressure requirements decrease. It is for this reason that governor pressure, which is a function of car speed, is directed to the area between the two different diameter spools at the outboard end of the valve. As the pressure increases, it creates an outward force on the modulator valve and in effect reduces the spring force of the modulator assembly. In addition, higher car speeds will produce a somewhat lower modulator and line pressure for any given vacuum, by virtue of the governor pressure acting on the modulator valve. Modulator pressure is then directd to the pressure regulator boost valve, the 1-2 shift control valve, the 2-3 shift control valve by way of the 3-2 control valve, the detent valve, the 1-2 acccumulator valve and the low speed downshift timing valve.

Accumulator Valve

The accumulator valve is used to establish a desired pressure to eventually control the rate of apply of the second clutch during the 1-2 upshift. Again, the regulating action is essentially the same as for the modulator valve or detent pressure regulating valve. The ports and spools operate in the same way. However, for increased engine torque, it is necessary to increase the accumulator pressure. This is done by introducing modulator pressure to the small end of the accumulator valve. As the modulator pressure increases, it adds to the spring force and increases the accumulator pressure. The accumulator pressure is fed to the bottom, spring-loaded side, of the accumulator piston.

Governor

The governor assembly is mounted on the output shaft. It contains two interconnected regulating valves. Its purpose is to supply an oil pressure that is a function of the output shaft speed and therefore, a function of the vehicle speed. Line pressure is supplied to the governor from the manual valve. The governor operates on the principle of centrifugal force. Line pressure is directed to the outer-most port of the secondary valve. The secondary spring holds the valve in an outward position so that the outer spool of the valve is open to line pressure. As the line pressure builds up between the spools, it exerts a force on the larger diameter inner spool to start counteracting the spring. When the hydraulic force is large enough, it moves the valve inward against the spring force until the outer spool closes the line port. If the pressure between the spools still creates a force larger than the spring force, the valve will continue to move inward until the excess pressure opens to the exhaust port. The valve then regulates between the line and the exhaust port. A fixed governor pressure in the secondary valve has now been established with no rotation of the output shaft. As the governor begins to rotate, the outward force (due to the weight of the secondary valve) is added to the force of the spring. Therefore, as the speed increases, the outward force and, in turn, the secondary valve pressure increases. The secondary valve pressure is directed to the feed port of the primary valve. With no rotation of the governor, the pressure acts against the large inner spool and forces it to open to the exhaust port. Since there is no spring force on the primary valve, it will continue to keep the feed port closed and the exhaust port open. The final governor pressure is then zero. As the governor begins to rotate, the weight of the primary valve creates an outward force working against the oil pressure. The pressure in the primary valve port now increases as a function of speed. This continues up to the speed where the outward force finally holds the primary valve outward, keeping the feed port open.

In summary, at zero speed, the governor pressure is zero. As the speed increases, the governor pressure will increase as dictated by the primary valve until the speed is great enough to hold the primary valve all the way out. At speeds above this point, governor pressure is established by the secondary valve. Governor pressure is then directed to the modulator valve, the 1-2 shift valve, the 2-3 shift valve, and the high speed downshift timing valve.

Manual Valve

The manual valve has a direct mechanical hook-up to the shift linkage. Its purpose is to direct the hydraulic pressure to the various circuits to establish the base hydraulic range of the transmission. Line pressure is fed to the manual valve. In the Park and Neutral positions, the valve seals line pressure from entering any of the circuits. At the same time all circuits are open to exhaust, so that the transmission remains in a neutral condition. In the Reverse position, line pressure is directed to the reverse clutch piston, the boost control valve and the reverse and manual control valves. All of the other manual control circuits are open to ex-

haust. In the Drive position, the manual valve directs oil to the governor, the 1-2 shift valve, the 1-2 accumulator valve and also to the apply side of the low servo piston by way of the high speed downshift timing valve. The Reverse, Second or Intermediate, and Low ports are exhausted. In Second or Intermediate, the Drive circuits remain pressurized. In addition, pressure is supplied to the boost control valve and to the 2-3 shift valve. The Reverse and Low ports are exhausted. In Low, the pressure is supplied to the 1-2 shift valve and to the reverse and manual control valve, in addition to the circuits already pressurized in Drive and Second. The Reverse port is exhausted.

Detent Valve

The purpose of the detent valve is to cause the transmission to shift to a lower gear for additional performance when the accelerator is depressed all the way down. The detent valve is connected to the throttle linkage mechanically. A spring holds the detent valve in the retracted position. Two pressures, "detent regulator" and "modulator" are supplied to the detent valve. In the retracted or part-throttle position, the detent valve directs modulator pressure to the 1-2 and the 2-3 shift control valves and to the 3-2 control valve. In the full throttle position, modulator pressure is blocked and the passages that were receiving modulator pressure now receive detent regulator pressure. In this position, detent regulator pressure is also supplied to additional ports of the 1-2 and 2-3 shift control valves and the 3-2 control valve.

1-2 Shift Valve

The 1-2 shift valve and the shift control valve both have the job of determining whether the transmission is in First or Second gear. With the shift valve bottomed in its bore, the valve blocks the Drive or line pressure and the second clutch is open to exhaust. The valve is held in this position by a spring as well as any modulator pressure that may be acting against the two spools of the 1-2 shift control valve. As the vehicle speed and the governor pressure increase, a force is developed on the end of the shift valve. When this force is great enough to overcome the spring and the force of the 1-2 shift control valve, the shift valve closes the exhaust and opens the line pressure port to the second clutch port.

To prevent a "hunting" condition of the shift valve, the modulator pressure supply to the second spool of the control valve is cut off as the shift valve opens line pressure to the second clutch. The oil in this pocket is exhausted out through the detent passage. An additional force keeping the valve in the upshifted position is obtained by line pressure acting on the larger diameter second spool of the shift valve. Because of this, even though the governor pressure might be maintained at a constant pressure after the valve upshifts, a higher modulator pressure is required to cause the valve to downshift. If the accelerator is depressed to the point where the detent spring force is felt, the vacuum will drop and the modulator pressure will increase. If the spring force plus the modulator pressure acting against the end spool of the shift valve are great enough to overcome the governor and line pressure acting on the shift valve, a "part throttle" forced downshift will occur. If not, the transmission will remain in the higher gear.

If the accelerator is pressed through the detent, the detent valve supplies detent regulator pressure to all three spools of the shift control valve, and a higher downshifting force is obtained as compared to the part throttle condition. Because of this, a through detent forced downshift can be obtained at a speed higher than for the part throttle condition. However, there is still a limiting speed at which a through detent forced downshift will occur. If the selector lever is placed in the Manual Low position, line pressure is supplied directly to the spring pocket between the valves. Since line pressure can never be less than governor pressure, the force established by the line pressure on the shift valve plus the spring force will move the shift valve to a downshifted position regardless of vehicle speed.

2-3 Shift Valve and 3-2 Control Valve

The function and operation of the 2-3 shift valve and the shift control valves are the same as that described above for the 1-2 shift valve, with the following exceptions. The downshifted position establishes Second or Intermediate gear, and the upshifted position establishes Third or High gear. Modulator pressure is supplied to the end spool of the 2-3 control valve through the 3-2 control valve. When the control valve moves to the upshifted position, line pressure is introduced to the third clutch circuit. The third clutch circuit also directs pressure to the end spool of the 3-2 control vlave. During light throttle conditions, third clutch pressure acting on the end of the 3-2 control valve moves the valve against the spring and the force established by the modulator pressure. This exhausts the modulator pressure from behind the end spool of the 2-3 control valve, and the spring is the only remaining force acting on the shift valve to produce a downshift. In this condition, it is not possible to obtain a "part throttle" forced downshift. If the accelerator is depressed far enough to cause a substantial drop in vacuum the increased modulator pressure on the 3-2 control valve plus the spring will overcome the force of the third clutch pressure. This feeds the modulator pressure back to the 2-3 control valve and a "part throttle" forced downshift will occur. As with the 1-2 shift valve, there is a limiting speed at which this can occur. When the selector lever is placed in Second or Intermediate, the line pressure is directed to the spring pocket between the 2-3 shift and shift control valves and the shift valve will be held in the downshifted or Second gear, regardless of vehicle speed.

Manual Low and Reverse Control Valve

The third clutch is applied in the manual Low position as well as the Reverse range to prevent a free wheeling condition. In Drive or Third gear, third clutch pressure is also directed to the release side of the Low servo. This is the pressure which causes the Low band to release during a 2-3 upshift. However, in manual Low the band must remain applied even though the third clutch is on. These conditions are achieved by routing third clutch pressure to the release side of the Low servo through the manual low and reverse control valve. In the Drive range, the spring holds the valve in its "bottomed" position and permits the third clutch pressure to be directed to the servo release circuit. When the selector lever is placed in the manual Low position, line pressure is introuced between the manual low and reverse control valves. This forces the low control valve over against the spring. In this position, third clutch pressure is cut off from the servo release and the servo release is opened to exhaust. The third clutch exhaust passage is now open to detent regulator pressure which applies the third clutch since the shift valve is in the downshifted position. Because the servo release passage is open to exhaust, the low band will remain applied. When the selector is placed in the Reverse position, the line pressure acts on the end of the reverse control valve and forces the low control valve into the same position as in the manual Low. This causes the third clutch to be applied.

Low Band Servo

The low band servo applies the band to provide engine braking in Second gear in the Low range. It is also used as an accumulator for the apply of the direct clutch and is used in conjunction with a series of check balls and the controlling orifices as a part of the timing for the release of the direct clutch. To keep from applying the front band in Neutral, Drive or Reverse the fluid is directed from the manual valve to the release side of the servo piston.

Planetary Gears

In this transmission, planetary gears are used as the basic means of multiplying the torque from the engine. Power flow through the planetary gear set is accomplished by applying power to one member, holding another member, thereby making it a reaction member, and obtaining the transmitted power from the third member. This results in any one of the following conditions.

1. The torque is increased along with a proportional decrease in output speed.
2. Speed can be increased with a proportional decrease of output torque.
3. It can act as a direct connection for direct drive.
4. It can reverse the direction of rotation.

The type of gear set that is used in the planetary utilizes two sets of planetary pinions in one carrier, two sun gears and one ring gear. The short pinions are in constant mesh with both the input (front) sun gear and the long pinions. The long pinions are in constant mesh with the reaction (rear) sun gear, the short pinions and the ring gear.

First Gear

In first gear, the reaction sun gear is held stationary. The input sun gear rotates in a clockwise direction (viewed from the front), turning the short pinions counterclockwise and the long planet pinions clockwise. The long planet pinions turn the ring gear clockwise and walk around the held reaction sun gear, driving the planet carrier and output shaft assembly in a clockwise direction.

Second Gear

In second gear, the reaction sun gear is again held stationary. The ring gear is the input and is driven in a clockwise direction turning the long planet pinions clockwise which walk around the stationary reaction sun gear, driving the planet carrier assembly and output shaft in a clockwise direction. The sprag allows the input sun gear to overrun.

Third Gear

In third gear, the ring gear is driven in a clockwise direction and the input sun gear is also driven in the same direction. The long and short planetary pinions cannot rotate on their shafts in this situation, thus causing the planetary carrier, output shaft and gears to rotate clockwise as a solid unit to provide direct drive.

Reverse

In reverse gear, the ring is held and the input sun gear is driven in a clockwise direction. This causes the short planet pinions to turn counterclockwise, turning the long planet pinions clockwise. The long planet pinions then walk around the inside of the stationary ring gear, driving the planet carrier assembly and output shaft in a counterclockwise direction.

DIAGNOSIS TESTS

Automatic transmission malfunctions may be caused by four general conditions: poor engine performance, improper adjustments, hydraulic malfunctions, or mechanical malfunctions.

The suggested sequence for diagnosis is as follows
1. Check and correct oil level
2. Check and adjust T.V. cable
3. Check and correct manual linkage
4. Check engine tune
5. Install oil pressure gauge and tachometer and check control pressure

6. Road test in all ranges, noting changes in operation and oil pressure

7. Attempt to isolate the unit involved in the malfunction

8. If road test indicates that the engine needs to be tuned up, do so before any corrective action is taken on the automatic transmission.

CONTROL PRESSURE TEST

1. Using a suitable transmission jack or support, take the weight off of the transmission crossmember.

2. Remove the crossmember bolts.

3. Carefully lower the transmission just enough to remove the pressure tap plug which is located on the left side of the transmission.

4. Install the pressure gauge and hose.

5. Install the crossmember.

6. Perform the pressure test.

NOTE: The pressures on the gauge are supplied by the servo apply pressure.

Test One

With the vehicle "coasting" at 30 MPH, the vacuum line still connected, and the accelerator closed (foot off throttle), minimum pressures should be:

Drive	65 psi.
L2	65 psi.
L1	80 psi.

Test Two

With the output shaft speed at zero, the vacuum line disconnected from the modulator and the engine speed adjusted to 1500 rpm, the maximum pressures should be:

Drive	120 psi.
L2	120 psi.
L1	160 psi.

NOTE: On replacing the pressure tap plug, torque to 5 to 7 foot-pounds.

STALL SPEED TEST

General Motors Corporation does not recommend the use of a stall speed test to determine whether a malfunction exists in the THM 180C automatic transmission. Excessive heat is generated by this test and could cause more transmission problems if not controlled.

ROAD TEST

When road testing a vehicle equipped with a THM 180C automatic transmission, be sure that the transmission is at the proper operating temperature. Operate the transmission in each position to check for slipping and any variation in the shifting pattern. Note whether the shifts are harsh or spongy. By utilizing the clutch and band application chart with any malfunctions noted, the defective unit or circuit can be found.

AIR PRESSURE TEST

The air pressure test can be used to determine if cross passages are present within the transmission case or the valve body of the THM 180C automatic transmission. Also, this test is used to determine if the clutch passages are open.

TORQUE CONVERTER STATOR OPERATION TEST

The torque converter stator assembly and its related roller clutch can possibly have one of two different type malfunctions.
1. The stator assembly freewheels in both directions.
2. The stator assembly remains locked up at all times.

CONDITION A

If the stator roller clutch becomes ineffective, the stator assembly freewheels at all times in both directions. With this condition, the vehicle will tend to have poor acceleration from a standstill. At speeds above 30-35 MPH, the vehicle may act normal. If poor acceleration problems are noted, it should first be determined that the exhaust system is not blocked, the engine is in good tune and the transmission is in gear when starting out.

If the engine will freely accelerate to high rpm in Neutral, it can be assumed that the engine and exhaust system are normal. Driving the vehicle in Reverse and checking for poor performance in forward gears will help determine if the stator is freewheeling at all times.

CONDITION B

If the stator assembly remains locked up at all times, the engine rpm and vehicle speed will tend to be limited or restricted at high speeds. The vehicle performance when accelerating from a standstill will be normal. Engine over-heating may be noted. Visual examination of the converter may reveal a blue color from the overheating that will result.

Under conditions A or B above, if the converter has been removed from the transmission, the stator roller clutch can be checked by inserting a finger into the splined inner race of the roller clutch and trying to turn the race in both directions. The inner race should turn freely in the clockwise direction, but not turn or be very difficult to turn in the counterclockwise direction.

VACUUM MODULATOR TESTING

A defective vacuum modulator can cause one or more of the following conditions:
  a. Soft up and down transmission shifts
  b. Harsh upshifts
  c. Engine burning automatic transmission fluid
  d. Automatic transmission overheating
  e. Delayed upshifts

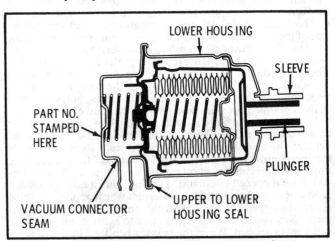

Modulator assembly—cross section (©General Motors Corp.)

Whenever a vacuum modulator is suspected of malfunctioning, a vacuum check should be made of the vacuum supply. If, the vacuum supply is found to be in proper order the vacuum modulator must be inspected. This inspection can be performed with the modulator either on or off the vehicle transmission.

Modulator Test (Modulator Installed In Vehicle)

1. Remove the vacuum line and attach a vacuum pump to the modulator connector pipe.

2. Apply 18 inches of vacuum to the modulator assembly. The vacuum should remain at 18 inches without leaking down.

3. If the vacuum reading drops sharply or will not remain at 18 inches, the diaphragm is leaking and the unit must be replaced.

4. If transmission fluid is present on the vacuum side of the diaphragm or in the vacuum hose, the diaphragm is leaking and the unit must be replaced.

NOTE: Gasoline or water vapors may settle on the vacuum side of the diaphragm. Do not diagnose as transmission fluid.

Modulator Test (Modulator Removed From Vehicle)

1. Remove the modulator assembly from the automatic transmission.

2. Attach a vacuum pump to the modulator connector pipe and apply 18 inches of vacuum.

3. The vacuum should hold at 18 inches, if the diaphragm is good and will drop to zero if the diaphram is leaking.

4. With the control rod in the transmission side of the vacuum modulator, apply vacuum to the connector pipe. The rod should move inward with light finger pressure applied to the end of the rod. When the vacuum is released, the rod will move outward by pressure from the internal spring.

Internal Spring Load Comparison Test

1. Install the known good vacuum modulator in one end of the gauge. Place the unit to be tested into the other end of the gauge.

2. While holding the assembly horizontally, force both vacuum modulator towards the center of the gauge, until one sleeve end touches the center line scribed on the gauge. The distance between the center line and the sleeve end on the opposite vacuum modulator should not exceed 1/16.

NOTE: Certain comparison gauges may use the control rod that is used with the vacuum modulator or a special rod. Follow the manufacturer's instructions.

3. Inspect the control rod for straightness and the vacuum modulator stem for being concentric to the can.

TORQUE CONVERTER CLUTCH

Converter Clutch Operation and Diagnosis

The THM 180C automatic transmission uses a torque converter clutch (TCC), which incorporates a unit inside the torque con-

verter with a friction material attached to a pressure plate and splined to the turbine assembly. When the clutch is applied, it presses against the converter cover. The result is a mechanical direct drive of the engine to the transmission. This eliminates slippage and improves fuel economy as well as reducing fluid temperature.

There are a number of controls to operate the torque converter clutch, all of which are determined by drive range selection. For example, the clutch is applied in direct drive above a certain preset minimum speed. At wider throttle openings the converter clutch will apply after the 2-3 shift. When the vehicle slows, or the transmission shifts out of direct drive, the fluid pressure is released, the converter clutch releases and the converter operates in the conventional manner.

The engaging of the converter clutch as well as the release is determined by the direction of the converter feed oil. The converter feed oil from the pressure regulator valve flows to the converter clutch apply valve. The position of the converter clutch apply valve controls the direction in which the converter feed oil flows to the converter.

A spring-loaded damper assembly is splined to the converter turbine assembly while the clutch pressure plate is attached to a pivoting unit on the damper assembly. The result is that the pressure plate is allowed to rotate independently of the damper assembly up to about 45 degrees. This rotation is controlled by springs in the damper assembly. The spring cushioning aids in reducing the effects felt when the converter clutch applies.

To help insure that the converter clutch applies and releases at the proper times, controls have been incorporated into the electrical system.

Troubleshooting The Torque Converter Clutch

Before diagnosing the torque converter clutch system as being at fault in the case of rough shifting or other transmission malfunctions, make sure that the engine is in at least a reasonable state of tune. Also, the following points should be checked:

1. Check the transmission fluid level and correct as necessary.
2. Check the manual linkage adjustment and correct as necessary.
3. Road test the vehicle to verify the complaint. Make sure that the vehicle is at normal operating temperature.

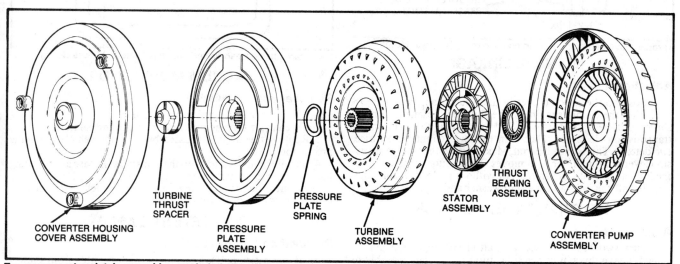

Torque converter clutch assembly—exploded view (©General Motors Corp.)

ON CAR SERVICES

Adjustments

T.V. AND DETENT CABLE

Adjustment

Before adjusting the cable the following steps should be taken:

    a. Check transmission fluid and correct as required.

    b. Be sure that the engine is operating and that the brakes are not dragging.

    c. Be sure that the correct cable has been installed in the vehicle that you are servicing.

    d. Check that the cable is connected at both ends.

1. To adjust, depress the re-adjust tab. Move the slider back through the fitting in the direction away from the throttle body until the slider stops against the fitting.

2. Release the re-adjust tab.

3. Open the carburetor or pump lever to the full stop throttle position to automatically adjust the cable. Release the carburetor or pump lever.

4. Check the cable for sticking or binding. Road test the vehicle as required.

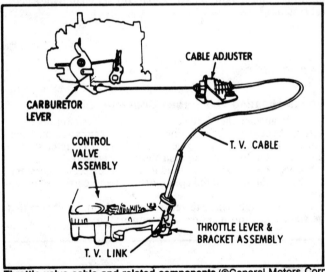

Throttle valve cable and related components (©General Motors Corp.)

MANUAL LINKAGE

Adjustment

1. Position the selector lever in the Neutral position.

2. With the connecting link of the shift rod loosened from the shifter, put the transmission in Park by moving the lever on the transmission clockwise. Then turn the lever counter-clockwise, two detents. This will be Neutral. Make sure neither the transmission lever nor the shifter lever moves. Then adjust the link so that the hole aligns with the shifter pin without preload or forcing.

3. Install the link, add the shimwasher and retainer.

NEUTRAL SAFETY SWITCH

Adjustment

1. New switches come with a small plastic alignment pin installed. Leave this pin in place. Position the shifter assembly in Neutral.

2. Remove the old switch and install the replacement, align the pin on the shifter with the slot in the switch, and fasten with the two screws.

3. Move the shifter from Neutral position. This shears the plastic alignment pin and frees the switch.

If the switch is to be adjusted, not replaced, insert a 3/32″ drill bit or similar size pin and align the hole and switch. Position switch, adjust as necessary. Remove the pin before shifting from Neutral.

LOW BAND

Adjustment

1. Raise the vehicle and support safely.

2. Remove the oil pan and valve body from the transmission.

3. Loosen the locknut and tighten the servo adjusting bolt to 40 in. lbs.

4. Back the adjusting bolt off exactly five turns and tighten the locknut while holding the adjusting bolt to prevent turning.

5. Reinstall the valve body and oil pan. Fill the transmission with fluid to specifications. Start the engine and recheck the level. Check for leakage. Lower the vehicle and roadtest.

Services

OIL PAN AND FILTER

Removal and Installation

1. Raise the vehicle and support it safely.

2. Remove the pan bolts from the transmission pan, and allow the fluid to drain.

3. Remove the transmission oil filter and discard the gasket.

4. Clean all gasket surfaces. Using a new gasket, install a new transmission filter to its mounting on the valve body assembly.

5. Install the transmission oil pan, using a new gasket, torque the retaining bolts to 7-10 ft. lbs.

6. Lower the vehicle, fill the transmission and check as required.

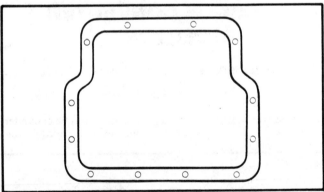

180C transmission pan identification (©General Motors Corp.)

VACUUM MODULATOR

Removal and Installation

1. Raise the vehicle and support it safely.

2. Disconnect the vacuum line from the modulator.

3. Remove the modulator assembly from the vehicle, using the proper tool.

4. Install the modulator into the transmission case after coating the O-ring with transmission fluid.

5. Connect the vacuum line. Lower the vehicle and check for proper operation of the modulator.

T.V./DETENT CABLE

Removal and Installation

1. If necessary, remove the air cleaner assembly from the vehicle.

2. Push in on the re-adjust tab and move the slider back through the fitting in the direction away from the throttle lever.

3. Disconnect the cable terminal from the throttle lever.

4. Compress the locking tabs and disconnect the cable assembly from the bracket.

5. Remove the routing clips or straps. Remove the screw and washer securing the cable to the automatic transmission assembly.

6. Disconnect the cable from the link and remove the cable from the vehicle.

7. Install a new seal into the automatic transmission case hole.

8. Connect and secure the transmission end of the cable to the transmission case using the bolt and washer. Torque the bolt to 8 ft. lbs.

9. Route the cable as removed and connect the straps or clips.

10. Position the cable through the bracket and engage the locking tabs of the cable on the bracket.

11. Connect the cable terminal to the throttle lever. Adjust the throttle cable as required.

12. If removed, install the air cleaner.

SERVO ASSEMBLY

Removal and Installation

1. Raise the vehicle and support it safely.

2. Remove the transmission oil pan, pan gasket and filter.

3. Remove the screw and the retainer securing the detent cable to the transmission. Disconnect the detent cable.

4. Remove the throttle lever and the bracket assembly.

5. Remove the manual detent roller and the spring assembly.

6. Remove the transfer plate reinforcement from the valve body assembly.

7. Remove the servo cover.

8. Remove the valve body attaching bolts and remove the valve body from the transmission.

9. Compress the servo piston, using the proper tool.

10. Remove the servo piston snap ring. Slowly loosen the servo piston compression tool and remove the servo piston, return spring and apply rod.

11. Installation is the reverse of removal. Be sure to adjust the apply rod by tightening the bolt to 40 in. lbs. Then back off the bolt exactly five turns. Tighten the locknut while holding the adjusting bolt firmly.

12. Lower the vehicle, fill the transmission with the proper grade and type automatic transmission fluid and road test as required.

SPEEDOMETER DRIVEN GEAR

Removal and Installation

1. Raise the vehicle and support it safely.

2. Disconnect the speedometer cable.

3. Remove the retainer bolt, retainer, speedometer driven gear and the O-ring seal.

4. Install the speedometer driven gear using a new O-ring seal coated with transmission fluid.

5. Continue the installation in the reverse order of removal. Adjust the transmission fluid level as required.

REAR OIL SEAL

Removal and Installation

1. Raise the vehicle and support it safely.

2. Remove the driveshaft. If equipped, remove the tunnel strap.

3. Pry the seal from the rear extension housing, using the proper tool.

4. Coat the new seal with transmission oil and drive it into place using a rear transmission seal installer tool, or equivalent.

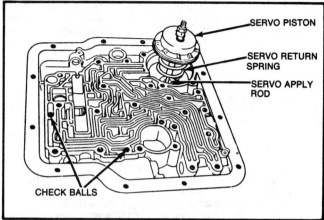

Servo and related components (©General Motors Corp.)

5. Continue the installation in the reverse order of the removal. Correct the transmission fluid level as required. Lower the vehicle and check for transmission fluid leaks.

VALVE BODY

Removal and Installation

1. Raise the vehicle and support it safely. Drain the transmission fluid.

2. Remove the oil screen and disconnect the detent cable by removing the capscrew and retainer on the outside of the case.

3. Remove the two bolts holding the manual detent roller and spring and remove the spring.

4. Remove the bolts from the transfer plate reinforcement and remove the reinforcement.

5. Remove the throttle lever and bracket assembly, being careful not to bend the throttle lever link.

6. Remove the servo cover and gasket.

7. Remove the remaining bolts attaching the valve body to the case. Remove valve body and transfer plate.

NOTE: The two check balls in the case may drop out when removing the valve body. Be careful that they do not become lost and, at reassembly, make sure they are installed in the proper position. Also, be careful that the manual valve and link are not dropped and damaged or lost.

8. Installation is the reverse of removal. Be sure to fill the automatic transmission with the proper grade and type transmission fluid. Lower the vehicle, and check for leaks. Correct as required.

GOVERNOR

Removal and Installation

1. Raise the vehicle and support it safely. Remove the driveshaft.

2. Support transmission with a suitable jack and remove the transmission mount. If necessary, remove the crossmember for clearance.

3. Remove the speedometer driven gear.

4. Remove the extension housing bolts and remove the housing and gasket.

5. Push down on the speedometer drive gear retaining clip and remove the speedometer drive gear.

6. Remove the four bolts from the governor body and remove governor and gasket.

7. If it is necessary to remove the governor hub use lock-ring pliers to remove the snap ring and slide governor hub off the shaft.

8. Installation is a reverse of the above. The governor-to-hub bolts are torqued to 6-7 foot-pounds. The extension housing bolts are torqued to 20-25 foot-pounds while the support-to-extension bolts are torqued to 29-36 foot-pounds.

TRANSMISSION REMOVAL

1. Disconnect the negative battery cable. Disconnect the detent cable at the carburetor or injection pump. Remove the air cleaner. Remove the dipstick.

2. On vehicles with air conditioning, remove the screws holding the heater core cover, disconnect the wire connector and, leaving the hoses attached, place the heater core cover out of the way.

3. Raise the vehicle on a hoist and support it safely. Drain the transmission fluid. Remove the driveshaft.

4. Remove speedometer cable and the cooler tubes from the transmission.

5. Disconnect the shift control linkages.

6. Raise the transmission slightly with a jack and remove the support bolts and converter bracket.

7. Disconnect the exhaust pipe at the rear of the catalytic converter and at the manifold and remove exhaust as an assembly.

8. Remove the pan under the torque converter and remove the three flexplate bolts. Mark the converter and flexplate so that they can be reassembled in the same manner.

9. Lower the transmission until there is access to the engine mounting bolts and remove them.

10. Raise the transmission to its normal position and support the engine with a jack. Slide transmission back and down.

NOTE: Hold on to the converter and/or keep the rear of the transmission lower than the front so it won't fall from the bell housing.

INSTALLATION

1. Transmission installation is the reverse of the removal procedure.

2. Make sure that the converter pump hub keyway is seated into the oil pump drive lugs. The distance from the face of the bell housing to the end of the hub should be .200 to .280 inches. Check that the converter has free movement.

3. Start all bolts in the converter finger tight so that the flexplate will not be distorted, and align the marks made at disassembly. Then torque the bolts to 20-30 foot-pounds. The transmission support-to-extension should be torqued to 29-36 foot-pounds and the exhaust system parts to 35 foot-pounds.

4. Lower the vehicle from the hoist. Fill the transmission with the proper grade and type automatic transmission fluid.

5. Make linkage and cable adjustments as required. Road test the vehicle as required.

Torque Converter Inspection

1. Make certain that the transmission is held securely.

2. The converter pulls out of the transmission. Be careful since the converter contains a large amount of oil. There is no drain plug on the converter so the converter should be drained through the hub.

3. If the oil in the converter is discolored but does not contain metal bits or particles, the converter is not damaged and need not be replaced. Remember that color is not longer a good indicator of transmission fluid condition.

4. If the oil in the converter contains metal particles, the converter is damaged internally and must be replaced. The oil may have an "aluminum paint" appearance.

5. If the cause of oil contamination was burned clutch plates or overheated oil, the converter is contaminated and should be replaced.

6. If the pump gears or cover show signs of damage, the converter will contain metal particles and must be replaced.

Before Disassembly

Before opening up the transmission, the outside of the unit should be thoroughly cleaned, preferably with high-pressure cleaning equipment. Dirt entering the transmission internal parts will negate all the effort and time spent on the overhaul. During inspection and reassembly, all parts should be thoroughly cleaned with solvent then dried with compressed air. Wiping cloths and rags should not be used to dry parts since lint will find its way into valve body passages. Lube the seals with Dexron® II and use ordinary unmedicated petroleum jelly to hold the thrust washers and to ease the assembly of seals. Do not use solvent on neoprene seals, friction plates or thrust washers. Be wary of nylon parts if the transmission failure was due to the cooling system. Nylon parts exposed to antifreeze solutions can swell and distort, so they must be replaced. Before installing bolts into aluminum parts dip the threads in clean oil. Anti-seize compound is also a good way to prevent the bolts from galling the aluminum and seizing.

Transmission Disassembly
EXTERNAL COMPONENTS

Removal

1. Remove the torque converter assembly from the transmission housing.

2. Mount the transmission assembly in the holding fixture.

3. Drain the automatic transmission through the rear extension housing.

4. Remove the oil pan retaining bolts. Remove the oil pan and discard the gasket.

5. Remove the manual detent roller and spring assembly. Remove the transmission filter and discard the filter gasket.

6. Disconnect the governor pressure switch electrical connector and the solenoid wiring harness.

7. Disconnect the governor pressure switch. Disconnect the torque converter clutch solenoid and the solenoid pipes. Be sure not to bend the solenoid pipes when removing them from their mounting place.

8. Remove the transfer plate reinforcment. Remove the servo cover and gasket.

9. Remove the valve body with its gasket and transfer plate. Care must be taken so that the manual valve and manual valve link are not damaged or lost during valve body removal. To prevent the manual link from falling into the transmission case, place a piece of paper toweling in the case void below it, before removing the valve body.

10. Compress the servo piston using the servo piston compressor tool. The offset of the tool must be positioned toward the rear of the transmission case.

11. Using needle nose pliers, remove the servo piston to transmission case snap ring.

12. Before removing the installation bolts on the tool, loosen the compression screw to relieve the tension on the servo spring.

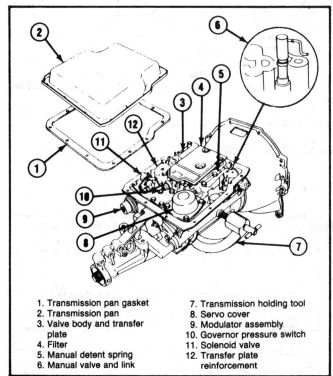

1. Transmission pan gasket
2. Transmission pan
3. Valve body and transfer plate
4. Filter
5. Manual detent spring
6. Manual valve and link
7. Transmission holding tool
8. Servo cover
9. Modulator assembly
10. Governor pressure switch
11. Solenoid valve
12. Transfer plate reinforcement

External components—exploded view (©General Motors Corp.)

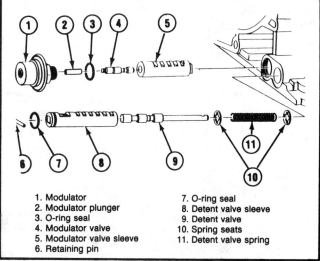

1. Modulator
2. Modulator plunger
3. O-ring seal
4. Modulator valve
5. Modulator valve sleeve
6. Retaining pin
7. O-ring seal
8. Detent valve sleeve
9. Detent valve
10. Spring seats
11. Detent valve spring

Modulator and detent valve with their related components
(©General Motors Corp.)

Remove the tool, the servo piston assembly, the return spring and the servo apply rod.

13. Remove the two check balls, which are located in the oil passage of the transmission case.

14. Remove the selector inner lever hex nut from the selector lever shaft. Place a piece of paper in the transmission case under the hex nut before removing it, this will catch the nut, should it drop.

15. Remove the selector inner lever from the selector lever shaft. Remove the selector lever shaft pin by pulling upwards with pliers. To prevent the spring from collapsing, insert a wire into the middle of the spring pin.

16. Install the outside attaching nut on the selector lever shaft. Place a suitable driver on the face of the nut and drive the selector lever shaft out of the transmission case.

17. Remove the selector lever shaft oil seal. Remove the electrical connector from the transmission case.

18. Remove the vacuum modulator and O-ring, using the proper tool. Care should be taken not to lose the modulator plug. Remove the modulator valve and sleeve from the transmission case.

19. Remove the detent valve retaining spring pin by pulling upward with pointed nose diagonal pliers. To prevent collapse of the spring pin, insert a wire into the middle of the spring pin. Remove the detent sleeve, valve, spring and spring seats.

EXTENSION HOUSING, SPEEDOMETER DRIVE GEAR AND GOVERNOR

Disassembly

1. Remove the speedometer driven gear housing retainer. Remove the gear assembly from the extension housing.

2. Remove the rear extension housing oil seal, using the proper seal removal tool. Remove the extension housing and gasket from the transmission.

3. Depress the speedometer drive gear retaining clip and remove the gear by sliding it off the output shaft.

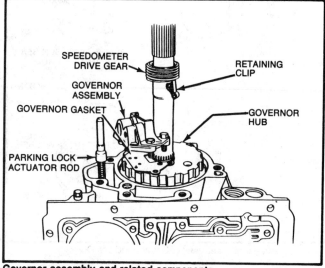

Governor assembly and related components
(©General Motors Corp.)

4. Remove the governor body and gasket. Remove the governor hub snap ring from the output shaft.

5. Slide the governor hub from the output shaft, be sure not to lose the governor hub oil screen.

INTERNAL COMPONENTS

Removal

1. Remove the torque converter housing oil seal, using the proper removal tool.

2. Remove the seven outer torque converter housing bolts. Loosen, but do not remove the five inner torque converter housing bolts.

3. Remove the O-ring seal from the input shaft.

NOTE: If the rubber seal is not removed, the second clutch and third clutch will come out with the converter housing. The rubber seal may shear, while you are holdng the parts, allowing the second clutch and third clutch to fall possibly causing personal injury.

4–141

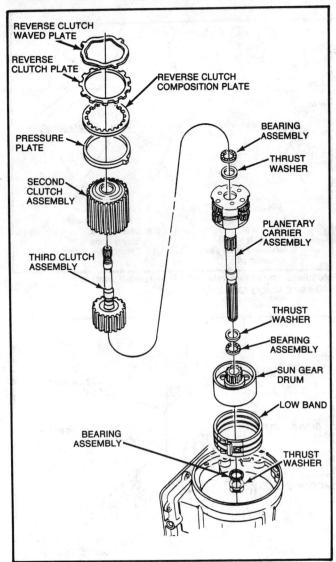

Internal parts—exploded view (©General Motors Corp.)

4. Remove the torque converter housing with the oil pump, oil pump flange gasket and the reverse clutch assemblies. Be sure not to loose the selective thrust washer between the oil pump hub and the second clutch drum.

5. Lift up on the input shaft and remove both the second and third clutch assemblies. Separate the assemblies from one another.

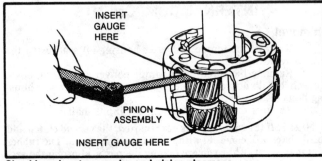

Checking planetary carrier and pinion clearance (©General Motors Corp.)

6. Remove the reverse clutch plates and the aluminum pressure plate from the automatic transmission case.

7. Remove the inside selector lever and the parking lock actuator rod from the automatic transmission case.

8. Remove the planetary carrier and the output shaft along with the two Torrington Bearings and thrust washer.

9. Remove the reaction sun gear and drum. Remove the low band.

10. If necessary, remove the case vent and install a new case vent upon reassembly. Do not use the old vent.

Unit Disassembly and Assembly

LOW BAND

1. Position the automatic transmission case so that the front of the transmission case is upward.

2. Inspect the low band for cracks, flaking, wear, looseness or burring. If any of these conditions exist the band must be replaced.

3. When installing the low band into the transmission case be sure to position it onto the anchor pins, which are located in the case.

REACTION SUN GEAR AND DRUM

1. Inspect the reaction gear assembly. Check for chipped or nicked teeth. Inspect the sun gear drum for scoring. Replace these components as required.

2. Check the reaction sun gear drum bushing. If necessary, replace.

3. When installing, place the thrust washer and the bearing into the transmission case. Use petroleum jelly to retain the thrust washer in place. The transmission case bushing acts as a guide to center the bearing.

4. Position the reaction sun gear and drum assembly into the transmission case. Be sure that the sun gear is facing upward.

5. Install the bearing and thrust washer onto the sun gear. Retain the thrust washer using petroleum jelly.

PLANETARY CARRIER

1. Inspect both the planetary carrier and the planetary pinions for excessive wear, damage or distortion. Replace as required.

2. Check the clearance of all planetary pinions using a feeler gauge. The clearance should be between .005-.035 inch. Replace the entire assembly if damage is evident.

3. When installing, insert the output shaft and the planetary carrier assembly from the front of the transmission case to spline with the reaction sun gear.

4. Install the thrust washer, then install the Torrington Bearing into the planetary carrier. Be sure to use petroleum jelly as a retainer.

THIRD CLUTCH ASSEMBLY

Disassembly

1. Position the third clutch assembly in a vise. Install special tool J-29351 or equivalent onto the third clutch drum. Position the five pins of the tool in the elongated slots. Do not put a pin into a slot, if the internal ring is not visible in the slot. Slide the compressing ring over the pin cage.

2. Pull on the pin cage handle or input sun gear. If the ring is hanging up on one side causing the clutch hub to cock, insert a punch into the slot, to compress the ring.

NOTE: Mounting the clutch pack at a ninety degree angle will prevent the ring from sliding back into its groove, during tool removal.

3. After the retaining ring has been pressed out of the clutch drum groove, remove the sprag assembly.

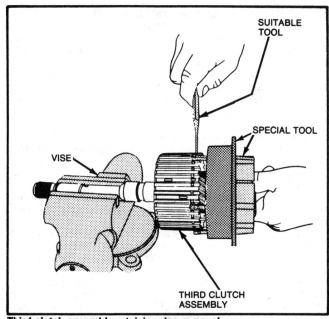

Third clutch assembly retaining ring removal
(©General Motors Corp.)

4. Remove the input shaft-to-input sun gear thrust washer and Torrington Bearing. This bearing and thrust washer may be staked together.

5. Remove the third clutch plates from the third clutch drum. The plates should be kept in the same sequence as they were installed in the clutch.

6. Remove the input sprag race and retainer assembly from the third clutch hub and input sun gear assembly.

7. Push the sprag assembly and retaining rings from the sprag race and retainer.

8. Using compressor tool J-23075 or equivalent, on the third clutch piston retaining seat, compress the third clutch piston return springs using an arbor press. Remove the snap ring. Use care not to let the retaining seat catch the snap ring groove. Remove the compressor tool.

9. Remove the retaining seat and the twelve return springs. Remove the third clutch piston from the third clutch drum.

Assembly

1. Before assembly, inspect all of the third clutch piston return springs for wear and defects. Inspect the check ball in the third clutch piston. Shake the piston and listen for movement of the check ball, which will indicate proper operation. If the ball is missing, falls out upon inspection or piston is damaged, replace the piston. Install a new lip seal on the piston, if necessary.

2. Remove the oil lip seal from the input shaft inside of the third clutch drum and install a new lip seal, if necessary, with the lip pointing downward.

3. Inspect, and if necessary replace the steel thrust washer on the front face of the third clutch drum. Air check the oil passages in the input shaft.

4. Install the third clutch piston into the third clutch drum. Use transmission fluid so that the seal is not damaged upon installation.

5. Install all of the third clutch piston return springs onto the piston. Install the retaining seat.

6. Use compressor tool J-23075 or equivalent, on the retaining seat and compress the piston return springs. Care must be taken so that the retaining seat does not catch in the snap ring groove and damage the retainer.

7. Install the snap ring. Remove the compressor tool. Inspect and replace, if necessary, the thick thrust washer and bearing.

8. Install the thrust washer and bearing onto the input shaft. The bearing will face the input sun gear if properly installed. Secure with petroleum jelly.

9. Inspect the sprag assembly for wear, damage, or sprags that freely fall out of the cage. Inspect the input sub gear for chipped or nicked teeth or abnormal wear. Replace parts as required.

10. Install the sprag cage and two retaining rings on the third clutch hub with the flared shoulder on the sprag cage outer diameter toward the input sun gear.

11. Install the outer sprag race and retainer assembly over the sprag assembly. Holding the input sun gear with the left hand, the sprag race and retainer assembly should hold firm when turned with your right hand in a clockwise direction and should rotate freely when turned counterclockwise.

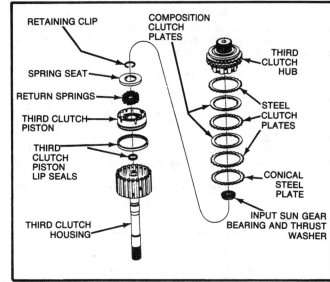

Third clutch assembly—exploded view (©General Motors Corp.)

NOTE: This procedure must be followed exactly, to be sure that the sprag has not been installed wrong.

12. Inspect the condition of the third clutch composition and steel plates, replace as required.

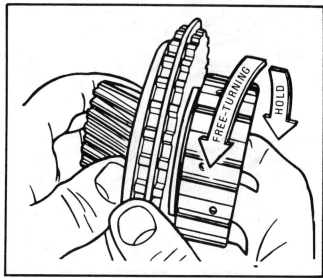

Correct sprag assembly test (©General Motors Corp.)

13. Install the third clutch plates on the third clutch hub in the following order: steel plate, composition plate, steel plate, composition plate, steel plate, conical steel plate. When installed correctly, the I.D. of the conical plate will touch the steel plate below it but, the O.D. will not.

14. Mount the clutch housing in a vise. Install the third clutch hub and sprag assembly into the third clutch housing until the sprag race rests on the third clutch drum.

15. Install tool J-29351 or equivalent, on the sprag race so that the pins in the tool fingers compress the retaining ring.

16. Slide the sprag assembly into the third clutch drum until the retaining ring is inside the drum.

NOTE: Do not push on the input sun gear while installing the sprag assembly or the clutch plates will slip off of their location and prevent assembly.

17. Remove the tool. Slide the sprag assembly into the third clutch drum until the retaining ring catches in the retaining ring groove. If necessary, use a small suitable tool to help compress the retaining ring while installing the sprag assembly.

SECOND CLUTCH ASSEMBLY

Disassembly

1. Remove the ring gear retaining ring from the second clutch drum. Remove the ring gear. Remove the second clutch spacer retaining ring. Remove the second clutch spacer.

2. Remove the second clutch steel and composition plates. The plates should be kept in the same sequence as they were installed in the clutch.

3. Remove the second clutch assembly to third clutch assembly thrust washer.

THRUST WASHER

RETAINING RING

PISTON RETURN SPRINGS (22)

SECOND CLUTCH

PISTON SEALS

SECOND CLUTCH HOUSING

RETAINING RING

RING GEAR

SPACER

STEEL CLUTCH PLATES

WAVED PLATE

Second clutch assembly (©General Motors Corp.)

4. Install clutch spring compressor tool J-23327 or its equivalent, on the second clutch piston return spring retainer and compress the second clutch piston return springs. Tool J-23327 must be adapted to fit in the second clutch piston, by using three sockets of the same height.

5. Remove the snap ring using the proper removal tool.

6. Remove the second clutch piston retaining seat and all of the piston return springs. Remove the second clutch piston.

Assembly

1. Before assembly, inspect the second clutch piston. If the piston is damaged or if the check falls out upon inspection, replace the piston. If necessary, install two new piston lip seals. One seal on the piston O.D. and one on the second clutch drum hub. Install with lips down.

2. Inspect the piston return springs. Replace as required. Inspect the second clutch hub bushing for wear or scoring, replace if necessary.

3. To install the second clutch piston into the second clutch drum, use tool J-23080 or its equivalent, to prevent damaging the outer lip seal. Use a liberal amount of transmission fluid for ease of installation and to prevent seal damage.

4. Remove the installation tool. Install all of the piston springs and the retaining seat on the second clutch piston.

5. Using spring compressor tool J-23327 or its equivalent, and three sockets on the retaining seat, compress the second clutch piston return springs. Care should be taken so that the retainer does not catch in the snap ring groove and damage the retainer.

6. Install the snap ring using the proper tool. Install the second to third clutch thrust washer so that the tang seats in the slot of the second clutch hub. Secure the assembly using petroleum jelly. Inspect the condition of the composition and steel plates, replace as required.

7. Install the second clutch plates into the second clutch drum with the waved clutch plate first, then a steel plate, a composition plate, a steel plate, etc. Use a liberal amount of transmission fluid.

8. Install the second clutch spacer plate into the second clutch drum with the wavy end toward the clutch plates.

9. Install the second clutch spacer retaining ring. Install the ring gear into the second clutch drum. Be sure that the grooved edge is facing up. Install the ring gear and the retaining ring.

SECOND AND THIRD CLUTCH ASSEMBLIES

1. When installing these two assemblies, align the tangs of the second clutch drive plates in the second clutch drum.

2. Align the tangs of the second clutch drive plates in the second clutch drum. Insert the third clutch drum and the input shaft through the top of the second clutch drum, seating the third clutch drum splines into the second clutch plate splines. Holding the second and third clutch assemblies by the input shaft, lower into the transmission case, indexing the ring gear in the second clutch drum with the long planetary pinion gear teeth.

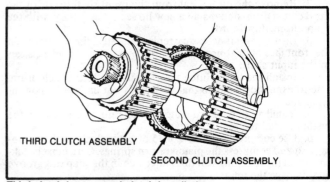

Third clutch into second clutch installation
(©General Motors Corp.)

SELECTIVE WASHER THICKNESS DETERMINATION

1. Install tool J-23085 or equivalent, on the transmission case flange and against the input shaft.

2. Loosen the thumb screw on the tool to allow the inner shaft of the tool to drop on the second clutch housing thrust face.

3. Tighten the thumb screw, then remove tool.

4. Compare the thickness of the selective washer removed earlier from the transmission to the protruding portion of the inner shaft of tool J-23085. The selective washer used in reassembly should be the thickest washer available without exceeding the dimension of the shaft protruding from tool J-23085. The dimension of the washer selected should be equal to or slightly less than the inner shaft for correct end play in the transmission.

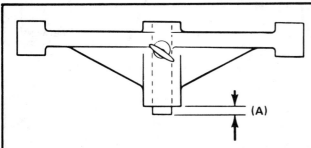

WHEN MEASURED GAP (A) IS:		USE WASHER PART NUMBER
INCHES	mm	
.069 – .074	1.78 – 1.88	5258202
.075 – .079	1.93 – 2.03	5258203
.080 – .084	2.06 – 2.16	5258204
.085 – .089	2.18 – 2.29	5258205
.090 – .094	2.31 – 2.41	5258206
.095 – .100	2.46 – 2.57	5258207

FOLLOWING THE PROCEDURE SHOULD RESULT IN FINAL END-PLAY FROM 0.36 mm TO 0.79 mm (.014 in. TO .031 in.)

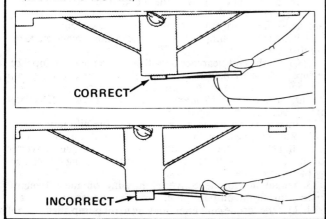

CORRECT

INCORRECT

Selective washer selection chart (©General Motors Corp.)

TORQUE CONVERTER HOUSING, OIL PUMP AND REVERSE CLUTCH ASSEMBLY

Dissembly

1. If not already removed, remove the selective washer from the oil pump shaft.

2. Remove the oil pump outer seal. Observe the position of the square cut seal.

3. Remove the bolts holding the torque converter housing to the oil pump. Separate the two assemblies. Remove the oil pump wear plate.

4. Check the torque converter pump hub for nicks, burrs or damage. Correct as required.

5. Mark the relative location of the oil pump gears and remove them.

6. Using compressor tool J-23327, or its equivalent, on the reverse clutch retaining seat, compress the clutch return springs. Remove the snap ring, using the proper removal tool.

7. Loosen compressor tool and remove the reverse clutch retaining ring and all of the reverse clutch springs. Use care not to let the spring retainer catch in the ring groove.

8. Remove the reverse clutch piston. The reverse clutch piston may be removed by blowing compressed air into the piston apply passage of the oil pump.

9. The valves, located in the oil pump, may be removed by using a pair of needle nose pliers to remove the retaining pin. However, it is not recommended that these valves be disassembled during overhaul, unless they were determined by oil pressure checks to have been malfunctioning.

NOTE: Use care when disassembling the valves as they are under spring pressure.

10. Remove from the pressure regulator bore the retaining pin, the pressure regulator boost valve sleeve, boost valve, the pressure regulator spring, the two spring seats, and the pressure regulator valve.

11. Remove from the converter clutch actuator bore the retaining pin, bore plug, spring and the converter clutch actuator valve.

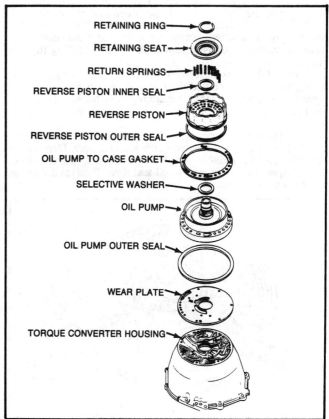

RETAINING RING
RETAINING SEAT
RETURN SPRINGS
REVERSE PISTON INNER SEAL
REVERSE PISTON
REVERSE PISTON OUTER SEAL
OIL PUMP TO CASE GASKET
SELECTIVE WASHER
OIL PUMP
OIL PUMP OUTER SEAL
WEAR PLATE
TORQUE CONVERTER HOUSING

Reverse clutch assembly—exploded view (©General Motors Corp.)

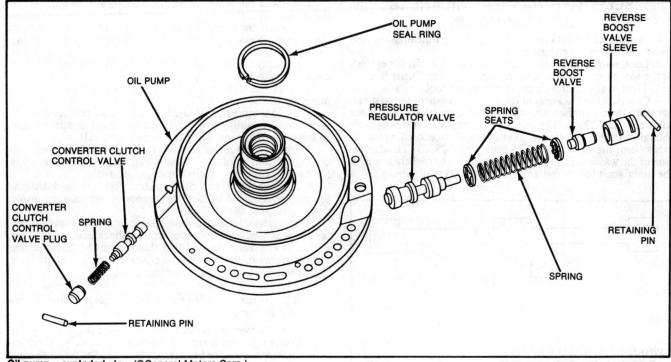

Oil pump—exploded view (©General Motors Corp.)

Assembly

1. Before assembly, inspect the pressure regulator boost valve, the pressure regulator valve and the torque converter clutch actuator valve for nicks or damage. Replace components as required.

2. Thoroughly clean the pressure regulator boost valve, the pressure regulator valve, and the converter clutch actuator valve. Immerse valves in transmission fluid before installing in their bores.

3. Install the pressure regulator valve, the two spring seats, the spring, the boost valve and the sleeve in the oil pump pressure regulator bore. Depress the pressure regulator boost valve sleeve until the back end lines up with the pin hole and insert the retaining pin to secure.

4. Install the converter clutch actuator valve, spring and bore plug in the oil pump converter clutch actuator bore. Depress the bore plug past the pinhole and insert the retaining pin to secure.

5. Inspect the oil pump hub oil seal rings. Replace if damage or side wear is found.

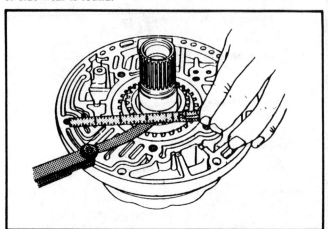

Oil pump clearance check (©General Motors Corp.)

6. Inspect the reverse clutch piston for damage. Replace as required. If necessary, install the two new oil seals on the reverse clutch piston. Install the reverse clutch piston onto the rear face of the oil pump. Be sure to use transmission fluid to aid in the installation.

7. Inspect the reverse clutch piston springs. Replace parts as required. Install all of the reverse clutch piston return springs. Set the retaining seat and the snap ring in place.

8. Compress the return springs using compressor tool J-23327 or equivalent. Care should be taken not to damage the retainer, should it catch in snap ring groove.

9. Install the snap ring using the proper tool. Do not air check the reverse clutch at this time, as the clutch is not complete and damage to the return spring retaining seat may occur.

10. Install the oil pump gears using the location mark made before disassembly.

11. Check the end clearance of both gears to the oil pump face. Be sure to measure between the face of the gears and the pump face, not between the crescent and the pump face. Use a straight edge and a feeler gage. Clearance should be between .0127-.0839 mm or .0005 to .00325 inch.

12. Install the oil pump wear plate onto the oil pump. Insert guide pins into the oil pump for alignment with the torque converter housing. Inspect the front face of the torque converter housing for oil leaks. Install a new converter housing oil seal, using the proper tool.

13. Install the torque converter housing on the oil pump. Loosely install the oil pump bolts into the converter housing.

14. Use the torque converter housing to oil pump aligning tool to align the two units. The tool should bottom on the oil pump gear.

NOTE: Failure to use this tool will cause pump damage when the transmission is operated after assembly.

15. Tighten the five inner torque converter housing bolts to 14 ft. lbs. Rotate the alignment tool to check for freeness. Remove the tool.

16. Install a new pump flange gasket and rubber seal.

17. Position the proper size selective thrust washer onto the oil pump shaft. Retain the washer using petroleum jelly.

18. Install the two guide pins in the transmission case and lower the torque converter housing and oil pump into the transmission case. Bolt the two units together and torque the retaining bolts to 25 ft. lbs.

19. Check for correct assembly by turning the input shaft. Install a new O-ring seal on the input shaft.

GOVERNOR HUB

Disassembly

1. Using the proper tool, remove the snap ring from the output shaft.

2. Remove the governor hub from the output shaft.

Assembly

1. Before installation, inspect the three seal rings on the governor hub. Replace as required.

2. Remove the governor hub oil screen. Use care not to damage or lose the oil screen. Inspect the screen, clean it with solvent and air dry. Replace, if necessary.

3. Install the oil screen flush with the governor hub.

4. Inspect the governor hub splines for cracks or chipped teeth. Replace the governor hub if required.

5. Turn the case so that the bottom of the transmission is facing upward.

6. Slide the governor hub along the output shaft and seat it into the case. Use a liberal amount of transmission fluid on the oil seal rings.

7. Install the snap ring over the output shaft, using the proper installation tool.

GOVERNOR BODY AND SPEEDOMETER DRIVE GEAR

Disassembly

1. Depress the secondary valve spring, using the proper tool. Remove the secondary valve spring retainer.

2. Remove the secondary valve spring, secondary valve and primary valve from the governor body.

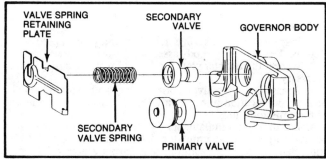

Governor assembly—exploded view (©General Motors Corp.)

3. Inspect the primary and secondary valve for nicks, burrs, etc. If necessary, use crocus cloth to remove any small burrs. Do not remove the sharp edges of the valve since these edges perform a cleaning action with the valve bore.

4. Inspect the secondary valve spring for breakage and distortion. Replace as required. Clean the assembly in solvent, air dry and blow out all oil passages. Inspect all oil passages and valve bores for nicks, burrs or varnish build up. Replace as necessary.

Assembly

1. Install the primary valve in the governor placing the small portion of the valve in first. Use a liberal amount of transmission fluid. There is no spring for the primary valve.

2. Install the secondary valve in the governor with the small spool portion of the valve first. Install the secondary valve spring.

3. Depress the secondary valve spring and install the retainer. Install a new governor body gasket.

4. Bolt the governor body to the governor hub. Torque to 6 ft. lbs. The two governor valves should move freely after the governor body is torqued.

5. Install the speedometer drive gear retaining clip onto the output shaft.

6. While depressing the retaining clip, slide the speedometer gear over the output shaft and install the gear and retaining clip.

SERVO PISTON

Disassembly

1. Remove the servo piston return spring and apply rod.

2. Hold the servo piston sleeve at the flat portion of the sleeve with a wrench, loosen the adjusting bolt locknut and remove.

3. Use an arbor press to depress the servo piston sleeve and remove the piston sleeve retaining ring.

4. Push the sleeve through the piston and remove the cushion spring and spring retainer.

5. Remove the servo piston ring.

6. Inspect the cushion spring, adjusting bolt and piston sleeve for damage. Inspect the piston for damage and the piston ring for side wear, replace if necessary.

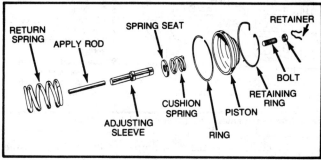

Servo piston and related components (©General Motors Corp.)

Assembly

1. Assemble the servo piston by reversing the disassembly procedure.

2. Install the servo apply rod, spring and piston into the case, using a liberal amount of transmission fluid.

3. Compress the servo piston spring using the piston compressor tool. Tap the servo piston with a rubber mallet while compressing, until the piston is seated, to avoid damage to the oil seal ring or case.

4. Install the servo retaining ring. Remove the tool.

5. Use a 3/16″ hex head wrench on the servo adjusting bolt and adjust the servo apply rod by tightening the adjusting bolt 40 in. lbs. Be certain that the locknut remains loose. Back off the bolt exactly five turns. Tighten the nut, holding the adjusting bolt and sleeve firmly.

VALVE BODY

Disassembly

1. Remove the manual valve and manual valve link from the valve body, then turn the unit over so that the transfer plate is facing upward. Remove the two bolts and take off the transfer plate and gasket.

2. Using caution to avoid damage to the valve body, use a small "C"-clamp to compress the accumulator piston.

3. Using the proper tool, remove the retaining ring.

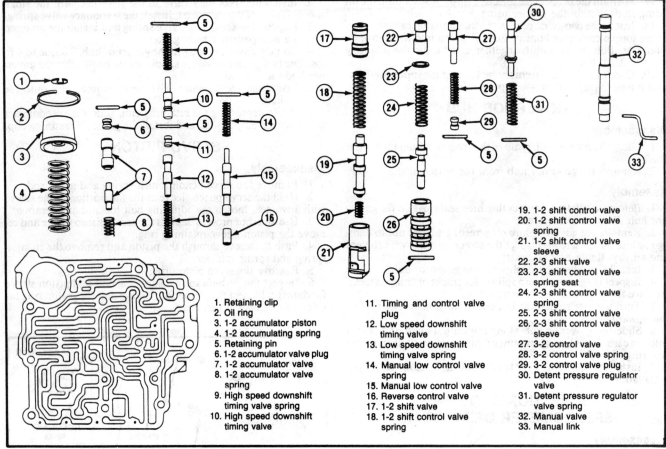

1. Retaining clip
2. Oil ring
3. 1-2 accumulator piston
4. 1-2 accumulating spring
5. Retaining pin
6. 1-2 accumulator valve plug
7. 1-2 accumulator valve
8. 1-2 accumulator valve spring
9. High speed downshift timing valve spring
10. High speed downshift timing valve

11. Timing and control valve plug
12. Low speed downshift timing valve
13. Low speed downshift timing valve spring
14. Manual low control valve spring
15. Manual low control valve
16. Reverse control valve
17. 1-2 shift valve
18. 1-2 shift control valve spring

19. 1-2 shift control valve
20. 1-2 shift control valve spring
21. 1-2 shift control valve sleeve
22. 2-3 shift valve
23. 2-3 shift control valve spring seat
24. 2-3 shift control valve spring
25. 2-3 shift control valve
26. 2-3 shift control valve sleeve
27. 3-2 control valve
28. 3-2 control valve spring
29. 3-2 control valve plug
30. Detent pressure regulator valve
31. Detent pressure regulator valve spring
32. Manual valve
33. Manual link

Valve body—exploded view (©General Motors Corp.)

4. Carefully loosen the "C"-clamp so that the spring pressure is gradually released.

5. Remove the accumulator piston, oil ring and spring.

6. Remove the 1-2 shift control valve retaining pin, valve sleeve, control valve, spring and valve. Check for burrs in the valve body bore made by the retaining pin before the other parts can be removed.

7. Similarly, remove the 2-3 shift control valve retaining pin and its related parts.

8. Remove the 3-2 control valve retaining pin and plug and its related parts.

9. Remove the detent pressure regulator valve retaining pin, its spring and detent pressure regulator valve.

10. Remove the high downshift timing valve retaining pins and its related parts.

11. Remove the manual low and reverse control valve retaining pin and its related parts.

12. Remove the manual low and reverse control valve retaining pin and its related parts.

13. Remove the 1-2 accumulator valve retaining pin and its related parts.

NOTE: It is extremely important to have a clean work area. Handle all parts carefully with clean hands. Tools should also be rinsed off since many valve failures are caused initially by dirt and other foreign matter preventing a valve from functioning properly. Use compressed air to blow out valve body passages. Check each valve for free movement in its bore. Crocus cloth may be used to remove small burrs. *DO NOT REMOVE THE SHARP EDGES OF THE VALVES. THESE SHARP EDGES ARE NECESSARY TO KEEP DIRT FROM STICKING IN THE BORE.*

The entire valve body should be inspected carefully and if any parts are damaged, the entire valve body should be replaced. Inspect the spring for distortion or collapsed coils.

Assembly

1. Inspect the valve springs for distortion or collapsed coils. Replace the entire valve body assembly if any parts are damaged.

2. Inspect the transfer plate for dents or distortion. Replace, if necessary.

3. Reassemble the valves, springs, plugs and retaining pins in their proper location into the valve body using a liberal amount of transmission fluid.

4. Install the accumulator spring and the piston into the valve body. Compress the accumulator piston using tool J-22269-01 or equivalent. Install the retaining ring. Install a new valve body gasket and transfer plate.

5. Bolt the transfer plate and gasket to the valve body. Torque the assembly to 7 ft. lbs. Install the steel check balls into the transmission case. Install a new transmission case to transfer plate gasket on the transmission case.

6. Install a guide pin in the transmission case for correct alignment of the valve body and transfer plate.

7. Install the manual valve into the valve body bore using liberal amount of transmission fluid. Install the long side of the manual valve link pin into manual valve and position the manual valve in its proper location.

8. Install the valve body onto the transmission case. Position the short side of the link pin into the inside selector lever.

9. Torque the valve body bolts to 14 ft. lbs. Start in the center of the valve body and work outward.

Transmission Assembly

If the automatic transmission was completely overhauled, rather than a specific component, follow the assembly procedure below.

1. Install a new seal for the selector lever shaft and carefully install the shaft from the outside. Be careful not to damage the seal. Lube with DEXRON® II.

2. Insert the spring pin to secure the selector lever shaft.

3. Install selector lever on shaft and tighten nut.

4. Insert the parking pawl actuating rod retaining ring.

5. With the transmission rotated so that the front of the case is upward, place the low band in the case and locate the band onto the anchor pins in the case.

6. Place the torrington bearing (reaction sun gear-to-case) into the case.

7. Install the reaction sun gear and drum into the band so that the reaction sun gear is facing upward. Then install the torrington bearing (output shaft-to-output sun gear) onto the sun gear. Petroleum jelly will help hold it in place.

8. Install the torrington bearing (sun gear-to-output shaft) and thrust washer (sun gear-to-output shaft) and planetary carrier assembly into the case so that it splines with the reaction sun gear.

9. Make sure the second clutch drive plates are aligned. Then, insert the third clutch drum, seating the third clutch drum splines into the second clutch spline plates.

10. Holding the assemblies together by grasping the input shafts, lower into the transmission case, indexing the ring gear in the second clutch drum with the long planetary pinion gear teeth.

11. Install the reverse clutch pressure plate with the flat side up. Make certain that the lug on the pressure plate fits into one of the narrow notches in the case. Then install a reverse clutch steel plate, composition plate, steel, composition, etc. into the case using plenty of DEXRON® II on the clutch surfaces.

12. Install the reverse clutch cushion plate (wave washer) making sure that all three of its lugs are engaged in the narrow notches in the case.

NOTE: At this point the end play must be measured. Jig up a dial indicator to index on the second clutch drum hub. Desired end play is .014 to .031 inch (0.36 mm to 0.79 mm). Compare the thickness of the selective washer (washer number 1, oil pump hub-to-second clutch). If this washer, in conjunction with the measured end play will not produce the desired end play, select a different washer from the chart.

SELECTIVE WASHER APPLICATION CHART
180C Automatic Transmission

Part Number	Thickness (inches)
5258202	.069-.074
5258203	.075-.079
5258204	.080-.084
5258205	.085-.089
5258206	.090-.094
5258207	.095-.100

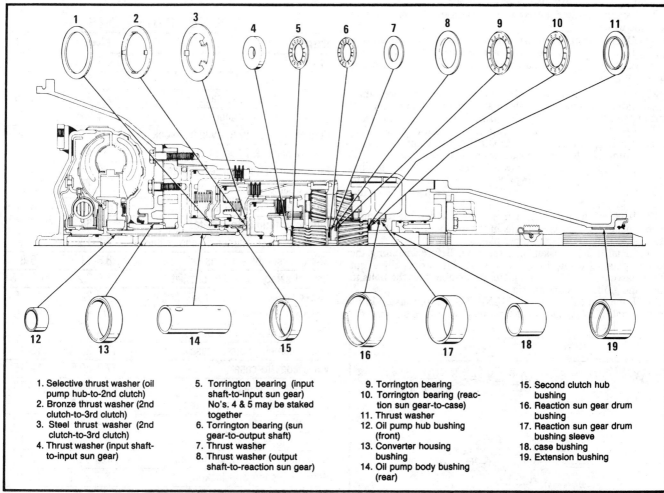

1. Selective thrust washer (oil pump hub-to-2nd clutch)
2. Bronze thrust washer (2nd clutch-to-3rd clutch)
3. Steel thrust washer (2nd clutch-to-3rd clutch)
4. Thrust washer (input shaft-to-input sun gear)
5. Torrington bearing (input shaft-to-input sun gear) No's. 4 & 5 may be staked together
6. Torrington bearing (sun gear-to-output shaft)
7. Thrust washer
8. Thrust washer (output shaft-to-reaction sun gear)
9. Torrington bearing
10. Torrington bearing (reaction sun gear-to-case)
11. Thrust washer
12. Oil pump hub bushing (front)
13. Converter housing bushing
14. Oil pump body bushing (rear)
15. Second clutch hub bushing
16. Reaction sun gear drum bushing
17. Reaction sun gear drum bushing sleeve
18. case bushing
19. Extension bushing

Thrust washer and bearing location guide (©General Motors Corp.)

13. After determining the proper selective washers, install on the oil pump shaft with petroleum jelly. Use two headless bolts as guide pins in the case to help guide the pump in place. After installing the guide pins, with a new pump flange gasket installed, lower the converter housing and oil pump assembly into the case.

14. Remove the guide pins; bolt the converter housing to the case. The bolts should be torqued to 22-26 foot-pounds. Make sure the input shaft turns as a check for correct assembly.

15. For governor installation, it is beneficial to turn the transmission so that the bottom (oil pan flange) is facing upward. Then slide the governor hub onto the output shaft and seat it in the case. Protect the seals on the hub with a liberal amount of DEXRON® II.

16. Install the snap ring over the output shaft.

17. Using a new governor body gasket, bolt the governor body to the hub; torque to 6-8 foot-pounds. The governor valves should move freely.

18. Install the speedometer drive gear retaining clip on the output shaft, then slide the gear on. Depress the clip and snap the gear into place.

19. With a new gasket, install the extension housing, making sure the parking pawl shaft will align. Torque the bolts to 20-30 foot-pounds.

20. Install the speedometer driven gear and the retainer into the slot in the speedometer housing. Torque the retainer bolt to 6-8 foot-pounds.

Install the detent valve, using a new seal and plenty of DEXRON® II. Make sure that the detent valve, sleeve, spring and spring seat are installed in proper order.

21. Depress the detent valve spring and insert the spring pin to hold the assembly.

NOTE: The detent valve sleeve contains slots which must face the oil pan. Make sure the spring pin is inserted into the groove provided in the sleeve and not into one of the oil passage slots in the sleeve.

22. Install the modulator valve and sleeve into the case, small end of valve first. Use a new "O-Ring" on the vacuum modulator and install plunger, and then thread modulator into the case.

23. Install the servo apply rod, spring and piston in the transmission case, again with DEXRON® II as lubricant.

24. Compress the servo piston spring being careful of the oil seal ring as the piston moves inward. Install the servo retaining ring. Adjust the servo by using a ³/₁₆" hex wrench on a socket to tighten the bolt to 40 inch-pounds, being certain that the lock nut remains loose.

25. Back off the bolt *exactly* five turns. Tighten the lock nut.

26. Install the steel check balls in the oil passages of the case. Make sure they are in the proper places

27. Install a new gasket onto the case. Before installing the transfer plate. Use bolts for guide pins to aid in correct alignment of the valve body and transfer plate. Remember that the bolts to make the guide pins are metric.

28. Lube manual valve with DEXRON® II and install in its bore. Install the *long side* of the manual valve link pin into the manual valve.

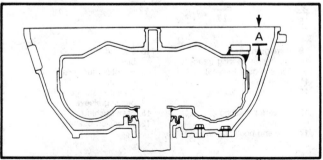

Torque converter installation (©General Motors Corp.)

29. Install the *short end* of the manual valve link into the selector lever. Install the valve body and transfer plate assembly over the guide pins; make sure valve link does not come off.

30. Install the selector lever roller spring and retainer. These bolts are torqued to 13-15 foot-pounds. The valve body bolts should be torqued starting from the center and working outward. These too should be torqued to 13-15 foot-pounds, as well as the reinforcement plate bolts which can be installed at this point.

31. Replacement of the oil strainer is recommended, as well as using a new gasket. Torque the screws to 13-15 foot-pounds.

32. With a new gasket, install the servo cover and torque the screws to 17-19 foot-pounds.

33. Install a new pan gasket and pan; torque the screws to 7-9 foot-pounds.

34. Slide the torque converter in place, making sure that the pump hub keyway is seated in the oil pump drive lugs. Check the distance from the case flange to the converter hub. It should be .83 to .91 inch. Turn the converter to check for free movement.

TORQUE LIMITATIONS

Item	N•m	Lb. Ft.
Oil pan-to-case	9.8-12.8	7-10
Modulator assembly	52	38
Extension housing-to-case	27.5-34.3	20-25
Oil pressure check plug	6.4-9.8	5-7
Converter housing-to-cylinder block	27-41	20-30
Transmission support-to-extension	39-48	29-36
Shift lever-to-selector lever shaft	20-34	15-25
Detent cable, retainer-to-case	7-10	5-7
Oil cooler fittings-to-case	11-16	8-12
Oil cooler fittings-to-radiator	20-34	15-25
Oil cooler hose clamps-to-cooler lines	0.8-1.2	.6-.9
Shifter assembly-to-console	8.5-11	6-8
Neutral safety switch-to-bracket	1.6-2.2	1.2-1.6
Lower cover-to-converter housing	18-22	13-16
Flexplate-to-converter	41-54	30-40
Transfer plate-to-valve body	7.8-10.8	6-8
Reinforcement plate-to-case	17.7-20.6	13-15
Valve body-to-case	17.7-20.6	13-15
Servo cover-to-case	22.6-25.5	17-19
Converter housing-to-oil pump	17.7-20.6	13-15
Converter housing-to-case	32.4-35.3	24-26
Selector lever locknut	10.8-14.7	8-11
Governor body-to-governor hub	7.8-9.8	6-7
Servo adjusting bolt locknut	16.7-20.6	12-15
Planetary carrier lock plate	27-48	20-35

SPECIAL TOOLS

TRANSMISSION HOLDING FIXTURE

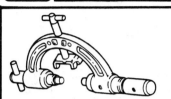

CLUTCH PISTON COMPRESSOR

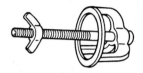

CONVERTER HOUSING BUSHING REMOVER/ INSTALLER

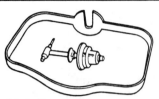

CONVERTER LEAK TEST FIXTURE

REACTION SUN GEAR DRUM BUSHING SLEEVE INSTALLER

2ND CLUTCH PISTON SEAL INSTALLER

CAPE CHISEL

BUSHING REMOVER

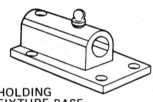

HOLDING FIXTURE BASE

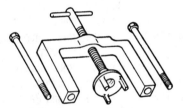

SERVO/3RD CLUTCH PISTON SPRING COMPRESSOR

REACTION SUN GEAR DRUM BUSHING INSTALLER

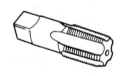

OIL PUMP BUSHING REMOVER (3/4 - 14 NPT)

REAR CASE BUSHING REMOVER/INSTALLER

CONVERTER-TO-OIL PUMP ALIGNMENT TOOL

DRIVER HANDLE

2ND CLUTCH DRUM BUSHING REMOVER/ INSTALLER

CONVERTER HOUSING OIL SEAL INSTALLER

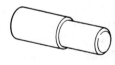

OIL PUMP BUSHING INSTALLER

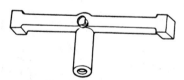

OIL PUMP-TO-2ND CLUTCH DRUM GAGING TOOL

SLIDE HAMMER

VACUUM MODULATOR WRENCH

EXTENSION HOUSING OIL SEAL INSTALLER

EXTENSION HOUSING BUSHING REMOVER/ INSTALLER

CONVERTER HOUSING SEAL REMOVER

INDEX

GENERAL MOTORS
TURBO HYDRA-MATIC 200-4R

APPLICATIONS

TRANSMISSION APPLICATION CHART

Year	Division	Transmission Model
1981 and Later	Buick	200-4R
1981 and Later	Cadillac	200-4R
1981 and Later	Oldsmobile	200-4R
1981 and Later	Pontiac	200-4R
1981 and Later	Chevrolet	200-4R

GENERAL DESCRIPTION

The T.H.M. 200-4R automatic transmission is a fully automatic unit consisting primarily of a three element hydraulic torque converter with a converter clutch, a compound planetary gear set and an overdrive unit. Five multiple-disc clutches, two roller clutches, and a band provide the friction elements required to obtain the desired function of the compound planetary gear set and the overdrive unit.

The torque converter couples the engine to the planetary gears and overdrive unit through oil and hydraulically provides additional torque multiplication when required. The combination of the compound planetary gear set and the overdrive unit provides four forward ratios and one reverse.

The torque converter consists of a converter clutch, a driving member, driven member and a reaction member known as the pressure plate and damper assembly, pump, turbine and stator.

Changing of the gear ratios is fully automatic in relation to vehicle speed and engine torque. Vehicle speed and engine torque signals are constantly fed to the automatic transmission to provide the proper gear ratio for maximum efficiency and performance at all throttle openings.

The shift quadrant has seven positions, which are indicated in the following order: P, R, N, D, 3, 2, 1.

Transmission and Torque Converter Identification

TRANSMISSION

The THM 200-4R automatic transmission can be identified by the serial number plate which is located on the right side of the transmission case. This identification tag contains information that is important when servicing the unit.

TORQUE CONVERTER

The torque converter is a welded unit and cannot be disassembled for service. Any internal malfunctions require the replacement of the converter assembly. The replacement converter must be matched to the model transmission through parts identification. No specific identification is available for matching the converter to the transmission for the average repair shop.

DIESEL ENGINES

Vehicles equipped with diesel engines use a different torque converter. To identify these units, examine the weld nuts. Most gas engine converters have their weld nuts spot welded onto the converter housing, usually in two spots. The diesel converters have the weld nuts completely welded around their entire circumference.

TURBOCHARGED V6 ENGINES

Vehicles equipped with turbocharged V6 engines use a different torque converter. These units have a high stall speed converter, allowing a stall speed of about 2800 rpm. These converters must not be replaced with a standard converter, otherwise performance will be sluggish and unsatisfactory. When ordering replacements for the turbocharged units, make certain to specify that the vehicle has turbocharging in order to obtain the proper replacement.

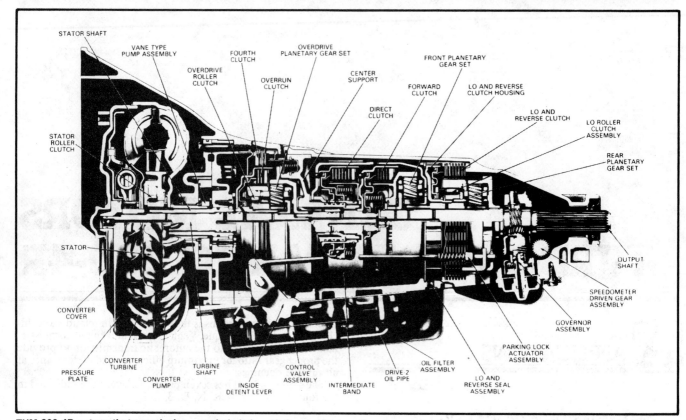

THM 200-4R automatic transmission—exploded view (© General Motors Corp.)

Metric Fasteners

Metric bolt sizes and thread pitches are used for all fasteners on the THM 200-4R automatic transmission. The use of metric tools is mandatory in the service of this transmission.

Do not attempt to interchange metric fasteners for inch system fasteners. Mismatched or incorrect fasteners can result in damage to the transmission unit through malfunctions, breakage or possible personal injury. Care should be taken to reuse the fasteners in the same locations as removed, whenever possible. Due to the large number of alloy parts used, torque specifications should be strictly observed. Before installing capscrews into aluminum parts, always dip screws into oil to prevent the screws from galling the aluminumn threads and to prevent seizing.

Fluid Capacities

The fluid capacities are approximate and the correct fluid level should be determined by the dipstick indicator. Use only Dexron® II automatic transmission fluid when adding or servicing the THM 200-4R automatic transmission.

FLUID CAPACITIES THM 200-4R

Division	Fluid Change (Pints)	Overhaul (Pints)
Buick	7	22①
Cadillac	7②	22
Oldsmobile	7	22
Pontiac	7	22
Chevrolet	7	22

①1983 models use 23 pints   ②1983 models use 10.6 pints

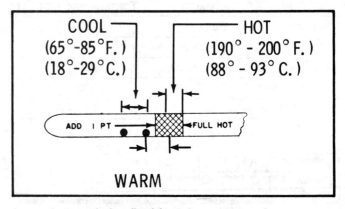

Automatic transmission dipstick (© General Motors Corp.)

Checking Fluid Level

The THM 200-4R transmission is designed to operate at the "FULL HOT" mark on the dipstick at normal operating temperatures, which range from 190° to 200° F. Automatic transmissions are frequently overfilled because the fluid level is checked when cold and the dipstick level reads low. However, as the fluid warms up, the level will rise, as much as ¾ of an inch. Note that if the transmission fluid is too hot, as it might be when operating under city traffic conditions, trailer towing or extended high speed driving, again an accurate fluid level cannot be determined until the fluid has cooled somewhat, perhaps 30 minutes after shutdown.

To determine proper fluid level under normal operating temperatures, proceed as follows:

1. Make sure vehicle is parked level.
2. Apply parking brake; move selector to Park.
3. Start engine but do not race engine. Allow to idle.
4. Move selector through each range, back to PARK, then check level. The fluid should read "FULL HOT" (transmission end of dipstick too hot to hold comfortably).

Do not overfill the transmission. Overfilling can cause foaming and loss of fluid from the vent. Overheating can also be a result of overfilling since heat will not transfer as readily. Notice the condition of the fluid and whether there seems to be a burnt smell or metal particles on the end of the dipstick. A milky appearance is a sign of water contamination, possibly from a damaged cooling system. All this can be a help in determining transmission problems and their source.

Fluid Drain Intervals

The main considerations in establishing fluid change intervals are the type of driving that is done and the heat levels that are generated by such driving. Normally, the fluid and strainer would be changed at 100,000 miles. However, the following conditions may be considered severe transmission service.
1. Heavy city traffic
2. Hot climates regularly reaching 90° F. or more
3. Mountainous areas
4. Frequent trailer pulling
5. Commercial use such as delivery service, taxi or police car use

If the vehicle is operated under any of these conditions, it is recommended that the fluid be changed and the filter screen serviced at 15,000 mile intervals. Again, be sure not to overfill the unit. Only one pint of fluid is required to bring the level from "ADD" to "FULL" when the transmission is hot.

Torque Converter Clutch Inoperative

Some 1983 vehicles equipped with the 200-4R automatic transmission may experience a condition of an inoperative torque converter clutch. On DFI equipped vehicles, a code 39 would also be present. This condition may be caused by the torque converter clutch harness locating clip grounding the wiring to the automatic transmission control valve body assembly.

For identification purposes, the automatic transmission serial number tag is located on the right rear of the transmission case. All 200-4R automatic transmissions with a date build code prior to 245 may have been assembled with the TCC harness locating clip positioned incorrectly.

To correct the inoperative TCC condition, replace the applicable torque converter clutch solenoid and harness assembly (P/N 8634960 for AA model transmissions and P/N 8634961 for OM model transmissions). Correctly position the clip by rotating it as far rearward as possible; then, bend it toward the depression in the control valve body assembly (approximately 60° clockwise from its original position.

HARSH AUTOMATIC TRANSMISSION UPSHIFTS

Some 1982 and 1983 "C" body vehicles, equipped with the HT4100 engine may experience delayed and/or harsher than normal automatic transmission upshifts, especially the 1-2 shift. This condition may be caused by a misrouted transmission TV cable, that results in an incorrect adjustment between the cable actuating lever on the throttle linkage and the TV link in the automatic transmission.

When diagnosing a vehicle exhibiting this condition, inspect the TV cable housing at the locator clip for any indication of the housing having been moved ¾ to 1 inch from its original mounting. If the cable has been moved, it will be necessary to reroute the cable to its correct position, before re-adjusting the TV cable. Replacement of the cable assembly is necessary only if the cable binds or sticks because of being kinked in the locator clip area. Be sure that the cable is routed around the oil pressure switch wiring harness, HEI distributor, air pipes and the manifold vacuum pipes.

"BUZZ" CONDITION ELIMINATION DUE TO NEW PUMP DESIGN

New design pump assemblies were introduced late in the 1982 model year. This new design eliminates a buzz condition that is caused by the pressure regulator valve. The design changes involve a new pump body, pump cover and the pressure regulator valve with its orifice.

When servicing these components on all 1981 and 1982 automatic transmissions, a complete pump assembly must be installed. See the pump assembly chart for the required part ordering information.

When replacing the pump, be sure not to mix the old pump halves (body and cover) with the new design pump halves as this may create a "no drive" condition, or may not eliminate the buzz condition at all.

AUTOMATIC TRANSMISSION OIL LEAK

Early production 1983 vehicles equipped with the TH4100 engine may experience an automatic transmission oil leak at the front edge of the automatic transmission oil pan. This condition can be caused by interference between the back edge of the flywheel cover and the front edge of the automatic transmission oil pan flange. A new design flywheel cover, part number 1627822, has been put into production to eliminate this interference.

Vehicles with this condition can be corrected by removing about 1/16-1/8 inch of material from the top, rear edge of the two radii and the back edge of the flywheel cover with a file. Before resealing the automatic transmission oil pan with RTV sealant make certain its flange is not deformed.

SHIFT BUSINESS WHILE IN CRUISE CONTROL—MODEL BY

Some 1982 vehicles equipped with the 4.1 Liter V6 engine may experience a transmission shift busyness at highway speeds with the cruise control engaged. This problem refers to the cycling of the automatic transmission between fourth and third gears. The condition is most noticeable while driving up inclines. To correct this condition, install transmission service package 8634987 into the control valve body assembly. This service package contains a 3-4 throttle valve, bushing and spring and a steel coiled pin.

PUMP ASSEMBLY CHART

Model Year	Transmission Model	Pump Assembly	Pressure Regulator
1981-82	BY,AA,AD	8639077	8637546
1982	OM	8639135	8637546
1982	AH	8639079	8637546

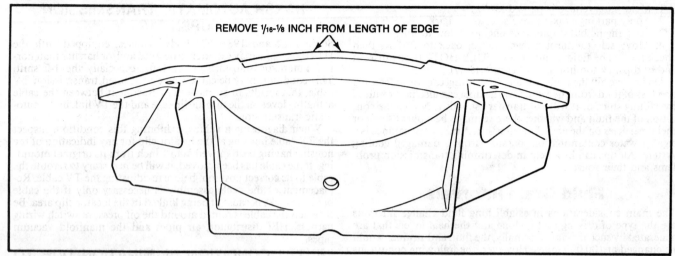

REMOVE ¹/₁₆-¹/₈ INCH FROM LENGTH OF EDGE

Grinding points on back edge of flywheel cover
(© General Motors Corp.)

PARKING LOCK ASSEMBLY—DESIGN CHANGE

Starting in October, 1982, the parking lock rod assembly in the 1983 model year automatic transmission was changed from a two piece unit to a one piece assembly.

When servicing the parking rod mechanism, it is important that the correct parts be used. The detent levers and rod assemblies are not interchangeable between the two designs. The two piece design part numbers are 8639110 and 8634184. The one piece design part number is 8634991.

NO FOURTH GEAR

Some 1982 and 1983 vehicles equipped with the THM 200-4R automatic transmission may experience a no fourth gear engagement condition. On DFI equipped vehicles, codes 29 and/or 39 could also be present. The no fourth gear condition may be caused by a broken hub weld on the overrun clutch housing.

When diagnosing vehicles exhibiting this condition, test drive the unit applying a light throttle in the drive range to about 30-35 MPH (third gear). Move the selector lever into manual second and release the throttle. The transmission should downshift and coastdown overrun braking should occur.

If this does not happen, remove the automatic transmission from the vehicle. Inspect the overrun clutch housing for a brake at the hub weld. If the hub weld is broken, the cause is due to misaligned pinions in the overdrive carrier. In all cases where the hub weld is broken on the overrun clutch housing, both the overrun clutch housing and the overdrive carrier must be replaced. Use part numbers 8634107 (overrun clutch housing) and 8634117 (overdrive carrier).

SELECTIVE WASHER—NEW DESIGN

Any time internal repairs are performed to the 200-4R automatic transmission, inspect the selective washer used to establish front (input) unit end play for wear or damage.

Beginning in late March 1983, a new design selective washer went into production and is now available for service on the forward clutch shaft to output shaft units. This new service part is made from a different material and can be identified by a raised portion on it's identification tab.

When servicing a transmission that necessitates the replacement of the front (input) unit selective washer due to wear or damage, use the following procedure:

1. Inspect both shafts for wear or damage on the selective washer mating surfaces. If damage or wear is found, replace the necessary components.

2. Select the proper selective washer from the selective washer chart.

3. Check the front end play to verify the proper washer selection. Front end play should be 0.022-0.051 in.

SELECTIVE WASHER CHART

ID Number	Color	Part Number	Thickness (inches)
1	—	8639291	0.065-0.070
2	—	8639292	0.070-0.075
3	Black	8639293	0.076-0.080
4	Light Green	8639294	0.081-0.085
5	Scarlet	8639295	0.086-0.090
6	Purple	8639296	0.091-0.095
7	Cocoa Brown	8639297	0.096-0.100
8	Orange	8639298	0.101-0.106
9	Yellow	8639299	0.106-0.111
10	Light Blue	8639300	0.111-0.116
11	Blue	8639301	0.117-0.121
12	—	8639302	0.122-0.126
13	Pink	8639303	0.127-0.131
14	Green	8639304	0.132-0.136
15	Gray	8639305	0.137-0.141

TROUBLE DIAGNOSIS

CLUTCH AND BAND APPLICATION CHART
Turbo Hydra-matic 200-4R

Selector	Converter Clutch	Overrun Clutch	Intermediate Band	Overdrive Roller Clutch	Direct Clutch	Low Reverse Clutch	Forward Clutch	Fourth Clutch	Low and Reverse Clutch
Park	Released	Released	Released	Holding	Released	Not Holding	Released	Released	Released
Neutral	Released	Released	Released	Holding	Released	Not Holding	Released	Released	Released
Drive (1st gear)	—	Holding	—	—	—	Holding	Applied	—	—
Drive (2nd gear)	Released	—	Applied	Holding	—	—	Applied	—	—
Drive (converter clutch applied)	Applied	—	Applied	Holding	—	—	Applied	—	—
Drive (3rd gear)	Applied	—	—	Holding	Applied	—	Applied	—	—
Drive (4th gear)	Applied	—	—	—	Applied	—	Applied	Applied	—
Reverse	—	—	—	Holding	Applied	Applied	—	—	—

CHILTONS THREE "C's" TRANSMISSION DIAGNOSIS CHART
Turbo Hydra-matic 200-4R

Condition	Cause	Correction
Transmission oil leak	a) Attaching pan bolts loose	a) Retorque pan bolts
	b) Filler pipe seal damaged	b) Replace filler pipe seal
	c) T.V. cable seal damaged, missing or improperly positioned.	c) Replace T.V. cable seal as required.
	d) Real transmission seal damaged	d) Replace rear transmission seal
	e) Speedometer drive O-ring damaged	e) Replace speedometer drive O-ring seal
	f) Line pressure tap plug	f) Replace or repair as necessary
	g) Fourth clutch pressure tap plug	g) Replace or repair as necessary
	h) Porous casting	h) Replace transmission case as required
	i) Intermediate servo O-ring damaged	i) Replace intermediate servo O-ring
	j) Front pump seal damaged or bolts loose	j) Replace pump seal or tighten pump bolts

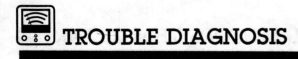

CHILTONS THREE "C's" TRANSMISSION DIAGNOSIS CHART

Condition	Cause	Correction
No drive in drive range	a) Low oil level	a) Adjust oil level
	b) Manual linkage misadjusted	b) Adjust manual linkage
	c) Low oil pressure due to plugged oil filter	c) Replace transmission oil filter
	d) Pump assembly pressure regulator valve stuck	d) Repair as required
	e) Pump rotor tangs damaged by torque converter	e) Correct the problem as required
	f) Overdrive unit springs missing in the roller clutch or rollers galled	f) Repair or replace defective components as required
	g) Forward clutch does not apply	g) Repair forward clutch assembly
	h) Cup plug leaking or missing in the rear of the forward clutch shaft	h) Replace cup plug
	i) Low and reverse roller clutch springs missing or rollers galled	i) Repair or replace low and reverse roller clutch assembly
High or low oil pressure	a) Throttle valve cable misadjusted, broken, binding or wrong part	a) Repair or replace throttle valve cable as required
	b) Throttle valve or plunger valve binding	b) Repair or replace defective component as required
	c) Pressure regulator valve binding	c) Free binding pressure regulator valve as required
	d) T.V. boost valve binding	d) Free binding T.V. boost valve
	e) T.V. boost valve not the right part number—low pressure only	e) Install the correct T.V. boost valve
	f) Reverse boost valve binding	f) Free binding reverse boost valve
	g) Manual valve unhooked or mispositioned	g) Repair or replace manual valve as required
	h) Pressure relief valve ball or spring missing	h) Repair or replace pressure relief valve components
	i) Pump slide seal stuck or seal damaged or missing	i) Replace seal as required
	j) Pump air bleed orifice missing or damaged	j) Replace as required
	k) T.V. limit valve binding	k) Correct T.V. limit valve as required
	l) Line bias valve binding in open position—high pressure only	l) Repair or replace valve as required
	m) Line bias valve binding in closed position—low pressure only	m) Repair or replace valve as required
1-2 shift—full throttle only	a) Throttle valve cable stuck, unhooked or broken	a) Repair or replace cable as required
	b) Throttle valve cable misadjusted	b) Adjust cable as required
	c) T.V. exhaust ball lifter or number 5 ball binding or mispositioned	c) Replace or reposition T.V. exhaust ball number 5 as required
	d) Throttle valve and plunger binding	d) Correct as required

CHILTONS THREE "C's" TRANSMISSION DIAGNOSIS CHART

Condition	Cause	Correction
1-2 Shift—Full throttle only	e) Valve body assembly gaskets leaking, damaged or incorrectly installed	e) Replace gaskets as necessary
	f) Transmission case porosity	f) Replace transmission case
First speed only—no 1-2 shift	a) Governor and governor feed passages plugged	a) Free blockage as necessary
	b) Governor ball or balls missing in governor assembly	b) Replace ball or balls as required
	c) Inner governor cover rubber O-ring seal missing or leaking	c) Replace rubber O-ring seal
	d) Governor seal shaft seal missing or damaged	d) Replace governor shaft seal
	e) Governor driven gear stripped	e) Replace governor driven gear
	f) Governor weights binding on pin	f) Correct problem as required
	g) Governor driven gear not engaged with governor shaft	g) Correct problem as necessary
	h) Control valve assembly 1-2 shift, Low 1st detent, or 1-2 throttle valve stuck in downshift position	h) Free stuck valve as required
	i) Control valve assembly spacer plate gaskets in wrong position	i) Replace spacer plate gaskets as required
	j) Intermediate band anchor pin missing or unhooked from band	j) Repair or replace band anchor pin
	k) Intermediate servo cover oil seal ring missing	k) Check and replace ring as required
	l) Wrong intermediate cover and piston	1) Check part and replace as necessary
	m) 1-2 accumulator housing bolts loose or housing face damaged	m) Repair or replace parts as required
	n) 1-2 accumulator plate missing or damaged	n) Replace spacer plate as required
Drive in Neutral	a) Manual linkage misadjusted	a) Adjust linkage
	b) Forward clutch does not release	b) Correct as necessary
	c) Forward clutch exhaust ball sticking	c) Free sticking exhaust ball as required
	d) Forward clutch plates burned together	d) Replace forward clutch assembly
	e) Case leakage at forward clutch passage (D4)	e) Correct case as required
First and second speed only—no 2-3 shift	a) 2-3 shift valve or 2-3 throttle valve stuck in the downshift position	a) Free the stuck valve or valves as required
	b) Valve body gaskets leaking mispositioned or incorrectly installed	b) Replace valve body spacer gaskets as required
	c) Reverse/3rd check ball not seating or damaged	c) Properly seat check ball or replace
	d) Transmission case porosity	d) Replace transmission case as needed

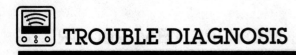

CHILTONS THREE "C's" TRANSMISSION DIAGNOSIS CHART
Turbo Hydra-matic 200-4R

Condition	Cause	Correction
First and second speeds only, no 2-3 shifts	e) Direct clutch feed passage in the center support plugged or not drilled through	e) Free blockage or drill passage through
	f) Direct clutch inner seal ring missing or damaged	f) Replace or install inner seal ring
	g) Center oil seal ring missing or damaged on direct clutch	g) Replace or install seal ring
	h) Check ball or retainer missing or damaged from direct clutch piston	h) Replace direct clutch ball
	i) Direct clutch plates damaged or missing	i) Replace damaged or missing plates as required
	j) Direct clutch backing plate snap ring out of groove	j) Correct as required
	k) Release spring guide mislocated, preventing piston check ball from seating in retainer	k) Correct as required
	l) Servo to case oil seal broken or missing on intermediate servo piston	l) Replace seal on piston
	m) Intermediate servo or capsule missing or damaged	m) Correct as required
	n) Exhaust hole in case between servo piston seal rings plugged or not drilled	n) Free blockage or drill passage
	o) Bleed orifice cup plug missing from intermediate servo pocket in case	o) Install orifice cup plug
Slips on 1-2 shift	a) Low fluid level	a) Correct fluid level
	b) Spacer plate and gaskets damaged or not installed properly	b) Correct as required
	c) Accumulator valve sticking in valve body or weak or missing spring	c) Free sticking valve or replace spring as required
	d) 1-2 accumulator piston seal leaking, spring missing or broken	d) Replace seal or spring as required
	e) Leak between 1-2 accumulator piston and pin	e) Correct as required
	f) 1-2 accumulator piston binding or piston bore damaged	f) Replace or repair accumulator piston as required
	g) Wrong intermediate band apply pin	g) Install correct pin
	h) Leakage between pin and case	h) Correct as required
	i) Apply pin feed hole not drilled completely	i) Drill hole through
	j) Porosity in intermediate servo piston	j) Replace piston
	k) Cover to intermediate servo seal ring missing or damaged	k) Replace seal ring
	l) Leak between intermediate servo apply pin and case	l) Find and correct leak
	m) T.V. cable not adjusted	m) Adjust T.V. cable
	n) T.V. limit valve binding	n) Correct as required

CHILTONS THREE "C's" TRANSMISSION DIAGNOSIS CHART
Turbo Hydra-matic 200-4R

Condition	Cause	Correction
Slips on 1-2 shift	o) Intermediate band worn or burned	o) Replace intermediate band
	p) Case porosity in 2nd clutch passage	p) Replace transmission case as required
Rough 1-2 shift	a) T.V. cable not adjusted or binding	a) Adjust T.V. cable or free binding as required
	b) Throttle valve or T.V. plunger binding	b) Correct binding as required
	c) T.V. limit valve, accumulator and line bias valve binding	c) Correct binding problem as required
	d) Wrong intermediate servo supply pin	d) Install correct pin
	e) Intermediate servo to case oil seal ring missing or damaged	e) Replace oil seal ring
	f) Bleed cup plug missing in case	f) Install bleed cup plug as required
	g) 1-2 accumulator oil ring damaged or piston stuck	g) Replace oil ring or free stuck piston
	h) 1-2 accumulator broken, spring missing or bore damaged	h) Repair or replace as required
	i) 1-2 shift check ball #8 missing or stuck	i) Free or replace check ball as required
Slips 2-3 shift	a) Low fluid level	a) Correct fluid level
	b) T.V. cable not adjusted	b) Adjust T.V. cable
	c) Throttle valve binding	c) Free binding throttle valve
	d) Direct clutch orifice partially blocked in spacer plate	d) Free blockage as required
	e) Intermediate servo to case oil seal ring missing or damaged	e) Replace or install new ring
	f) Intermediate servo or piston bore damaged	f) Replace intermediate servo
	g) Intermediate servo orifice bleed cup plug missing in case	g) Install cup plug as required
	h) Direct clutch piston or housing cracked	h) Replace piston or housing as required
	i) Direct clutch piston seals cut or missing	i) Replace or install new seals
	j) Check ball or capsule damaged in direct clutch	j) Correct as required
	k) Center support seal rings damaged or missing	k) Replace or install new center support seal rings
Rough 2-3 shift	a) T.V. cable mispositioned	a) Reposition T.V. cable
	b) Throttle valve and plunger binding	b) Correct binding as required
	c) T.V. limit valve binding	c) Correct T.V. limit valve binding
	d) Exhaust hole undrilled or plugged between intermediate servo piston seals, not allowing the servo piston to complete its stroke	d) Unplug or drill hole as required

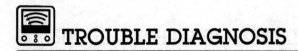 **TROUBLE DIAGNOSIS**

CHILTONS THREE "C's" TRANSMISSION DIAGNOSIS CHART
Turbo Hydra-matic 200-4R

Condition	Cause	Correction
Rough 2-3 shift	e) 3rd accumulator check ball number 2 missing or stuck	e) Free blockage or replace ball as required
	f) 3-2 exhaust check ball number 4 missing or stuck	f) Free blockage or replace ball as required
Slips 3-4 shift	a) Low fluid level	a) Correct fluid level
	b) Control valve assembly spacer plate or gasket mispositioned or damaged	b) Correct gasket or spacer plate as required
	c) Accumulator valve sticking causing low 3-4 accumulator pressure	c) Free sticking accumulator valve as required
	d) Weak or missing accumulator valve spring	d) Install or replace spring
	e) 3-4 accumulator piston stuck, bore damaged or oil ring damaged	e) Replace defective component as required
	f) Center support bolts loose	f) Tighten bolts
	g) Fourth clutch piston surface damaged or seals damaged	g) Replace piston or seals as required
	h) Improper clutch plate usage	h) Check and replace plates as needed
	i) Fourth clutch plates burned	i) Replace fourth clutch plates
Rough 3-4 shift	a) T.V. cable mispositioned	a) Reposition cable as required
	b) Throttle valve and plunger binding	b) Free binding problem as required
	c) T.V. limit valve binding	c) Free binding T.V. limit valve
	d) 3-4 accumulator piston stuck or bore damaged	c) Free piston or correct bore damage as required
	e) Fourth clutch piston stuck	d) Free fourth clutch piston
First, second and third speed only—no 3-4 shift	a) 3-4 shift valve or 3-4 throttle valve stuck	a) Free stuck valve as required
	b) Orifice in spacer plate blocked	b) Free blockage as required
	c) Center support oil passages plugged or not drilled	c) Correct as required
	d) Center support bolts loose	d) Tighten bolts
	e) Fourth clutch piston cracked or damaged	e) Replace piston as required
	f) Fourth clutch piston seals damaged, missing or not installed properly	f) Correct as required
	g) Fourth clutch burned	g) Replace fourth clutch
	h) Overrun clutch plates burned	h) Replace clutch plates
No engine braking in manual low—first gear	a) Manual linkage misadjusted	a) Adjust manual linkage
	b) D-3 orifice in spacer plate plugged	b) Free blockage in spacer plate
	c) Valve body gaskets leaking, damaged or installed wrong	c) Correct as required
	d) D-2 oil pipe leaking or out of position	d) Correct leaking as required
	e) Low overrun clutch valve binding in valve body	e) Free binding valve

No engine braking in manual low—first gear	f) Low/reverse check ball number 10 missing or defective	f) Install or replace check ball
	g) Low/detent check ball number 9 missing or defective	g) Install or replace check ball
	h) Low/reverse overrun clutch orifice in spacer plate plugged	h) Unplug or replace spacer plate as required
	i) PT/D-3 check ball number 3 check ball missing or defective	i) Install or replace check ball
	j) D-3 oil passage plugged or not drilled in turbine shaft or overrun clutch	j) Drill or unplug oil passage
	k) Oil seals missing or damaged in the overrun clutch piston	k) Replace seals as required
	l) Overrun clutch seals burned	l) Replace as required
	m) Overrun clutch backing plate snap ring out of groove	m) Replace as required
	n) Low reverse clutch piston seals missing or damaged	n) Replace seals as required
	o) Cup plug or seal missing or damaged	o) Install or replace cup plug
No engine braking in manual 2nd—2nd gear	a) Manual linkage misadjusted	a) Adjust manual linkage
	b) Valve body gaskets leaking or damaged	b) Replace gaskets as required
	c) D-2 oil pipe leaking or out of position	c) Reposition or replace pipe
	d) D-3 orifice in spacer plate plugged	d) Unplug orifice in spacer plate
	e) PT/D-3 check ball number 3 missing or defective	c) Install or replace check ball as required
	f) Intermediate servo cover to case oil seal ring missing or damaged	f) Install or replace ring as required
	g) Intermediate band off its anchor pin, broken or burned	g) Repair as required
	h) D-3 oil passage not drilled through in overrun clutch or turbine shaft	h) Drill passage through
	i) Oil seals missing or damaged in overrun clutch piston	i) Replace or install seals
	j) Overrun clutches burned	j) Replace overrun clutch
	k) Overrun clutch backing plate snap ring out of groove	k) Replace components as required
No engine braking in manual 3rd—3rd gear	a) Valve body gaskets leaking or damaged	a) Replace gaskets as required
	b) D-2 oil pipe leaking or out of position	b) Reposition or replace pipe
	c) D-3 orifice in spacer plate plugged	c) Unplug orifice in spacer plate
	d) PT/D-3 check ball number 3/ missing or defective	d) Install or replace check ball as required
	e) D-3 oil passage not drilled through in overrun clutch or turbine shaft	e) Drill passage through
	f) Oil seals missing or damaged in overrun clutch piston	f) Replace or install seals
	g) Overrun clutches burned	g) Replace overrun clutch
	h) Overrun clutch backing plate snap ring out of groove	h) Replace components as required

CHILTONS THREE "C's" TRANSMISSION DIAGNOSIS CHART
Turbo Hydra-matic 200-4R

Condition	Cause	Correction
Will not hold in park	a) Manual linkage misadjusted	a) Adjust manual linkage
	b) Parking pawl binding in case	b) Correct binding as required
	c) Actuator rod, spring, or plunger damaged	c) Correct as required
	d) Parking pawl broken, loose or damaged	d) Correct as required
	e) Manual shaft to case pin missing or loose	e) Correct as required
	f) Inside detent lever and pin nut loose	f) Replace or repair as required
	g) Manual detent roller and spring assembly bolt loose	g) Tighten bolt as required
	h) Manual detent pin or roller damaged, mispositioned or missing	h) Correct as required

Hydraulic Control System
THROTTLE VALVE SYSTEM

The THM 200-4R automatic transmission uses a throttle valve (T.V.) cable between the carburetor and the control valve assembly. The T.V. cable controls automatic transmission line pressure, shift points, shift feel, part throttle downshift and detent downshifts. The cable operates the throttle lever and the bracket assembly. This cable and bracket assembly serve two basic functions.

1. The first duty of this assembly is to transfer the movement of the carburetor throttle plate to the T.V. plunger in the control valve assembly. Thus the T.V. pressure and line pressure can increase according to throttle opening; it also controls part throttle and detent downshifts. The proper adjustment of the T.V. cable is therefore critical, and is based on the T.V. plunger being fully depressed to flush with the T.V. bushing at wide open throttle.

2. The second function of the assembly involves the T.V. exhaust valve lifter rod, spring and T.V. exhaust ball. The function of this system is to prevent the transmission from operating at low (idle position) pressures, should the cable break or become disconnected. As long as the cable is properly connected, not broken or stretched, the T.V. lifter rod will not move from its normal, spring-loaded "up" position which holds the T.V. exhaust check ball off its seat. The T.V. lifter rod will drop down to allow the T.V. exhaust ball to seat only if the cable breaks or becomes disconnected and out of adjustment. With the transmission pan removed, it should be possible to pull down on the T.V. exhaust valve lifter rod and the spring should return the rod to its normal "up" position. If the throttle lever and bracket assembly and/or lifter rod binds or sticks so that the T.V. lifter rod cannot lift the exhaust ball off its seat, high line pressures and delayed upshifts will result. The shape of the throttle valve lifter rod is critical, especially the 90° (right angle) bend. *It must not be bent to any other angle or it will not work properly.*

The importance of the throttle valve cable and its associated parts becomes clear since, if the T.V. cable is broken, sticky, misadjusted or if an incorrect part for the car model is fitted, the car may exhibit various malfunctions which could be attributed to internal component failure. Sticking or binding T.V. linkage can result in delayed or full throttle shifts. The T.V. cable must be free to travel to the full throttle position and return to the closed throttle position without binding or sticking.

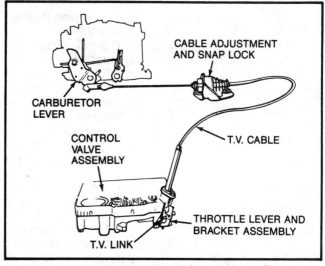

T.V. cable and related components (© General Motors Corp.)

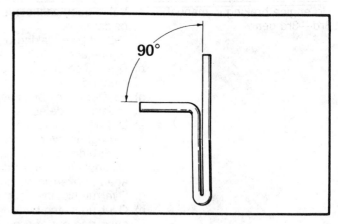

Throttle valve lifter rod (© General Motors Corp.)

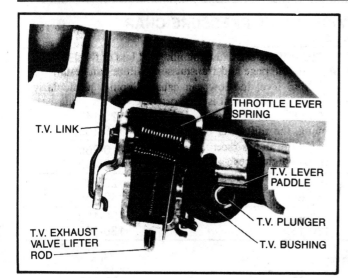

Throttle valve and bracket assembly (© General Motors Corp.)

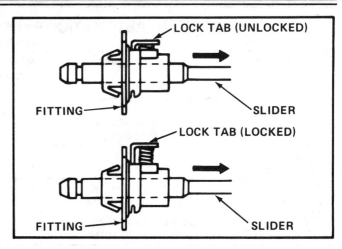

T.V. cable adjustment—typical (© General Motors Corp.)

Checking the throttle valve linkage for sticking or binding should be done with the engine running at idle, the selector in Neutral and the brakes set. The reason for this is that some binding or sticking may occur only when the engine is running and may not be noted or obtained with the engine turned off.

Checking T.V. Cable

To check the T.V. cable with the engine running, pull the cable to full travel beyond the carburetor lever pin that it attaches to and then release the cable. It should return to the closed throttle position against the carburetor lever pin.

If the T.V. cable sticks, it will remian ahead of the carburetor lever pin.

Sticking or Incorrectly Adjusted Cable Problems

NOTE: Check the throttle valve cable with the engine running in Neutral, not Park. Adjust the cable with the engine off.

A stuck T.V. cable could be caused by one or more of the following problems:
1. A damaged cable housing or a housing with a kink or sharp bend. Re-route or replace if required.
2. A sharp burr on the T.V. link dragging in the cable housing. Correct with a fine file, smoothing the end.
3. The T.V. link could be bent. Straighten or replace.
4. Misalignment of the throttle lever and bracket assembly, at the coiled pin in the control valve assembly.
5. The throttle lever and bracket assembly could possibly be binding or damaged or the throttle lever spring unhooked. Correct as necessary.

Length adjustment is also important to the T.V. cable. If it is adjusted too long, it may result in:
1. Early shifts or slipping shifts and/or no detent downshifts.
2. Delayed or full throttle shifts caused by forcing the transmission to operate in the high pressure mode. The complaint could be described as no upshifts and the closed throttle 3-2 shift may automatically occur as high as 45 mph. The transmission senses a malfunction and to prevent burning the clutches and band because of low line pressures, it will go into the high pressure mode. Line pressures and test procedures are described below. If, on the other hand, the T.V. cable is adjusted too short or not adjusted at all, it will result in raising the line pressure and shift points. Also, the carburetor will not be able to reach full throttle opening.

Diagnosis Tests

CONTROL PRESSURE TEST

NOTE: Before making the control pressure test, check the transmission fluid level, adjust the T.V. cable, check and adjust the manual linkage and be sure that the engine is not at fault, rather than the automatic transmission.

1. Install the oil pressure gauge to the automatic transmission. Connect a tachometer to the engine.
2. Raise and support the vehicle safely. Be sure that the brakes are applied at all times.
3. Total running time must not exceed two minutes.
4. Minimum line pressure check: set the T.V. cable to specification. Apply the brakes and take the reading in the ranges and rpm's that are indicated in the automatic transmission oil pressure chart.
5. Full line pressure check: hold the T.V. cable to the full extent of its travel. Apply the brakes and take the reading in the ranges and rpm's that are indicated in the automatic transmission oil pressure chart.
6. Record all readings and compare them with the data in the chart.

AIR PRESSURE TEST

Air pressure testing should be done in moderation to avoid excessive fluid spray and damage to the internal parts during disassembly and assembly, through partial retention of units.

STALL SPEED TEST

Stall speed testing is not recommended by General Motors Transmission Division. Extreme overheating of the transmission unit can occur, causing further internal damages. By pressure testing and road testing, the malfunction can be determined and by consulting the diagnosis chart, the cause and correction can normally be found.

ROAD TEST

1. Road test using all selective ranges, noting when discrepancies in operation or oil pressure occur.
2. Attempt to isolate the unit or circuit involved in the malfunction.
3. If engine performance indicates an engine tune-up is required, this should be performed before road testing is completed or transmission correction attempted. Poor engine performance can result in rough shifting or other malfunctions.

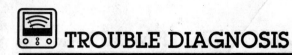

TROUBLE DIAGNOSIS

THM 200-4R AUTOMATIC TRANSMISSION OIL PRESSURE CHART

1981 MODELS

Transmission	Selector Lever	NORMAL OIL PRESSURE AT MINIMUM T.V. (P.S.I.)	NORMAL OIL PRESSURE AT FULL T.V. (P.S.I.)
BM,BY,EG, OG,CU,PH,	Park and Neutral at 1000 RPM	55-65 55-65	120-145 140-165
BM,BY,EG, OG,CU,PH,	Reverse at 1000 RPM	100-110 100-110	210-250 240-285
BM,BY,EG, OG,CU,PH,	Drive and Manual Third at 1000 RPM	55-65 55-65	120-145 140-165
BM,BY,EG, OG,CU,PH,	Manual Second and Low at 1000 RPM	130-150 130-150	130-150 130-150

1982 MODELS

Transmission	Selector Lever	NORMAL OIL PRESSURE AT MINIMUM T.V. (P.S.I.)	NORMAL OIL PRESSURE AT FULL T.V. (P.S.I.)
CQ,AA,HG,OG BY,CR	Park and Neutral at 1000 RPM	55-65 55-65	120-145 130-155
CQ,AA,HG,OG BY,CR	Reverse at 1000 RPM	100-110 100-110	210-250 215-260
CQ,AA,HG,OG BY,CR,	Drive and Manual Third at 1000 RPM	55-65 55-65	120-145 130-155
CQ,AA,HG,OG BY,CR	Manual Second and Low at 1000 RPM	130-150 130-150	130-150 130-150

1983 MODELS

Transmission	Selector Lever	NORMAL OIL PRESSURE AT MINIMUM T.V. (P.S.I.)	NORMAL OIL PRESSURE AT FULL T.V. (P.S.I.)
OM OG,AA,AP,BY,HE AH BR ZR	Park and Neutral at 1000 RPM	55-65 55-65 55-65 55-65 65-75	110-127 123-140 140-160 145-170 160-190
OM OG,AA,AP,BY,HE AH BR ZR	Reverse at 1000 RPM	105-120 105-120 105-120 105-120 120-140	220-235 230-260 260-297 275-315 293-350
OM OG,AA,AP,BY,HE AH BR ZR	Drive and Manual Third at 1000 RPM	55-65 55-65 55-65 55-65 65-75	110-127 123-140 140-160 145-170 160-190
OM OG,AA,AP,BY,HE AH BR ZR	Manual Second and Low at 1000 RPM	122-140 122-140 122-140 122-140 140-160	122-140 122-140 122-140 122-140 140-160

1984 MODELS

Transmission	Selector Lever	NORMAL OIL PRESSURE AT MINIMUM T.V. (P.S.I.)	NORMAL OIL PRESSURE AT FULL T.V. (P.S.I.)
AA,AP BQ BT,BY,CH,CR,OG,OJ CQ HE,HG OF OZ OM	PARK & NEUTRAL @ 1000 RPM	50-60 55-65 55-65 55-62 55-65 55-65 70-90 55-65	120-135 155-175 117-132 165-190 125-140 105-122 175-210 110-127

THM 200-4R AUTOMATIC TRANSMISSION OIL PRESSURE CHART
1981 MODELS

Transmission	Selector Lever	NORMAL OIL PRESSURE AT MINIMUM T.V. (P.S.I.)	NORMAL OIL PRESSURE AT FULL T.V. (P.S.I.)
1984 MODELS			
AA,AP		95-110	220-250
BQ		105-120	285-325
BT,BY,CH,CR,OG,OJ		105-120	220-245
CQ	REVERSE	80-90	235-265
HE,HG	@ 1000 RPM	105-120	230-260
OF		105-120	200-225
OZ		130-167	325-392
OM		105-120	207-235
AA,AP		50-60	120-135
BQ		55-65	155-175
BT,BY,CH,CR,OG,OJ	DRIVE (D4)	55-65	117-132
CQ	& MANUAL	55-62	165-190
HE,HG	THIRD (D3)	55-65	125-140
OF	@ 1000 RPM	55-65	105-122
OZ		70-90	175-210
OM		55-65	110-127
AA,AP	*MANUAL	112-127	112-127
BQ,BT,BY,CH,CR, HG,HE,OG,OF,OJ, OM	SECOND (D2) & LOW (D1) @ 1000 RPM	122-137	122-137
CQ		115-130	115-130
OZ		152-196	152-196

CONVERTER STATOR OPERATION

The torque converter stator assembly and its related roller clutch can possibly have one of two different type malfunctions.
A. The stator assembly freewheels in both directions.
B. The stator assembly remains locked up at all times.

Condition A

If the stator roller clutch becomes ineffective, the stator assembly freewheels at all times in both directions. With this condition, the vehicle will tend to have poor acceleration from a standstill. At speeds above 30-35 MPH, the vehicle may act normal. If poor acceleration problems are noted, it should first be determined that the exhaust system is not blocked, the engine is in good tune and the transmission is in gear when starting out.

If the engine will freely accelerate to high rpm in neutral, it can be assumed that the engine and exhaust system are normal. Driving the vehicle in reverse and checking for poor performance will help determine if the stator is freewheeling at all times.

Condition B

If the stator assembly remains locked up at all times, the engine rpm and vehicle speed will tend to be limited or restricted at high speeds. The vehicle performance when accelerating from a standstill will be normal. Engine over-heating may be noted. Visual examination of the converter may reveal a blue color from the over-heating that will result.

Under conditions A or B above, if the converter has been removed from the transmission, the stator roller clutch can be checked by inserting a finger into the splined inner race of the roller clutch and trying to turn the race in both directions. The inner race should turn freely in the clockwise direction, but not turn or be very difficult to turn in the counterclockwise direction.

Converter Clutch Operation and Diagnosis

TORQUE CONVERTER CLUTCH

The GM Torque Converter Clutch (TCC) incorporates a unit inside the torque converter with a friction material attached to a pressure plate and splined to the turbine assembly. When the clutch is applied, it presses against the converter cover. The result is a mechanical direct drive of the engine to the transmission. This eliminates slippage and improves fuel economy as well as reducing fluid temperature.

There are a number of controls to operate the torque converter clutch, all determined by drive range selection. For example, the clutch is applied in direct drive above a certain preset minimum speed. At wider throttle openings the converter clutch will apply after the 2-3 shift. When the vehicle slows, or the transmission shifts out of direct drive, the fluid pressure is released, the converter clutch releases and the converter operates in the conventional manner.

The engaging of the converter clutch as well as the release is determined by the direction of the converter feed oil. The converter feed oil from the pressure regulator valve flows to the converter clutch apply valve. The position of the converter clutch apply valve controls the direction in which the converter feed oil flows to the converter.

A spring-loaded damper assembly is splined to the converter turbine assembly while the clutch pressure plate is attached to a pivoting unit on the damper assembly. The result is that the pressure plate is allowed to rotate independently of the damper assembly up to about 45 degrees. This rotation is controlled by springs in the damper assembly. The spring cushioning aids in reducing the effects felt when the converter clutch applies.

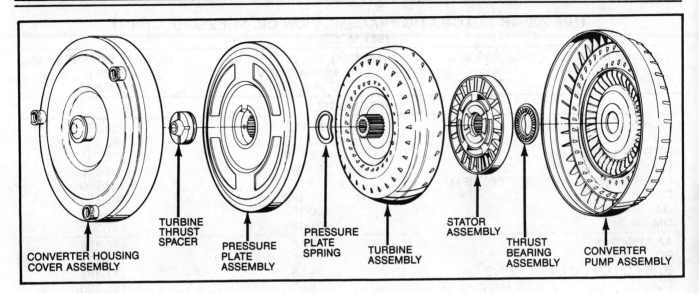

Torque converter clutch—exploded view (© General Motors Corp.)

To help insure that the converter clutch applies and releases at the proper times, controls have been incorporated into the electrical system. The converter clutch is applied when all of the conditions listed below exist:

1. The brake pedal is released.
2. The engine vacuum is above 2.5 inches of vacuum.
3. The engine coolant is above 130°F.
4. The vehicle speed is above 24 to 30 mph.
5. The transmission is in 3rd gear.

Converter clutch units are different for V6 or V8 engine applications and should not be interchanged. Poor engine and/or transmission operation will occur.

TROUBLESHOOTING THE TORQUE CONVERTER CLUTCH

Before diagnosing the TCC system as being at fault in the case of rough shifting or other malfunctions, make sure that the engine is in at least a reasonable state of tune. Also, the following points should be checked:

1. Check the transmission fluid level and correct as necessary.
2. Check the manual linkage adjustment and correct as necessary.
3. Road test the vehicle to verify the complaint. Make sure that the vehicle is at normal operating temperature.

G.M. TORQUE CONVERTER CLUTCH (TCCC)

Condition	Cause	Correction
Clutch applied in all ranges (engine stalls when put in gear)	a) Converter clutch valve stuck in apply position	a) Free stuck valve as required
Clutch does not apply: applies erratically or at at wrong speeds	a) Electrical malfunction in most instances	a) Follow troubleshooting procedure to determine if problem is internal or external to isolate defect
Clutch applies erratically; shudder and jerking felt	a) Vacuum hose leak b) Vacuum switch faulty c) Governor pressure switch malfunction d) Solenoid loose or damaged e) Converter malfunction; clutch plate warped	a) Repair hose as needed b) Replace switch c) Replace switch d) Service or replace e) Replace converter
Clutch applies at a very low or high 3rd gear	a) Governor switch shorted to ground b) Governor malfunction c) High line pressure d) Solenoid inoperative or shorted to case	a) Replace switch b) Service or replace governor c) Service pressure regulator d) Replace solenoid

CAUTION: When inspecting the stator and turbine of the torque converter clutch unit, a slight drag is normal when turned in the direction of freewheel rotation because of the pressure exerted by the waved spring washer, located between the turbine and the pressure plate.

If it has been determined that there is a problem with the TCC system, the next step is to determine if the problem is internal or external. The following procedure can be used:

1. Disconnect the electrical connector at the transmission case.
2. Raise and safely support vehicle.
3. Start engine and adjust the speed to 2000 rpm, gear selector in Neutral.
4. Test for 12 volts at the connector using a volt/ohm meter or a test light. If 12 volts are present, the problem is internal. If no voltage (or low voltage according to the meter) is present at the connector, the problem is external.

If the problem is internal (12 volts at the connector) the following steps can be taken:

1. With the wire to the transmission case disconnected, take a 12 volt test light and connect it to the female connector and ground to the male transmission connector.
2. Start the engine and adjust the speed to 2000 rpm, gear selector in park.
3. If the test light comes on, the governor switch or the internal wiring is shorted to ground. The oil pan will have to be removed, the wiring checked and/or the governor switch replaced.
4. If the test light does not light, make sure that the vehicle is off the ground, and run in Drive until the transmission shifts to 3rd gear. Keep the engine speed to 2000 rpm.

5. If the test light now comes on, the internal hydraulic/mechanical controls will have to be checked.
6. If the test light still does not light, there is a problem with the solenoid or governor switch.

To test for solenoid or governor switch electrical malfunction, the following steps can be used:

1. Drain the transmission fluid and remove the oil pan.
2. Using an external 12 volt source, (self-powered test light or small lantern battery, etc.) connect a positive lead to the case connector. Remove the lead wire from the governor pressure switch and connect it to the ground lead of the external 12 volt source.

CAUTION

Do not reverse the leads or the solenoid diode will be destroyed by the reverse voltage. Do not use an automobile battery for this test. A self-powered test light is best for these tests.

---

3. If the solenoid clicks, it can be considered serviceable; replace the governor switch.
4. If the solenoid does not click, check the wiring. If the wiring appears to be good, replace the solenoid and recheck.
5. For more information refer to the section on torque converter clutch operation, which is located in the front of this manual.

 ON CAR SERVICES

 ON CAR
SERVICES

Adjustments

SHIFT INDICATOR

1. With the engine off, position the selector lever in neutral.
2. If the pointer does not align with the "N" indicator position, move the clip on the shift bowl until alignment has been achieved.

NOTE: The manual linkage must be adjusted correctly before this adjustment can be made properly.

NEUTRAL SAFETY SWITCH

The neutral safety switch is not adjustable on vehicles equipped with this type of automatic transmission. The switch is located in the steering column, and can be serviced only after removing the steering wheel.

T.V. AND DETENT CABLE

NOTE: Before any adjustment is made to the T.V. or detent cable be sure that the engine is operating properly, the transmission fluid level is correct, the brakes are not dragging, the correct cable has been installed and that the cable is connected at both ends.

Diesel Engine

1. Stop the engine. If equipped, remove the cruise control rod.
2. Disconnect the transmission T.V. cable terminal from the cable actuating lever.
3. Loosen the lock nut on the pump rod and shorten it several turns.
4. Rotate the lever assembly to the full throttle position. Hold this position.
5. Lengthen pump rod until the injection pump lever contacts the full throttle stop.
6. Release the lever assembly and tighten pump rod lock nut. Remove the pump rod from the lever assembly.
7. Reconnect the transmission T.V. cable terminal to cable actuating lever.
8. Depress and hold the metal re-adjust tab on the cable upper end. Move the slider through the fitting in the direction away from the lever assembly until the slider stops against the fitting.
9. Release the re-adjust tab. Rotate the cable actuating lever assembly to the full throttle stop and release the cable actuating lever assembly. The cable slider should adjust out of the cable fitting toward the cable actuating lever.
10. Reconnect the pump rod and cruise control throttle rod if so equipped.
11. If equipped with cruise control, adjust the servo throttle rod to minimum slack (engine off) then put clip in first free hole closest to the bellcrank but within the servo bail.

Gas Engine

1. Stop engine.
2. Depress and hold the metal re-adjust tab on the cable upper end. Move the slider through the fitting in the direction away from the lever assembly until the slider stops against the fitting. Release re-adjust tab.
3. Rotate the cable actuating lever assembly to the full throttle stop and release the cable actuating lever assembly. The cable slider should adjust out of the cable fitting toward the cable actuating lever. Check cable for sticking and binding.
4. Road test the vehicle. If the condition still exists, remove the transmission oil pan and inspect the throttle lever and bracket assembly on the valve body.

5. Check to see that the T.V. exhaust valve lifter rod is not distorted or binding in the valve body assembly or spacer plate.
6. The T.V. exhaust check ball must move up and down as the lifter does. Be sure that the lifter spring holds the lifter rod up against the bottom of the valve body.
7. Make sure that the T.V. plunger is not stuck. Inspect the transmission for correct throttle lever to cable link.

MANUAL LINKAGE

There are many variations of the manual linkage used with the THM 200-4R, due to its wide vehicle application. A general adjustment procedure is outlined for the convenience of the repairman.

Loosen the shift rod adjusting swivel clamp at the cross shaft. Place the transmission detent in the neutral position and the column selector in the neutral position. Tighten the adjustment swivel clamp and shift the selector lever through all ranges, confirming the transmission detent corresponds to the shift lever positions and the vehicle will only start in the park or neutral positions.

INTERMEDIATE BAND

The intermediate band is adjusted by a selective intermediate servo apply pin. Because of possible inaccessibility to the intermediate servo piston assembly on the transmission/vehicle application, due to the clearance restrictions, the special tools needed for the pin selection cannot be installed and used. Should the band adjustment be necessary and the adjusting clearance is not available, the transmission would have to be removed from the vehicle.

The servo apply pin selection procedure is as follows.

1. Remove the servo cover and retaining ring from the transmission case with a special depressing tool.
2. Pull the servo cover from the case and remove the seal.
3. Remove the intermediate servo piston and band apply pin assembly.
4. Install the special tool into the intermediate servo bore and retain with the servo cover retaining ring.
5. Install the special pin, (part of the special tool), into the tool base, making sure the tapered pin end is properly positioned against the band apply lug. Be sure the band anchor pin is properly located in the case and the band anchor lug.
6. Install a dial indicator gauge on top of the special tool zero post and adjust the gauge to zero.

NOTE: The special tool base should be squarely against the servo cover retaining ring.

7. Align the special tool and the special indicator pin. The band selection pin should register between the high and low limits as marked on the tool. If not, possible trouble can exist with the intermediate band, direct clutch or the transmission case.
8. If the band selection pin registers in its proper zone, apply 100 in. lbs. of torque to the hex nut on the special tool. Measure the amount of travel on the dial indicator gauge. Remove the special tools.
9. Using the accompanying chart, install the proper band apply pin into the transmission and reassemble the intermediate servo into the transmission case.

INTERMEDIATE BAND APPLY PIN SELECTION CHART

Dial Indicator Travel	Apply Pin Identification
0.0-0.029 inch	one ring
0.029-0.057 inch	two rings
0.057-0.086 inch	three rings
0.086-0.114 inch	wide ring

NOTE: The apply pin identification ring is located on the band end of the pin.

Services

FLUID CHANGE AND OIL PAN

Removal

1. Raise and safely support car.
2. Place drain pan under transmission oil pan. Remove the oil pan attaching bolts from the front and side of the pan.
3. Loosen, but do not remove the rear pan bolts, then bump the pan loose and allow fluid to drain.
4. Remove the remaining bolts and remove pan.
5. Clean pan in solvent and dry with compressed air.
6. Remove the two bolts holding the filter screen to the valve body. Discard gasket.

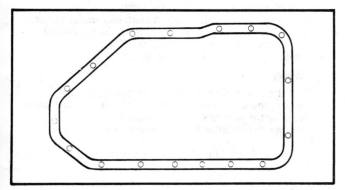

Transmission pan identification (© General Motors Corp.)

Installation

1. Clean screen in solvent and blow dry, or replace as required.
2. Install screen with new gasket; torque bolts to 6-10 foot-pounds.
3. Install new gasket on pan and torque bolts to 10-13 foot-pounds.
4. Lower car and add 3 quarts of Dexron® II.
5. With selector in Park start engine and idle. Apply parking brake. *Do not race engine.*
6. Moe selector through all ranges and end in Park. Check fluid level. Add if necessary to bring the level between the dimples on the dipstick.

VALVE BODY

Removal

1. Raise the vehicle on a hoist and support it safely.
2. Drain the transmission fluid. Remove the fluid pan. Discard the gasket or R.T.V. sealant.
3. Remove the two bolts retaining the filter. Remove the filter from the valve body.
4. Remove bolt and washer securing T.V. cable to transmission and disconnect T.V. cable.
5. Remove throttle lever and bracket assembly. Do not bend throttle lever link.
6. Disconnect the wire leads on the electrical connector at the transmission case. Remove the connectors at the 4-3 pressure switch and the fourth clutch pressure switch. Remove the solenoid attaching bolts, clips and solenoid assembly.
7. Remove the manual detent roller and spring assembly.
8. Support the valve body assembly. Remove the valve body retaining bolts.

9. Holding the manual valve, remove the valve body assembly. Be sure to use care in order to prevent loss of the three check balls located in the valve body assembly.
10. Position the valve body assembly down with the spacer plate side up.

Installation

1. Using new gaskets, as required, position the valve body assembly against the transmission. Install the valve body retaining bolts and torque them 7-10 ft. lbs.
2. Continue the installation in the reverse order of the removal procedure.
3. Use a new gasket when installing the oil pan. If the transmission uses R.T.V. sealant be sure that both mating surfaces are clean before applying new sealant.
4. Lower the vehicle. Fill the transmission to specification using the proper grade and type automatic transmission fluid.
5. Start the engine and check for leaks. Check to see that the transmission is properly filled once the vehicle has reached operating temperature.

INTERMEDIATE SERVO

Removal

1. Raise the vehicle and support it safely.
2. Remove the four catalytic converter heat shield bolts. Slide the heat shield away from the catalytic converter.
3. Using the special intermediate servo removal tool, depress the servo cover and remove the retaining ring.
4. Remove the cover and the seal ring. The seal ring may be in the transmission case.
5. Remove the intermediate servo piston and bore apply pin assembly.

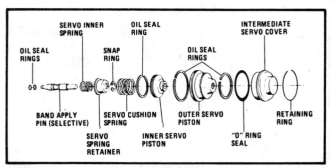

Intermediate servo—exploded view (© General Motors Corp.)

Installation

1. Position the bore apply pin assembly and the intermediate servo piston to its mounting in the transmission case.
2. Using a new seal ring install the servo cover. Depress the servo cover and install the retainer ring.
3. Continue the installation in the reverse order of the removal procedure.
4. Lower the vehicle. Check the fluid level and correct as required.
5. Start the engine and check for leaks. Once the vehicle has reached operating temperature recheck the fluid level.

ACCUMULATOR

Removal

1. Raise the vehicle and support it safely.
2. Drain the transmission fluid. Remove the fluid pan, gasket and filter.

3. Remove the valve body assembly.

4. Support the 1-2 accumulator housing. Remove the four 1-2 accumulator housing retaining bolts, housing and gasket.

5. Lay the 1-2 accumulator housing down with the plate side facing upward.

6. Support the valve body spacer plate, gaskets and accumulator plate to prevent dropping the eight check balls and the 3-4 accumulator spring, piston and pin which are located in the transmission case.

7. Remove the remaining retaining bolt on the accumulator plate. After the removal of the spacer plate and gaskets the band anchor pin may come out of its bore.

Installation

NOTE: The intermediate band anchor pin must be located on the intermediate band or damage to the transmission will result.

1. Installation is the reverse of the removal procedure.

2. Be sure to retain the eight check balls, using petroleum jelly or equivalent.

3. Replace all gaskets as required. Lower the vehicle. Fill with the proper grade and type transmission fluid. Start the engine and check for leaks.

GOVERNOR

Removal

1. Raise the vehicle and support it safely.

2. Drain the transmission fluid. Remove the fluid pan. Discard the gasket and screen.

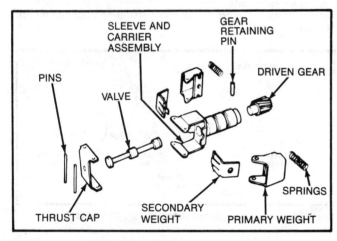

Governor assembly (© General Motors Corp.)

3. Remove the governor cover attaching bolts.

4. Remove the governor cover and gasket. Discard the gasket.

5. The governor assembly may come out along with the cover. It may also be necessary to rotate the output shaft counterclockwise while removing the governor. Do not use any type of pliers to remove the governor from its bore.

Installation

1. To install, position the governor assembly in its bore within the transmission case.

2. Install the governor cover using a new gasket. Torque the retaining bolts to 18 ft. lbs.

3. Continue the installation in the reverse order of the removal procedure.

4. Lower the vehicle. Fill the transmission with the proper grade and type automatic transmission fluid.

5. Start the engine and check for leaks. Once the engine has reached operating temperature recheck the fluid.

SPEEDOMETER DRIVEN GEAR

Removal

1. Raise the vehicle and support it safely.

2. Disconnect the speedometer cable.

3. Remove the retainer bolt, retainer, speedometer driven gear and O-ring seal.

Installation

1. Coat the new O-ring seal with transmission fluid and install it along with the other components.

2. Install the speedometer cable.

3. Lower the vehicle. Replace any transmission that may have been lost.

REAR OIL SEAL

Removal

1. Raise the vehicle and support it safely.

2. Remove the driveshaft. Place a container under the driveshaft to catch any fluid that may leak from the transmission.

3. If equipped, remove the tunnel strap.

4. Pry out the old seal using a suitable tool.

Installation

1. Coat the new seal with clean transmission fluid.

2. Position the seal on the rear extension housing and drive it in place using the proper seal installation tools.

3. Continue the installation in the reverse order of the removal procedure.

4. Lower the vehicle. Fill the transmission with clean oil, as required.

 REMOVAL & INSTALLATION

REMOVAL

1. Disconnect the negative battery cable.

2. Remove the air cleaner assembly. Disconnect the T.V./detent cable at its upper end.

3. Remove the transmission dipstick. Remove the bolt holding the dipstick tube, if accessible.

4. Raise the vehicle and support it safely.

5. Drain the transmission. Remove the driveshaft. Matchmark the driveshaft to aid in installation.

6. Disconnect the speedometer cable at the transmission. Disconnect the shift linkage at the transmission.

7. Disconnect all electrical leads at the transmission. Disconnect any clips that retain these leads.

8. Remove the inspection cover. Mark the flexplate and torque converter, to aid in installation.

9. Remove the torque converter to flex plate retaining bolts. Remove the catalytic converter support bracket, if necessary.

10. Position the transmission jack under the transmission and remove the rear transmission mount.

11. Remove the floor pan reinforcement. Remove the crossmember retaining bolts. Move the crossmember out of the way.

12. Remove the transmission to engine bolt on the left side. This is the bolt that retains the ground strap.

13. Disconnect and plug the transmission oil cooler lines. It may be necessary to lower the transmission in order to gain access to the cooler lines.

14. With the transmission still lowered, disconnect the T.V. cable.

15. Support the engine, using the proper tools and remove the remaining engine to transmission bolts.

16. Carefully disengage the transmission assembly from the engine. Lower the transmission from the transmission jack.

Installation

1. To install, reverse the removal procedures.
2. Before installing the flexplate to torque converter bolts, make sure that the weld nuts on the converter are flush with the flexplate and that the torque converter rotates freely by hand in this position.
3. Be sure that the flexplate and torque converter marks made during the removal procedure line up with each other.
4. Adjust the shift linkage and the T.V. cable as required.
5. Lower the vehicle and fill the transmission with the proper grade and type automatic transmission fluid.
6. Start the engine and check for leaks. Once the vehicle has reached operating temperature, recheck the fluid level.

BENCH OVERHAUL

Before Disassembly

Clean the exterior of the transmission assembly before any attempt is made to disassemble, to prevent dirt or other foreign materials from entering the transmission assembly or its internal parts.

NOTE: If steam cleaning is done to the exterior of the transmission, immediate disassembly should be done to avoid rusting from condensation of the internal parts.

Take note of thrust washer locations. It is most important that thrust washers and bearings be installed in their original positions. Handle all transmission parts carefully to avoid damage.

Ring groove wear on governor supports, input shaft, pump housings and other internal parts should be checked using new rings installed in the grooves.

Converter Inspection

The converter cannot be disassembled for service.

CHECKING CONVERTER END PLAY

1. Place the converter on a flat surface with the flywheel side down.
2. Attach a dial indicator so that the end play can be determined by moving the turbine shaft in relation to the converter hub. End play for the Turbo Hydra-Matic 200-4R should be no more than .0-.050 of an inch. If the clearance is greater than this, replace the converter.

CHECKING CONVERTER ONE-WAY CLUTCH

1. If the one-way clutch fails, then the stator assembly freewheels at all times in both directions. The car will have poor low speed acceleration.
2. If the stator assembly remains locked up at all times, the car speed will be limited or restricted at high speeds. The low speed operation may appear normal, although engine overheating may occur.
3. To check one-way clutch operation, insert a finger into the splined inner race of the roller clutch and try to turn the race in both the directions. The inner race should turn freely in the clockwise direction, but it should not turn in the counterclockwise direction.

NOTE: Inspect the converter hub for burrs or jagged metal to avoid personal injury.

4. If the one-way clutch does not operate properly, the converter must be replaced.

VISUAL INSPECTION OF THE CONVERTER

Before installation of the converter, check it carefully for damage, stripped bolt holes or signs of heat damage. Check for loose or missing balance weights, broken converter pilot, or leaks, all of which are cause for replacement. Inspect the converter pump drive hub for nicks or burrs that could damage the pump oil seal during installation.

Transmission Disassembly

TORQUE CONVERTER AND OIL PAN

1. Remove the torque converter from the transmission by pulling it straight out of the transmission housing.
2. Position the transmission assembly in a suitable holding fixture.
3. Remove the transmission fluid pan bolts. Discard the pan gasket, or R.T.V. sealant. Remove and discard the fluid screen.

VALVE BODY

1. Disconnect the wire leads at the electrical connector, which is located in the transmission case and the pressure switches on the valve body assembly.
2. Remove the electrical connector and O-ring seal from the transmission case.
3. Remove the solenoid assembly attaching bolts. Remove the clips and the solenoid.
4. Remove the throttle lever and bracket assembly. Do not bend the throttle lever link. The T.V. exhaust valve lifter and spring may separate from the throttle lever and bracket assembly.
5. Remove the manual detent roller and spring assembly, the signal pipe retaining clip and the signal pipe.
6. On non-C3 models, remove the 4-3 pressure switch retaining bolt.
7. Remove the remaining valve body retaining bolts. Do not drop the manual valve, as damage to the valve may occur.
8. Hold the manual valve in its bore. Remove the valve body assembly. Care must be taken as three check balls will be exposed on top of the spacer plate to valve body gasket. Remove the exposed check balls.
9. Lay the valve body assembly down with the spacer plate gasket side up.

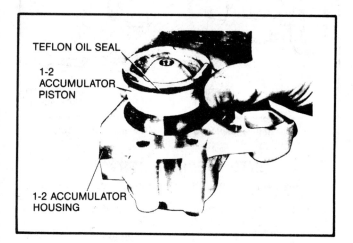

Accumulator housing and related components
(© General Motors Corp.)

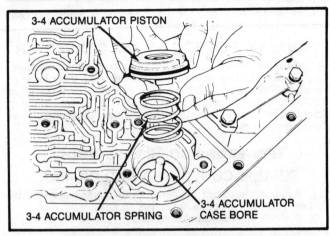

3-4 accumulator assembly removal (© General Motors Corp.)

1-2 ACCUMULATOR

1. Remove the 1-2 accumulator housing retaining bolts. Remove the accumulator housing, spring, plate and gasket.
2. Remove the 1-2 accumulator piston from the housing. It may be necessary to apply low air pressure (about 3 PSI) to the orifice in the accumulator housing passage in order to remove the 1-2 accumulator piston.
3. Remove the valve body gaskets and spacer plate. Remove the 3-4 accumulator spring, piston and pin from the transmission case. Remove the eight check balls from the core passages in the transmission case.

GOVERNOR

1. Remove the governor cover attaching bolts. Remove the governor cover and gasket.
2. Remove the governor assembly from the transmission case.
3. It may be necessary to rotate the output shaft counterclockwise while removing the governor. Do not use any type of pliers to remove the governor from its bore in the transmission case.

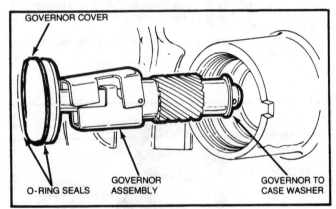

Governor assembly—exploded view (© General Motors Corp.)

INTERMEDIATE SERVO

1. Using a suitable tool, remove the intermediate servo cover retaining ring.
2. Remove the servo cover. Discard the seal ring. The cover seal ring may be located in the transmission case.
3. Remove the intermediate servo piston and the band apply pin assembly.
4. Install the special tool into the intermediate servo bore and retain with the servo cover retaining ring.

5. Install the special pin, (part of the special tool), into the tool base, making sure the tapered pin end is properly positioned against the band apply lug. Be sure the band anchor pin is properly located in the case and the band anchor lug.
6. Install a dial indicator gauge on top of the special tool zero post and adjust the gauge to zero.

NOTE: The special tool base should be squarely against the servo cover retaining ring.

7. Align the special tool and the special indicator pin. The band selection pin should register between the high and low limits as marked on the tool. If not, possible trouble can exist with the intermediate band, direct clutch or the transmission case.
8. If the band selection pin registers in its proper zone, apply 100 in. lbs. of torque to the hex nut on the special tool. Measure and record the amount of travel on the dial indicator gauge. Select a new apply pin as required. Remove the special tools.

INTERMEDIATE BAND APPLY PIN SELECTION CHART

Dial Indicator Travel	Apply Pin Identification
0.0-0.029 inch	one ring
0.029-0.057 inch	two rings
0.057-0.086 inch	three rings
0.086-0.114 inch	wide ring

3-4 ACCUMULATOR

1. Inspect the 3rd accumulator check valve for missing check ball, check ball binding or stuck in the tube, oil feed slot in tube missing or restricted, improperly assembled, loose fitting or not fully seated in the case and damaged or missing.
2. If the 3rd accumulator check valve assembly requires replacement, use a 6.3 mm (#4) easy out and remove the check valve assembly from the case by turning and pulling straight out.
3. Install new check valve assembly, small end first, into the case. Position the oil feed slot in tube so it faces the servo cover.
4. Using a ⅜″ diameter metal rod and hammer, drive the check valve assembly until it is seated in the 3rd accumulator case hole.

FRONT OIL PUMP

1. Remove the front pump seal.
2. Remove the front pump retaining bolts and washers.
3. Using a suitable tool, remove the front pump assembly and gasket from the transmission case.
4. Remove the front pump oil deflector plate.

OVERDRIVE UNIT

1. Check the overdrive unit end play by installing the rear unit support tool (J-25013-1) on the sleeve of the output shaft. Then bolt the output shaft loading fixture adapter tool (J-29332) to the end of the transmission case.
2. Position the transmission accordingly and remove the pump-to-case bolt and washer. Install an eleven inch long bolt and locking nut.
3. Install the oil pump end play checking fixture adapter (J-25022) on to the oil pump remover tool (J-24773-5) and secure both tools on the end of the turbine shaft.
4. Position the dial indicator gauge and clamp the assembly on the bolt positioning indicator point cap nut on top of the oil pump removal tool.
5. Lift up on the oil pump removal tool with approximately three lbs. of upward force and while holding the upward force, set the indicator to zero.

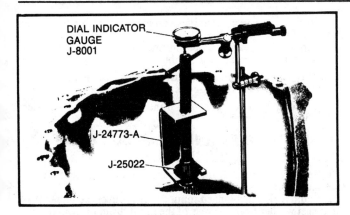

Overdrive unit end play check (© General Motors Corp.)

NOTE: The above procedure must be performed to eliminate the tolerance difference between the turbine shaft snap ring and the overdrive carrier.

6. With the indicator gauge set at zero, increase the force to about twenty lbs. Read the dial indicator and record the reading. Overdrive unit end play should be .004-.027 in.

7. The selective washer controlling this end play is located between the pump cover and the overrun clutch housing. Replace the thrust washer as required to bring the end play within specification.

8. Remove the dial indicator gauge and the special tools once the washer has been replaced.

OVERDRIVE UNIT END PLAY SELECTIVE WASHER CHART

Identification	Thickness (inches)
Zero—Scarlet	0.167-0.171
One—White	0.172-0.176
Two—Cocoa Brown	0.177-0.180
Three—Gray	0.181-0.185
Four—Yellow	0.186-0.190
Five—Light Blue	0.191-0.195
Six—Purple	0.196-0.200
Seven—Orange	0.201-0.204
Eight—Green	0.205-0.209

FOURTH CLUTCH ASSEMBLY

1. Remove the front pump using the required pump removal tool.

2. Remove the fourth clutch plate-to-transmission case snap ring.

3. Grasp the turbine shaft and lift out the overdrive unit assembly and the fourth clutch plate assembly.

4. Remove the fourth clutch plate assembly from the overdrive unit assembly. Be sure to remove the one remaining steel plate from the inside of the transmission case.

5. Remove the overdrive internal gear-to-carrier thrust washer. Remove the overdrive internal gear. Remove the internal gear-to-support thrust washer.

6. Install the fourth clutch compressor and center support tool (J-29334-1) on the fourth clutch spring and retainer assembly. Install the rest of the tool to the transmission case housing using two governor cover bolts.

Overdrive assembly and fourth clutch plate assembly (© General Motors Corp.)

7. Compress fourth clutch spring and retainer assembly; then remove support to fourth clutch spring snap ring and spring and retainer assembly.

8. Remove the fourth clutch compressor and center support tool.

9. Remove the fourth clutch piston from inside the transmission assembly.

FRONT UNIT PARTS

1. To check the front unit end play, push the forward clutch shaft downward. Install the forward and direct clutch unit fixture tool (J-29337) in the end of the forward clutch shaft.

2. Mount the dial indicator gauge and clamp the assembly on the bolt positioning indicator point on top of the forward and direct clutch unit fixture tool.

3. Move the output shaft upward by turning the adjusting screw on the output shaft loading fixture adapter tool (J-29332) until the white line on the sleeve of the rear unit support tool (J-25013-1) begins to disappear. Set the dial indicator gauge to zero.

4. Pull the forward and direct clutch unit fixture tool upward and read the end play. Record the result. The end play should be .022-.051 in. If the end play is not within the specification given, replace the front unit end play selective washer. The selective washer controlling this end play is located between the forward clutch shaft and the output shaft.

5. Remove the dial indicator and clamp assembly. Do not remove the output shaft loading fixture adapter tool or the rear unit support tool.

CENTER SUPPORT

1. Remove the two center support-to-transmission case bolts.

2. Remove the center support-to-transmission case beveled snap ring.

3. Install the fourth clutch compressor and center support remover tool (J-29334-1) on the center support. Retain the tool in place using the center support-to-fourth clutch spring snap ring.

4. Using the fourth clutch compressor and center support removal tool and a slide hammer, remove the center support. Be sure that the center support bolts located on the valve body assembly have been removed.

5. Remove the support-to-direct clutch thrust washer. The thrust washer may be stuck on the back of the direct clutch.

FRONT UNIT END PLAY SELECTIVE WASHER CHART

Identification	Thickness (inches)
One— —	0.065-0.070
Two— —	0.070-0.075
Three—Black	0.076-0.080
Four—Light Green	0.081-0.085
Five—Scarlet	0.086-0.090
Six—Purple	0.091-0.095
Seven—Cocoa Brown	0.096-0.100
Eight—Orange	0.101-0.106
Nine—Yellow	0.106-0.111
Ten—Light Blue	0.111-0.116
Eleven— —	0.117-0.121
Twelve— —	0.122-0.126
Thirteen—Pink	0.127-0.131
Fourteen—Green	0.132-0.136
Fifteen—Gray	0.137-0.141

FORWARD AND DIRECT CLUTCH ASSEMBLY

1. Install the forward and direct clutch unit fixture tool (J-29337) in the end of the forward clutch shaft. Grasp the tool and remove the direct and forward clutch assemblies.
2. Lift the direct clutch assembly off of the forward clutch assembly.

NOTE: The direct forward clutch thrust washer may stick to the end of the direct clutch housing when it is removed from the forward clutch housing.

3. Remove the intermediate band assembly.
4. Remove the band anchor pin.

FRONT GEAR PARTS

1. Remove the output shaft-to-forward clutch shaft front selective washer. The washer may be stuck to the end of the forward clutch shaft.
2. To check the rear unit end play, loosen the output shaft loading fixture adapter tool (J-29332) adjusting screw on the output shaft. Push the output shaft downward.
3. Install the dial indicator gauge and plunger extension (J-7057). Position the extension end against the output shaft and set the dial indicator gauge to zero.
4. Move the output shaft upward by turning the adjusting screw on the tool until the white line on the sleeve of the tool begins to disappear. Read and record the rear unit end play. It should be .004-.025 in. If the end play is not within specification, replace the rear unit selective washer. The selective washer is located between the front internal gear thrust washer and the output shaft snap ring.
5. Remove the dial indicator and clamp assembly. Do not remove the rear unit support tool or the output shaft loading fixture adapter tool.
6. Using snapring pliers, remove the output shaft-to-selective washer snap ring.

FRONT INTERNAL GEAR

1. Remove the front internal gear, rear selective washer and the thrust washer from inside the transmission case.

2. Remove the rear selective washer and the thrust washer from the internal gear.
3. Remove front carrier assembly and the front internal gear to front carrier thrust bearing assembly.

NOTE: The front sun gear to front carrier thrust bearing assembly and race may come out as the front carrier is removed.

4. Remove front sun gear and front sun gear to front carrier thrust bearing assembly.

INPUT DRUM AND REAR SUN GEAR ASSEMBLY

1. Remove the input drum and rear sun gear assembly from inside the transmission case.
2. Remove the four-tanged input drum-to-reverse clutch housing thrust washer from the rear of the input drum or from the reverse clutch housing.

LOW AND REVERSE CLUTCH HOUSING ASSEMBLY

1. Grind approximately ¾ in. from end of 6.3 mm (#4) easy out to remove cup plug. Remove the housing to case cup plug assembly by turning the easy out 2 or 3 turns and pulling it straight out. Do not reuse the plug and seal assembly.
2. Remove the low and reverse clutch housing-to-transmission case beveled snap ring.

NOTE: The flat side of the snap ring should have been against the low and reverse clutch housing with the beveled side up.

3. Using the reverse clutch housing installer and remover tool (J-28542), remove the low and reverse clutch housing assembly.
4. Remove the low and reverse clutch housing-to-transmission case spacer ring.

REAR GEAR PARTS

1. Be sure that the governor assembly has been removed from the transmission case.
2. Grasp the output shaft and lift out the remainder of the rear unit gear parts. Lay the assembly down in a horizontal position.

ROLLER CLUTCH AND REAR CARRIER ASSEMBLY

1. Remove the roller clutch and rear carrier assembly from the output shaft.
2. Remove the four-tanged rear carrier-to-rear internal gear thrust washer from the end of the rear carrier, or from the inside of the rear internal gear.
3. Remove the low and reverse clutch plates from the output shaft.

REAR INTERNAL GEAR

1. Remove the rear internal gear-to-rear sun gear thrust bearing assembly from the rear internal gear.
2. Remove the rear internal gear from the output shaft.
3. Position the transmission so that the rear unit support tool and the output shaft loading fixture adapter tool can be removed from the transmission case.
4. If necessary, remove the rear oil seal.

MANUAL SHAFT AND PARKING PAWL PARTS

1. Position the transmission with the oil pan side up.
2. Remove the hex nut which holds the inside detent lever to the manual shaft.
3. Remove the parking lock actuator rod and the inside detent lever assembly.
4. Remove the manual shaft retaining pin from the transmission case and slide the manual shaft out of the case.

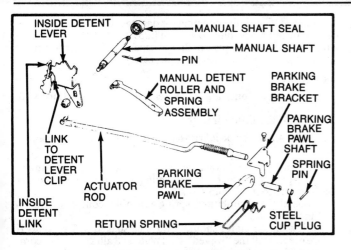

Manual shaft and parking pawl components
(© General Motors Corp.)

5. Inspect the manual shaft to case oil seal for damage, replace as required.

6. Remove the parking lock bracket. Remove the parking lock pawl shaft retaining pin.

7. Using a 6.3 mm (#4) Easy-Out®, remove the parking lock pawl cup plug. You will have to grind about ¾ inch from the end of the Easy-Out in order to remove the cup plug.

8. Using a 4 mm (#3) Easy-Out®, remove the parking pawl shaft. Remove the parking pawl and the return spring.

Unit Disassembly and Assembly

FRONT UNIT

Disassembly

1. Disassemble the front unit (direct clutch) by removing the snap ring that fits into the circumference of the direct clutch housing.

2. Take out the backing plate and the clutch plates.

NOTE: The number of clutch plates to be found will vary depending on the model of the unit. Take care when disassembling the clutch packs so that the right number of clutches will be installed and will be in their proper order. Also, due to differences in their composition material, do not mix the plates of the direct clutch with those of the forward clutch.

3. With a suitable spring compressor, compress the retainer and spring assembly and then remove the snap ring.

4. Remove the retainer and spring assembly.

5. Remove the release spring guide and the clutch piston.

NOTE: It is not necessary to remove the clutch apply ring from the piston unless the piston or apply ring requires replacement. These apply rings will vary between models. There will be an identifying number on the apply ring which will be needed if replacement is required.

Inspection

Clean and inspect all parts. Look for cracks, wear, burrs or any damage. Blow parts dry with compressed air. Discard all seals that are removed. Do not overlook the center seal in the direct clutch housing. Check for free operation of the check balls and see that the snap ring grooves are not damaged.

Assembly

1. If the clutch apply ring has been removed, install it back on the piston.

2. Install the inner and outer seals on the piston. Lube them with Dexron® II.

NOTE: The lips on the seals should face away from the clutch apply ring side.

3. Install the new center seal in the clutch housing, again with the lip in the proper position and well lubricated.

4. Install the piston into the housing. The seals should be well lubricated with transmission fluid.

NOTE: A tool can be fabricated from wire to help ease installation. This can be slipped down between the seal and the housing to help compress the lip of the seal while pushing down on the piston.

5. Install the release spring guide onto the piston, indexing the guide notch onto the check ball assembly. Install the retainer and spring assembly.

6. Compress the assembly and install the snap ring.

7. New clutch plates should always be soaked in Dexron® II for at least thirty minutes before use. Install a flat steel plate and alternate with composition plates, following the sequence of disassembly.

8. Install the backing plate, chamfered side up, and the snap ring.

9. Double-check to make sure that the composition plates turn freely.

10. Set the unit aside until the transmission is ready for assembly.

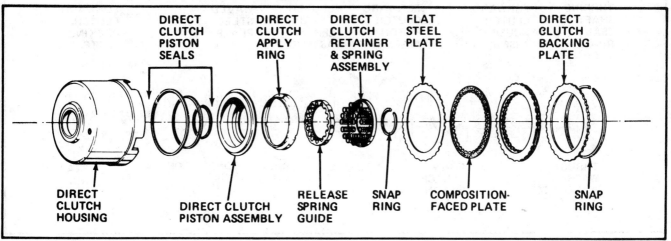

Direct clutch assembly—exploded view (© General Motors Corp.)

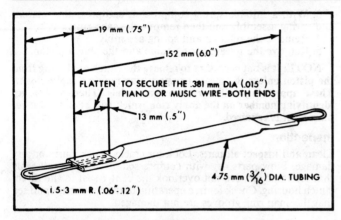

Piston installation tool (© General Motors Corp.)

INTERMEDIATE UNIT

Disassembly

1. Inspect the teflon seals on the shaft of the forward clutch assembly. They should not be removed unless they are going to be replaced. Original equipment seals will be teflon while service replacements are usually cast iron. Test the teflon seals to see that they rotate freely in the grooves. If they are to be replaced, cut them off with a sharp knife, then check the grooves for nicks or burrs. Lubricate any new seals with petroleum jelly.

2. Remove the thrust washer from the forward clutch and check it for wear or damage.

3. Turn the clutch over with the shaft facing down. Remove the snap ring and the clutch backing plate.

4. Remove the clutch plates, taking note of the number of plates and their order so that they can be reassembled properly. With a compressor, push against the retainer and remove the snap ring. Remove the retainer and spring assembly from the housing.

5. Remove the forward clutch piston.

Inspection

Clean and inspect all parts well. Look for cracks, wear, burrs or any damage. Blow parts dry with compressed air. Discard all seals that are removed. Check the housing for cracks and see that the lube hole is open, the check ball is free and the snap ring groove is

undamaged. Check the cup plug for damage. If replacement is required, install a new plug to 1.0mm (.039") below the surface. As with the direct clutch piston, do not separate the piston from the apply ring unless replacement is necessary.

Assembly

1. Install new inner and other seals on the piston. Lube with transmission fluid.

2. Install the piston into the housing.

NOTE: A tool can be fabricated from wire to help ease installation.

This can be slipped down between the seal and the housing to help compress the lip of the seal while pushing down on the piston. Use the tool on both seals.

3. Install the retainer and spring assembly, compress and install the snap ring.

4. New clutch plates should always be soaked in Dexron® II before use. Install the "waved" steel plate and alternate with composition plates referring to the proper sequence at disassembly.

5. Install the backing plate, with the chamfered side up, then the snap ring.

NOTE: Double-check to make certain that the composition plates turn freely.

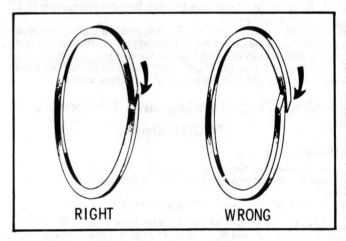

Teflon seal installation (© General Motors Corp.)

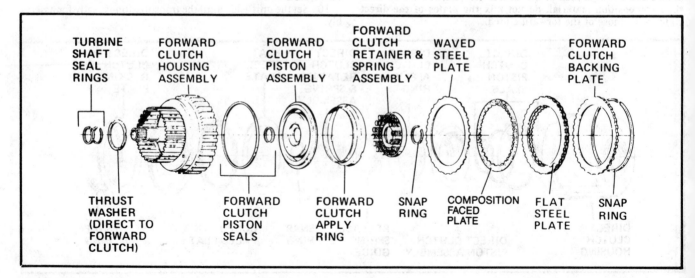

Forward clutch assembly—exploded view (© General Motors Corp.)

6. If the teflon rings on the the turbine are to be replaced, cut off the original rings and install new ones, well lubricated with transmission fluid.

NOTE: Service replacement rings may be teflon or cast iron. If they are teflon, make certain that they are installed properly, and that the ends do not overlap. Also, new teflon rings may sometimes look like they are distorted after they have been installed. However, once they have been exposed to the oil and the normal operating temperatures of the transmission, the new seals will regain their proper shape and fit freely in their grooves. A free fit of a used teflon ring does not indicate leakage in operation.

7. Install the thrust washer onto the turbine shaft.
8. Set the unit aside until the transmission is ready for assembly.

REAR UNIT

Disassembly

1. The low/reverse clutch pack is removed after the low/reverse clutch housing is withdrawn from the case.
2. Disassemble the clutch housing by compressing the clutch spring retainer and removing the snap ring. Remove the retainer.
3. Remove the wave release spring and then the piston.

Inspection

Clean and inspect all parts well. Look for cracks, wear, burrs or any damage. Blow parts dry with compressed air. Discard all seals that are removed. Check the housing for cracks and see that the oil feed hole is clear. Check the bushing for excessive wear.

Assembly

1. Install new inner and outer seals on the piston. Lube with transmission fluid.
2. Be certain that the lips of the seals are away from the clutch apply ring side. Install the piston into the housing. The seals should be well lubricated with transmission fluid.
3. Install the waved release spring, then the retainer with the cupped side down.
4. Compress the assembly and install the snap ring.
5. Set the unit aside until the transmission is ready for assembly.

CENTER SUPPORT

Disassembly

1. Remove the fourth clutch inner and outer seals.
2. Check the condition of the cast iron oil rings. Remove them from the center support if necessary.

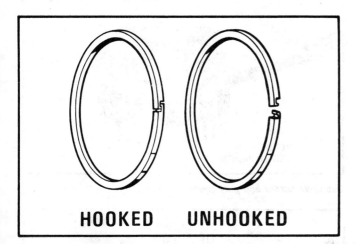

HOOKED   UNHOOKED

Cast iron seal ring installation (© General Motors Corp.)

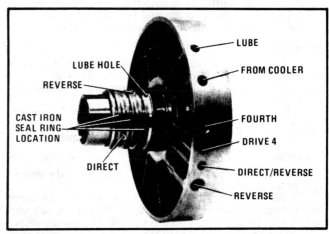

Center support and seal rings (© General Motors Corp.)

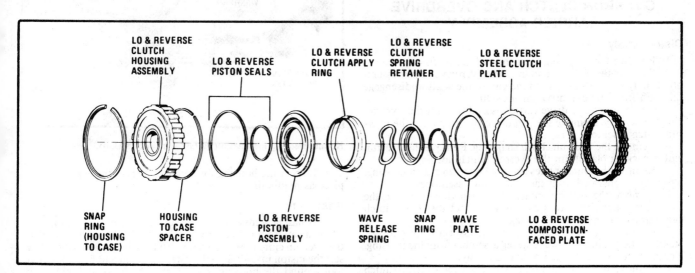

Low and reverse clutch piston assembly—exploded view (© General Motors Corp.)

Inspection

1. Inspect the bushings for scoring, wear or galling. Check the oil ring grooves and the oil rings for nicks and other damage.
2. Air check the oil passages to be sure that they are open and not interconnected.
3. Inspect the piston sealing surfaces for scoring and scratches.
4. Inspect the center support for cracks or porosity.
5. Inspect the center support for burrs or raised edges. If present, remove them with fine abrasive paper.

Assembly

1. If removed, install the cast iron oil seal rings on the center support.

NOTE: When installing cast iron oil seal rings, make sure the ends overlap and interlock. Make sure ends are flush with each other when interlocked, and oil seal rings are seated in ring grooves to prevent damage to rings during reassembly of mating parts over rings. Coat the rings with clean transmission fluid.

2. Set the assembly aside until the transmission is ready to be assembled.

FOURTH CLUTCH ASSEMBLY

Disassembly

1. Remove the fourth clutch snap rings.
2. Separate the release springs and the retainer.
3. Remove the composition faced and steel plates.

Inspection

1. Inspect the fourth clutch snap rings. Inspect the fourth clutch piston for cracks, wear or other visable damage.
2. Inspect the release springs and retainer assembly for distortion or damage.
3. Inspect the composition faced and steel plates for damage, wear or burning.
4. Inspect the backing plate for scratches or damage.

Assembly

1. Assemble the fourth clutch piston assembly in the reverse order of the disassembly procedure.
2. Install the fourth clutch inner and outer seals on the center support assembly with the lips facing down. Retain the seals with transmission fluid. The inner seal is identified by a white stripe.
3. Set the assembly aside until the transmission is to be reassembled.

OVERRUN CLUTCH AND OVERDRIVE CARRIER ASSEMBLY

Disassembly

1. Remove the snap ring from the turbine shaft.
2. Remove the turbine shaft from the overdrive carrier assembly. It may be necessary to tap the end of the shaft to disengage the shaft from the overdrive carrier splines.
3. Remove the overdrive carrier assembly from the overrun clutch assembly. Remove the overdrive sun gear from the overrun clutch assembly.
4. Remove the overrun clutch snap ring. Remove the overrun clutch backing plate from the overrun clutch housing.
5. Remove the clutch plates from the overrun clutch housing. Keep them separated from the other plate assemblies.
6. Remove the overrun clutch hub snap ring. Remove the overdrive roller clutch cam assembly. Inspect the roller clutch cam ramps for any signs of damage.
7. Remove the roller clutch assembly. Inspect it for damage. Remove the retainer and wave spring assembly from the housing. Remove the overrun clutch piston assembly.
8. Remove the inner and outer seal from the overrun clutch piston. Inspect the piston for damage.

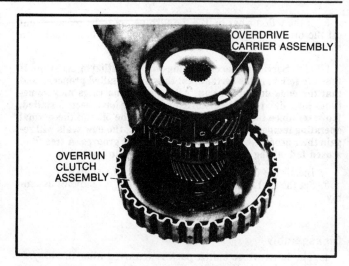

Overrun clutch and overdrive carrier assembly
(© General Motors Corp.)

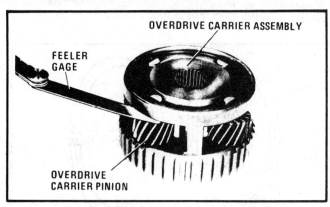

Overdrive carrier end play check (© General Motors Corp.)

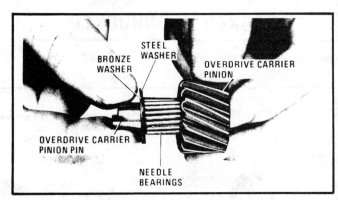

Needle bearing installation (© General Motors Corp.)

9. Inspect the housing for damage. Inspect the snap rings, replace as required.

Assembly

1. Install new inner and outer seals on the piston with the lips facing away from clutch apply ring side. Oil the seals and install the overrun clutch piston. To make the piston easier to install, insert the piston installing tool between the seal and housing; rotate tool around the housing to compress the lip of the seal, while pushing down slightly on the piston.

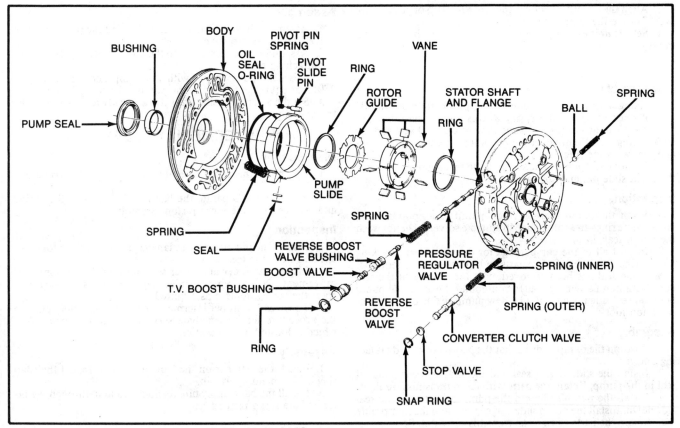

Front pump—exploded view (© General Motors Corp.)

2. Install the overrun clutch waved release spring. Install the overrun clutch waved spring retainer with the cupped face down.

3. Install the roller clutch cam on the roller clutch assembly. The locating tangs on the roller clutch must set on the roller clutch cam.

4. Install the roller clutch assembly on the overrun clutch hub.

5. Compress the retainer by pushing down on the roller clutch assembly. Install the narrow snap ring.

6. Oil and install the overrun clutch plates into the overrun clutch housing, starting with a flat steel and alternating composition-faced and flat steel clutch plates.

7. Install the backing plate with the chamfered side up. Install the snap ring. Be sure that the composition faced plates turn freely.

8. Set the unit aside until you are ready to reassemble the transmission.

OVERDRIVE CARRIER ASSEMBLY

Disassembly

1. Before disassembly, inspect the locating splines for damage. Inspect the roller clutch race for damage, scratches or wear. Inspect the carrier housing for cracks or wear. Inspect the pinions for damage, rough bearings or tilt.

2. Check the pinion end play. It should be .009-.024 in. If it is not within specification, repair or replace the defective component as required.

3. Remove the overdrive carrier snap ring. Remove the overdrive carrier pinions. Remove the pinions, thrust washer and roller needle bearings.

4. Inspect the pinion pocket thrust faces for burrs. Remove them if they are present.

5. Remove the overdrive sun gear to overdrive carrier thrust bearing assembly. Inspect the unit.

Assembly

1. Install thrust bearing assembly by placing the small diameter race down. Hold in place with transmission fluid.

2. Install 19 needle bearings into each pinion, thumb lock needle bearings to hold them in place.

3. Place a bronze and steel thrust washer on each side so steel washer is against pinion. Hold them in place with transmission fluid.

4. Place the pinion assembly in position in the carrier and use a pilot shaft to align parts in place.

5. Push the pinion pin in place while rotating the pinion from side to side. Install the overdrive carrier snap ring to retain the pinion pins.

6. Set the assembly aside until you are ready to assemble the transmission.

FRONT PUMP

Disassembly

1. Remove the front pump-to-transmission case seal ring. Inspect the ring, replace as required.

2. Position the pump so that the front pump cover side is up.

3. Remove the pump cover-to-pump body attaching bolts. Separate the front pump cover from the front pump body.

4. Remove the stator shaft to overrun selective thrust washer. Inspect the components for damage and replace as required.

Assembly

1. Assemble the pump cover to the pump body. Retain the components in place using the attaching bolts.

2. Align the cover and body using tool #J-25015 or equivalent.

3. Torque the retaining bolts to 18 ft. lbs.

4. Set the unit aside until you are ready to assemble the transmission.

PUMP BODY

Disassembly

1. Using a suitable tool, remove the pump slide spring. Be careful when removing this spring as it is under high pressure.

2. Remove the pump slide, slide to wear plate oil seal and the back O-ring seal. Remove the rotor, rotor guide, the seven vanes and the two vane rings.

3. Remove the pump slide seal and the seal support. Remove the pivot slide pin and spring.

Inspection.

1. Wash the pump body, springs, pump slide, pump rotor, vanes, vane rings and rotor guide in suitable solvent. Do not wash the pump seals in the solvent.

2. Inspect all of the components for damage, wear, cracking and scoring. Replace defective parts and components as required.

3. If the seal has been removed, coat the outside of the seal body with non-hardening sealing compound. Position the pump body oil seal side up and install a new pump seal using a seal installation tool.

Assembly

1. Position the pump body so that the pump pocket side is facing upward.

2. Install the slide O-ring seal and the slide-to-wear plate oil seal in the pump. Retain the parts with clean transmission fluid.

3. Install the pump slide into the pump pocket with the seal side down. Install the pump slide seal support and the pump slide seal. Retain the parts using clean transmission fluid. Be sure that this seal is installed correctly.

4. Install the pivot pin spring and the pivot slide pin. Install the vane ring in the pump pocket. Install the rotor guide in the pump rotor.

5. Install the pump rotor into the pump pocket. Center and seat the pump rotor on the rotor guide so that the rotor is flush with the pump slide.

6. Install the seven vanes into the pump. Be sure that the vane pattern is installed against the vane ring. Install the top vane ring. Install the pump slide spring.

PUMP COVER

Disassembly

1. Using a suitable tool, remove the converter clutch valve from the pump cover.

2. Using a punch, remove the pressure relief spring retaining sleeve. Remove the pressure relief valve assembly.

Inspection

1. Inspect both the converter clutch valve and the pressure relief valve for damage. Replace defective components as required.

2. Inspect the pump cover to be sure that all the oil passages are open.

3. Inspect the four cup plugs. If a plug is missing, install a new cup plug to 1/32 in. below the top of the hole. Use a 9/32 in. rod on the two smaller cup plugs a 5/16 in. rod on the line to case plug and a 7/16 in. rod on the larger plug when installing new plugs.

4. Inspect the orifice plugs. If they require replacement, position the new plug, orifice end first, into the plug hole from the rough casting side.

5. Drive the plug flush to 0.10 in. below the top of the hole. Stake the top of the hole in two places to retain the plug.

Assembly

1. Assemble the converter clutch valve and the pressure relief valve into the pump cover.

2. Assemble the pump cover to the pump body. Finger tighten the attaching bolts.

3. Align the pump cover with the pump body using tool J-25015 or equivalent. Torque the pump bolts to 18 ft. lbs.

4. Set the assembly aside until you are ready to reassemble the transmission.

INTERMEDIATE SERVO

Disassembly

1. Compress the servo assembly and remove the retaining ring.

2. Remove and separate the band apply pin, and the spring and washer from the servo piston assembly.

Inspection

1. Inspect the pin oil seals for damage. Inspect the pin for damage and fit to the case bore.

2. Inspect the inner and outer seal rings for damage and free movement. Do not remove unless replacement is required. Inspect the spring, replace as required.

3. To check for proper intermediate servo band apply pin; see the procedure for intermediate servo removal and installation which is located in this section.

Assembly

1. Install the retainer on the band apply pin. Install the snap ring on the band apply pin.

2. Install the band apply pin, retainer end first, through the intermediate servo pistons.

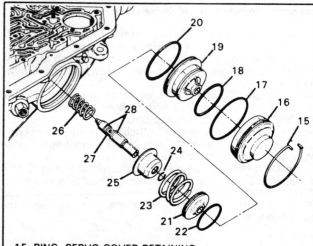

15 RING, SERVO COVER RETAINING
16 COVER, INTERMEDIATE SERVO
17 SEAL, "O" RING (INTERMEDIATE SERVO COVER)
18 RING, OIL SEAL (INNER)
19 PISTON, INTERMEDIATE SERVO (OUTER)
20 RING, OIL SEAL (OUTER)
21 PISTON, INTERMEDIATE SERVO (INNER)
22 RING, OIL SEAL PISTON (INNER)
23 SPRING, INTERMEDIATE SERVO CUSHION
24 RING, SNAP (APPLY PIN/RETAINER)
25 RETAINER, SERVO SPRING
26 SPRING, INTERMEDIATE SERVO (INNER)
27 PIN, INTERMEDIATE BAND APPLY (SELECTIVE)
28 RING, OIL SEAL (INTERMEDIATE BAND APPLY PIN)

Intermediate servo assembly—exploded view

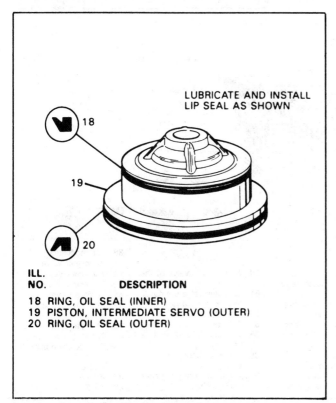

LUBRICATE AND INSTALL
LIP SEAL AS SHOWN

ILL. NO.	DESCRIPTION
18	RING, OIL SEAL (INNER)
19	PISTON, INTERMEDIATE SERVO (OUTER)
20	RING, OIL SEAL (OUTER)

Intermediate Servo Piston Assembly

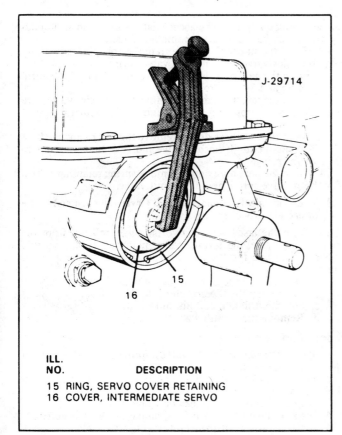

ILL. NO.	DESCRIPTION
15	RING, SERVO COVER RETAINING
16	COVER, INTERMEDIATE SERVO

Installing Intermediate Servo Cover

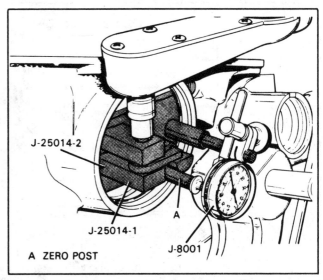

J-25014-2
J-25014-1
A
J-8001
A ZERO POST

Intermediate Band Apply Pin Check

INTERMEDIATE BAND APPLY PIN SELECTION CHART

DIAL INDICATOR TRAVEL		APPLY PIN IDENTIFICATION
.0 - .72mm	(.0″ - .029″)	1 GROOVE
.72 - 1.44mm	(.029″ - .057″)	2 GROOVES
1.44 - 2.16mm	(.057″ - .086″)	3 GROOVES
2.16 - 2.88mm	(.086″ - .114″)	NONE

Intermediate Band Apply Pin Selection Chart

3. If removed, install new intermediate servo inner and outer seal rings. Be sure that the cut ends are assembled in the same manner as they were cut. Be sure that the rings are seated in the grooves to prevent damage to them. Retain the rings using transmission fluid.

4. Install a new seal ring on the servo cover.

5. Set the unit aside until you are ready to reassemble the transmission.

VALVE BODY

Disassembly

NOTE: As each valve train component is removed, place the individual valve train in the order that it is removed and in a separate location relative to its position in the valve body. None of the valves, bushings or springs are interchangeable; some roll pins are interchangeable. Remove all roll pins and spring retaining sleeve by pushing through from the rough case surface side of the control valve pump assembly, except for the blind hole roll pins.

1. Lay the control valve assembly machined face up, with the manual valve at the top.

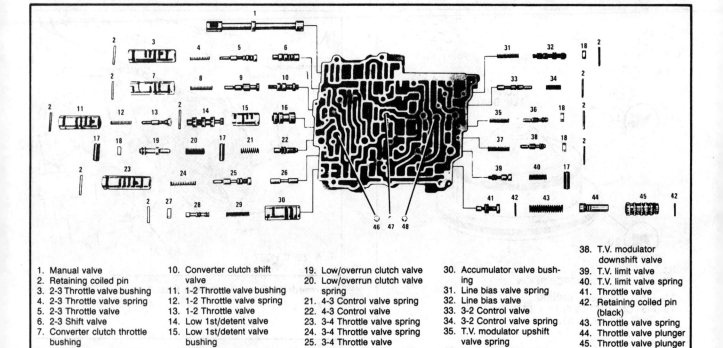

1. Manual valve
2. Retaining coiled pin
3. 2-3 Throttle valve bushing
4. 2-3 Throttle valve spring
5. 2-3 Throttle valve
6. 2-3 Shift valve
7. Converter clutch throttle bushing
8. Converter clutch throttle valve spring
9. Converter clutch throttle valve
10. Converter clutch shift valve
11. 1-2 Throttle valve bushing
12. 1-2 Throttle valve spring
13. 1-2 Throttle valve
14. Low 1st/detent valve
15. Low 1st/detent valve bushing
16. 1-2 Shift valve
17. Spring retaining sleeve
18. Valve bore plug [12.50mm (.500 in.)]
19. Low/overrun clutch valve
20. Low/overrun clutch valve spring
21. 4-3 Control valve spring
22. 4-3 Control valve
23. 3-4 Throttle valve spring
24. 3-4 Throttle valve spring
25. 3-4 Throttle valve
26. 3-4 Shift valve
27. Valve bore plug
28. Accumulator valve
29. Accumulator valve spring
30. Accumulator valve bushing
31. Line bias valve spring
32. Line bias valve
33. 3-2 Control valve
34. 3-2 Control valve spring
35. T.V. modulator upshift valve spring
36. T.V. modulator upshift valve
37. T.V. modulator downshift valve spring
38. T.V. modulator downshift valve
39. T.V. limit valve
40. T.V. limit valve spring
41. Throttle valve
42. Retaining coiled pin (black)
43. Throttle valve spring
44. Throttle valve plunger
45. Throttle valve plunger bushing
46. #11 D3
47. #8 1-2 shift valve
48. #1-2 low 1st

Valve body assembly—exploded view (© General Motors Corp.)

2. Remove the 3 check balls if still intact.
3. Remove manual valve from upper bore.

NOTE: **Some of the roll pins in the valve body assembly have pressure against them. Hold a shop towel over the bore while removing the pin, to help prevent possibly losing a bore plug, spring, etc.**

4. Remove roll pin from upper left bore. Remove 2-3 throttle valve bushing, 2-3 throttle valve spring, 2-3 throttle valve and 2-3 shift valve. The 2-3 throttle valve spring and 2-3 throttle valve may be inside the 2-3 throttle valve bushing.
5. From the next bore down, remove the roll pin. Remove the converter clutch throttle bushing, converter clutch throttle valve spring, converter clutch throttle valve, and converter clutch shift valve. The converter clutch throttle valve spring and the converter clutch throttle valve may be inside the converter clutch throttle bushing.
6. From the next bore down, remove the outer roll pin. Remove the 1-2 throttle valve bushing, 1-2 throttle valve spring, 1-2 throttle valve and low 1st/detent valve: The 1-2 throttle valve spring and the 1-2 throttle valve may be inside the 1-2 throttle valve bushing. Remove the inner roll pin. Remove the low 1st/detent valve bushing and 1-2 shift valve.
7. From the next bore down, remove the outer spring retaining sleeve. This spring is under load so be careful when removing it. Remove the bore plug, 4-3 control valve and spring. This spring is also under extreme pressure. Remove the low overrun clutch valve spring and the low overrun clutch valve.
8. From the next bore down, remove the roll pin. Remove the 3-4 throttle valve bushing, 3-4 throttle valve spring, 3-4 throttle valve and 3-4 shift valve. The 3-4 throttle valve and spring may be inside the 3-4 throttle valve bushing.
9. From the last bore down, remove the roll pin and bore plug. Remove the accumulator valve bushing, accumulator valve

and accumulator spring. The accumulator spring and accumulator valve may be inside the accumulator bushing.
10. From the upper right bore, remove roll pin, valve bore plug, line bias valve and line bias valve spring.
11. From the next bore down, remove the roll pin, 3-2 control valve spring and 3-2 control valve.
12. From the next bore down, remove the roll pin, valve bore plug, T.V. modulator upshift valve and T.V. modulator upshift valve spring.
13. From the next bore down, remove the roll pin, valve bore plug, T.V. modulator downshift valve and T.V. modulator downshift valve spring.
14. From the next bore down, remove spring retaining sleeve, T.V. limit valve spring and T.V. limit valve.

NOTE: **This sleeve is under a load. Cover the open end of the bore to prevent loss of spring.**

15. From the last bore, remove the outer roll pin from the rough casting side, the throttle valve plunger bushing, throttle valve plunger and throttle valve spring. Remove the inner roll pin as follows:
    a. Grind a taper to one end of a #49 drill.
    b. Lightly tap the tapered end into the roll pin.
    c. Pull the drill and coil pin out.
    d. Remove the throttle valve.

Inspection

1. Wash control valve body, valves, springs, and other parts in clean solvent and air dry.
2. Inspect valves for scoring, cracks and free movement in their bores.
3. Inspect bushings for cracks or scored bores.
4. Inspect valve body for cracks, damage or scored bores.
5. Inspect springs for distortion or collapsed coils.
6. Inspect bore plugs for damage.

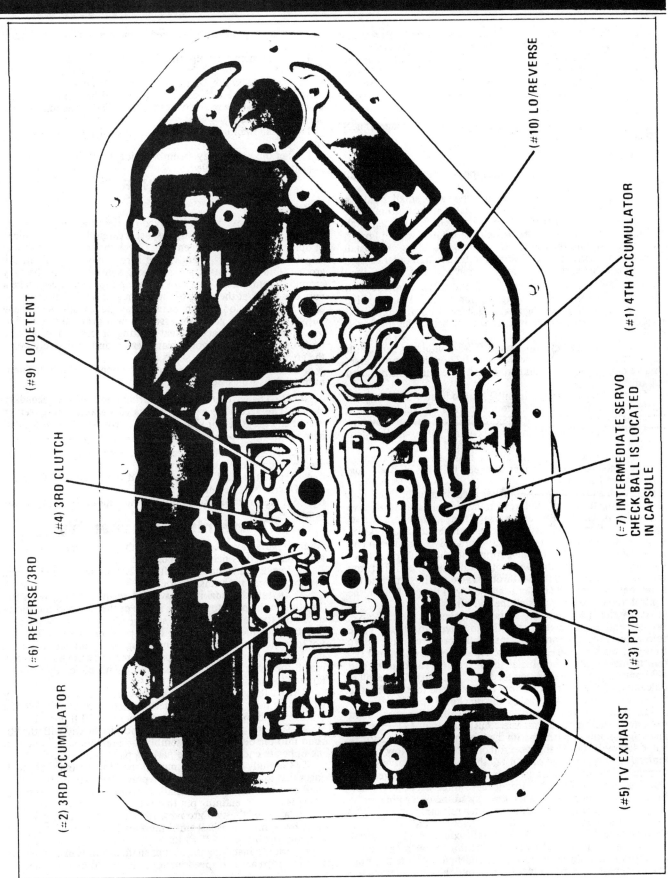

(#9) LO/DETENT

(#10) LO/REVERSE

(#1) 4TH ACCUMULATOR

(#4) 3RD CLUTCH

(#6) REVERSE/3RD

(#2) 3RD ACCUMULATOR

(#7) INTERMEDIATE SERVO CHECK BALL IS LOCATED IN CAPSULE

(#3) PT/D3

(#5) TV EXHAUST

200-4R-230 Check Ball Location in Bottom of Case

NOTE: Install all flared coiled pins (zinc coated), flared end out, and from the machined face of the control valve assembly. Install the two (2) tapered coiled pins (black finish) that retain the throttle valve and throttle valve bushing, tapered end first. Coiled pins do not fit flush on rough casting face. Make sure that all coiled pins are flush at machined face or damage to the transmission will occur.

7. Inspect the fourth clutch pressure switch for damage. If necessary replace pressure switch.

Assembly

1. Position the valve body assembly so that you can install the fourth clutch pressure switch.

2. Install into lower right bore, throttle valve, smaller outside diameter land first, make sure valve is seated at the bottom of the bore. Install inner roll pin between the lands of this valve. Install the T.V. spring into the bore. Install the throttle valve plunger, stem end first, into the throttle valve plunger bushing and install these two parts into the bore, valve end first. Install outer roll pin from rough cast surface side, aligning pin with slot in bushing.

3. In the next bore up, install T.V. limit valve, stem end first. Install the T.V. limit valve spring. Compress the T.V. limit valve spring and install the spring retaining sleeve from the machined face. Make sure the sleeve is level with or below the machined surface.

4. In the next bore up, install T.V. modulator downshift valve spring then install the T.V. modulator downshift valve, smaller chamfered stem end first. Install bore plug, hole out, and roll pin.

5. In the next bore up, install T.V. modulator upshift valve spring then install the T.V. modulator upshift valve, smaller chamfered stem end first. Install bore plug, hole out, and roll pin.

6. In the next bore up, install 3-2 control valve, smaller stem end out. Install 3-2 control valve spring and roll pin.

7. In the next bore up, install line bias valve spring then install the line bias valve, smaller stem end first. Install bore plug, hole out, and roll pin.

8. In the lower left bore, install accumulator bushing into bore, aligning the pin slot in line with the pin hole in control valve assembly. Install accumulator spring then accumulator valve, smaller end first. Next, install bore plug, hole out, and roll pin.

9. In the next bore up, install 3-4 shift valve, chamfered end first. Install the 3-4 throttle valve spring into the 3-4 throttle valve bushing. Next, install the 3-4 throttle valve, stem end first into the 3-4 bushing. Install the 3-4 throttle valve and bushing into the bore, making sure the pin slot is aligned with the pin hole in control valve assembly. Install the roll pin.

10. In the next bore up, install low/overrun clutch valve, smaller end first. Next, install the low/overrun clutch valve spring; compress the spring and install the spring retaining sleeve from the machined face. Make sure the sleeve is level with or below the machined surface. Install the 4-3 control valve spring and 4-3 control valve smaller stemmed end first. Install bore plug, hole out and retaining sleeve from the machined face. Make sure the sleeve is level with or below the machined surface.

11. In the next bore up, install the 1-2 shift valve, small stem outward. Install the 1-2 shift valve bushing, small I.D. first and aligning pin hole in bushing with the inner pin hole in the control valve assembly. Install the inner roll pin. Next, install low 1st/detent valve long stem end out. Install the 1-2 throttle valve spring into the 1-2 throttle valve bushing and the 1-2 throttle valve, stem end first, into the bushing. Install these three parts, valve end first, into the bore, aligning the bushing so the outer pin can be installed in the pin slot. Install the outer roll pin.

12. In the next bore up, install the converter clutch shift valve, short stemmed end first. Next, install the converter clutch throttle valve spring into the converter clutch throttle bushing and the converter clutch throttle valve, stem end first, into the bushing. Install these three parts, valve end first, into the bore, aligning the

bushing so the pin can be installed in the pin slot. Install the roll pin.

13. In the next bore up, install the 2-3 shift valve, large end first. Next, install the 2-3 throttle valve spring into the 2-3 throttle valve bushing and the 2-3 throttle valve, stem end first, into the bushing. Install these three parts, valve end first, into the bore, aligning the bushing so the pin can be installed in the pin slot. Install the roll pin.

14. Install manual valve with the inside detent lever pin hole first.

15. Set the valve body assembly aside, until the transmission is ready to be reassembled.

GOVERNOR

Inspection

1. Once the governor is removed (as described in the Transmission Disassembly section), inspect it carefully. Look for nicks or other damage to the driven gear. Check the governor shaft seal ring for cuts or damage and make sure it has a free fit in the groove. Often the gears are stripped, sometimes all the way around. Assuming the governor is free and turns properly, take a very close look at the edges of the worm gear teeth that drive the governor. If they look rough, polish away the roughness or a new output gear and/or governor assembly will promptly be ruined too. Sometimes the factory heat treating process leaves a scale build up on the output worm gear oversize and excessively wears the governor gear or even strips it.

2. Note that different transmission models will use different numbers of springs, some one, some two.

NOTE: The governor with one spring must use the secondary spring. The secondary spring is located between the governor weight and the governor shaft, while the primary spring is located between the governor weight and the governor gear.

3. Make sure the two check balls are in place and that the washer is not worn excessively.

4. Renew the seals as necessary and remember to lube them with Dexron® II or petroleum jelly.

5. Set the governor aside until the transmission is ready for assembly.

Transmission Assembly

After all parts have been cleaned and the subassemblies checked and overhauled as necessary, begin reassembly.

1. Install a new manual shaft seal. Make sure the lip is facing inward, toward the transmission.

2. Install the parking pawl and return spring, with the "tooth" on the pawl toward the inside of the case and the spring under the tooth. Make sure the spring ends push against the case pad and slide the pawl shaft in. The tapered end goes in first. Do not forget to install a new shaft cup plug, open end out, driving it in with a rod ⅜ in. in diameter, past the retaining pin hole. Finally install the retaining pin, by positioning with pliers, then gently tapping the pin in place.

3. Install the parking lock bracket. The pawl will probably have to be held toward the center of the transmission while the bracket is put into place. Torque the bolts to 20 ft. lbs.

4. Take the parking actuator rod and slip the end with the 90° bend into the detent plate assembly, on the "pin side," and slip the opposite end between the parking pawl and lock bracket.

5. Carefully examine the manual shaft for any burrs or rough edges that might damage the seal. Apply some Dexron® II to the shaft and seal and slide the shaft, threaded end first, into place. There is a small retaining pin that is installed. It retains the shaft in the larger of the two grooves. Install the detent lever on the shaft by turning the shaft until the flats line up. Install the hex nut, and tighten to 20-25 ft. lbs.

6. Prior to installing the output shaft, carefully examine it for damage or wear. The speedometer gear should be examined for wear.

7. If the output shaft is determined to be re-usable, install the rear internal gear onto the shaft, hub end first, if it has not been assembled yet. The roller clutch and rear pinion carrier should have already been assembled. See the section "Unit Disassembly and Assembly." Slide the rear carrier into the rear internal gear, and, holding the output shaft from its front end, lower the assembly into the case, until the parking pawl lugs on the rear gear align flush with the parking pawl tooth. At this point, install the speedometer driven gear as a double check that the drive gear is positioned on the proper journal. The attaching bolt is torqued to 6-10 ft. lbs.

8. Make certain that the low/reverse clutch selective washer is in place, then install the clutches, well oiled with Dexron® II, alternating friction and steel plates in the order determined at disassembly. Install the housing-case spacer ring. The low-reverse clutch housing assembly is installed next. Note that there is a feed hole in the housing that must be aligned with the feed hole to the reverse clutch case feed passage. Make sure that the low-reverse clutch housing assembly is fully seated past the case snap ring groove.

9. Install the snap ring, flat side against the housing, beveled side up. Position the ring gap 180° from the parking rod.

10. Install the rear sun gear and input drum. Double-check for chipped teeth, damaged splines or other damage. Install the four-tanged thrust washer on the input drum. Retain the components in place using transmission fluid.

11. Inspect the front sun gear for damage or wear. Note that there is an identification mark (either a groove or a drill spot) which goes against the input drum snap ring. Follow this with the thrust bearing and race assembly. The needle bearings go against the gear.

12. The front planetary carrier should have been cleaned, inspected and end play checked. If the planetary is within specifications, install in the case. The thrust bearing which goes on the front of the corner is installed with its small diameter race against the carrier. Transmission fluid will hold it in place.

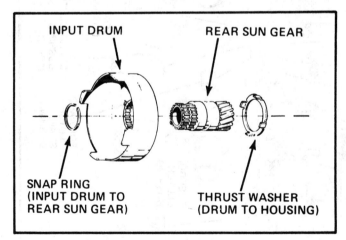

INPUT DRUM **REAR SUN GEAR**

SNAP RING (INPUT DRUM TO REAR SUN GEAR) **THRUST WASHER (DRUM TO HOUSING)**

Rear sun gear and input drum (© General Motors Corp.)

13. After the front internal gear has been inspected, install along with tanged thrust washer, using transmission fluid to hold it. If the *rear* selective washer had been determined to give proper end play, reuse it. Then install the snap ring, making sure it is seated in its groove.

NOTE: The rear selective thrust washer must be installed with the identification number toward the front of the transmission.

14. Install the *front* selective washer, using transmission fluid and with the identification number to the front.

15. Install a new intermediate band with the apply lug and anchor pin lug located in the case slot. Install the band anchor, pin stem end first, seating the stem in the band hole lug.

16. The direct clutch and forward clutch assemblies are now ready for installation. It would be convenient to have a hole in the work bench or other arrangement to receive the turbine shaft. First position the direct clutch assembly over the hole, clutch plate end up. It will help if the clutch plate teeth are aligned to make the forward clutch assembly easier to install.

17. Install the forward clutch assembly, shaft first into the direct clutch; rotate back and forth until the forward clutch is seated.

18. To check that the forward clutch is in fact, seated, measure with a steel ruler from the tang end of the direct clutch housing to the end of the forward clutch drum. It should be approximately 5/8 in.

19. Holding both clutch assemblies so that they do not separate, lower into the case. The direct clutch housing will be approximately 4 1/8 in. from the pump face in the case if it is correctly seated.

20. Install the overdrive unit assembly at this time. As you are installing the overdrive unit assembly, it may be necessary to rotate the assembly in order to get the assembly seat properly.

21. Install the front pump assembly. One suggestion is to take two 8mm x 1.25 bolts several inches long and by sawing the heads off, make two guide pins to help with the pump assembly. Before installing the pump, confirm that the intermediate band anchor pin lug is aligned with the band anchor pin hole in the case and that the stem of the anchor pin locates in the hole of the band lug. Apply transmission fluid to the direct clutch thrust washer, install on the back of the pump and turn the pump over. With a new gasket and the outer seal well lubricated, lower the pump gently onto the guide pins and into the case.

22. Use new washers on the pump bolts and torque them 15-20 ft. lbs. Try the turbine shaft to make sure that it can be rotated as the pump is being tightened. If not, the forward and direct clutch housings have not been properly installed, and are not indexing with all the clutch plates. This must be corrected before the pump can be installed completely. The turbine shaft must rotate freely.

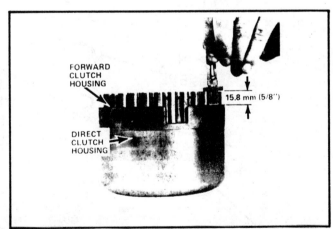

FORWARD CLUTCH HOUSING

15.8 mm (5/8")

DIRECT CLUTCH HOUSING

Forward clutch installation (© General Motors Corp.)

23. Front unit end play should have been checked at disassembly to determine if a thrust washer change is necessary. It can be rechecked by mounting a dial indicator to rest on the turbine shaft. This end play is controlled by the selective washer between the output shaft and turbine shaft. At this point, the transmission fixture, if used, should be rotated so that the transmission is horizontal, oil pan side up.

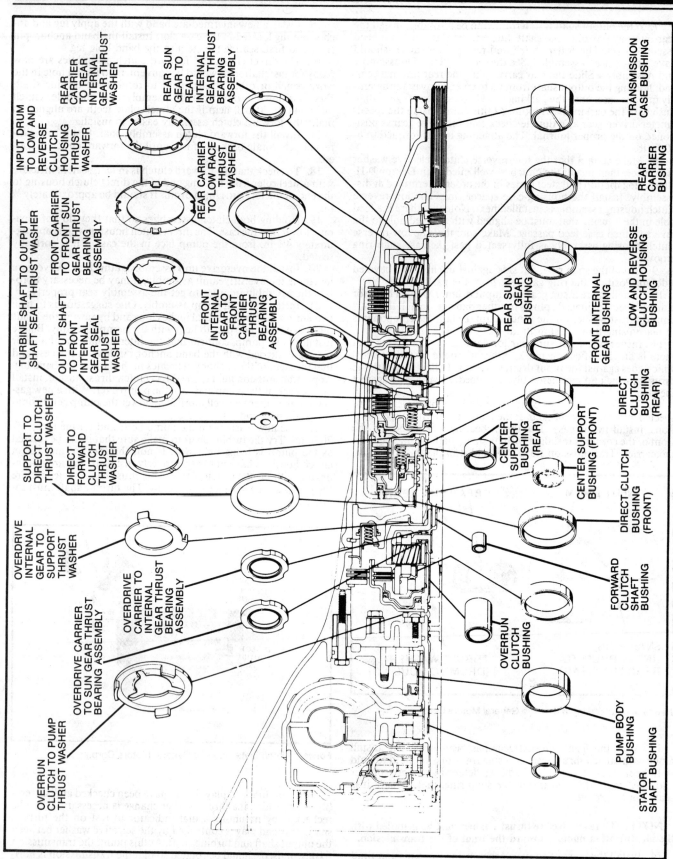

Automatic transmission bushing locations (© General Motors Corp.)

CLUTCH PLATE AND APPLY RING USAGE CHART

Overrun Clutch				Fourth Clutch				Direct Clutch					
Flat Steel Plate		Composition Faced Plate		Flat Steel Plate		Composition Faced Plate		Flat Steel Plate		Composition Faced Plate		Apply Ring	
No.	Thickness	No.		No.	Thickness	No.		No.	Thickness	No.		ID.	Width
2	0.077 in.	2		3	0.077 in	2		6	0.091 in	6		9	0.429 in.

CLUTCH PLATE AND APPLY RING USAGE CHART

Forward Clutch							Low and Reverse Clutch						
Wave Plate		Flat Steel Plate		Composition Faced Plate	Apply Ring		Wave Plate		Flat Steel Plate		Composition Faced Plate	Apply Ring	
No.	Thickness	No.	Thickness	No.	ID.	Width	No.	Thickness	No.	Thickness	No.	ID.	Width
1	0.062 in.	3	0.077 in.	4	8	0.0492 in.	1	0.077 in.	7	0.077 in.	6	0	0.0516 in.

24. Take the clean and already inspected governor assembly and place the governor to case washer on the end, holding it with transmission fluid. Install the governor cover onto the governor, making certain that the small seal on the end has been renewed.

NOTE: If the seal is slightly damaged, a pressure leak will result. For the same reason, examine carefully the governor cover. A seal that had previously leaked or that deteriorated would have allowed enough end play to develop to have badly worn the governor cover. If in doubt, replace the governor cover.

CAUTION

Do not use any type of hammer to install the governor assembly into the case. Lube the seals with Dexron II® to help ease installation. Expect the last 1/16 in. of travel to be a tight fit as the governor cover end seal is being compressed, but do not hammer on the governor cover.

25. Install the intermediate servo piston into its cover and install the assembly into the case, tapping lightly if necessary. Make sure the tapered end of the band apply pin is properly located against the band apply lug. Install the retaining ring.
26. The low and reverse clutch housing to case cup plug and seal is installed next. Make sure the seal seats against the housing. Use a ⅜″ rod to drive the plug in until it seats against the seal.
27. Install the valve body. Note the check balls that belong in the valve body and case. There should be four in the valve body and one in the case. Begin by installing the 5th check ball in the case. It is the throttle valve exhaust check ball.

28. It would be helpful to make two guide pins from some 6.3 x 1.0 metric bolts. Screw these into the case on each side of the valve body location. Install four check balls into the ball seat pockets in the control valve body and retain with transmission fluid. Install the valve body spacer plate gasket.
29. Put the spacer plate on the gasket and then the spacer plate to case gasket.
30. Put two of the attaching bolts through the assembly, then install using a finger to position the manual valve into the detent lever pin.
31. Hold the assembly tightly as it is lowered onto the guide pins so that the check balls, 1-2 accumulator piston and manual valve do not fall out. Remove the guide pins and start the bolts into place. Replace the manual detent roller and spring assembly.
32. Install the throttle lever and bracket assembly seeing that the slot fits into the roll pin, aligning the lifter through the valve body bore and link through the T.V. linkage case bore. Install the bolt, then torque all control valve assembly bolts to 12 ft. lbs.
33. Install the oil screen using a new gasket and torque to 12 ft. lbs.
34. Install the oil pan with a new gasket and torque the retaining bolts evenly to 10-13 ft. lbs. The transmission is now ready for the torque converter and then to be moved to the transmission jack for installation.

S SPECIFICATIONS

TORQUE SPECIFICATIONS—THM 200-4R

Item	Ft. Lbs.	N•m
Pump cover bolts	18	24
Pump to case attaching bolts	18	24
Parking pawl bracket bolts	18	24
Control valve body bolts	11	15
Oil screen retaining bolts	11	15
Bottom pan attaching bolts	12	16
Converter to flywheel bolts	35	48

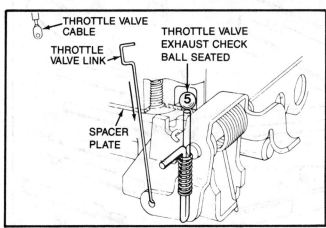

Throttle lever assembly (© General Motors Corp.)

TORQUE SPECIFICATIONS—THM 200-4R

Item	Ft. Lbs.	N•m
Transmission to engine mounting bolts	25	34
Converter dust shield screws	8	11
Manual shaft nut	23	31
Speedometer driven gear attaching bolts	8	11
Detent cable attaching screw	6	9
Oil cooler line to transmission connector	25	37
Oil cooler line to radiator connector	20	27
Linkage swivel clamp nut	30	41

TORQUE SPECIFICATIONS—THM 200-4R

Item	Ft. Lbs.	N•m
Shifter assembly to sheet metal screws	8	10
Converter bracket to adapter nuts	13	17
Catalytic converter to rear exhaust pipe nuts	17	23
Exhaust pipe to manifold nuts	12	16
Rear transmission support bolts	40	54
Mounting assembly to support nuts	21	29
Mounting assembly to support center nut	33	44
Adapter to transmission bolts	33	44

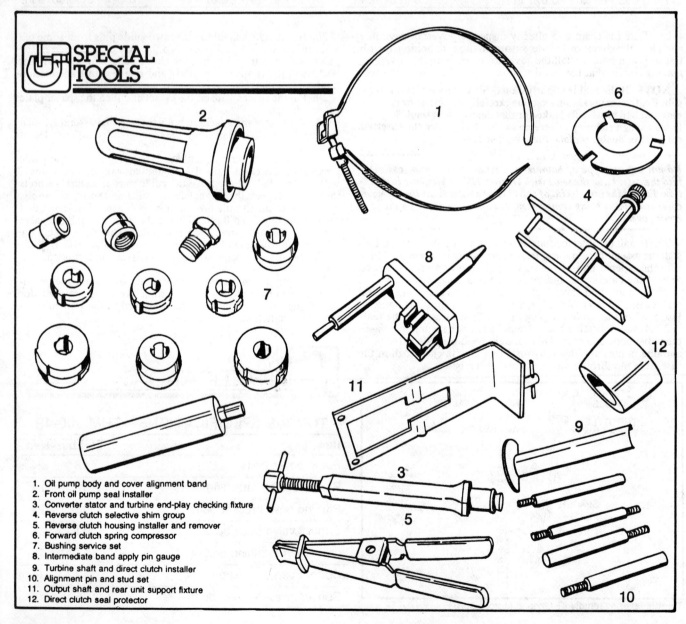

SPECIAL TOOLS

1. Oil pump body and cover alignment band
2. Front oil pump seal installer
3. Converter stator and turbine end-play checking fixture
4. Reverse clutch selective shim group
5. Reverse clutch housing installer and remover
6. Forward clutch spring compressor
7. Bushing service set
8. Intermediate band apply pin gauge
9. Turbine shaft and direct clutch installer
10. Alignment pin and stud set
11. Output shaft and rear unit support fixture
12. Direct clutch seal protector

SPECIAL TOOLS

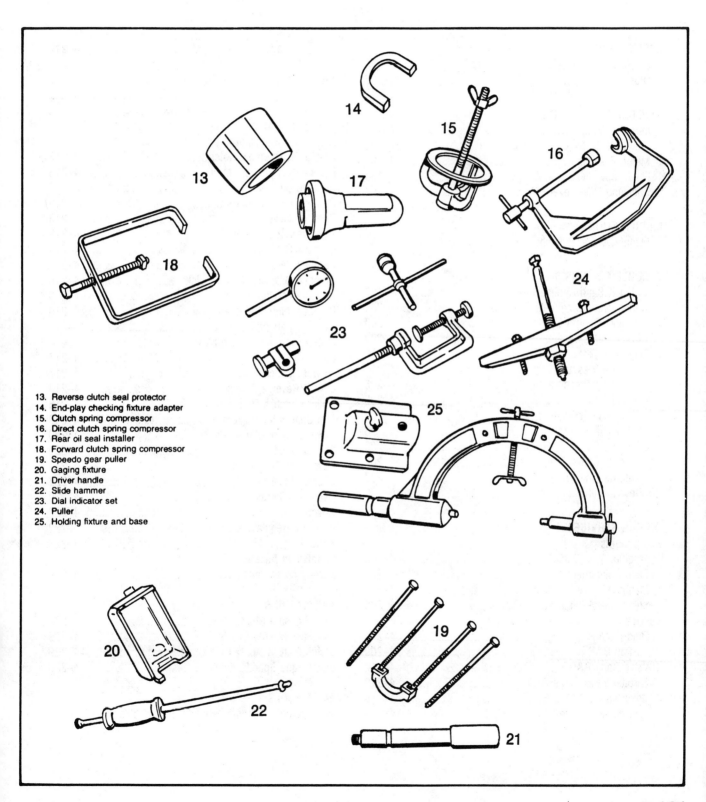

13. Reverse clutch seal protector
14. End-play checking fixture adapter
15. Clutch spring compressor
16. Direct clutch spring compressor
17. Rear oil seal installer
18. Forward clutch spring compressor
19. Speedo gear puller
20. Gaging fixture
21. Driver handle
22. Slide hammer
23. Dial indicator set
24. Puller
25. Holding fixture and base

INDEX

GENERAL MOTORS
TURBO HYDRA-MATIC 325-4L
Automatic Transaxle

APPLICATIONS

TRANSAXLE APPLICATION CHART

Year	Transmission Type	Model Identification
1982 and Later	325-4L	Riviera
1982 and Later	325-4L	Seville, Eldorado
1982 and Later	325-4L	Toronado

GENERAL DESCRIPTION

The THM 325-4L automatic transaxle is a fully automatic front wheel drive transaxle. This unit, consists primarily of a four element hydraulic torque converter with a torque converter clutch, three compound planetry gear sets and an overdrive unit. Five multiple-disc clutches, two roller clutches and a band provide the friction elements required to obtain the desired function of the compound planetary gear set and the overdrive unit.

The torque converter smoothly couples the engine to the overdrive unit and planetary gears through oil and hydraulically provides additional torque multiplication as required. The combination of the compound planetary gear set and the overdrive unit provides four forward ratios and one reverse. Changing of the gear ratios is fully automatic in relation to vehicle speed and engine torque. Vehicle speed and engine torque signals are constantly fed to the automatic transaxle to provide the proper gear ratios for maximum efficiency and performance at all throttle openings.

A hydraulic system, pressurized by a gear type pump, provides the working pressure required to operate the friction elements and automatic controls.

Transaxle and Torque Converter Identification

TRANSAXLE

The THM 325-4L automatic transaxle can be identified by locating the unit number plate, which is positioned on the left side of the torque converter housing.

TORQUE CONVERTER

Torque converter usage differs with the type engine and the area in which the vehicle is to be operated. No specific physical markings are designated so that the repairman can identify the converter. It is most important to obtain the transaxle model and serial number, know the specific operating area and the vehicle VIN number before replacement of the converter unit is attempted.

Metric Fasteners

Metric bolts sizes and thread pitches are used for the THM 325-4L. The metric fastener dimensions are very close to the dimensions of the familiar inch system fasteners.

For this reason, replacement fasteners must have the same measurement and strength as those removed.

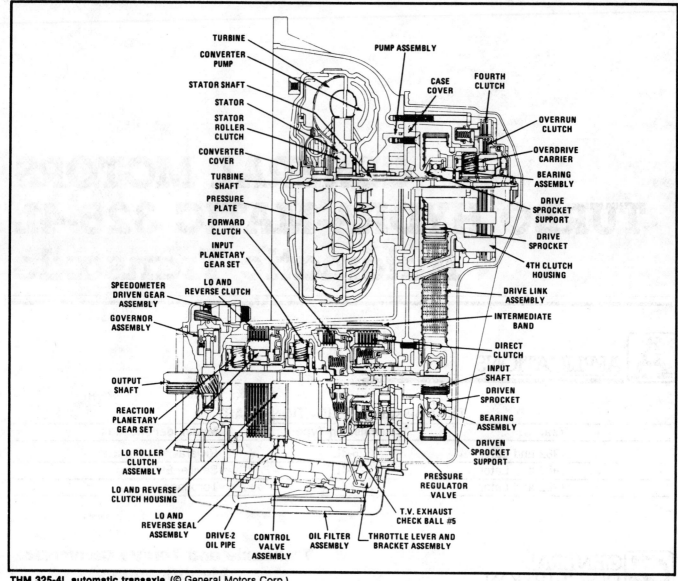

TURBINE

CONVERTER PUMP

STATOR SHAFT

STATOR

STATOR ROLLER CLUTCH

CONVERTER COVER

TURBINE SHAFT

PRESSURE PLATE

FORWARD CLUTCH

INPUT PLANETARY GEAR SET

SPEEDOMETER DRIVEN GEAR ASSEMBLY

GOVERNOR ASSEMBLY

LO AND REVERSE CLUTCH

OUTPUT SHAFT

REACTION PLANETARY GEAR SET

LO ROLLER CLUTCH ASSEMBLY

LO AND REVERSE CLUTCH HOUSING

LO AND REVERSE SEAL ASSEMBLY

DRIVE-2 OIL PIPE

CONTROL VALVE ASSEMBLY

OIL FILTER ASSEMBLY

THROTTLE LEVER AND BRACKET ASSEMBLY

T.V. EXHAUST CHECK BALL #5

PRESSURE REGULATOR VALVE

DRIVEN SPROCKET SUPPORT

BEARING ASSEMBLY

DRIVEN SPROCKET

INPUT SHAFT

DIRECT CLUTCH

INTERMEDIATE BAND

DRIVE LINK ASSEMBLY

4TH CLUTCH HOUSING

DRIVE SPROCKET

DRIVE SPROCKET SUPPORT

BEARING ASSEMBLY

OVERDRIVE CARRIER

OVERRUN CLUTCH

FOURTH CLUTCH

CASE COVER

PUMP ASSEMBLY

THM 325-4L automatic transaxle (© General Motors Corp.)

Do not attempt to interchange metric fasteners for inch system fasteners. Mismatched or incorrect fasteners can result in damage to the unit through malfunctions, breakage or possible personal injury.

Care should be taken to re-use the fasteners in the same locations as removed, whenever possible.

Fluid Capacities

The fluid level should be checked at each engine oil change. Use only Dexron® II or equivalent transmission fluid. Under normal operating conditions, the fluid may be expected to last 100,000 miles. However, if the vehicle is used in what is considered severe service, then the automatic transaxle fluid and filter should be changed every 18,000-24,000 miles or a time interval of 18-24 months. Severe service would include heavy city traffic, high temperature areas (regularly over 90°F), mountainous areas or frequent trailer towing. Commercial service is also to be considered severe service, so more frequent service to the automatic transaxle will be required.

If any service work is performed on the transaxle, the following capacities should be followed.

a) pan removal—five quarts of fluid

b) overhaul—twelve quarts of fluid

The fluid level should be established by the dipstick rather than by using a specific number of quarts.

Checking Fluid Levels

1. Verify that the transaxle is at normal operating temperature. At this temperature, the end of the dipstick will be too hot to hold in the hand. Make sure that the vehicle is level.

2. With the selector in Park, allow the engine to idle. *Do not race engine.* Move the selector through each range and back to Park.

3. Immediately check the fluid level, engine still running. Fluid level on the dipstick should be at the "FULL HOT" mark.

4. If the fluid level is low, add fluid as required. One pint will bring the fluid level from "ADD" to "FULL."

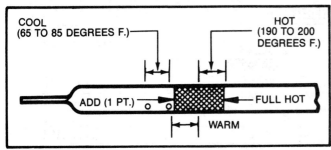

THM 325-4L automatic transaxle dipstick (© General Motors Corp.)

Often it is necessary to check the fluid level when there is no time or opportunity to run the vehicle to warm the fluid to operating temperature. In this case, the fluid should be around room temperature (70°). The following steps can be used:

1. Place the selector in Park and start engine. *Do not race engine.* Move the selector through each range and back to Park.

2. Immediately check the fluid level, engine still running, off fast idle. Fluid level should be between the two dimples on the dipstick, approximately ¼" *below* the "ADD" mark on the dipstick.

3. If the fluid level is low, add fluid as required, to bring the fluid level to between the two dimples on the dipstick. *Do not overfill.*

M MODIFICATIONS

Automatic Transaxle Fluid Leak in Final Drive Area

Some 1982 vehicles equipped with the THM 325-4L automatic transaxle may experience a transaxle oil leak in the final drive to transaxle case mating area and/or an erratic upshift caused by a loose or missing governor cup plug in the transaxle case passage.

Automatic transaxle with the fluid leak problem are, model AJ serial number 58856, model AL serial number 8327 and model AM serial number 5587. These identification numbers are located on the transaxle identification tag, which is mounted on the left side of the bell housing.

In order to correct an automatic transaxle with this problem the following procedure must be followed.

1. Remove the final drive assembly.

2. Remove the original governor cup plug.

3. Install a new governor cup plug, part number 8620318. Use Loctite 240® or equivalent to seal the cup plug to the transaxle case. Staking the plug cup into the transaxle case will not assure an adequate seal.

4. Install the final drive assembly. Be sure to use a new gasket.

Hydraulic Noise in Park and/or Neutral

Some vehicles equipped with the THM 325-4L automatic transaxle may experience a hydraulic buzzing sound in the park and/or neutral shift selector positions. This condition may be caused by an orifice cup plug missing from the pressure regulator valve.

In order to correct this condition, the following procedure must be performed.

1. Remove the automatic transaxle fluid pan. Discard the pan gasket. Remove the pressure regulator valve.

2. Inspect the valve for a missing orifice cup plug, which is located at the end of the valve.

3. If the cup plug is missing, inspect the pressure regulator valve bushing as well as the transaxle case for the original plug. If found, remove and discard.

4. Install the new pressure regulator valve, part number 8623422. Be sure that the new valve is equipped with an orifice cup plug.

5. Install the transaxle oil pan, using a new pan gasket. Fill the transaxle as required with the proper grade and type automatic transmission fluid.

No Drive or Slipping

Some 1983 vehicles may experience a no drive or slipping condition. This condition, while not noticed in reverse, could be caused by a loose or missing solid cup plug in the input shaft of the forward clutch assembly.

When servicing any 1983 THM 325-4L automatic transaxle for this condition, and the forward clutch assembly shows signs of burning, inspect the housing end of the input shaft which is part of the forward clutch assembly. If the solid cup plug is loose or missing, apply pressure to the forward clutch piston will be lost. This pressure loss will cause the no drive or slipping condition. If the solid cup plug is missing from the feed passage, it must be located before the transaxle is reassembled. It could possibly be in the open end of the output shaft. Tap the end of the output shaft on a table and remove the solid cup plug.

The solid cup plug, part number 8628145, is serviced separately. To install the cup plug, use a ⁵⁄₁₆ inch punch. Apply loctite 290® or equivalent to the plug before installation. Drive the plug into the larger of the two holes until it is .039 inch below the surface.

New Style Selective Washer for Input to Output Shaft

Beginning at the end of March, 1983, a new style selective washer went into production. This new style selective washer is now available when servicing the input to output shaft assemblies.

This new style selective washer is made from a different material and can be identified by a raised portion on the identification tab.

When servicing an automatic transaxle that necessitates the replacement of this selective washer due to wear or damage, the following procedure must be followed.

1. Inspect the two shafts for wear or damage on the selective

SELECTIVE WASHER CHART

Identification Number	Color	Part Number	Thickness (inches)
1	----	8639291	.065-.070
2	----	8639292	.070-.075
3	Black	8639293	.076-.080
4	Light Green	8639294	.081-.085
5	Scarlet	8639295	.086-.090

SELECTIVE WASHER CHART

Identification Number	Color	Part Number	Thickness (inches)
6	Purple	8639296	.091-.095
7	Cocoa Brown	8639297	.096-.100
8	Orange	8639298	.101-.106
9	Yellow	8639299	.106-.111
10	Light Blue	8639300	.111-.116
11	Blue	8639301	.117-.121
12	----	8639302	.122-.126
13	Pink	8639303	.127-.131
14	Green	8639304	.132-.136
15	Gray	8639305	.137-.141

washer mating surfaces. If wear or damage is found, replace the components.

2. Use the proper selective washer from the selective washer chart.

3. Check the front end play to verify the proper selective washer selection. Front end play for the THM 325-4L automatic transaxle should be .022 to .051 inch.

Automatic Transaxle Bottom Oil Pan Leak

Some 1982 and 1983 vehicles equipped with the THM 325-4L automatic transaxle may exhibit a transaxle oil pan leak. This leak is caused by an interference between the 1-2 and 3-4 accumulator housing and the bottom oil pan.

Starting in April, 1983, all THM 325-4L automatic transaxles are being produced using a new design accumulator housing. This new design housing will correct the interference problem.

When servicing any THM 325-4L transaxle for a bottom oil leak problem, be sure to inspect for interference with the accumulator housing when the oil pan is removed. If evidence of interfer-

ence exists, remove the housing and grind off a portion of the casting boss. When performing this operation use caution to avoid damaging the cup plug.

4-2 Downshift Hitch (Model AJ)

Some vehicles equipped with transaxle model AJ, may experience a 4-2 downshift as a two shift feel with a slight hesitation and a bump in between. This condition is most noticeable during a full throttle detent downshift.

Starting in February, 1983, all automatic transaxles beginning with transaxle serial number 83-AJ-30723 are being built using a new control valve assembly. This new control valve assembly will correct the 2-4 downshift condition.

When servicing any 1983 model AJ transaxle for this condition use service package 8635949, which contains the proper components to correct the problem.

No Fourth Gear Engagement

Some 1982 and 1983 vehicles equipped with the THM 325-4L automatic transaxle may not engage into fourth gear (overdrive).

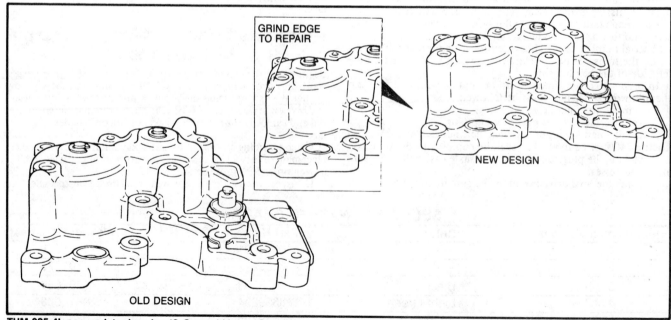

GRIND EDGE TO REPAIR

NEW DESIGN

OLD DESIGN

THM 325-4L accumulator housing (© General Motors Corp.)

If this condition exists, the following diagnostic procedure will aid in correction of the problem.

Test drive the vehicle applying light throttle in drive range (third gear) to about 30-35 mph. Move the selector lever into manual second "2" and release the throttle. The automatic transaxle should downshift and coastdown overrun braking should occur.

If coastdown overrun braking does not occur, remove the transaxle from the vehicle. Inspect the overrun clutch housing for a break in the hub weld. If the hub weld is broken, the cause may be due to misaligned pinions in the overdrive carrier.

In all instances where the weld is broken on the overrun clutch housing, both the overrun clutch housing, part number 8635941, and the overdrive carrier, part number 8635036, must be replaced.

Shift Busyness, Cruise Control Engaged

Some 1981 and 1982 vehicles equipped with the THM 325-4L automatic transaxle model AM and the 4.1 Liter engine, may experience a transaxle shift busyness at highway speeds with the cruise control engaged. Shift busyness refers to the cycling of the transaxle between fourth and third gears. This condition is noticeable while driving up inclines. In order to correct this problem, install service package 8635944 into the control valve assembly.

Delayed Engagement and/or Loss of Engine Power

Some 1982 and 1983 vehicles equipped with the THM 325-4L automatic transaxle may exhibit either delayed engagement when moving the selector lever from Park to Reverse or an apparent loss of engine power resulting from improper low and reverse clutch operation.

Delayed engagement (hot or cold) when moving the automatic

TROUBLE DIAGNOSIS

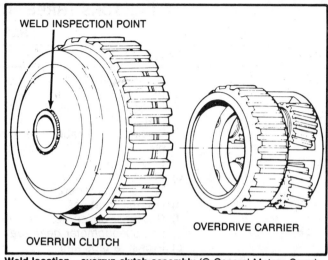

Weld location—overrun clutch assembly (© General Motors Corp.)

transaxle selector lever from Park to Reverse after a complete overhaul can be caused by an insufficient number of flat steel plates installed in the low and reverse clutch assembly. The low and reverse clutch assembly should have one waved steel plate, seven flat steel plates and six composition faced plates installed in it.

An apparent loss of engine power or automatic transaxle operation in two gears at one time on any THM 325-4L automatic transaxle can be caused by the incorrect application of the low and reverse clutch assembly. The incorrect apply of the low and reverse clutch can be caused by a missing locator pin securing the low first detent valve bushing in the control valve assembly. This missing pin will permit the bushing to rotate in its control valve bore and apply the low and reverse clutch through the use of "detent" fluid. Whenever the throttle angle is above sixty degrees of travel, "detent" fluid is available.

To correct this condition, reroute the valve body, orient the bushing correctly in its bore and install a locator pin, part number 8628400.

CLUTCH AND BAND APPLICATION CHART
THM 325-4L

Range/Gear		Low-Reverse Clutch	Low-Roller Clutch	Forward Clutch	Intermediate Band	Direct Clutch	Fourth Clutch	Overrun Clutch	Overdrive Roller Clutch
Park		—	—	—	—	—	—	—	—
Reverse		Applied	—	—	—	Applied	—	—	Holding
Neutral		—	—	—	—	—	—	—	Holding
Drive 4	First	—	Holding	Applied	—	—	—	—	—
	Second	—	—	Applied	Applied	—	—	—	Holding
	Third	—	—	Applied	—	Applied	—	—	Holding
	Fourth	—	—	Applied	—	Applied	Applied	—	—
Drive 3	First	—	Holding	Applied	—	—	—	Applied	—
	Second	—	—	Applied	Applied	—	—	Applied	—
	Third	—	—	Applied	—	Applied	—	Applied	—
Drive 2	First	—	Holding	Applied	—	—	—	Applied	—
	Second	—	—	Applied	Applied	—	—	Applied	—
Low	First	Applied	—	Applied	—	—	—	Applied	—

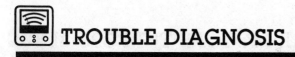

CHILTONS THREE "C's" DIAGNOSIS CHART
THM 325-4L

Condition	Cause	Correction
Oil leak	a) Sprocket cover and bottom oil pan	a) Retorque attaching bolts
	b) Cover to pan sealant bead broken	b) Reseal assemblies as required
	c) Throttle valve cable, filler pipe, electrical connector missing or damaged seals.	c) Replace missing or damaged seal, as required
	d) Engine mount and T.V. cable interference	d) Correct interference between T.V. cable and engine mount
	e) Manual shaft seal	e) Seal assembly damaged, replace as required
	f) Governor cover, servo cover and speedometer gear seal	f) Repair damaged O-ring seal
	g) Pressure taps and cooler fittings	g) Retorque fittings
	h) Transaxle case at final drive	h) Repair damaged gasket Retorque attaching bolts
	i) Torque converter	i) Weld seam leaking, repair or replace
	j) Pump seal damaged	j) Repair as required Retorque attaching bolts
	k) Governor oil cup plug not seated properly	k) Repair or replace as required
Oil out the vent	a) Oil level too high	a) Correct as required
	b) Drive sprocket support holes blocked	b) Clear blockage
	c) Water in transaxle fluid	c) Correct as required
High or low oil pressure	a) Oil level high or low	a) Correct as required
	b) T.V. cable broken or misadjusted	b) Repair cable or replace as required
	c) T.V. lever and bracket assembly roll pin binding	c) Correct as required
	d) Throttle valve plunger stuck or binding	d) Repair or replace as required
	e) Pressure regulator valve stuck, binding or misassembled	e) Repair or replace as required
	f) Reverse boost valve and bushing stuck or binding	f) Repair or replace as required
	g) M.T.V. valve and bushing stuck or binding	g) Repair or replace as required
	h) Pressure relief valve ball missing or spring distorted	h) Repair or replace as required
	i) T.V. limit valve or line bias valve stuck or mising	i) Repair or replace as required
	j) Oil filter plugged, pick up tube cracked, O-ring damaged	j) Replace filter, replace pick up tube replace damaged O-ring
	k) Oil pump gears damaged	k) Replace oil pump assembly
No drive in drive range	a) Low oil level	a) Correct oil level
	b) Low oil pressure	b) Correct as required
	c) Manual linkage misadjusted	c) Adjust
	d) Torque converter stator roller clutch broken	d) Replace components as required
	e) Overdrive roller clutch springs or rollers damaged	e) Repair or replace roller clutch assembly

CHILTONS THREE "C's" DIAGNOSIS CHART
THM 325-4L

Condition	Cause	Correction
No Drive in drive range	f) Overdrive carrier assembly pinions, sun gear or internal gear damage	f) Repair or replace components as required
	g) Damaged oil pump gears	g) Replace pump
	h) Broken drive link assembly sprockets or sprocket bearings	h) Replace components as required
	i) Forward clutch assembly damaged, or components within the unit defective	i) Repair or replace assembly or components as required
	j) Damaged low roller clutch springs or rollers	j) Repair or replace components as necessary
	k) Forward clutch input shaft broken	k) Replace components as required
No engine braking in manual ranges	a) Manual linkage misadjusted	a) Adjust linkage
	b) Spacer plate or gaskets misadjusted, damaged or plugged	b) Replace defective components as required
	c) 4-3 control valve-manual third only binding or stuck	c) Replace 4-3 control valve
	d) D2 signal pipe-manual second only leaking	d) Repair as required
	e) Number three check ball leaking or missing	e) Replace check ball
	f) Number nine check ball leaking or missing	f) Replace check ball as required
	g) Overrun clutch assembly damaged piston seals or clutch plates	g) Replace or repair damaged components
	h) Blocked turbine shaft oil passages	h) Remove blockage
No part throttle	a) T.V. valve or plunger stuck or binding	a) Repair or replace valve or plunger
	b) Spacer plate or gaskets mispositioned or blocked	b) Replace components as required or remove blockage
	c) T.V. cable misadjusted	c) Readjust cable
	d) T.V. downshift modulator valve	d) Correct as required
Won't hold in park	a) Manual linkage misadjusted	a) Adjust linkage
	b) Internal linkage damaged, mispositioned or mis-assembled	b) Correct as required
Shifts points high or low	a) T.V. cable misadjusted or damaged	a) correct as required
	b) T.V. limit valve stuck or binding	b) Correct as required
	c) Throttle valve or plunger stuck or binding	c) Repair or replace as required
	d) T.V. modulator upshift or downshift valves stuck or binding	d) Repair or replace valves as needed

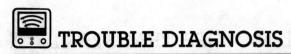

CHILTONS THREE "C's" DIAGNOSIS CHART
THM 325-4L

Condition	Cause	Correction
Shifts points high or low	e) Mispositioned or binding spacer plate or gaskets	e) Replace spacer plate or gasket
	f) T.V. lever and bracket assembly unhooked, mispositioned or mis-assembled	f) Correct as required
	g) Governor assembly or cover damaged or leaking	g) Repair or replace as required
First speed only, no 1-2 shift	a) Internal governor assembly damaged	a) Repair or replace unit, as required
	b) Control valve assembly 1-2 shift, low/first detent or 1-2 throttle valve stuck	b) Repair or replace defective components as necessary
	c) Intermediate band assembly burned or damaged.	c) Replace band assembly
	d) Intermediate band assembly band anchor plug missing	d) Install band anchor plug
	e) Intermediate servo assembly band apply pin too short, wrong piston cover combination or seal damage	e) Correct defective component as necessary
	f) 1-2, 3-4 accumulator housing spacer plate blocked or mispositioned, gasket damage or housing porosity	f) Correct the problem as required
1-2 Shift—full throttle only	a) T.V. cable misadjusted, bent or disconnected	a) Readjust, connect or replace as required
	b) T.V. lever and bracket assembly roll pin wedged against the valve body casting	b) Correct as required
	c) Throttle valve or plunger stuck or binding	c) Free component as required
	d) Spacer plate or gaskets blocked or mispositioned	d) Free blockage or reposition components
First and second speeds only—no 2-3 shift	a) Control valve assembly 2-3 shift valve or throttle valve stuck or binding	a) Correct components as necessary
	b) Spacer plate and gaskets blocked or mispositioned	b) Free blockage or reposition spacer plate and gaskets
	c) Transaxle case interconnected passages or porosity	c) Replace transaxle case
	d) Transaxle case cover number 7 pellet and spring not seating	d) Replace number 7 pellet and spring
	e) Transaxle case cover third/reverse cup plug leaking	e) Replace third/reverse cup plug
	f) Transaxle case cover gasket damaged	f) Replace case cover gasket
	g) Transaxle case cover porosity or interconnected passages	g) replace transaxle case cover
	h) Driven sprocket support oil seal rings leaking or sprocket support gasket damaged	h) Replace oil seal rings or sprocket support gasket
	i) Number 6 check ball missing or leaking	i) Replace number 6 check ball

CHILTONS THREE "C's" DIAGNOSIS CHART

THM 325-4L

Condition	Cause	Correction
First and second speeds only—no 2-3 shift	j) Direct clutch assembly piston seals or ball capsule damaged k) Internal governor assembly malfunction	j) Replace components or assembly as required k) Repair or replace governor assembly as required
No reverse or slips in reverse	a) Manual linkage misadjusted b) Low and reverse clutch piston, seals or cup plug seal assembly damaged c) Direct clutch misassembled, piston or seals or ball capsule damaged d) Reverse boost valve binding or stuck e) Reverse oil pipe leaking f) Number 6 check ball missing or leaking g) Number 10 check ball missing or leaking h) Spacer plate mispositioned or damaged i) Transaxle case cover, case or valve body reverse passage leak	a) Adjust linkage b) Replace piston, seals or cup plug seal assembly as required c) replace defective component d) Repair or replace reverse boost valve e) Replace reverse oil pipe f) Replace number 6 check ball g) Replace number 10 check ball h) Replace spacer plate i) Replace defective component as necessary
Drive in neutral range	a) Manual linkage misadjusted b) Transaxle case or case cover leak into D4 oil passage c) Forward clutch plates fused or ball capsule won't release	a) Adjust linkage b) Replace defective component as required c) Replace forward clutch plates or free ball capsule
Slipping 1-2 shift	a) Low fluid level b) Mispositioned or blocked spacer plate/gaskets c) Accumulator housing bolts loose or housing porosity d) 1-2 accumulator piston or seal damaged, stuck or binding e) Intermediate band damaged or burnt lining f) Intermediate servo assembly porosity, damaged piston or seal or wrong apply pin g) Pressure regulator valve sticking or binding h) Line bias valve, throttle valve or plunger binding or stuck i) T.V. cable misadjusted j) Transaxle case or case cover leak in second oil passage	a) Correct as required b) Correct as required c) Tighten bolts or replace housing d) Correct as required e) Replace intermediate band or lining f) Repair or replace defective component as required g) Repair or replace pressure regulator valve h) Correct the required component as necessary i) Adjust cable j) Replace component as required
Rough 1-2 shift	a) T.V. cable misadjusted b) Throttle valve or plunger binding or stuck	a) Adjust cable b) Correct as required

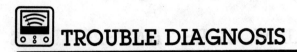

CHILTONS THREE "C's" DIAGNOSIS CHART
THM 325-4L

Condition	Cause	Correction
Rough 1-2 shift	c) T.V. limit valve, line bias valve or accumulator stuck or binding	c) Repair or replace components as required
	d) Intermediate servo assembly piston bleed orifice missing or wrong apply pin	d) Correct as required
	e) 1-2 accumulator piston or seal damaged, binding or stuck	e) Repair or replace defective components as required
	f) Number 8 check ball missing or blocked	f) Free blockage or install number 8 check ball
	g) Servo bleed orifice missing or blocked	g) Repair as required
Slipping 2-3 shift	a) Low fluid level	a) Correct as required
	b) T.V. cable misadjusted	b) Adjust cable
	c) Mispositioned or blocked spacer plate/gaskets	c) Free blockage or replace spacer plate/gaskets
	d) Leaking number 7 pellet and spring	d) Replace number 7 pellet and spring
	e) Servo piston seal rings damaged	e) Replace servo piston or seal rings
	f) Servo bleed orifice missing or damaged	f) Correct as required
	g) Driven sprocket support case, case cover leak in the third clutch passage	g) Correct the defective component as required
	h) Direct clutch damaged ball capsule, piston or housing	h) Repair or replace defective component
	i) Direct clutch piston or seal damaged	i) Repair or replace component as necessary
Rough 2-3 shift	a) T.V. cable misadjusted	a) Adjust cable
	b) Throttle valve or plunger binding or stuck	b) Correct defective parts as necessary
	c) T.V. limit or bias valve stuck or binding	c) Correct defective component as required
	d) M.T.V up valve stuck or binding	d) Correct defective component as required
No 3-4 shift	a) Shift valve or 3-4 throttle valve stuck or binding	a) Repair or replace defective parts as necessary
	b) Fourth clutch assembly piston or seal damaged	b) Replace piston or seals
	c) Fourth clutch assembly fourth feed O-ring seal missing or damaged	c) Replace O-ring seal
	d) Accumulator housing leak in the 3-4 accumulator circuit	d) Repair or replace as required
	e) Mispositioned or blocked spacer plate/gaskets	e) Free blockage or replace spacer plate/gaskets
Slipping 3-4 shift	a) Accumulator housing piston or seal damaged	a) Replace or repair piston or seal
	b) Accumulator valve binding or stuck	b) Replace component as required

CHILTONS THREE "C's" DIAGNOSIS CHART
THM 325-4L

Condition	Cause	Correction
Slipping 3-4 shift	c) Fourth clutch internal leak, damaged piston seal, housing bolts loose or feed O-ring damaged	c) Correct and replace components as necessary
Rough 3-4 shift	a) T.V. cable misadjusted b) Throttle valve or plunger binding or stuck c) 3-4 accumulator piston and seal damaged, stuck or spring weak or missing	a) Adjust cable b) Correct as necessary c) Repair or replace defective component

Hydraulic Control System

The major components incorporated within the hydraulic control system are: Oil pump assembly, Pressure regulator assembly, Throttle valve, T.V. limit valve, T.V. exhaust check ball, Line bias valve and governor assembly.

OIL PUMP ASSEMBLY

The hydraulic pressure system requires a supply of transmission fluid and a pump to pressurize the fluid. This Transaxle uses an internal-external gear type pump with its oil intake connected to a screen assembly. The screen intake draws oil from the transaxle bottom pan or sump. The pump drive gear is keyed to the converter pump hub and therefore turns whenever the engine is operating. As the drive gear turns, it also turns the driven gear, causing oil to be lifted from the sump. As the gears turn, the oil is carried past the crescent section of the pump. Beyond the crescent, the gear teeth begin to come together causing the oil to be pressurized as it is squeezed from between the gear teeth. At this point, the oil is delivered through the pump outlet to the pressure regulator.

PRESSURE REGULATOR

Oil pressure is controlled by the pressure regulator valve. Oil is directed to the line passage and through an orifice to the top of the pressure regulator valve. When the desired line pressure is reached, the line pressure at the top of the valve moves the valve against the pressure regulator spring, thus opening a passage to feed the torque converter.

THROTTLE VALVE

Throttle valve pressure is related to carburetor opening which is related to engine torque. The system is a mechanical type with a fixed, direct relation between the carburetor throttle plate opening and the transaxle throttle plunger movement. As the accelerator pedal is depressed and the carburetor opened, the mechanical linkage (T.V. cable) relays the movement to the throttle plunger and increases the force of the T.V. spring against the throttle valve, increasing T.V. pressure which can regulate from 0 to 90 psi. T.V. oil is directed through the T.V. plunger to provide a hy-

draulic assist reducing the pedal effort necessary to actuate the plunger.

THROTTLE VALVE LIMIT VALVE

The pressure requirements of the transaxle for apply of the band and clutches vary with engine torque and throttle opening. Under heavy throttle operation, the 60 psi line pressure is not sufficient to hold the band and clutches on without slipping. To provide higher line pressure with greater throttle opening, a variable oil pressure related to throttle opening is desired. The throttle valve regulates line pressure in relation to carburetor opening. The T.V. limit valve limits this variable pressure to avoid excessive line pressure. The T.V. limit valve feeds the throttle valve and receives oil directly from the oil pump. As the pressure in the line leading from the T.V. limit valve and feeding the T.V. valve exceeds 90 psi, the pressure will push against the T.V. limit valve spring, bleeding off excess pressure. This limits T.V. feed pressure to a maximum of approximately 90 psi.

THROTTLE VALVE EXHAUST CHECK BALL

The throttle valve exhaust check ball feature has been included in the T.V. system. This feature will prevent the transaxle from becoming burned or damaged due to low line pressure in the event the T.V. cable becomes disconnected or broken. This check ball is located in the transaxle case pad at the T.V. regulating exhaust port. This check ball (5) is held off its seat by the T.V. exhaust lifter rod, when the T.V. cable is properly connected. This allows T.V. pressure to regulate normally. If the cable becomes disconnected, or is not properly adjusted, the T.V. exhaust lifter rod will drop down and allow the ball to close the exhaust port. This will block the T.V. exhaust port and keep T.V. pressure at 90 psi and line pressure at approximately 140 psi to prevent operating the transaxle with 60 psi line pressure at increased throttle openings.

LINE BIAS VALVE

The capacity of a band or clutch to resist slipping is directly related to the pressure supplied to its apply piston. In order to prevent the band or clutches from slipping when an increase in engine torque is applied, oil pressure supplied to the apply piston must be increased. However, excessive pressure will cause a harsh shift. It is then desirable to vary the pressure supplied to each apply piston at the same rate at which engine torque is varied.

The line bias valve is fed by throttle valve fluid and performs the function of regulating throttle valve pressure as engine torque varies. This oil pressure, called modulated throttle valve oil (M.T.V.), is fed to the T.V. boost valve in the bottom of the pressure regulator valve, to increase line pressure as throttle opening or engine torque increases.

In summary, the T.V. plunger controls T.V. pressure which becomes M.T.V. pressure at the line bias valve, and the line bias valve regulates line pressure at the pressure regulator valve. It is this line pressure the manual valve routes to the shift valves and to the band and clutches. Line pressure is varied in relation to engine torque in such a manner as to apply the band or clutch with just enough pressure to hold against engine torque plus a safety factor, but not so much pressure that the shifts are harsh.

GOVERNOR ASSEMBLY

The vehicle speed signal for the shift is supplied by the transaxle governor which is driven by the output shaft. The governor assembly consists of a governor shaft, a driven gear, two check balls, a primary weight, a secondary weight, primary and secondary spring, one governor weight pin and an oil seal ring. The check balls seat in two pockets directly opposite each other in the governor shaft. The weights are so arranged that the primary weight, assisted by the primary spring, acts on one check ball and the secondary weight, assisted by the secondary spring, act on the other ball. As the governor turns, the weights are moved outward by the centrifugal force. This force is relayed to the check balls, seating them; and as the speed of the governor increases, so does the force acting on the check balls, which tends to keep them closed.

THROTTLE VALVE SYSTEM

The T H M 325-4L transaxle makes use of a throttle valve cable system, rather than the vacuum modulator system. This cable system controls line pressure, shift points, shift feel, part throttle downshifts, and detent downshifts. Do not think of the T.V. cable in the same way as a conventional detent cable. The function of the T.V. cable system more closely resembles the combined functions of both a detent cable and a vacuum modulator system. Since so many functions of the transaxle are dependent on the T.V. system, it can be seen that proper adjustment is critical to the satisfactory performance of the unit. Many transaxle problems or malfunctions could be solved by careful attention to, and adjustment of, the T.V. system.

The throttle valve cable operates the throttle lever and bracket assembly. This lever and bracket assembly serves two basic functions:

1. The first duty of this assembly is to transfer the movement of the carburetor throttle plate to the T.V. plunger in the control valve assembly. Thus, the T.V. pressure and line pressure can increase according to the throttle opening. It also controls part

throttle and detent downshifts. The proper adjustment of the T.V. cable is, then, very important and is based on the T.V. plunger being fully depressed to flush with the T.V. bushing at wide open throttle.

2. The second function of the assembly involves the T.V. exhaust valve lifter rod, spring and T.V. exhaust ball. The function of this system is to prevent the transmission from operating at low (idle position) pressures, should the cable break or become disconnected. As long as the cable is properly connected, not broken or stretched, the T.V. lifter rod will not move from its normal, spring loaded "UP" position which holds the T.V. exhaust check ball off its seat. The T.V. lifter rod will drop down to allow the exhaust ball to seat only if the cable breaks or becomes disconnected or out of adjustment. With the transaxle pan removed, it should be possible to pull down on the T.V. exhaust valve lifter rod and the spring should return the rod to its normal "UP" position. If the throttle lever and the bracket assembly and/or the lifter rod binds or sticks so that the T.V. lifter rod cannot lift the exhaust ball off its seat, high line pressures and delayed upshifts will be the result. The shape of the throttle valve lifter rod is critical, especially the 90° bend.

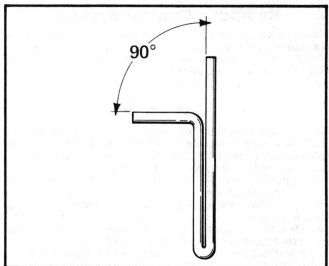

Throttle valve lifter rod (© General Motors Corp.)

If the T.V. cable is broken, sticky, mis-adjusted or if an incorrect part for the model is fitted, the vehicle may exhibit various malfunctions which could be attributed to internal component failure. Sticking or binding T.V. linkage can result in delayed or full throttle shifts. The T.V. cable must be free to travel to the full throttle position and return to the closed throttle position without binding or sticking.

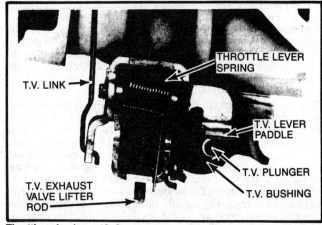

Throttle valve lever (© General Motors Corp.)

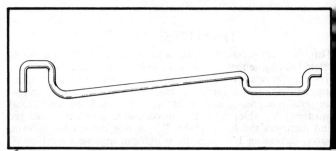

Throttle lever to cable link identification (© General Motors Corp.)

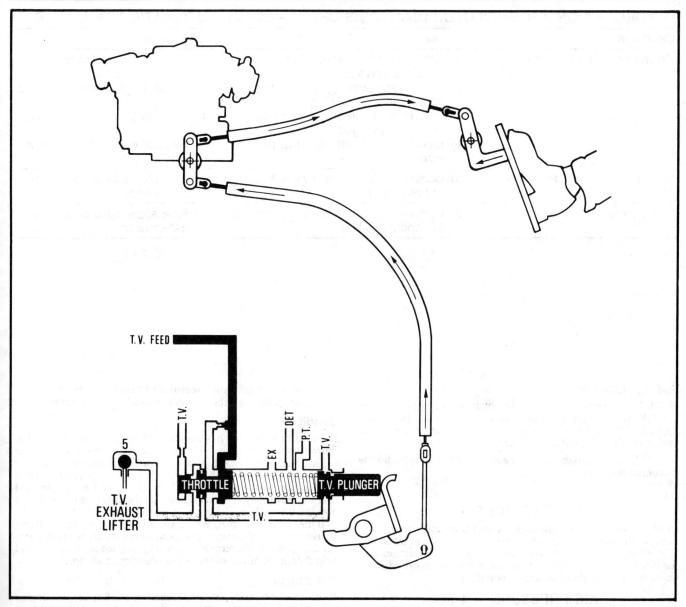

T.V. FEED

5

T.V. EXHAUST LIFTER

THROTTLE

T.V. PLUNGER

Accelerator, carburetor and T.V. linkage. (General Motors Co.)

Diagnosis Tests

Automatic transaxle malfunctions may be caused by four general conditions: poor engine performance, improper adjustments, hydraulic malfunctions, or mechanical malfunctions.

The suggested sequence for diagnosis is as follows:
1. Check and correct oil level.
2. Check and adjust T.V. cable.
3. Check and correct manual linkage.
4. Check engine tune.
5. Install oil pressure gauge and tachometer and check control pressure.
6. Road test in all ranges, noting change in operation and oil pressure.
7. Attempt to isolate the unit involved in the malfunction.

CONTROL PRESSURE TEST

1. Check transaxle oil level and correct as required. Check and adjust T.V. cable. Check and correct outside manual linkage. Check engine tune.
2. Install an oil pressure gauge to line pressure and fourth clutch pressure taps.
3. Connect the tachometer to the engine.
4. Perform the control pressure test in the following manner.

Minimum T.V. Line Pressure Test

Adjust the T.V. cable to specifications. With the brakes applied, take the line pressure readings in the ranges and at the engine rpm's indicated in the chart.

TORQUE CONVERTER CLUTCH DIAGNOSIS CHART—325-4L AUTOMATIC TRANSAXLE

Condition	Cause	Correction
Converter clutch will not apply	a) Converter clutch shift valve stuck b) Solenoid stays open c) Converter clutch apply valve and spring stuck or binding d) Converter clutch signal pipe leaks or screen plugged e) Electrical connector short or pinched wire	a) Correct as necessary b) Replace solenoid c) Repair or replace as required d) Replace pipe or free blockage e) Replace connector or repair pinched wire
Rough converter clutch apply	a) Converter clutch apply valve bushing orifice plugged	a) Free blockage or replace component
No converter clutch release	a) Converter clutch apply valve stuck in the apply position	a) Free apply valve or replace defective unit

Full T.V. Line Pressure Test

Full T.V. line pressure readings are obtained by tying or holding the T.V. cable to the full extend of its travel. With the brakes applied, take the line pressure readings in the ranges and at the engine rpm's as indicated in the chart.

NOTE: Total testing time must not exceed two minutes for the combinations.

ROAD TEST

In addition to the oil pressure tests, a road test can be performed, using all the selective ranges, to help isolate the unit or circuit that is involved in the malfunction. Comparing the malfunction with the element that is known to be in use, the problem can often be quickly narrowed down to one or two units.

AIR PRESSURE TESTS

The positioning of the transaxle in the vehicle and the valve body location will cause the air pressure tests to be very difficult. It is advisable to make any air pressure tests during the disassembly and assembly of the transaxle to ascertain if a unit is operating.

STALL SPEED TEST

General Motors Corporation does not recommend performing a stall test because of the excessive heat that is generated within the transaxle by the converter during the tests.

Recommendations are to perform the control pressure test and road test to determine and localize any transaxle malfunctions.

CONVERTER STATOR OPERATION DIAGNOSIS

The torque converter stator assembly and its related roller clutch can possibly have one of two different type malfunctions.

1. The stator assembly freewheels in both directions.
2. The stator assembly remains locked up at all times.

Condition A

If the stator roller clutch becomes ineffective, the stator assembly freewheels at all times in both directions. With this condition, the vehicle will tend to have poor acceleration from a standstill. At speeds above 30-35 mph, the vehicle may act normal. If poor acceleration problems are noted, it should first be determined that the exhaust system is not blocked, the engine is in good tune and the transaxle is in First gear when starting out.

If the engine will freely accelerate to high rpm in Neutral, it can be assumed that the engine and exhaust system are normal. Driving the vehicle in Reverse and checking for poor performance will help determine if the stator is freewheeling at all times.

Condition B

If the stator assembly remains locked up at all times, the engine rpm and vehicle speed will tend to be limited or restricted at high speeds. The vehicle performance when accelerating from a standstill will be normal. Engine over-heating may be noted. Visual examination of the converter may reveal a blue color from the over-heating that will result.

Torque Converter Clutch Operation and Diagnosis

TORQUE CONVERTER CLUTCH

The GM Torque Converter Clutch (TCC) incorporates a unit inside the torque converter with a friction material attached to a pressure plate and splined to the turbine assembly. When the clutch is applied, it presses against the converter cover. The result is a mechanical direct drive of the engine to the transaxle. This eliminates slippage and improves fuel economy as well as reducing fluid temperature.

There are a number of controls to operate the torque converter clutch, all determined by drive range selection. When the vehicle slows, or the transaxle shifts out of direct drive, the fluid pressure

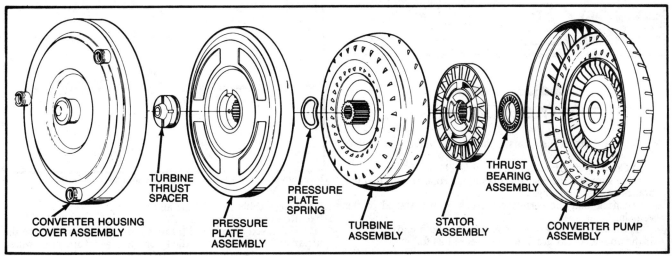

Torque converter clutch assembly (© General Motors Corp.)

is released, the converter clutch releases and the converter operates in the conventional manner.

The engaging of the converter clutch as well as the release is determined by the direction of the converter feed oil. The converter feed oil from the pressure regulator valve flows to the converter clutch apply valve. The position of the converter clutch apply valve controls the direction in which the converter feed oil flows to the converter.

A spring-loaded damper assembly is splined to the converter turbine assembly while the clutch pressure plate is attached to a pivoting unit on the damper assembly. The result is that the pressure plate is allowed to rotate independently of the damper as-

sembly up to about 45 degrees. This rotation is controlled by springs in the damper assembly. The spring cushioning aids in reducing the effects felt when the converter clutch applies.

To help insure that the converter clutch applies and releases at the proper times, controls have been incorporated into the electrical system. The converter clutch is applied when all of the conditions listed below exist:
1. The brake pedal is released.
2. The engine vacuum is above 2.5 inches of vacuum.
3. The engine coolant is above 130°F.
4. The vehicle speed is above 24 to 30 mph.
5. The transaxle is in High gear.

TRANSAXLE LINE PRESSURE CHART
THM 325-4L

Model	Range-Zero Throttle (1000 RPM's)			Range-Full Throttle		
	N, D4, D3	D2, D1	Reverse	N, D4, D3	D2, D1	Reverse
AJ	64-77	126-153	91-109	129-155	126-153	184-220
AL, OK	64-77	126-153	91-109	129-155	126-153	184-220
AM, BE	64-77	148-179	126-152	113-135	148-179	224-268
OE	55-65	128-152	109-129	121-130	128-152	239-284
BJ	55-65	128-152	109-129	108-129	128-152	229-254

 **ON CAR SERVICES**

Adjustments

THROTTLE VALVE CABLE

Before making any adjustments to the throttle valve cable, check the following:

    a) Check and correct transaxle oil level

    b) Be sure that the engine is operating properly, and that the brakes are not dragging.

    c) Check that the correct cable has been installed in the vehicle.

    d) Check that the cable is connected at both ends.

If the above checks are in order, proceed as follows.

DIESEL ENGINE

1. Stop engine. If equipped, remove cruise control rod.
2. Disconnect the transaxle T.V. cable terminal from the cable actuating lever.
3. Loosen the lock nut on the pump rod and shorten it several turns.
4. Rotate the lever assembly to the full throttle position and hold.
5. Lengthen the pump rod until the injection pump lever contacts the full throttle stop.
6. Release the lever assembly and tighten the pump rod lock nut. Remove the pump rod from the lever assembly.
7. Reconnect the transaxle T.V. cable terminal to the cable actuating lever.
8. Depress and hold the metal re-adjust tab on the cable upper end. Move the slider through the fitting in the direction away from the lever assembly until the slider stops against the fitting.
9. Release the re-adjust tab. Rotate the cable actuating lever assembly to the full throttle stop and release the cable actuating lever assembly. The cable slider should adjust (or ratchet) out of the cable fitting toward the cable actuating lever.
10. Reconnect the pump rod, and cruise control throttle rod if so equipped.
11. If equipped with cruise control, adjust the servo throttle rod to minimum slack (engine off) then put clip in first free hole closest to the bellcrank but within the servo bail.
12. Road test the vehicle.
13. If delayed or only full throttle shifts still occur, remove the oil pan and inspect the throttle lever and bracket assembly which is located on the control valve assembly.
14. Check that the T.V. exhaust valve lifter rod is not distorted or binding in the control valve assembly or spacer plate. The T.V. exhaust check ball must move up and down the same way that the lifter does.
15. Be sure that the lifter spring holds the lifter rod up against the bottom of the control valve assembly. Make sure that the T.V. plunger is not stuck. Inspect the transaxle for correct throttle lever to cable link positioning.

GAS ENGINE

1. Stop engine.
2. Depress and hold the metal re-adjust tab on the cable upper end. Move the slider through the fitting in the direction away from the lever assembly until the slider stops against the fitting.
3. Release re-adjust tab.
4. Rotate the cable actuating lever assembly to the full throttle stop and release the cable actuating lever assembly. The cable slider should adjust (or ratchet) out of the cable fitting toward the cable actuating lever.

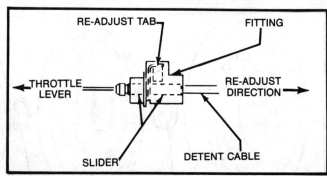

Throttle valve cable adjustment (© General Motors Corp.)

5. Check cable for sticking and binding.
6. Road test the vehicle.
7. If delayed or only full throttle shifts still occur, remove the oil pan and inspect the throttle lever and bracket assembly which is located on the control valve assembly.
8. Check that the T.V. exhaust valve lifter rod is not distorted or binding in the control valve assembly or spacer plate. The T.V. exhaust check ball must move up and down the same way that the lifter does.
9. Be sure that the lifter spring holds the lifter rod up against the bottom of the control valve assembly. Make sure that the T.V. plunger is not stuck. Inspect the transaxle for correct throttle lever to cable link positioning.

BAND ADJUSTMENT

The THM 325-4L automatic transaxle uses an intermediate band and intermediate servo assembly with a selective band apply pin. The use of special tools are needed to select the proper length apply pin and because of the transaxle location in the vehicle body, only the servo cover and piston assembly removal should be attempted. Band apply pin measurements should be accomplished with the transaxle removed from the vehicle.

MANUAL LINKAGE ADJUSTMENT

The THM 325-4L does not use a vacuum modulator system. In addition to the T.V. cable, the manual linkage is the only other major adjustment that can be made with the transaxle in the vehicle. This will vary somewhat with the vehicle. However, the following steps should be applicable to all.

1. Place the selector in Park.
2. Check that the selector pointer on the indicator quadrant lines up properly with the range indicator in Park.
3. Apply the brakes and start the vehicle. If the vehicle starts in Park, shut the engine off and place the selector in Neutral. Again, the pointer should line up with the range indicator. Try to start the vehicle. If it starts in Neutral as well as Park, then the manual linkage should be in proper adjustment.
4. A further check can be made by verifying that the back-up lamps only come on in Reverse. Also, set the parking brake, apply the service brakes firmly and attempt to start the vehicle in gear. The engine must not start in any range except Park and Neutral.
5. If adjustment is needed, begin by loosening the clamp screw on the transaxle control shift rod. See that the rod is free to slide through the clamp. Lubricate if necessary.
6. Place the upper shift lever against the Neutral stop in the steering column. A detent will hold it there.
7. Set the transaxle outer lever in the Neutral position. Tighten the clamp screw to 20 foot-pounds torque.
8. Check the operation by the above tests. In addition, be sure that the key cannot be removed when the steering wheel is not locked, and the key in the Run position. With the key in Lock and the shift lever in Park, be sure that the key can be removed. When

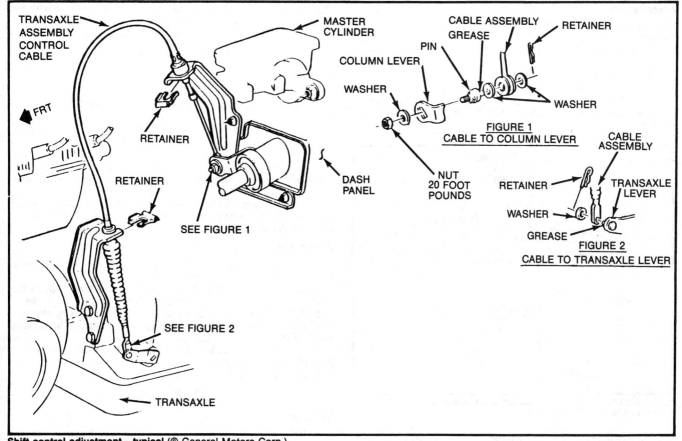

Shift control adjustment—typical (© General Motors Corp.)

the steering wheel is locked, verify that the shift lever cannot be moved from Park. Check for proper operation of the back-up lamps.

SHIFT CONTROL CABLE ADJUSTMENT

1. Position the steering column shift lever in the neutral position.
2. Set the transaxle lever in the neutral position.
3. Assemble the transaxle control cable. Install the pin, washer and retainer to the cable.
4. Properly install the shift cable. Do not hammer on the retainers.
5. Be sure that the shoulder on the pin is engaged in the column lever. Move the pin to give a free pin fit in the column lever. Tighten the attaching nut to the proper torque.

Services

FLUID CHANGE

General Motors does not recommend periodic fluid changes because of normal maintenance and lubrication requirements performed on the automatic transmission by mileage or time intervals.

If the vehicle is used in continuous service or driven under severe conditions (police or taxi type operation), the transaxle should be drained and refilled at mileage intervals of 18,000 to 24,000 miles or a time interval of 18 to 24 months.

NOTE: The miles or time intervals given are average. Each vehicle operated under severe conditions should be treated individually.

Automatic transmission fluids, (Dexron® II), meeting General Motors specifications, should be used in the THM 325-4L automatic transaxle. Failure to use the proper grade and type automatic transmission fluid could result in internal transaxle damage.

When the transaxle has to be removed for major repairs, the unit should be drained completely. The converter, cooler and cooler lines should be flushed to remove any particles or dirt that may have entered the components as a result of the malfunction or failure.

OIL PAN

Removal

1. Raise the vehicle on a hoist and support it safely.
2. With a drain pan placed under the transaxle oil pan, remove the pan attaching bolts from the front and side of the transaxle pan.
3. Loosen the rear pan attaching bolts approximately four turns.
4. Carefully pry the transaxle oil pan loose, using a suitable tool. Allow the transaxle fluid to drain.
5. Remove the remaining pan bolts and remove the pan from the vehicle.
6. Discard the pan gasket. If R.T.V. sealant has been used in place of the pan gasket, be sure to properly clean the pan in order to remove the old sealant.
7. If required, remove the screen to valve body bolts. Remove the screen from the valve body. Replace or clean the screen as necessary.

Installation

1. Install the screen using a new gasket or O-ring onto the valve body assembly.

2. Install the transaxle oil pan, using a new transaxle pan gasket. If, R.T.V. sealant is used as a sealer, apply a $1/16''$ bead of R.T.V. Sealant to the part flange and assemble wet. The bead of R.T.V. should be applied around the inside of the bolt holes. If the part flange has depressed stiffening ribs, the bead of R.T.V. must be installed on the high portion of the surface, not in the groove.

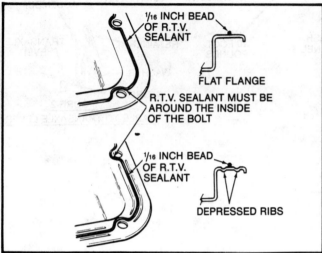

R.T.V. sealant application on special transaxle pans
(© General Motors Corp.)

3. Lower the vehicle from the hoist. Add the proper amount of automatic transmission fluid.

4. With selector lever in PARK position, apply parking brake, start engine and let idle. DO NOT RACE ENGINE.

5. Move selector lever through each range and, with selector lever in PARK range, check fluid level.

6. Correct the fluid level as required to bring the level between the dimples on the dipstick (transaxle cool).

VALVE BODY

Removal

1. Raise the vehicle on a hoist and support it safely. Drain the fluid.

2. Remove the transaxle oil pan and gasket. If R.T.V. sealant has been used, properly clean the mating surfaces.

3. Remove the oil screen and rubber O-ring.

4. Remove the screw and washer holding the cable to the transaxle and disconnect the T.V. cable.

5. Remove the throttle lever and bracket assembly. Do not bend the throttle lever link.

6. Disconnect the torque converter clutch electrical wiring.

7. Remove the oil transfer pipes and hold down brackets.

8. Support the valve body assembly. Remove the valve body retaining bolts.

9. Position the valve body assembly down and note the location of the check ball in the valve body.

NOTE: If the accumulator housing is to be removed, support the spacer plate during removal. Note the location of the check balls and then remove the spacer plate. Also, note the location of the check ball in the accumulator housing.

Installation

1. Installation is the reverse of removal. Be sure to torque the valve body retaining bolts to 8 ft. lbs.

2. Lower the vehicle from the hoist. Fill the transaxle with the proper grade and type automatic transmission fluid.

3. Adjust the fluid level as required. Road test the vehicle.

THROTTLE VALVE CABLE

Removal

1. If necessary, remove the air cleaner.

2. Depress and hold the metal re-adjust tab on the cable upper end.

3. Move the slider through the fitting in the direction away from the cable actuating lever until the slider stops against the fitting. This is in the re-adjust position. Release the re-adjust tab.

4. Disconnect the cable terminal from the cable actuating lever. Compress the locking tangs and disconnect the cable assembly from the bracket.

5. Remove the routing clips or straps, if used. Remove the screw and washer securing the cable to the transaxle. Disconnect the cable from the link.

Installation

1. Install a new seal into the transaxle case hole before inserting the cable into the transaxle case.

2. Connect and secure the transaxle end of the cable to the transaxle case, using the bolt and washer. Torque to 8 ft. lbs.

3. Route the cable as removed. Connect the clips or straps.

4. Pass the cable through the bracket and engage the locking tangs of the cable on to the bracket.

5. Connect the cable terminal to the cable actuating lever. Adjust the cable as required.

6. If removed, install the air cleaner.

GOVERNOR

Removal

1. Raise the vehicle on a hoist and support it safely.

2. Remove the speedometer cable at the transaxle. Be sure to catch fluid in a drain pan.

3. Remove the screws securing the governor cover. Remove the cover.

4. Remove the drive gear and the governor assembly from the transaxle case.

Installation

1. Install the drive gear and governor assembly into the transaxle case.

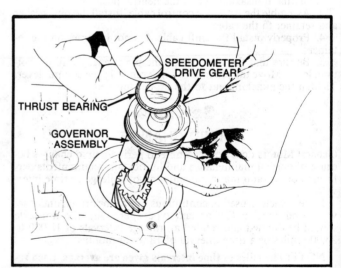

Governor assembly and related components
(© General Motors Corp.)

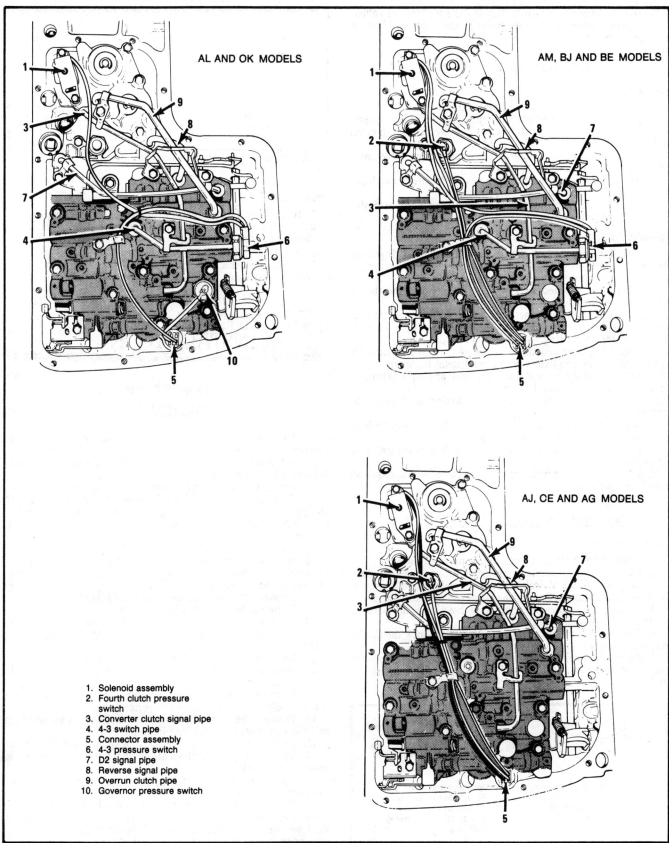

AL AND OK MODELS

AM, BJ AND BE MODELS

AJ, CE AND AG MODELS

1. Solenoid assembly
2. Fourth clutch pressure switch
3. Converter clutch signal pipe
4. 4-3 switch pipe
5. Connector assembly
6. 4-3 pressure switch
7. D2 signal pipe
8. Reverse signal pipe
9. Overrun clutch pipe
10. Governor pressure switch

Torque converter clutch electrical control connections (© General Motors Corp.)

2. Install the governor cover. Install the speedometer cable.

3. Lower the vehicle from the hoist. Correct the fluid level as required. Road test the vehicle.

INTERMEDIATE SERVO

Removal

1. Raise the vehicle on a hoist and support it safely.

2. Install special tool J-28493 or equivalent, on to the transaxle case. Tighten the bolt to remove the intermediate servo cover bolt.

3. Using a suitable tool, remove the intermediate servo cover retaining ring. Remove the special tool.

4. Remove the intermediate servo cover. Be sure to catch any excess fluid in a drain pan.

5. Remove the servo piston and the band apply pin assembly from the transaxle case.

Installation

1. Install the new inner and outer servo piston, seal rings and/or apply pin seal rings. Make sure that the rings are seated in the grooves to prevent damage to the rings.

2. Install the intermediate servo piston assembly into the intermediate servo cover.

3. Lubricate and install a new seal ring on the intermediate servo cover. This seal ring must be well lubricated to prevent damage or cutting of the ring.

4. Install the intermediate servo assembly into the transaxle case. If necessary, tap lightly with a non metal hammer.

5. Be sure that the tapered end of the band apply pin is properly located against the band apply lug.

6. Install the special tool on to the transaxle case. Tighten the bolt to depress the servo cover.

7. Install the servo retaining ring. Align the ring gap with an end showing in the transaxle case slot.

8. Remove the special tool from its mounting on the transaxle assembly.

9. Lower the vehicle from the hoist. Replace fluid as required. Road test the vehicle.

PRESSURE REGULATOR VALVE

Removal

1. Raise the vehicle on a hoist and support it safely.

2. Drain the fluid from the transaxle. Remove the oil pan. Discard the gasket. If R.T.V. sealant has been used, be sure to properly clean the mating surfaces of the transaxle case and the pan.

3. Pushing on the pressure regulator valve with a suitable tool, compress the pressure regulator spring.

4. Remove the retaining ring and slowly withdraw the tool to release spring tension.

5. Remove the drive to boost valve and bushing, reverse boost plunger and bushing, spacer, and the pressure regulator spring and valve.

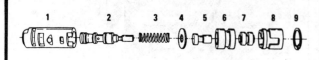

1. Pressure regulator bushing
2. Pressure regulator valve
3. Pressure regulator spring
4. Valve bushing spacer
5. Reverse boost valve
6. Reverse boost valve bushing
7. MTV boost valve
8. MTV boost valve bushing
9. Snap ring

Pressure regulator valve—exploded view (© General Motors Corp.)

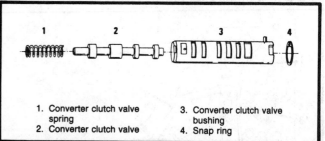

1. Converter clutch valve spring
2. Converter clutch valve
3. Converter clutch valve bushing
4. Snap ring

Torque converter clutch valve—exploded view (© General Motors Corp.)

Installation

1. Installation is the reverse of removal. Be sure to push the sleeve assembly past the retaining ring groove by compressing the pressure regulator valve spring, using tool J-24684 or equivalent.

2. Install the retaining ring.

3. Lower the vehicle from the hoist. Fill the transaxle, using the proper grade and type fluid. Road test as required.

 REMOVAL & INSTALLATION

Transaxle
REMOVAL

1. Disconnect the negative battery cable. Remove the air cleaner.

2. If the vehicle is equipped with a diesel engine, install tool J-26996-1 or equivalent.

3. Disconnect the speedometer cable at the transaxle.

4. Disconnect the T.V. cable from the bellcrank, diesel equipped, or the throttle lever, gasoline equipped, at its upper end. Disconnect the T.V. cable from its bracket by depressing both retainers and sliding the cable out.

5. Remove the center final drive bolt from the assembly.

6. Remove the two nuts on the left exhaust pipe to manifold connection. Remove the transaxle to engine bolts and stud from the top of the assembly.

7. Remove the fuel line to transaxle bracket.

8. Raise the vehicle on a hoist and support it safely.

9. Remove the starter. Remove the TCC electrical connector.

10. Disconnect the transaxle oil cooler lines. Plug the openings.

11. Remove the transaxle inspection cover.

12. Remove the two bolts on the left exhaust pipe. Remove the exhaust pipe.

13. Remove the three nuts on the right exhaust pipe. Remove the two bolts from the catalytic converter hanger.

14. Remove the four bolts from the torque box cross-member, and the two bolts from the fuel lines.

15. Remove the torque converter to flexplate bolts. Be sure to mark the position of the flexplate to the torque converter for proper reassembly.

16. Disconnect the shift linkage at the transaxle. Remove the right hand transaxle mount.

17. Position the transmission jack under the transaxle.

18. Remove the left transaxle mount through bolt. Remove the lower bracket to transaxle bolt. Raise the transaxle assembly about two inches in order to gain access to the remaining upper bracket to transaxle bolts.

19. Remove the remaining engine to transaxle bolt and bracket.

20. Lower the transaxle carefully while disengaging the final drive unit.

21. Install a torque converter holding tool and remove the automatic transaxle from the vehicle.

INSTALLATION

1. Installation is the reverse of removal. The following points should be observed. Always replace the final drive to transaxle gasket.

2. Use care to see that the final drive splines engage the transaxle and then loosely install two final drive to transaxle lower attaching bolts.

3. After the final drive and transaxle are mated, align the bellhousing with the engine and install the remaining attaching bolts.

4. Before installing the converter to flywheel bolts, make sure that the weld nuts on the torque converter are flush with the flywheel and that the converter rotates freely in this position. Hand start the bolts and tighten all three finger tight so that the converter and flywheel will be in the proper alignment. Then torque the bolts evenly to specification.

5. Be sure to adjust the T.V. cable.

6. Refill the transaxle to the proper level with Dexron® II.

BENCH OVERHAUL

Before Disassembly

Before opening up the transaxle, the outside of the unit should be thoroughly cleaned, preferably with high-pressure cleaning equipment such as a car wash spray unit. Dirt entering the transaxle internal parts will negate all the effort and time spent on the overhaul. During inspection and reassembly, all parts should be thoroughly cleaned with solvent, then dried with compressed air. Wiping cloths and rags should not be used to dry parts since lint will find its way into valve body passages. Lube seals with Dexron® II and use ordinary unmedicated petroleum jelly to hold thrust washers to ease assembly of seals, since it will not leave a harmful residue as grease often will. Do not use solvent on neoprene seals, friction plates or thrust washers. Be wary of nylon parts if the transaxle failure was due to a failure of the cooling system. Nylon parts exposed to antifreeze solutions can swell and distort and so must be replaced. Before installing bolts into aluminum parts, always dip the threads into clean oil. Anti-seize compound is also a good way to prevent the bolts from galling the aluminum and seizing. Always use a torque wrench to keep from stripping the threads. Take care of the seals when installing them, especially the smaller O-rings. The internal snap rings should be expanded and the external snap rings should be compressed, if they are to be reused. This will help insure proper seating when installed.

Torque Converter

Removal and Inspection

1. Make certain that the transaxle is held securely.

2. The converter pulls out of the transaxle. Be careful since the converter contains a large amount of oil. There is no drain plug on the converter so the converter should be draind through the hub.

The fluid drained from the torque converter can help diagnose transaxle problems.

1. If the oil in the converter is discolored but does not contain metal bits or particles, the converter is not damaged and need not be replaced. Remember that color is no longer a good indicator of transmission fluid condition. In the past, dark color was associated with overheated transmission fluid. It is not a positive sign of transaxle failure with the newer fluids like Dexron® II.

2. If the oil in the converter contains metal particles, the converter is damaged internally and must be replaced. The oil may have an "aluminum paint" appearance.

3. If the cause of oil contamination was due to burned clutch plates or overheated oil, the converter is contaminated and should be replaced.

Transaxle Disassembly

OIL PAN AND FILTER

Removal

1. Remove the oil pan attaching bolts.

2. Separate the pan from the transaxle assembly.

3. Discard the pan gasket. If R.T.V. sealant has been used, properly clean both mating surfaces.

4. Remove the transaxle oil filter and O-ring.

EXTERNAL PARTS

Removal

1. Disconnect the wire leads at the transaxle case electrical connector and the pressure switches.

2. Remove the electrical connector and the O-ring seal from the transaxle case. Using a suitable tool, depress the connector tangs while pushing out on the connector.

3. Remove the solenoid assembly retaining bolts. Remove the solenoid and the O-ring seal.

4. Remove the pressure regulator assembly retaining snap ring.

5. Remove the T.V. boost valve bushing and valve, reverse boost valve bushing and valve, pressure regulator spring and the pressure regulator bushing and valve.

NOTE: If replacement of the valve bushing is necessary, you will have to determine if the bushings are oversized. The bushings and the transaxle case will both be stamped with "O.S", if in fact they are oversized.

6. Remove the torque converter clutch apply valve retaining snap ring.

7. Remove the torque converter clutch valve bushing, valve and spring.

8. Remove the D2 signal pipe, reverse signal pipe and the overrun pipe retainers. Remove the remaining oil pipes and solenoid assembly.

VALVE BODY

Removal

1. Remove the throttle lever and bracket assembly.

2. Remove the valve body assembly attaching bolts. Disconnect the manual valve.

3. Do not drop the manual valve, as damage to this component may result.

4. Remove the valve body assembly from its mounting in the transaxle. Note the location of the check ball on the spacer plate. Remove the check ball.

1-2, 3-4 ACCUMULATOR ASSEMBLY

Removal

1. Remove the 1-2, 3-4 accumulator housing attaching bolts. Remove the accumulator housing, and 1-2 spring.

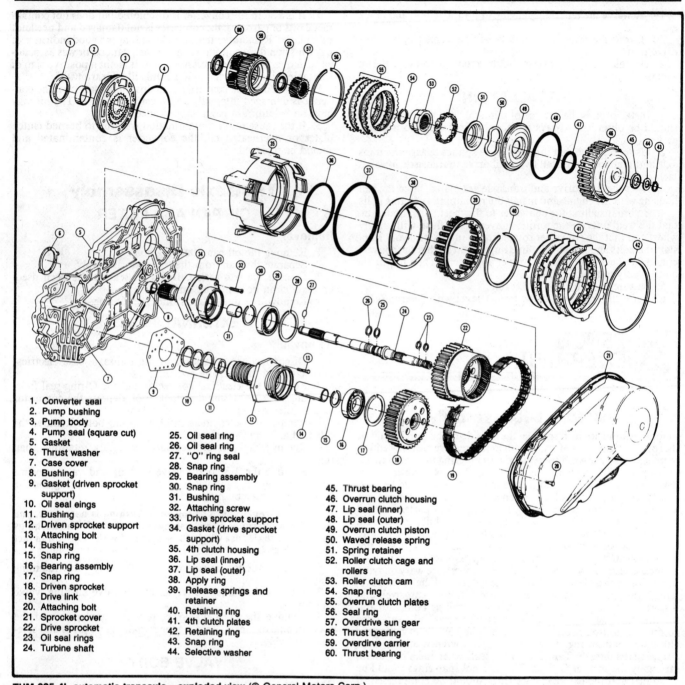

1. Converter seal
2. Pump bushing
3. Pump body
4. Pump seal (square cut)
5. Gasket
6. Thrust washer
7. Case cover
8. Bushing
9. Gasket (driven sprocket support)
10. Oil seal eings
11. Bushing
12. Driven sprocket support
13. Attaching bolt
14. Bushing
15. Snap ring
16. Bearing assembly
17. Snap ring
18. Driven sprocket
19. Drive link
20. Attaching bolt
21. Sprocket cover
22. Drive sprocket
23. Oil seal rings
24. Turbine shaft

25. Oil seal ring
26. Oil seal ring
27. "O" ring seal
28. Snap ring
29. Bearing assembly
30. Snap ring
31. Bushing
32. Attaching screw
33. Drive sprocket support
34. Gasket (drive sprocket support)
35. 4th clutch housing
36. Lip seal (inner)
37. Lip seal (outer)
38. Apply ring
39. Release springs and retainer
40. Retaining ring
41. 4th clutch plates
42. Retaining ring
43. Snap ring
44. Selective washer

45. Thrust bearing
46. Overrun clutch housing
47. Lip seal (inner)
48. Lip seal (outer)
49. Overrun clutch piston
50. Waved release spring
51. Spring retainer
52. Roller clutch cage and rollers
53. Roller clutch cam
54. Snap ring
55. Overrun clutch plates
56. Seal ring
57. Overdrive sun gear
58. Thrust bearing
59. Overdirve carrier
60. Thrust bearing

THM 325-4L automatic transaxle—exploded view (© General Motors Corp.)

2. One check ball will be exposed on top of the spacer plate to valve body gasket, remove it.

3. Remove the accumulator housing assembly gasket and spacer plate.

4. Remove the nine check balls from the transaxle case passages.

GOVERNOR AND SPEEDOMETER DRIVE GEAR ASSEMBLY

Removal

1. Remove the speedometer driven gear retaining bolt and retainer clip.

2. Remove the speedometer driven gear assembly from the governor cover. Remove the "O" ring seal from the speedometer driven gear assembly.

3. Remove the governor cover bolts.

4. Remove the governor cover and the "O" ring seal.

5. Remove the governor, bearing assembly and speedometer drive gear assembly from the transaxle case. The governor bearing assembly may be in the governor cover.

INTERMEDIATE SERVO ASSEMBLY

Removal

1. Install tool J-28493, or equivalent on to the transaxle case.

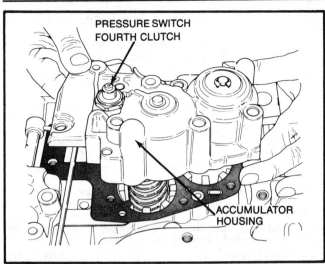

1-2, 3-4 accumulator housing (© General Motors Corp.)

Tighten the bolt to depress the servo cover.

2. Using a suitable tool, remove the intermediate servo cover retaining ring. Remove tool J-28493.

3. Using a pliers, remove the intermediate servo cover and O-ring seal. The cover O-ring seal may be located in the transaxle case bore.

4. If the intermediate servo cover and seal assembly can not be removed easily, apply air pressure into the intermediate servo exhaust port.

5. Remove the intermediate servo piston and band apply pin assembly.

OVERDRIVE UNIT AND DRIVE LINK ASSEMBLY

Removal

1. Rotate the transaxle to the sprocket cover side up.
2. Remove the sprocket cover retaining bolts.
3. Remove the cover and the R.T.V. sealant. Clean the excess sealant from the cover. Do not hit the cover to remove it, use a putty knife to break the transaxle case to case cover seal.
4. Check the overdrive end play by mounting a dial indicator gauge against the end of the turbine shaft. Set the dial indicator to zero.
5. Move the turbine shaft upward by pushing up from the converter side. The overdrive unit end play should be .004"-.029".
6. The selective washer controlling this end play is located between the turbine shaft retaining ring and the thrust bearing assembly. If more or less washer thickness is required to bring the end play within specifications, select the proper washer.
7. Remove the fourth clutch snap ring. Remove the fourth clutch plates. Remove the turbine shaft snap ring.
8. Remove the turbine shaft thrust washer, thrust bearing, and overdrive unit. The overdrive carrier may have to be removed separately.
9. Remove the overdrive carrier to drive sprocket thrust bearing assembly.
10. Remove the fourth clutch housing retaining bolts. Remove the housing.
11. Remove the fourth clutch housing to case cover O-ring seal. The O-ring seal may be located in the fourth clutch housing.
12. Using a snap ring pliers, remove the sprocket bearing retaining snap rings which are located under the drive and driven sprockets. The snap ring grooves are located in the sprocket supports.

13. To remove the drive and driven sprockets, drive link bearings and turbine shaft pull alternately on the drive and driven sprockets.

NOTE: If the sprockets are difficult to remove, place a small piece of fiberboard between the sprocket and a pry bar and alternately pry under each sprocket.

14. Remove the drive link from the drive and driven sprockets. Remove the turbine shaft.
15. Remove the oil pump by first removing the two opposite pump attaching flat head screws from the drive sprocket support using tool J-25359-5.
16. Install two 5/16x4 inch guide pins into the assembly. Remove the remaining pump flat head screws from the drive sprocket support.
17. Steady the pump from underneath with one hand then gently tap the guide pins until the pump is removed from the transaxle case.

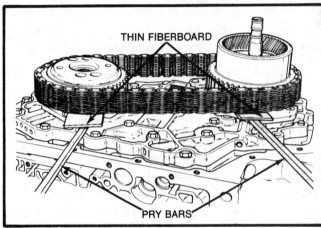

Sprocket removal (© General Motors Corp.)

18. To check the input end play, install tool J-26958 to the output end of the transaxle.
19. Remove the driven sprocket support to case bolt and install the dial indicator gauge.
20. Push the input shaft down. Raise the input shaft by pushing the handle down on the dial indicator gauge. The input unit end play should be .022-.051 inch.
21. The selective washer controlling this end play is located between the output shaft and the input shaft. If a different washer thickness is required to bring the end play within specifications, refer to the selective washer chart for the proper washer.
22. Remove the dial indicator gauge and tool J-25025-7. Do not remove tool J-26958.
23. To remove the case cover, remove the cover retaining bolts. Do not remove the driven sprocket support bolts. Remove the case cover and gasket.
24. Remove the thrust washer from the hub of the driven sprocket support. The thrust washer may be on the direct clutch housing.
25. Remove the pressure regulator screen from the case cover side of the case.

FORWARD AND DIRECT CLUTCH

Removal

1. Remove the forward and direct clutch assemblies from the transaxle case, using tool J-26959 or equivalent.
2. Remove the intermediate band.
3. Remove the direct clutch assembly from the forward clutch assembly.

4. The direct clutch to the forward clutch thrust washer may be on the end of the direct clutch housing when it is removed from the forward clutch housing.

INTERNAL GEAR PARTS

Removal

1. Remove the output shaft to the input shaft selective washer.

2. Check the reaction unit end play by loosening the adjusting screw on tool J-26958, which should still be installed on the output shaft. Push the output shaft downward.

3. Install the dial indicator gauge onto the transaxle case. Position the extension, of the tool, against the end of the output shaft. Set the dial indicator to zero.

4. Move the output shaft upward by turning the adjusting screw on tool J-26958 until it stops. Read and record the end play. The reaction unit end play should be .004-.025 inches.

5. The selective washer controlling this end play is located between the input internal gear thrust washer and the output shaft snap ring.

6. If more or less washer thickness is required, select the proper thrust washer from the reaction unit end play selective washer chart.

7. Remove the dial indicator gauge. Do not remove tool J-26958.

8. Using a snap ring pliers, remove the output shaft to the selective washer snap ring. Tighten the adjusting screw on tool J-26958, which is still located on the output shaft, to remove the snap ring.

9. Remove the input internal gear, reaction selective washer, and tanged thrust washer. Remove the reaction selective washer, and thrust washer from the input internal gear.

10. Remove the input carrier assembly. Remove the input internal gear to the input carrier thrust bearing assembly.

NOTE: The input sun gear, to the input carrier, thrust bearing assembly and race may come out as the input carrier is removed.

11. Remove the input sun gear to the input carrier thrust bearing assembly, and input sun gear. This thrust bearing requires only one thrust race.

12. Remove the input drum and the reaction sun gear assembly.

13. Remove the four tanged input drum to the low and reverse clutch housing thrust washer from the rear of the input drum or from the low and reverse clutch housing.

14. To remove the low and reverse clutch housing assembly, grind about ¾ inch from the end of a 6.3mm (#4) screw extractor to remove the cup plug. Remove the low and reverse cup plug assembly by turning the easy out two or three turns and pulling it straight out.

15. Remove the low and reverse clutch housing to transaxle case beveled snap ring. Using tool J-28542, remove the low and reverse clutch housing assembly and spacer ring.

16. Be sure that the governor assembly has been removed at this time.

17. Grasp the output shaft and lift out the rest of the reaction unit parts.

18. Remove the roller clutch and reaction assembly from the output shaft.

19. Remove the four tanged reaction carrier to reaction internal gear thrust washer, from the end of the reaction carrier or from the inside of the reaction internal gear.

20. Remove the low and reverse clutch plates from the output shaft. Take note that the top clutch plate is waved.

21. Remove the reaction internal gear to reaction sun gear thrust bearing assembly from the reaction internal gear. Remove the reaction internal gear from the output shaft.

22. Remove the governor assembly drive gear.

MANUAL SHAFT AND PARKING PAWL

Removal

1. Position the transaxle so that the oil pan is facing upward. Remove tool J-26958 from the transaxle case.

2. Remove the manual detent roller and spring assembly. Remove the parking strut shaft retaining ring.

3. Use a 6.3mm (#4) screw extractor to remove the parking lock cup plug by grinding about ¾ inch from the end of the Easy Out.

4. Use a 4mm (#4) screw extractor to remove the parking strut shaft. Remove the parking lock spring, strut and lever.

5. Remove the parking lock cam. Remove the inside detent lever retaining nut.

6. Remove the manual shaft and the inside detent lever assembly.

7. Inspect the manual shaft and seal for damage. If necessary, pry out the manual shaft seal using the proper seal removal tool.

8. Using a 4mm (#3) screw extractor, remove the parking pawl shaft.

9. Remove the parking pawl and the return spring.

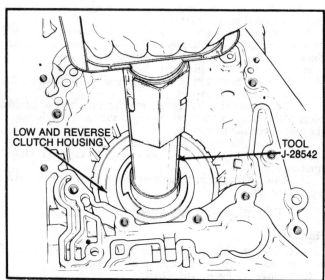

Low and reverse clutch housing assembly removal
(© General Motors Corp.)

LOW AND REVERSE CLUTCH HOUSING

TOOL J-28542

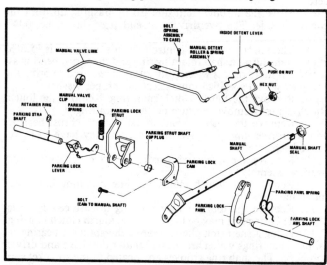

Manual shaft and related components (© General Motors Corp.)

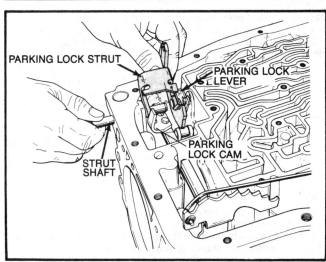

Parking strut shaft and related components
(© General Motors Corp.)

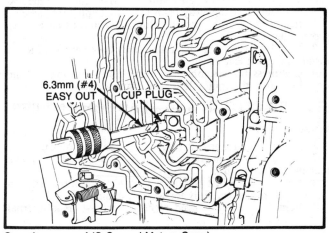

Cup plug removal (© General Motors Corp.)

Unit Disassembly and Assembly
DIRECT CLUTCH

Disassembly

1. Pry the snap ring from the direct clutch housing and remove the snap ring along with the backing plate.

2. Remove the clutch pack from the housing. Take note of the number of clutch plates and the order in which they are removed. Be sure to keep the direct clutch plates separate from the forward clutch plates.

3. Compress the assembly so that the spring retainer snap ring can be removed. Remove the retainer and spring assembly from the clutch housing.

4. Remove the release spring guide and the clutch piston. The seals should be discarded. Do not remove the apply ring from the piston unless the piston is damaged and replacement is required.

Inspection

1. Clean all parts thoroughly in solvent and blow dry with compressed air.

2. Check the piston for cracks or damage. See that the check ball is free.

Assembly

1. With all parts clean and dry, carefully install new seals on the piston. The seals should be lubricated with Dexron® II or petroleum jelly.

2. Install a new center seal in the clutch housing. Lubricate the seal to ease installation.

3. With the seals well lubricated, install the piston into the clutch housing using care so the seal will not be torn or damaged by the sharp edge on the snap ring groove. Rotate the piston to help with the installation. A tool made from wire can be fabricated to help ease the lip of the seal into place.

4. Install the spring guide making sure that the notch, or omitted rib, is aligned with the check ball in the piston. Then install the spring and retainer assembly. Compress the unit and install the snap ring.

5. Install the clutch pack. The composition plates should be well-oiled with Dexron® II preferably soaking in the oil for at least 20 minutes prior to assembly. Install the plates in the order determined at removal, staring with a steel plate.

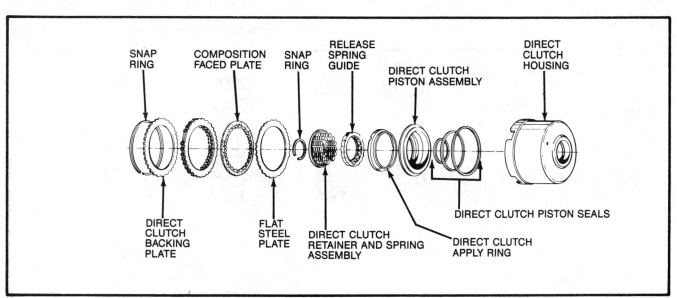

Direct clutch assembly—exploded view (© General Motors Corp.)

6. Install the backing plate with the chamfered side up. Install the snap ring and check the clutch pack for proper assembly by verifying that the composition plates turn freely. Set the direct clutch aside until the transaxle is ready for assembly.

FORWARD CLUTCH

Disassembly

1. Pry the snap ring from the forward clutch housing and remove the snap ring along with the backing plate.

2. Remove the clutch pack from the housing. Take note of the number of clutch plates and the order in which they are removed. Be sure to keep the forward clutch plates separate from the direct clutch plates.

3. Compress the assembly so that the spring retainer snap ring can be removed. Remove the retainer and spring assembly from the clutch housing.

4. Remove the forward clutch piston. The seals should be discarded. Do not remove the apply ring from the piston unless the piston is damaged and requires replacement.

Inspection

1. Clean all parts thoroughly in solvent and blow dry with compressed air.

2. Check the piston for cracks or damage. See that the check ball is free.

3. Inspect the housing snap-ring groove for burrs or damage. Check the input shaft for open and clean oil passages on both ends of the shaft. Inspect the cup plug for damage. If it needs to be replaced, a #4 Easy-Out can be used for removal. Install the replacement cup plug to 1mm (.039") below the surface.

Assembly

1. With all parts clean and dry, carefully install new seals on the piston. The seals should be lubricated with Dexron® II or petroleum jelly.

2. Install a new center seal in the clutch housing. Lubricate the seal to ease installation.

3. With the seals well lubricated, install the piston into the clutch housing using care so the seal will not be torn or damaged by the sharp edge on the snap ring groove. Rotate the piston to help with the installation.

4. Install the spring retainer assembly. Be careful that the retainer does not hang-up in the groove for the snap ring. Compress the assembly and install the snap ring.

5. Install the clutch pack. The composition plates should be well-oiled with Dexron® II, preferably soaking in the oil for at least 20 minutes prior to assembly. Install the plates in the order determined at removal, starting with the waved steel plate.

6. Install the backing plate with the chamfered side up. Install the snap ring and check the clutch pack for proper assembly by verifying that the composition plates turn freely. Set the forward clutch aside until the transaxle is ready for assembly.

LOW AND REVERSE CLUTCH

Disassembly

1. Compress the spring retainer and remove the snap ring from the clutch housing.

2. Remove the waved spring and the clutch piston. Remove the seals and discard.

Inspection

1. Check the feed holes in the clutch housing. Inspect the housing for scoring or other damage.

2. Remove any burrs from the splines on the snap ring groove. Check the piston for cracks or other damage.

Assembly

1. Replace the seals on the piston. The seals should be lubricated with Dexron® II or petroleum jelly. Make sure that the seal lips are installed facing away from the clutch apply ring side.

2. Install the piston into the clutch housing using care to protect the seal.

3. Install the waved spring, followed by the retainer, which is installed cupped face down.

4. Compress the assembly and install the snap ring. Set the unit aside until the transaxle is ready for assembly.

ROLLER CLUTCH AND REACTION CARRIER

Disassembly

1. Remove the roller clutch race from the reaction carrier.

2. Remove the roller clutch assembly. Be careful of the rollers and energizing springs.

3. Remove the thrust bearing from the inside of the reaction carrier and inspect carefully for wear or damage.

4. Remove the four-tanged reaction carrier-to-rear internal gear thrust washer.

Inspection

1. Examine the thrust washers to determine if they can be re-used. If slight scoring or scuff marks are noted, replacement is not necessary. Replacement is necessary only when the thrust washer

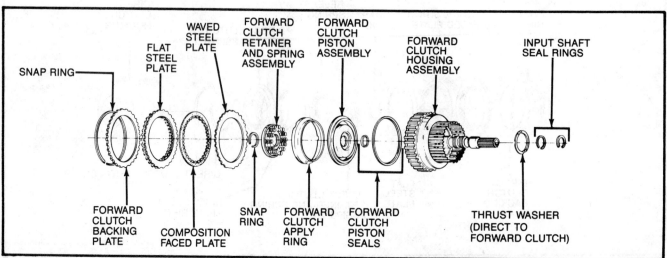

Forward clutch assembly—exploded view (© General Motors Corp.)

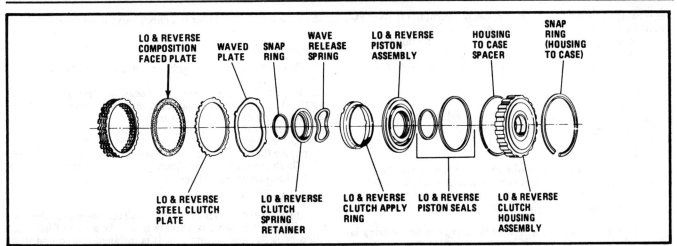

Low and reverse clutch assembly—exploded view (© General Motors Corp.)

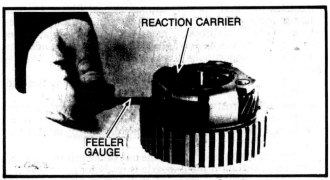

Checking pinion end play (© General Motors Corp.)

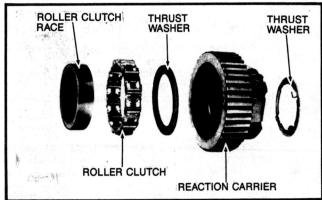

Roller clutch and reaction carrier assembly (© General Motors Corp.)

indicates material pick-up on the thrust surface or when the surface that the washer protects is deeply cut or scored.

2. Check the rollers for wear or damage. If a roller should come out of the cage assembly, the roller should be replaced from the outside of the cage to keep from damaging the energizing spring.

3. The pinion end-play on the planetary carrier should be checked. A feeler gauge can be used. End-play should be between .009″ to .027″.

Assembly

1. Install the roller clutch-to-reaction carrier thrust washer into the reaction carrier.

2. Install the roller clutch cage into the reaction carrier.

3. Install the roller clutch race into the roller clutch cage, spline side out, turning as needed to get the race into position.

4. Install the four-tanged thrust washer onto the reaction carrier on the pinion end. Align the slots on the carrier with the tangs on the washer. Use petroleum jelly to hold the washers in place.

5. The reaction internal gear should still be on the output shaft. If the output shaft does not need to be disassembled, that is, the governor gear does not have to be removed and the shaft is not damaged, then install the reaction carrier assembly onto the output shaft and into the reaction internal gear. Set the unit aside until the transaxle is ready for assembly.

REACTION SUN GEAR AND INPUT DRUM

Disassembly

1. Remove the input drum-to-reaction sun gear snap-ring.

2. Remove the reaction sun gear from the input drum.

3. Remove the four-tanged thrust washer from the end of the reaction sun gear. This thrust washer protects the input drum and the low and reverse clutch housing.

Inspection

1. Clean all parts well in solvent and blow dry.

2. Carefully inspect the reaction sun gear for broken teeth, splits, or cracks. Check that the lubrication holes are not plugged. Inspect the bushing for scoring or damage.

3. Make sure that the snap ring is not distorted or bent. If necessary, replace this snap ring. Carefully examine the four-tanged thrust washer for excessive wear. Expect some light marks or polishing of the surface, but deep gouges or signs of metal transfer means that the washer must be replaced.

Assembly

1. Insert the reaction sun gear into the input drum, making

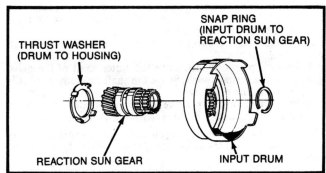

Reaction sun gear and input drum assembly
(© General Motors Corp.)

sure that the spline side goes in first. Being careful not to over-expand the snap ring, install the ring on the sun gear.

2. Coat the four-tanged thrust washer with petroleum jelly and install it on the sun gear. Align the tangs on the washer with the slots in the input drum. Set the assembly aside until the transaxle is ready for assembly.

INPUT SUN GEAR AND INPUT CARRIER ASSEMBLY

Disassembly

1. Once the input sun gear and input carrier have both been removed, there is no further disassembly to be done. Take note that the planetary carrier is a stamped unit and that the pinion pins are not removable. If there is damage to the carrier, it must be replaced.

2. If the input internal gear-to-input carrier thrust bearing is still on the carrier, remove it for cleaning. Also, the input sun gear-to-front carrier thrust bearing should be removed.

Inspection

1. Clean all parts thoroughly in solvent and blow dry with compressed air.

2. Thoroughly clean the input sun gear-to-input carrier thrust bearing for pitted or rough surfaces.

3. Inspect the input carrier for chipped teeth on the pinions, or excessive play. The pinion end-play should be checked. A feeler gauge can be used to check this clearance. Pinion end-play should be between .009" and .027".

Assembly

1. Install the input carrier-to-input internal gear thrust bearing onto the input carrier. Make sure that the smaller diameter race goes against the carrier. A liberal amount of petroleum jelly will help hold the bearing in place.

2. The input sun gear will be installed at the time of assembly, along with the input carrier, transaxle input internal gear and the selective thrust washer. Set these components aside until the transaxle is ready for assembly.

CASE COVER, DRIVE AND DRIVEN SPROCKETS

Disassembly

1. Normally, the drive and driven sprocket support assemblies are not to be removed. The support assemblies are pressed into the case cover and special care is needed to remove them. The case cover and the sprocket supports should be carefully examined for damage or any sign of leakage. The drive sprocket support stator shaft splines should be checked carefully *before* they are disassembled. Generally, unless the transaxle has seen abuse or is otherwise heavily damaged, the supports should not be removed. If it has been determined that the sprocket supports must be removed, first verify that all sprocket support-to-case cover attaching bolts have been removed.

2. Set the cover on two wood blocks or otherwise support it, stator shaft side up.

3. Using a plastic mallet, drive the stator shaft of the drive sprocket support downward. Strike the shaft carefully until it is removed from the case cover.

4. In the same manner, strike the hub of the driven sprocket support downward until it too is free from its bore.

NOTE: **Use care when driving out these parts. Avoid damaging or distorting the stator shaft or the ring grooves in the hub of the driven sprocket support.**

5. Discard the gaskets that will be found between the support and case cover.

6. Remove the converter out check valve from the pump cover.

Inspection

1. Clean all parts thoroughly in solvent and blow dry with compressed air.

2. Inspect the case cover for cracks or damage. A metal straightedge can be used to check the cover for flatness. Check the cup plugs for tightness, or signs of leakage.

3. Check for the presence of a check ball in the case cover.

4. Carefully examine the sprocket supports for damage. Do not remove the oil seal rings unless they are going to be replaced.

5. Check the case cover oil passages for cracks or any damage that could cause cross-leakage.

Assembly

1. Install the converter out check valve into its bore in the case cover. This check valve is actually more like a coil of flat steel with a hook or tang on one end. It is installed tanged end first, into the oil passages. Then coil the remainder of the valve within itself.

2. Install the drive sprocket support gasket. Two guide pins should be made to help keep the sprocket support straight. Carefully install the drive sprocket support and use a plastic mallet as needed to seat the support in the housing. Remove the guide pins and install the support bolts. These are to be torqued to 18 foot-pounds.

3. In the same manner, install the driven support gasket and support.

4. Install the thrust washer on the hub of the driven sprocket support. Note that there is a tab on the bearing that must be located in the case cover locating hole. Use petroleum jelly to hold it in place.

5. Replace the oil seal rings on the hub of the driven sprocket. Set the unit aside until the transaxle is ready for assembly.

OIL PUMP

Disassembly

1. Remove the drive gear from the oil pump body.
2. Remove the driven gear from the oil pump body.
3. Remove and discard the oil pump body to case O-ring seal.

Inspection

1. Clean all parts thoroughly in solvent and blow dry with compressed air.

2. Check the gear pocket for wear, nicks, burrs or other damage.

3. Check the gears in the same manner for any damage or excessive wear.

4. Pry out the front seal and check the bore for damage.

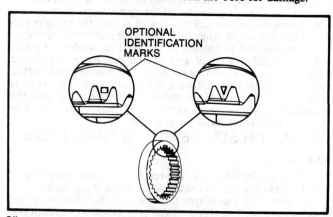

Oil pump gear identification marks (© General Motors Corp.)

5. The pump body can be installed on the torque converter hub to check for an out-of-round condition or excessive play.

6. Gear clearance should also be checked. Put the gears in the pocket of the pump aligning the match-marks made earlier. With a feeler gauge, check the pump body face-to-gear clearance. The clearance should be from .0013″ to .0035″. Note that the larger, driven gear should have a factory identification mark on it. This mark must be down when the gear is installed. The pump drive gear, which is the smaller of the two and has the square tang on the inside diameter, has its identification mark on the tang and it must be installed with the mark facing up.

Assembly

1. Apply a thin coat of non-hardening sealer to the outer edge of the pump seal and install the seal to the pump.

2. With all parts clean and dry, install the pump gears into the pump pocket matching up the marks made at disassembly. Be sure that the driven gear's identification mark is down and the drive gear's mark on the tang faces up.

3. Lubricate the large outer O-ring that goes around the circumference of the pump with petroleum jelly and install the seal. The pump spacer plate and the pump can be set aside until the transaxle is ready for assembly.

FOURTH CLUTCH ASSEMBLY

Disassembly

1. Using a press and tool J-29334-1, or equivalent, compress the fourth clutch spring and retainer assembly. Remove the snap ring and the retainer assembly.

2. Remove the fourth clutch piston. Remove the fourth clutch inner and outer seals and discard them.

Inspection

1. Inspect the fourth clutch housing for cracks, or porosity. Also inspect for burrs or raised edges. If present, remove them with a fine stone or fine abrasive paper.

2. Inspect the piston sealing surface for scratches.

3. Air check the fourth clutch oil passage making certain it is open and not interconnected.

4. Inspect the snap ring for damage.

5. Inspect the release spring, and retainer assembly for distortion or damage.

6. Inspect the composition-faced and steel clutch plates for signs for wear or burning.

7. Inspect the backing plate for scratches or damage.

Assembly

1. Install the fourth clutch outer and inner seals on the fourth clutch housing with the seal lips facing downward. Use petroleum jelly to aid in the installation of the seals. The fourth clutch housing inner seal is identified by a white stripe.

2. Install the fourth clutch piston assembly into the fourth clutch piston housing.

3. Install the fourth clutch spring and retainer assembly. Using the press and tool J-29334-1, compress the spring and retainer assembly and install the fourth clutch housing to spring assembly snap ring.

4. Set the assembly aside until the transaxle is ready for reassembly.

OVERRUN CLUTCH ASSEMBLY

Disassembly

1. Separate the overdrive carrier assembly from the overrun clutch assembly. Remove the overdrive sun gear from the overrun clutch assembly.

2. Remove the overrun clutch snap ring using the proper tool. Remove the overrun clutch backing plate from the overrun clutch housing.

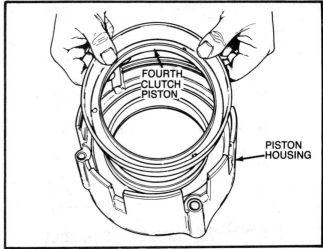

Fourth clutch piston assembly (© General Motors Corp.)

3. Remove the clutch plates from the overrun clutch housing. Be sure to keep them separated from the other plate assemblies.

4. Remove the overrun clutch hub snap ring, using a snap ring pliers.

5. Remove the overdrive roller clutch cam assembly. Remove the roller clutch assembly.

6. Remove the retainer and wave spring assembly from the housing. Remove the overrun clutch piston assembly.

7. Remove both the inner and outer seal from the overrun clutch piston.

Inspection

1. Inspect both the composition plates and the steel plates for wear and heat damage. Replace as required.

2. Inspect both the overrun clutch and the clutch housing for scoring, distortion, cracks, damage, wear and open oil passages. Replace components as required.

Assembly

1. Install new inner and outer seals on the piston with the lips facing away from the clutch apply ring side.

2. Install seal protector tool J-29335. Lubricate the seals and install the overrun clutch piston. To make the piston easier to install, insert the piston installing tool between the seal and the

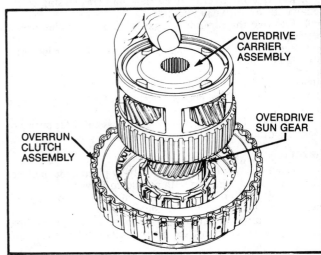

Overrun clutch assembly and related components
(© General Motors Corp.)

housing. Rotate the tool around the housing to compress the lip of the seal, while pushing down slightly on the piston. Remove tool J-29335.

3. Install the overrun clutch waved release spring. Install the overrun clutch waved spring retainer. Be sure that the cupped face is positioned downward.

4. Install the roller clutch cam on the roller clutch assembly. The locating tangs on the roller clutch must set on the roller clutch cam.

5. Install the roller clutch assembly on the overrun clutch hub. Install the narrow snap ring by pushing the roller clutch assembly down.

6. Lubricate and install the overrun clutch plates as they were removed, starting with a flat steel plate.

7. Install the backing plate, micro finish side down. Install the snap ring. Be sure that the composition plates turn freely.

8. Set the unit aside until you are ready to reassemble the transaxle.

OVERDRIVE CARRIER ASSEMBLY

Disassembly

1. Before disassembly, inspect the locating splines for damage, the roller clutch race for scratches or wear and the carrier housing for cracks and wear. Also, inspect the pinions for damage, rough bearings or tilt.

2. Check the pinion end play, it should be .009"-.024".

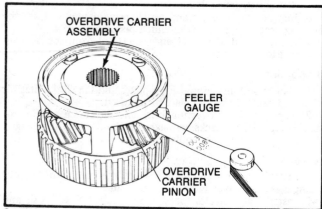

Overdrive carrier pinion end play check (© General Motors Corp.)

3. Remove the overdrive carrier snap ring. Using pliers, remove the overdrive pinion pins.

4. Remove the pinions, thrust washer and roller needle bearings.

5. Remove the overdrive sun gear to overdrive carrier thrust bearing assembly.

Inspection

1. Inspect the pinion pocket thrust faces for burrs. Remove if present.

2. Clean and inspect the thrust bearing assembly for pitting or rough conditions.

Assembly

1. Install the thrust bearing assembly by placing the diameter race down. Retain the part with petroleum jelly.

2. Install nineteen needle bearings into each pinion. Thumb lock the needle bearings to hold them in place.

3. Place a bronze and steel thrust washer on each side so that the steel washer is against the pinion. Retain the washer using petroleum jelly.

4. Place the pinion assembly in position inside the carrier. Use a pilot shaft to align the parts in place.

5. Push the on pin into place while rotating the pinion from the side.

6. Install the overdrive carrier snap ring in order to retain the pinion pins. Set the assembly aside until the transaxle is ready for assembly.

INTERMEDIATE SERVO PISTON

Disassembly

1. Once the intermediate servo assembly is out of the case, there is little to disassemble on the unit. It should be noted that there is an inner and an outer servo piston. These will separate by pulling them apart.

2. Be careful of the oil seal rings on the servo pistons, both inner and outer. They should not be removed unless they are to be replaced.

3. If the apply pin is to be removed, pry the snap ring from its groove and separate the apply pin from the retainer.

Inspection

1. Check the pin carefully for wear or damage, as well as the fit in the case bore. Do not remove the small oil seal rings from the apply pin unless they are to be replaced.

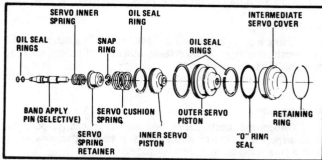

Intermediate servo—exploded view (© General Motors Corp.)

2. Inspect the inner and outer piston seal rings. These too should not be removed unless they are to be replaced.

3. Checking for proper pin application, since the pin is a selective fit, requires special tools that, in the case of the factory type units, only give relative readings. That is, dial indicator readings are valid only when used in conjunction with these special tools. However, in nearly all cases, the apply pin will not have to be changed unless the case itself is defective and needs to be replaced. In the course of a normal rebuild or seal job, the pin would not have to be replaced or even removed from the rest of the servo parts.

Assembly

1. If any of the oil seal rings have been removed, lubricate the replacement rings with Dexron® II or petroleum jelly. Make certain that the rings are fully seated in their grooves so that they will not be damaged on assembly.

2. Install the intermediate servo inner piston into the outer piston taking care to see that the seals are not torn or displaced as the inner servo goes into place.

3. Again using plenty of lubricant, replace the O-ring on the servo cover.

4. Assemble the apply pin retainer and the apply pin. Install the snap-ring to secure the assembly. Install this assembly into the servo inner piston along with the cushion spring. Set the unit aside until the transaxle is ready for assembly.

GOVERNOR

Disassembly

1. Remove the speedometer drive gear from the top of the

governor. There should be a thrust washer on the top of the drive gear to protect the assembly from the governor cover. This too should be removed.

2. Remove and discard the O-ring seal from the governor cover.

Inspection

1. Clean all parts in solvent and dry with compressed air.

2. Inspect the governor cover for wear or damage. Inspect the governor shaft seal ring for cuts or other signs of damage. If this seal is removed, it must be replaced.

3. Make sure that the two check balls are in place and that the weight springs are not mispositioned or damaged.

Assembly

1. If the shaft seal is to be replaced, it can be cut from the shaft. Be careful not to damage the groove. Lubricate the replacement seal with Dexron® II or petroleum jelly. Install the seal on the shaft and then carefully place the seal end of the shaft into the pilot hole in the case to size the seal to the shaft.

2. Install the speedometer drive gear onto the governor shaft on the weight end and make sure that the slot on the gear aligns with the pin in the shaft. Install the thrust washer on top of the drive gear.

3. Lubricate the replacement O-ring seal for the governor cover and install it into the groove in the cover. Set these parts aside until the transaxle is ready for assembly.

PRESSURE REGULATOR

Disassembly

1. The pressure regulator can be found in the transaxle case. It sits in a bore next to the hole for the oil pipe. With snap ring pliers, remove the internal snap ring.

2. Remove the pressure regulator bushing. Ordinary slip-joint pliers can be used to carefully graps the regulator bushing and pull it from the case.

3. Compress the regulator valve into the bushing and remove the inner roll pin. *Be careful* since the valve is under strong spring pressure. Release the valve slowly.

4. Remove the pressure regulator valve, the guide, and the spring from the bushing.

Inspection

1. Clean all parts well in solvent and blow dry with compressed air.

2. Check the fit of the pressure regulator valve in the bushing. It must move freely in the bore of the bushing.

3. Check the spring for distortion.

4. Examine the outer pin in the bushing for damage, such as being bent. This roll pin should not be removed unless it is to be replaced.

Assembly

1. With all parts clean and dry, begin by installing the spring and guide into the bushing.

2. Install the regulator valve into the bushing, stem end last. Retain with the roll pin.

3. If the outer roll pin has been removed, replace it with a new one. Set these parts aside until the transaxle is ready for reassembly.

VALVE BODY

Disassembly

NOTE: As each valve train is removed, place the individual valve train in the order that it is removed and in a separate location relative to its position in the valve body assembly. None of the valves, bushings or springs are interchangeable; some roll pins

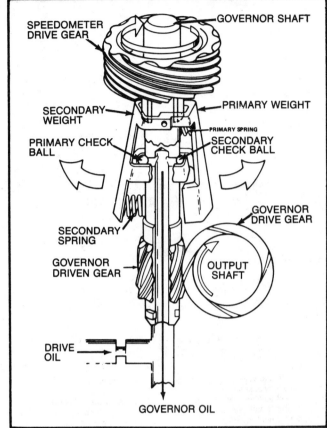

Governor assembly—cross section (© General Motors Corp.)

are interchangeable. Remove all the roll pins and spring retaining sleeves by pushing through from the rough case surface side of the control valve pump assembly, except for the blind hole pins.

1. Lay the control valve assembly machined face up, with the manual valve at the top.

2. If, the number one check ball is still in place remove it.

3. Some of the roll pins in the control valve assembly have pressure against them. Hold a towel over the bore while removing the pin to help prevent possibly losing a bore plug, spring, etc.

4. Remove the roll pin from the upper left bore. Remove the 2-3 throttle valve bushing, 2-3 throttle valve spring, 2-3 throttle valve and the 2-3 shift valve. The 2-3 throttle valve spring and 2-3 throttle valve may be inside the 2-3 throttle valve bushing.

5. From the next bore down, remove the spring retaining sleeve, valve bore plug, T.V. modulator downshift valve and the T.V. modulator downshift valve spring.

6. From the next bore down, remove the outer roll pin. Remove the 1-2 throttle valve bushing, 1-2 throttle valve spring, 1-2 throttle valve and the low 1st/detent valve. The 1-2 throttle valve spring and the 1-2 throttle valve may be inside the 1-2 throttle valve bushing. Remove the inner roll pin. Remove the low 1st/detent valve bushing and 1-2 shift valve.

7. From the next bore down, remove the spring retaining sleeve, valve bore plug, T.V. modulator upshift valve and the T.V. modulator upshift valve spring.

8. From the next bore down, remove the roll pin. Remove the converter clutch throttle bushing, converter clutch throttle valve spring, converter clutch throttle valve, and the converter clutch shift valve. The converter clutch throttle valve spring and the converter clutch throttle valve may be inside the converter clutch throttle bushing.

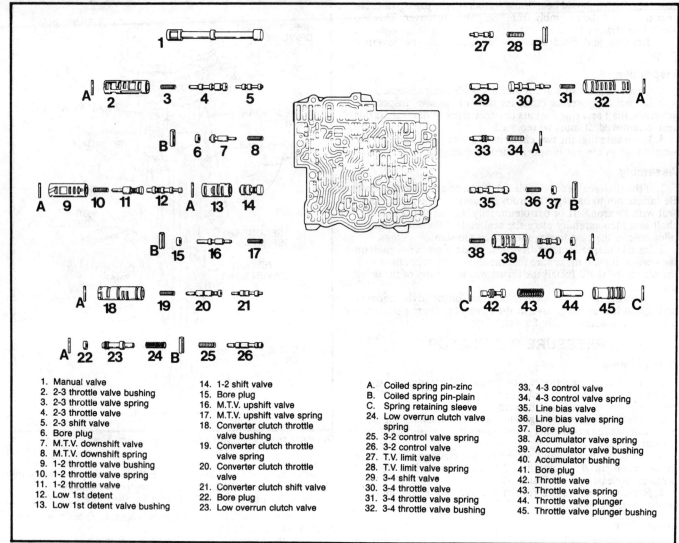

1. Manual valve
2. 2-3 throttle valve bushing
3. 2-3 throttle valve spring
4. 2-3 throttle valve
5. 2-3 shift valve
6. Bore plug
7. M.T.V. downshift valve
8. M.T.V. downshift spring
9. 1-2 throttle valve bushing
10. 1-2 throttle valve spring
11. 1-2 throttle valve
12. Low 1st detent
13. Low 1st detent valve bushing

14. 1-2 shift valve
15. Bore plug
16. M.T.V. upshift valve
17. M.T.V. upshift valve spring
18. Converter clutch throttle
    valve bushing
19. Converter clutch throttle
    valve spring
20. Converter clutch throttle
    valve
21. Converter clutch shift valve
22. Bore plug
23. Low overrun clutch valve

A. Coiled spring pin-zinc
B. Coiled spring pin-plain
C. Spring retaining sleeve
24. Low overrun clutch valve
    spring
25. 3-2 control valve spring
26. 3-2 control valve
27. T.V. limit valve
28. T.V. limit valve spring
29. 3-4 shift valve
30. 3-4 throttle valve
31. 3-4 throttle valve spring
32. 3-4 throttle valve bushing

33. 4-3 control valve
34. 4-3 control valve spring
35. Line bias valve
36. Line bias valve spring
37. Bore plug
38. Accumulator valve spring
39. Accumulator valve bushing
40. Accumulator bushing
41. Bore plug
42. Throttle valve
43. Throttle valve spring
44. Throttle valve plunger
45. Throttle valve plunger bushing

Valve body assembly—exploded view (© General Motors Corp.)

9. From the last bore down, remove the roll pin. Remove the bore plug, low/overrun clutch valve and the low/overrun clutch valve spring. From the same bore, remove the inner spring retaining sleeve. Remove the 3-2 control valve spring and valve.

10. From the upper right bore, remove the spring retaining sleeve. Remove the T.V. limit spring and valve.

11. From the next bore down, remove the roll pin. Remove the 3-4 throttle valve bushing, 3-4 throttle valve spring, 3-4 throttle valve and the 3-4 shift valve. The 3-4 throttle valve and spring may be inside the 3-4 throttle valve bushing.

12. From the next bore down, remove the roll pin. Remove the 4-3 control valve spring and 4-3 control valve.

13. From the next bore down remove the spring retaining sleeve, valve bore plug, line bias valve and the line bias valve spring.

14. From the next bore down, remove the roll pin and the bore plug. Remove the accumulator valve bushing, accumulator valve and the accumulator spring. The accumulator spring and accumulator valve may be inside the accumulator bushing.

15. From the last bore, remove the outer roll pin from the rough casting side, the throttle valve plunger bushing, throttle valve plunger and the throttle valve spring.

16. Remove the inner roll pin by grinding a taper to one end of

a #49 drill bit. Insert the tapered end into the roll pin. Pull the drill bit and the coil retaining pin out. Remove the throttle valve.

Inspection

1. Wash the control valve body, valves, springs, and other parts in clean solvent and air dry.

2. Inspect the valves for scoring, cracks and free movement in their bores.

3. Inspect the bushings for cracks or scored bores.

4. Inspect the valve body for cracks, damaged or scored bores.

5. Inspect the springs for distortion or collapsed coils.

6. Inspect the bore plugs for damage.

7. Install all of the flared coiled pins (zinc coated), flared end out, and from the machined face of the valve body assembly.

8. Install the two tapered coiled pins (black finish) that retain the throttle valve and throttle bushing, tapered end first.

9. The coiled pins do not fit flush on the rough casting face. Make sure that all of the coiled pins are flush against the machined face or damage to the transaxle will occur.

Assembly

1. Be sure that the control valve body is clean before starting the assembly procedure.

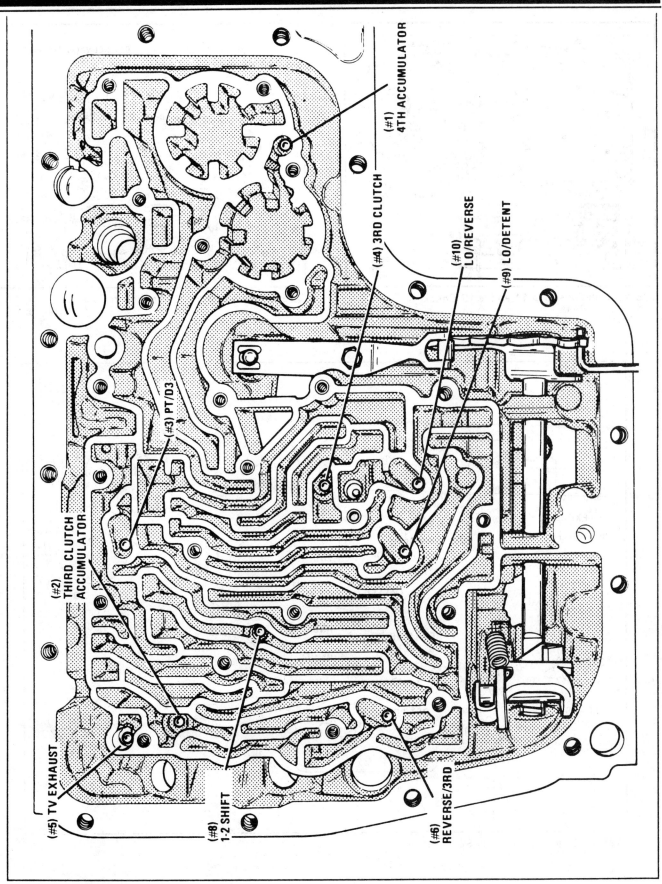

(#1) 4TH ACCUMULATOR

(#4) 3RD CLUTCH

(#10) LO/REVERSE

(#9) LO/DETENT

(#3) PT/D3

(#2) THIRD CLUTCH ACCUMULATOR

(#5) TV EXHAUST

(#8) 1-2 SHIFT

(#6) REVERSE/3RD

Check Ball Locations

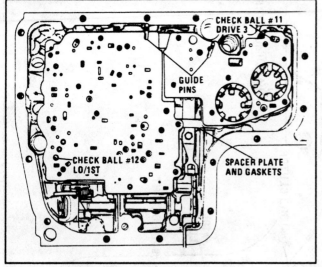

325-4L- Locations of #11 and #12 Check Balls

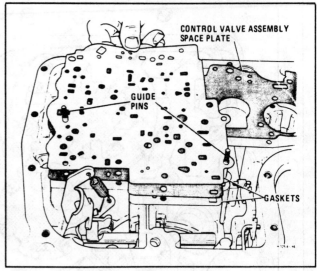

325-4L- Installing the Guide Pins in Case

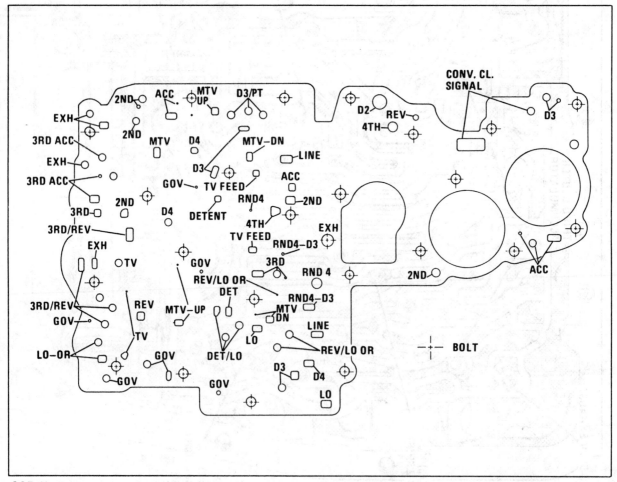

325-4L- Typical Valve Body Spacer Plate

2. Install into the lower right bore, throttle valve, the smaller outside diameter land first, make sure the valve is seated at the bottom of the bore. Install the inner roll pin between the lands of this valve. Install the T.V. spring into the bore. Install the throttle valve plunger, stem end first, into the throttle valve plunger bushing. Install these two parts into the bore, valve end first. Install the outer roll pin from the rough cast surface side, aligning the pin with the slot in the bushing.

3. In the next bore up, install the accumulator bushing into the bore. Align the pin slot in line with the pin hole in the control valve assembly. Install the accumulator spring, then accumulator valve, the smaller end first. Next, install the bore plug, hole out, and roll pin.

4. In the next bore up, install the line bias valve spring, then install the line bias valve, smaller stem end first. Install the bore plug, the hole out, and the spring retaining sleeve.

5. In the next bore up, install the 4-3 control valve, the smaller end first, then the 4-3 control valve spring. Compress the spring and install the roll pin.

6. In the next bore up, install the 3-4 shift valve, chamfered end first. Install the 3-4 throttle valve spring into the 3-4 throttle valve bushing. Next, install the 3-4 throttle valve, the stem end first into the 3-4 bushing. Install the 3-4 throttle valve and bushing into the bore, making sure the pin slot is aligned with the pin hole in the control valve assembly. Install the roll pin.

7. In the last bore up, install the T.V. limit valve, the stem end first. Install the T.V. limit valve spring. Compress the T.V. limit valve spring and install the spring retaining sleeve from the machined face side. Make sure the sleeve is level with the machined surface.

8. In the lower left bore, install the 3-2 control valve, the smaller stem end out. Install the 3-2 control valve spring. Compress the spring and install the spring retaining sleeve. Be sure that the spring retaining sleeve does not damage the coiled spring. In the same bore install the low/overrun clutch valve spring and then the low/overrun clutch valve, the smaller end first. Next, install the bore plug, the hole out, and the roll pin.

9. In the next bore up, install the converter clutch shift valve, the short stemmed end first. Next, install the converter clutch throttle valve spring into the converter clutch throttle bushing and the converter clutch throttle valve, the stem end first, into the bushing. Install these three parts, the valve end first, into the bore, aligning the bushing so the pin can be installed in the pin slot. Install the roll pin.

10. In the next bore up, install the T.V. modulator upshift valve spring then install the T.V. modulator upshift valve, the smaller chamfered stem end first. Install the bore plug, the hole out and the spring retaining sleeve. Make sure the sleeve is level with or below the machined surface.

11. In the next bore up, install the 1-2 shift valve, the small stem outward. Install the 1-2 shift valve, the small stem outward. Install the 1-2 shift valve bushing, the small I.D. first, and align pin hole in bushing with the inner pin hole in the control valve assembly. Install the inner roll pin. Next, install the low 1st/detent valve, the long stem end out. Install the low 1st/detent valve, the long stem end out. Install the 1-2 throttle valve spring into the 1-2 throttle valve bushing and the 1-2 throttle valve, the stem end first, into the bushing. Install these three parts, the valve end first, into the bore, aligning the bushing so the outer pin can be installed in the pin slot. Install the outer roll pin.

12. In the next bore up, install the T.V. modulator downshift valve spring then install the T.V. modulator downshift valve, the smaller chamfered stem end first. Install the bore plug, the hole out, and the spring retaining sleeve. Make sure the sleeve is level with or below the machined surface.

13. In the next bore up, install the 2-3 shift valve, the large end first. Next, install the 2-3 throttle valve spring into the 2-3 throttle valve bushing, the 2-3 throttle valve stem end first, into the bushing. Install these three parts, the valve end first, into the bore,

aligning the bushing so the pin can be installed in the pin slot. Install the roll pin.

14. Install the manual valve with the inside detent lever rod slot first.

15. Set the control valve body aside, until ready to reassemble the transaxle.

Transaxle Assembly

1. Verify that the transaxle case has been thoroughly cleaned and that all bolt holes have been checked for damage or stripped threads. The vent should not be remove unless it is broken and must be replaced. It should be retained with Loctite®. Any cooler line connectors that may have been removed should have their threads coated with a good sealer. They should be torqued to 28 foot-pounds.

2. After it has been determined that all of the parking linkage is serviceable and that the manual linkage is not excessively worn, begin assembly by installing a new manual shift seal. Make sure that the lip of the seal faces inward toward the transaxle case.

3. Install the parking pawl and the return spring with the tooth of the pawl toward the center of the case and the spring under the tooth of the pawl. The ends of the spring should be toward the inside of the case, with the ends located in the slots in the case.

4. Install the parking pawl shaft with the tapered end going in first.

5. install the manual shaft and inside detent lever assembly making certain that the shaft fits into the flats in the hole in the detent lever. If the manual link (long thin rod) was disassembled from the detent lever, install the end back into the lever and retain with a new push nut.

6. Install the hex nut on the manual shaft and with a 15mm socket, torque to 23 foot-pounds.

7. Install the park lock cam onto the side of the manual shaft and with an 8mm socket, tighten to a torque of 8 foot-pounds.

8. Install the parking strut shaft (different from the parking pawl shaft) into the case while aligning the lock strut and lock lever. See that the lower strut arm is positioned between the lever tangs. Some units may have a washer that belongs between the parking lock lever and the case. Install the washer if so equipped. Finally, install the retaining ring.

9. Install the parking lock spring.

10. Using a piece of ⅜" rod for a driver, install a new parking strut shaft cup plug with the open end out. Drive the cup in until it is flush with the face of the case.

11. Install the manual lever detent roller and spring assembly. With a 10mm socket torque the retaining bolt to 11 foot-pounds. Check the operation of the park lock by working the linkage by hand.

12. The output shaft should be inspected for plugged lubrication passages and the splines checked for damage. Some early production output shafts were made with a snap ring while others have a shoulder against which the governor drive gear rests. If the output shaft has a shoulder, *do not* try to use a snap ring on this shaft. Install the reaction internal gear on the output shaft hub end first.

13. Install the reaction internal gear-to-reaction sun gear thrust bearing on the output shaft with the small diameter race first. This will place the inside diameter of the race toward the carrier assembly.

14. The roller clutch should have been installed at this point, into the reaction planetary carrier, and the thrust washers should be in place. Install the roller clutch and reaction carrier assembly into the reaction internal gear.

15 A special fitting is normally used to align the output shaft and aid in the installation of the reaction unit parts. The purpose of this assembly tool is also to help adjust the height of the reaction internal gear parking pawl lugs to align flush with the top of the parking pawl tooth. The main point is to make sure the parking pawl lugs align with the top of the pawl tool and to hold the

assembly in this relationship while the transaxle is being assembled. Sight through the parking pawl slot to line things up. With the reaction unit installed and aligned properly, the low and reverse clutch components are installed next.

16. The low and reverse clutch plates should be well-oiled with Dextron® II before use. It is normally recommended that clutch plates be soaked for at least 20 minutes before assembly. Install this clutch pack by starting with a flat steel plate and then alternate with the composition plates. Install the low and reverse clutch housing-to-case spacer into the transaxle case. The low and reverse clutch housing and piston should already be overhauled at this point, and it should be installed in the case aligning the feed hole to the clutch and the case passage. A large spring-loaded plier-like tool is used to lower the low and reverse clutch housing since it would be difficult to handle the clutch housing. Make sure that the clutch housing is seated past the case snap ring groove. If the clutch housing does not seat past the snap ring groove, take the reaction sun gear and input drum assembly, and using them as a tool, install them into the transaxle case. Rotate the reaction sun gear and drum back and forth tapping lightly to align the roller clutch race and the low and reverse clutch hub splines. Remove the sun gear and drum and check to see if the low and reverse clutch housing is now seated below the snap ring groove. If not, repeat the above process. Install the housing-to-case snap ring with the flat side against the housing which puts the ring's beveled side up.

17. Check to make sure that the four-tanged thrust washer is properly installed on the input drum. Petroleum jelly will help hold it in place. Install the reaction gear and input drum assembly.

18. Install the input sun gear with the identification mark on the gear against the input drum. These identification marks could be either a groove or a drill spot depending on the gear.

19. Lubricate and install the input sun gear-to-front carrier thrust bearing and race assembly. Install it so that the needle bearings are against the input sun gear.

20. Install the input carrier-to-input internal gear thrust bearing assembly on the input carrier. Make sure that the inside diameter race goes against the gear. Use plenty of petroleum jelly to hold it in place. Install the input carrier assembly onto the input sun gear.

21. Install the reaction selective washer and thrust washer on the input internal gear and use petroleum jelly to hold them in place. The selective washer must be installed with the identification number facing the snap-ring side.

22. Install the input internal gear.

23. Install the snap ring on the output shaft, making certain that it is fully seated in the groove. When completed, the output end-play can be checked with a dial indicator. End-play should be from .004" to .025".

24. Install the output shaft-to-input shaft selective thrust washer, indexing it into the output shaft with plenty of petroleum jelly to hold it in place.

25. Install the intermediate band into the case making sure that the anchor pin lug on the band as well as the apply lug is properly located in the case slot.

26. Position the direct clutch assembly with the clutch plate end up, over a hole in the workbench. Align the composition clutch plates so that the teeth in the plates are lined up with one another. This will make installation of the forward clutch easier. Install the forward clutch assembly with the input end first into the direct clutch. It may be helpful to hold the direct clutch housing and rotate the forward clutch until it is seated. As a check to make sure that the forward clutch is fully seated, it will be approximately ⅝" from the tang end of the direct clutch housing to the end of the forward clutch drum. Grasp the direct and forward clutch assemblies together so that they will not separate and turn them so that the input shaft is facing up. Install these assemblies into the transaxle case, turning them as necessary to get them to seat. As a check to make sure that the clutches are fully seated, it

will be approximately ⁷/₁₆" from the case cover face to the direct clutch housing when the clutches are fully installed, and correctly seated.

27. Install a new case cover gasket and carefully lay the case cover on the transaxle. Install the retaining bolts and with a 13mm socket, snug down the bolts. Make sure the bolts are in their proper locations. While the bolts are being tightened, rotate the input shaft. If the shaft cannot be rotated, then the forward or direct clutch housings have not been installed properly and the clutch plates are not indexed. At this point, the condition must be corrected before the case cover can be fully installed.

28. Install the overhauled oil pump next with plenty of lubricant on the outer O-ring. It will be helpful to make two guide pins. These particular screws are not metric, but a standard ⁵/₁₆-18 thread. The guide pins should be about 4 inches long. First install the two guide pins in the pump attaching screw holes. Then install the pump spacer plate onto the oil pump face, taking care to align it properly. Align the guide pins with matching holes in the case cover and insert the flat-head screws into the open holes in the pump body. Tighten the flat-head screws, remove the guide pins and put in the last two screws. Tighten all of the screws to 18 foot-pounds.

29. Place the drive link chain around the drive and driven sprockets so that the links engage the teeth of the sprockets. Look for one of the links to be of a different color and to have etched numbers on it. This side should face the sprocket cover. Place the drive link chain and sprockets in position on the sprocket supports at the same time. A plastic mallet will probably be needed to gently seat the sprocket bearings into the sprocket supports.

30. Reach through the access holes in the sprockets with a pair of snap ring pliers and install the snap rings. Make sure that they are seated in their grooves.

31. Install a new sprocket cover gasket and then install the cover. The attaching bolts are torqued to 8 foot-pounds.

32. Install the overdrive carrier assembly and the overrun clutch unit.

33. Install the governor assembly into the transaxle case. Lubricate the O-ring and install the cover. Torque the retaining bolts to 8 ft. lbs. Install the speedometer driven gear assembly, retainer clip and bolt.

34. Install the intermediate servo piston assembly. Take care when the rings enter the transaxle case so that they will not be torn or damaged. The servo cover ring must also be well lubricated to prevent damage or cutting of the ring. A plastic hammer may be needed to gently tap the assembly into the case. Be sure that the tapered end of the apply pin is properly located against the band apply lug. Depress the cover and install the retaining ring. Align the ring gap of the ring with the slot in the case to aid future removal.

35. Position the transaxle with the oil pan side up. Install a new low and reverse clutch housing-to-case cup plug. The rubber end goes in first, into the hole in the case. A ⅜" diameter rod can be used as a driver. Tap gently with a plastic mallet on the rod until the plug seats against the low and reverse clutch housing.

36. Install the 1-2 accumulator spring into the transaxle case.

37. Install the #5 and the #6 check balls into their proper locations in the case. Two guide pins should be made up to aid with the installation of the control valve body. These guide pins should be made from metric bolts, 6mm in diameter and several inches in length. Install the remaining four check balls in their proper locations in the valve body. Since the valve body will be inverted to install it, use some petroleum jelly to hold them in place.

38. Place the valve body-to-spacer plate gasket in place on the valve body. The two gaskets that are used in valve body assembly are usually very similar. They are often marked to help prevent confusion. Look for a "VB" either printed or stamped into the gasket. Install the spacer plate on top of this gasket. Place the spacer plate-to-case gasket on spacer plate. This gasket should be marked with a "CB".

39. Install two of the valve body attaching screws through the assembly. Carefully grasp the entire assembly and invert, and install on the case. Be very careful to hold all these parts together so that the check balls and the accumulator parts do not fall out of place. Also, hold one finger on the manual valve so that it will not fall from the valve body. Take note that two of the valve body bolts are ⅜″ longer than the rest and must not be interchanged. Start the valve body bolts into the case with the exception of the throttle lever and bracket assembly and the oil screen retainer bolts. Remove the guide pins and replace with bolts.

40. Install the link onto the manual valve and install the valve clip.

41. Install the throttle lever and bracket assembly, making sure to locate the slot in the bracket with the roll pin, and aligning the lifter through the valve body opening and the link through the T.V. linkage case bore. Install the retaining bolt. Carefully torque all the control valve assembly bolts to 11 foot-pounds.

42. The pressure regulator can be installed next. Its bore is next to the opening for the oil suction pipe. Install the assembly into the case, and carefully fit the snap ring into the groove.

43. The oil screen should have been carefully and thoroughly cleaned in solvent and blown dry with compressed air. A new O-ring seal should be installed on the intake pipe. Lubricate this ring well with petroleum jelly, and install the screen assembly.

44. The oil pan should also have been cleaned well and all traces of debris removed. Check the pan for straightness and make sure that none of the bolt holes have been dished in from over-torque. A block of wood and a rubber mallet can be used to straighten the gasket flanges on the pan. Install the pan with a new gasket. The oil pan bolts are to be tightened to 12 foot-pounds of torque.

NOTE: One particular point to watch when installing the oil pan is a possible interference between the oil pan and the sprocket cover, resulting in oil leaks. Due to this interference, the pan will not seat tightly against the gasket and leaks will result. The interference is caused by the outside diameter of the washer head bolts squeezing the outer pan flange against the sprocket cover. A second design bolt with a smaller washer head has been used in later models. The smaller washer head will prevent this interference. The part number for the new bolt is GM #11502670.

SPECIFICATIONS

OVERDRIVE UNIT END PLAY SELECTIVE WASHER CHART

Identification Number	Color	Thickness (inches)
One	Gray	.063-.067
Two	Dark Green	.070-.074
Three	Pink	.077-.081
Four	Brown	.084-.088
Five	Light Blue	.091-.095
Six	White	.098-.102
Seven	Yellow	.105-.109
Eight	Light Green	.112-.116
Nine	Orange	.119-.123
Ten	Violet	.126-.130
Eleven	Red	.133-.137
Twelve	Dark Blue	.140-.144

INPUT UNIT END PLAY SELECTIVE WASHER CHART

Identification Number	Color	Thickness (inches)
One	—	.065-.070
Two	—	.070-.075
Three	Black	.076-.080
Four	Light Green	.081-.085
Five	Scarlet	.086-.090
Six	Purple	.091-.095
Seven	Cocoa Brown	.096-.100
Eight	Orange	.101-.206
Nine	Yellow	.106-.111
Ten	Light Blue	.111-.116
Eleven	—	.117-.121
Twelve	—	.122-.126
Thirteen	Pink	.127-.131
Fourteen	Green	.132-.136
Fifteen	Gray	.137-.141

REACTION UNIT END PLAY SELECTIVE WASHER CHART

Identification Number	Color	Thickness (inches)
One	Orange	.114-.119
Two	White	.021-.126
Three	Yellow	.128-.133
Four	Blue	.135-.140
Five	Red	.143-.147
Six	Brown	.150-.154
Seven	Green	.157-.161
Eight	Black	.164-.168
Nine	Purple	.171-.175

FORWARD CLUTCH PLATE AND APPLY RING USAGE CHART

	Wave Plate	Flat Steel Plate	Composition Faced Plate	Apply Ring
Number	One	Three	Four	—
Thickness	.062 Inch	.077 Inch	—	—
Identification	—	—	—	Eight
Width	—	—	—	.492 Inch

S SPECIFICATIONS

DIRECT CLUTCH PLATE AND APPLY RING USAGE CHART

	Flat Steel Plate	Composition Faced Plate	Apply Ring
Number	Six	Six	—
Thickness	.91 Inch	—	—
Identification	—	—	Nine
Width	—	—	.492 Inch

LOW AND REVERSE CLUTCH PLATE AND APPLY RING USAGE CHART

	Wave Plate	Flat Steel Plate	Composition Faced Plate	Apply Ring
Number	One	Seven	Six	—
Thickness	.077 Inch	.077 Inch	—	—
Identification	—	—	—	Zero
Width	—	—	—	.516 Inch

TORQUE SPECIFICATIONS

Description of Usage	Torque
Valve Body Assembly to Case	9-12 ft.-lbs.
Accumulator Housing to Case	9-12 ft.-lbs.
Manual Detent Spring Assembly to Case.	9-12 ft.-lbs.
Governor Cover to Case	6-10 ft.-lbs.
Sprocket Cover to Case	9 ft.-lbs.
Case Cover to Pump Body (Flat Head)	15-20 ft.-lbs.
Driven Support to Case Cover	15-20 ft.-lbs.
Case Cover to Case	15-20 ft.-lbs.
Fourth Clutch Housing to Case	15-20 ft.-lbs.
Oil Pan to Case	7-10 ft.-lbs.
Cam to Manual Shaft	6-9 ft.-lbs.
Manual Shaft to Inside Detent Lever (Nut)	20-25 ft.-lbs.
Cooler Connector	26-30 ft.-lbs.
Line Pressure Take-Off	5-10 ft.-lbs.
Third Accumulator Take-Off	5-10 ft.-lbs.
Governor Pressure Take-Off	5-10 ft.-lbs.
Fourth Pressure Take-Off	5-10 ft.-lbs.
Pressure Switch	5-10 ft.-lbs.

SPECIAL TOOLS

	J 21369-B	Converter Leak Test Fixture		J 26744	Clutch Seal Installer
	J 21465-17	Pump Bushing Remover and Installer		J 26900-9	Thickness Gauge Set
	J 21465-3	Drive Sprocket Support Bushing Installer		J 26900-27	Flexible Steel Scale (150mm)
	J 21465-15	Drive Sprocket Support Bushing Remover		J 26958	Output Shaft Alignment and Loading Tool
	J 23327	Direct & Forward Clutch Spring Compressor		J 28492	Holding Fixture
	J 23456	Clutch Pack Compressor		J 28493	Servo Cover Depressor

	J 25010	Direct Clutch Seal Protector		J 28494	Input Shaft Lifter
	J 25011	Reverse Clutch Seal Protector		J 28542	Low & Reverse Clutch Housing Installer & Remover
	J 25014	Intermediate Band Apply Pin Gauge		J 28585	Snap Ring Remover and Installer
	J 25018-A	Adapter Forward Clutch Spring Compressor		J 28667	Dial Indicator Extension
	J 25019	Bushing Service Set		J 29060	Converter End Play Checking Fixture
	J 25025-A	Pump and Valve Guide Pin and Indicator Stand Set		J 29334-1	Fourth Clutch Spring Compressor
	J 25359-5	#40 Drive Bit		J 29846	Turbine Shaft Seal Installer

INDEX

GENERAL MOTORS TURBO HYDRA-MATIC 440 T4 Automatic Transaxle

 APPLICATIONS

1984½ General Motors "C" Cars

 GENERAL DESCRIPTION

Transmission and Converter Identification

TRANSMISSION

The THM 440-T4 (ME9) automatic transaxle is a fully automatic unit, consisting of four multiple disc clutches, a roller clutch, a sprag and two bands, requiring hydraulic and mechanical applications to obtain the desired gear ratios from the compound planetary gears. The transaxle identification can be located on one of three areas of the unit. An identification plate on the side of the case, a stamped number on the governor housing or an ink stamp on the bell housing.

CONVERTER

Two types of converters are used in the varied vehicle applications. The first type is the three element torque converter combined with a lock-up converter clutch. The second type is the three element torque converter combined with a viscous lock-up converter clutch that has silicone fluid sealed between the cover and the body of the clutch assembly. Identification of the torque converters are either ink stamp marks or a stamped number on the shell of the converter.

Metric Fasteners

The metric fastener dimensions are very close to the dimensions of the familiar inch system fasteners, and for this reason, replacement fasteners must have the same measurement and strength as those removed.

Do not attempt to inter-change metric fasteners for inch system fasteners. Mismatched or incorrect fasteners can result in damage to the transmission unit through malfunctions, breakage or possible personal injury.

Care should be taken to re-use the fasteners in the same locations as removed.

Capacities

Fluid Checking Procedure

1. Verify that the transmission is at normal operating temperature. At this temperature, the end of the dipstick will be too hot to hold in the hand. Make sure that the vehicle is level.

2. With the selector in Park, allow the engine to idle. *Do not race engine.* Move the selector through each range and back to Park.

3. Immediately check the fluid level, engine still running. Fluid level on the dipstick should be at the "FULL HOT" mark.

4. If the fluid level is low, add fluid as required, remembering that only one pint will bring the fluid level from "ADD" to "FULL".

Often it is necessary to check the fluid level when there is no time or opportunity to run the vehicle to warm the fluid to operating temperature. In this case, the fluid should be around room temperature (70°). The following steps can be used:

1. Place the selector in Park and start engine. *Do not race engine.* Move the selector through each range and back to Park.

2. Immediately check the fluid level, engine still running, off fast idle. Fluid level should be between the two dimples on the dipstick, approximately ¼ in. *below* the "ADD" mark on the dipstick.

3. If the fluid level is low, add fluid as required, to bring the fluid level to between the two dimples on the dipstick. *Do not overfill.* The reason for maintaining the low fluid level at room tem-

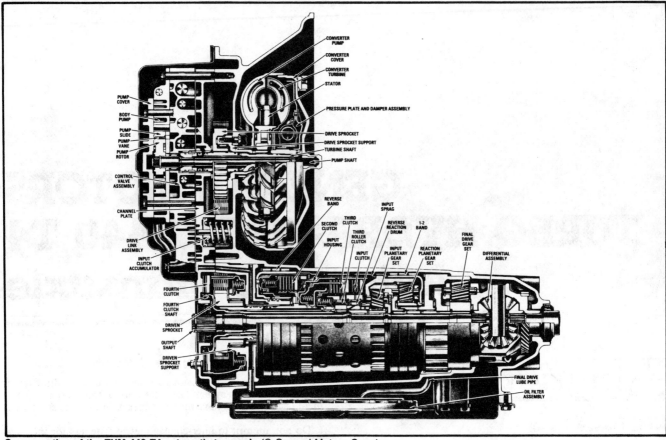

Cross section of the THM 440-T4 automatic transaxle (© General Motors Corp.)

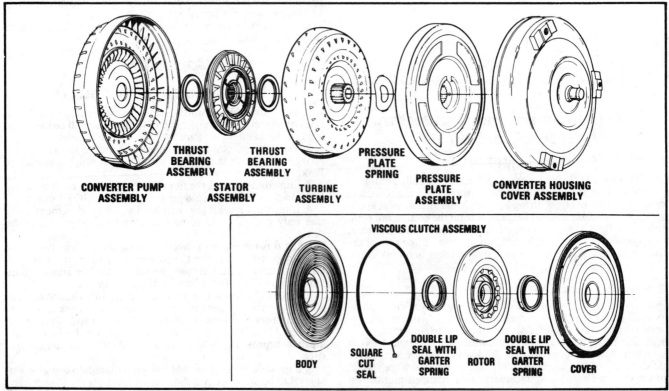

Two types of converters used with the THM 440-T4 transaxle (© General Motors Corp.)

perature is that the transmission fluid level will rise as the unit heats up. If too much fluid is added when cold, then the fluid will rise to the point where it will be forced out of the vent and overheating can occur.

If the fluid level is *correctly* established at 70°F as outlined above, it will appear at the "FULL" mark when the transmission reaches operating temperature.

When checking the fluid level it is a good idea to notice the condition of the fluid. If there is a burnt smell, or if there are metal particles on the end of the dipstick, these signs can help in diagnosing transmission problems. A milky appearance is a sign of water contamination, possibly from a damaged cooling system. Note if there are any air bubbles on the end of the dipstick. Oil with air bubbles indicates that there is an air leak in the suction lines which can cause erratic operation and slippage. All of these signs can help in determining transmission problems and their source.

FLUID SPECIFICATIONS

Dexron® II automatic transmission fluid or its equivalent is the only recommended automatic transmission fluid to be used in this unit. Make certain that only good quality, clean fluid is used when servicing this transmission. The use of any other grade of fluid can lead to unsatisfactory performance or complete unit failure.

Modification publications not available at time of printing.

CLUTCH AND BAND APPLICATION CHART
THM 440-T4 Automatic Transaxle

Range		4th Clutch	Reverse Band	2nd Clutch	3rd Clutch	3rd Roller Clutch	Input Sprag	Input Clutch	1-2 Band
NEUTRAL PARK							①	①	
DRIVE	1						HOLD	ON	ON
	2			ON			OVER-RUNNING	①	ON
	3			ON	ON	HOLD			
	4	ON		ON	①	OVER-RUNNING			
MANUAL	3			ON	ON	HOLD	HOLD	ON	
	2			ON			OVER-RUNNING	①	ON
	1				ON	HOLD	HOLD	ON	ON
REVERSE		ON					HOLD	ON	

①APPLIED BUT NOT EFFECTIVE

CHILTON'S THREE "C's" DIAGNOSIS CHART
THM 440-T4 Automatic Transaxle

Condition	Cause	Correction
Oil Leakage	a) Side cover, bottom pan and gaskets, loose bolts	a) Repair or replace cover, gasket and torque bolts
	b) Damaged seal at T.V. cable, fill tube or electrical connector	b) Replace seal as required
	c) Damaged seal assembly on manual shaft	c) Replace seal assembly as required

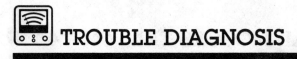 TROUBLE DIAGNOSIS

CHILTON'S THREE "C's" DIAGNOSIS CHART
THM 440-T4 Automatic Transaxle

Condition	Cause	Correction
Oil Leakage	d) Leakage at governor cover, servo covers, modulator, parking plunger guide, speedometer driven gear sleeve	d) Replace damaged "O" ring seals as required
	e) Converter or converter seal leaking	e) Replace converter and seal as required
	f) Axle seals leaking	f) Remove axles and replace seals as required
	g) Pressure ports or cooler line fittings leaking	g) Tighten or repair stripped threads
Fluid foaming or blowing out the vent	a) Fluid level high	a) Correct fluid level
	b) Fluid foaming due to contaminates or over-heating of fluid	b) Determine cause of contamination or overheating and repair
	c) Drive sprocket support has plugged drain back holes	c) Open drain back holes in sprocket support
	d) Thermo element not closing when hot	d) Replace thermo element
	e) Fluid filter "O" ring damaged	e) Replace fluid filter "O" ring
High or low fluid pressure, verified by pressure gauge	a) Fluid level high or low	a) Correct fluid level as required
	b) Vacuum modulator or hose leaking	b) Repair or replace hose or modulator
	c) Modulator valve, pressure regulator valve, pressure relief valve nicked, scored or damaged. Spring or ball checks missing or damaged	c) Repair or replace necessary components
	d) Oil pump or components damaged, parts missing	d) Repair or replace oil pump assembly
No drive in DRIVE range	a) Fluid level low	a) Correct fluid level
	b) Fluid pressure low	b) Refer to low fluid pressure causes
	c) Manual linkage mis-adjusted or disconnected	c) Repair or adjust manual linkage
	d) Torque converter loose on flex plate or internal converter damage	d) Verify malfunction and repair as required
	e) Oil pump or drive shaft damaged	e) Repair or replace oil pump and/or drive shaft
	f) Number 13 check ball mis-assembled or missing	f) Correct or install number 13 check ball in its proper location
	g) Damaged drive link chain, sprocket or bearings	g) Replace damaged components
	h) Burned or missing clutch plates, damaged piston seals or piston, Housing check ball damaged, input shaft seals or feed passages blocked or damaged	h) Repair and/or replace damaged input clutch assembly components

CHILTON'S THREE "C's" DIAGNOSIS CHART
THM 440-T4 Automatic Transaxle

Condition	Cause	Correction
No drive in DRIVE range	i) Input sprag and/or input sun gear assembly improperly assembled or sprag damaged	i) Correctly assemble or replace input sprag and input sun gear assembly
	j) Pinions, sun gear or internal gears damaged on input and reaction carrier assemblies	j) Repair or replace carrier assemblies as required
	k) 1-2 band or servo burned or damaged. Band apply pin incorrect in length	k) Repair or replace band and/or servo components as required
	l) 1-2 servo oil pipes leaking fluid	l) Correct oil tubes to prevent leakage
	m) Final drive assembly broken or damaged	m) Repair or replace necessary components of final drive
	n) Parking pawl spring broken	n) Replace parking pawl spring
	o) Output shaft damage, broken or misassembled	o) Repair, replace or re-assemble output shaft
First speed only, no 1-2 shift	a) Governor assembly defective	a) Repair or replace governor
	b) Number 14 check ball missing	b) Install number 14 check ball
	c) 1-2 shift valve sticking or binding	c) Repair or clean valve and bore
	d) Accumulator and/or pipes	d) Repair/renew components
	e) 2nd clutch assembly damaged	e) Repair/renew components
	f) Oil seal rings damaged on driven sprocket support	f) Replace oil seal rings
	g) Splines damaged or parts missing from reverse reaction drum	g) Repair or replace damaged components
Harsh or soft 1-2 shift	a) Fluid pressure	a) Check pressure and correct
	b) Defective accumulator assembly	b) Repair or replace accumulator assembly
	c) Accumulator valve stuck	c) Repair or clean valve and bore
	d) Missing or mislocated number 8 check ball	d) Install or re-locate number 8 check ball
High or low 1-2 shift speed	a) Disconnected or misadjusted T.V. cable	a) Connect and/or adjust T.V. cable
	b) Bent or damaged T.V. link, lever and bracket assembly	b) Repair or replace damaged components
	c) Stuck or binding T.V. valve and plunger	c) Correct binding condition or remove stuck valve and plunger
	d) Incorrect governor pressure	d) Correct governor pressure
No 2-3 upshift, 1st and 2nd speeds only	a) Defective 1-2 servo or components	a) Repair or replace 1-2 servo assembly
	b) Defective number 7 check ball and capsule assembly	b) Check, repair or replace check ball and capsule
	c) Number 11 check ball not seating	c) Check, repair or replace check ball
	d) 2-3 shift valve stuck in control valve assembly	d) Remove stuck 1-2 shift valve and repair
	e) Seals damaged or blocked passages on input shaft	e) Replace seals and open passages
	f) Defective third clutch assembly	f) Overhaul third clutch assembly

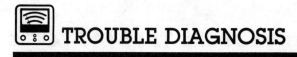

CHILTON'S THREE "C's" DIAGNOSIS CHART
THM 440-T4 Automatic Transaxle

Condition	Cause	Correction
No 2-3 upshift, 1st and 2nd speeds only	g) Defective third roller clutch assembly h) Numbers 5, 6 and/or accumulator valve stuck	g) Inspect, repair or replace necessary components h) Repair as required
Harsh or soft 2-3 shift	a) Fluid pressure b) Mislocated number 12 check ball	a) Test and correct fluid pressure b) Correct check ball location
High or low 2-3 shift speed	a) Disconnected or misadjusted T.V. cable b) Bent or damaged T.V. link, lever and bracket assembly c) Stuck or binding T.V. valve and plunger d) Incorrect governor pressure	a) Connect and/or adjust T.V. cable b) Repair or replace damaged components c) Correct binding condition or remove stuck valve and plunger d) Correct governor pressure
No 3-4 shift	a) Incorrect governor pressure b) 3-4 shift valve stuck in control valve assembly c) Defective 4th clutch assembly d) Spline damage to 4th clutch shaft	a) Correct governor pressure b) Free 3-4 shift valve and repair control valve assembly c) Overhaul 4th clutch assembly d) Replace 4th clutch shaft
Harsh or soft 3-4 shift	a) Fluid pressure b) Defective accumulator assembly c) Mislocated number 1 check ball	a) Test and correct fluid pressure b) Repair or replace accumulator assembly c) Correct check ball location
High or low 3-4 shift	a) Disconnected or misadjusted T.V. cable b) Bent or damaged T.V. link, lever and bracket assembly c) Stuck or binding T.V. valve and plunger d) Incorrect governor pressure	a) Connect and/or adjust T.V. cable b) Repair or replace damaged components c) Correct binding condition or remove stuck valve and plunger d) Correct governor pressure
No converter clutch apply (Vehicles equipped with E.C.M.)	a) Improper E.C.M. operation b) Electrical system of transaxle malfunctioning c) Converter clutch apply valve stuck d) Number 10 check ball missing e) Converter clutch blow-off check ball not seating or damaged f) Seals damaged on turbine shaft g) Damaged seal on oil pump drive shaft	a) Verify proper E.C.M. operation b) Test and correct electrical malfunction c) Free converter clutch apply valve and repair d) Install missing check ball e) Inspect channel plate and check ball. Repair as required f) Replace seals and inspect shaft g) Replace seal on oil pump drive shaft
No converter clutch apply (vehicles not equipped with E.C.M.)	a) Electrical system of transaxle malfunctioning b) Converter clutch shift and/or apply valves stuck c) Number 10 check ball missing	a) Test and correct electrical malfunction b) Free converter clutch shift and/or apply valves c) Install missing check ball

CHILTON'S THREE "C's" DIAGNOSIS CHART
THM 440-T4 Automatic Transaxle

Condition	Cause	Correction
No converter clutch apply (vehicles not equipped with E.C.M.)	d) Converter clutch blow-off check ball not seated or damaged e) Seals damaged on turbine shaft f) Damaged oil seal on seal pump drive shaft	d) Inspect channel plate and check ball. Repair as required e) Replace seals and inspect shaft f) Replace seal on oil pump drive shaft
Converter clutch does not release	a) Converter clutch apply valve stuck in the apply position	a) Free apply valve for converter clutch and repair as required
Rough converter clutch apply	a) Converter clutch regulator valve stuck b) Converter clutch accumulator piston or seal damaged c) Seals damaged on turbine shaft	a) Free converter clutch regulator valve and repair as required b) Replace seal or piston as required Check accumulator spring c) Replace seals on turbine shaft
Harsh 4-3 downshift	a) Number 1 check ball missing in control valve assembly	a) Install number 1 check ball in control valve assembly
Harsh 3-2 downshift	a) 1-2 servo control valve stuck b) Number 12 check ball missing c) 3-2 control valve stuck d) Number 4 check ball missing e) 3-2 coast valve stuck f) Input clutch accumulator piston or seal damaged	a) Free 1-2 servo control valve b) Install number 12 check ball c) Free 3-2 control valve d) Install number 4 check ball e) Free 3-2 coast valve f) Replace input clutch accumulator piston and/or seal
Harsh 2-1 downshift	a) Number 8 check ball missing	a) Install number 8 check ball
No reverse	a) Fluid pressure b) Defective oil pump c) Broken, stripped or defective drive link assembly d) Reverse band burned or damaged e) Defective input clutch f) Defective input sprag g) Piston or seal damaged, pin selection incorrect for rear servo assembly h) Defective input and reaction carriers	a) Test and correct fluid pressure b) Test and correct oil pump malfunction c) Repair, replace as required d) Replace rear band as required e) Repair, replace defective input clutch components f) Replace defective sprag g) Repair rear servo as required and install correct pin if needed h) Replace input and reaction carriers as required
No park range	a) Parking pawl, spring or parking gear damaged b) Manual linkage broken or out of adjustment	a) Repair as required b) Repair linkage or adjust as required
Harsh shift from Neutral to Drive or from Neutral to Reverse	a) Number 9 check ball missing b) Number 12 check ball missing c) Thermal elements not closing when warm	a) Install number 9 check ball b) Install number 12 check ball c) Replace thermal elements
No viscous clutch apply (Vehicles with E.C.M.)	a) Improper E.C.M. operation b) Damaged thermister c) Damaged temperature switch	a) Verify E.C.M. operation and repair as required b) Replace thermister c) Replace temperature switch

PRELIMINARY CHECK PROCEDURE

CHECK TRANSMISSION OIL LEVEL
CHECK AND ADJUST T.V. CABLE
CHECK OUTSIDE MANUAL LINKAGE AND CORRECT
CHECK ENGINE TUNE
INSTALL OIL PRESSURE GAGE*
CONNECT TACHOMETER TO ENGINE

CHECK OIL PRESSURES IN THE FOLLOWING MANNER:

Minimum T.V. Line Pressure Check
Set the T.V. cable to specification; and with the brakes applied, take the line pressure readings in the ranges and at the engine r.p.m.'s indicated in the chart below.

Full T.V. Line Pressure Check
Full T.V. line pressure readings are obtained by tying or holding the T.V. cable to the full extent of its travel; and with the brakes applied, take the line pressure readings in the ranges and at the engine r.p.m.'s indicated in the chart below.

NOTICE	Total running time for this combination not to exceed 2 minutes.

CAUTION	Brakes must be applied at all times.

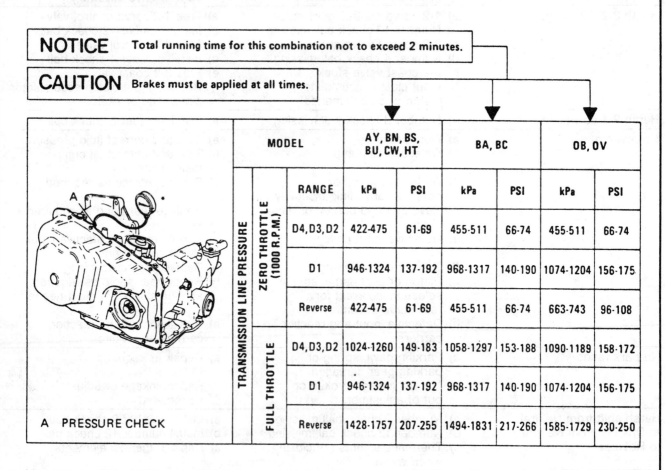

A PRESSURE CHECK

	MODEL		AY, BN, BS, BU, CW, HT		BA, BC		OB, OV	
		RANGE	kPa	PSI	kPa	PSI	kPa	PSI
ZERO THROTTLE (1000 R.P.M.)		D4,D3,D2	422-475	61-69	455-511	66-74	455-511	66-74
		D1	946-1324	137-192	968-1317	140-190	1074-1204	156-175
		Reverse	422-475	61-69	455-511	66-74	663-743	96-108
FULL THROTTLE		D4,D3,D2	1024-1260	149-183	1058-1297	153-188	1090-1189	158-172
		D1	946-1324	137-192	968-1317	140-190	1074-1204	156-175
		Reverse	1428-1757	207-255	1494-1831	217-266	1585-1729	230-250

(left axis label: TRANSMISSION LINE PRESSURE)

Line pressure is basically controlled by pump output and the pressure regulator valve. In addition, line pressure is boosted in Reverse, Intermediate and Lo by the reverse boost valve.

Also, in the Neutral, Drive and Reverse positions of the selector lever, the line pressure should increase with throttle opening because of the T.V. system. The T.V. system is controlled by the T.V. cable, the throttle lever and bracket assembly and the T.V. link, as well as the control valve pump assembly.

Operation Principles

The torque converter smoothly couples the engine to the planetary gears and the overdrive unit through fluid and hydraulically/mechanically provides additional torque multiplication when required. The combination of the compound planetary gear set provides four forward gear ratios and one reverse. The changing of the gear ratios is fully automatic in relation to the vehicle speed and engine torque. Signals of vehicle speed and engine torque are constantly being directed to the transaxle control valve assembly to provide the proper gear ratio for maximum engine efficiency and performance at all throttle openings.

Quadrant Positions

The quadrant has seven positions indicated in the following order: P, R, N, Ⓓ, D, 2, 1.

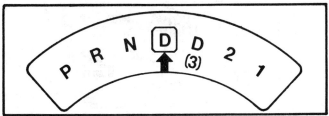

Quadrant for shift selector (© General Motors Corp.)

P—Park position enables the transaxle output shaft to be held, thus preventing the vehicle from rolling either forward or backward. (For safety reasons, the vehicle parking brake should be used in addition to the transaxle "Park" position). Because the output shaft is mechanically locked, the park position should not be selected until the vehicle has come to a stop. The engine may be started in the Park position.

R—Reverse enables the vehicle to be operated in a rearward direction.

N—Neutral position enables the engine to be started and operated without driving the vehicle. If necessary, this position must be selected if the engine has to be restarted while the vehicle is moving.

Ⓓ—Drive Range is used for most highway driving conditions and maximum economy. Drive Range has four gear ratios, from the starting ratio, through direct drive to overdrive. Downshifts to a higher ratio are available for safe passing by depressing the accelerator.

D—Manual Third can be used for conditions where it is desired to use only three gears. This range is also useful for braking when descending slight grades. Upshifts and downshifts are the same as in Drive Range for first, second and third gears, but the transaxle will not shift to fourth gear.

2—Manual Second adds more performance. It has the same starting ratio as Manual Third Range, but prevents the transaxle from shifting above second gear, thus retaining second gear for acceleration or engine braking as desired. Manual Second can be selected at any vehicle speed. If the transaxle is in third or fourth gear it will immediately shift to Second Gear.

1—Manual Lo can be selected at any vehicle speed. The transaxle will shift to second gear if it is in third or fourth gear, until it slows below approximately 40 mph. (64 km/h), at which time it will downshift to first gear. This is particularly beneficial for maintaining maximum engine braking when descending steep grades.

Torque Converter

The torque converter assembly serves three primary functions. First, it acts as a fluid coupling to smoothly connect engine power through oil to the transaxle gear train. Second, it multiplies the torque or twisting effort from the engine when additional performance is desired. Thirdly, it provides direct drive through the torque converter.

The torque converter assembly consists of a three-element torque converter combined with a friction clutch. The three elements are the pump (driving member), the turbine (driven or output member), and the stator (reaction member). The converter cover is welded to the pump to seal all three members in an oil filled housing. The converter cover is bolted to the engine flexplate which is bolted directly to the engine crankshaft. The converter pump is therefore mechanically connected to the engine and turns at engine speed whenever the engine is operating.

The stator is located between the pump and turbine and is mounted on a one-way roller clutch which allows it to rotate clockwise but not counterclockwise.

The purpose of the stator is to redirect the oil returning from the turbine and change its direction of rotation back to that of the pump member. The energy in the oil is then used to assist the engine in turning the pump. This increases the force of the oil driving the turbine; and as a result, multiples the torque or twisting force of the engine.

The force of the oil flowing from the turbine to the blades of the stator tends to rotate the stator counterclockwise, but the roller clutch prevents it from turning.

With the engine operating at full throttle, transaxle in gear, and the vehicle standing still, the converter is capable of multiplying engine torque by approximately 2.0:1.

As turbine speed and vehicle speed increases, the direction of the oil leaving the turbine changes. The oil flows against the rear side of the stator vanes in a clockwise direction. Since the stator is now impeding the smooth flow of oil in the clockwise direction, its roller clutch automatically releases and the stator revolves freely on its shaft.

CONVERTER CLUTCH

The converter clutch mechanically connects the engine to the drive train and eliminates the hydraulic slip between the pump and turbine. When accelerating, the stator gives the converter the capability to multiply engine torque (maximum torque multiplication is 2.0:1) and improve vehicle acceleration. At this time the converter clutch is released. When the vehicle reaches cruising speed, the stator becomes inactive and there is no multiplication of engine torque (torque multiplication is 1.0:1). At this time the converter is a fluid coupling with the turbine turning at almost the same speed as the converter pump. The converter clutch can now be applied to eliminate the hydraulic slip between the pump and turbine and improve fuel economy.

The converter clutch cannot apply in Park, Reverse, Neutral, and Drive Range—First Gear.

On those vehicles equipped with computer command control, converter clutch operation is controlled by the solenoid. The solenoid is controlled by the brake switch, third clutch pressure switch, and the computer command control system.

On those vehicles not equipped with computer command control, converter clutch operation is controlled by the converter clutch shift valve and the solenoid together. When the converter clutch shift valve is used, it is used in place of the converter clutch plug.

Converter Clutch Applied

The converter clutch is applied when oil pressure is exhausted between the converter pressure plate and the converter cover and pressure is applied to push the pressure plate against the converter cover.

When the solenoid is energized converter signal oil, from the converter clutch plug, pushes the converter clutch apply valve to the left. Converter clutch release oil then exhausts, and converter clutch apply oil from the converter clutch regulator valve pushes the converter pressure plate against the converter cover to apply the converter clutch.

Converter Clutch Released

The converter clutch is released when oil pressure is applied between the converter cover and the converter pressure plate.

Converter feed pressure from the pressure regulator valve passes through the converter clutch apply valve into the release passage. The release oil feeds between the pump shaft and the turbine shaft to push the pressure plate away from the converter cover to release the converter clutch.

Converter Clutch Apply Feel

Converter clutch apply feel is controlled by the converter clutch regulator valve and the converter cutch accumulator.

The converter clutch regulator valve is controlled by T.V. and controls the pressure that applies the converter clutch. The converter clutch accumulator absorbs converter feed oil as converter release oil is exhausting. Less oil is then fed to the converter clutch regulator valve and the apply side of the converter clutch. This gives a cushion to the converter clutch apply (Figure 53).

VISCOUS CONVERTER CLUTCH

Viscous Converter Clutch Applied

The viscous converter clutch performs the same function as the torque converter clutch that is explained in this section. The primary difference between the converter clutch and the viscous converter clutch is the silicone fluid that is sealed between the cover and the body of the clutch assembly. The viscous silicone fluid provides a smooth apply of the clutch assembly when it engages with the converter cover.

When the viscous clutch is applied there is a constant but minimal amount of slippage between the rotor and the body. Despite this slippage (approx. 40 rpm @ 60 mph) good fuel economy is attained at highway torque. The viscous converter clutch is controlled by the solenoid and through the ECM which monitors: vehicle speed; throttle angle; transmission gear; transmission fluid temperature; engine coolant temperature; outside temperature and barometric pressure.

Viscous Converter Clutch Apply

The viscous converter clutch is capable of applying at approximately 25 mph providing that the transmission is in second gear and the transmission oil temperature is below 93.3° C (200° F). (This temperature is monitored by the ECM through the thermistor.) When transmission oil temperatures are above 93.3° C (200° F) but below 157° C (315° F) the viscous clutch will not apply until approximately 38 mph. If transmission oil temperature exceeds 157° C (315° F) a temperature switch located in the channel plate will open and release the viscous clutch to protect the transmission from overheating.

Description of Hydraulic Components

MANUAL VALVE: Mechanically connected to the shift selector. It is fed by line pressure from the pump and directs pressure according to which range the driver has selected.

1-2 SERVO: A hydraulic piston and pin that mechanically applies the 1-2 band in first and second gear. Also absorbs third clutch oil to act as an accumulator for the 2-3 shift.

REVERSE SERVO: A hydraulic piston and pin that mechanically applies the reverse band when reverse range is selected by the driver.

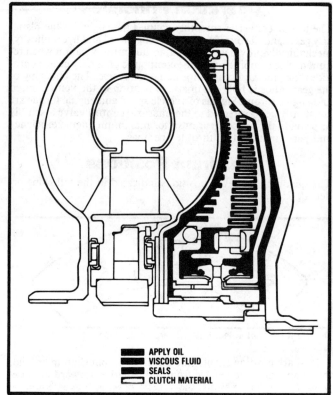

Viscous converter clutch applied (© General Motors Corp.)

Legend:
APPLY OIL
VISCOUS FLUID
SEALS
CLUTCH MATERIAL

MODULATOR VALVE: Is controlled by the vacuum modulator assembly and regulates line pressure, into a modulator pressure that is proportional to engine vacuum (engine torque).

MODULATOR ASSEMBLY: By sensing engine vacuum, it causes the modulator valve to regulate modulator pressure that is proportional to engine torque (inversely proportional to engine vacuum).

1-2 ACCUMULATOR PISTON: Absorbs second clutch oil to provide a cushion for the second clutch apply. The firmness of the cushion is controlled by the 1-2 accumulator valve.

3-4 ACCUMULATOR PISTON: Absorbs fourth clutch oil to provide a cushion for the fourth clutch apply. The firmness of the cushion is controlled by the 1-2 accumulator valve.

1-2 SERVO THERMO ELEMENTS: When cold, it opens another orifice to the servo, to provide less of a restriction for a quick servo apply. When warm, it blocks one of the two orifices to the servo and slow the flow of oil and provide a good neutral/drive shift feel.

INPUT CLUTCH ACCUMULATOR: Absorbs input clutch apply oil to cushion the input clutch apply.

CONVERTER CLUTCH ACCUMULATOR: Cushions the converter clutch apply by absorbing converter clutch feed oil and slowing the amount of oil feeding into the converter clutch apply passage.

3-2 DOWNSHIFT VALVE: Controlled by 2nd clutch oil that opens the valve when line pressure exceeds 110 psi and allows T.V. oil to enter the 3-2 downshift passage.

THERMO ELEMENT: Maintains a level of transmission fluid in the side cover that is needed for the operation of the hydraulic pressure system. The thermo element allows fluid levels to increase or decrease with the increase or decrease of fluid temperature.

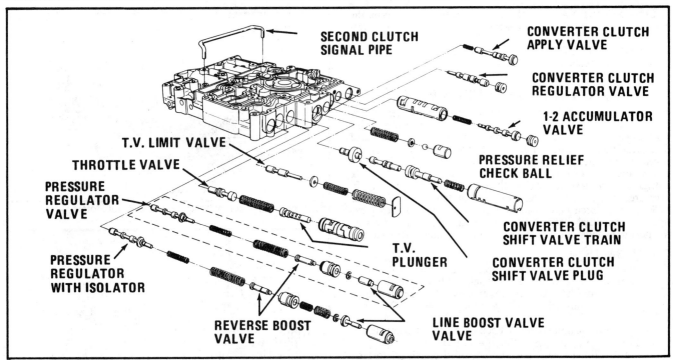

Exploded view of the second side of the valve body assembly (© General Motors Corp.)

1-2 SHIFT VALVE TRAIN: Shifts the transaxle from 1st to 2nd gear or 2nd to 1st gear, depending on governor, T.V., detent, or low oil pressures.

3-4 M.T.V. VALVE: Modulates T.V. pressure going to the 3-4 throttle valve to a lower pressure so that a light throttle 3-4 up-shift will not be delayed.

2-3 ACCUMULATOR VALVE: Receives line pressure from the manual valve and controlled by modulator pressure. The 2-3 accumulator valve, in third gear and overdrive, varies 1-2 servo. Apply (2-3 accumulator) oil pressure in proportion to changes in modulator pressure (engine torque).

3-2 CONTROL VALVE: Controlled by governor oil, it controls the 3-2 downshift timing by regulating the rate at which the third clutch releases and the 1-2 band applies.

2-3 SHIFT VALVE TRAIN: Shifts the transaxle from 2nd to 3rd gear or 3rd to 1st gear, depending on governor T.V., detent or drive 2 oil pressures.

3-4 SHIFT VALVE TRAIN: Shifts the transaxle from 3rd to 4th gear or 4th to 3rd gear, depending on governor, 3-4 M.T.V., 4-3 M.T.V., part throttle, or drive 3 oil pressures.

4-3 M.T.V. VALVE: Modulates T.V. pressure going to the 3-4 throttle valve to a lower pressure to prevent an early downshift at light to medium throttle.

REVERSE SERVO BOOST VALVE: Under hard acceleration the higher line pressure will open the valve to provide a quick feed to the reverse servo and prevent the reverse band from slipping during application.

1-2 SERVO CONTROL VALVE: Closed by second oil during a drive range 3-2 downshift, it slows down the 1-2 servo apply.

1-2 SERVO BOOST VALVE: Under hard acceleration the higher line pressure will open the valve to provide a quick feed to the 1-2 servo and prevent the 1-2 band from slipping during application.

CONVERTER CLUTCH APPLY VALVE: Controlled by the converter clutch solenoid, it directs oil to either the release or the apply side of the converter clutch.

CONVERTER CLUTCH REGULATOR VALVE: Controlled by T.V. pressure and fed by converter clutch feed pressure it regulates converter clutch apply pressure.

1-2 ACCUMULATOR VALVE: Receives line pressure from the manual valve and controlled by modulator pressure. It varies 1-2 and 3-4 accumulator pressure in proportion to changes in modulator pressure (engine torque).

PRESSURE RELIEF CHECK BALL: Prevents line pressure from exceeding 245-360 psi.

CONVERTER CLUTCH SHIFT VALVE PLUG: Allows second oil to feed into the converter clutch signal passage. The plug is used on models with vehicles equipped with computer command control.

CONVERTER CLUTCH SHIFT VALVE TRAIN (NON C3 SYSTEMS): Sends signal oil to the converter clutch apply valve and together with the converter clutch solenoid determines whether the converter clutch should be released or applied. It is controlled by governor, T.V. and detent oil.

T.V. LIMIT VALVE: Limits the line pressure fed to the throttle valve to 90 psi.

THROTTLE VALVE: A regulating valve that increases T.V. pressure as the accelerator pedal is depressed and is controlled by T.V. plunger movement.

T.V. PLUNGER: Controlled by the throttle lever and bracket assembly and linked to the accelerator pedal. When accelerating, this valve compresses the throttle valve spring causing the throttle valve to increase T.V. pressure. It also controls the opening of the part throttle and detent ports.

PRESSURE REGULATOR VALVE: Controls line pressure by regulating pump output and is controlled by the pressure regulator spring, the reverse boost valve, and the line boost valve.

PRESSURE REGULATOR VALVE WITH ISOLATOR: Same function as pressure regulator valve except isolator system assists in stabilizing the pressure regulator system.

REVERSE BOOST VALVE: Boosts line pressure by pushing the pressure regulator valve up when acted on by Park, Reverse, Neutral (PRN) oil or lo oil pressure.

LINE BOOST VALVE: Boosts line pressure by pushing the pressure regulator valve up when acted on by modulator oil pressure.

SECOND CLUTCH SIGNAL PIPE: Directs second clutch oil to apply or release the 1-2 control valve.

CLUTCH EXHAUST CHECK BALLS

To complete the exhaust of apply oil when the input, second, or third clutch is released, an exhaust check ball assembly is installed near the outer diameter of the clutch housings. Centrifugal force, resulting from the spinning clutch housings, working on the residual oil in the clutch piston cavity would give a partial apply of the clutch plates if it were not exhausted. The exhaust check ball assembly is designed to close the exhaust port by clutch apply pressure seating the check ball when the clutch is being applied.

When the clutch is released and clutch apply oil is being exhausted, centrifugl force on the check ball unseats it and opens the port to exhaust the residual oil from the clutch piston cavity.

CHECK BALLS

1. Fourth Clutch Check Ball: Forces fourth clutch oil to feed through one orifice and exhaust through a different orifice.

2. 3-2 Control Check Ball: Forces exhausting 1-2 servo release oil to either flow through an orifice or the regulating 3-2 control valve.

3. Part Throttle and Drive 3 Check Ball: Separates part throttle and drive 3 oil passages to the 3-4 shift valve.

4. Third Clutch Check Ball: Forces third clutch oil to feed through one orifice and exhaust through a different orifice.

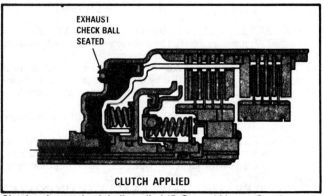

Clutch exhaust check ball applied (© General Motors Corp.)

5. 2-3 Accumulator Feed Check Ball: In third gear and fourth gear forces D4 oil to be orificed into the 1-2 servo (2-3 accumulator).

6. 2-3 Accumulator Exhaust Check Ball: In first gear, allows the 2-3 accumulator exhaust passage to feed and apply the 1-2 servo unrestricted. In third gear, forces exhausting 1-2 servo oil to either flow through an orifice or the regulating 2-3 accumulator valve.

7. Third Clutch Accumulator Ball and Spring: In third gear, it closes the 1-2 servo release passage exhaust. On a 3-2 down shift after 1-2 servo release oil has dropped to a low pressure, the spring will unseat the check ball and allow the oil to exhaust completely.

8. Second Clutch Check Ball: Forces second clutch oil to feed through one orifice and exhaust through a different orifice.

9. Reverse Servo Feed Check Ball: Forces oil feeding the reverse servo to orifice, but allows the oil to exhaust freely.

10. Converter Clutch Release/Apply Check Ball: Separates converter clutch release and converter clutch apply passages to the converter clutch blow-off ball.

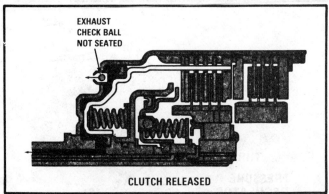

Clutch exhaust check ball released (© General Motors Corp.)

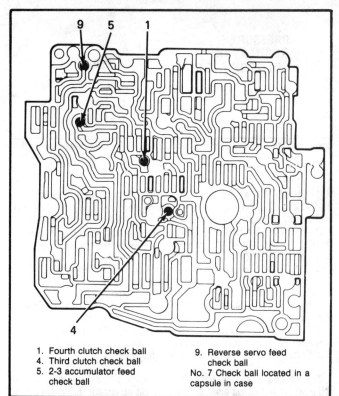

1. Fourth clutch check ball
4. Third clutch check ball
5. 2-3 accumulator feed check ball
9. Reverse servo feed check ball
No. 7 Check ball located in a capsule in case

Check ball location in valve body (© General Motors Corp.)

11. Third/Lo-1st Check Ball: Separates the third clutch and lo-1st passages to the third clutch.

12. 1-2 Servo Feed Check Ball: Forces oil feeding the apply side of the 1-2 servo to orifice but allows the oil to exhaust unrestricted.

13. Input Clutch/Reverse Check Ball: Minimizes neutral—drive and neutral—reverse apply time by allowing the park, reverse, neutral (PRN) circuit to feed and apply the input clutch quicker.

14. Detent/Modulator Check Ball: Allows detent oil to apply force to the pressure regulator system during part or full throttle detent and when driving at high altitude.

15. Converter Clutch Blow Off Check Ball: Prevents converter clutch release or apply pressure from exceeding 100 psi.

16. Low Blow Off Check Ball: Prevent lo-1st pressure to the third clutch from exceeding 70 psi in manual lo.

17. Cooler Check Ball: When the engine is shut off the spring seats the ball to prevent converter drainback.

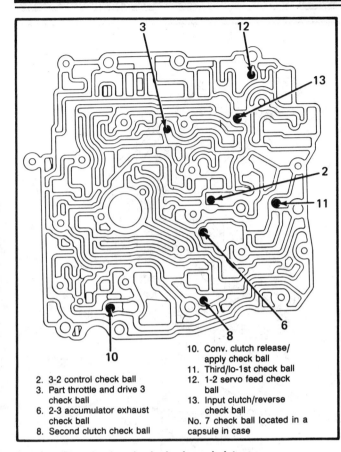

Check ball location in valve body channel plate
(© General Motors Corp.)

2. 3-2 control check ball
3. Part throttle and drive 3 check ball
6. 2-3 accumulator exhaust check ball
8. Second clutch check ball
10. Conv. clutch release/ apply check ball
11. Third/lo-1st check ball
12. 1-2 servo feed check ball
13. Input clutch/reverse check ball
No. 7 check ball located in a capsule in case

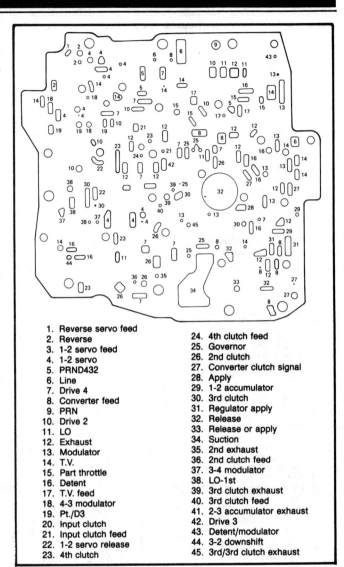

1. Reverse servo feed
2. Reverse
3. 1-2 servo feed
4. 1-2 servo
5. PRND432
6. Line
7. Drive 4
8. Converter feed
9. PRN
10. Drive 2
11. LO
12. Exhaust
13. Modulator
14. T.V.
15. Part throttle
16. Detent
17. T.V. feed
18. 4-3 modulator
19. Pt./D3
20. Input clutch
21. Input clutch feed
22. 1-2 servo release
23. 4th clutch
24. 4th clutch feed
25. Governor
26. 2nd clutch
27. Converter clutch signal
28. Apply
29. 1-2 accumulator
30. 3rd clutch
31. Regulator apply
32. Release
33. Release or apply
34. Suction
35. 2nd exhaust
36. 2nd clutch feed
37. 3-4 modulator
38. LO-1st
39. 3rd clutch exhaust
40. 3rd clutch feed
41. 2-3 accumulator exhaust
42. Drive 3
43. Detent/modulator
44. 3-2 downshift
45. 3rd/3rd clutch exhaust

Identification of spacer plate passages (© General Motors Corp.)

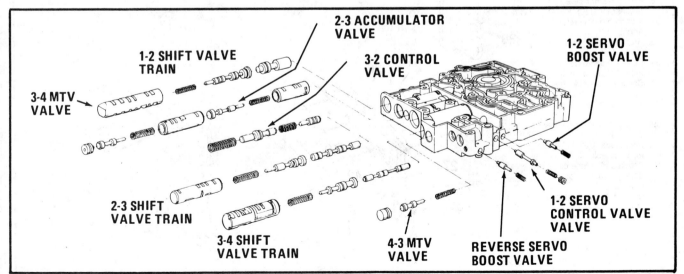

Exploded view one side of the valve body assembly (© General Motors Corp.)

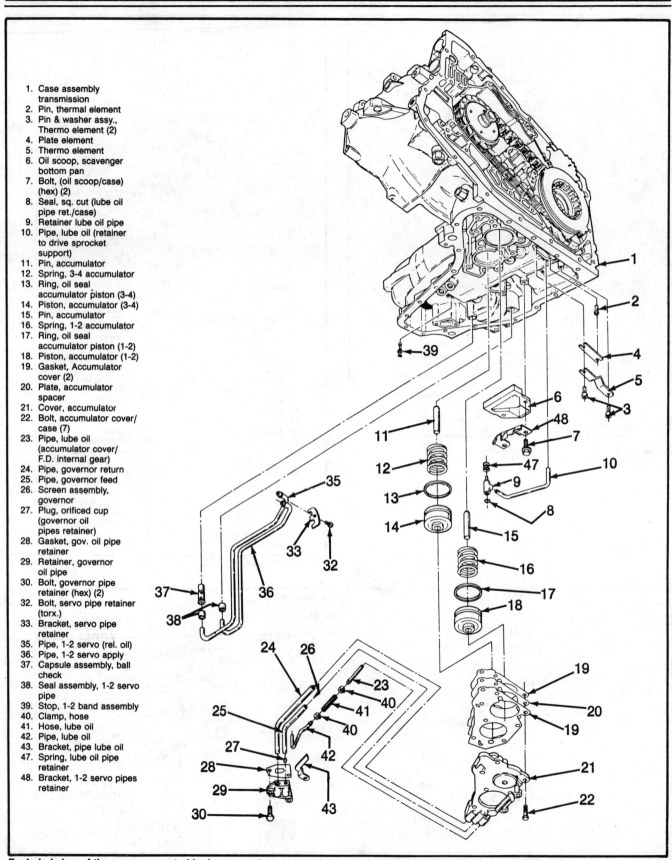

1. Case assembly transmission
2. Pin, thermal element
3. Pin & washer assy., Thermo element (2)
4. Plate element
5. Thermo element
6. Oil scoop, scavenger bottom pan
7. Bolt, (oil scoop/case) (hex) (2)
8. Seal, sq. cut (lube oil pipe ret./case)
9. Retainer lube oil pipe
10. Pipe, lube oil (retainer to drive sprocket support)
11. Pin, accumulator
12. Spring, 3-4 accumulator
13. Ring, oil seal accumulator piston (3-4)
14. Piston, accumulator (3-4)
15. Pin, accumulator
16. Spring, 1-2 accumulator
17. Ring, oil seal accumulator piston (1-2)
18. Piston, accumulator (1-2)
19. Gasket, Accumulator cover (2)
20. Plate, accumulator spacer
21. Cover, accumulator
22. Bolt, accumulator cover/ case (7)
23. Pipe, lube oil (accumulator cover/ F.D. internal gear)
24. Pipe, governor return
25. Pipe, governor feed
26. Screen assembly, governor
27. Plug, orificed cup (governor oil pipes retainer)
28. Gasket, gov. oil pipe retainer
29. Retainer, governor oil pipe
30. Bolt, governor pipe retainer (hex) (2)
32. Bolt, servo pipe retainer (torx.)
33. Bracket, servo pipe retainer
35. Pipe, 1-2 servo (rel. oil)
36. Pipe, 1-2 servo apply
37. Capsule assembly, ball check
38. Seal assembly, 1-2 servo pipe
39. Stop, 1-2 band assembly
40. Clamp, hose
41. Hose, lube oil
42. Pipe, lube oil
43. Bracket, pipe lube oil
47. Spring, lube oil pipe retainer
48. Bracket, 1-2 servo pipes retainer

Exploded view of the governor control body, accumulator cover, pistons and component parts (© General Motors Corp.)

1. Plate assembly, 4th clutch
2. Plate, 4th clutch apply
3. Bearing assembly, 4th clutch hub/channel plate
4. Bearing assembly, 4th clutch shaft
5. Hub & shaft assembly, 4th clutch
6. Washer, thrust (4th clutch hub/driven sprocket)
7. Ring, oil seal (shaft/sleeve) (2)
8. Washer, thrust (drive sprocket/channel plate)
9. Ring, snap (turbine shaft/drive sprocket)
10. Link assembly, drive
11. Sprocket, drive
12. Washer, thrust (drive sprocket/sprocket support)
13. Shaft, turbine
14. Ring, oil seal (turbine shaft/support)
15. Seal, "O" ring (turbine shaft/turbine hub)
16. Bearing assembly, drawn cup
17. Support, drive sprocket
18. Bushing, drive sprocket support
19. Sprocket, driven
20. Washer, thrust (driven sprocket/sprocket support)
21. Ring, snap (4th clutch ret. spring)
22. Spring asm., 4th clutch piston return
23. Piston, 4th clutch
24. Seals, 4th clutch piston
25. Case, transmission
26. Connector, cooler (1)
27. Pin, dowel
28. Vent assembly
29. Plug, pipe (line pressure)
32. Pin, anchor (1-2 band) (2)
33. Pin, anchor (reverse band) (2)
34. Plug, case servo (orifice)
35. Plug, cup (park lock-out)
36. Screw, nameplate
37. Nameplate
38. Plug, pipe (governor pressure)
39. Bushing, case
40. Seal assembly, axle oil
41. Helix seal assembly, (converter oil)
42. Screw, button head (4) case/drum sprocket)
43. Ring, servo cover retaining
44. Cover, servo (reverse)
45. Seal, "O" ring (cover to case)
46. Ring, snap (band apply pin)
47. Ring, oil seal piston
48. Piston, reverse servo
49. Spring, reverse servo cushion
50. Retainer, servo cushion spring
51. Pin, reverse apply
52. Spring, servo return
53. Ring, servo cover retaining
54. Cover, servo (1-2)
55. Seal, "O" ring (cover to case)
56. Ring, snap (band apply pin)
57. Ring, oil seal piston
58. Piston, 1-2 servo
59. Spring, 1-2 servo cushion
60. Retainer, servo cushion spring
61. Pin, 1-2 band apply
62. Spring, servo return
63. Spring, reverse servo curved
64. Ring, retaining (output shaft)
65. Shaft, output
66. Bearing, input sun gear/output shaft
67. Ring, snap (output shaft/differential)
68. Retainer, servo spring
69. Connector, cooler
70. Ball, connector cooler
71. Spring, connector cooler
72. Connector, inverted flared

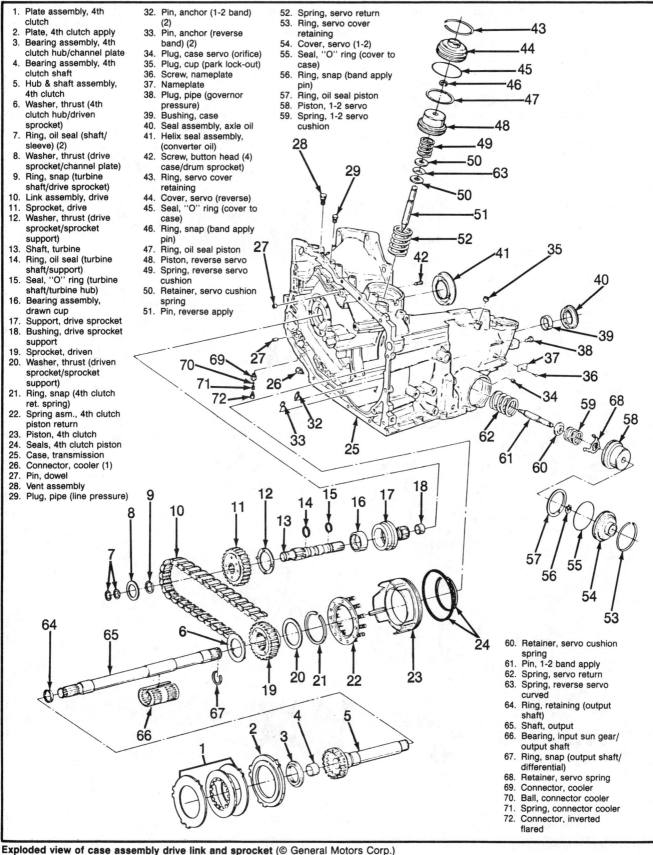

Exploded view of case assembly drive link and sprocket (© General Motors Corp.)

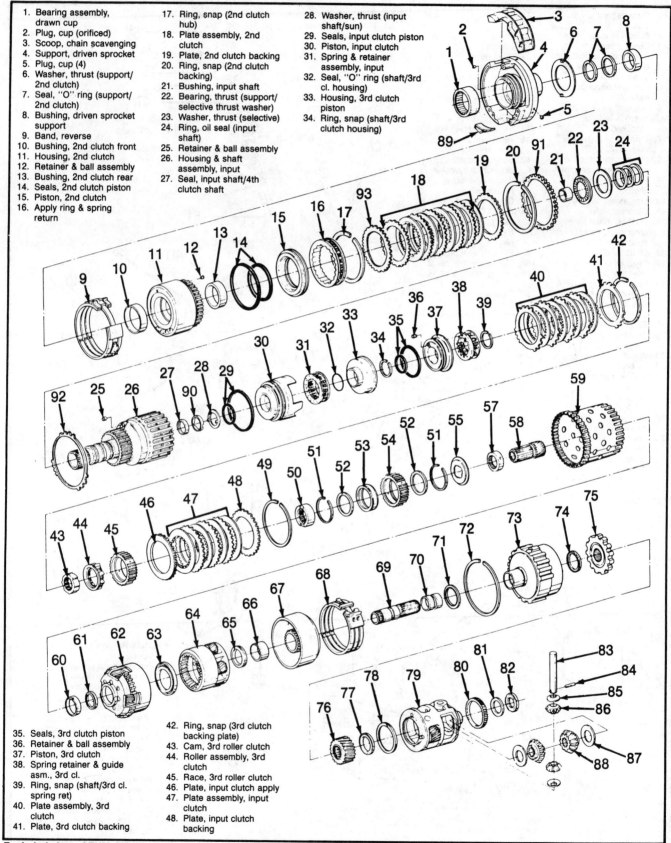

1. Bearing assembly, drawn cup
2. Plug, cup (orificed)
3. Scoop, chain scavenging
4. Support, driven sprocket
5. Plug, cup (4)
6. Washer, thrust (support/ 2nd clutch)
7. Seal, "O" ring (support/ 2nd clutch)
8. Bushing, driven sprocket support
9. Band, reverse
10. Bushing, 2nd clutch front
11. Housing, 2nd clutch
12. Retainer & ball assembly
13. Bushing, 2nd clutch rear
14. Seals, 2nd clutch piston
15. Piston, 2nd clutch
16. Apply ring & spring return

17. Ring, snap (2nd clutch hub)
18. Plate assembly, 2nd clutch
19. Plate, 2nd clutch backing
20. Ring, snap (2nd clutch backing)
21. Bushing, input shaft
22. Bearing, thrust (support/ selective thrust washer)
23. Washer, thrust (selective)
24. Ring, oil seal (input shaft)
25. Retainer & ball assembly
26. Housing & shaft assembly, input
27. Seal, input shaft/4th clutch shaft

28. Washer, thrust (input shaft/sun)
29. Seals, input clutch piston
30. Piston, input clutch
31. Spring & retainer assembly, input
32. Seal, "O" ring (shaft/3rd cl. housing)
33. Housing, 3rd clutch piston
34. Ring, snap (shaft/3rd clutch housing)

35. Seals, 3rd clutch piston
36. Retainer & ball assembly
37. Piston, 3rd clutch
38. Spring retainer & guide asm., 3rd cl.
39. Ring, snap (shaft/3rd cl. spring ret)
40. Plate assembly, 3rd clutch
41. Plate, 3rd clutch backing

42. Ring, snap (3rd clutch backing plate)
43. Cam, 3rd roller clutch
44. Roller assembly, 3rd clutch
45. Race, 3rd roller clutch
46. Plate, input clutch apply
47. Plate assembly, input clutch
48. Plate, input clutch backing

Exploded view of THM 440-T4 transaxle internal components (© General Motors Corp.)

49. Ring, snap (input clutch backing plate)
50. Race, input sprag inner
51. Ring, snap (sprag)
52. Wear plate, input sprag
53. Sprag assembly, input clutch
54. Race, input sprag outer
55. Retainer, input sprag
57. Spacer, input sun gear
58. Gear, input sun
59. Drum, reverse reaction
60. Bushing, reaction internal gear
61. Bearing assembly, (input sun/carrier)
62. Carrier assemby, input
63. Bearing asm., (input/reaction carrier)
64. Carrier assembly, reaction
65. Bearing assembly, (reaction carrier/sun gear)

66. Bushing, reaction sun
67. Gear & drum asm., reaction sun
68. Band, 1-2
69. Shaft, final drive sun gear
70. Bushing, final drive internal
71. Bearing assembly, reaction sun gear/internal gear
72. Ring, snap (internal gear/case)
73. Gear, final drive internal
74. Bearing asm., (int. gear/park gear)
75. Gear parking
76. Gear, final drive sun
77. Bearing, thrust (sun gear/carrier)
78. Ring, snap (final drive carrier)

79. Carrier, final drive
80. Gear, governor drive
81. Washer, carrier/case selective
82. Bearing asm., (selective washer/case)
83. Shaft, differential pinion
84. Pinion, differential pinion shaft ret.
85. Washer, pinion thrust
86. Pinion, differential
87. Washer, differential side gear thrust
88. Gear, differential side
89. Weir, oil reservoir
90. Sleeve, lock up
91. Support, 2nd clutch housing
92. Plate, reverse reaction drum
93. Plate, 2nd clutch waved

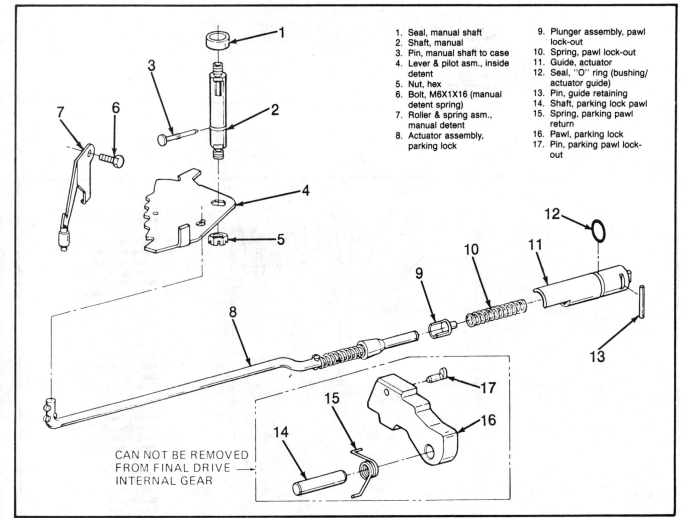

1. Seal, manual shaft
2. Shaft, manual
3. Pin, manual shaft to case
4. Lever & pilot asm., inside detent
5. Nut, hex
6. Bolt, M6X1X16 (manual detent spring)
7. Roller & spring asm., manual detent
8. Actuator assembly, parking lock
9. Plunger assembly, pawl lock-out
10. Spring, pawl lock-out
11. Guide, actuator
12. Seal, "O" ring (bushing/actuator guide)
13. Pin, guide retaining
14. Shaft, parking lock pawl
15. Spring, parking pawl return
16. Pawl, parking lock
17. Pin, parking pawl lock-out

CAN NOT BE REMOVED FROM FINAL DRIVE INTERNAL GEAR

Exploded view of manual linkage assembly (© General Motors Corp.)

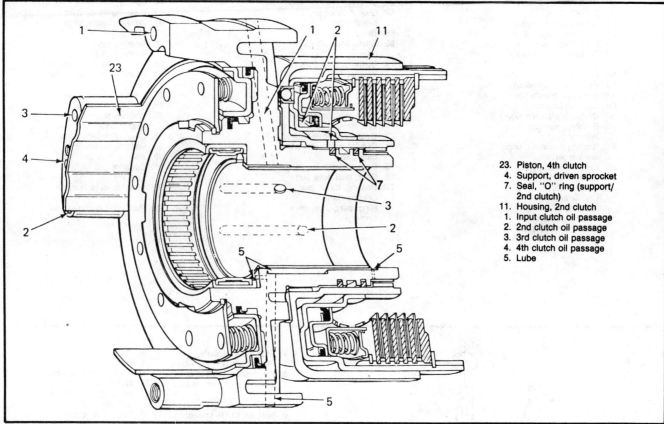

23. Piston, 4th clutch
4. Support, driven sprocket
7. Seal, "O" ring (support/
   2nd clutch)
11. Housing, 2nd clutch
1. Input clutch oil passage
2. 2nd clutch oil passage
3. 3rd clutch oil passage
4. 4th clutch oil passage
5. Lube

Cross section of the driven sprocket support assembly (© General Motors Corp.)

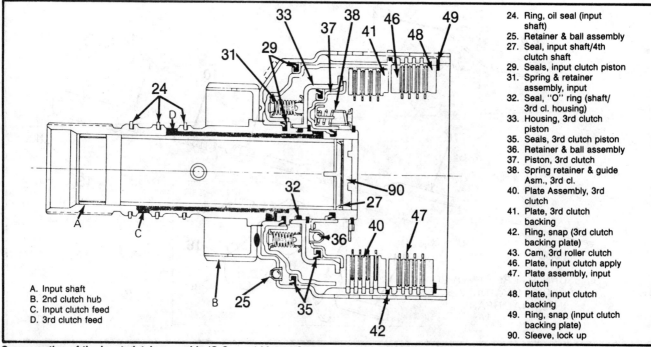

A. Input shaft
B. 2nd clutch hub
C. Input clutch feed
D. 3rd clutch feed

24. Ring, oil seal (input
   shaft)
25. Retainer & ball assembly
27. Seal, input shaft/4th
   clutch shaft
29. Seals, input clutch piston
31. Spring & retainer
   assembly, input
32. Seal, "O" ring (shaft/
   3rd cl. housing)
33. Housing, 3rd clutch
   piston
35. Seals, 3rd clutch piston
36. Retainer & ball assembly
37. Piston, 3rd clutch
38. Spring retainer & guide
   Asm., 3rd cl.
40. Plate Assembly, 3rd
   clutch
41. Plate, 3rd clutch
   backing
42. Ring, snap (3rd clutch
   backing plate)
43. Cam, 3rd roller clutch
46. Plate, input clutch apply
47. Plate assembly, input
   clutch
48. Plate, input clutch
   backing
49. Ring, snap (input clutch
   backing plate)
90. Sleeve, lock up

Cross section of the input clutch assembly (© General Motors Corp.)

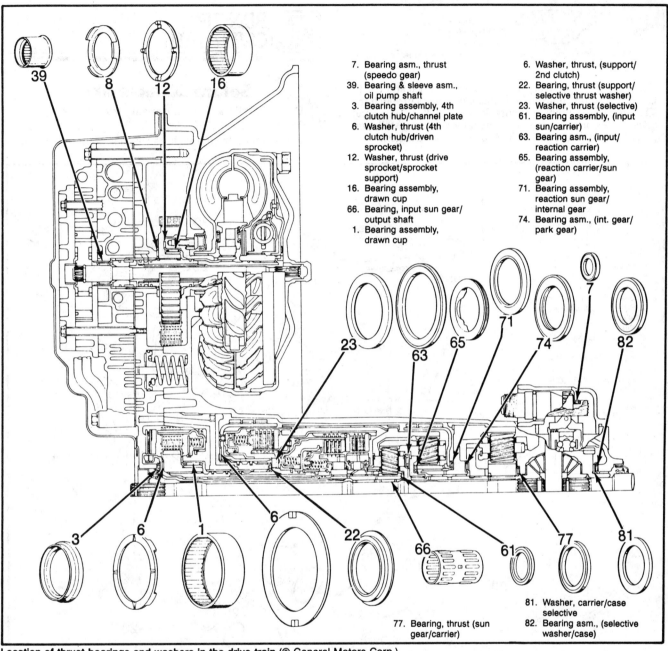

7. Bearing asm., thrust (speedo gear)
39. Bearing & sleeve asm., oil pump shaft
3. Bearing assembly, 4th clutch hub/channel plate
6. Washer, thrust (4th clutch hub/driven sprocket)
12. Washer, thrust (drive sprocket/sprocket support)
16. Bearing assembly, drawn cup
66. Bearing, input sun gear/output shaft
1. Bearing assembly, drawn cup

6. Washer, thrust, (support/2nd clutch)
22. Bearing, thrust (support/selective thrust washer)
23. Washer, thrust (selective)
61. Bearing assembly, (input sun/carrier)
63. Bearing asm., (input/reaction carrier)
65. Bearing assembly, (reaction carrier/sun gear)
71. Bearing assembly, reaction sun gear/internal gear
74. Bearing asm., (int. gear/park gear)

77. Bearing, thrust (sun gear/carrier)

81. Washer, carrier/case selective
82. Bearing asm., (selective washer/case)

Location of thrust bearings and washers in the drive train (© General Motors Corp.)

There has been no factory recommended removal and installation procedure published affecting the vehicles in which the THM 440-T4 automatic transaxle are installed. A general guide can be used by following the removal and installation procedures for the THM 125C and automatic transaxle.

Before Disassembly

NOTE: Cleanliness is an important factor in the overhaul of the transaxle.

Before opening up the transaxle, the outside of the unit should be

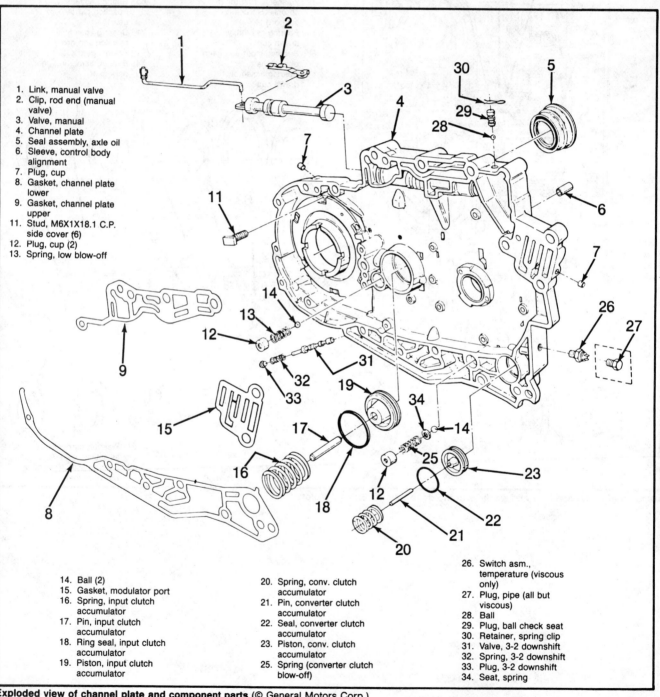

1. Link, manual valve
2. Clip, rod end (manual valve)
3. Valve, manual
4. Channel plate
5. Seal assembly, axle oil
6. Sleeve, control body alignment
7. Plug, cup
8. Gasket, channel plate lower
9. Gasket, channel plate upper
11. Stud, M6X1X18.1 C.P. side cover (6)
12. Plug, cup (2)
13. Spring, low blow-off

14. Ball (2)
15. Gasket, modulator port
16. Spring, input clutch accumulator
17. Pin, input clutch accumulator
18. Ring seal, input clutch accumulator
19. Piston, input clutch accumulator

20. Spring, conv. clutch accumulator
21. Pin, converter clutch accumulator
22. Seal, converter clutch accumulator
23. Piston, conv. clutch accumulator
25. Spring (converter clutch blow-off)

26. Switch asm., temperature (viscous only)
27. Plug, pipe (all but viscous)
28. Ball
29. Plug, ball check seat
30. Retainer, spring clip
31. Valve, 3-2 downshift
32. Spring, 3-2 downshift
33. Plug, 3-2 downshift
34. Seat, spring

Exploded view of channel plate and component parts (© General Motors Corp.)

1. Converter assembly
2. Bushing, converter pump
3. Sleeve, governor shaft
4. Ring, oil seal (governor shaft)
5. Governor assembly
6. Gear, speedometer drive
7. Bearing asm., thrust (speedo gear)
8. Seal, "O" ring (governor cover)
9. Cover, governor
10. Screw, governor cover/case
11. Gear, speedo driven
12. Seal, "O" ring
13. Sleeve, speedo driven gear
14. Retainer, speedo driven gear
15. Bolt, speedo gear retaining
16. Case assembly
17. Valve, modulator
18. Seal, "O" ring
19. Modulator assembly
20. Retainer, modulator
21. Bolt, (modulator)
23. Electrical connector
24. Channel, plate assembly
25. Ball, check valve (7)
26. Bolt, channel plate to case (5)
27. Bolt, channel plate to driven support (6)
28. Ring, oil seal (oil pump)
29. Shaft, oil pump drive
30. Gasket, spacer plate/channel plate
31. Plate, valve body spacer
32. Gasket, spacer plate/valve body
33. Screen asm., conv. clutch solenoid
34. Ball, check valve (5)
35. Valve assembly, control
38. Screen assembly, oil pump pressure
39. Bearing & sleeve asm., oil pump shaft
40. Bolt, V.B. to C.P. (torque head) (6)
41. Bolt, valve body to C.P. (hex) (1)
42. Bolt, V.B. to driven support (torque) (2)
43. Bolt, valve body to case (hex) (3)
44. Pump assembly
45. Bolt, pump body to case (hex) (2)
46. Bolt, pump cover to C.P. (hex) (10)
47. Bolt, pump cover to valve body (hex) (1)
48. Harness, wiring
49. Link throttle lever to cable
50. Lever & bracket assembly, throttle
51. Pan, case side cover
52. Screw, special M8x1.25x16.0
53. Nut, flanged hex (M6x1.0)
54. Bolt, M6x1.0x35 LG. P.B./C.P. hex (1)
55. Bolt, M6x1.0x45 LG. V.B./C.P. (2)
56. Washer, conical
57. Seal assembly, oil filter
58. Filter assembly, oil
59. Pan, transmission oil
60. Screw, special M8x1.25x16.0)
61. Wire conduit
62. Clip, two wire
63. Gasket, transmission oil pan
64. Gasket, side cover to case
65. Gasket, side cover to channel plate
66. Magnet, chip collector

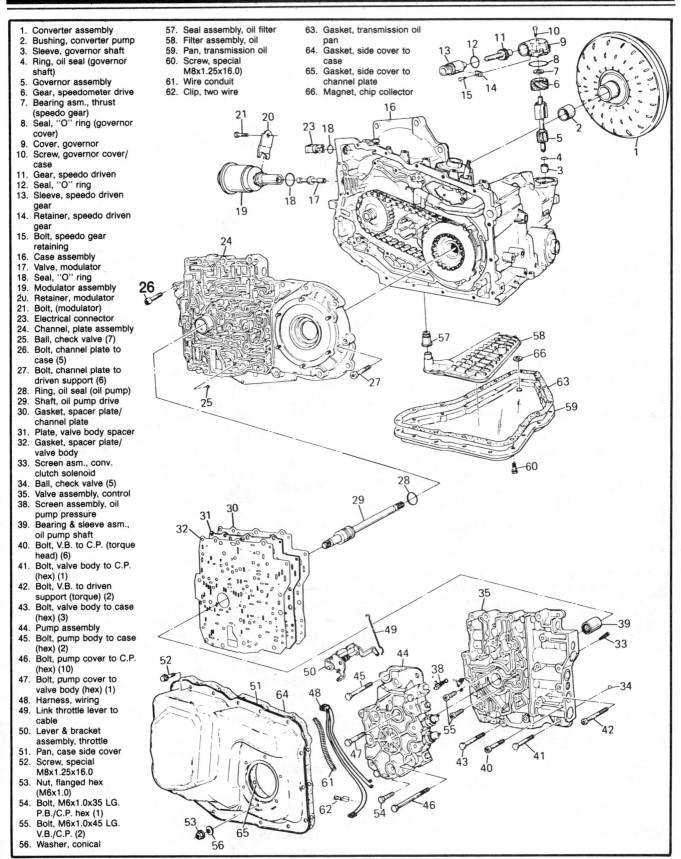

Exploded view of the THM 440-T4 automatic transaxle (© General Motors Corp.)

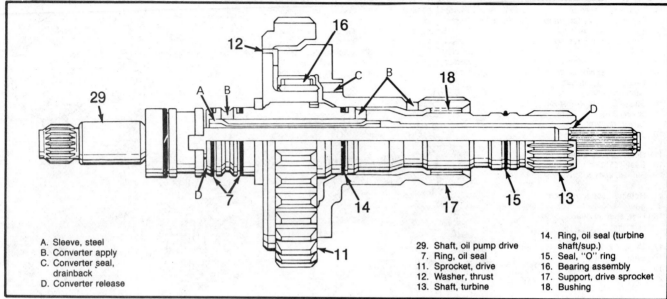

A. Sleeve, steel
B. Converter apply
C. Converter seal, drainback
D. Converter release

29. Shaft, oil pump drive
7. Ring, oil seal
11. Sprocket, drive
12. Washer, thrust
13. Shaft, turbine

14. Ring, oil seal (turbine shaft/sup.)
15. Seal, "O" ring
16. Bearing assembly
17. Support, drive sprocket
18. Bushing

Cross section of the pump drive shaft and drive sprocket support (© General Motors Corp.)

thoroughly cleaned, preferably with high-pressure cleaning equipment such as a car wash spray unit. Dirt entering the transaxle internal parts will negate all the effort and time spent on the overhaul. During inspection and reassembly, all parts should be thoroughly cleaned with solvent, then dried with compressed air. Wiping cloths and rags should not be used to dry parts since lint will find its way into valve body passages. Wheel bearing grease, long used to secure thrust washers and to lube parts should not be used. Lube seals with Dexron® II and use ordinary unmedicated petroleum jelly to hold thrust washers to ease assembly of seals, since it will not leave a harmful residue as grease often will. DO NOT use solvent on neoprene seals, friction plates or thrust washers. Be wary of nylon parts if the transaxle failure was due to a failure of the cooling system. Nylon parts exposed to antifreeze solutions can swell and distort and so must be replaced (Speedo gears, some thrust washers, etc.). Before installing bolts into aluminum parts, always dip the threads into clean oil. Anti-seize compound is also a good way to prevent the bolts from galling the aluminum and seizing. Always use a torque wrench to keep from stripping the threads. Take care of the seals when installing them, especially the smaller O-rings. The internal snap rings should be expanded and the external snap rings should be compressed, if they are to be reused. This will help insure proper seating when installed.

Transaxle Disassembly

TORQUE CONVERTER

Removal

1. Make certain that the transaxle is held securely.
2. Pull the converter straight out of the transaxle. Be careful since the converter contains a large amount of oil. There is no drain plug on the converter so the converter should be drained through the hub.

NOTE: The transaxle fluid that is drained from the converter can help diagnose transaxle problems.

1. If the oil in the converter is discolored but does not contain metal bits or particles, the converter is not damaged and need not be replaced. Remember that color is no longer a good indicator of transaxle fluid condition. In the past, dark color was associated with overheated transaxle fluid. It is not a positive sign of transaxle failure with the newer fluids like Dexron® II.
2. If the oil in the converter contains metal particles, the converter is damaged internally and must be replaced. The oil may have an "aluminum paint" appearance.
3. If the cause of oil contamination was due to burned clutch plates or overheated oil, the converter is contaminated and should be replaced.

OIL PAN

Removal

1. Remove the oil pan attaching bolts. Carefully bump the pan with a rubber mallet to free the pan. If the pan is pried loose instead, be very careful not to damage the gasket surfaces. Discard the pan gasket.
2. Inspect the bottom of the pan for debris that can give an indication of the nature of the transaxle failure.
3. Check the pan for distorted gasket flanges, especially around the bolt holes, since these are often dished-in due to over-torque. They can be straightened with a block of wood and a rubber mallet if necessary.

OIL FILTER

Removal

1. The oil filter is retained by a clip as well as the interference fit of the oil intake tube and O-ring. Move the clip out of the way and pull the filter from the case.
2. Discard the O-ring from the intake pipe. Often the O-ring will stick to its bore in the case.

3. The oil pressure regulator is in the bore next to the oil filter intake pipe opening. If it is to be removed, use snap ring pliers to remove the retaining ring and then pull the pressure regulator bushing assembly from the case.

GOVERNOR

Removal

1. Since the speedometer drive gear is attached to the governor assembly, first remove the speedometer driven gear attaching bolt and the retaining clip.

2. Remove the speedometer driven gear from the governor cover. Remove the governor cover and discard the O-ring.

3. Remove the governor assembly along with the speedometer drive gear thrust bearing.

4. Remove the modulator retainer and lift out the modulator. Discard the O-ring. Lift out the modulator valve using a magnet.

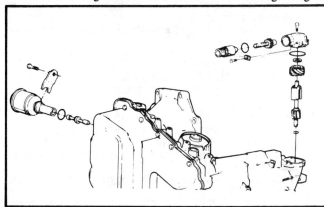

Governor and modulator assemblies—exploded view
(© General Motors Corporation)

INTERMEDIATE SERVO

Removal

1. To remove the intermediate servo assembly, the cover will have to be depressed to relieve the spring pressure on the snap ring that retains the cover. The factory type tool for this is similar to a clamp that hooks to the case and applies with a screw operated arm. The object is to release the spring holding pressure the cover has on the snap ring.

2. Pliers can be used to grasp the servo cover to remove it. Discard the seals. Make sure that the cover seal ring is not stuck in the case groove where it might be overlooked.

3. Remove the servo assembly and the servo return springs. Remove the apply pin from the servo assembly.

NOTE: The servo assemblies are not interchangeable due to the reverse servo pin being longer than the 1-2 pin. Keep the servo assemblies separate.

OIL PUMP

Removal

1. Disconnect the side cover attaching nuts and bolts and remove the side cover. Discard the gaskets.

2. Detach the solenoid wiring harness from the case connector and pressure switch. Remove the throttle valve assembly from the valve body.

3. Remove the oil pump bolts and lift the oil pump assembly from the valve body.

VALVE BODY

Removal

1. Remove the valve body retaining bolts. Remove the valve body from the channel plate.

2. Remove the check balls from the spacer plate. Detach the spacer plate and discard the gaskets.

3. Remove the check balls from the channel plate.

OIL PUMP SHAFT AND CHANNEL PLATE

Removal

1. Remove the oil pump shaft by sliding it out of the channel plate assembly. Place the detent lever in the park position and remove the manual valve clip.

2. Remove the channel plate attaching bolts and lift the channel plate from the transaxle case.

3. Remove the accumulator piston and the converter clutch accumulator piston and spring. Discard all gaskets.

FOURTH CLUTCH (C4) AND OUTPUT SHAFT

Removal

1. Remove the three (3) clutch plates along with the apply plate. Remove the thrust bearing.

NOTE: The thrust bearing may still be on the case cover from earlier disassembly.

2. Remove the front clutch hub and shaft assembly. This will expose the output shaft.

3. Rotate the output shaft until the output shaft C ring is visible. Remove the C-ring.

4. Pull the output shaft from the transaxle being careful not to damage it.

DRIVE LINK ASSEMBLY

Removal

1. Remove the turbine shaft O-ring located at the front of the unit.

2. Reach through the access holes in the sprockets and slip the snap rings from their grooves.

3. Remove the sprockets and the chain as an assembly. It will require alternately pulling on the sprockets until the bearings come out of their support housings.

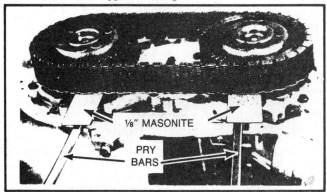

Removing tight sprockets (© General Motors Corporation)

NOTE: If the sprockets are difficult to remove, use two small pieces of masonite or similar material to act both as wedges and as pads for a pry bar. Do not pry on the chain or the case.

4. After removing the drivelink assembly, take note as to the position of the colored link on the chain. It should be facing out. Remove the thrust washers.

5. Lift out the drive sprocket support and remove the thrust washer located on the casing.

CHECK INPUT UNIT END-PLAY

The factory tools for checking the end-play on this unit are somewhat elaborate and it is unlikely that every shop will be equipped in the same manner. However, the object is to pre-load the output shaft to remove the clearance. The factory tool mounts on the output end of the transaxle and a knob and screw arrangement can be tightened, forcing the output shaft upwards. A dial indicator is mounted on the end of the input shaft. By raising the input shaft with a suitable bar (again, the factory tool is special and grips the input shaft splines) the input unit end-play can be measured. Input shaft end-play should be 0.020-0.042 in.

INPUT HOUSING AND SHAFT ASSEMBLY

Removal

1. Clamp the clutch and final drive tool to the second clutch housing. Lift the second clutch housing and the input shaft assembly out of the unit.
2. Remove the reverse band and the reverse reaction drum. Remove the input carrier assembly.

NOTE: The reverse band assembly may lift out with the second clutch housing.

3. Remove the thrust bearing and the reaction carrier. The thrust bearing is located at one end of the reaction carrier.
4. Remove the thrust bearing which may be stuck to the reaction carrier.
5. Remove the reaction sun gear and the drum assembly. Remove the 1-2 band.

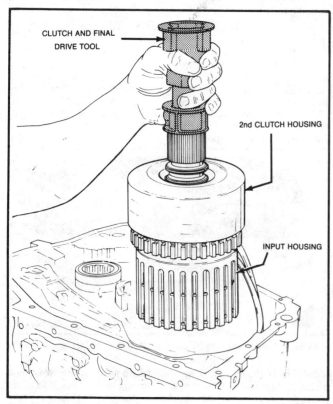

Input housing and shaft removal (© General Motors Corporation)

NOTE: The 1-2 band assembly should not be cleaned in cleaning solvent.

6. Remove the bearing ring and the sun gear shaft. A final drive internal bushing will be found on the sun gear shaft.

FINAL DRIVE ASSEMBLY

Removal

1. With a suitable tool, remove the snap ring at the head of the internal final drive gear.
2. Clamp the clutch and final drive tool to the final drive internal gear and lift it out.
3. The final drive carrier will be removed with the final drive internal gear.

MANUAL SHAFT/DETENT LEVER AND ACTUATOR ROD

Removal

1. It is not necessary to disassemble the manual shaft, detent lever or the actuator rod unless replacement is needed.
2. Remove the manual shaft and detent lever retaining bolts and remove the shaft and lever.
3. Remove the actuator rod assembly and check for wear or damage.
4. Remove the actuator guide assembly and check for damage.

MANUAL LINKAGE/ACTUATOR REPLACEMENT

Removal

1. Remove the manual shaft lock nut and pin and lift out the manual shaft with the detent lever.
2. Remove the retaining pin from the case and detach the actuator guide assembly.
3. Remove the O-ring from the actuator guide and discard the O-ring.

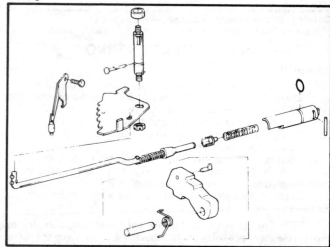

Manual linkage—exploded view (© General Motors Corporation)

4. The parking lock pawl assembly cannot be removed from the final drive internal gear.

NOTE: If the manual shaft seal is needed to be replaced, remove the axle oil seal along with the converter seal and then remove the manual shaft seal from its mounting in the transaxle case.

Unit Disassembly and Assembly

CASE

1. Clean the case well and inspect carefully for cracks. Make certain that all passages are clean and that all bores and snap ring

grooves are clean and free from damage. Check for stripped bolt holes. Check the case bushings for damage.

2. A new manual shaft seal can be installed at this time, along with a new axle oil seal and converter seal. Tap the seals into place with a suitable tool. The seal lips must face into the case.

3. Check the drive sprocket support assembly for damage. If it requires replacement, a slide hammer type puller can be used to pull the bearing from the sprocket support. Once the bearing is out, inspect the bore for wear or damage. The new bearing should be drive in straight and with care so as not to damage it.

4. If the parking pawl and related parts are to be removed, begin by turning the transaxle to the oil pan side up. Use a punch to remove the cup plug. Remove the parking pawl shaft retainer, then the shaft, pawl and return spring. Check the pawl for cracks.

FINAL DRIVE UNIT

Disassembly

1. Remove the final drive gear snap ring and lift out the final drive gear unit. Lift out the bearing assembly and the parking gear.

2. Remove the final drive sun gear along with the final drive carrier and the governor drive gear.

3. Remove the carrier washer and the bearing assembly.

4. Place the final drive carrier on its side and remove the differential pinion shaft by tapping out the pinion shaft retaining pin.

5. Remove the pinion thrust washer and the differential pinion. Remove the two side gear thrust washers and the two differential side gears.

Inspection

1. Clean all parts well and check for damage or excessive wear. Check the gears for burrs and cracks.

2. Inspect the washers and replace any that appear damaged or warped.

3. Check the bearing assemblies for any mutilation and replace as needed.

Assembly

1. Install the differential side gears and washers into the final drive carrier.

2. Install the pinion thrust washers onto the differential pinions. A small amount of petroleum jelly can be used to hold the washers in place.

3. Place the pinions and washers into the final drive carrier.

4. Insert the differential pinion shaft into the final drive carrier to check alignment of the pinions, then remove.

5. If the pinions are out of alignment, correct and reinstall the differential pinion shaft. Tap the pinion shaft retaining pin into position.

6. Assemble the sun gear into the final drive carrier with the stepped side facing out. Install the parking gear onto the sun gear.

7. Install the thrust bearing assembly into the final drive internal gear and place the unit onto the final drive carrier.

8. Install the thrust washer and the thrust bearing onto the carrier hub. Install the snap ring making sure it seats properly in its groove.

FINAL DRIVE SUN GEAR SHAFT

Disassembly

1. Lift off the reverse reaction drum and the input carrier assembly.

2. Remove the reaction carrier bearing, which may be stuck to the input carrier assembly.

3. Remove the internal gear bearing and the final drive sun gear.

4. Lift out the 1-2 band from the reaction sun gear and drum assembly.

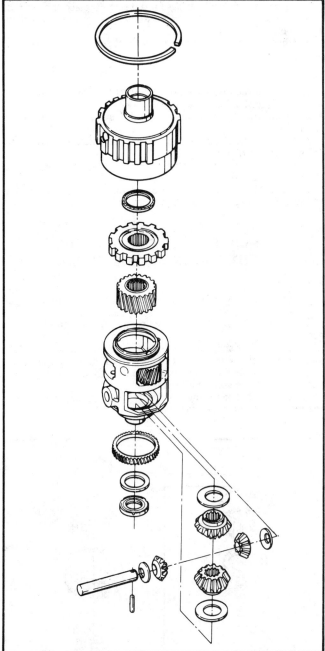

Final drive assembly—exploded view
(© General Motors Corporation)

5. Remove the bearing assembly from the reaction sun gear and drum assembly.

Inspection

1. Inspect the final drive sun gear shaft for damaged splines. Replace if necessary.

2. Inspect the 1-2 band assembly for damage from heat or excessive wear. Check the band assembly for lining separation or lining cracks.

NOTE: The 1-2 band assembly is presoaked in a friction solution and should not be washed in a cleaning solvent.

3. Inspect the sun gear/drum assembly for any scoring or damaged teeth.

4. Inspect the thrust bearings for damage. Replace as required.

5. Check the reaction carrier assembly for pinion end play. Pinion end play should be 0.23-0.61mm.

6. Check for pinion damage or internal gear damage. Replace as needed.

Assembly

1. Position the inside race of the thrust bearing against the final drive internal gear.

2. Install the reaction sun gear and drum assembly into the case.

3. Position the thrust bearing inside the reaction carrier and retain with petroleum jelly.

4. Install the reaction carrier and rotate until all the pinions engage with the sun gear.

5. Install the reverse reaction drum making sure that all the spline teeth engage with the input carrier.

THIRD ROLLER CLUTCH AND INPUT SUN GEAR AND SPRAG

Disassembly

1. Remove the input sprag and 3rd roller clutch from the input sun gear.

2. Remove the input sun gear spacer and retainer from the sun gear.

3. Remove the snap ring and lift off the input sprag wear plate and the 3rd roller clutch race and cam from the roller assembly.

4. Disassemble the input sprag assembly by removing the inner race from the sprag assembly.

5. Remove the snap ring that holds the wear plate and lift out the wear plate and sprag assembly.

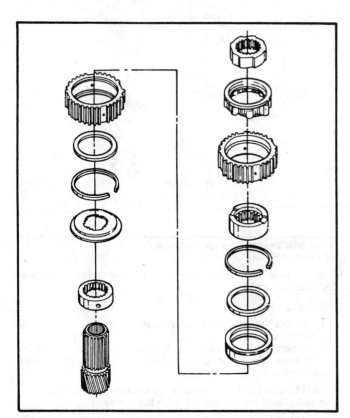

Third roller clutch assembly (© General Motors Corporation)

Inspection

1. Clean all parts in cleaning solvent and blow dry using compressed air.

2. Inspect the outer race and roller cam for any cracks or scoring. Replace as required.

3. Inspect the roller assembly for damaged rollers and springs. Replace any loose rollers by depressing the spring and inserting the roller.

4. Inspect the sprag assembly for damaged sprags or cages and replace as required.

5. Inspect the inner race and wear plates for any scoring or damage. Replace as required.

Assembly

1. Position one wear plate against the snap ring and hold in position with petroleum jelly.

2. Insert the wear plate with the snap ring against the sprag assembly.

3. Install the spacer on the input sun gear and place the sprag retainer over the spacer.

4. Install the sprag snap ring. Make sure the snap ring seats properly.

5. Install the sprag assembly and the roller clutch onto the sun gear.

INPUT CLUTCH ASSEMBLY

Disassembly

1. Remove the input clutch snap ring and remove the input clutch backing plate.

2. Remove the steel and compostion clutch plates, along with the input clutch apply plate.

3. Remove the third clutch snap ring and remove the third clutch backing plate.

4. Remove the steel and composition clutch plates, along with the third clutch waved plate.

5 Remove the snap ring from the spring retainer, and lift out the third clutch piston from its housing.

6. Remove the third clutch piston inner seal from the shaft.

7. Compress the third clutch piston housing and remove the snap ring. Remove the third clutch piston housing.

8. Remove the O-ring seal and take out the spring retainer. Remove the input clutch piston and inner seal.

Inspection

1. Wash all parts in cleaning solvent and blow dry using compressed air.

2. Inspect all parts for scoring, wear or damage.

3. Inspect the input clutch housing for damaged or worn bushings.

4. Check the fourth clutch shaft seal and replace if damaged or cut.

5. Repair or replace any parts found to be defective.

Assembly

1. Lubricate all parts with automatic transmission fluid prior to assembling.

2. Install the input clutch piston seal with the piston seal protector. Position the input piston in the housing.

3. Install the O-ring on the input shaft making sure it seats properly. Install the spring retainer in the piston.

4. Install the third clutch piston housing into the input housing. Compress the third clutch housing with a clutch spring compressor and install the snap ring.

5. Install the third clutch inner seal on the third clutch piston and install the third clutch piston.

6. Install the third clutch spring retainer and compress the spring retainer using the clutch spring compressor and install the snap ring.

7. Install the third clutch plates and the third clutch backing plate. Install the snap ring.

NOTE: When installing the third clutch plates, start with a steel plate and alternate between composition and steel plates. When installing the third clutch backing plate, make sure that the stepped side is facing up.

8. Install the input clutch apply plate with the notched side facing the snap ring. Install the input clutch plates.

NOTE: When installing the input clutch plates, start with a composition plate and alternate between steel and composition plates.

9. Install the input clutch backing plate and secure with a snap ring.

RETAINER AND BALL ASSEMBLY

Replacement (optional)

1. Remove the retainer and ball assembly from the housing with a ⅜ in. (9.5mm) drift.
2. Tap in a new retainer using a ⅜ in. (9.5mm) drift.

PISTON SEAL

Replacement (optional)

1. Remove the input clutch piston seal and the third clutch piston seal.
2. Inspect the clutch piston for any remaining seals and remove.
3. Install a new input clutch piston seal and a new third clutch piston seal. Lubricate with transmission fluid.

FOURTH CLUTCH SHAFT SEAL

Replacement (optional)

1. Remove the lock up sleeve using a suitable tool and remove the oil seal.
2. Install the new oil seal into the input shaft making sure that the seal tab aligns with the slot in the shaft.
3. Install the lock up sleeve in the shaft using a bench press.

INPUT SHAFT SEAL

Replacement (optional)

1. Remove the seal rings from the input shaft with a suitable tool.
2. Adjust the seal protector so that the bottom matches the seal ring groove.
3. Lubricate the oil seal ring and place it on the seal protector.
4. Slide the seal into position with the seal pusher over the seal protector.
5. Size the seal with a seal sizer and gently work the tool over the seal with a twisting motion.

SECOND CLUTCH HOUSING

Disassembly

1. Remove the second clutch hub snap ring and lift out the second clutch wave plate.
2. Remove the second clutch plate assembly and the second clutch backing plate.
3. Remove the next snap ring and remove the second clutch housing support.
4. Remove the thrust bearing and thrust washer. Remove the reverse band.
5. Remove the second clutch housing. Remove the second clutch piston seals which may be stuck on the second clutch housing.

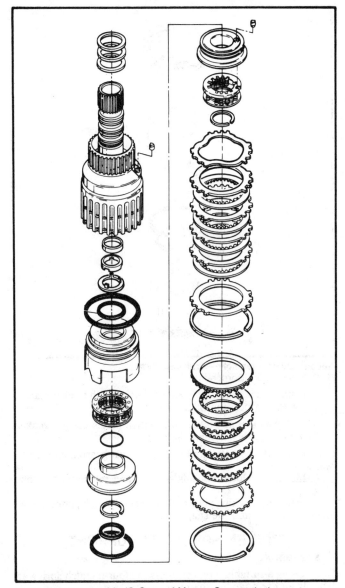

Input clutch assembly (© General Motors Corporation)

6. Remove the second clutch piston and the spring return apply ring.

Inspection

1. Wash all parts in cleaning solvent and blow dry using compressed air.
2. Inspect all parts for scoring, wear or damage.
3. Inspect the second clutch piston and seal for damage or warping.
4. Repair or replace damaged parts as required.

Assembly

1. Lubricate a new piston seal with automatic transmission fluid and install in the second clutch piston.
2. Install a new retainer and ball assembly into the second clutch housing using a ⅜ in. (9.5mm) drift.
3. Install the second clutch piston into the second clutch housing. Make sure the seals are not damaged.
4. Install the spring return apply ring into the second clutch housing. Using a spring compressor, compress the spring return apply ring and insert the snap ring.

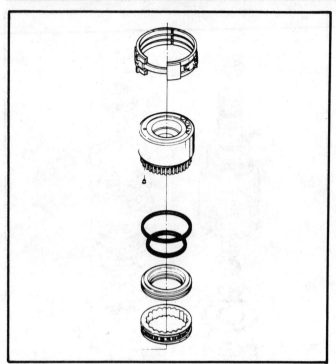

Second clutch assembly—exploded view
(© General Motors Corporation)

5. Install the second clutch plates starting with steel and alternating with composition plates.

NOTE: **The second clutch composition plates are pre-soaked and do not require soaking in solvent.**

6. Install the reverse band and the thrust bearing and thrust washers.

7. Install the clutch housing support and secure with a snap ring.

8. Install the backing plate and the second clutch plates. Install the waved plate and the snap ring.

DRIVEN SPROCKET SUPPORT

Disassembly

1. Using a suitable tool, compress the fourth clutch spring assembly and remove the snap ring.

2. Remove the fourth clutch piston and the piston seals. Discard the piston seals.

3. Remove the bearing assembly from the driven sprocket support.

4. Remove the chain scavening scoop and the oil reservoir weir.

5. Remove the thrust washer and the O-ring seals.

Inspection

1. Clean all parts thoroughly in solvent and blow dry using compressed air.

2. Inspect the driven sprocket support for cracks or damage.

3. Inspect the seals and pistons for damage. Replace as necessary.

4. Check the spring retainer for distorted or damaged springs.

5. Inspect the oil reservoir weir and the chain scavenging scoop for cracks or damage.

6. Repair or replace any damaged parts as required.

Assembly

1. Lubricate new O-rings and install them behind the thrust washer on the driven sprocket support.

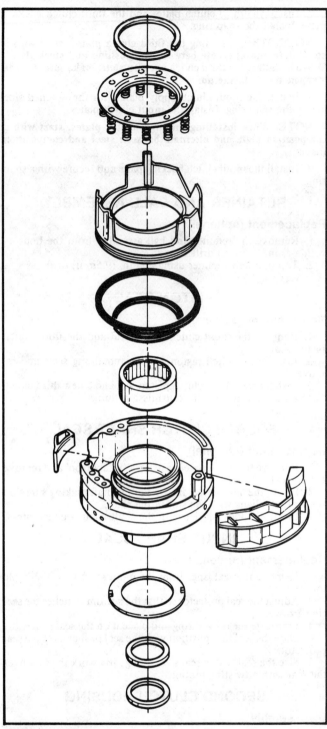

Drive sprocket support—exploded view
(© General Motors Corporation)

2. Install the oil reservoir weir and the chain scavenging scoop on the driven sprocket support.

3. Install the bearing assembly on the driven sprocket support, between the oil reservoir weir and the chain scavenging scoop.

4. Install new seals on the fourth clutch piston and install the fourth clutch piston.

5. Install the fourth clutch spring assembly and using a suitable tool, compress the springs and insert the snap ring.

OIL PUMP

Disassembly

1. Remove the pump bolts from the cover. Lift off the pump cover. Remove the pump cover sleeve and the vane ring.
2. Remove the pump rotor and the bottom vane ring. Remove the oil seal ring and the O-ring from the oil pump slide.
3. Remove the oil pump slide. Enclosed in the oil pump slide are the pump slide seal and support along with the inner and outer pump priming springs.

NOTE: The pump rotor, vane rings and oil pump slide are factory matched units. Therefore, if any parts need replacing, all parts must be replaced.

4. Remove the pivot pin and the roll pin from the oil pump body.
5. Remove the 3-2 coast down valve, spring and bore plug from the oil pump body. Remove the oil pressure switches.

Inspection

1. Clean all parts thoroughly in solvent and blow dry using compressed air.
2. Check the pump body for warping or cracks. Make sure that the oil passages are free of debris.
3. Check the oil pump slide and springs for excessive wear.
4. Check the rotor and vanes for any damage. Inspect the pump slide seal and support for any cracks.
5. Repair or replace damaged parts as needed.

Assembly

1. Install the 3-2 coast down valve along with the spring and bore plug into the oil pump body.
2. Install the lower vane ring into the pump pocket. Install the pump slide into the pump pocket being careful not to dislodge the lower vane ring.
3. Install the pump slide seal and support into the oil pump slide. Install the pump slide seal and support into the pump slide.
4. Insert the inner priming spring into the outer priming spring. Press the springs into the pump body.
5. Install the O-ring seal onto the oil pump slide. Install the oil seal ring onto the oil pump slide.
6. Install the oil pump rotor onto the oil pump body. Insert the pump vanes into the oil pump rotor. Install the upper vane ring onto the oil pump rotor.

NOTE: The pump vanes must be installed flush with the top of the oil pump rotor.

7. Install the pump cover onto the oil pump body. Install the cover bolts.
8. Install the oil pressure switches into the oil pump body.

CONTROL VALVE

Disassembly

NOTE: As each part of the valve train is removed, place the pieces in order and in a position that is relative to the position on the valve body to lessen the chances for error in assembly. None of the valves, springs or bushings are interchangeable.

1. Lay the valve body on a clean work bench with the machined side up and the line boost valve at the top. The line boost valve should be checked for proper operation before its removal. If it is necessary to remove the boost valve, grind the end of a #49 drill to a taper (a small Allen wrench can sometimes be substituted) and lightly tap the drill into the roll pin. Push the line boost valve out of the top of the valve body.
2. The throttle valve should be checked for proper operation before removing it, by pushing the valve against the spring. If it is necessary to remove the throttle valve, first remove the roll pin holding the T.V. plunger bushing and pull out the plunger and

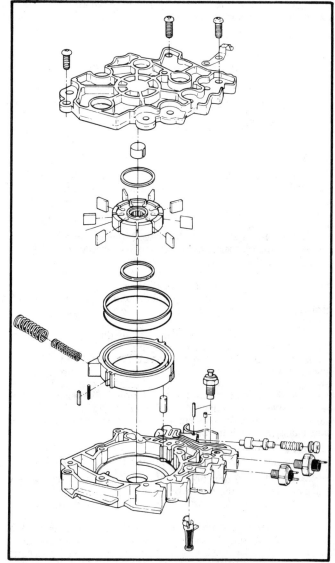

Oil pump—exploded view (© General Motors Corporation)

bushing. Remove the throttle valve spring. Remove the blind hole roll pin using the drill method described earlier.
3. On the same side of the valve body, move to the next bore down. Remove the straight pin and remove the reverse boost valve and bushing, as well as reverse boost spring and the pressure regulator assembly.

NOTE: Models equipped with a diesel engine will not have a reverse boost spring or a pressure regulator modulator spring.

4. Move to the other side of the valve body and from the top bore remove the 1-2 throttle valve bushing retainer and lift out the throttle valve bushing and the 1-2 shift valve assembly.
5. Move to the next bore and remove the spring pin and lift out the 2-3 accumulator bushing assembly.
6. At the next bore down, remove the 3-2 control sleeve and the 3-2 control valve assembly.
7. Move to the next bore and remove the spring pin. Remove the 2-3 throttle valve bushing and its components.
8. Move to the next bore and remove the coiled spring pin. Remove the 3-4 throttle valve assembly.
9. The two remaining bores contain the 1-2 servo pipe lip seals, which are removed.

10. At the side of the valve body, are the servo assemblies. Remove the coiled spring pins and detach the servo valves and springs.

Inspection

1. Wash all bushings, springs and valves in solvent. Blow dry using compressed air.
2. Inspect all valves and bushings for any scoring or scratches.
3. Check the springs for collapsed coils and bore plugs for damage.
4. Repair or replace any defective parts as needed.

Assembly

1. Install the reverse servo boost valve and spring into its proper location in the valve body. Install the 1-2 servo control valve and the 1-2 servo boost valve into the valve body. Install the correct springs behind each valve and insert the coiled spring pins.
2. Install new servo pipe lip seals. Move to the bore next to the servo pipe lip seals, and install the 3-4 throttle assembly. Install the coiled spring pin.
3. Move to the next bore and install the 2-3 throttle valve bushing and its components. Install the spring pin.
4. Move to the next bore and in the following order install 3-2 isolator valve and spring, 3-2 control valve and spring and the 3-2 control sleeve.
5. Move to the next bore and install the 2-3 accumulator bushing assembly. Install the spring pin.
6. Move to the last bore on the side and install in the following order the 1-2 shift valve and throttle valve, the throttle valve spring and the 1-2 throttle valve bushing. Insert the 1-2 throttle valve bushing retainer into the 1-2 throttle valve bushing.
7. Move to the other side of the valve body and install the pressure regulator valve assembly. Install the reverse boost valve and bushing. Install the straight pin.

NOTE: The reverse boost spring and a pressure regulator modulator spring will only be found on models with a gas engine.

8. Install in the next bore the throttle valve assembly. Secure the assembly with the spring pin.
9. Install the line boost valve assembly. Insert the retainer clip.
10. Move to the next bore and install throttle valve feed valve assembly. Make sure the valve stop plate is in its proper position.
11. Move to the next bore and install the converter clutch shift valve assembly securing it with the bushing and a coiled spring pin.
12. The next bore is the pump pressure valve and bushing locations. Secure these units with a spring pin.
13. Moving to the next bore, install the 1-2 accumulator assembly. Make sure that the valve bore plug is in place over the 1-2 accumulator valve.
14. Assemble the converter clutch valves into the next two bore holes.
15. Install the second clutch pipes in the top of the valve body. Install the release cover plate gasket and cover.

CHANNEL PLATE

Disassembly

1. Remove the modulator port gasket and the upper channel plate gasket. Remove the lower channel plate gasket and discard all gaskets.
2. Remove the input clutch accumulator piston and spring. Discard the input clutch ring seal.
3. Remove the converter clutch accumulator piston and the converter clutch spring. Discard the converter clutch seal. Remove the axle oil seal.
4. Remove the manual valve assembly. Detach the manual

valve clip. Remove the channel plate stud from the case side of the channel plate.

Inspection

1. Wash all parts in solvent and blow dry using compressed air.
2. Check all parts for excessive wear or damage. Check the channel plate for cracks or warping.
3. Repair or replace any defective parts as required.

Assembly

1. Install the channel plate stud to the channel plate. Attach the manual valve clip to the manual valve assembly. Install the manual valve assembly into the channel plate.
2. Press in a new axle oil seal using a suitable tool. Install a new converter clutch seal. Place the converter clutch piston on its shaft with a new seal. Install the converter clutch spring.
3. Install the input clutch piston with a new input clutch seal. Install the input clutch spring.

Transaxle Assembly

Before the assembly of the transaxle begins, make certain that the case and all other parts are clean that all parts are serviceable or have been overhauled. Inspect the case carefully for cracks or any signs of porosity. Inspect the vents to make sure that they are open. Check the case lugs, the intermediate servo bore and snap ring grooves for damage. Check the bearings that are in the case and replace if they appear to be worn.

NOTE: If the bearings are replaced, they must be installed with the bearing identification facing up.

The converter seal should be replaced. After all these preliminary checks have been made, begin the reassembly of the transaxle. Make certain that the transaxle is held securely in the proper support fixture.

1. If the manual shaft seals were removed during disassembly, replace as required. Install a new O-ring on the actuator guide assembly and install the actuator rod onto the detent lever.
2. Install the manual shaft into the transaxle case and insert the detent lever onto the manual shaft. Install the lock nut and torque to 25 ft. lbs. (34 N•m). Insert the retaining pin into the case.
3. Turn the case so it faces up and using the clutch and final drive tool, install the final drive internal gear assembly. Install the final drive snap ring.

NOTE: The clutch and final drive tool will clamp to the end of the final drive internal gear assembly.

4. The input housing and shaft assembly should be installed at this point. It should be overhauled as outlined earlier in the section "Unit Disassembly and Assembly". Carefully lower the input housing into the case.
5. Clamp the clutch and final drive tool to the second clutch housing. Lower the second clutch housing into the transaxle case, on top of the input housing and shaft assembly.
6. Install the already assembled driven sprocket support onto the input shaft. Install the drawn cup bearing assembly to the front of the driven sprocket.
7. The drive link assembly is to be installed next. Make sure that the gears and related parts are in satisfactory condition. Install the thrust washers using petroleum jelly to hold them in place.
8. Install the sprockets into the link assembly and locate the colored guide link so that it will be assembled facing the case cover. With the assembly oriented properly, install into the case.
9. Install the thrust washer on the driven sprocket with the outer race against the sprocket.
10. The output shaft should be installed next. Insert the output shaft into the driven sprocket and install the output shaft differ-

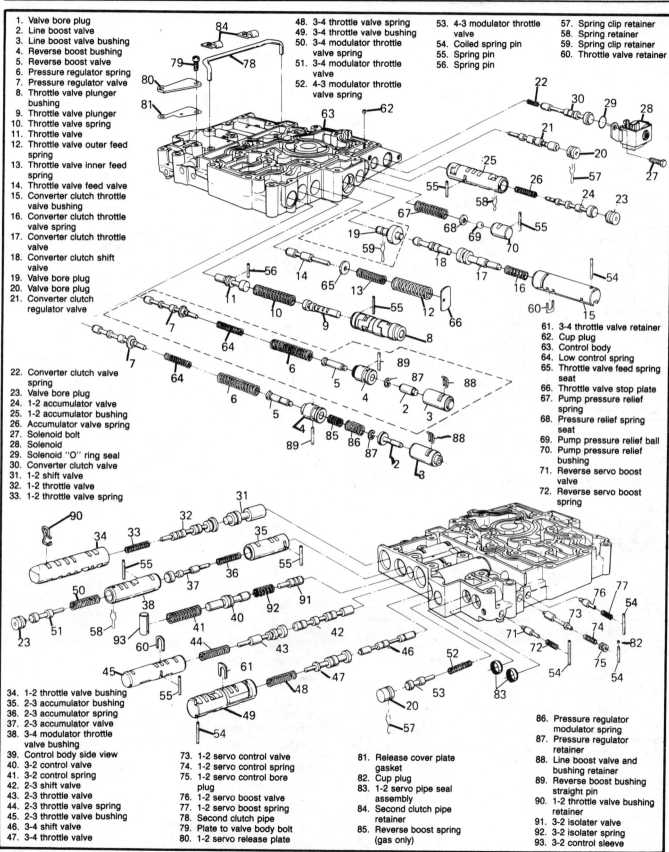

1. Valve bore plug
2. Line boost valve
3. Line boost valve bushing
4. Reverse boost bushing
5. Reverse boost valve
6. Pressure regulator spring
7. Pressure regulator valve
8. Throttle valve plunger bushing
9. Throttle valve plunger
10. Throttle valve spring
11. Throttle valve
12. Throttle valve outer feed spring
13. Throttle valve inner feed spring
14. Throttle valve feed valve
15. Converter clutch throttle valve bushing
16. Converter clutch throttle valve spring
17. Converter clutch throttle valve
18. Converter clutch shift valve
19. Valve bore plug
20. Valve bore plug
21. Converter clutch regulator valve
22. Converter clutch valve spring
23. Valve bore plug
24. 1-2 accumulator valve
25. 1-2 accumulator bushing
26. Accumulator valve spring
27. Solenoid bolt
28. Solenoid
29. Solenoid "O" ring seal
30. Converter clutch valve
31. 1-2 shift valve
32. 1-2 throttle valve
33. 1-2 throttle valve spring

48. 3-4 throttle valve spring
49. 3-4 throttle valve bushing
50. 3-4 modulator throttle valve spring
51. 3-4 modulator throttle valve
52. 4-3 modulator throttle valve spring
53. 4-3 modulator throttle valve
54. Coiled spring pin
55. Spring pin
56. Spring pin

57. Spring clip retainer
58. Spring retainer
59. Spring clip retainer
60. Throttle valve retainer

61. 3-4 throttle valve retainer
62. Cup plug
63. Control body
64. Low control spring
65. Throttle valve feed spring seat
66. Throttle valve stop plate
67. Pump pressure relief spring
68. Pressure relief spring seat
69. Pump pressure relief ball
70. Pump pressure relief bushing
71. Reverse servo boost valve
72. Reverse servo boost spring

34. 1-2 throttle valve bushing
35. 2-3 accumulator bushing
36. 2-3 accumulator spring
37. 2-3 accumulator valve
38. 3-4 modulator throttle valve bushing
39. Control body side view
40. 3-2 control valve
41. 3-2 control spring
42. 2-3 shift valve
43. 2-3 throttle valve
44. 2-3 throttle valve spring
45. 2-3 throttle valve bushing
46. 3-4 shift valve
47. 3-4 throttle valve

73. 1-2 servo control valve
74. 1-2 servo control spring
75. 1-2 servo control bore plug
76. 1-2 servo boost valve
77. 1-2 servo boost spring
78. Second clutch pipe
79. Plate to valve body bolt
80. 1-2 servo release plate

81. Release cover plate gasket
82. Cup plug
83. 1-2 servo pipe seal assembly
84. Second clutch pipe retainer
85. Reverse boost spring (gas only)

86. Pressure regulator modulator spring
87. Pressure regulator retainer
88. Line boost valve and bushing retainer
89. Reverse boost bushing straight pin
90. 1-2 throttle valve bushing retainer
91. 3-2 isolater valve
92. 3-2 isolater spring
93. 3-2 control sleeve

Valve body assembly (© General Motors Corporation)

ential snap ring. The snap ring location is at the opposite end of the case.

11. Install the fourth clutch shaft onto the output shaft. Insert the fourth clutch bearing assembly into the exposed end of the fourth clutch shaft. Install the fourth clutch apply plate. Install the fourth clutch plate assembly.

12. Install the channel plate assembly to the front of the case. Torque the bolts to 10 ft. lbs. (14 N•m).

NOTE: When installing the channel plate to the case, petroleum jelly can be used to hold the gaskets in place.

13. The oil pump drive shaft should be installed next. Install a new oil seal ring on the oil pump drive shaft. Insert the drive shaft into its proper position and rotate to be sure of the spline engagement in the drive sprocket.

14. Install the eight check balls in the channel plate. Install the spacer plate/channel plate gasket, the valve body spacer plate and the spacer plate/valve body gasket.

15. Install the check balls in the spacer plate. Attach the valve body assembly to the channel plate assembly. Torque the bolts to 10 ft. lbs. (14 N•m).

16. The oil pump assembly is installed next. Bolt the oil pump to the valve body. Torque to 10 ft. lbs. (14 N•m).

17. Install the throttle lever on the pump assembly. Attach the throttle lever to the throttle lever cable link.

18. Position the wiring harness and secure using the two wire clip and the five wire clip.

19. Install the side cover gaskets. These can be held in place with a small amount of petroleum jelly. Install the side cover case pan. Torque to 10 ft. lbs. (14 N•m).

20. Turn the transaxle in its fixture so as that the top of the unit faces up. Install the servo return spring and the reverse apply pin. Slide the servo cushion spring retainer with the reverse servo spring and the other retainer onto the reverse apply pin.

21. Install the reverse servo piston with the oil seal ring and secure with the snap ring. Install the reverse servo cover and retaining ring.

22. On the side of the transaxle install the servo return spring and the 1-2 band apply pin. Install the retainer. Assemble the 1-2

servo piston and install behind the 1-2 band apply pin. Install the snap ring.

23. Rotate the transaxle assembly in its fixture so as to gain access to the bottom of the unit. Attach the scavenging scoop and bolts.

24. Install the accumulator assemblies making sure not to intermix the accumulator springs. Install the drive sprocket support oil lube pipe and retainer.

25. Install the cover gasket and accumulator spacer plate.

NOTE: At this point the thermo element assembly can be installed, if removed earlier. Make certain that the element plates are not bent during installation.

26. Assemble the oil pipes, cover, body and gasket as a unit and install with their gaskets into the transaxle.

27. Install the governor control body bolts and torque to 20 ft. lbs. (27 N•m). Install the accumulator cover bolts. No torque is needed on these bolts.

28. Install a new oil filter with a new oil filter seal. Install a new pan gasket on the pan and install the pan on the transaxle. Torque to 10 ft. lbs. (14 N•m).

29. Using a magnet, install the modulator valve. Place a new O-ring on the modulator and insert into its proper position. Install the modulator retainer and bolt. Torque to 20 ft. lbs. (27 N•m).

30. Install a new oil seal ring on the governor assembly. Install the governor into the transaxle case.

31. Install the speedometer drive gear onto the governor assembly. Install the speedometer gear thrust bearing onto the speedometer gear.

32. Install a new O-ring seal in the governor cover. Install the speedometer driven gear into the governor cover.

33. Install the O-ring seal on the speedometer driven gear sleeve. Assemble the gear sleeve to the governor cover. Install the unit to the case and secure with the speedometer gear retainer and bolt. Install the governor cover case screw.

34. The transaxle is now ready for the torque converter and then to be moved to the transaxle jack for installation.

SPECIFICATIONS

TORQUE SPECIFICATIONS
THM 440-T4

Item	Foot Pounds	N•m
Cooler Fitting Connector	30	41
Modulator to Case	20	27
Pump Cover to Channel Plate	10	14
Pump Cover to Pump Body	20	27
Pump Cover to Pump Body (Torx Head)	20	27
Pipe Plug	10	14
Case to Drive Sprocket Support	20	27
Manifold to Valve Body	10	14
Governor to Case	20	27
Pressure Switch	10	14
Solenoid to Valve Body	10	14
Detent Spring to Valve Body	10	14

TORQUE SPECIFICATIONS
THM 440-T4

Case Side Cover to Channel Plate	10	14
Pump Cover to Valve Body	10	14
Pump Cover to Channel Plate	10	14
Valve Body to Case (Torx Head)	20	27
Valve Body to Case	20	27
Pump Body to Case	20	27
Valve Body to Channel Plate	10	14
Valve Body to Channel Plate (Torx Head)	10	14
Channel Plate to Case (Torx Head)	20	27
Channel Plate to Driven Sprocket Support (Torx Head)	20	27
Side Cover to Case	10	14
Accumulator Cover to Case	20	27
Oil Scoop to Case	10	14
Governor Control Body Retainer	20	27
Transmission Oil Pan to Case	10	14
Manual Shaft to Inside Detent Lever (Nut)	25	34

THRUST WASHER GUIDE
THM 440-T4

I.D. Number	Dimension (in.)	Color
1	2.90-3.00	Orange/Green
2	3.05-3.15	Orange/Black
3	3.20-3.30	Orange
4	3.35-3.45	White
5	3.50-3.60	Blue
6	3.65-3.75	Pink
7	3.80-3.90	Brown
8	3.95-4.05	Green
9	4.10-4.20	Black
10	4.25-4.35	Purple
11	4.40-4.50	Purple/White
12	4.55-4.65	Purple/Blue

THRUST WASHER GUIDE
THM 440-T4

I.D. Number	Dimension (in.)	Color
13	4.70-4.80	Purple/Pink
14	4.85-4.95	Purple/Brown
15	5.00-5.10	Purple/Green

FINAL DRIVE END PLAY
THM 440-T4

I.D. Number	Thickness
1	0.059-0.062 inches (1.50-1.60mm)
2	0.062-0.066 inches (1.60-1.70mm)
3	0.066-0.070 inches (1.70-1.80mm)
4	0.070-0.074 inches (1.80-1.90mm)
5	0.074-0.078 inches (1.90-2.00mm)
6	0.078-0.082 inches (2.00-2.10mm)

SPECIAL TOOLS

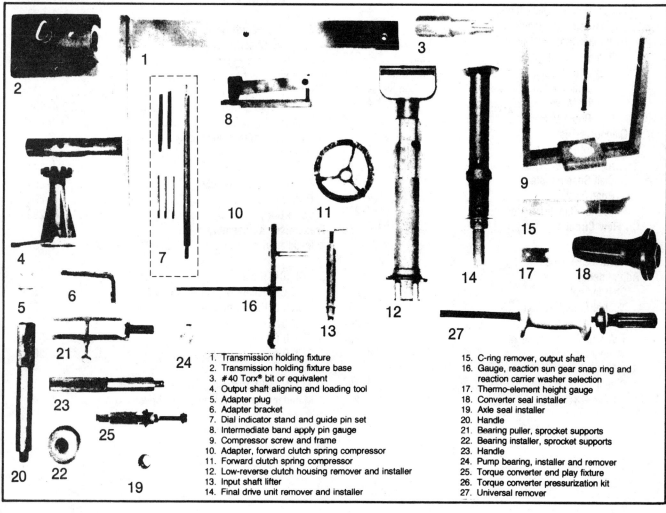

1. Transmission holding fixture
2. Transmission holding fixture base
3. #40 Torx® bit or equivalent
4. Output shaft aligning and loading tool
5. Adapter plug
6. Adapter bracket
7. Dial indicator stand and guide pin set
8. Intermediate band apply pin gauge
9. Compressor screw and frame
10. Adapter, forward clutch spring compressor
11. Forward clutch spring compressor
12. Low-reverse clutch housing remover and installer
13. Input shaft lifter
14. Final drive unit remover and installer
15. C-ring remover, output shaft
16. Gauge, reaction sun gear snap ring and reaction carrier washer selection
17. Thermo-element height gauge
18. Converter seal installer
19. Axle seal installer
20. Handle
21. Bearing puller, sprocket supports
22. Bearing installer, sprocket supports
23. Handle
24. Pump bearing, installer and remover
25. Torque converter end play fixture
26. Torque converter pressurization kit
27. Universal remover

INDEX

GENERAL MOTORS
TURBO HYDRA-MATIC 700-R4

APPLICATIONS

TRANSMISSION APPLICATION CHART

Year	Model	Transmission Model
1982 and later	Corvette	700-R4
1982 and later	Impala, Caprice	700-R4
1983 and later	Camaro, Firebird	700-R4
1983 and later	Parisienne	700-R4
1982 and later	Chevrolet and GMC	700-R4

GENERAL DESCRIPTION

The T.H.M. 700-R4 is a fully automatic transmission consisting of a three element hydraulic torque converter with the addition of a torque converter clutch.

This automatic transmission also consists of two planetary gear sets, five multiple disc type clutches, two roller or one way clutches and a band which are used in order to provide the friction elements to produce four forward speeds, the last of which is an overdrive speed.

The torque converter, through oil, couples the engine power to the gear sets and hydraulically provides additional torque multiplication when required. Also, through the converter clutch, the converter drive and driven members operate as one unit when applied providing mechanical drive from the engine through the transmission.

The gear ratio changes are fully automatic in relation to vehicle speed and engine torque. Vehicle speed and engine torque are directed to the transmission providing the proper gear ratios for maximum efficiency and performance at all throttle openings.

A hydraulic system pressurized by a variable capacity vane type pump provides the operating pressure required for the operation of the friction elements and automatic controls.

Transmission and Torque Converter Identification

TRANSMISSION

The THM 700-R4 automatic transmission can be identified by the serial number plate which is located on the right side of the transmission case. This identification tag contains information that is important when servicing the unit.

TORQUE CONVERTER

The torque converter is a welded unit and cannot be disassembled for service. Any internal malfunctions require the replacement of the converter assembly. The replacement converter must be matched to the model transmission through parts identification. No specific identification is available for matching the converter to the transmission for the average repair shop.

DIESEL ENGINE

Vehicles equipped with diesel engines use a different torque converter. To identify these units, examine the weld nuts. Most gas engine converters have their weld nuts spot welded onto the converter housing, usually in two spots. The diesel converters have the weld nuts completely welded around their entire circumference.

TURBOCHARGED V-6 ENGINE

Vehicles equipped with turbo-charged V-6 engines use a different torque converter. These units have a high stall speed converter, allowing a stall speed of about 2800 rpm. These converters must not be replaced with a standard converter, otherwise performance will be sluggish and unsatisfactory. When ordering replacements for the turbo-charged units, make certain to specify that the vehicle has turbo-charging in order to obtain the proper replacement.

Metric Fasteners

Metric fasteners are very close to the dimensions of the familiar

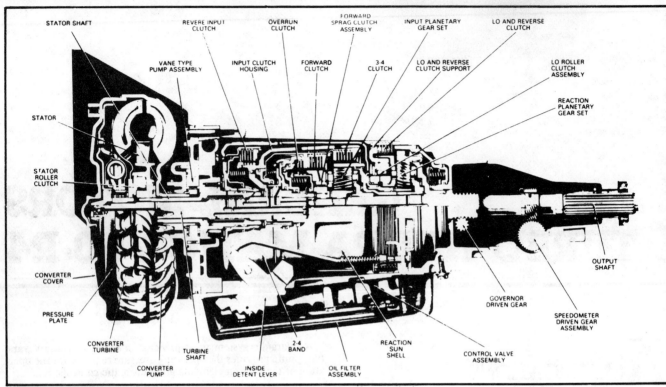

T.H.M. 700-R4 automatic transmission—exploded view (© General Motors Corporation)

inch system fasteners. For this reason, replacement fasteners must have the same measurement and strength as those removed.

NOTE: Do not attempt to interchange metric fasteners for inch system fasteners. Mismatched or incorrect fasteners can result in damage to the transmission unit through malfunctions or breakage, or possible personal injury.

Care should be taken to reuse the fasteners in the same locations as removed. Keep them in a safe place when removed, for if lost, replacement could be inconvenient. The following locations could have metric fasteners.

Fluid Capacities

The fluid capacities are approximate and the correct fluid level should be determined by the dipstick indicator. Use only Dexron® II automatic transmission fluid when adding, or servicing the THM 700-R4 automatic transmission.

When a complete transmission overhaul is done the transmission will take about 23 pints of transmission fluid. When a fluid change is done the transmission will take about 10 pints of transmission fluid.

Checking Fluid Level

The THM 700-R4 transmission is designed to operate at the "FULL HOT" mark on the dipstick at normal operating temperatures, which range from 190° to 200° F. Automatic transmissions are frequently overfilled because the fluid level is checked when cold and the dipstick level reads low. However as the fluid warms up the level will rise, as much as ¾ of an inch. Note that if the transmission fluid is too hot, as it might be when operating under city traffic conditions, trailer towing or extended high speed driving, again an accurate fluid level cannot be determined until the fluid has cooled somewhat, perhaps 30 minutes after shutdown.

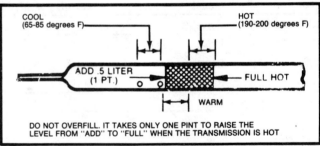

T.H.M. 700-R4 dipstick (© General Motors Corporation)

To determine proper fluid level under normal operating temperatures, proceed as follows:
1. Make sure vehicle is parked level.
2. Apply parking brake; move selector to Park.
3. Start engine but do not race engine. Allow to idle.
4. Move selector through each range, back to PARK, then check level. The fluid should read "FULL HOT" (transmission end of dipstick too hot to hold comfortably.)

Do not overfill the transmission. Overfilling can cause foaming and loss of fluid from the vent. Overheating can also be a result of overfilling since heat will not transfer as readily. Notice the condition of the fluid and whether there seems to be a burnt smell or metal particles on the end of the dipstick. A milky appearance is a sign of water contamination, possibly from a damaged cooling system. All this can be a help in determining transmission problems and their source.

Fluid Drain Intervals

The main considerations in establishing fluid change intervals are the type of driving that is done and the heat levels that are generated by such driving. Normally, the fluid and strainer would be changed at 100,000 miles. However, the following conditions may be considered severe transmission service.

1. Heavy city traffic
2. Hot climates regularly reaching 90°F. or more
3. Mountainous areas
4. Frequent trailer pulling
5. Commercial use such as delivery service, taxi or police car use

If the vehicle is operated under any of these conditions, it is recommended that the fluid be changed and the filter screen serviced at 15,000 mile intervals. Again, be sure not to overfill the unit. Only one pint of fluid is required to bring the level from "ADD" to "FULL" when the transmission is hot.

MODIFICATIONS

VALVE BODY SPACER PLATE HOLE MISSING

On some early T.H.M. 700-R4 automatic transmissions the valve body spacer plate gasket may be missing a passage hole. If this is the case replace the valve body spacer gasket with part number 8647065.

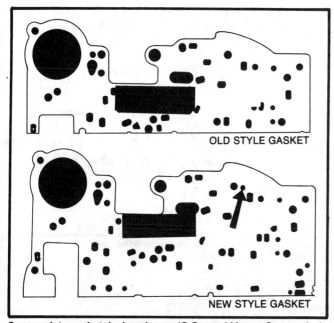

OLD STYLE GASKET

NEW STYLE GASKET

Spacer plate gasket design change (© General Motors Corporation)

TRANSMISSION OIL FILTER INSTALLATION

When installing the oil filter assembly to the valve body of the T.H.M. 700-R4 automatic transmission, be very careful not to cut the O-ring. Press the O-ring into its bore with firmness, but do not strike the filter. A cut O-ring will show up as a valve buzz, low line pressure, soft shifts or a converter clutch chatter.

FRONT TORRINGTON BEARING INSTALLATION

On some T.H.M. 700-R4 transmissions, when installing the front Torrington bearing that goes between the pump and the input housing, be sure that the black surface of the bearing is against the pump. If the black surface is not against the pump damage to the unit can occur.

VALVE BODY BORE PLUGS

Valve body bore plugs that have a recess on one side should always be assembled with the recess towards the outside of the valve body. Failure to do so can result in mispositioning of the valves and springs.

R.T.V. SEALANT—TRANSMISSION OIL PAN

Do not use R.T.V. sealant on the fluid pan of the T.H.M. 700-R4 automatic transmission. Use of this sealant could cause a possible block to the servo exhaust port. Use only a fluid pan gasket, when servicing this transmission.

LATE OR ERRATIC UPSHIFTS

On some T.H.M. 700-R4 automatic transmissions late or erratic upshifts may occur, if this happens, especially with changes in the temperature, remove the valve body and inspect the T.V. sleeve. The sleeve can rotate in the valve body and cause the passages to become blocked. To correct the problem, remove the sleeve and the valve and clean thoroughly. When reinstalling the sleeve, align it so that all of the ports are open and not restricted. Insert the pin and make sure it is seated all the way down to help prevent rotation of the sleeve.

TORQUE CONVERTER RATTLE— DIESEL ENGINE

On some vehicles equipped with a diesel engine, a rattle may exist coming from the torque converter area when the torque converter clutch is applied. If this problem exists, you should check the clutch throttle line up. If it is found to be defective, it should be replaced with part number 8642970. The new part will raise the lock up torque converter function to forty miles per hour, which is in fourth gear only. If the rattle still persists, the torque converter should be checked and replaced with part number 8647323 or 8642964.

BURNT 3-4 CLUTCH AND BAND ASSEMBLY

If, when servicing a T.H.M. 700-R4 automatic transmission you encounter a burnt 3-4 clutch and band, inspect the third accumulator check valve. The accumulator valve is located behind servo assembly, which is located in the transmission case. To inspect the third accumulator, pour some solvent into the capsule. The solvent should only go into the servo area. If there is any trace of solvent inside the case barrel the capsule assembly should be replaced. Follow the procedure outlined below when making this repair.

1. Remove the old capsule using a #4 easy-out.
2. Install new capsule, small end first so that one of the four holes will align with the passage into the servo when the capsule is fully installed.
3. Use a ⅜ in. rod to drive the capsule into the case, far enough that the feed hole in the capsule is completely open into the servo bore. This can be approximated by marking the rod at 1⅝ in. from the end that goes into the case. Again recheck with solvent.

ACCUMULATOR HOUSING PLATE ELIMINATION

In 1982 a change was made to eliminate the plate that was located under the accumulator housing in all T.H.M. 700-R4 automatic transmissions. To do this the accumulator housing and gaskets have been changed.

If you are rebuilding the old style unit, update it by ordering service kit 8642967. This kit contains all the necessary parts to eliminate the accumulator housing plate.

PARK TO REVERSE—HARSH SHIFT

Changes have been made to the T.H.M. 700-R4 automatic transmission to improve the park to reverse shift feel. If a park to reverse harsh shift feel exists, it can be corrected by installing a new style valve body spacer gasket, case gasket and spacer plate. Also a new style oil pump cover and waved reverse clutch plate must be installed.

The new design parts are being used beginning with transmission serial number 9M6184D. All earlier units can be updated by using these new parts. Refer to the park/reverse change chart for the proper information when ordering the required parts.

PARK/REVERSE CHANGE CHART

Component	Old Part Number	New Part Number
Valve body to spacer gasket	8642920	8642952
Spacer-to-case gasket	8642920	8642952
Spacer plate	8642655-A	8647066-B
Oil pump cover (TN,TP,M6,MH units)	8647043	8647190
Oil pump cover (all other units)	8647042	8647189
Waved reverse clutch plate	8642060	8647067

DETENT LEVER TO MANUAL VALVE LINK CHANGE

Beginning July 1983 all T.H.M. 700-R4 automatic transmissions are being built with a new detent lever to manual valve link. This design change provides a new retention of the detent lever to the manual valve link without the retaining clip.

When servicing a transmission using the old style assembly, replace it with the new design components. The new design component is part number 8654113. The old design part numbers are 8642244 and 8654037.

BURNED THREE-FOUR CLUTCH DIAGNOSIS-MODEL 3YH

When diagnosing a 700-R4 transmission for a burned three-four clutch check the spacer plate for correct identification. If the plate is identified by the letter "C" change the plate. Use part number 8654083 which is identified with the letter "E". The new spacer plate has a larger third clutch feed hole.

NEW DESIGN OIL PUMP COVER

Beginning in the middle of July 1983, all transmissions are being fitted with a new oil pump cover and a new oil pump to transmission case gasket. This new design prevents the transmission oil from exhausting out of the breather during harsh deceleration or on sharp turns.

When servicing any 1983 and later automatic transmissions be sure to use the new design components. The new pump to case gasket cannot be used with the past oil pump cover. The past pump to case gasket can be used with the new oil pump cover, but will continue to exhaust oil out of the breather.

GRINDING NOISE IN ALL RANGES

Starting in January, 1984, all transmissions are being built using a new reaction internal gear. This new gear has a change in the parking lugs and internal gear teeth area. The new gear is the same dimensionally as the old gear. Identification marks have been added to the new gear. Two types of identifying marks are being used. The first is an ink stamp on the top chamfer of the gear. The second is a broached line on three parking lugs equally spaced around the internal reaction gear. When replacing the internal reaction gear order part number 8654161.

NEW STYLE SERVO COVER AND SEAL

Starting in the middle of April 1983, transmissions are being built using a new style 2-4 servo cover and seal. This design change provides a tighter fit between the case bore and the 2-4 servo cover.

The new servo cover seal is part number 8647351 and can be identified with a green stripe. The old servo cover seal is part number 8642112 which is still used on the second apply piston. The old servo seal cannot be used on the new design servo cover. If it is used the cover will leak transmission oil.

Forward Sprag Clutch Modification

During the 1982 model year, a new front sprag was introduced in the T.H.M. 700-R4 automatic transmission. This new design front sprag assembly is replacing the front roller clutch assembly. To update the older units to the new design sprag assembly, order service kit 8642947.

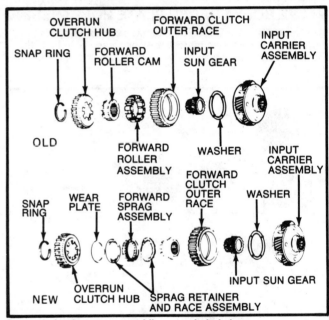

Forward clutch sprag assemblies—exploded view
(© General Motors Corporation)

TROUBLE DIAGNOSIS

Hydraulic Control System

To provide the working pressure within the transmission, a gear-type pump is used to operate the elements to provide the friction function as well as the automatic controls.

The hydraulic control system directs the path and pressure of the fluid so that the proper element can be applied as needed. The valve body directs the pressure to the proper clutch or band servo.

PUMP ASSEMBLY

A hydraulic pressure system requires a source of clean hydraulic fluid and a pump to pressurize the fluid. The THM 700-R4 uses a variable capacity vane type pump. The pump rotor is keyed to the converter pump hub and therefore turns whenever the engine is

CLUTCH AND BAND APPLICATION CHART

Selector Range	2-4 Band	Reverse Input Clutch	Overrun Clutch	Forward Clutch	Forward Sprag Clutch Assembly	3-4 Clutch	Low Roller Clutch	Low Reverse Clutch
First DR4				Applied	Applied		Applied	
Second DR4	Applied			Applied	Applied			
Third DR4				Applied	Applied	Applied		
Fourth DR4	Applied			Applied		Applied		
Third DR3			Applied	Applied	Applied	Applied		
Second DR2	Applied		Applied	Applied	Applied			
First Low			Applied	Applied	Applied		Applied	Applied
Reverse		Applied						Applied

CHILTON'S THREE "C's" AUTOMATIC TRANSMISSION DIAGNOSIS CHART

Condition	Cause	Correction
High or low shift points	a) T.V. cable binding or out of adjustment b) Improper external linkage travel c) Binding throttle valve or plunger in valve body d) T.V. modulator up or down valve sticking in valve body e) Valve body gaskets or spacer plate damaged f) T.V. limit valve sticking in valve body g) Pressure regulator valve to T.V. boost valve sticking in pump assembly h) Front pump slide sticking	a) Correct as required b) Correct linkage as required c) Free binding as required d) Free sticking modulator valve e) Replace gaskets or spacer plate as required f) Free sticking T.V. limit valve g) Repair or replace defective component as required h) Correct as required
First speed only— no upshift	a) Governor valve sticking b) Governor driven gear damaged, retaining pin missing, wrong retaining pin, weights and springs damaged c) Burrs on governor output shaft, sleeve or case d) 1-2 shift valve sticking in valve body e) Valve body gaskets or spacer plate damaged f) Case porosity or restricted fluid passages g) 2-4 servo apply passages blocked, damaged servo piston seals or damaged apply pin h) 2-4 band assembly burned, band anchor pin missing	a) Correct as required b) Repair or replace defective component as necessary c) Correct defective part as required d) Correct 1-2 shift valve as required e) Replace gaskets or spacer plate required f) Correct as required g) Repair or replace defective component as required h) Repair or replace defective component as required

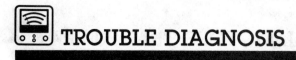
CHILTON'S THREE "C's" AUTOMATIC TRANSMISSION DIAGNOSIS CHART

Condition	Cause	Correction
Slips in first gear	a) Forward clutch plates burned	a) Replace forward clutch plates as required
	b) Forward clutch piston seals cut or damaged	b) Replace piston seals as required
	c) Damaged forward clutch housing or check ball	c) Replace defective part as required
	d) Forward clutch low oil or oil pressure	d) Correct as required
	e) Valve body accumulator valve sticking	e) Free sticking valve
	f) Valve body gaskets or spacer plate damaged	f) Replace gaskets or spacer plate as required
	g) Internal T.V. linkage binding	g) Correct internal linkage problem as required
	h) 1-2 accumulator piston seals cut or damaged	h) Replace damaged seals as required
	i) Leak between 1-2 accumulator piston and pin	i) Repair or replace as required
	j) Missing or broken 1-2 accumulator spring	j) Replace spring as required
1-2 full throttle shifts only	a) T.V. cable not adjusted	a) Adjust cable as required
	b) Throttle lever and bracket assembly missing, damaged or missing exhaust check ball	b) Repair or replace defective component as required
	c) Throttle link not connected, damaged or hanging on upper sleeve	c) Repair throttle link as required
	d) Throttle valve or plunger sticking in open position	d) Free sticking plunger as required
	e) Blockage of case interconnecting passages	e) Free blockage as required
1-2 slip or rough shift	a) Throttle lever and bracket assembly damaged or not installed properly	a) Correct lever and bracket assembly as required
	b) Valve body throttle valve or bushing sticking	b) Correct as required
	c) Sticking 1-2 shift valve in valve body	c) Free sticking valve
	d) Valve body gaskets or spacer plate damaged	d) Replace damaged gaskets or spacer plate as required
	e) Sticking line bias valve, accumulator valve or T.V. limit valve in valve body	e) Repair or replace defective component as required
	f) 2-4 servo incorrect apply pin	f) Install proper apply pin
	g) Damaged 2-4 servo oil seal rings, bore or restricted oil passages	g) Repair or replace as required
	h) Damaged second accumulator piston seal, bore or porus piston	h) Replace defective part as required
	i) Missing second accumulator spring or restricted oil passages	i) Replace missing spring or free restricted oil passages
	j) Burned 2-4 band	j) Replace 2-4 band
2-3 slip or rough shift	a) 2-3 shift valve sticking in valve body	a) Free sticking valve
	b) Accumulator valve sticking in valve body	b) Free sticking accumulator valve
	c) Valve body gaskets or spacer plate damaged	c) Replace spacer plate or gaskets
	d) Throttle valve sticking or T.V. limit valve sticking in valve body	d) Free sticking valve as required

CHILTON'S THREE "C's" AUTOMATIC TRANSMISSION DIAGNOSIS CHART

Condition	Cause	Correction
2-3 slip or rough shift	e) 3-4 clutch plates burned, piston seals damaged, case porosity, exhaust ball open, restricted apply passages	e) Repair or replace defective component as required
	f) 3-4 clutch check ball #7 damaged, excessive clutch plate travel	f) Repair or replace defective component as required
3-4 slip or rough shift	a) 2-3 shift valve sticking in valve body	a) Free sticking valve
	b) Accumulator valve sticking in valve body	b) Free sticking accumulator valve
	c) Valve body gaskets or spacer plate damaged	c) Replace spacer plate or gaskets
	d) Throttle valve sticking or T.V. limit valve sticking in valve body	d) Free sticking valve as required
	e) Servo damaged or missing piston seals, piston bores, piston porosity or incorrect band apply pin	e) Repair or replace defective component as required
	f) 3-4 clutch burned	f) Replace defective clutch as required
	g) 2-4 band burned	g) Replace defective band as required
	h) 3-4 accumulator spring missing, piston porosity, bore damaged, restricted feed passage or broken oil seal ring	h) Repair or replace defective component as required
No reverse, or slips in reverse	a) Forward clutch will not release	a) Correct
	b) Manual linkage out of adjustment	b) Adjust manual linkage as required
	c) Reverse boost valve sticking in front pump assembly	c) Correct stuck valve as required
	d) Valve body spacer plate or gaskets damaged	d) Replace defective gaskets or spacer plate as required
	e) Low reverse clutch piston seals damaged, plates burned, restricted apply passages, or cover plate loose or gasket damaged	e) Repair or replace defective component as required
	f) Reverse input clutch plates burned, seals damaged or restricted apply passages	f) Repair or replace defective component as required
	g) Reverse input housing exhaust ball and capsule damaged	g) Repair or replace as required
No part throttle downshifts	a) Binding external or internal linkage	a) Adjust linkage as required
	b) T.V. modulator downshift valve or throttle valve binding in valve body	b) Replace defective component as required
	c) Throttle valve bushing or feed hole restricted	c) Correct as required
	d) Check ball #3 mispositioned in valve body	d) Replace or reposition check ball as required
No overrun braking—manual 3-2-1	a) External manual linkage not adjusted	a) Adjust linkage as required
	b) Overrun clutch plates burned or piston seals damaged	b) Replace clutch plates or seals as required
	c) Overrun clutch piston exhaust check ball sticks	c) Free sticking ball as required
	d) Valve body gaskets or spacer plate damaged	d) Replace damaged gaskets or spacer plate as required
	e) 4-3 sequence valve sticking in valve body	e) Free sticking valve as required

CHILTON'S THREE "C's" AUTOMATIC TRANSMISSION DIAGNOSIS CHART

Condition	Cause	Correction
No overrun braking—manual 3-2-1	f) Valve body check balls 3, 9 or 10 mispositioned g) Turbine shaft oil feed passage restricted, teflon seal rings damaged or plug missing	f) Replace or reposition check balls as required g) Repair or replace components as required
Drive in neutral	a) Forward clutch burned b) Manual linkage or manual valve disconnected or out of adjustment	a) Replace forward clutch as required b) Connect or adjust components as required
Second speed start—drive range	a) Governor assembly b) T.V. cable	a) Repair as required b) Adjust or repair as required
No park or will not hold in park	a) Actuator rod assembly bent or damaged b) Parking lock pawl return spring damaged c) Parking brake bracket, detent roller or parking pawl damaged d) Parking lock pawl interference with low reverse piston	a) Repair or replace parts as needed b) Correct as required c) Replace parts as required d) Correct interference

operating. A slide is fitted around the rotor and vane which automatically regulates pump output, according to the needs of the transmission. Maximum pump output is obtained when the priming spring has been fully extended and has the slide held against the side of the body. As the slide moves toward the center, the pump output is reduced until minimum output is reached.

PRESSURE REGULATOR

As the pump rotor rotates, the pump output is directed to the pressure regulator valve. The pressure regulator valve is held closed by the pressure regulator valve spring. As the pump pressure increases, the pressure regulator valve is opened, directing oil from the pressure regulator to a cavity on the side of the pump opposite the priming spring. This oil pressure acts against the priming spring and moves the slide, decreasing the pump output to a steady 65 psi. With the engine off, the slide is held in a maximum output position by the priming spring.

T.V. LIMIT VALVE

The pressure requirements of the transmission for apply of the band and clutches vary with engine torque and throttle opening. Under heavy throttle operation, 65 psi line pressure is not sufficient to hold the band and clutches on without slipping. To provide higher line pressure with greater throttle opening, a variable oil pressure related to throttle opening is desired. The throttle valve regulates line pressure in relation to carburetor opening. The T.V. limit valve limits this variable pressure to avoid excessive line pressure. The T.V. limit valve feeds the throttle valve and receives oil directly from the oil pump. As the pressure in the line leading from the T.V. limit valve and feeding the T.V. valve exceeds 90 psi, the pressure will push against the T.V. limit valve spring, bleeding off excess pressure. This limits T.V. feed pressure to a maximum of approximately 90 psi.

THROTTLE VALVE

Throttle valve pressure is related to carburetor opening which is related to engine torque. The system is a mechanical type with a direct, straight line relation between the carburetor throttle plate opening and the transmission throttle plunger movement.

As the accelerator pedal is depressed and the carburetor opened, the mechanical linkage (T.V. Cable) relays the movement to the throttle plunger and increases the force of the T.V. spring against the throttle valve, increasing T.V. pressure which can regulate to 90 psi. T.V. oil is directed through the T.V. plunger to provide a hydraulic assist reducing the pedal effort necessary to actuate the plunger.

LINE BOOST VALVE

A feature has been included in the T.V. system that will prevent the transmission from being operated with low or minimum line pressure in the event the T.V. cable is disconnected or broken. This feature is the line boost valve which is located in the control valve and pump assembly at the T.V. regulating exhaust port.

The line boost valve is held off its seat by the throttle lever and bracket assembly (this allows T.V. oil to regulate normally) when the T.V. cable is properly adjusted. If the T.V. cable becomes disconnected or is not adjusted properly, the line boost valve will close the T.V. exhaust port and keep T.V. and line pressure at full line pressure.

Diagnosis Tests

CONTROL PRESSURE TEST

NOTE: Before making the control pressure test, check the transmission fluid level, adjust the T.V. cable, check and adjust the manual linkage and be sure that the engine is not at fault, rather that the automatic transmission.

1. Install the oil pressure gauge to the automatic transmission. Connect a tachometer to the engine.
2. Raise and support the vehicle safely. Be sure that the brakes are applied at all times.
3. Total running time must not exceed two minutes.
4. Minimum line pressure check: set the T.V. cable to specification. Apply the brakes and take the reading in the ranges and rpm's that are indicated in the automatic transmission oil pressure chart.
5. Full line pressure check; hold the T.V. cable the full extent of its travel. Apply the brakes and take the reading in the ranges and rpm's that are indicated in the automatic transmission oil pressure chart.
6. Record all readings and compare them with the data in the chart.

AUTOMATIC TRANSMISSION OIL PRESSURES

Range	Model	Normal Oil Pressure At Minimum T.V. psi	Normal Oil Pressure At Full T.V. psi
Park & Neutral @ 1000 rpm	TC, MB, MC, MJ, VN	55–65	130–170
	TE, TH, TK, MD, ME, MK, MW, VH	55–65	130–170
	VA, ML, T7, MP, MS, PQ, YN, YK, YP, YG, YF, T2	55–65	130–170
	T8, TZ, TP, TS, MH, VJ, TL	65–75	130–170
	YH	65–75	140–180
	Y9	55–65	140–180
Reverse @ 1000 rpm	TC, MB, MC, MJ, VN	90–105	210–285
	TE, TH, TK, MD, ME, MK, MW, VH	90–105	210–285
	VA, ML, T7, MP, MS, PQ, YN, YK, YP, YG, YF, T2	90–105	210–285
	T8, TZ, TP, TS, MH, VJ, TL	110–120	210–285
	YH	110–120	225–300
	Y9	90–105	225–300
Drive & Manual Third @ 1000 rpm	TC, MB, MC, MJ, VN	55–65	130–170
	TE, TH, TK, MD, ME, MK, MW, VH	55–65	130–170
	VA, ML, T7, MP, MS, PQ, YN, YK, YP, YG, YF, T2	55–65	130–170
	T8, TZ, TP, TS, MH, VJ, TL	65–75	130–170
	YH	65–75	140–180
	Y9	55–65	140–180
Manual Second & LO @ 1000 rpm	TC, MB, MC, MJ, VN	100–120	100–120
	TE, TH, TK, MD, ME, MK, MW, VH	100–120	100–120
	VA, ML, T7, MP, MS, PQ, YN, YK, YP, YG, YF, T2	100–120	100–120
	T8, TZ, TP, TS, MH, VJ, TL	100–120	100–120
	YH	100–120	100–120
	Y9	100–120	100–120

AIR PRESSURE TEST

Air pressure testing should be done in moderation to avoid excessive fluid spray and damage to the internal parts during disassembly and assembly, through partial retention of units.

STALL SPEED TEST

Stall speed testing is not recommended by General Motors Transmission Division. Extreme overheating of the transmission unit can occur, causing further internal damages. By pressure testing and road testing, the malfunction can be determined and by consulting the diagnosis chart, the cause and correction can normally be found.

ROAD TEST

1. Road test using all selective ranges, noting when discrepancies in operation or oil pressure occur.
2. Attempt to isolate the unit or circuit involved in the malfunction.
3. If engine performance indicates and engine tune-up is required, this should be performed before road testing is completed or transmission correction attempted. Poor engine performance can result in rough shifting or other malfunctions.

Converter Stator Operation

The torque converter stator assembly and its related roller clutch can possibly have one of two different type malfunctions.
1. The stator assembly freewheels in both directions.
2. The stator assembly remains locked up at all times.

CONDITION A

If the stator roller clutch becomes ineffective, the stator assembly freewheels at all times in both directions. With this condition, the vehicle will tend to have poor acceleration from a standstill. At speeds above 30-35 mph, the vehicle may act normal. If poor acceleration problems are noted, it should first be determined that the exhaust system is not blocked, the engine is in good tune and the transmission is in gear when starting out.

If the engine will freely accelerate to high rpm in neutral, it can be assumed that the engine and exhaust system are normal. Driving the vehicle in reverse and checking for poor performance will help determine if the stator is freewheeling at all times.

CONDITION B

If the stator assembly remains locked up at all times, the engine rpm and vehicle speed will tend to be limited or restricted at high speeds. The vehicle performance when accelerating from a standstill will be normal. Engine over-heating may be noted. Visual examination of the converter may reveal a blue color form the overheating that will result.

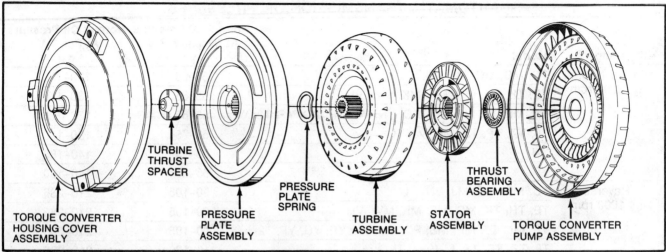

Torque converter assembly—exploded view (© General Motors Corporation)

Under conditions A or B above, if the converter has been removed from the transmission, the stator roller clutch can be checked by inserting a finger into the splined inner race of the roller clutch and trying to turn the race in both directions. The inner race should turn freely in the clockwise direction, bu not turn or be very difficult to turn in the counterclockwise direction.

Converter Clutch Operation and Diagnosis

TORQUE CONVERTER CLUTCH

The torque converter clutch assembly consists of a three-element

torque converter with the addition of a converter clutch. The converter clutch is splined to the turbine assembly, and when operated, applies against the converter cover providing a mechanical direct drive coupling of the engine to the planetary gears.

Converter clutch operation is determined by a series of controls and by drive range selection. The transmission must be in drive range, and the car must have obtained a preset speed depending on engine and transmission combination.

For more detailed information regarding the torque converter clutch, refer to the general section in this manual.

CHILTON'S THREE "C's" TRANSMISSION DIAGNOSIS CHART— G.M. TORQUE CONVERTER CLUTCH (TCCC)

Condition	Cause	Correction
Clutch applied in all ranges (engine stalls when put in gear)	a) Converter clutch valve stuck in apply position	a) R&R oil pump and clean valve
Clutch does not apply: applies erratically or at wrong speeds	a) Electrical malfunction in most instances	a) Follow troubleshooting procedure to determine if problem is internal or external to isolate defect
Clutch applies erratically; shudder and jerking felt	a) Vacuum hose leak b) Vacuum switch faulty c) Governor pressure switch malfunction d) Solenoid loose or damaged e) Converter malfunction; clutch plate warped	a) Repair hose as needed b) Replace switch c) Replace switch d) Service or replace e) Replace converter
Clutch applies at a very low or high 3rd gear	a) Governor switch shorted to ground b) Governor malfunction c) High line pressure d) Solenoid inoperative or shorted to case	a) Replace switch b) Service or replace governor c) Service pressure regulator d) Replace solenoid

CAUTION: When inspecting the stator and turbine of the torque converter clutch unit, a slight drag is normal when turned in the direction of freewheel rotation because of the pressure exerted by the waved spring washer, located between the turbine and the pressure plate.

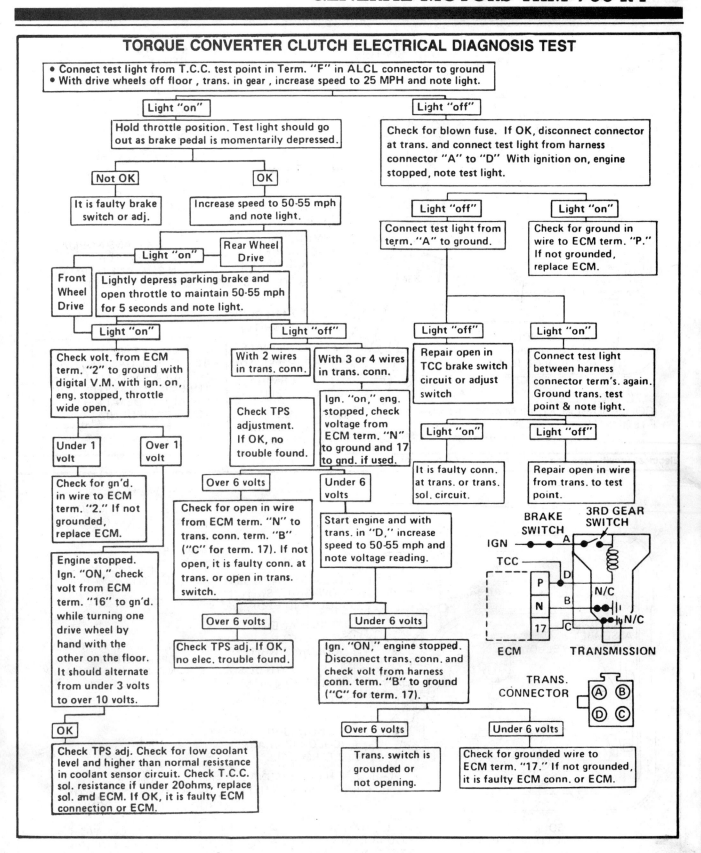

TORQUE CONVERTER CLUTCH ELECTRICAL DIAGNOSIS TEST

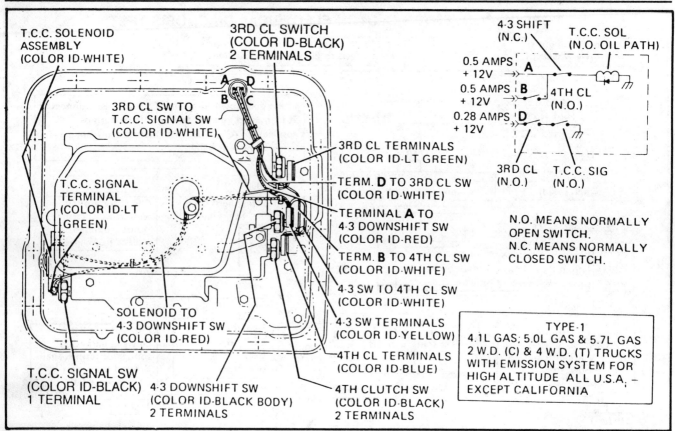

Electrical circuit—type one (© General Motors Corporation)

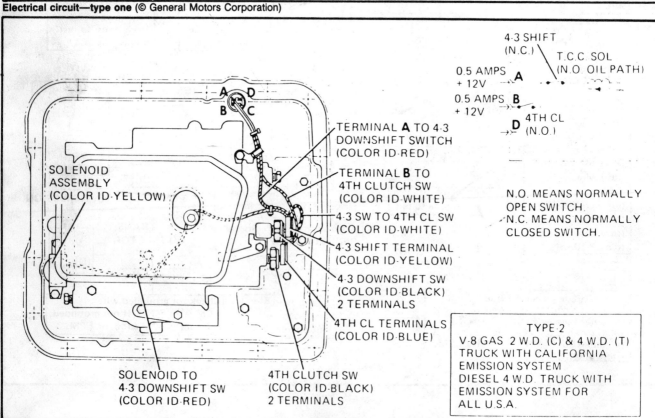

Electrical circuit—type two (© General Motors Corporation)

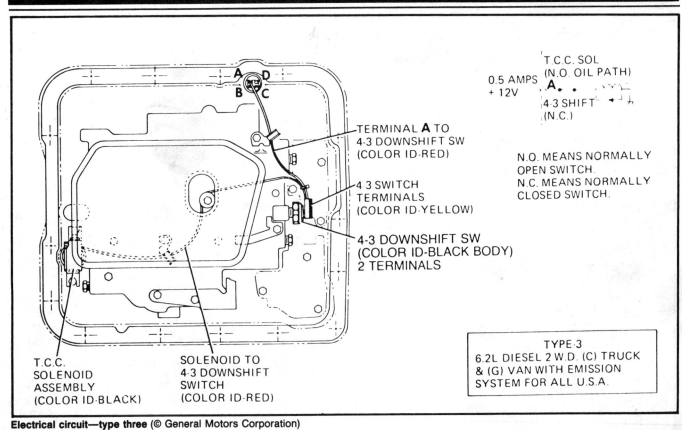

Electrical circuit—type three (© General Motors Corporation)

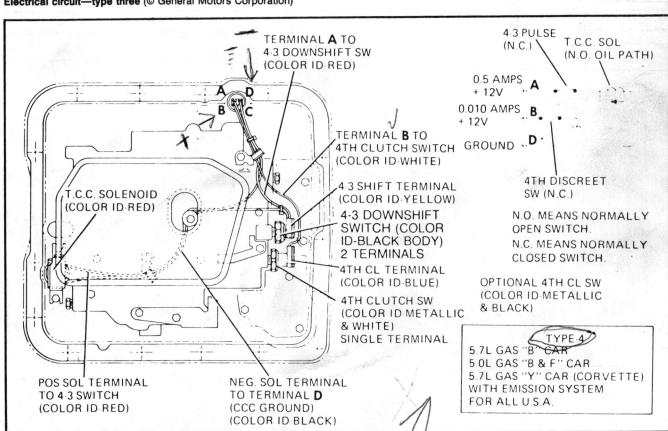

Electrical circuit—type four (© General Motors Corporation)

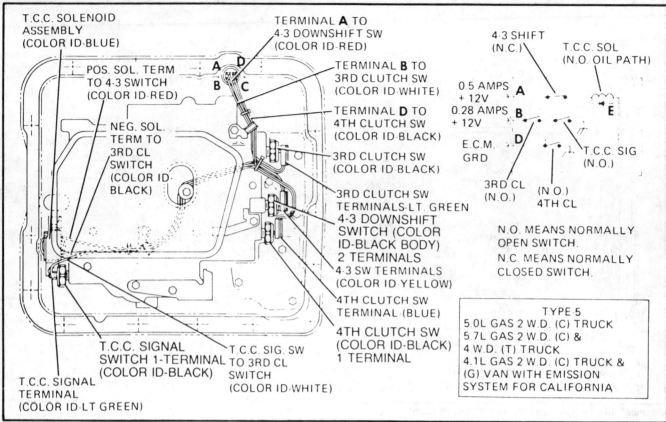

Electrical circuit—type five (© General Motors Corporation)

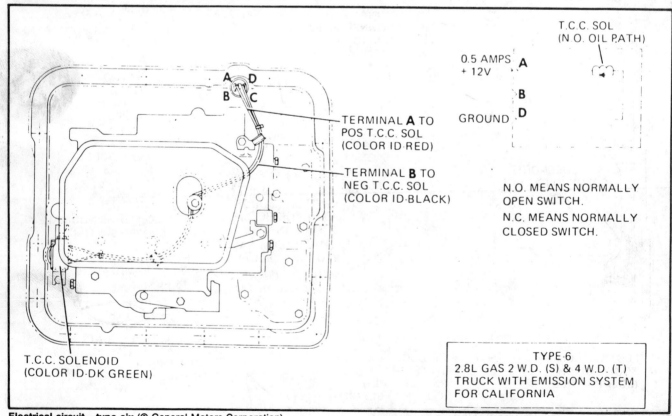

Electrical circuit—type six (© General Motors Corporation)

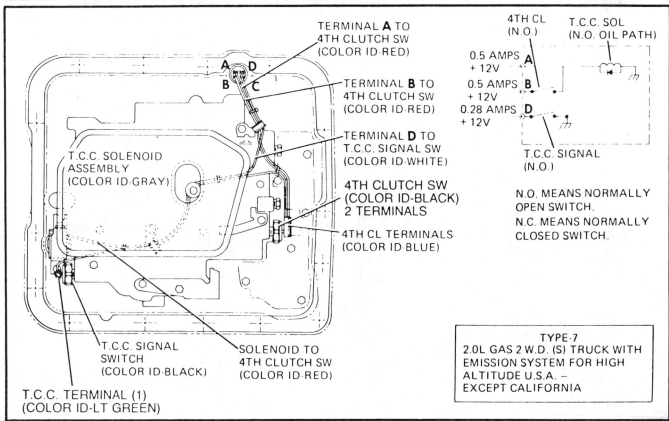

Electrical circuit—type seven (© General Motors Corporation)

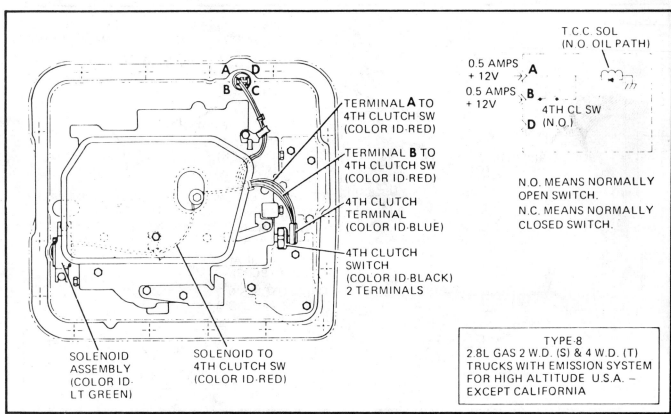

Electrical circuit—type eight (© General Motors Corporation)

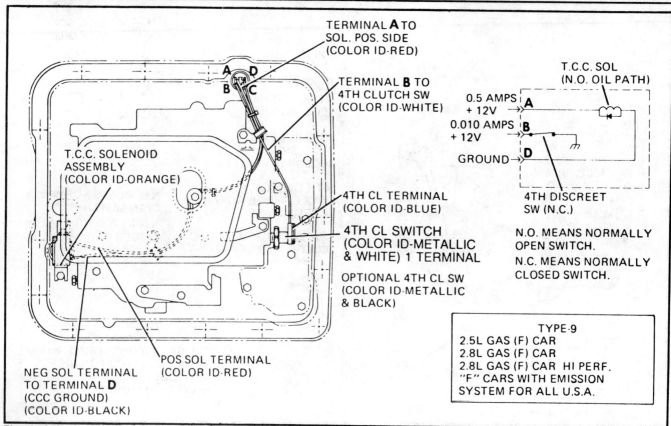

Electrical circuit—type nine (© General Motors Corporation)

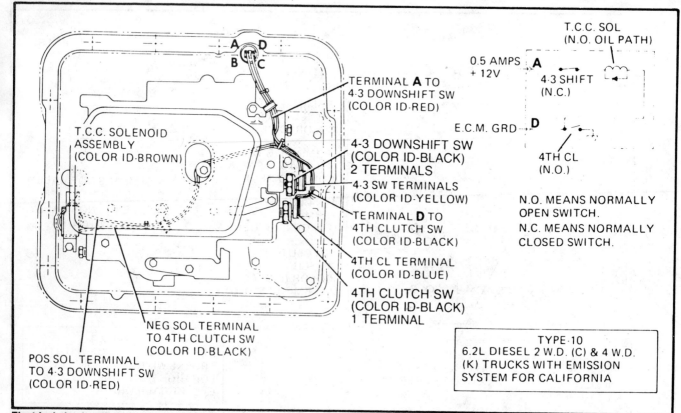

Electrical circuit—type ten (© General Motors Corporation)

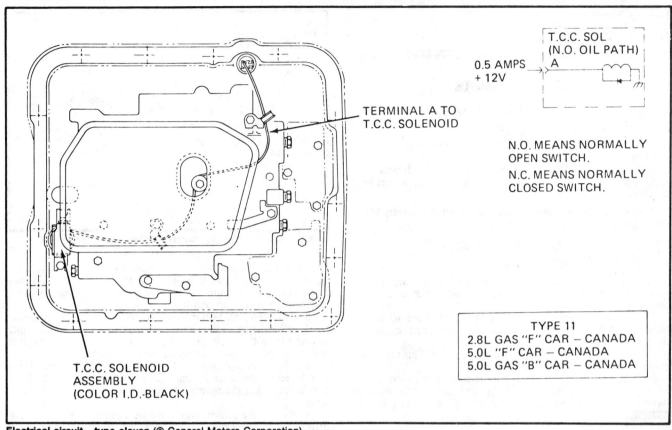

TERMINAL A TO
T.C.C. SOLENOID

T.C.C. SOL
(N.O. OIL PATH)
A

0.5 AMPS
+ 12V

N.O. MEANS NORMALLY
OPEN SWITCH.

N.C. MEANS NORMALLY
CLOSED SWITCH.

TYPE 11
2.8L GAS "F" CAR — CANADA
5.0L "F" CAR — CANADA
5.0L GAS "B" CAR — CANADA

T.C.C. SOLENOID
ASSEMBLY
(COLOR I.D.-BLACK)

Electrical circuit—type eleven (© General Motors Corporation)

ON CAR SERVICES

Adjustments

SHIFT INDICATOR

COLUMN MOUNTED SHIFTER

1. With the engine off, position the selector lever in the neutral position.
2. If the pointer does not align with the "N" indicator position, move the clip on the shift bowl until alignment has been achieved.

NOTE: The manual linkage must be adjusted correctly before this adjustment can be made properly.

NEUTRAL SAFETY SWITCH

FLOOR MOUNTED SHIFTER

1. New switches comes with a small plastic alignment pin installed. Leave this pin in place. Position the shifter assembly in neutral.
2. Remove the old switch and install the replacement, align the pin on the shifter with the slot in the switch, and fasten with the two screws.
3. Move the shifter from the neutral position. This shears the plastic alignment pin and frees the switch.

If the switch is to be adjusted, not replaced, insert a 3/32 in. drill bit or similar size pin and align the hole and switch. Position switch, adjust as necessary. Remove the pin before shifting from neutral.

COLUMN MOUNTED SHIFTER

1. Remove wire connectors from the combination back-up and neutral safety switch.
2. Remove two screws attaching the switch to the steering column.
3. Installation is the reverse of removal. To adjust a new switch:

    a. Position the shift lever in neutral.
    b. Loosen the attaching screws. Install a 0.090 in. gauge pin into the outer hole in the switch cover.
    c. Rotate the switch until the pin goes into the alignment hole in the inner plastic slide.
    d. Tighten the switch to column attaching screws and remove the gauge pin. Torque the screws to 20 in. lbs. maximum.
    c. Make sure that the engine starts only in the park and neutral positions.

MECHANICAL NEUTRAL START SYSTEM

Vehicles with this system use a mechanical block rather than an electrical neutral start system. The system only allows the lock cylinder to rotate to the start position when the shift lever is in neutral or park.

PARK LOCK CABLE

FLOOR MOUNTED SHIFTER

1. Remove the floor shift knob. Remove the console trim panel. Remove the hush panel.
2. Snap the park lock cable (column end) into the steering column bracket.
3. Snap the cable to the steering column sliding pin. The steering column sliding pin must have the ignition lock cylinder in the lock position.
4. Position the selector lever in the park position.
5. Install the park lock cable terminal to the shifter park lock

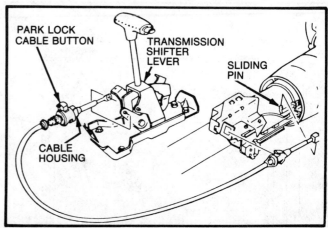

Park lock cable—floor shift models (© General Motors Corporation)

lever pin. Install the retainer pin. Install the park lock cable to the shifter mounting bracket by pushing the lock button housing against the adjusting spring. Drop the cable through the slot in the mounting bracket and seat the housing to shifter.

6. Push the lock button down to complete the adjustment.
7. With the ignition key in the lock position, depress the button on the shift handle. Button travel must not be sufficient to move the shifter lever out of the park position.
8. Move the ignition key to the run position. The shifter should select the gears by the handle rotating rearward. Check and be sure that the ignition lock cylinder cannot be turned to lock.
9. Return the shifter lever to the park position. Return the ignition key to the lock position. Check that the ignition key can be removed from the ignition switch.
10. If the ignition key is difficult to remove, pry up on the park lock cable button. Pull the cable housing rearward one notch and push the lock button down.
11. If the ignition key is still difficult to remove repeat the above step.

NOTE: If the park lock cable housing is moved to far rearward, the shift lever may come out of the park position when the ignition is locked with the key removed.

T.V. AND DETENT CABLE

NOTE: Before any adjustment is made to the T.V. or detent cable be sure that the engine is operating properly, the transmission fluid level is correct, the brakes are not dragging, the correct cable has been installed and that the cable is connected at both ends.

DIESEL ENGINE

1. Stop the engine. If equipped, remove the cruise control rod.
2. Disconnect the transmission T.V. cable terminal from the cable actuating lever.
3. Loosen the lock nut on the pump rod and shorten it several turns.
4. Rotate the lever assembly to the full throttle position. Hold this position.
5. Lengthen pump rod until the injection pump lever contacts the full throttle stop.
6. Release the lever assembly and tighten pump rod lock nut. Remove the pump rod from the lever assembly.
7. Reconnect the transmission T.V. cable terminal to cable actuating lever.
8. Depress and hold the metal re-adjust tab on the cable upper end. Move the slider through the fitting in the direction away from the lever assembly until the slider stops against the fitting.

9. Release the re-adjust tab. Rotate the cable actuating lever assembly to the full throttle stop and release the cable actuating lever assembly. The cable slider should adjust out of the cable fitting toward the cable actuating lever.

10. Reconnect the pump rod and cruise control throttle rod if so equipped.

11. If equipped with cruise control, adjust the servo throttle rod to minimum slack (engine off) then put clip in first free hole closest to the bellcrank but within the servo bail.

GAS ENGINE

1. Stop engine.

2. Depress and hold the metal re-adjust tab on the cable upper end. Move the slider through the fitting in the direction away from the lever assembly until the slider stops against the fitting. Release readjust tab.

3. Rotate the cable actuating lever assembly to the full throttle stop and release the cable actuating lever assembly. The cable slider should adjust out of the cable fitting toward the cable actuating lever. Check cable for sticking and binding.

4. Road test the vehicle. If the condition still exists, remove the transmission oil pan and inspect the throttle lever and bracket assembly on the valve body.

5. Check to see that the T.V. exhaust valve lifter rod is not distorted or binding in the valve body assembly or spacer plate.

6. The T.V. exhaust check ball must move up and down as the lifter does. Be sure that the lifter spring holds the lifter rod up against the bottom of the valve body.

7. Make sure that the T.V. plunger is not stuck. Inspect the transmission for correct throttle lever to cable link.

MANUAL LINKAGE

There are many variations of the manual linkage used with the THM 700-R4 due to its wide vehicle application. A general adjustment procedure is outlined for the convenience of the repairman.

Loosen the shift rod adjusting swivel clamp at the cross shaft. Place the transmission detent in the neutral position and the column selector in the neutral position. Tighten the adjustment swivel clamp and shift the selector lever through all ranges, confirming the transmission detent corresponds to the shift lever positions and the vehicle will only start in the park or neutral positions.

Services

DETENT CABLE

Removal

1. If required, remove the air cleaner.

2. Push in on the readjust tab and move the slider back through the fitting in the direction away from the throttle lever.

3. Disconnect the cable terminal from the throttle lever. Compress the locking tab and disconnect the cable assembly from the bracket.

4. Remove the routing clips or straps. Remove the screw and washer securing the cable to the transmission and disconnect the cable from the link.

Installation

1. Install a new oil seal in the transmission case hole. Position and install the cable at the transmission case. Torque the retaining bolt to 8 ft. lbs.

2. Route the cable as removed and connect the clip or strap. Pass the cable through the bracket and engage the locking tabs of the cable on the bracket.

3. Connect the cable terminal to the throttle lever. Adjust the cable as required.

4. Install the air cleaner, if removed.

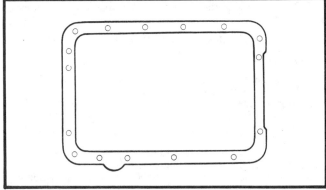

Oil pan identification (© General Motors Corporation)

FLUID CHANGE AND OIL PAN

Removal

1. Raise and safely support the vehicle.

2. Place drain pan under transmission oil pan. Remove the oil pan attaching bolts from the front and side of the pan.

3. Loosen, but do not remove the rear pan bolts, then bump the pan loose and allow fluid to drain.

4. Remove the remaining bolts and remove pan.

5. Clean pan in solvent and dry with compressed air.

6. Remove the two bolts holding the filter screen to the valve body. Discard gasket.

Installation

1. Install the filter assembly using a new gasket. Torque the retaining bolts to 6-10 ft. lbs.

2. Install new gasket on pan and torque bolts to 10-13 ft. lbs.

3. Lower the vehicle and add 3 quarts of Dexron® II.

4. With selector in park start engine and idle. Apply parking brake. *Do not race engine.*

Move selector through all ranges and end in park. Check fluid level. Add if necessary to bring the level between the dimples on the dipstick.

VALVE BODY

Removal

1. Raise the vehicle on a hoist and support it safely.

2. Drain the transmission fluid. Remove the fluid pan. Discard the gasket.

3. Remove the two bolts retaining the filter. Remove the filter from the valve body.

4. Disconnect the electrical connectors at the valve body assembly.

5. Remove the detent spring and roller assembly from the valve body. Remove the valve body to transmission case retaining bolts.

6. Remove the valve body assembly while disconnecting the manual control valve link from the range selector inner lever. Remove the throttle lever bracket from the T.V. link.

7. Position the valve body assembly down, with the spacer plate side up.

Installation

1. Using new gaskets, as required, position the valve body assembly against the transmission. Install the valve body retaining bolts and torque them 7-10 ft. lbs.

2. Continue the installation in the reverse order of the removal procedure.

3. Use a new gasket when installing the oil pan.

4. Lower the vehicle. Fill the transmission to specification using the proper grade and type automatic transmission fluid.

ON CAR SERVICES

5. Start the engine and check for leaks. Road test the vehicle as required.

SERVO ASSEMBLY

Removal

1. Raise the vehicle and support safely.
2. Drain the transmission fluid, as required.
3. Remove the two oil pan bolts and install tool #J-29714 or equivalent. Using the tool, depress the servo cover.
4. With a suitable tool, remove the servo cover retaining ring. Remove the tool.
5. Remove the servo cover and the seal ring. The seal ring may be stuck inside the transmission case.
6. Remove the servo piston and the band apply pin assembly.

Installation

1. Position the band apply pin assembly and the servo piston in its mounting within the transmission case.
2. Using a new seal ring install the servo cover.
3. Continue the installation in the reverse order of the removal procedure.
4. Lower the vehicle. Check and correct the fluid level, as required. Start the engine and check for leaks.

SPEEDOMETER DRIVEN GEAR

Removal

1. Raise the vehicle and support it safely.
2. Disconnect the speedometer cable.
3. Remove the retainer bolt, retainer, speedometer driven gear and O-ring seal.

Installation

1. Coat the new O-ring seal with transmission fluid and install it along with the other components.
2. Install the speedometer cable.
3. Lower the vehicle. Replace any transmission that may have been lost.

REAR OIL SEAL

Removal

1. Raise the vehicle and support it safely.
2. Remove the driveshaft. Place a container under the driveshaft to catch any fluid that may leak from the transmission.
3. If equipped, remove the tunnel strap.
4. Pry out the old seal using a suitable tool.

Installation

1. Coat the new seal with clean transmission fluid.
2. Position the seal on the rear extension housing and drive it in place using the proper seal installation tools.
3. Continue the installation in the reverse order of the removal procedure.
4. Lower the vehicle. Fill the transmission with clean oil, as required.

GOVERNOR

Removal

1. Raise the vehicle and support it safely. Drain the transmission fluid as required.
2. If the vehicle is equipped with four wheel drive, disconnect the shift rod at the shifter. Remove the transfer case shifter bolts at the transmission case and move the unit aside.
3. Remove the governor cover from the side of the transmission case. Be sure not to damage the cover.
4. Remove the governor from its mounting inside the transmission case.

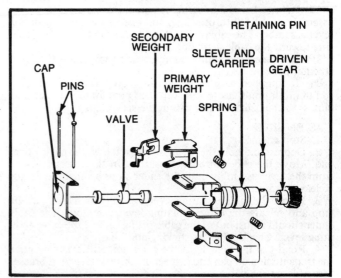

Governor assembly—exploded view
(© General Motors Corporation)

Installation

1. Position the governor-assembly inside the transmission case.
2. Install the governor cover using a new O-ring coated with clean transmission fluid.
3. Lower the vehicle. Check and correct the transmission fluid level as required. Start the engine and check for leaks.

REMOVAL & INSTALLATION

ALL EXCEPT TRUCKS

Removal

1. Disconnect the negative battery cable.
2. Remove the air cleaner assembly. Disconnect the T.V./detent cable at its upper end.
3. Remove the transmission dipstick. Remove the bolt holding the dipstick tube, if accessible.
4. Raise the vehicle and support it safely. Remove the torque arm to transmission retaining bolts, on F Series vehicles.

NOTE: Rear spring force will cause torque arm to move toward floor pan when arm is disconnected from transmission. Carefully place a piece of wood between the floor pan and torque arm when disconnecting the arm to avoid possible damage to floor pan and to avoid possible injury to hand or fingers.

5. Drain the transmission. Remove the driveshaft. Matchmark the driveshaft to aid in installation.
6. Disconnect the speedometer cable at the transmission. Disconnect the shift linkage at the transmission.
7. Disconnect all electrical leads at the transmission. Disconnect any clips that retain these leads.
8. Remove the inspection cover. Mark the flexplate and torque converter, to aid in installation.
9. Remove the torque converter to flexplate retaining bolts. Remove the catalytic converter support bracket, if necessary.
10. Position the transmission jack under the transmission and remove the rear transmission mount.
11. Remove the floor pan reinforcement. Remove the crossmember retaining bolts. Move the crossmember out of the way.
12. Remove the transmission to engine bolt on the left side.

4–286

This is the bolt that retains the ground strap.

13. Disconnect and plug the transmission oil cooler lines. It may be necessary to lower the transmission in order to gain access to the cooler lines.

14. With the transmission still lowered, disconnect the T.V. cable.

15. Support the engine, using the proper tools and remove the remaining engine to transmission bolts.

16. Carefully disengage the transmission assembly from the engine. Lower the transmission from the transmission jack.

Installation

1. To install, reverse the removal procedures.

2. Before installing the flexplate to torque converter bolts, make sure that the weld nuts on the converter are flush with the flexplate and that the torque converter rotates freely by hand in this position.

3. Be sure that the flexplate and torque converter marks made during the removal procedure line up with each other.

4. Adjust the shift linkage and the T.V. cable as required.

5. Lower the vehicle and fill the transmission with the proper grade and type automatic transmission fluid.

6. Start the engine and check for leaks. Once the vehicle has reached operating temperature recheck the fluid level.

TRANSMISSION REMOVAL

Removal
TRUCK MODELS

NOTE: If the vehicle is equipped with four wheel drive the transfer case must first be removed.

1. Disconnect the negative battery cable. Remove the air cleaner assembly.

2. Disconnect the T.V. cable at the upper end. Raise the vehicle and support it safely.

3. Remove the driveshaft. Match mark the assembly to aid in installation.

4. Disconnect the speedometer cable at the transmission. Disconnect the shift linkage at the transmission.

5. Disconnect all electrical leads at the transmission and remove any clips that retain the leads to the transmission.

6. Remove the brake line to crossmember clips and remove the crossmember, four wheel drive vehicles only.

7. Remove the transmission support brace attaching bolts at the torque converter, if equipped.

8. Remove the exhaust crossover pipe and catalytic converter attaching bolts. Remove the components as an assembly.

9. Remove the inspection cover and mark the torque converter in relation to the flywheel. This will assure proper line up on installation.

10. Remove the torque converter to flywheel retaining bolts. Disconnect the catalytic converter support bracket.

11. Position the transmission jack under the transmission and raise the unit slightly.

12. Remove the transmission support retaining bolts.

13. Remove the left body mounting bolts and loosen the radiator support mounting bolt.

14. Raise the vehicle on the left side to gain the necessary clearance to remove the upper attaching bolt. Support the vehicle by placing a block of wood between the frame and the first body mount.

15. Slide the transmission support toward the rear of the vehicle.

16. Lower the transmission jack in order to gain access to the oil cooler lines. Disconnect and plug the oil cooler lines.

17. Support the engine using the proper tool and remove the transmission to engine retaining bolts.

18. Slide the transmission back and lower it from the vehicle. Install the torque converter retaining tool as required.

Installation

1. To install, reverse the removal procedures.

2. Before installing the flexplate to torque converter bolts, make sure that the weld nuts on the converter are flush with the flexplate and that the torque converter rotates freely by hand in this position.

3. Be sure that the flexplate and torque converter marks made during the removal procedure line up with each other.

4. Adjust the shift linkage and the T.V. cable as required.

5. Lower the vehicle and fill the transmission with the proper grade and type automatic transmission fluid.

6. Start the engine and check for leaks. Once the vehicle has reached operating temperature recheck the fluid level.

TRANSFER CASE

Removal

1. Shift the transfer case into the 4HI position.

2. Disconnect the negative battery cable. Raise the vehicle and support it safely.

3. Drain the transfer case.

4. Match mark the transfer case driveshafts to aid in reinstallation. Remove and match mark the rear driveshaft.

5. Disconnect the speedometer cable and the vacuum harness at the transfer case.

6. Remove the catalytic converter hanger bolts at the converter.

7. Raise the transmission and transfer case and remove the transmission mount retaining bolts. Remove the mount and the catalytic converter hanger. Lower the transmission and the transfer case.

8. Support the transfer case and remove the case attaching bolts. Remove the shift lever mounting bolts, in order to remove the upper left transfer case retaining bolt.

9. Separate the transfer case from the adapter plate and remove the assembly from the vehicle.

Installation

1. Installation is the reverse of the removal procedure.

2. Fill the transfer case with the proper grade and type fluid.

3. Lower the vehicle. Start the engine and check for leaks. Roadtest as required.

Before Disassembly

Clean the exterior of the transmission assembly before any attempt is made to disassemble, to prevent dirt or other foreign materials from entering the transmission assembly or its internal parts.

NOTE: If steam cleaning is done to the exterior of the transmission, immediate disassembly should be done to avoid rusting from condensation of the internal parts.

Take note of thrust washer locations. It is most important that thrust washers and bearings be installed in their original positions. Handle all transmission parts carefully to avoid damage.

Ring groove wear on governor supports, input shaft, pump housings and other internal parts should be checked using new rings installed in the grooves.

Converter Inspection

The converter cannot be disassembled for service.

Checking Converter End Play

1. Place the converter on a flat surface with the flywheel side down.
2. Atttach a dial indicator so that the end play can be determined by moving the turbine shaft in relation to the converter hub. End play for the Turbo Hydra-Matic 700-R4 should be no more than 0.0-0.050 of an inch. If the clearance is greater than this, replace the converter.

Checking Converter One-Way Clutch

1. If the one-way clutch fails, then the stator assembly freewheels at all times in both directions. The car will have poor low speed acceleration.
2. If the stator assembly remains locked up at all times, the car speed will be limited or restricted at high speeds. The low speed operation may appear normal, although engine overheating may occur.
3. To check one-way clutch operation, insert a finger into the splined inner race of the roller clutch and try to turn the race in both directions. The inner race should turn freely in the clockwise direction, but it should not turn in the counterclockwise direction.

NOTE: Inspect the converter hub for burrs or jagged metal to avoid personal injury.

4. If the one-way clutch does not operate properly, the converter must be replaced.

Visual Inspection of the Converter

Before installation of the converter, check it carefully for damage, stripped bolt holes or signs of heat damage. Check for loose or missing balance weights, broken converter pilot, or leaks, all of which are cause for replacement. Inspect the converter pump drive hub for nicks or burrs that could damage the pump oil seal during installation.

Transmission Disassembly

TORQUE CONVERTER AND OIL PAN

1. Remove the torque converter from the transmission by pulling it straight out of the transmission housing.
2. Position the transmission assembly in a suitable holding fixture.
3. Remove the transmission fluid pan bolts. Discard the pan gasket, or R.T.V. sealant. Remove and discard the fluid screen.

VALVE BODY

1. Remove the transmission fluid pan retaining bolts. Remove the transmission pan from the transmission assembly.
2. Remove and discard the pan gasket. Remove the fluid filter and O-ring. Discard these components.
3. Disconnect the inner harness connector at the outside connector located in the transmission case, by bending the locking tab outward and pulling the connector upwards.
4. Remove the outside electrical connector and O-ring seal from the transmission case by bending the inner tab inward with a suitable tool and pushing downward.
5. Remove the solenoid and attaching bolts, and O-ring from the case and pump. If the solenoid is not free, gently pry up with a suitable tool.
6. Disconnect all wire leads at the pressure switches and remove the complete wiring harness and solenoid assembly.
7. Remove the 1-2 accumulator housing, attaching bolts, 1-2 accumulator spring, and piston, gasket and plate.
8. Remove the oil passage cover retaining bolts from the transmission case. Remove the cover.
9. Remove the manual detent roller assembly. Remove the harness wire retaining clips and retaining bolts.

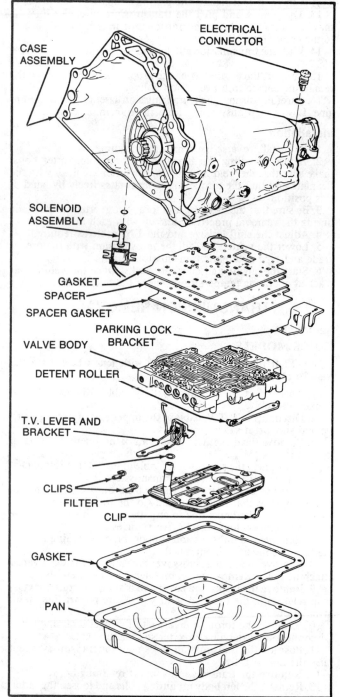

Valve body component location (© General Motors Corporation)

10. Remove the throttle lever bracket assembly and the T.V. link.
11. Remove the remaining valve body retaining bolts.
12. Unhook the manual valve retaining clip at the inside detent lever and remove the valve body, spacer plate and gaskets.
13. Be sure not to loose the three check balls located in the valve body.
14. Remove the 3-4 accumulator spring, piston and pin from the transmission case.
15. Remove the five check balls and the check valve from the transmission case passages.

16. Remove the converter clutch and governor screens from the transmission case.

GOVERNOR

1. Position the transmission assembly in a suitable holding fixture.
2. Remove the governor cover from the side of the transmission case. Be sure not to damage the cover.
3. Remove and discard the governor cover seal ring.

SECOND AND FOURTH SERVO

1. Install the servo removal tool on the servo cover. Depress the cover and remove the snap ring using a suitable tool.
2. Remove the servo cover and the O-ring. Remove the fourth gear apply piston and its O-ring.
3. Remove the second servo piston assembly.
4. Remove the inner servo piston assembly and the oil seal ring. Release the spring.
5. To check the 2-4 servo apply pin, proceed as follows.
6. Install the servo apply pin checking tool into the servo bore and secure it in place using a snap ring.
7. Disassemble the 2-4 servo using the proper tools and insert the servo pin into the tool, locating the end of the pin on the band anchor lug.
8. Apply 100 inch lbs. of torque using a torque wrench.
9. If any part of the white line appears in the window, the pin is the correct length. If the white line does not appear select another pin until the line does appear.

SERVO PIN SELECTION CHART

Pin Identification	Pin Length (inches)
Two rings	2.61-2.62
Three rings	2.67-2.68
Wide band	2.72-2.73

END PLAY CHECK

1. Position the transmission vertically, with the front pump side upward.
2. Remove the front pump to transmission case bolt and washer. Install an eleven inch bolt and locking nut.
3. Install the transmission end play checking tool on the end of the turbine shaft. Mount the dial indicator gauge on the bolt positioning the indicator point cap nut on top of the tool. With the dial indicator set at zero, pull upwards. The end play should be 0.005-0.036 inch. The selective washer controlling transmission end play is located between the input housing and the roller bearing on the pump hub. If more or less end clearance is required to bring the transmission within specifications, select the proper washer.

TRANSMISSION END PLAY— SELECTIVE WASHER CHART

Identification	Thickness (inches)
67	0.074–0.078
68	0.080–0.084
69	0.087–0.091
70	0.094–0.098
71	0.100–0.104
72	0.107–0.111
73	0.113–0.118
74	0.120–0.124

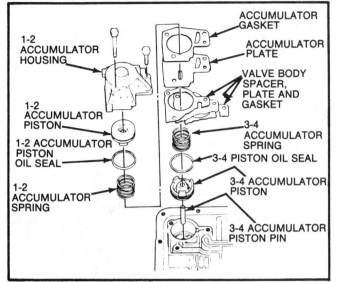

1-2 and 3-4 accumulator assembly—exploded view
(© General Motors Corporation)

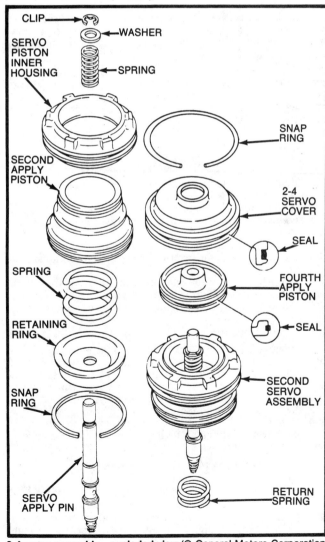

2-4 servo assembly—exploded view (© General Motors Corporation)

FRONT PUMP

1. Remove and discard the front pump seal.
2. Remove the front pump to transmission case retaining bolts and washers.
3. Using the pump removal tool, remove the front pump from the transmission case.

NOTE: The filter and the solenoid must be removed from the transmission assembly before the front pump can be removed.

REVERSE AND INPUT CLUTCHES

1. Remove the front pump assembly from the transmission case.
2. Remove the reverse input drum washer from the pump assembly.
3. Remove the reverse and input clutch assemblies from the transmission by lifting them out with the turbine shaft.
4. Do not remove the teflon seals on the turbine shaft unless required.

2-4 BAND AND INPUT GEAR SET

1. Remove the 2-4 band assembly from the transmission case. Remove the band anchor pin from the transmission case.
2. Remove the input gear assembly from the transmission case.

NOTE: The output shaft and the reaction internal gear assembly are assembled using loctite, which holds the parts together as one unit to aid in installation. During transmission operation these parts may separate.

3. Install the output shaft holding tool to the rear of the transmission case in order to prevent the output shaft from falling.
4. Remove the input carrier to output shaft snap ring.
5. Remove the output shaft tool and the output shaft.
6. Remove the input carrier and the thrust washer.
7. Remove the reaction shaft thrust bearing from the input internal gear assembly.

REACTION GEAR ASSEMBLY

1. Remove the reaction shaft to reaction sun gear washer. Remove the reaction shell.
2. Remove the reaction shell to inner race washer. Remove the low and reverse support to transmission case retaining ring and support spring.
3. Remove the reaction sun gear assembly.
4. Remove the low and reverse clutch plate assemblies and the low and reverse clutch plates.
5. Remove the reaction internal gear. Remove the output shaft if not removed and the bearing assembly.
6. Remove the support bearing assembly from the transmission case hub.

LOW AND REVERSE CLUTCH COMPONENTS

1. Remove the parking lock bracket and the retaining bolts, if not yet removed. Position the parking lock pawl inwards.
2. Using the clutch spring compressor tool, compress the low and reverse clutch spring retainer. Remove the spring retaining ring and the low and reverse spring assembly.
3. Remove the special tools.
4. Remove the low and reverse clutch piston. It may be necessary to apply compressed air to the transmission case passage to remove the component.

MANUAL LINKAGE (INNER)

NOTE: It is not necessary to remove the inner manual linkage unless service to these components is required.

1. Rotate the transmission to the horizontal position and loosen the manual shaft retaining nut and move inboard on the manual shaft.

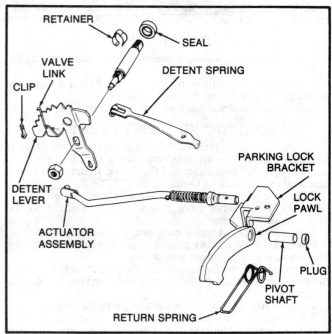

Parking lock linkage—exploded view (© General Motors Corporation)

2. Move the inner detent lever and connected actuator rod assembly and manual shaft retainer inboard.
3. Gently tap the manual shaft outboard until the retaining nut is free. If necessary install a retaining nut on the outside end of the manual shaft. With a suitable tool remove the manual shaft retainer and connect inner detent lever and actuator rod.
4. Remove the manual shaft and nut.
5. If necessary remove the inside detent lever from the actuator rod assembly by rotating the rod and indexing notches in the rod with the hole in the lever.
6. Inspect the manual shaft seal in the case for damage, and proper location and if necessary replace the seal.
7. Be sure that the new seal is bottomed square against the bottom of the transmission case bore and with the opened or sealing surface facing inward.
8. Inspect all components for wear, scoring or damage. Repair or replace defective components as required.

PARKING PAWL AND RETURN SPRING

1. Using an easy out, remove the parking lock shaft retaining plug from the outside of the transmission case.
2. Remove the parking lock shaft, pawl and return spring.
3. Inspect all components for wear, damage or scoring. Repair or replace defective components as required.

TRANSMISSION CASE AND RELATED COMPONENTS

1. Inspect the case for damage, cracks, porosity or interconnected oil passages.
2. Inspect the valve body case pad for flatness or land damage. The oil passage lands are considered acceptable if a void exists and is not in excess of fifty percent of the land. The condition of the valve body pad can also be checked by inspecting the mating face of the spacer plate to the case gasket for a proper case land impression.
3. Air check the case passages for restrictions or blockage.
4. Inspect the case internal clutch plate lugs for damage, or wear.
5. Inspect the speedometer, servo and accumulator bores for damage and clearance relative to the mating parts.

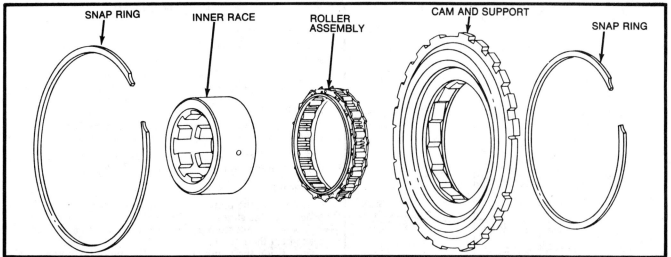

Low reverse clutch support assembly—exploded view (© General Motors Corporation)

6. Inspect all bolt holes for damaged threads. Helo-coils® can be used for repair.

7. Inspect the cooler line connectors for damage and proper torque.

8. Inspect all snap ring grooves for damage.

9. Inspect the governor locating pin for proper length. A incorrect length results in a damaged governor gear.

10. Inspect the third accumulator check valve, if replacement is needed, use a number four easy out and remove the check valve assembly from the transmission case.

11. Install a new check valve assembly into transmission case with the small end first. Position the oil feed slot in the sleeve so that it aligns with the servo passage.

12. Seat the check valve assembly using a ⅜ in. metal rod. Drive the check valve assembly into the case approximately 42.0mm.

Unit Disassembly and Assembly
GOVERNOR

Disassembly

1. Cut off the end of the governor weight pin. Remove the pins, outer weight and the secondary weight.

2. Remove the governor valve from the governor sleeve.

3. To replace the governor drive gear, remove the governor gear retaining pin.

4. Support the governor assembly on plates installed in the exhaust slots of the governor sleeve. Position the assembly in a press. Press the gear out of the sleeve.

5. Clean and inspect all parts of the governor assembly. Repair or replace defective components as required.

Assembly

1. Support the governor assembly on the plates which are installed in the exhaust slots of the sleeve.

2. Position a new gear on the sleeve and press it in place until it is seated against the shoulder.

3. A new pin hole must be drilled through the sleeve and gear. Locate the new hole position 90 degrees from the existing hole. While supporting the governor in press, center punch and drill a new hole through the sleeve and gear using a standard ⅛ in. drill.

4. Install the retaining pin. Stake the assembly in two locations.

5. Set the assembly aside until the transmission is ready to be reassembled.

LOW REVERSE SUPPORT ASSEMBLY
Disassembly

1. Remove the low reverse inner race.

2. Remove the snap ring retaining the roller assembly to the low reverse support.

3. Remove the roller assembly from its mounting.

4. Clean and inspect all components for wear, damage or scoring. Repair or replace defective components as required.

Assembly

1. Position the low reverse support on the bench so that the chamfered side is up.

2. Install the low reverse roller assembly into the support with the oil lube hole down or rearward.

NOTE: **Care should be taken to insure that the rollers and spring are not damaged and that the rollers do not become dislodged.**

3. Install the low reverse inner race into the roller assembly by rotating clockwise.

4. When installed, the inner race should rotate in the clockwise direction and lock up in the counterclockwise direction.

5. Set the assembly aside until the transmission is ready to be reassembled.

THROTTLE LEVER AND BRACKET ASSEMBLY
Disassembly

1. Unhook and remove the line boost spring. Remove the retaining nut from the pin.

2. Remove the pin, torsion lever spring, line boost lever, throttle lever and the bracket.

3. Clean and inspect all parts for wear, damage or scoring. Repair or replace defective components as required.

Assembly

1. Assembly of the throttle lever and bracket is the reverse of the disassembly procedure.

2. Set the assembly aside until you are ready to reassemble the transmission.

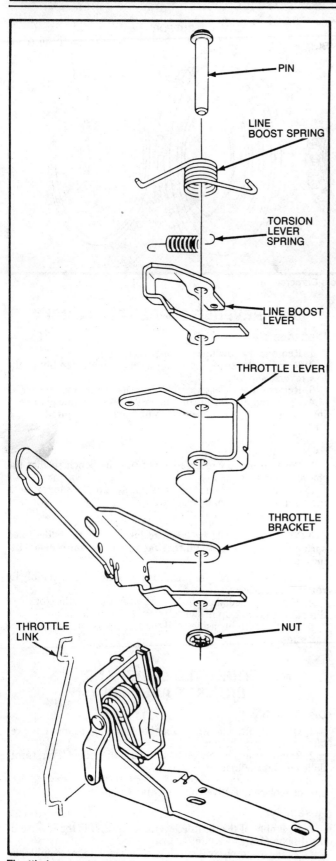

PIN

LINE
BOOST SPRING

TORSION
LEVER
SPRING

LINE BOOST
LEVER

THROTTLE LEVER

THROTTLE
BRACKET

THROTTLE
LINK

NUT

Throttle lever and bracket assembly (© General Motors Corporation)

FRONT PUMP

Disassembly

1. Remove the pump to drum washer from the front pump assembly.
2. Remove the front pump cover to transmission case gasket from the front pump cover.
3. Remove the front pump to case oil seal ring from the front pump assembly.

NOTE: Do not remove the two teflon oil seal rings from the front pump hub unless they require replacement.

4. Remove the front pump cover and separate the front pump cover from the front pump body.

Assembly

1. Position the pump body assembly over a hole in the work bench with the stator shaft facing down.
2. Assemble the front pump cover to the front pump body. Finger tighten the retaining bolts.
3. Position the pump cover and the pump body using tool J-21368 or equivalent. Place a holding bolt or a suitable tool through a pump to transmission case bolt hole and the bench hole.
4. Remove the tool and torque the retaining cover bolts to 18 ft. lbs.
5. Install a new pump to case gasket on the pump, use transmission fluid to retain the gasket in place.
6. If removed install the teflon oil seal rings on the stator hub.
7. Install the pump to case oil seal on the pump cover. Be sure not to twist the seal and make certain that the seal is seated properly.
8. Install the front pump cover to transmission case gasket aligning holes. Retain them using clean transmission fluid.
9. Install the front pump to drum thrust washer.

FRONT PUMP BODY

Disassembly

1. Remove the pump slide spring by compressing the spring and pulling it straight out.

NOTE: The spring is retained under pressure. Before removing it place a covering over the assembly to prevent injury.

2. From the pump pocket remove the pump guide rings, the pump vanes, the pump rotor and the rotor guide.
3. Remove the slide from the pump pocket. Remove the slide seal and support seal from the pump body pocket or slide.
4. Remove the pivot slide pin and spring from the pocket.
5. Remove the slide seal ring. Remove the slide backup seal from the slide pump pocket.
6. Clean and inspect all parts for wear, damage or scoring. Repair or replace defective components as required.

Assembly

1. Install the pump slide O-ring and flat steel ring into the groove on the back side of the slide.
2. Install the small pivot pin and spring into the small hole located in the pump pocket.
3. Install the pump slide into the pump, position the assembly so that the notch in the slide is indexed with the pivot pin hole and the flat oil seal ring facing downward into the pocket.
4. Install the pump slide seal and support into the slide adjacent to the rotor.

NOTE: The pump slide seal (composition) must be positioned against the other diameter of the pump pocket.

5. Install a vane ring into the pump pocket centering on the stator hole.
6. Install the composition rotor guide into the deep pocket of

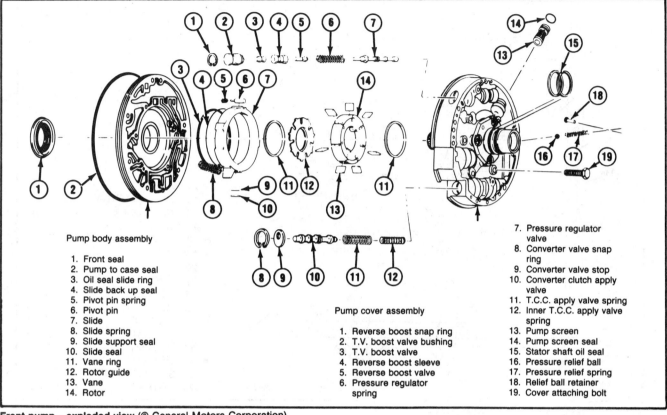

Pump body assembly

1. Front seal
2. Pump to case seal
3. Oil seal slide ring
4. Slide back up seal
5. Pivot pin spring
6. Pivot pin
7. Slide
8. Slide spring
9. Slide support seal
10. Slide seal
11. Vane ring
12. Rotor guide
13. Vane
14. Rotor

Pump cover assembly

1. Reverse boost snap ring
2. T.V. boost valve bushing
3. T.V. boost valve
4. Reverse boost sleeve
5. Reverse boost valve
6. Pressure regulator spring
7. Pressure regulator valve
8. Converter valve snap ring
9. Converter valve stop
10. Converter clutch apply valve
11. T.C.C. apply valve spring
12. Inner T.C.C. apply valve spring
13. Pump screen
14. Pump screen seal
15. Stator shaft oil seal
16. Pressure relief ball
17. Pressure relief spring
18. Relief ball retainer
19. Cover attaching bolt

Front pump—exploded view (© General Motors Corporation)

the rotor indexing the notches, and retain with transmission fluid.

7. Install the rotor and guide into the pump pocket with the guide positioned downward.

8. Install the vanes into the rotor positioning so they are flush with the rotor and with the full wear pattern against the slide.

9. Install the vane guide ring into the rotor.

10. Compress the pump slide spring and install into pump pocket.

FRONT PUMP COVER

Disassembly

1. Using a suitable tool, push on the retainer to compress the converter clutch apply valve spring and remove the snap ring. Release the valve spring tension. Remove the retainer and the converter clutch apply valve and spring.

2. Remove the pressure relief spring retaining rivet. The rivet is under strong spring pressure. Remove the pressure relief spring and ball. If required, apply air pressure to remove the ball.

3. Remove the oil screen and O-ring from the pump cover.

4. Position the pump assembly with the stator shaft facing downward. Secure the assembly to the work bench using a pump holding bolt.

5. Clean and inspect all components for wear, scoring and damage. Repair or replace defective components as required.

Assembly

1. Position the O-ring on the oil pump screen and install it in the bore of the front pump with the seal end last.

2. Install the pressure relief ball and spring. Be sure to install the ball first. Install the retaining rivet.

3. Install the converter clutch apply valve spring on the long

end of the converter clutch apply valve. Retain the components in place using transmission fluid.

4. Install the converter clutch apply valve and spring into the bore of the front pump. Install the retainer and the snap ring.

5. Position the pump cover so that the pressure regulator bore is located in the vertical position.

6. Install the pressure regulator valve into the bore of the pump cover with the large land and the orifice hole first, positioning with a magnet into the bottom of the bore.

7. Install the pressure regulator valve spring into the bore.

8. Install the T.V. boost valve into the T.V. bushing. Long land of valve into the large hole of the bushing—retain the valve with transmission fluid.

9. Install the reverse boost valve into the reverse sleeve, small end of valve first. Retain with transmission fluid.

10. Using a small magnet install the T.V. boost valve and bushing into the cover with the small hole in the bushing rearward.

11. With the T.V. boost valve and bushing compressed and with the snap ring groove visible, install the snap ring.

VALVE BODY

Disassembly

As each part of the valve train is removed, place the individual part in the order that it was removed and in the same relative location as its true position in the valve body. All parts must be reassembled in the same location as they were removed.

Remove all outside roll pins by pushing them through from the rough casting side of the valve body assembly. Removal of the inner roll pins can be accomplished by grinding a taper to one end of a number 49 ($\frac{1}{16}$ in.) drill bit. Lightly tap the tapered end of the drill bit into the roll pin and then pull the drill bit and the pin out of its bore.

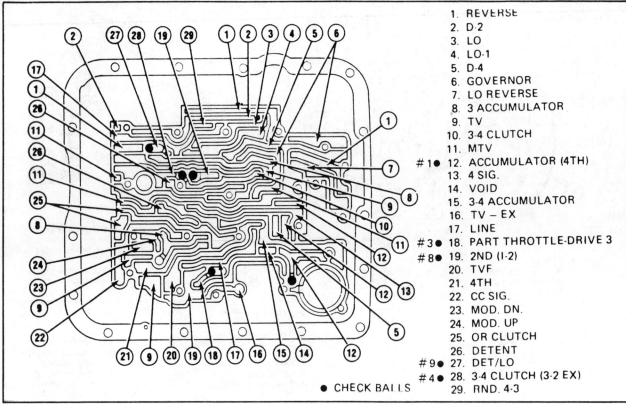

Case Oil Passages and Checkball Locations

1. REVERSE
2. D-2
3. LO
4. LO-1
5. D-4
6. GOVERNOR
7. LO REVERSE
8. 3 ACCUMULATOR
9. TV
10. 3-4 CLUTCH
11. MTV
#1● 12. ACCUMULATOR (4TH)
13. 4 SIG.
14. VOID
15. 3-4 ACCUMULATOR
16. TV — EX
17. LINE
#3● 18. PART THROTTLE-DRIVE 3
#8● 19. 2ND (1-2)
20. TVF
21. 4TH
22. CC SIG.
23. MOD. DN.
24. MOD. UP
25. OR CLUTCH
26. DETENT
#9● 27. DET/LO
#4● 28. 3-4 CLUTCH (3-2 EX)
29. RND. 4-3

● CHECK BALLS

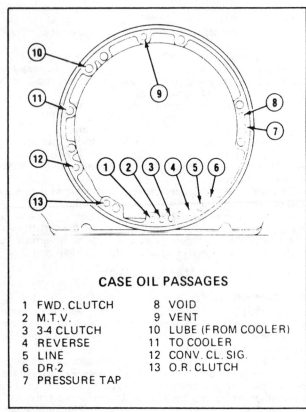

CASE OIL PASSAGES

1 FWD. CLUTCH	8 VOID
2 M.T.V.	9 VENT
3 3-4 CLUTCH	10 LUBE (FROM COOLER)
4 REVERSE	11 TO COOLER
5 LINE	12 CONV. CL. SIG.
6 DR-2	13 O.R. CLUTCH
7 PRESSURE TAP	

Pump to Case Oil Passages

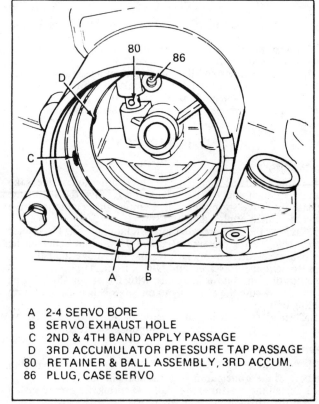

A 2-4 SERVO BORE
B SERVO EXHAUST HOLE
C 2ND & 4TH BAND APPLY PASSAGE
D 3RD ACCUMULATOR PRESSURE TAP PASSAGE
80 RETAINER & BALL ASSEMBLY, 3RD ACCUM.
86 PLUG, CASE SERVO

Servo Oil Passages

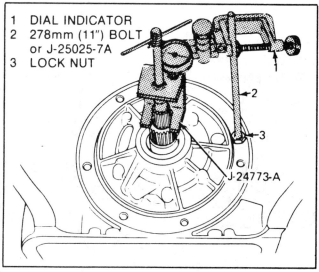

1 DIAL INDICATOR
2 278mm (11") BOLT
  or J-25025-7A
3 LOCK NUT

J-24773-A

End Play Check

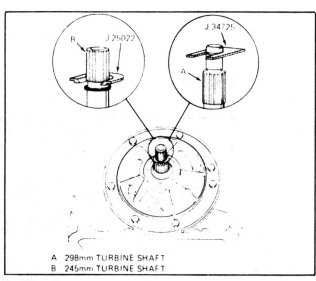

A 298mm TURBINE SHAFT
B 245mm TURBINE SHAFT

End Play Tool

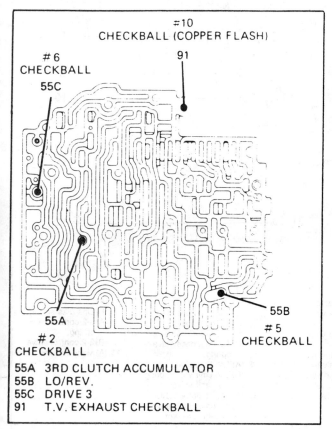

#10
CHECKBALL (COPPER FLASH)

91

#6
CHECKBALL

55C

#2
CHECKBALL

55A

55B

#5
CHECKBALL

55A 3RD CLUTCH ACCUMULATOR
55B LO/REV.
55C DRIVE 3
91 T.V. EXHAUST CHECKBALL

Valve Body Checkballs

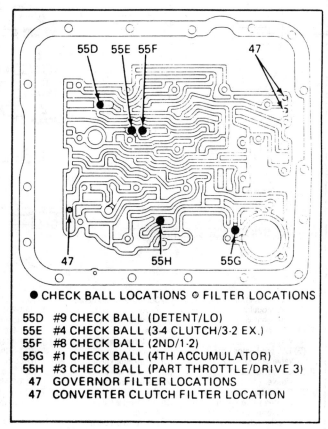

55D   55E 55F   47

47   55H   55G

● CHECK BALL LOCATIONS ⊘ FILTER LOCATIONS

55D #9 CHECK BALL (DETENT/LO)
55E #4 CHECK BALL (3-4 CLUTCH/3-2 EX.)
55F #8 CHECK BALL (2ND/1-2)
55G #1 CHECK BALL (4TH ACCUMULATOR)
55H #3 CHECK BALL (PART THROTTLE/DRIVE 3)
47   GOVERNOR FILTER LOCATIONS
47   CONVERTER CLUTCH FILTER LOCATION

Case Checkballs and Filters

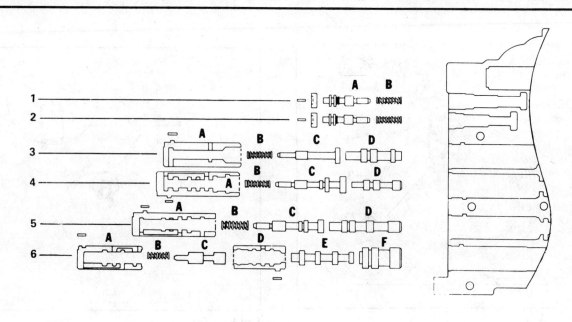

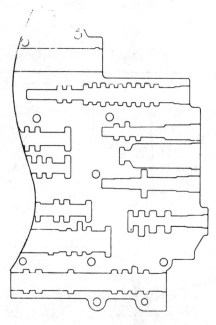

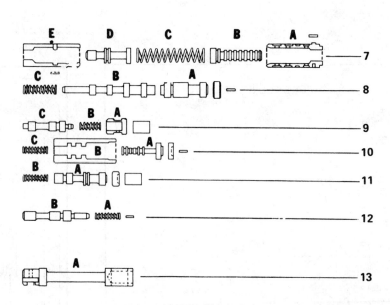

1. (A) T.V. modulator.
   (B) T.V. modulator downshift valve spring.
2. (A) T.V. modulator upshift valve.
   (B) T.V. modulator upshift valve spring.
3. (A) Converter clutch valve sleeve.
   (B) Converter clutch throttle valve spring.
   (C) Converter clutch throttle valve.

(D) Converter clutch shift valve.
4. (A) 3-4 throttle valve sleeve.
   (B) 3-4 throttle valve spring.
   (C) 3-4 throttle valve.
   (E) 3-4 shift valve.
5. (A) 2-3 throttle valve sleeve.
   (B) 2-3 throttle valve spring.
   (C) 2-3 throttle valve.

(D) 2-3 shift valve.
6. (A) 1-2 throttle valve sleeve.
   (B) 1-2 throttle valve spring.
   (C) 1-2 throttle valve.
   (D) Lo range sleeve.
   (E) 1-2 lo range valve.
   (F) 1-2 shift valve.
7. (A) Throttle valve plunger sleeve.
   (B) Throttle valve plunger.
   (C) Throttle valve spring.

(D) Throttle valve.
   (E) Throttle valve sleeve.
8. (A) 3-4 relay valve.
   (B) 4-3 sequence valve.
   (C) 4-3 sequence valve spring.
9. (A) T.V. limit plug.
   (B) T.V. limit valve spring.
   (C) T.V. limit valve.
10. (A) Accumulator valve.
    (B) Accumulator sleeve.
    (C) Accumulator spring.

11. (A) Line bias valve.
    (B) Line bias spring.
12. (A) 3-2 control valve spring.
    (B) 3-2 control valve.
13. (A) Manual valve.

Valve body assembly—bore location (© General Motors Corporation)

The spring retaining sleeves can be removed by compressing with needle-nose pliers and moving upward through the exposed hole.

Some of the roll pins have applied pressure against them. When removing, care should be taken to prevent the possible loss of parts.

Do not remove the pressure switches unless they require replacement.

Remove the three check balls from the passage side of the body—if present.

Position the valve body machined side up: positioning the manual valve lower right and remove the link and retaining clip, if attached.

1. From the No. 1 bore, remove the retaining pin, valve bore plug, T.V. modulator downshift valve and T.V. modulator downshift valve spring.

2. From bore No. 2, remove the retaining pin, valve bore sleeve, T.V. modulator upshift valve and T.V. modulator upshift valve spring.

3. From bore No. 3, remove the retaining pin. Remove the converter clutch throttle sleeve, converter clutch throttle valve spring and valve, and the converter clutch shift valve.

4. From bore No. 4, remove the retaining pin, 3-4 throttle valve sleeve, 3-4 throttle valve spring, 3-4 throttle valve and 3-4 shift valve.

5. From bore No. 5, remove the retaining pin. Remove the 2-3 throttle valve sleeve and 2-3 throttle valve spring, 2-3 throttle valve and 2-3 shift valve.

6. From bore No. 6, remove the outer roll pin. Remove the 1-2 throttle valve sleeve, 1-2 throttle valve spring, 1-2 throttle valve and lo range valve. Remove the inner retaining pin and remove the lo range valve sleeve and 1-2 shift valve.

7. From bore No. 7, remove the outer roll pin from the rough casting side, the throttle valve plunger sleeve, throttle valve plunger and throttle valve spring. Remove the inner roll pin and valve.

8. From bore No. 8, remove the retaining roll pin and plug. Remove the 3-4 relay valve, 4-3 sequence valve and spring.

9. From bore No. 9, using needle nose pliers, compress and remove the spring retainer. Remove the T.V. limit plug and spring valve.

10. From bore No. 10, remove the retaining roll pin and plug. Remove the 1-2 accumulator valve, spring and sleeve.

11. From bore No. 11, using needle nose pliers, compress the line bias valve spring retainer and remove the plug, line bias and spring.

12. From bore No. 12, remove the roll pin, 3-2 control valve spring and 3-2 control valve.

13. Remove the manual valve from bore No. 13.

14. Clean and inspect all components for wear, scoring and damage. Repair or replace defective components as required.

Assembly

Install all parts in the reverse order as they were removed. Assemble all bore plugs against the retaining pins with the recessed holes outboard. All the roll pins must be installed so they do not extend above the flat machined face of the valve body pad. Install all flared coiled pins with the flared end out.

Make certain all retaining or roll pins are installed into the proper locating slots in the sleeves, not in the oil passage holes. The bushing for the 1-2 accumulator valve train must be assembled with the small hole for the roll pin facing the rough casting side of the valve body.

Transmission Assembly

NOTE: Some assembly and disassembly of components will be done as the transmission is being assembled.

1. Position the transmission on the transmission stand in the vertical position. The parking pawl must be removed from the case before installing the low reverse clutch piston.

2. Lubricate and install the inner, center, and outer seals on the low reverse clutch piston if removed. (Large aluminum).

3. Install the low reverse clutch piston in the transmission case indexing the piston with the notch at the bottom of the case with the hub facing downward, making certain the piston is fully seated in the downward position and the parking pawl will index into the opening in the piston wall.

4. Install the low reverse clutch spring retainer assembly into the case with the flat side of the retainer upwards.

5. Install tool J-23327 or equivalent and compress the springs, indexing the tool retaining plate so that the tool is free to slide over the case hub.

6. Install the low reverse clutch snap ring. Remove all of the tools.

7. Install the reaction internal gear support bearing on the case hub so that the longer inside "L" race is positioned downward.

8. Install the reaction internal gear and output shaft on the bearing assembly in the case. When the gear is properly seated it will be centered with the long open slot in the case and the parking lock pawl can be engaged with the external teeth of the internal gear.

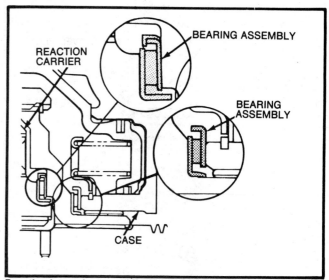

Reaction internal gear—"L" race positioning
(© General Motors Corporation)

NOTE: If the reaction internal gear and output shaft were removed as one unit, install them into the case at this time, otherwise assemble only the reaction internal gear.

9. Install the reaction carrier to support thrust bearing on the internal gear support so that the longer outside "L" race is positioned downwards.

10. Install the reaction carrier (with large outside hub) locating the reverse hub upwards.

11. Install the low and reverse clutch plates, starting with steel and alternating with composition indexing with the splines of the reaction carrier and case, aligning the steel plates.

NOTE: Some low reverse clutch assemblies are equipped with four composition and four steel plates, while others are equipped with five steel and five composition plates. The thickness for the composition faced plates is 0.088 in. The thickness for the steel plates is 0.069 in.

12. Remove the low reverse inner race and install the low reverse support and roller assembly with the chamfered side up in the case, indexing with the case splines.

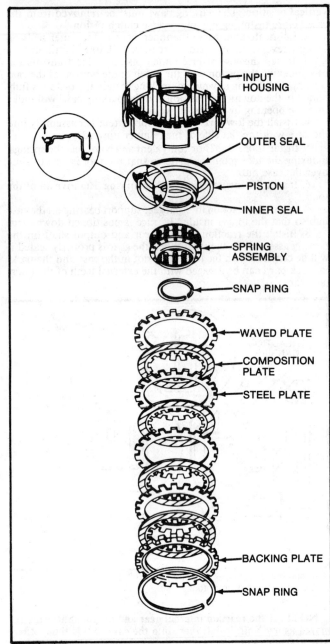

Reverse input clutch—exploded view
(© General Motors Corporation)

the tangs in the shell. (Bronze thrust washer with wide thrust face).

5. Install the input internal gear and shaft positioning the shaft end first.

NOTE: If the output shaft and the reaction gear were removed as separate parts during disassembly position the output shaft into the transmission with all components. Install the output shaft support tool J-29837 or equivalent and adjust it so that the output shaft is positioned up as far as possible.

6. Install the input carrier to reaction shaft thrust bearing with the "L" race on the outside.

7. Install the input carrier assembly, with the hub end down.

8. Install a new snap ring on the output shaft.

9. Install the input sun gear, indexing the gear end with the input carrier pinions.

10. Install the input carrier thrust washer on the input carrier.

REVERSE INPUT CLUTCH

1. Separate the reverse input clutch from the input clutch assembly. Remove the stator bearing and the selective washer from the input housing.

2. Remove the snap ring from the reverse input housing. Remove the reverse input clutch backing plate.

3. Remove the reverse input steel and composition clutch plates.

4. Using tool J-23327 or equivalent compress the reverse input spring assembly and remove the snap ring.

5. Remove all tools.

6. Remove the reverse input clutch release spring assembly.

7. Remove the reverse input clutch piston. Remove both the inner and outer seals.

8. Clean and inspect all components for wear, damage and scoring. Repair or replace defective components as required.

9. Lubricate and install the inner and outer seals on the reverse input clutch piston with the seal lips facing away from the hub.

10. Install the reverse clutch piston into the reverse input housing with the hub facing up. Use a feeler gauge to position the seal.

11. Install the reverse input clutch spring assembly with the large opening first.

12. Install tools J-23327 and J-25018-A or equivalent on the spring retainer. Compress the spring retainer and install the retaining snap ring. Remove all tools. Install the waved steel reverse plate. The waved plate is thinner and will show some high burnished spots. No indexing or alignment is required. Install a composition plate, then install the balance of the plates, alternating composition faced and flat steel. The reverse clutch plates are the largest plates with equally spaced tangs.

13. Install the reverse input backing plate with the chamfered side facing upward.

14. Install the backing plate snap ring.

INPUT CLUTCH ASSEMBLY

1. Position the input clutch assembly on the bench with the turbine shaft located in a bench hole and resting on the turbine shaft housing.

2. Remove the snap ring retaining the 3-4 clutch backing plate.

3. Remove the 3-4 clutch backing plate.

4. Remove the 3-4 clutch plates (composition and steel).

5. Remove the 3-4 clutch apply plate.

6. Remove the 3-4 clutch retaining apply ring.

7. Remove the forward clutch backing plate snap ring.

8. Remove the forward clutch backing plate.

9. Remove the forward clutch cam assembly and outer race by pulling up.

13. Install the low reverse inner race into the roller assembly and rotate until the internal splines are engaged, then push downward for full engagement. The bottom tanges will be flush with the reaction hub when seated.

14. Install the low reverse snap ring and support spring into the transmission case.

REACTION AND INPUT GEAR SETS

1. Install the snap ring on the reaction sun gear if removed, and install into the reaction carrier indexing the pinions.

2. Install the nylon low reverse thrust washer with 4 locating ears on the low reverse inner race.

3. Install the reaction sun gear shell (large housing with end slots and holes) engaging the splines of the shell onto the sun gear.

4. Install the reaction shaft to shell thrust washer indexing.

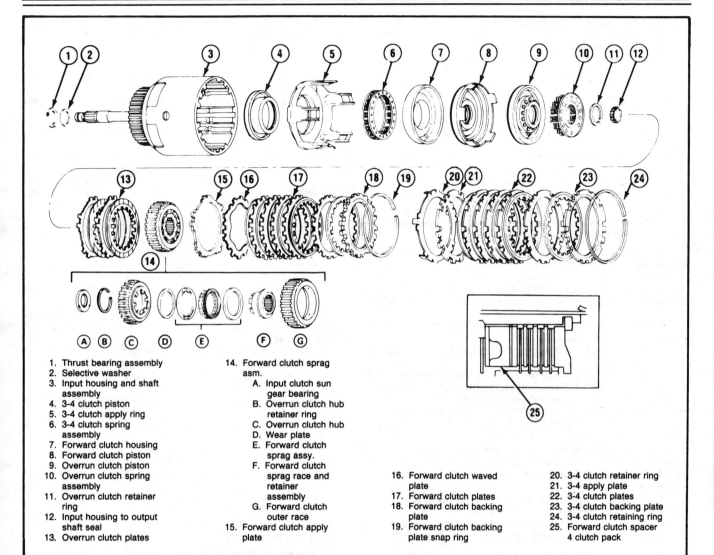

1. Thrust bearing assembly
2. Selective washer
3. Input housing and shaft assembly
4. 3-4 clutch piston
5. 3-4 clutch apply ring
6. 3-4 clutch spring assembly
7. Forward clutch housing
8. Forward clutch piston
9. Overrun clutch piston
10. Overrun clutch spring assembly
11. Overrun clutch retainer ring
12. Input housing to output shaft seal
13. Overrun clutch plates
14. Forward clutch sprag asm.
  A. Input clutch sun gear bearing
  B. Overrun clutch hub retainer ring
  C. Overrun clutch hub
  D. Wear plate
  E. Forward clutch sprag assy.
  F. Forward clutch sprag race and retainer assembly
  G. Forward clutch outer race
15. Forward clutch apply plate
16. Forward clutch waved plate
17. Forward clutch plates
18. Forward clutch backing plate
19. Forward clutch backing plate snap ring
20. 3-4 clutch retainer ring
21. 3-4 apply plate
22. 3-4 clutch plates
23. 3-4 clutch backing plate
24. 3-4 clutch retaining ring
25. Forward clutch spacer 4 clutch pack

Input clutch—exploded view (© General Motors Corporation)

10. Remove the input sun gear bearing (it can be located on the back side of the input inner race).

11. Remove the input housing to output shaft nylon seal.

12. Remove the forward clutch plates (steel and composition).

13. Remove the forward clutch apply plate and spacer if used.

14. Remove the overrun clutch plates (2 steel and 2 composition).

15. Install tools J-23456 and J-25018 or equivalent and compress the overrun clutch spring retainer.

16. Remove the overrun clutch retaining snap ring.

17. Remove all tools.

18. Remove the overrun clutch spring assembly.

19. Remove the overrun clutch piston assembly.

20. Remove the inner and outer seals from the overrun clutch piston.

21. Remove the forward clutch piston assembly.

22. Remove the inner and outer seals from the forward clutch piston assembly.

NOTE: Apply air pressure to the 3-4 feed hole in the turbine shaft—3rd hole from the shaft end. If unable to remove parts strike housing on a soft surface squarely on the open end.

23. Remove the forward clutch housing assembly.

24. Remove the 3-4 clutch spring assembly.

25. Remove the 3-4 apply ring and piston.

26. Remove the O-ring from the input housing assembly.

27. Inspect the four teflon oil seal rings on the turbine shaft for damage or distortion. Replace only if necessary.

28. Clean and inspect all components for wear, damage and scoring.

29. Repair or replace defective components as required. Do not replace the teflon seals unless required.

30. If removed, install the four teflon oil seal rings on the turbine shaft. Assemble with a long edge to a long edge and the large O-ring on the inside of the housing.

31. If removed, install the small O-ring on the end of the turbine shaft.

32. Position the input clutch housing over the bench hole with the turbine shaft in a hole pointing downwards.

33. Install the inner and outer seals on the 3-4 piston (smallest aluminum piston) with the lips facing away from the hub.

34. Install the 3-4 piston in the input housing, rotate and gently push downward making certain the piston is properly seated.

35. Install the inner and outer seals on the forward clutch piston (largest aluminum) with the lips facing away from the tangs.

36. Lubricate and install the forward clutch piston (largest remaining aluminum) into the forward clutch housing (large steel).

37. Install the 3-4 spring retainer into the 3-4 clutch apply ring.

38. Install the assembled forward clutch housing and piston on the spring retainer in the 3-4 apply ring.

39. The notches of the forward clutch piston must be indexed with the long apply tangs of the 3-4 apply ring.

40. Install seal protector tool J-29883 or equivalent on the input housing shaft.

41. Hold the 3-4 apply ring and assembled parts by the tangs and install into the input clutch housing and firmly seat the forward clutch piston. Remove tools. Use care to insure the pistons do not separate.

42. Install overrun clutch seal protector tool J-29882 or equivalent on the input housing shaft and install the overrun clutch piston with the hub facing upward and remove tool.

NOTE: When properly positioned the overrun piston will be approximately 3/16 inch below the snap ring groove on the input housing hub if not seated install the clutch spring compressor tool and gently tap until all parts are fully seated.

43. Install the overrun clutch spring retainer on the overrun clutch piston locating the release springs on the piston tabs.

44. Position tools J-23456 and J-25018 or equivalent overrun on the overrun spring assembly and compress the spring retainer. Do not over compress springs as distortion to the retainer can occur.

45. With the springs compressed, install the retaining snap ring and remove all tools.

46. Install the splined nylon output shaft seal, with the seal lip facing up.

47. On the forward clutch piston in the input housing, install the four overrun clutch plates. Starting with a steel plate and positioning so that the long recessed slot is indexed with a wide notch in the housing, install the remaining clutch plates alternating steel and composition. The overrun plates will be the smallest of the three clutch plate sets in the input clutch assembly.

48. Install the input sun gear bearing assembly on the input clutch hub on top of the nylon seal positioning the outside "L" race in the downward position. Make certain the bearing is properly centered.

49. With a suitable tool, align and center the inside drive tangs of the two overrun clutch plates (composition).

50. Install the assembled forward clutch cam assembly and outer race clutch hub indexing the overrun clutch plates.

51. Install the forward clutch spacer (thick steel) into the input clutch housing, indexing the lug on the spacer with the large slot in the input housing.

NOTE: A five clutch plate forward clutch, will use a single thick apply plate. A four clutch forward clutch, will use a thick spacer and a thin apply plate, and must be assembled with the thin apply plate first and the spacer with the holes facing the thin apply plate.

52. Install the waved steel forward clutch plate into the input housing. Index the wide slot with two small ears with the wide notch in the housing. The waved steel plate will show high burnish marks.

53. Install a forward clutch plate assembly (composition) on the forward clutch hub. In the same manner, install the remaining forward clutch plates alternating composition and steel. The last plate installed will be composition.

54. Install the forward clutch backing plate into the input housing, with the chamfered side up.

55. Install the forward clutch backing plate snap ring into the input clutch housing (smaller ring with the larger gap).

56. Install the 3-4 gear ring retaining plate (flat plate with legs) into the clutch housing indexing each apply lug with the ends of the 3-4 gear apply ring.

57. Install the 3-4 gear apply plate (thick steel) into the input housing indexing the long wide gear of the plate with the wide slot in the housing.

58. Install a 3-4 plate assembly (composition) then install the remaining plates alternating steel and composition indexing the long wide ear of the plate with the wide slot in the housing. The last plate installed will be composition.

59. Install the 3-4 gear backing plate with the chamfered side up.

60. Install the 3-4 gear retaining ring into the input housing assembly.

FORWARD CLUTCH CAM ASSEMBLY

1. Remove the overrun clutch hub snap ring, clutch hub and clutch thrust washer.

2. Remove the forward clutch sprag cam and the sprag assembly from the forward clutch cam.

3. Clean and inspect all components for wear, damage and scoring. Repair or replace defective components as required.

THIRD AND FOURTH CLUTCH INFORMATION CHART

Model	Flat Steel		Composition Faced		Apply Plate		Backing Plate		Apply Ring	
	No.	Thickness	No.	Thickness	No.	Thickness	No.	Thickness	No.	Length
TC, MB, MC, MJ, VN, T2, VA, ML, MP, MS, T7, YF, PQ	4	0.077 in. (1.97mm)	5	0.079 in. (2.03mm)	1	0.128 in. (3.30mm)	1	selective	1	3.71 in. (94.13mm)
All Others	5	0.077 in. (1.97mm)	6	0.079 in. (2.03mm)	1	0.128 in. (3.30mm)	1	selective	1	3.71 in. (94.13mm)

FORWARD CLUTCH INFORMATION CHART

Model	Flat Steel		Composition Faced		Apply Plate		Spacer		Waved Steel	
	No.	Thickness	No.	Thickness	No.	Thickness	No.	Thickness	No.	Thickness
TC, MB, MC, MJ, VN, T2, VA, ML, MP, MS, T7, YH, YF, PQ	3	0.077 in. (1.97mm)	4	0.079 in. (2.03mm)	1	0.251 in. (6.44mm)	1	0.330 in. (0.835mm)	1	0.079 in. (2.03mm)
All Others	4	0.077 in. (1.97mm)	5	0.079 in. (2.03mm)	1	0.251 in. (6.44mm)	none	—	1	0.079 in. (2.03mm)

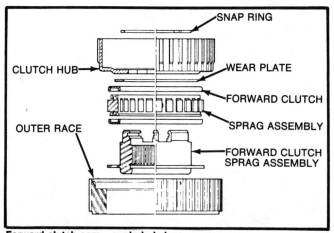

Forward clutch cam—exploded view
(© General Motors Corporation)

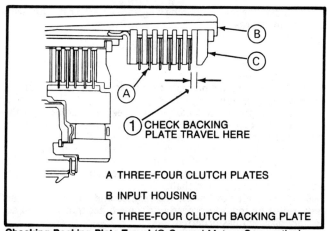

A THREE-FOUR CLUTCH PLATES

B INPUT HOUSING

C THREE-FOUR CLUTCH BACKING PLATE

Checking Backing Plate Travel (© General Motors Corporation)

4. Position the forward clutch cam with the retainer side down.

5. Install the sprag assembly with the two retainer rings over the cam hub. The lipped edge of the sprag must face the cam retainer.

6. Place the overrun hub thrust washer on the overrun hub and place this unit over the forward clutch cam. Install the overrun clutch hub snap ring in the I.D. of the cam.

7. Install the assembled hub, cam and sprag inner race assembly into the forward clutch outer race by rotating clockwise. When holding the outer race in your left hand, the overrun clutch hub must rotate freely, in a clockwise direction and must hold in a counterclockwise direction.

THREE-FOUR CLUTCH PISTON TRAVEL CHECK

1. Using a feeler gauge measure the end clearance between the backing plate and the first composition plate.

2. If the clearance is not within specification, select the proper backing plate from the three-four backing plate court.

INTERNAL COMPONENTS

1. Install the reverse and input clutch assemblies as a complete unit into the transmission case indexing the 3-4 clutch plates of the input assembly with the input internal gear. The complete assembly will be properly seated when the reverse housing is located just below the pump face of the case. Be sure that all clutch plates are fully seated.

2. Install the 2-4 band into the case indexing the band anchor pin end with the case pin hole. Install the band anchor pin in the case, indexing with the 2-4 band end.

SERVO ASSEMBLY

1. Remove the fourth apply piston from the second apply piston assembly.

2. Remove the servo apply pin spring from the servo apply pin.

3. Disassemble the second servo apply piston assembly using J-22269-01 and separate the second apply piston, spring and retainer.

4. Remove the retaining 'E' ring washer and spring from the servo apply pin, and remove the pin.

5. Remove all oil seal rings.

6. Clean and inspect all components for wear, scoring and damage. Repair or replace defective components as required.

7. Assemble the components in the reverse order of disassembly. Assemble all flat edged seals with the flat edge to the flat edge and coat them with clean transmission fluid.

8. Install the complete 2-4 servo assembly into the case indexing the servo apply pin on the 2-4 band end and check for proper engagements.

9. Recheck for the correct apply pin length if any of the servo parts, the 2-4 band or the input housing have been replaced.

10. Install the servo cover and O-Ring into the case.

11. Install the servo compressing tool and compress the servo cover.

THREE—FOUR BACKING PLATE SELECTION CHART

Model	Backing Plate Travel	Backing Plate ①	
		DIM	I.D.
TC, MB, MC, MJ, VN, T2, VA, ML, MP, MS, T7, YF, PQ	0.055-0.109 in. (1.39-2.78mm)	0.278 in. (7.125mm)	1
		0.239 in. (6.125mm)	2
All Others	0.049-0.113 in. (1.25-2.87mm)	0.200 in. (5.125mm)	3
		0.161 in. (4.125mm)	4

① Use backing plate which gives correct travel

12. Install the servo retaining ring, indexing the ring ends with the slot in the case.

FRONT PUMP

1. Place the transmission in the vertical position in order to install the front pump.
2. Place the aligning pins into the transmission case. Install the front pump assembly.
3. Align the filter and the pressure regulator holes with the transmission case holes. Retain the pump in place using the pump attaching bolts. Torque the bolts to 18 ft. lbs.
4. Rotate the transmission assembly to a horizontal position and rotate the output shaft by hand.
5. If the shaft will not rotate, loosen the pump retaining bolts and attempt to rotate the shaft.
6. If the shaft now turns, the reverse and input assemblies have not been indexed properly or some other transmission assembly problem has occurred.
7. Rotate the transmission vertically and correct the problem before proceeding.

VALVE BODY AND RELATED COMPONENTS

1. Install the 1-2 accumulator pin into the transmission case. Install the accumulator piston and seal over the pin with the lug end up. Install the spring on the piston.
2. Install the governor and the converter clutch oil screens into the transmission case.
3. Install the five check balls into the transmission case. Install the valve body alignment pins.
4. Install the spacer plate to case gasket with the small "C" on the transmission case.
5. Install the valve body spacer on the guide pins. Align the holes as required.
6. Install the valve body to spacer gasket with the "V" on the spacer plate.
7. Install the three exhaust balls and check valve into the valve body. Retain the components using transmission fluid.
8. Install the valve body and connect the link to the inside detent lever.
9. Remove the aligning pins. Install the valve body retaining bolts. Torque the bolts to 8 ft. lbs.
10. Attach the retaining clip to the link and the inside detent lever.
11. Install the throttle lever and bracket, and T.V. link locating the slot in the bracket with the roll pin on the valve body top face, aligning the link through the T.V. linkage case bore. Attach with two valve body to case bolts and torque to 8 ft. lbs.
12. Install the valve body attaching bolt and harness clip and torque to 8 ft. lbs.

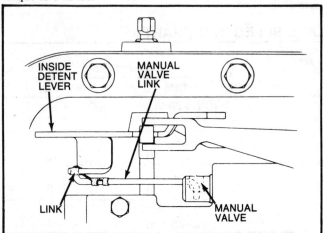

Manual valve link and clips (© General Motors Corporation)

13. Install the parking pawl bracket and torque to 18 ft. lbs.
14. Install the manual detent spring and roller assembly with an attaching bolt and torque to 11 N•m (8ft. lbs.).
15. Install the O-Ring on the solenoid assembly and install into the pump, locating the attaching wire harness toward the transmission, attaching with two bolts and torque to 18 ft. lbs.
16. Install the wiring harness and connect to all pressure switches.

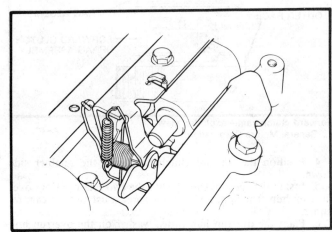

T.V. bracket and lever assembly (© General Motors Corporation)

NOTE: Each pressure switch will be color coded for switch and wire connector identification. When connecting the wire connectors to the pressure switches, always connect the same colors. It is not necessary to connect the same wire connector to the same pressure switch terminals as they are reversible.

17. Install the O-Ring on the outside electrical connector and install into the case by compressing the inside tang or gently taping in board, locating the tab with the case notch.
18. Attach the inside electrical connector to the outside electrical connector.
19. Install the oil passage cover to the transmission with the three attaching bolts. Torque the component to 8 ft. lbs.
20. Install the 3-4 accumulator piston into the accumulator housing with the lug end up. Install the piston spring into the housing on the piston.
21. Position the 3-4 accumulator plate and gasket on the transmission with the gasket on top. Install the housing, spring and piston on the transmission case. Secure the assembly in place with the retaining bolts. Torque the bolts 8 ft. lbs.
22. Install the speedo gear and retaining clip on the output shaft, positioning the large notch on the speedo gear rearward. On shafts with 2 speedo clip holes, use the hole nearest the yoke end of the shaft for Corvettes only.
23. Install the output shaft seal in the output shaft sleeve and install on the output shaft with J25016 or equivalent. Install the oil seal ring on the case extension and install on case. Positioning so the speedo hole is located on the same side as the governor. Torque bolts to 23 ft. lbs.
24. Install the governor assembly. Apply a sealant to the edge of the cover, then install cover.
25. Install the O-Ring on the filter and install in the transmission.
26. Install a new oil pan gasket on the transmission case and install the attaching bolts and torque to 8 ft. lbs.
27. Install all remaining outside connectors such as driven speedo gear and adaptor, outside manual lever and nut.
28. Remove the transmission from the holding fixture and install the torque converter.

S SPECIFICATIONS

TORQUE SPECIFICATIONS

Location	Quantity	Size	Torque
Accumulator cover to case	2	1.0x30.3	11 N•m (8 ft. lb.)
Accumulator cover to case	1	1.0x60.0	11 N•m (8 ft. lb.)
Detent spring to valve body	1	1.75x20.0	22 N•m (18 ft. lb.)
Valve body to case	15	1.0x50.0	11 N•m (8 ft. lb.)
Oil passage cover to case	3	1.0x16.0	11 N•m (8 ft. lb.)
Solenoid assembly to pump	2	1.0x12.0	11 N•m (8 ft. lb.)
Transmission oil pan to case	16	1.25x16	24 N•m (18 ft. lb.)
Pressure switches	1-3	1/8-27	11 N•m (8 ft. lb.)
Park brake bracket to case	2	1.25-20.00	41 N•m (31 ft. lb.)
Pump cover to body	5	1.25-40.00	22 N•m (18 ft. lb.)
Pump assembly to case	7	1.25-60	22 N•m (18 ft. lb.)
Case extension to case	4	1.50-30.0	31 N•m (23 ft. lb.)
Manual shaft to inside det. lever	1	1.50 Nut	31 N•m (23 ft. lb.)
Pressure plugs	3	1/8-27	11 N•m (8 ft. lb.)
Connector cooler pipe	2	1/4-18	38 N•m (28 ft. lb.)

SPECIAL TOOLS

Holding Fixture & Base

Oil Pump Body & Cover Alignment Band
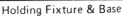

Rear Seal Installer

Pump Oil Seal Installer

Piston Compressor

Bushing Remover

Clutch Spring Compressor

Clutch Spring Compressor Adaptor

Clutch Spring Compressor Press

Universal Remover
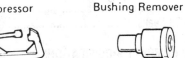

Oil Pump Remover & End Play Checking Fixture
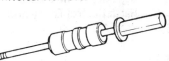

End Play Checking Fixture Adaptor

Bushing & Universal Remover Set

Bushing Remover

Servo Cover Compressor

Output Shaft Support Fixture

Inner Overrun Clutch Seal Protector

Inner Forward Clutch Seal Protector

2-4 Band Apply Pin Tools

INDEX

MODIFICATIONS

CHANGES AND/OR MODIFICATION SECTION

This section includes the known and available change or modification information, pertaining to the automatic transmissions/transaxles covered in the first Chilton's Automatic Transmission Manual, and remaining in use from 1980 to present.

Included in this section are numerous late changes and/or modifications for the transmissions/transaxles covered in this second Chilton's Automatic Transmission Manual, that could not be added to the individual sections before the manual publication date. Although many changes and modifications are covered, others have been published by the manufacturer that have not reached our segment of the transmission industry. Therefore, it behooves the repairman to continually update his knowledge through attendance at the numerous technical seminars, sponsored by Transmission Associations, Parts and Supply Firms and other Vocational and Trade groups. Many monthly publications are available to update the rebuilding process of particular units, while Technical and Trade Associations have visual and/or sound products, both for rent or sale.

CHRYSLER CORPORATION
AMERICAN MOTORS CORPORATION/JEEP

Torque Command, TorqueFlite A-904, A-998, A-999, A-727

1980 Changes and/or Modifications

REAR CLUTCH RETAINER (A-904 FAMILY)

Snap ring groove diameter increased by .110 snap ring free diameter also increased by .110 and radial section decreased to provide clearance with wide ratio annulus gear. 1979 retainers will be serviced with packages including the new retainers and snap rings.

	1980 P/N	1979 P/N
Rear Clutch Retainer (4-Plate)	4130765	4130195
Rear Clutch Retainer (3-Plate)	4202046	4130461
Snap Ring (thin)	4130761	1942421
Snap Ring (medium)	4130762	1942423
Snap Ring (thick)	4130763	2538617

KICKDOWN BAND ASSEMBLIES (A-904 FAMILY)

New design flex-band P/N 4058863 replaces standard band 2204792 and the A-999 wide band 3681921. New part will service prior models.

VALVE BODY ASSEMBLIES (A-904 & A-727)

The following changes to the valve body assemblies were effective at the start of 1980 model year.

The lockup valve was changed to a two-diameter valve to provide a snap action valve in order to eliminate lockup valve buzz and to decrease lockup shudder. The lockup speed was also increased for most applications in order to eliminate lockup shudder and vibrations.

The part throttle kickdown limit for A-904 6-cylinder applications was increased approximately 10 mph for better driveability.

The switch valve was changed to eliminate switch valve buzz when in reverse. A .005 deep x .06 wide step on one land was added. This change was effective at S/N 6595-0000. The new valve body components will be used to service prior models.

GOVERNORS (A-904 & A-727)

The outer governor weights were increased 0.070 on the O.D. This was done in order to create a new governor weight for wide ratio transmissions and to maintain common tooling for all governor bodies. The new weights and body assemblies can be used together to service prior models.

AMC-JEEP 4-WHEEL DRIVE TRANSMISSIONS

Three new transmissions were introduced for 4-wheel drive usage—A-998 transmission for 258 AMC passenger car, A-999 for CJ-7 Jeep and A-727 for the Senior Jeep. New parts involved are two (2) output shafts, output shaft bearing, bearing snap ring, adapter seal and three (3) adapters. The transfer case hole mounting pattern on the two Jeep adapters is rotated 3° counterclockwise from the passenger car version.

AMC 2.5 LITRE 4-CYLINDER (A-904)

New case with revised bell housing configuration to fit the Pontiac 4-cylinder engine. Not interchangeable with prior models.

TRUCK 4-WHEEL DRIVE TRANSMISSIONS (A-727)

New die cast aluminum adapter (P/N 4130675) to replace cast iron adapter 3743285. Over-all length of new adapter was increased by 1.84", which requires a new output shaft P/N 4130677. Lockup feature was also added to most 4-wheel drive units. New parts and transmissions are not interchangeable with prior model.

MEDIUM TRUCK EXTENSION (A-727)

Casting and machining of the brake attaching boss was modified to accept 9 x 3 parking brake for 18,000# GVW motor home. New extension will service prior model year.

DODGE TRUCK WITH 446-IHC ENGINE (A-727)

New transmission similar to 1979 440-engine except uses case P/N 3743105. Adapter plate is used to mate transmission bell with back of block of the International Harvester engine.

MMC TRANSMISSION (Colt & Arrow)

New shift quality package developed for the 1980 MMC transmissions. Modifications include V-8 type valve body with provisions for rear servo feed check ball, elimination of bleed hole in case accumulator bore, longer accumulator spring and 2-disc front clutch with thicker pressure plate for use with 1.6 and 2.0 litre engines. New parts are not interchangeable with prior models.

Effective with transmission serial #4202084-6604-9994, the 1.6 and 2.0 litre front clutch packs were modified to improve clutch life. The separator plate (P/N 1942403) between the two friction discs was replaced by 2801969 and the thick pressure plate P/N 4202090) was replaced by the standard plate 2801969. Transmissions built before date code "6604" should be repaired with the revised clutch pack.

1980 Model Year Running Changes

VALVE BODY ASSEMBLIES (A-998 & A-999)

Valve body assemblies for 318 wide ratio and 360 engine applications were revised by adding an orifice in the steel plate to improve 1-2 shift quality and adding a check ball in the rear clutch circuit to improve neutral to drive shift. Kickdown spring loads were increased and the top accumulators spring removed in these transmissions at the same time.

Part	New P/N	Old P/N
Valve Body Assy. (318 Wide Ratio)	4203667	4202215
Valve Body Assy. (360)	4202668	4202213
Steel Plate	4202567	4202071
Kickdown Spring	4058885	4058781
Top Accumulator Spring	—	3515114

New valve body assemblies will service the old by changing springs as noted above.

VALVE BODY ASSEMBLIES (LOCK-UP)

Lockup speed was increased to 40 MPH for Chrysler 6-cylinder applications by using a 6.2# lockup spring. New valve body assembly will service the old.

Part	New P/N	Old P/N
Valve Body Assembly	4202780	4202214
Lockup Spring	4202672	4202642

Lockup speeds were increased during the model year on all other transmissions by increasing the lockup spring loads.

1981 Changes and/or Modifications

VALVE BODY ASSEMBLIES (A-904 AND A-727)

Part throttle 3-2 kickdown limits were increased by either increasing load of limit valve spring or by eliminating the limit body. Elimination of the limit body required changes to the steel plates and transfer plates.

Check ball added to the rear servo circuit of A-904 6-cylinder to improve neutral to reverse shift quality. Can be used to service prior models if counterbore is added to Rear Servo Feed passage of Case (½").

TRANSMISSION CASE (A-904)

Counterbore added to rear servo feed passage to allow use of check ball in this circuit to improve neutral to reverse shift quality. New case will service prior models.

	New P/N	Supersedes
Case (Chrysler 6-Cylinder)	4202774	3743111
Case Assy. (Chrysler 6-Cylinder)	4202777	3743125
Case (AMC 6-Cylinder Pass.)	4202776	3743113
Case Assy. (AMC 6-Cylinder Pass.)	4202778	3743127

TRANSMISSION CASE (A-904T)

New case added for 6-cylinder truck usage incorporates double-wrap reverse band and A-999 servo piston bores. Also used in heavy-duty 6-cylinder passenger car applications. 4130663 case-4202047 case assembly.

TRANSMISSION CASE (A-727 JEEP)

Casting revised to provide clearance with lockup valve body. New case will service prior models.

	New P/N	Supersedes
Case	4202577	3743105
Case Assembly	4202576	3743120

1981 Model Year Running Changes
INPUT SHAFT (A-904 FOUR-CYLINDER)

The turbine hub spline was lengthened by .540" to eliminate possible strip-out of hub spline. The new shaft should be used to service all 1979 through 1981 transmissions used with 2.0 and 2.6 litre MMC engines and 2.0 and 2.5 litre AMC and AM General engines. The old shaft with ground pilot will continue to be used with the 1.6 litre MMC engine.

NEW P/N: 4269140
OLD P/N: 4130788

OVERFILLING AUTOMATIC TRANSMISSIONS 1981 AND LATER

Automatic transmissions are frequently overfilled because the fluid level is checked when it is not "HOT" and the dipstick indicates that fluid should be added. However, the low reading is normal since the level will rise as the fluid temperature increases. Therefore, as a running change in the 1981 model year, all automatic transmission dipsticks have been revised with improved labeling and dimples added for checking oil level warm (85°F-125°F). These new dipstick part numbers are:

New P/N	Old P/N	Models Affected
4202951	2466302	J,E,T and Truck
4202952	4117465	Truck
4202953	4028908	B,F,G,X,S, and Y
4202971	[5224068] [4207049]	M,Z,P, and D

NOTE: Fluid level should not be checked when it is cold to the touch. "Warm" is fluid between 85°-125°F (29°-52°C).

If the fluid level checks low, add sufficient fluid to bring the level to within the marks indicated for the appropriate temperature.

——————— CAUTION ———————
DO NOT OVERFILL THE TRANSMISSION.

Overfilling may cause leakage out the RWD vent which may be misdiagnosed as a pump seal leak. In addition, overfilling causes aeration or foaming due to the oil being picked up by the rotating gear train. This significantly reduces the life of the oil and may cause a transmission failure.

——————— CAUTION ———————
On front wheel drive vehicles, the dipstick is located just behind the radiator electric cooling fan. Caution should be taken so as not to allow your hand or fingers to be caught in the fan.

All Chrysler built transmissions use DEXRON® or DEXRON® II Automatic Transmission Fluid.

1982 Changes and/or Modifications
REAR SERVO PISTON SEAL (A-904)

New Viton seal 4202864 released for fleet usage to replace 2464462 on designated assemblies that require high temperature endurance. New seal can also be used in all other applications.

WIDE RATIO GEAR SET (A-904 FAMILY)

AMC 6 and 8-cylinder and MMC 4-cylinder transmissions were changed to the wide ratio gear set, making wide ratio common across the board for 1982. New case, valve body assemblies, governors, kickdown lever and springs, accumulator springs, and some front clutch assemblies are required in the above transmissions to match shift quality with the ratio change.

TRANSMISSION ASSEMBLY (A-900 TRUCK)

The A-904 transmission was introduced for some 1981 model trucks equipped with 225 and 318 engines up to 6000 GVW. The usage has been expanded up to 8500 GVW for the 1982 Model Year.

EXTENSION SEAL (A-904)

New Vamac extension seal with improved high temperature material, P/N 4058047. This seal must be used in 1982 B3 vans and wagons equipped with A-900 transmissions. The new seal can be used in place of old part 3515384 in all other applications.

CASE (A-904 MMC)

The MMC case has an added orifice in the accumulator servo bore. New case assembly 4269141 is not interchangeable with 1981 part 4202116. The new assembly can be used in place of 3878521 for 1974 through 1979 Model Years.

VALVE BODY ASSEMBLY (A-904)

The MMC valve body has an added drilled orifice. Various springs were also changed to match shift quality to the wide ratio gear set. Miscellaneous valve body changes were also made for AMC and Jeep six-cylinder usages.

1983 Changes and/or Modifications
INNER OIL PUMP ROTOR (A-904 FOUR-CYLINDER)

To improve wear and brinelling on the inner rotor lugs in all 4-cylinder applications, the lug shape becomes dovetailed. The mating broached slot in the torque converter is also revised. Rotors are not interchangeable. The new 1983 torque converters can be used with prior model transmissions. The new 1983 transmissions with revised rotor lugs *cannot* be used with prior model torque converters. New service oil pump assemblies are also required for 4-cylinder applications:

Part	1983	1982 & Prior
Oil pump assembly w/rotors	4269948	4130192
Oil pump and support Assy. w/rotors	4269935	4202371

4X4 ADAPTERS (AMC AND JEEP)

The bearing shoulder area in the adapters was increased which reduced the seal bore by .123 in. The seals are not interchangeable.

Past model units can be serviced by using the new 1983 adapters in conjunction with the new seal. Part numbers involved are:

Part	1983 P/N	1982 P/N
Adapter (AMC Pass)	4269952	4130540
Adapter (A-999 Jeep)	4269953	4130730
Adapter (A-727 Jeep)	4269957	4130734
Seal-Adapter	4269956	4130539

MMC 4-WHEEL DRIVE (A-904)

A new transmission is being introduced for Mitsubishi built 4X4 trucks. It is similar to that used in 2.6 litre passenger car except for transfer case adapter and related parts. A 4-disc rear clutch will also be used in place of 3-disc for durability.

REAR SERVO PISTON SEAL (A-904 MMC)

The material of the rear servo seal is changed to VITON to improve resistance to high temperature. This seal is common with A-404 family and can be used to service prior models.

STEADY DRIVING SURGE AT 30-40 MPH FOLLOWING CONVERTER LOCK-UP

1981-83 TRUCKS AND VANS, w/318 2V AND LIGHT DUTY FEDERAL EMISSIONS ASPIRATOR WITH LOCK-UP CONVERTER.

Road test the vehicle and verify the condition. If surge is still present at steady speeds between 30 and 40 mph with the engine fully warmed up and after all warm driveability diagnostic procedures have been followed, do the following:

Repair Procedure

Parts Required:	1-Spring	PN 4202672
	1-Lock-Up Body	PN 4202209

1. Remove transmission oil pan
2. Remove valve body and inspect the lock-up valve body.
   a. If lock-up valve body is the *old* style lock-up body, replace it with new style Lock-Up Body PN 4202209 *and* replace the present lock-up spring with a new Lock-Up Spring PN 4202672. Reuse the existing lock-up valve, fail safe valve and spring, cover, and screws.
   b. If the lock-up valve body is the *new* style lock-up body install only the new Lock-Up Spring PN 4202672.
3. Install the valve body and torque screws to 35 inch pounds (4 N•m).
4. Install the transmission oil pan and fill with ATF fluid.
5. Adjust shift and throttle linkage as necessary.

CHRYSLER CORPORATION, AMERICAN MOTORS CORPORATION/JEEP

Updated Modifications From 1978 to 1980

NOTE: Because of the importance of the modifications for the affected transmissions, the following updates are included.

RECURRING TRANSMISSION PROBLEMS FROM A CLOGGED FLUID COOLER

Recurring transmission problems may be a direct result of a clogged transmission cooler. A clogged transmission cooler may result in overheated transmission fluid and a loss or reduction in lubrication resulting in a transmission failure.

The transmission oil cooler can be clogged by particles of the lock-up clutch friction material or other foreign material, as the transmission oil cooler has an inner mesh design to help distribute the heat to the radiator lower tank coolant.

When a transmission repair requires transmission and/or torque converter removal, cooler line flow at the transmission must be checked. This is accomplished by:

1. Disconnect the fluid return line at the rear of the transmission and place a collecting container under the disconnected line.
2. Run the engine at *curb idle speed*, with the transmission in neutral.

If fluid flow is intermittent or if it takes longer than 20 seconds to collect a quart of ATF, the cooler lines and radiator cooler must be reverse flushed.

--- CAUTION ---

With transmission fluid level at specification, fluid collection should not exceed one (1) quart of automatic transmission fluid (ATF), or internal transmission damage may occur.

Whenever friction material is found in the transmission cooler inlet, it is necessary to replace the torque converter and clean and repair the transmission as required.

When reverse flushing of the cooler system fails to clear all the obstructions from the system, the radiator transmission cooler must be repaired by an approved repair facility or the radiator replaced.

REVERSE FLUSH PROCEDURE

When reverse flushing an automatic transmission fluid cooler, the following procedure should be used:

1. Disconnect the cooler lines at the transmission.
2. Disconnect the cooler lines at the radiator and remove the inlet fitting from the oil cooler connection on the radiator. In all operations use a back-up wrench to prevent damage.
3. Carefully dislodge any material that may be collected at the inlet of the cooler with a small screwdriver or other suitable tool and intermittent spurts of compressed air in the outlet of the cooler.
4. Using a hand suction gun filled with mineral spirits, reverse flush the cooler by pumping in mineral spirits and clearing with intermittent spurts of compressed air.
5. Using the method in Step 4, flush the cooler lines separately to ensure they are free flowing.
6. Reinstall the brass fitting in the cooler and reinstall the cooler lines.
7. To remove all remaining mineral spirits from the cooler and lines, one (1) quart of ATF should be pumped through the cooler and cooler lines prior to connecting the lines to the transmission.

When the reverse flushing of the cooler is completed, fill the transmission to the specified level and check the system for leaks.

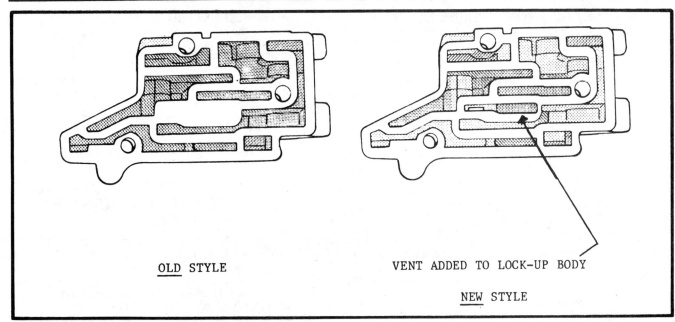

OLD STYLE

VENT ADDED TO LOCK-UP BODY

NEW STYLE

Comparison of old style and new style lock-up converter valve bodies (© Chrysler Corp.)

If reverse flushing of the cooler system fails to clear all obstructions from the system, the radiator cooler assembly must be replaced.

TORQUE CONVERTER LOCK-UP CLUTCH DRAG IN REVERSE GEAR AND IMPROVED TRANSMISSION PERFORMANCE—1978-79 A-904 and LA-904 (A-998 and A-999)

Some 1978 and 1979 vehicles equipped with a lock-up transmission may exhibit a condition of the torque converter lock-up clutch engaging when the vehicle transmission is shifted into reverse.

Torque converter lock-up clutch engagement can be easily diagnosed when one or more of the following conditions exist:

a. The engine hesitates and/or feels extremely sluggish when accelerating in reverse.

b. The torque converter stall speed is lower in reverse than in drive by 200 rpm or more.

In normal reverse operation, the torque converter lock-up clutch should not be applied. A lock-up clutch drag condition may be caused by a hydraulic pressure build up due to a restricted transmission oil cooler or by oil leakage into the lock-up circuit at the inner face of the pump assembly.

To correct this condition, the transmission oil cooler system must first be reversed flushed to ensure proper transmission oil cooling and fluid flow. If the transmission continues to exhibit the lock-up drag condition after the cooler has been reverse flushed and fluid flow checked, it will be necessary to replace the pump assembly with a new service pump assembly which includes:

1. Pump housing
2. Select Fit Pump Gears
3. Reaction Shaft Support Assembly
4. Pump-Torque Converter Oil Seal
5. All Steel Seal Rings

Part Number	Application
4202089	Lock-Up Transmission
4202298	Non-Lock-Up Transmission

These pumps will provide improved transmission performance.

These pump assemblies have select fit components and must not be disassembled or modified.

The pump assemblies are not interchangeable, and should only be used on the transmission for which they were released. The seal and gasket package for both pump assemblies is P/N 4131042, which includes:

1. Torque Converter Impeller Hub Seal
2. Pump Housing Seal
3. Reaction Shaft Support Gasket
4. Reaction Shaft Support Seal Rings (2)
5. Front Input Shaft Seal Ring
6. Rear Clutch Retainer Seal Ring

It is important gear selector and throttle linkage be adjusted properly and transmission fluid level is correct to ensure proper transmission performance.

A-727 OVERRUNNING CLUTCH FAILURES

Overrunning clutch failure may be the result of poor retention of the output shaft bearing by the bearing snap ring under a high load condition. This results in rearward movement of the output shaft which may cause the overrunning clutch to bear against the case and carry the full gear train thrust load. This in turn causes excessive wear of the race into the case and overrunning clutch failure.

The output shaft bearing may have an improperly machined snap ring groove in the outside diameter of the bearing outer race, which allows the snap ring to spread and release during heavy throttle operation.

All known overrunning clutch failures have resulted from bearings date coded with "□". If the transmission output shaft bearing has the date code shown; the bearing, snap ring, and case must be replaced.

Part	Part Number
Output Shaft Bearing	2466224
Output Shaft Bearing Snap Ring	2400320

MODIFICATIONS

LOW MILEAGE FRONT WHEEL BEARING FAILURES—1978, 1979, AND 1980 OMNI AND HORIZON MODELS

Low mileage front wheel bearing failures may be the direct result of an improperly grounded engine to battery (negative terminal).

A high pitch whining sound from a bearing with this type of wear pattern (electrostatic) is characteristic.

The ball bearings on the outboard race will be black from the electrostatic arcing on the race.

Diagnosis and Repair

A bearing failure due to poor engine ground can be easily diagnosed by looking for the following conditions:

• Removal of the noisy wheel bearing from the steering knuckle and disassembly of the bearing will show a pattern on the outer race of the wheel bearing assembly.

Check engine for proper installation of ground cable and straps.

• Check for proper attachment of the negative battery cable to the transmission mounting bolt.

• Check for proper attachment of the negative battery cable to body ground.

• Check for proper installation of the braided ground strap on the right side engine mount.

• Check for proper installation of the braided ground strap on the right rear side of the engine to the firewall.

If all ground cables and straps are properly installed, a check of the battery ground cable may be necessary.

To check the ground cable, use the following procedure:

1. Disconnect the negative battery cable from the transmission.

2. Using a test lamp, connect one lead to the positive battery terminal and the other end to the battery ground cable eyelet from the transmission. Firmly tug (not yanking), straighten out the cable and move the cable up and down and sideways. If the test light does not stay lit, the cable is defective and should be replaced.

3. Repeat the procedure in Step 2 and test the body ground cable from the negative battery terminal.

4. Clean cables and connections of all corrosion.

5. Reconnect the battery cable to the transmission and torque to 70 ft. lbs. (95 N•m). Also, reconnect the body ground wire, torque to 65 inch lbs. (7 N•m).

Replace the defective wheel bearing.

NO DRIVE IN ANY GEAR CONDITION—A-904 (1978-79)

In many instances, an automatic transmission failure, reported as *A NO DRIVE CONDITION,* may be mis-diagnosed. To ensure proper diagnosis and repair, the following procedure should be used when a "No Drive in any Gear" condition exists:

1. Check to verify the "No Drive in any Gear" condition.

NOTE: To save time and possibly unnecessary work; first, perform those operations that do not require transmission removal.

2. If after performing all diagnostic in-vehicle tests, the condition still exists, remove the transmission and torque converter from the vehicle.

3. Drain the transmission and torque converter as best as possible.

NOTE: Remove as much oil as possible from the torque converter. This will enable you to inspect the torque converter turbine hub inner-drive splines.

4. Before you begin any major disassembly, disassemble the transmission only to the point of removing the input shaft.

5. Insert the input shaft into the torque converter turbine hub spline until the shaft bottoms against the front of the torque converter and then pull the shaft back a ½ in. Rotate the input shaft slowly and firmly so as to drive the torque converter turbine. This will enable you to check the turbine hub spline for a stripped condition by noting the resistance to turning. If strippage is found, it will be necessary to replace the torque converter and input shaft, as well as clean and flush the transmission cooler and lines.

6. If torque converter turbine hub strippage is not found, it will be necessary to disassemble the transmission, clean and inspect the transmission for worn or failed parts. Replace all worn or failed parts as required.

7. Reassemble the transmission as necessary. Install the transmission. Check the transmission cooler flow. Check and set fluid level and check and adjust throttle and shift linkage.

8. Road test vehicle for proper transmission operation.

TORSIONAL VIBRATION AND LOCK-UP TORQUE CONVERTER TRANSMISSION SHUDDER

All 1978-79 225, 318, 360 FEDERAL ENGINES, CAR AND TRUCK MODELS EQUIPPED WITH LOCK-UP TORQUE CONVERTER

A slight amount of vibration is possible in a lock-up torque converter equipped vehicle under certain operating conditions (i.e. direct gear engine lugging, climbing hills under light throttle, or accelerating in direct gear, etc.). This occurs when the hydraulic drive through the torque converter is changed to a direct drive when the lock-up clutch is applied. In these instances, no repairs should be attempted. When an abnormal driveline disturbance is experienced on these vehicles, it can best be described as either torsional vibration or lock-up shudder. These conditions can be identified and corrected as follows:

Torsional Vibration

A continuous drumming or groaning sound emitted by the drive train, especially at low lock-up speeds. This condition may occur at the minimum lock-up speed up through approximately 45 mph. This condition can best be described as being similar to lugging a manual transmission vehicle in high gear.

If the condition experienced is abnormal torsional vibration, a new lock-up spring should be installed. This will increase the minimum lock-up speed.

Before installing a new lock-up spring, advise the customer that raising the lock-up speed may slightly reduce fuel economy and the torsional vibration is not harmful to the transmission. Once properly informed the customer may decide *not* to have the lock-up spring changed.

To ensure the proper lock-up spring and valve body end cover are used with the proper transmission application, check the list below:

1978 Transmissions

1. Transmission with a serial number up to 5911-xxxx, install Lock-up Valve Spring and Cover Package, P/N 4186185.

2. Transmissions with a serial number 5911-xxxx to 6093-xxxx, install lock-up spring from Package P/N 4186185, and retain the original valve body end cover.

3. Transmissions with a serial number after 6093-xxxx, install Lock-up Spring, P/N 4130478 (orange color code) and retain the original valve body end cover.

1979 Transmissions

1. 1979 transmissions, install Lock-up Spring, P/N 4130478 (orange color code) and retain the original valve body end cover.

Lock-Up Shudder

A vertical shaking of the instrument panel and steering column induced by faulty operation of the lock-up clutch, similar to driving over toll booth speed warning strips.

These disturbances may be caused by any one or a combination of the following items:
1. Rough Road Conditions
2. Poor Engine Performance
3. Poor Driveability
4. Defective Torque Converter
5. Defective Transmission or Pump

A road test should be performed, preferably with the owner, as he can best verify the condition when it occurs. Many times poor engine performance, roughness, cold engine bucking, or surge are misdiagnosed as lock-up shudder and/or torsional vibration.

In some instances, a new engine (under 300 miles) may affect engine and transmission performance. If a lock-up shudder or vibration condition is encountered on a new engine vehicle, no transmission repairs should be attempted until the vehicle has accumulated a minimum of 300 miles. The condition may clear up with mileage.

If the condition experienced is lock-up shudder, the following procedure should be used:

To ensure proper diagnosis and repair, if engine performance is suspected, check engine tune (timing, propane idle adjustment, and scope check). Also, check transmission fluid level and throttle linkage adjustment.

If after ensuring the above items are satisfactory and lock-up shudder still exists, proceed as follows:
1. Transmission oil cooler flow should be checked. Inadequate flow is an intermittent flow of less than 1 quart in 20 seconds (in neutral, transmission at operating temperature, engine at curb idle). If inadequate flow is found, inspect the cooler lines for kinks and sharp bends, then disconnect the transmission cooler line at the cooler inlet and remove the brass fitting. Inspect the inlet of the cooler for lock-up clutch friction material. If friction material is found or cooler flow is low, the transmission cooling system should be reverse flushed as described in the same Technical Service Bulletin. When reverse flushing fails to clear any obstruction from the system and flow remains inadequate, the radiator should be repaired or replaced.

Whenever friction material is found in the cooler inlet, it will be necessary to replace the torque converter and clean and repair the transmission as necessary.

If no friction material is found and cooler flow is adequate, proceed to Step 2.
2. Replace the lock-up valve body assembly with Service Stepped Lock-up Valve Body, P/N 4202219. This lock-up valve body is designed to provide a more positive lock-up shift. If the shudder condition persists, proceed to Step 3.
3. Replace the torque converter with a new Service Unit, P/N 4058489 for 318 CID and P/N 4058287 for 360 CID A-904-LA Torqueflite transmissions. These torque converters are easily identified by yellow paint on the torque converter ring gear. The transmission pump assembly on all A-904 series transmissions should also be replaced with Service Pump Assembly, P/N 4202089. When replacing a torque converter in an A-904 225 CID application or all A-727 applications, use a standard service unit.

When replacing the pump assembly, inspect the input shaft seal rings for cracking or sticking and lube the rings prior to pump installation. Input shaft end-play should also be checked after the new pump assembly is installed to ensure proper input shaft end-play.

After the replacement of these components is completed, a short break-in period may be necessary to achieve maximum improvement. No further major repairs to improve transmission performance, other than adjustment, should be attempted until at least 300 miles have been accumulated on these new components.

This information pertains to ALL 1978 225, 318, and 360 engine applications and 1979 225, 318, and 360 Federal engine applications *only*, and not for use on 1979 California engine applications.

OIL PUMP NOISES, REVERSE MOAN, BUZZ, ENGINE RUNAWAY, OIL BURPING FROM DIPSTICK OR VENT
1978-79 A-904 TORQUEFLITE TRANSMISSIONS

Some hydraulic-mechanical noise (low pitch moan) is normal when shifting a transmission into "Reverse," especially after parking overnight. Noise is caused by air being purged from the transmission.

When transmission performance is suspect, first inspect transmission oil level and correct if required. At this time, inspect the transmission and cooler lines for any leaks and/or obstructions. Also, check for proper adjustment and operation of the throttle and gearshift linkage.

Pump Noises—"Moan" or "Buzzing" in Reverse Gear

When shifting a vehicle into reverse gear (primarily cold) or when the shift lever is between "Park" and "Reverse" position, a loud moan or buzz may be heard from the transmission for a time period longer than 10 seconds.

If the noise continues for more than 10 seconds, the following diagnostic procedure should be used:
1. Set the parking brake and apply the service brake with your left foot. Shift the transmission into "Reverse."
2. With your right foot depress the accelerator pedal slowly, raising the engine rpm until the "Moan" or "Buzzing" noise is heard. Hold the engine rpm at the point of the loudest moan for approximately 1-2 minutes, return to idle, shift into "Park," and immediately check the transmission oil dipstick for foamy oil.

──────────── CAUTION ────────────

Do not hold engine rpm at or near torque converter stall speed for more than five seconds.

a. If foaming is not present, replace the transmission valve body switch valve with a new Valve, P/N 4202474. This new valve has an undercut land to eliminate "buzz" or "moan."
b. If foam is present, replace the transmission pump assembly.

NOTE: If the buzzing noise is still present after replacement of the valve and foaming is not present, an inspection of the transmission oil cooler is required. A plugged cooler may cause switch valve buzz.

Intermittent Engine Runaway and/or Transmission Slippage

Engine runaway, especially when stopping after initial start, or transmission slippage that is not consistent and does not occur during heavy throttle application may be another indication of a transmission pump deficiency.

When either/or both conditions occur, check the transmission oil for aeration. If aeration is present, replace the transmission pump assembly. In all repairs, follow the procedures for transmission and torque converter removal and installation and transmission pump replacement.

Oil Burping from Transmission Vent and/or Dipstick Tube Due to Pump Aeration

Oil burping from the vent on the transmission pump may be misdiagnosed as a front pump seal leak. Also, in more severe cases, oil may leak out the dipstick tube. Both of these conditions may also be due to a transmission pump defect causing oil aeration. If either of these conditions are present:
1. Check the transmission fluid level and condition.
a. If the oil level is above specification, correct the fluid level and retest

After every repair, make certain the following are correct:
a. Transmission Fluid Level
b. Shift Linkage Adjustment
c. Throttle Linkage Adjustment

MODIFICATIONS

THM125/125C Transaxles

CASE COVER CHANGE

Starting mid October 1980, a new design transaxle case cover went into production for some THM 125 automatic transaxles. Both the new and old design transaxle case covers are being used in current production.

The new design case cover has a larger manual valve plug bore. Also, a section of the transaxle case casting has been removed. The changes made in the new design transaxle case cover require a larger manual bore plug and a six lobe socket head screw.

Regardless of design, when servicing the transaxle case cover order service package number 8631957. The old design transaxle case cover uses manual bore plug part number 8631165 and retaining bolt number 11503562. The new design transaxle case cover uses manual bore plug part number 8637091 and retaining bolt part number 8637087. Also, the new design six lobe socket head screw requires a number 40 internal six lobe socket head bit and holder.

DRIVEN SPROCKET WASHER

A new service driven sprocket to driven sprocket support thrust washer has been released for the first design driven sprocket support assembly. This revision is incorporated in all THM 125 automatic transaxles now being built. This design change also affects the transaxle case cover gaskets.

When reassembling the transaxle unit with the first design driven sprocket support, use the second design gaskets and the first design service driven sprocket to·driven sprocket support thrust washer. The part number is 8631981 and the color of the required component is red.

NOTE: The first design washer is white. The second design is black. The first design service washer is red.

The second design driven sprocket support and case to case cover gaskets must be used, as the first design driven sprocket support and case to case cover gaskets are not available for service.

Incorrect combination of these parts may result in gasket distortion, oil leaks, loss of bolt torque or needle bearing damage.

DRIVEN SPROCKET SUPPORT

When diagnosing a THM 125 automatic transaxle for lack of drive, slipping shifts or erratic shift points check the driven sprocket support for wear.

If present, the wear pattern is caused by the bearing sleeve on the outside hub of the driven sprocket moving and wearing into the driven sprocket support. As this wear condition reaches an advanced state the oil feed passages in the driven sprocket support are exposed, allowing an oil pressure cross leak.

The bearing sleeve on the driven sprocket may, after causing the described wear pattern groove, move back to its proper location and appear to be normal. If this condition is suspected, be

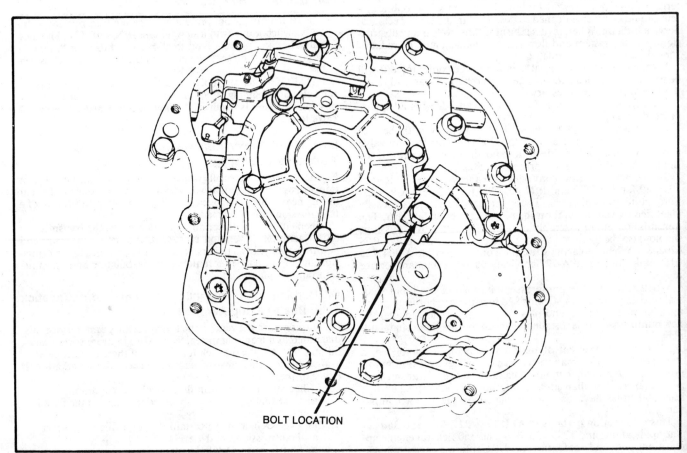

BOLT LOCATION

Pump cover bolt location THM 125 automatic transaxle (© General Motors Corp.)

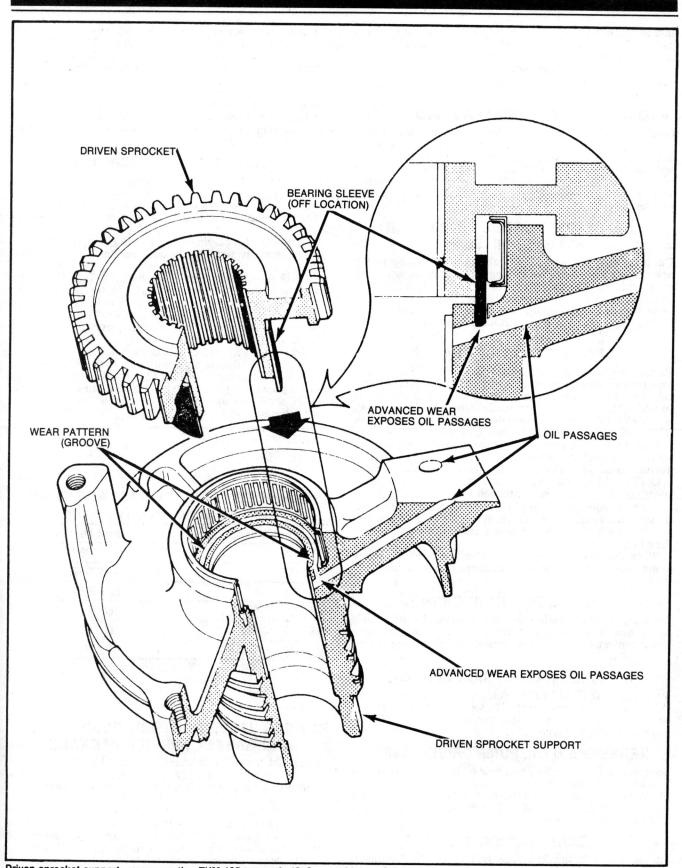

DRIVEN SPROCKET

BEARING SLEEVE
(OFF LOCATION)

ADVANCED WEAR
EXPOSES OIL PASSAGES

OIL PASSAGES

WEAR PATTERN
(GROOVE)

ADVANCED WEAR EXPOSES OIL PASSAGES

DRIVEN SPROCKET SUPPORT

Driven sprocket support—cross section THM 125 transaxle (© General Motors Corp.)

sure to check the driven sprocket support for the wear pattern groove. If this condition exists, replace the driven sprocket, driven sprocket support and any other damaged parts.

Check the direct clutch assembly and the intermediate band for wear. Repair or replace as required. Check the torque converter and flush if necessary.

PRODUCTION TRIAL RUN BACKING PLATE

On all THM 125 automatic transaxles built around the middle of July 1981, a production trial run was started using a new design direct clutch backing plate, part number 8631764. The new design backing plate has a raised side for identification, and a flat side, which when assembled, should face toward the clutch plates. When servicing the transaxle use the current backing plate, part number 8631027.

NEW DESIGN TORQUE CONVERTER HOUSING OIL SEAL

Due to the usage of a new turbine shaft and oil pump drive shaft a new design torque converter housing oil seal is being used in all units produced after December 1981. The new oil seal can be identified by the part number 8637420 stamped on the front face of the seal. The old design seal will have either the part number 8631158 stamped on it, or no part identification at all.

PUMP COVER BOLT CHANGE

Starting at the end of May, 1981, some THM 125 automatic transaxles were being built with a substitute pump cover to transaxle case bolt. The standard pump cover to transaxle case bolt is an M8X1.25X105 bolt, part number 8637266. The substitute bolt is an M8X1.25X110 bolt, part number 11501048 and is used with two standard steel washers that are 1.5mm thick.

If the substitute pump cover to transaxle case bolt and washers are removed from any transaxle, be sure to use the two washers during reassembly. Failure to use the two washers on the M8X1.25X110 bolt may result in interference between the converter and the pump cover to case bolt.

After reassembly a THM 125 automatic transaxle containing the substitute pump cover to case bolt, install the torque converter and check for interference between the torque converter and the pump cover to case bolts.

When installing the pump cover to transaxle case bolt, be sure to use thread sealer on the bolt threads and torque the bolt to 15 ft. lbs.

TORQUE CONVERTER CHANGE

Beginning in 1982 all THM 125 automatic transaxles will be built with a new design 245 millimeter torque converter. This new torque converter will replace the present 254mm torque converter.

TURBINE SHAFT DESIGN CHANGE

Starting in 1982 all THM 125 automatic transaxles will be built using a new design turbine shaft. The spline of the new shaft will be 5.1mm longer than the old style turbine shaft, and will incorporate an O-ring seal groove.

REDESIGNED OIL PUMP DRIVE SHAFT

Beginning in January 1982 a new style oil pump drive shaft will be used in all THM 125 automatic transaxles. The new shaft is 23.7mm longer and has 15 teeth on the oil pump spline rather than the old shaft which has 20 teeth on the oil pump spline and is 23.7mm shorter in length.

BAND ANCHOR PLUG

Starting around the middle of June, 1981, all THM 125 automatic transaxles are being built with a new design band anchor plug.

This new anchor plug has a tab which holds the part in place. All automatic transaxles built prior to mid June 1981 have the old design style band anchor plug. A staking operation was required to hold the old plug in place. This staking operation is not required with the new design band anchor plug, as due to the tab extension, it is held in place by the reverse oil pipe.

When replacing the new design band anchor plug the reverse oil pipe must be removed first. When repairing automatic transaxles prior to mid June 1981 use the new style band anchor plug, part number 8637640.

VALVE BUZZING DIAGNOSIS

Some X body vehicles, Citation, Skylark, Omega and Phoenix, equipped with the THM 125 automatic transaxle may experience a buzzing noise which can only be detected at curb idle.

When diagnosing a buzzing noise problem, follow the procedure below:

1. Be sure that the buzzing noise is coming from the automatic transaxle assembly, not from the engine or any other part of the vehicle.

2. Check for engine coolant in the transaxle fluid. Check and correct the transaxle fluid level. Recheck for the buzzing noise.

3. If the buzzing noise is not corrected, remove the oil pan and check for a plugged or damaged oil strainer. Also, check for a missing or damaged O-ring seal. Replace parts as required. Recheck for the buzzing noise.

4. If the noise is still present, check the automatic transaxle serial number. If the transaxle was built prior to the break point serial numbers in the chart below, the buzzing noise may be coming from the pressure regulator valve, which is located in the valve body. To correct the noise replace the valve body unit, according to the valve body service package chart.

VALVE BODY SERVICE PACKAGE CHART
THM 125 Automatic Transaxle

Break point serial number	Part Number
1980 CV	8637958
1980 PZ	8637957
1981 CD	8637961
1981 CV	8637960
1981 CT	8637960
1981 PZ	8637959

REVISED VALVE CONTROL SPACER PLATE AND GASKET SERVICE PACKAGE

Starting with the month of March 1982, all THM 125 automatic transaxles are being built with a revised version of the valve control spacer plate and gasket. It is important that the proper spacer plate and gasket be used when servicing the unit.

The revised design spacer plate package will include the proper gasket, and can be identified with a yellow stripe. Be sure to use the old design gasket with the old design spacer plate and the revised design gasket with the revised spacer plate, as these parts are not interchangeable. Use the following charts to determine proper usage.

BURNT BAND AND DIRECT CLUTCH ASSEMBLY CONDITION

Some THM 125 automatic transaxle equipped vehicles may experience a burnt band and direct clutch condition. A possible cause of the burnt band and direct drive condition might be the third accumulator check valve not seating properly. This condition allows the intermediate band to drag while the direct clutch is applied causing excessive friction. If the third accumulator is found to be defective order service package part number 8643964, which contains a new dual land third accumulator check valve and a conical spring. Refer to the following procedure to replace the accumulator assembly:

1. Remove the intermediate servo cover and gasket.
2. Remove the third accumulator check valve and spring. Inspect the third accumulator valve bore for wear and damage to the valve seat and also for the presence of the valve seat.
3. Plug both the feed and exhaust holes in the bore using petroleum jelly.
4. Replace the third accumulator check valve with the new dual land check valve. Center the valve to be sure that it is seated properly.
5. Leak test the valve seat by pouring solvent into the accumulator check valve bore. Check for a leak on the inside of the case. A small amount of leakage is acceptable.
6. If the valve leaks tap the assembly with a brass drift and rubber mallet in order to try and reseat the valve.
7. Repeat the leak test procedure. If the valve still leaks it may be necessary to replace the transaxle case.
8. If the valve does not leak, remove the check valve and install the new conical valve spring onto the valve. Be sure that the small end goes on first. Install the valve into the transaxle case bore.
9. Using a new gasket, install the servo cover.

OIL WEIR USAGE

During the month of September 1981 all THM 125 automatic transaxles were assembled with a new part called an oil weir. This part was used by General Motors for one month as a production trial run. The oil weir is located in the rear case oil pan area. Its function is to revise the lubrication flow around the differential assembly.

The oil weir part number is 8637836 and is held in position by a retaining clip part number 8637837. Both of these parts are available for service. Automatic transaxles built without an oil weir do not require the addition of the part during service.

NO REVERSE WHEN HOT

When diagnosing a THM 125 automatic transaxle for a no reverse when hot condition, first check the line pressure. If after checking the line pressure it is okay, carefully check the following list of possible causes in addition to those listed in the Chilton Three C's diagnostic section in the proper Chilton manual.

1. Be sure that the transaxle case to low and reverse clutch housing cup plug assembly is fully seated and not restricted.
2. Check for a damaged or missing low and reverse pipe washer and O-ring.
3. Check and retorque all driven sprocket support bolts to 18 ft. lbs. If the bolt torque is found to be low, use Loctite® on the bolt threads.
4. Check the transaxle case cover (valve body and driven sprocket mating surfaces) and the driven sprocket support sealing surface for possible damage, porosity, leaks and flatness.
5. Check for the correct transaxle case to transaxle case cover gasket and the transaxle case to transaxle cover center gasket.
6. Make certain that the low and reverse clutch inner and outer piston lip seals are installed correctly. The lip seals must be installed with the lip facing down and away from the clutch apply ring. With the low and reverse clutch piston installed into the low

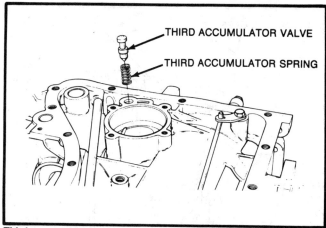

Third accumulator assembly—THM 125 automatic transaxle (© General Motors Corp.)

& reverse clutch housing, the lip of the lip seal must contact the walls of the low and reverse clutch housing to form a proper seal. Installation of these seals in any other manner may cause the no reverse condition.

LOW-REVERSE CLUTCH ASSEMBLY— DESIGN CHANGE

Starting around the middle of July, 1981, some 1982 vehicles produced with the THM 125 automatic transaxle were assembled with a new design low and reverse clutch assembly. This design change produces a more desirable neutral to reverse shift. The new design assembly consists of a modified low and reverse piston which eliminates the need for an apply ring. A smaller low and reverse clutch housing feed orifice is also used, as is a waved steel clutch plate located next to the low and reverse piston. The new waved steel clutch plate eliminates the one flat steel clutch plate.

INTERMEDIATE BAND/DIRECT CLUTCH HOUSING ASSEMBLY SERVICE PACKAGE

A new intermediate band and direct clutch housing service package has been assembled and released to repair all THM 125 automatic transaxles. This new service package, part number 8643941 consists of a direct clutch housing and drum assembly, and an intermediate band assembly. Both of these parts must be used together as a complete unit. These two items are no longer available individually and can only be obtained as a set.

PRESSURE REGULATOR VALVE RETAINING PIN

When diagnosing the 125C automatic transaxle for no drive or harsh shifts (high line pressure), check the control valve assembly for a worn or missing pressure regulator valve retaining pin.

If after checking the pressure regulator valve retaining pin it is found to be either missing or worn the following repair must be performed.

1. Position the control valve and oil pump assembly with the machined portion face down. Be certain that the machined face is protected in order to prevent damage to its surface.
2. Using a ⅜ in. drift punch and hammer close the pressure regulator valve retaining pin hole. Close the pin hole only enough to hold the new retaining pin in place after assembly.
3. Reassemble the pressure regulator and reverse boost valve train assembly.

MODIFICATIONS

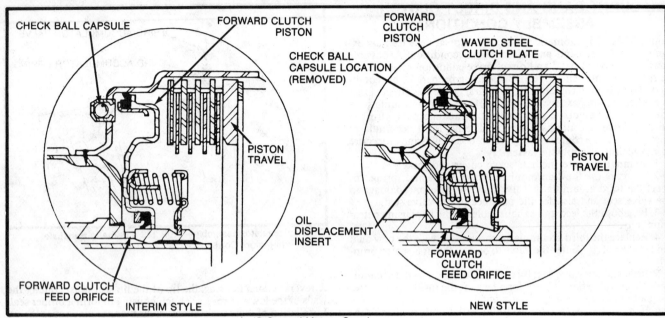

Forward clutch assembly—125 automatic transaxle (© General Motors Corp.)

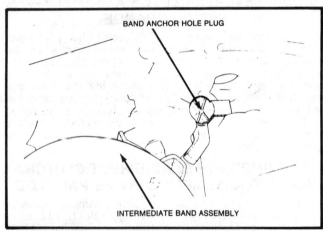

Band anchor plug location—THM 125 automatic transaxle
(© General Motors Corp.)

4. Retain the valve train with a new steel retaining pin, part number 112496. Be sure that the new pin is inserted from and flush with the machined face of the control valve and oil pump assembly.

FORWARD CLUTCH CHANGES

Beginning July, 1981 some changes were made in the forward clutch assembly to produce a more desirable neutral to drive engagement and eliminate the check ball capsule. These changes were accomplished in two phases. Phase one began in July of 1981 and phase two in September of the same year.

PHASE ONE—INTERIM STYLE

The forward clutch was redesigned, eliminating the apply ring. The backing plate was made selective to control piston travel.

PHASE TWO—NEW STYLE

The forward clutch feed orifice was reduced in size and an oil displacement insert was added. These two changes will produce a more desirable neutral to drive engagement.

The check ball capsule was removed and two exhaust holes were added to the new forward clutch piston. When the piston is applied the exhaust holes will seal against the waved steel clutch plate. When the piston is released, apply oil will exhaust between the waved steel clutch plate and the piston. Also, a selective backing plate is used to control piston travel.

FORWARD CLUTCH SELECTION PROCEDURE

When servicing the individual parts that make up the forward clutch assembly in the THM 125 automatic transaxle, first identify the forward clutch assembly style and then select only those parts specified in the chart below for each style.

When replacing the selective backing plate, check the end play in the following manner.

1. Insert a feeler gauge between the selective backing plate and the snap ring.
2. The end play should be 0.040–0.070 in., without compressing the waved clutch plate.
3. If the end play is not within specification, select the proper backing plate from the forward clutch end play chart. For positive identification measure the thickness of the selective backing plate.
4. Use of a new style forward clutch housing without proper end play will cause excessive clutch plate wear.

FRONT BAND CHANGE

THM 125 automatic transaxles produced late in 1981, were being built with a wider intermediate band which conforms with the wider band surface on the direct clutch drums. The new wider band cannot be used with the early model THM 125 automatic transaxle equipped with the narrow band area. The early design band will fit the late model transaxle, but this is not recommended. The early design band is 1.49 inches wide. The new design band is 1.74 inches wide.

LATE OR DELAYED UPSHIFTS

Some THM 125 automatic transaxles may experience late or delayed upshifts (1-2 shift 30-35 mph, 2-3 shift 50-55 mph) this will occur if the line boost lever cannot lift the line boost valve off its

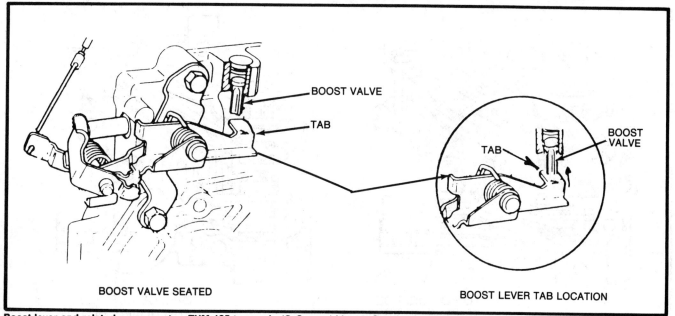

BOOST VALVE

TAB

BOOST VALVE

TAB

BOOST VALVE SEATED

BOOST LEVER TAB LOCATION

Boost lever and related components—THM 125 transaxle (© General Motors Corp.)

seat. To correct the problem, check and adjust the T.V. linkage. If the linkage adjustment does not correct the condition, connect a pressure gauge and check the transaxle line pressure. If the line pressure is high, remove the valve body cover and check the position of the line boost lever. With the T.V. linkage set and the throttle linkage on the idle stop. The boost lever should be holding the boost valve off its seat. If not, and the linkage is correctly adjusted and the linkage springs properly positioned, bend the tab on the boost lever to hold the boost valve completely off its seat.

GOVERNOR COVER AND BEARING ASSEMBLY CHANGE

Beginning late March, 1981, some THM 125 automatic transaxles were built with a new design governor cover bearing in place of the old design thrust washer. This new bearing will be in either one of two designs. The first design is a three piece thrust bearing assembly and the second design is a one piece bearing. For service, only the first design bearing will be used.

A new design governor cover was also put into production. The height of the governor hub was reduced to allow clearance for the increased thickness of the bearing.

This design change was made to reduce friction between the speedometer gear and governor cover during high speed, long distance driving.

ROUGH NEUTRAL-TO-DRIVE GEAR ENGAGEMENT 1982 THROUGH 1984 THM 125/125C TRANSAXLES

Some 1982 through 1984 THM 125/125C transaxles may exhibit a rough neutral to drive gear engagement condition. A new forward clutch wave plate went into production, beginning in December of 1983, to prevent this condition. When servicing any 1982 through 1984 THM 125/125C transaxles for this condition, order and install part number 8652126, forward clutch wave plate.

NEW AUXILIARY VALVE BODY AND AUXILIARY VALVE BODY GASKET ALL THM 125C TRANSAXLES

During September, 1983 all THM 125C transmissions were built with a new auxiliary valve body and valve body gasket. The new auxiliary valve body does not have an orifice plug hole. The auxiliary valve body gasket is made out of a new material and has three holes made larger.

The changes to the gasket eliminate a no torque converter clutch release (TCC) or no TCC apply condition that resulted from incorrect cup plug, mispositioned cup plug or eroding gasket.

NOTE: The only interchangeable part is the auxiliary valve body cover. The new auxiliary valve body gasket must be used with the new auxiliary valve body. The past auxiliary valve body gasket must be used with the past auxiliary valve body.

When servicing any model THM 125C transmission auxiliary valve body, refer to the following procedure:

1. Identify the auxiliary valve body being serviced (past or new).

2. If the past auxiliary valve body is being serviced, refer to Chart 1. If the new auxiliary valve body is being serviced, refer to Chart 2.

CHART 1

Part Being Replaced	Part Number
Auxiliary Valve Body Cover Gasket	8653947
Auxiliary Valve Body Cover	8643645
Auxiliary Valve Body Orifice Plug	8623796
Auxiliary Valve Body (past design not serviced)	New auxiliary valve body must be used, order Service Package 8653946.

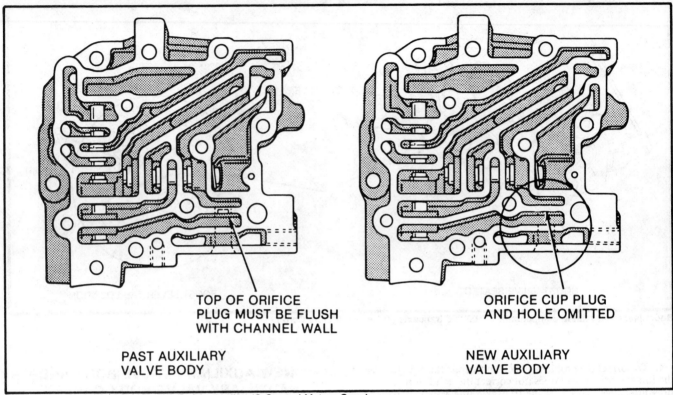

TOP OF ORIFICE
PLUG MUST BE FLUSH
WITH CHANNEL WALL

ORIFICE CUP PLUG
AND HOLE OMITTED

PAST AUXILIARY
VALVE BODY

NEW AUXILIARY
VALVE BODY

Comparison of past and new auxiliary valve bodies (© General Motors Corp.)

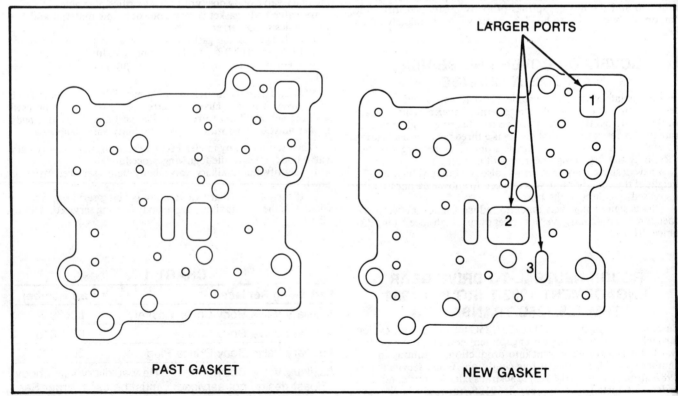

LARGER PORTS

PAST GASKET

NEW GASKET

Comparison of past and new gaskets (© General Motors Corp.)

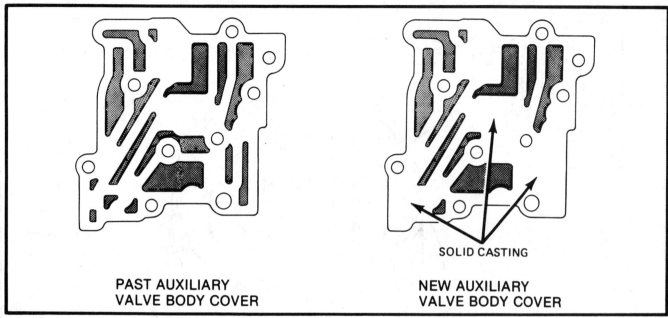

PAST AUXILIARY
VALVE BODY COVER

NEW AUXILIARY
VALVE BODY COVER

SOLID CASTING

Comparison of past and new auxiliary valve body covers (© General Motors Corp.)

CHART 2

Part Being Replaced	Part Number
Auxiliary Valve Body Cover Gasket	8643863
Auxiliary Valve Body Cover	8643645
Auxiliary Valve Body Models BF, BL, CE, CA, CB, CF, CL, CK, CT, CC, PE, PD, PG, PJ, PW, PF	8653946 Service Package
EM, EN, EW, HC, HS, HV, HY, OP	8653945 Service Package

Service Package 8653946 contains:

1	8643865	Auxiliary Valve Body
1	8643863	Gasket Aux. Valve Body
1	8643645	Cover Aux. Valve Body

Service Package 8653945 contains:

1	8643864	Body Auxiliary Valve
1	8643863	Gasket Aux. Valve Body Cover
1	8643645	Cover Aux. Valve Body

Only the Auxiliary Valve Body Cover, Auxiliary Valve Body Cover Gasket, and Auxiliary Valve Body Orifice Plug can be ordered separately.

When servicing any auxiliary valve body inspect it for erosion around the hole number 1. If the gasket is eroded, then replace the auxiliary valve body cover along with the required gasket.

If the auxiliary valve body being serviced needs the orifice plug replaced due to damage or missing follow the procedure listed below.

    a. If the orifice plug is still in the auxiliary valve body, remove it with a $7/32$ in. screw extractor.

    b. Install a new orifice plug (Part Number 8623796) using a $5/16$ in. punch to start the orifice plug in the hole. Then use a $3/8$ in. punch and seat the plug flush with the wall.

WHINING NOISE AND SLOW ENGAGEMENT THM 180C AUTOMATIC TRANSMISSIONS (CHEVETTE AND 1000 MODELS)

The vehicles referenced may exhibit the following condition:

Whining noise which increases with engine rpm in all ranges and slow or slipping engagement. The whining noise is similar to noise caused from the steering wheel being turned to its maximum travel.

If the above condition exists check the transmission oil line pressure. If the line pressure is low, refer to the following service procedure.

    1. Remove the oil filter and visually inspect for restriction.

    2. If the oil filter is restricted remove the transmission from vehicle, and remove the pump.

    3. After removing the pump inspect the selective end-play washer for wear especially on the inside diameter of the washer. If the washer is worn replacement is necessary.

    4. Remove any burrs that may be present around the lube hole on the Pump Tower.

    5. Inspect washer thrust surface for damage. Remove any high spot that may be present by stoning the thrust surface.

    6. Thoroughly wash the Pump Body with solvent before reassembly.

    7. Select the correct end-play washer and assemble pump to the transmission, replace oil screen and install transmission into vehicle.

NEW THIRD OIL PRESSURE SWITCH— THM 200-C 01 MODEL TRANSMISSION 5.0L BONNEVILLE AND GRAND PRIX

Beginning November 1, 1983 a new third oil pressure switch and bracket assembly went into production for all 1984 THM 200-C 01 model transmissions.

When servicing any 1984 THM 200-C 01 model transmission that necessitates that the third oil pressure switch be replaced

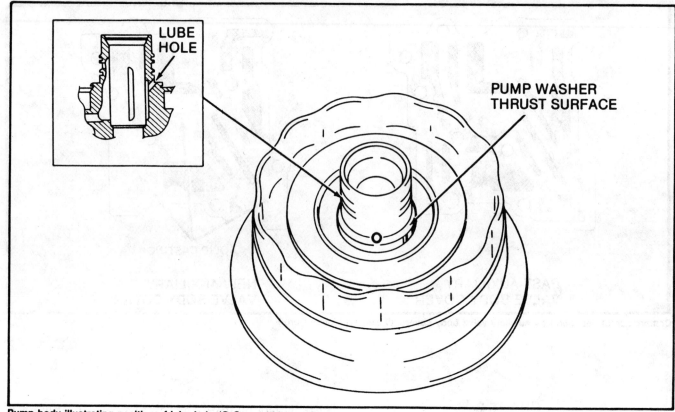

Pump body illustrating position of lube hole (© General Motors Corp.)

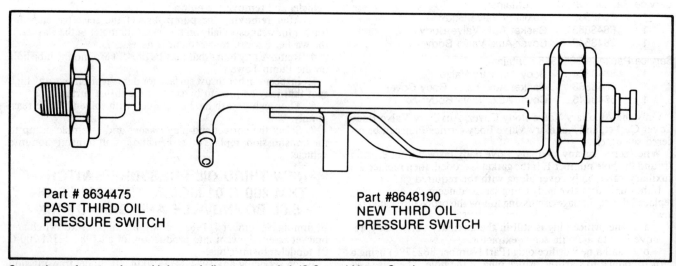

Part # 8634475
PAST THIRD OIL
PRESSURE SWITCH

Part #8648190
NEW THIRD OIL
PRESSURE SWITCH

Comparison of past and new third speed oil pressure switch (© General Motors Corp.)

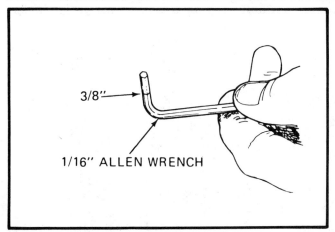

Marking of Allen wrench (© General Motors Corp.)

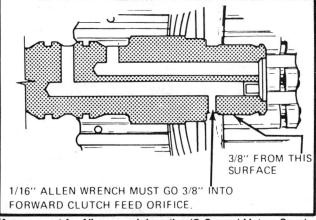

3/8'' FROM THIS SURFACE

1/16'' ALLEN WRENCH MUST GO 3/8'' INTO FORWARD CLUTCH FEED ORIFICE.

Measurement for Allen wrench insertion (© General Motors Corp.)

compare the oil pressure switch being replaced to the past and new oil switches.

If the oil switch required is the past switch, order Part Number 8634475. If the oil switch required is the new switch, order Part Number 8648190.

THM 200C LATE ENGAGEMENT OR SLIPPING IN DRIVE

Some early 1984 model THM 200C transmissions may experience a late engagement or slipping in drive range. This condition is usually detected while the transmission is cold, and could be caused by a blocked forward clutch feed orifice hole.

Other causes for late engagement or slipping in drive are low transmission oil level, cut or nicked forward clutch piston seals and/or cut or nicked Teflon® turbine shaft seals.

Refer to the list below and verify serial and model numbers. If the serial number is prior to the number on the chart then follow this procedure:

MODEL SERIAL NUMBER

BH	54628
OI	21738
OR	1119
OU	1727

SERVICE PROCEDURE

1. Check transmission oil level and correct if required.
2. Visually inspect the forward clutch piston seals and turbine shaft Teflon® seals for cuts and nicks.
3. Use a 1/16 in. Allen wrench marked on the short end as illustrated in Fig. 8. Using the Allen wrench as a gage, insert the small end into the forward clutch feed orifice hole and make sure the Allen wrench goes into the orifice hole (without any drag) until the mark on the wrench is flush with the ground diameter.

The part numbers for the Housing Assembly Forward Clutch is 8638944 and 8628924 for the Forward Clutch Piston Seals. The Turbine Shaft Teflon® Seals are part number 8628090.

PRESSURE REGULATOR BUSHING PIN DESIGN CHANGE—THM 325-4L

Beginning early in April, 1983 production the THM 325-4L transmission case was changed along with the coiled spring pin used to retain the pressure regulator valve bushing in the case. A larger solid steel pin is now used. This change was made to eliminate the possibility of case face distortion during assembly. Due to different sizes, these two pins cannot be used interchangeably. Furthermore, the new solid pin is a slip fit in the transmission case and may fall out if the transmission is rolled over; loss of this pin would enable the pressure regulator bushing to lose proper orientation, resulting in the loss of proper pressure regulator valve operation.

If replacement of either pin is required, use the following numbers to order the correct pin.

Coiled Spring   P/N 9437415
Solid Pin   P/N 141202

HIGH SHIFT POINTS—1982 "C" CARS WITH 5.7L DIESEL AND 350C TRANSMISSION

1982 "C" cars equipped with the 5.7 liter diesel engine and 350 C transmission may experience higher than normal or delayed transmission shift points. This condition may be caused by low vacuum at the modulator due to a disconnected or damaged EGR valve vacuum hose on the underside of the air cleaner.

Modifications and Changes Made to More Than One THM Model Transmission/Transaxle Assemblies

SERVICING OF THE TORQUE CONVERTER CLUTCH VALVE TRAIN IN THM 200-4R AND 325-4L TRANSMISSIONS

Early production 1982 Cadillacs equipped with gasoline engines and THM 200-4R or 325-4L transmissions were built with a Torque Converter Clutch (TCC) valve train installed in the control valve assembly of the transmission. This original design control valve assembly contains a hollow bushing as well as converter clutch throttle and shift valves and a calibration spring.

Beginning late in 1982 production, the TCC valve train was eliminated from the control valve assembly. This second design control valve assembly contains only a bushing to direct TCC signal oil flow. The converter clutch apply valve and solenoid continues to be used as the control for TCC operation.

When servicing the THM 200-4R or 325-4L Transmission, it is essential that the proper gasket and spacer plate be used to assure proper transmission operation. On THM 200-4R applications, incorrect parts can result in poor fuel economy, and a hard diagnostic code 39 on HT4100 equipped vehicles. On THM 325-4L applications, incorrect parts can result in poor shift quality.

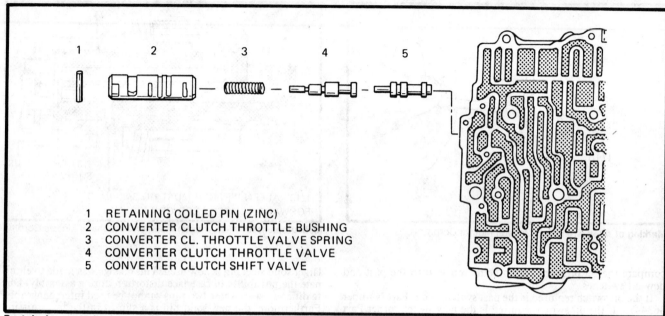

Past design control valve assembly with T.C.C. Valve train (© General Motors Corp.)

1 RETAINING COILED PIN (ZINC)
2 CONVERTER CLUTCH THROTTLE BUSHING
3 CONVERTER CL. THROTTLE VALVE SPRING
4 CONVERTER CLUTCH THROTTLE VALVE
5 CONVERTER CLUTCH SHIFT VALVE

THM 200-4R (DEVILLE AND BROUGHAM SERIES VEHICLES)

1. Service Information

a. First and second design control valve assemblies and spacer plates are not interchangeable. Care must be taken to ensure the correct spacer plate is used with the correct control valve assembly. Second design gaskets can be used to service both first and second design components.

b. First design parts are no longer available through GMWDD. Second design parts must be used to service these applications. Refer to the following parts information chart for the proper second design part numbers to be used when servicing first design applications:

THM 325-4L (ELDORADO AND SEVILLE)

2. Service Information

a. On 4.1L V-6 applications, the complete service package must be used when replacing a first design control valve assembly. The first design spacer plate is currently available but must not be used on the second design control valve assembly. On HT4100 applications, the second design control valve assembly is compatible with the existing first design spacer plate. The gasket package will service both first and second applications.

b. First design control valve assemblies are no longer available through GMWDD. Second design parts must be used to service these applications. Refer to the following parts information chart for the proper second design part numbers to be used when servicing first design applications.

THM 325-4L, FIRST DESIGN TCC VALVE TRAIN SERVICE PARTS INFORMATION

Engine	Trans Code	Control Valve Assembly	Gasket Package	Spacer Plate	Service Package
4.1L V-6	AM	N/A	8635919	8635648①	8635938②
HT4100	AJ	8635955	8635919	8635640	N/A

① First design spacer plate, not to be used on second design control valve assembly.
② Contains second design control valve assembly, spacer plate and gasket package.

THM 200-4R, FIRST DESIGN TCC VALVE TRAIN SERVICE PARTS INFORMATION

Engine	Trans Code	Control Valve Assembly	Gasket Package	Spacer Plate	Service Package
4.1L V-6	BY	N/A	8634969	N/A	8639133①
HT4100	AA & AP	N/A	8634969	8639027②	8634981①

① Contains second design control valve assembly, spacer plate, and gasket package.
② Second design spacer plate, not to be used on first design control valve assembly.

NOTE: If disassembly of a second design control valve assembly on either model transmission is performed, location of the solid bushing during reassembly is important. Locate the bushing so that the coiled pin fits into the groove in the bushing. Failure to do so could result in a loss of TCC operation.

NO REVERSE, FIRST GEAR ONLY IN FORWARD
THM 200, 200C, 200-4R, 325 and 325-4L TRANSMISSIONS/TRANSAXLES 1979-84 CADILLACS

1979-1984 Cadillacs equipped with THM 200, 200C, 200-4R, 325 and 325-4L transmissions may experience a lack of engagement in reverse, second, third and fourth gear while operating properly in first gear. This condition can be caused by stripped input drum to rear sun gear splines and does not affect line pressure readings or cause worn clutch plates or bands.

Inform technicians performing overhaul procedures for this condition that careful inspection of the input drum splines and planetary gear assemblies is necessary. Spline wear may not be completely obvious during disassembly since the splines may hold with hand torque but not with engine torque.

To inspect the splines properly, remove the snap ring used to retain the input drum to the sum gear shaft and remove the input drum from the sun gear. If any wear is present, replace the sun gear shaft along with the input drum.

DIAGNOSING OIL LEAKS ON 125C, 200C, 325-4L, AND 200-4R TRANSMISSIONS

When any THM transmission oil leak is detected, it is very important that they are diagnosed properly, as to the location and type of leak.

The following procedure will help in diagnosing the type and location of the leak and will suggest a correction method.

DIAGNOSIS PROCEDURE: (ALL MODELS)
The following procedure can be performed on car.

1. Clean all residual oil from the transmission, concentrating on the transmission oil pan to case mating areas (all models), TV cable connector (all models), transmission oil pan to sprocket cover interface area (325-4L model), final drive to case connection (325-4L model), and valve body cover to transmission oil pan interface area (125C model) (CRC's electramotive cleaner, or equivalent is recommended).

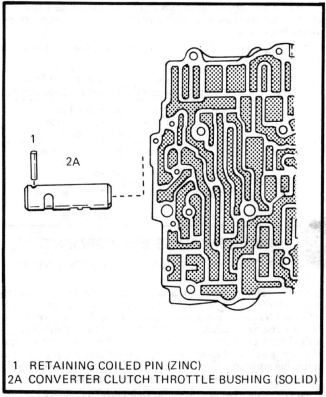

| 1 | RETAINING COILED PIN (ZINC) |
| 2A | CONVERTER CLUTCH THROTTLE BUSHING (SOLID) |

New Design control valve assembly with T.C.C. Valve train eliminated (© General Motors Corp.)

2. Dust the entire transmission with a white powder (e.g., Leak Tracing Powder or Foot Powder Spray).

3. After normal operating temperature (170°F) has been obtained, let the transmission stand for at least (30) thirty minutes.

4. Inspect the transmission for leakage by following red oil traces left in the powder. (If residual oil is left on the transmission oil pan or sprocket cover, it will be necessary for the cleaning and dusting procedure to be repeated.)

5. Once the leak has been diagnosed as to location and type (wetness, moist, damp, etc.), refer to the following repair procedure:

MODIFICATIONS

Ford Motor Company

OVERSENSITIVE 3-2 DOWNSHIFT

1982 CAR LINES EQUIPPED WITH C-3 A/T AND 2.3L ENGINE

If an oversensitive 3-2 downshift is encountered prior to servicing the main control valve assembly, inspect the modulator vacuum line for engine fuel.

If gasoline is found in the line or modulator, drain the line and replace the modulator. Route the vacuum hose from the vacuum tree on the fire wall to a higher elevation than the vacuum tree. Use a clamp and a self-tapping screw to hold the hose in place. If no fuel is encountered, proceed with the normal valve body repair.

CHANGE OF THREAD SPECIFICATIONS ON INTERMEDIATE BAND ADJUSTING SCREW AND LOCKNUT

1981-82 CAR LINES USING C-4 AND EARLY C-5 AUTOMATIC TRANSMISSIONS

During 1981 model C4 transmission production, a change of intermediate band adjustment screw and nut was incorporated—from coarse to fine thread.

The adjustment specifications are different for the intermediate band on C4 transmissions depending on the thread of the adjustment screw and nut.

Prior to any intermediate band adjustment or service, always examine the threads of the adjustment screw to determine the type of threads—fine or coarse.

MODIFICATIONS

NOTE: Enough of the thread can be seen from the outside of the transmission case so the screw does not have to be removed for visual inspection.

The following chart denotes Intermediate Band Adjustment Specifications:

Transmission Type	Pitch on Thread Adjustment Screw & Nut	Intermediate Band Adjustment Specification
C4	Fine pitch thread (C4 trans W/C5 case)	Back-off 3 turns. Locknut torque 35-45 lbs.-ft
C4	Coarse pitch thread	Back-off 1¾ turns. Locknut torque 35-45 lbs.-ft.
C5	Fine pitch thread only	Back-off 4¼ turns. Locknut torque 35-45 lbs.-ft.

HARSH REVERSE ENGAGEMENT

1981 CAR LINES EQUIPPED WITH C-4 (PEN) AND 3.3L ENGINE

A harsh neutral to reverse idle engagement may be encountered on the above 1981 3.3L passenger car applications equipped with a C-4 (PEN) model transmission. To service this concern, check the model identification tag on the transmission to verify that it is a PEN model C-4 transmission. If it is a PEN model, verify that the idle speed is set to specifications, if the concern persists, the following are instructions to rework the main control valve body assembly and the low-reverse servo piston assembly:

NOTE: This rework may cause slight delay in reverse engagement.

1. Remove the low-reverse servo assembly. Mustang and Capri vehicle applications require lowering the transmission to remove the low-reverse servo. To lower the transmission, remove the engine support-to-crossmember nuts, loosen the right-hand crossmember nut, remove the left hand crossmember bolt and nut, and then swing the crossmember to the right side. The transmission can be lowered with a jack sufficiently to gain access to the low-reverse servo cover.

2. After removing the low-reverse servo piston, drill a 0.031 ± 0.002 in. (#68 Drill) diameter hole through the piston top surface. This can be facilitated by inserting the servo in a vise. Allow the top of the servo to be supported by the vise not clamped in the jaws. Clean the metal shavings from the drilled hole and from the piston.

3. Reinstall the piston into the case as outlined in the low-reverse servo shop manual procedure. Follow Steps 5-7. No special orientation of the hole is necessary when installing the servo assembly into the case.

4. Adjust the low-reverse band.

Step II Main Control Valve Body Assembly Modification

1. Remove the main control valve body assembly from the transmission.

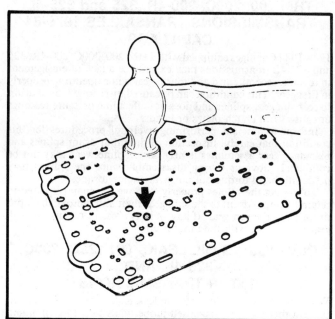

Seating ball bearing in separator plate orifice (© Ford Motor Co.)

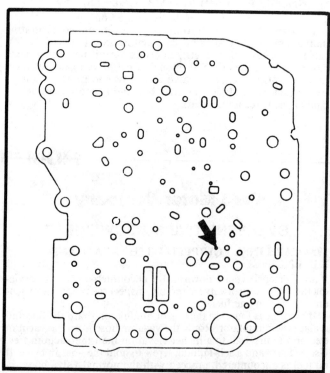

Location of reworked orifice in separator plate (© Ford Motor Co.)

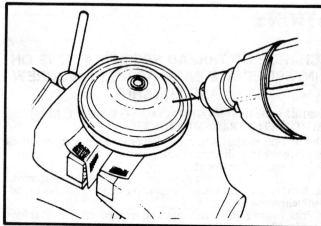

Drilling Servo piston (© Ford Motor Co.)

5-22

2. Remove the seperator plate from the main control valve body be careful not to lose the upper valve body shuttle valve and check valve when seperating the upper and lower valve bodies.

3. Rework the seperator plate as follows: Place the seperator plate on a hard flat surface such as a steel block. Using a steel ball approximately ¼ in. diameter, rework hole #76 by placing the ball in the hole and striking the ball with a hammer until the hole is reduced to a diameter of slightly less than 0.040 in. where a #60 drill will not quite pass through the hole. Then size the hole to 0.040 in. by drilling with a #60 drill. Clean-up metal burrs from the hole after the operation.

4. Reassemble the main control valve body assembly.

5. Reinstall the main control valve body assembly in the transmission.

INTERMEDIATE BAND ADJUSTMENT SPECIFICATION CHANGE

1981 C-4 A/T WITH FINE THREAD ADJUSTING SCREWS

All C-4 automatic transmissions built from January 5, 1981 and after are equipped with a new intermediate band adjuster screw and nut using a fine pitch ½-28 thread.

Whenever a band adjustment is required, the following new procedure is required:

Intermediate Band Adjustment

1. Remove and discard the lock nut.
2. Install a new lock nut on the adjusting screw. With the special tool, tighten the adjusting screw until the handle clicks. The tool is a pre-set torque wrench which clicks and breaks when the torque on the adjusting screw reaches 10 ft. lbs.
3. Back off the adjusting screw exactly 3 turns.
4. Hold the adjusting screw from turning and tighten the lock nut to 40 ft. lbs.

This change is being incorporated concurrently with new cases bearing a yellow trademark. The fine adjusting screw can be identified by an identification rib on the strut end.

Transmissions built prior to the above date will continue to use the old band adjustment procedure.

New Model Numbers Affected

PEJ-	AC4	PEB-	N11	Pen-	A1	PEM-	W1
	AD4		P9		B1		C6
			Z1				D6
			U5				E6
PEA-	CP1	PEE-	FL6				AC1
	CE10		GB1				AD1
							AE1
							AL1
							AM1
							AN1
							AK1

SHIFT HUNTING CONDITION DURING 1-2 UP-SHIFT

1980 AND EARLIER C-4 MODELS

To service the condition of shift hunting (rapid progressional first-to-second, second-to-first gear shift) occurring during the 1-2 shift period at minimum throttle (approximately 10 mph), replace the production primary valve with service primary valve part no. D70Z-7C054-A as follows:

1. Remove the transmission extension.
2. Remove the governor from the transmission output shaft.
3. Remove the governor primary valve snap ring, washer and spring. Remove the governor primary valve from the body and discard.

4. Install the new service primary valve reinstall the spring washer and snap ring in the governor body.
5. Reinstall the governor to the oil collector body.
6. Reinstall the extension assembly.
7. Lower the vehicle and fill the transmission to proper level.
8. Test drive vehicle.

PART NUMBER	PART NAME
D70Z-7C054-A	Primary Valve

NOTE: C4 automatic transmission use transmission fluid meeting Ford Specification ESP-M2C 138-CJ or Dexron® II, Series D.

HARSH 1-2 UPSHIFTS AND/OR 3-2 DOWNSHIFTS

1981 CAR LINES EQUIPPED WITH 3.3L ENGINES AND C-4 A/T, MODELS PEN-A, PEN-B, PEN-M and PEN-N ONLY

Customer concerns of harsh 1-2 upshifting and/or 3-2 downshifts, may be resolved by reinstalling the following shift feel revision kit—E1DZ-7D371-A.

Shift feel revision kit consists of:
1-Intermediate servo cover
1-Main intermediate servo cover gasket
1-Intermediate servo piston
1-Intermediate servo piston seal (small)
1-Intermediate servo piston seal (large)
1-Intermediate servo piston return spring (orange)
1-Main control intermediate servo accumulator spring (yellow)

DESCRIPTION OF CHANGES AND INTERCHANGEABILITY

1984 CAR AND TRUCK LINES EQUIPPED WITH C-5 AUTOMATIC TRANSMISSIONS

NOTE: Information arrived to late to include with the C-5 testing and overhaul section.

C-5 transmission will have the following changes incorporated for 1984 model vehicles.

Approximately eight (8) models of C-5 automatic transmissions are currently released for the 1984 five (5) passenger car models and three (3) truck models. The truck models are two (2) F-150 applications and one (1) Ranger/Bronco II 2.8L 4x4 application.

Vehicle and/or engine combinations dropped for 1984 include the 3.3L (200 CID) engine with the C-5 transmission; the 2.3L Ranger with the C-5 transmission; and the 3.8L F-100 C-5 truck model. the 2.3L Ranger will be released with the C-3 transmission. New to C-5 will be the F-150 4.9L and 5.0L truck applications.

MAJOR DESIGN CHANGES FOR INCORPOATION OF THE 4.9L AND THE 5.0L F-150 APPLICATIONS

1. Micro-finish change to the forward planet assembly for durability. This planet will be the service replacement for all C-5 model transmissions.
2. The reverse clutch pack will have one more steel plate (3 to 4) and one more friction plate (3 to 4).
3. The converter is a 12 in. non-converter clutch 110 type converter. It is as deep as the converter clutch type converter because the clutch assembly has been replaced by a spacer.
4. Two new flywheels (flywheels were used on other vehicle/engine applications).
5. Transmission fluid capacity is 11.5 U.S. quarts.

C-5 PASSENGER CAR HIGHLIGHTS

1. All passsenger car applications have 12 in. converter clutch type converters.

2. All converters will have an orange special balance mark on a converter stud. If a flywheel bolt has a special balance mark, align the marks to avoid powertrain imbalance.

3. Fleet applications with the auxiliary cooler option will have an "H" fitting that bypasses some of the auxiliary cooler flow, especially during cold weather operation to ensure lubrication flow to the transmission.

INTERCHANGEABILITY:

The following 1984 transmission and main control are carryover 1983 and can be used on the 1983 model if service replacement is necessary:

For 1983 Ranger/Bronco II 2.8L 4x4:
Use 1984:
PEJ-AJ E37P-7000-EA
Main Control E37P-7A100-AA

NO OR DELAYED UPSHIFTS/DOWNSHIFTS

1982 ECONOLINE AND BRONCO WITH C-6 A/T WITH BUILD DATE OF 10-1-81 to 2-28-82

No or delayed upshift or downshift may be caused by forward and reverse planet assembly bearing spacer and thrust washer deterioration.

No or delayed upshift or downshift in vehicles with transmissions built between 10-1-81 and 2-28-82 may be caused by forward and reverse planet assembly bearing spacer and thrust washer deterioration. The transmission build date can be determined from the metal I.D. tag that is bolted to the intermediate servo cover.

This condition is characterized by fine scoring of the pump gear cavities and/or converter impeller hub and excessive magnetic metallic contamination in the main control, oil pan, and oil filter screen.

When servicing a transmission built between 10-1-81 and 2-28-82 that exhibits no or delayed upshift or downshift, remove the transmission oil pan to inspect for excessive magnetic contamination. If such contamination is present in the oil pan or oil pickup screen, remove the transmission and replace the forward and reverse planet assemblies. Replace other components as necessary. Replacement of the planet assemblies is required because deterioration of the spacer and thrust washer is not readily apparent by visual inspection.

SLIPPING 3RD GEAR PERFORMANCE AND/ OR NO 4-3 DOWNSHIFT

1981-82 CAR LINES EQUIPPED WITH AOD TRANSMISSION

Slipping 3rd gear, lack of 3rd gear performance, or no 4-3 downshift may be caused by a blocked hydraulic passageway in the transmission case.

The blockage of this particular transmission case passageway feeds both the overdrive servo for release of the overdrive band and forward clutch circuit in 3rd gear. Therefore, if this passageway is blocked, the overdrive band may drag resulting in a lack of 3rd gear performance. Eventually, the overdrive band material may deteriorate and require replacement.

To identify cases that have a portion of the case hydraulic passageway blocked, visually check the die numbers located on the top of the case in the converter housing area. With the transmission in the vehicle, it is necessary to use a long-handled flexible mirror with a flashlight to read the numbers.

If the case is identified with the die numbers 31 or 32, remove the oil pan, filter, and main control valve body. Identify the passageway and the portion that maybe blocked. If the passageway is blocked, replace the case.

HARSH 2-3 SHIFT

1980-81 CAR LINES EQUIPPED WITH AOD TRANSMISSION

The following is a procedure to resolve harsh 2-3 shift concerns.

Check for a Missing Check Ball

1. Remove the valve body from the transmission.
2. Disassemble the valve body.
3. Refer to the illustration for the proper location of the number 3 check ball.
4. If the check ball is present, proceed to the check the 2-3 capacity modulator valve.

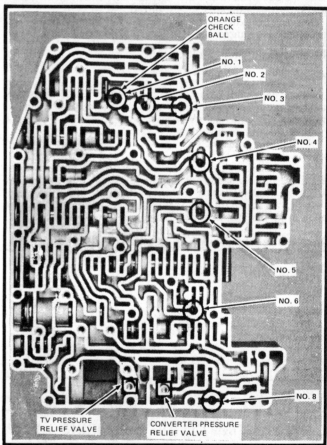

Location of Check balls. Check ball number seven was eliminated in later production years (© Ford Motor Co.)

5. If the check ball is not present, install the check ball and reassemble the valve body.
6. Reassemble the transmission fluid level following the "Fluid Level Check" procedure.

Check the 2-3 Capacity Modulator Valve

1. Follow the valve body disassembly procedure for both the location of the 2-3 capacity modulator valve and for the removal and installation of the valve.

2. If the valve does not move freely, remove any obstructions. If necessary, replace the main control valve body.

3. If the valve moves freely, reassemble the valve body and proceed to the check for a blocked 2-3 accumulator case feed passage.

4. When reassembling the valve body, follow the assembly instructions which includes replacing the gaskets, using the valve body alignment pins when installing the separator plate on the valve body and when installing the valve body on the case. Also, properly torque the valve body to avoid repeat servicing.

Check the 2-3 Accumulator Case Feed Passage

1. With the transmission in the vehicle and the main control valve body removed, air pressure test the 2-3 accumulator piston. Apply air to the 2-3 accumulator passage. Use service tool T80L-77030-B or equivalent in order to seal the apply passage. Check for a free flow of air to the accumulator bore.

2. If there is not free flow air, drill the passage free by using a 6 in. long ¼ in. diameter (.250) drill bit. Drill into the passage 3 in. To indicate when a 3 in. depth has been reached, wrap a thin piece of black electrical tape around the drill 3 in. from the drill tip. Drill into the passage until the black tape meets the case valve body surface. After this operation, it is important to remove all metal particles and shavings from the passageway and the bore. Use regulated air pressure and blow into both the passageway and accumulator bore to remove all particles thoroughly.

3. Check with air to insure the passageway is clear. If there is still no free flow of air to the accumulator bore, replace the case.

4. If there is free flow of air to the accumulator bore, reinstall the 2-3 accumulator assembly into the case. Reinstall the valve body using the alignment pins. Finish assembling the transmission.

5. Check the transmission fluid level.

ERRATIC 4-3 AND 3-4 SHIFTS

1981-82 CAR LINES EQUIPPED WITH AOD TRANSMISSION

Erratic 4-3 and 3-4 shifts at vehicle speeds of 45 mph or more may be the result of the governor body restricting line pressure to the governor valve and/or areas in the transmission allowing leakage in the governor pressure circuit.

This condition may be recognized by a 4-3 downshift as the throttle is released followed by a 3-4 upshift as the throttle is applied even though the vehicle speed remains nearly constant.

To service, first verify the transmission build date. On 1981 and 1982 vehicles with transmissions built prior to November 16, 1981, check the governor casting vendor symbol located on the casting back below the body casting number. If the vendor symbol   appears, replace the governor.

If the transmission has a build date after November 16, 1981, or if the governor body casting does not have the specified vendor symbol, or if the vehicle is not corrected by replacing the governor, check for areas that may cause leakage in the governor pressure circuit. Such areas include: a broken or unseated governor retaining ring, worn or damaged output shaft seal rings, a worn or damaged case bore at the output shaft seal rings, an oversize counterweight bore or an undersized output shaft at the counterweight bore, case porosity, leakage between the main control and case due to out of flatness or loose main control valve body bolts.

REVISED OVERDRIVE BAND PIN

1980-81 CAR LINE EQUIPPED WITH AOD TRANSMISSIONS

Overdrive premature band wear may be caused by movement of the overdrive anchor pin. For anchor pin movement inboard more than ³/32″ (2.38mm) from machined face of the case boss, replace the anchor pin with the revised anchor pin—(E2AZ-7F295-A).

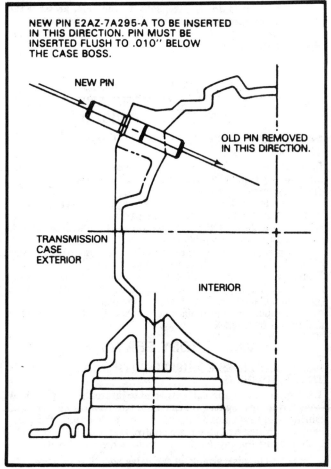

Installation of new band anchor pin (© Ford Motor Co.)

NEW PIN E2AZ-7A295-A TO BE INSERTED IN THIS DIRECTION. PIN MUST BE INSERTED FLUSH TO .010″ BELOW THE CASE BOSS.

NEW PIN

OLD PIN REMOVED IN THIS DIRECTION.

TRANSMISSION CASE EXTERIOR

INTERIOR

NOTE: During transmission overhaul verify anchor pin movement-replace if movement is in excess of ³/32″ (2.38mm).

── CAUTION ──

Exercise care not to cock the pin in the case. By driving the old pin out with new pin, the old pin will help to align the service pin to prevent the service pin from cocking in the transmission case hole. If the pin digs into the case, leakage, case cranks or metal contamination may result.

WHISTLING NOISE ON ACCELERATION

1981-82 CAR LINES EQUIPPED WITH AOD TRANSMISSION

A high frequency "whistle" noise may be emitted on vehicles equipped with the AOD transmission. This noise may be caused by converter pressure relief valve oscillation.

This "whistle" type noise is reported to occur in first, second and reverse gears during light throttle with a cold transmission and/or heavier throttle with a warm transmission. The "whistle" noise is approximately the same pitch in all three gears. Usually, the "whistle" stops one or two seconds before the 2-3 shift occurs. Because of the throttle openings involved, it may not be possible to find the "whistle" noise in reverse.

If the "whistle" noise is identified, change, only the converter pressure relief valve with the new design valve (E2AZ-7E217-A) and not the throttle pressure relief valve.

MODIFICATIONS

NOTE: Prior to this change, both the converter pressure relief valve and the throttle pressure relief valve were identical. For any main control valve body service involving disassembly, do not mix the relief valves.

NEW PLANETARY CARRIER ASSEMBLY, DIRECT CLUTCH CYLINDER AND DIRECT CLUTCH HUB REVISED

1980-81 CARLINES EQUIPPED WITH AOD TRANSMISSIONS

In the event internal servicing of an automatic overdrive transmission is required, it is essential to identify the design of the planetary carrier assembly.

In order to identify transmission's design planet carrier, visually check the direct clutch hub. If the part numbers stamped on it are EOAZ-7F236-AB, the transmission was built with the new design planet carrier assembly.

If the planet carrier assembly must be replaced, no replacement of the direct clutch cylinder and hub is required.

However, if the part numbers stamped on the direct clutch hub are EOAZ-7F236-AA, the transmission was built with the previous design planet carrier assembly. If the planet carrier assembly must be replaced, then replacement of the direct clutch cylinder and hub is required.

VALVE BODY SERVICING

1980-81 CARLINES EQUIPPED WITH AOD TRANSMISSIONS

The No. 7 ball was originally installed to insure the 3-4 shift, however, further testing has indicated the No. 7 ball is not required. During AOD transmission production in May 1981, the No. 7 ball was omitted from the main control (valve body). If the valve body is apart for any servicing—remove the No. 7 ball.

NOTE: The disassembly of the valve body just to remove the No. 7 ball is not recommended.

ASSEMBLY INTERCHANGE, CHANGES AND MODIFICATIONS

1981-83 CAR LINES EQUIPPED WITH AOD TRANSMISSION 1981-82

In the event an entire automatic overdrive transmission assembly is to be replaced in a 1981 vehicle, the following chart lists the 1982 AOD transmissions that can be used to service the 1981 models/applications:

NOTE: 1981 models not listed to be serviced by individual components only.

1982 Models	1981 Models	1982 Model Applicatons
PKA-AS5 E2AP-7000-KA	PKA-AS E1AP-7000-ACA	5.8L Ford/Mercury Police 49S/Can. 2.73 axle
PKA-M13 E2VP-7000-AA	PKA-M8 E1VP-7000-BA	5.9L Lincoln/Mark 50S/Unq. Can. 3.08 axle
PKA-AG5 E2AP-7000-CA	PKA-AL E1AP-7000-JA	5.0L Ford/Mercury (Sedan) and Ford/Mercury Police W/O Low Gear Lockout 50S/Unq. Can. 3.08 axle
PKA-AG5 E2AP-7000-CA	PKA-AG E1AP-7000-HA	5.0L Ford/Mercury (SW) 50S/ Unq. Can. 3.08 axle
PKA-AU5 E2AP-7000-HA	PKA-AU E1AP-7000-AEA	5.0L Ford/Mercury Police 50S 3.08 axle

1982 Models	1981 Models	1982 Model Applicatons
PKA-AH5 E2SP-7000-AA	PKA-AH E1SP-7000-BA	4.2L T'Bird/XR-7 50S/Can. 3.08 axle
PKA-AF5 E2AP-7000-AA	PKA-AF E1AP-7000-GA	4.2L Ford/Mercury Sedan and Ford/Mercury Police W/O Low Gear Lockout 50S/Can: 3.08 axle
PKA-AT5 E2AP-7000-CA	PKA-AT E1AP-7000-ADA	4.2L Ford/Mercury Police 50S/ Can. 3.08 axle
PKB-A6 E2TP-7000-AA	PKB-A1 E1TP-7000-AAC	5.0L F-100/150 3.25 axle, F-250 3.54/3.73 axle 49S/Can.
PKB-A6 E2TP-7000-AA	PKB-A E1TP-7000-AAA	5.0L F-100/150 3.25 axle, F-250 3.54/3.73 alxe 49S/Can.

1982-83

Fifteen (15) models are currently released for the 1983 Automatic Overdrive Transmission (AOD) program; eleven (11) passenger car models and four (4) truck models.

New to AOD for 1983 is the 4.9L engine, the LTD/Marquis, and the Econoline Series applications. There are two (2) new transmission calibrations for the 4.9L applications and one (1) new transmission for the 5.0L.

C. New cable TV linkage. The TV linkage adjustment can be checked and reset, but there is no fine adjustment capability as with the TV rod type linkage (4.9L E and F Series application).

INTERCHANGEABILITY:

The following chart lists the 1983 AOD transmissions that can be used to service the 1982 models and applications listed:

1983 Models	Replaces	1982 Models	1982 Model Applications
PASSENGER CARS			
PKA-AU17 ① E3AP-7000-CA	—	PKA-AU7 E2AP-7000-HA	5.0L Ford/Mercury Police 3.08 A/R
PKA-AS17 ① E3AP-7000-BA	—	PKA-AS7 E2AP-7000-KA	5.8L H.O. Ford/Mercury Police 2.73 A/R
PKA-AG17 ① E3AP-7000-AA	—	PKA-AG7 E2AP-7000-CA	5.0L Ford/Mercury and Police 3.08 A/R with Low Gear Lockout Delete
PKA-AY12 ① E3AP-7000-DA	—	PKA-AY2 E2AP-7000-DA	5.0L Ford/Mercury Altitude and Police 3.42 A/R with Low Gear Lockout Delete
PKA-BC5 ① E3VP-7000-BA	—	PKA-BC3① E2VP-7000-BB	5.0L Lincoln/Mark 3.42 A/R
PKA-BD12 ① E35P-7000-CB	—	PKA-BD3① E25P-7000-AB	5.0L Continental 3.08 A/R
PKA-BB12 ① E35AP-7000-EA	—	PKA-BB2 E2AP-7000-JA	5.0L Ford/Mercury Police 3.42 A/R
PKA-C25 ① E3AP-7000-FA	—	PKA-C15 E2AP-7000-NA	5.8L Ford/Mercury Police 2.73 A/R
PKA-M25 ① E3VP-7000-AB	—	PKA-M16① E2VP-7000-AB	5.0L Lincoln/Mark 3.08 A/R

Truck			
PKB-A20	—	PKB-A10	5.0L F100/150/250
E3TP-7000-MA		E2TP-7000-AB	3.25, 3.54, 3.73, A/R

① Push connect cooler line fittings. Replace with ¼ in. pipe fitting where applicable for 1982 usage.

The following 1982-83 main control valve body assembly can be used to service the 1982 main control listed:

Main Control	Replaces	1982 Main Control
E25P-7A100-AA (PVG)	—	E2SP-7A100-CA (SWG, SVG)
E2AP-7A100-DA (OVH)	—	E2AP-7A100-FA (RWH, RVH)
E2AP-7A100-HA (OCH)	—	E2AP-7A100-JA (RBH, RCH)
E3SP-7A100-CA (UVR)	—	E2SP-7A100-DB, DA (NWM, NVM)

There are four (4) new main control assemblies for 1983.

USE OF ANAEROBIC SEALANT ON EXTENSION HOUSING BOLTS

1983 CAR LINES EQUIPPED WITH AOD TRANSMISSIONS

The #3, #4 and #5 extension housing bolts are now pre-coated with Dri-Loc Anaerobic Sealant in production. Revised service procedures are provided for reinstalling these extension housing to main case bolts—removal and recoating of thread sealant.

In the event service requires removing and reinstalling the extension housing bolts (#3, #4, and #5 bolts).

The following is the revised procedure:
1. Wire brush the bolts and case bolt holes.
2. Remove as much loose sealant as possible.
3. Recoat the bolt with either Threadlock and Sealer (EOAZ-19554-A), Pipe Sealant with Teflon (D8AZ-19554-A), or Teflon Tape.
4. Torque the bolts to specifications (16-20 ft. lbs. or 22-27 N•m).

FLUID LEAKAGE AT LEFT SIDE OF PAN GASKET AREA

1983 CAR LINES EQUIPPED WITH AOD TRANSMISSION

Fluid leakage in the left (driver's) side area of the oil pan may be caused by an undertorqued pipe plugs(s).

Some vehicles built before February 7, 1983, may exhibit a fluid leakage at the left side area of transmission oil pan gasket. Prior to servicing the suspect oil pan gasket:
1. Clean off all traces of transmission fluid.
2. Test drive vehicle and then observe the concern area.
3. If fluid leakage is occurring from a pipe plug(s), the following service procedure should be performed:
    a. Remove the suspect plug(s).
    b. Clean plug threads and reinstall to proper torque specifications (8-16 N•m—6-12 ft. lbs.) and road test.

NEW 3-4 SHIFT VALVE AND SPRING

1983½ CAR LINES EQUIPPED WITH AOD TRANSMISSION

A new 3-4 shift valve configuration and spring was incorporated in production mid-March, 1983 which moves the 3-4 shift to a

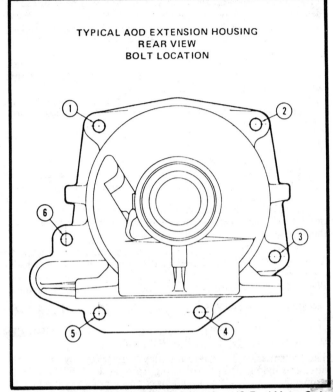

Rear extension housing retaining bolt locations (© Ford Motor Cc.)

higher vehicle speed and provides for 4-3 downshifts with less throttle and at higher speeds.

The new shift valve is comprised of two parts and the bore plug has been replaced by a sleeve.

The new shift valve configuration will change the 3-4/4-3 shift pattern. The new 3-4 shift will feel delayed if compared directly with the old design. This condition may be more evident prior to engine break-in. The old design would upshift to fourth gear even during fairly heavy throttle acceleration. The new design will not upshift to fourth gear until the vehicle approaches the desired speed and the driver eases up on the accelerator pedal.

The new 4-3 downshift pattern feels more responsive than the old design. At moderate to heavy throttle, the transmission will downshift to third gear easier and at higher speeds.

Easier downshifts or delayed 3-4 shifts is the intent of the new shift pattern and it should not be interpreted as a transmission concern. Proper explanation to the customer of this as normal vehicle operation is required.

REVISED FRONT OIL PUMP GASKET

1983½ CAR LINES EQUIPPED WITH AOD TRANSMISSION

A new front pump gasket was incorporated in May, 1983 on all AOD transmissions. The pump gasket is notched matching a recessed area in the pump body. This will allow transmission fluid to drainback at a higher point into the case, alleviating some transmission pump case leakage concerns.

The AOD transmission case will also be changed removing the drainback hole in the case casting. Refer to Figure 17 for the location of the case drainback hole. The hole allowed transmission fluid in the pump area to drainback to the oil pan. Instead, the notched gasket will provide a path for the fluid to vent at a higher point into the case.

For service, the transmission service case will be changed to the new level which deletes the drainback hole. The service case is interchangeable for all model years and all AOD applications.

The new notched pump gasket can be used on both type transmission cases. However, the original design gasket cannot be used on the new level case which has the drainback hole deleted since there is no provision for drainback of transmission fluid with this combination of parts.

Service stock with the original design front pump gasket may be used on a case with no drainback hole; modify the gasket as follows:

1. Using the new notched gasket as a template, lay it on top of the original design gasket.
2. Match all feed passages and bolt holes.
3. Trace the notched portion on the original design level gasket.
4. Remove the notched gasket and cut out the portion traced on the original design level gasket.
5. Verify that the modification was done correctly by comparing the modified gasket with the notched gasket that was used as the template.

NOTE: Do not use sealer or adhesive on pump case or gasket-to-pump surface. If the pump gasket notch is blocked, no fluid drainback is possible. This could result in the fluid pressure unseating the front pump O-ring or fluid coming out the filler tube.

ERRATIC SHIFTING AND/OR NO 4TH GEAR

1983½ THUNDERBIRD, COUGAR CAR LINES EQUIPPED WITH 5.0L ENGINE (CFI 3-22W-ROO)

The transmission TV rod return may be restricted by the rubber hose and nylon fittings protruding from the back of the rubber connector block (—9E455—). This connector block is attached to the 3-port PVS valve, at the rear left hand side of the intake manifold. Potential kinking of the canister purge hose is also possible due to incorrect orientation of the nylon elbow on the rubber block connector.

On a customer concern basis, the following procedure is recommended:

1. Remove the rubber connector block from the 3-port PVS.
2. With an appropriate deep well socket (slotted), rotate the 3-port PVS (located on rear left hand side of intake manifold) clockwise, until the ports are aimed toward the upper rear rocker cover attaching bolt (approximately 15-20° from original position).
3. Rotate the middle and lower nylon elbow 180° from original position. The middle elbow should face the left side of the engine and the lower elbow should face the right side of the engine.
   a. Route the middle vacuum hose from the connector block in front of the PVS valve and under the kickdown (TV-rod) rod. The upper and middle hoses should be tucked down.
   b. Route the lower vacuum hose to the purge control valve under the kickdown rod.
4. Ensure that any other hoses that are routed in area of kickdown rod (TV rod) are placed under or away from rod to eliminate any potential interference.

REVISED THROTTLE VALVE (TV) LINKAGE ADJUSTMENT PROCEDURE USING TV CONTROL PRESSURE

1980-84 CAR AND TRUCK LINES EQUIPPED WITH AOD TRANSMISSION

This revised method of setting TV linkage on all AOD transmissions is the only way to adjust the linkage to the middle of the specification curve. This method sets the linkage to the most sensitive point.

The new procedure attached uses a TV control pressure gage block service tool D84P-70332-A. If that tool is not available, alternates are listed in the procedure. Highlights of the new procedure include setting the TV linkage at idle with the gage block installed to 35 ± 5 PSI.

The "TV pressure method for adjusting the TV linkage" procedure is recommended for transmission shift concerns while the "Linkage Adjusted at the Carburetor" procedure is recommended when the idle speed is changed by 50 rpm.

LINKAGE ADJUSTMENT USING TV CONTROL PRESSURE TV ROD SYSTEMS ONLY

The following procedure may be used to check and/or adjust the throttle valve (TV) rod linkage using the TV control pressure.

1. Check/adjust the engine curb idle speed to specification required.
2. Attach a 0-100 PSI pressure gage, T73L-6600-A, with the adapter fitting D80L-77001-A, or equivalent to the TV port on the transmission with sufficient flexible hose to make gage accessible while operating the engine.
3. Obtain a TV control pressure gage block, service tool no. D84P-70332-A, or fabricate a block .397 ± .007 inch thick. The following drill bit shanks may also be used in order of preference: Letter X (.397 in.), 10mm (.3937 in.) or 25/64 (.3906 in.).
4. Operate the engine until normal operating temperature is reached and the throttle lever is off fast idle or the Idle Speed Control plunger (if equipped) is at its normal idle position. The transmission fluid temperature should be approximately 100-150°F. Do not make pressure check if transmission fluid is cold or too hot to touch.
5. Set parking brake, place shift selector in N (neutral), remove air cleaner, shut off air conditioner. If equipped with a Vacuum Operated Throttle Modulator, disconnect and plug the vacuum line to this unit. If equipped with a Throttle Solenoid Positioner or an Idle Speed Control, do not disconnect either of these units.

--- CAUTION ---

Do not make pressure check in Park.

6. With engine idling in neutral, and no accessory load on engine, insert gage block (or drill shank) between the carburetor throttle lever and adjustment screw on the TV linkage lever at the carburetor. The TV pressure should be 35 psi ± 5 psi. For best transmission function, use the adjusting screw to set the pressure as close as possible to 35 psi. Turning the screw in will raise the pressure 1.5 psi per turn. Backing out the screw will lower the pressure. If equipped with Idle Speed Control, some "hunting" may occur and an average pressure reading will have to be determined. If the adjusting screw does not have enough adjustment range to bring TV pressure within specification, first adjust rod at the transmission.
7. Remove gage block, allowing TV lever to return to idle. With engine still idling in neutral, TV pressure must be at or near zero (less than 5 psi). If not, back out adjusting screw until TV pressure is less than 5 psi. Reinstall gage block and check that TV pressure is still 35 psi ± 5 psi.

DESCRIPTION OF CHANGES AND INTERCHANGEABILITY FOR 1984

1984 CAR AND TRUCK LINES EQUIPPED WITH AOD TRANSMISSIONS

AOD transmissions will have the following changes incorporated for 1983½ and 1984 model vehicles.

Twenty (20) models are currently released for the 1984 Automatic Overdrive Transmission (AOD) Program: Sixteen (16) passenger car models and four (4) F-Series/E-Series models. This is an increase of three (3) transmission models from the 17 models in 1983.

New to AOD for 1984 is the Mustang/Capri vehicle line equipped with either the 3.8L or the 5.0L H.O. engine both scheduled as a running change after Job #1. The 5.0L H.O. engine is a new power plant application with the AOD transmission. The

Mark VII for 1984 will have the console floor shift (cable type) transmission shift controls. The F-100 truck has been replaced by the Range application.

MAJOR DESIGN CHANGES FOR INCORPORATION OF THE 3.8L AND 5.0L H.O. MUSTANG/CAPRI APPLICATION:

1. Two (2) new transmission models.
2. One (1) new main control valve body calibration.
3. New manual lever (oriented down).
4. Both 3.8L and 5.0L H.O. packages will have the cable type floor shift with the shift brackets installed on the rear of the transmission on two extension housing stud bolts.
5. 13mm hex head extension housing bolts will be used in holes No. 1 and No. 6 (maybe No. 2 also) to provide bearing surface for the floor shift bracket. 10mm hex head bolts will be used in the remainder of the extension housing holes.
6. TV levers will be curved similar to truck lever.
7. Converter has a higher stall ratio (165K) for improved performance.

MARK VII APPLICATIONS:

1. 13mm hex head extension housing bolts will be used in holes No. 1 and No. 6 (maybe No. 2 also) to provide bearing surface for the floor shift bracket. 10mm hex head bolts will be used in the remainder of the extension housing holes.

AOD DESIGN CHANGES AFFECTING ALL PASSENGER CAR AND TRUCK APPLICATIONS:

1. The dipstick full mark for operating temperature (hot) fluid fill has been changed to a cross-hatched area instead of arrows. Do not add fluid if level is in the cross-hatched area.
2. The following AOD hardware changes were incorporated on all model AOD transmissions with build dates of July 19, 1983, and later. These hardware changes will affect past model serviceability.

   a. Common forward and reverse clutch friction plates and pressure plates which result in new internal snap ring locations for both the reverse clutch drum and the forward clutch cylinder. (The forward clutch cylinder friction plate usage will be expanded to the reverse clutch drum and the reverse clutch pressure plate usage will be expanded to the forward clutch cylinder.)

   b. A new external retaining ring for the intermediate one-way clutch which results in a deeper snap ring groove on the reverse clutch drum.

SERVICEABILITY REVISIONS:

The new reverse clutch drums (no change to number of plates; passenger car—3 plate drum, 4 plate drum and truck drum; and no change to clutch pack clearance specification) will have an internal identification groove adjacent to the clutch pack snap ring groove. This internal identification groove indicates that the 7E311 friction plates and the new intermediate one-way clutch snap ring must be used with this drum.

When service stock is exhausted on the original reverse clutch drum, the new reverse clutch drum will be the replacement drum for all past model service.

The new intermediate one-way clutch retaining ring will replace all stock in the depot and will be used on all past model service, but must be used with the new level reverse clutch drum. Therefore, whenever the original design level drum is replaced on past model applications, the intermediate one-way clutch retaining ring must be replaced with the new level retaining ring.

The new forward clutch cylinder (same number of plates as before—4 plate cylinder and 5 plate cylinder; and no change to clutch pack clearance specification) will be identified by a reduced chamfer ($1/32$ in. x 45° O.D.) vs. the previous forward clutch cylinder ($1/8$ in. x 45° O.D.). The reduced chamfer identification indicates that 7F278 pressure plate must be used.

1. All converters will have an orange special balance mark on a converter stud. If a flywheel bolt hole has a special balance mark, align the marks to avoid powertrain imbalance.
2. Passenger cars with auxiliary oil coolers will have an "H" fitting that bypasses some of the auxiliary cooler flow, especially during cold weather operation to ensure lubrication flow to the transmission.

CARBURETOR DESIGN CHANGES AFFECTING TV LINKAGE ADJUSTMENT PROCEDURE:

Passenger Cars with 3.8L CFI

1. A new idle speed control motor (ISC) will be incorporated on the carburetor replacing other conventional throttle positioners. The ISC will have the capability of varying the throttle plate opening via signals from the EEC IV system. Before TV linkage is checked and adjusted, the ISC motor plunger must be retracted.

E and F-Series 4.9L YFA-FBC

1. A new idle speed control motor (ISC) (same as 3.8L CFI passenger car) will be incorporated on the carburetor replacing other throttle positioners. The ISC motor plunger must be retracted before checking and adjusting the TV linkage. But, in addition, there is a fast idle cam lever and that must be de-cammed before the TV linkage is adjusted.

INTERCHANGEABILITY

The following chart lists the 1984 AOD transmissions that can be used to service the 1983 models and applications listed.

1984 Model	Replaces	1983 Models	1983 Model Applications
		PASSENGER CARS	
PKA-AS23 E4AP-7000-JA	—	PKA-AS21 E3AP-7000-BA	5.8L H.O. Ford police 2.73 Axle with low gear lockout.
PKA-C31 E4AP-7000-KA	—	PKA-C29 E3AP-7000-FA	5.8L H.O. Ford/Mercury (Sedan) Canada, Ford P (50S/Can/Alt) 2.73 Axle without low gear lock
PKA-BD18 E45P-7000-DA	—	PKA-BD16 E35P-7000-CB	5.0L Continental (50S/Can) 3.08 Axle, (Alt/ 3.27 Axle.
PKA-AG23 E4AP-7000-HA	—	PKA-AG21 E3AP-7000-AA	5.0L Ford/Mercury (Passenger Car) F (50S/Unique Canada) 3.08 Axle with lockout.

THM 200C

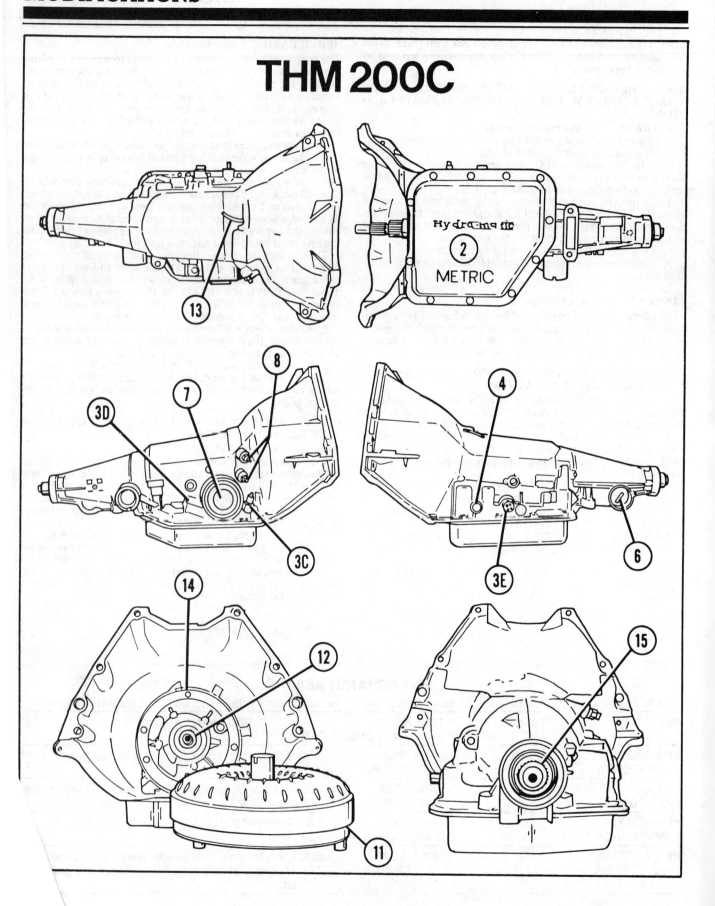

THM 125C

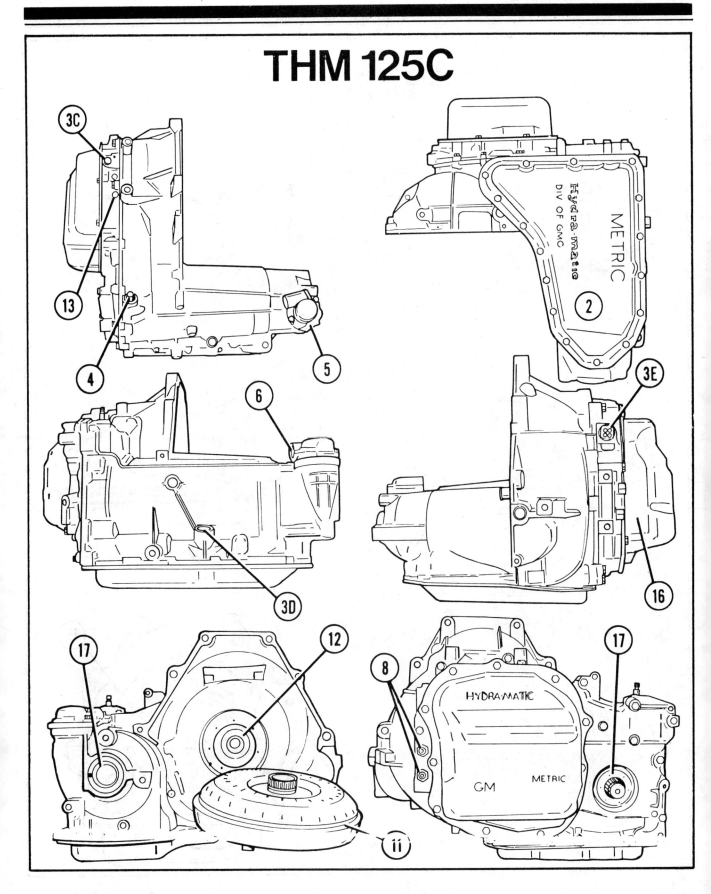

THM 200 4R

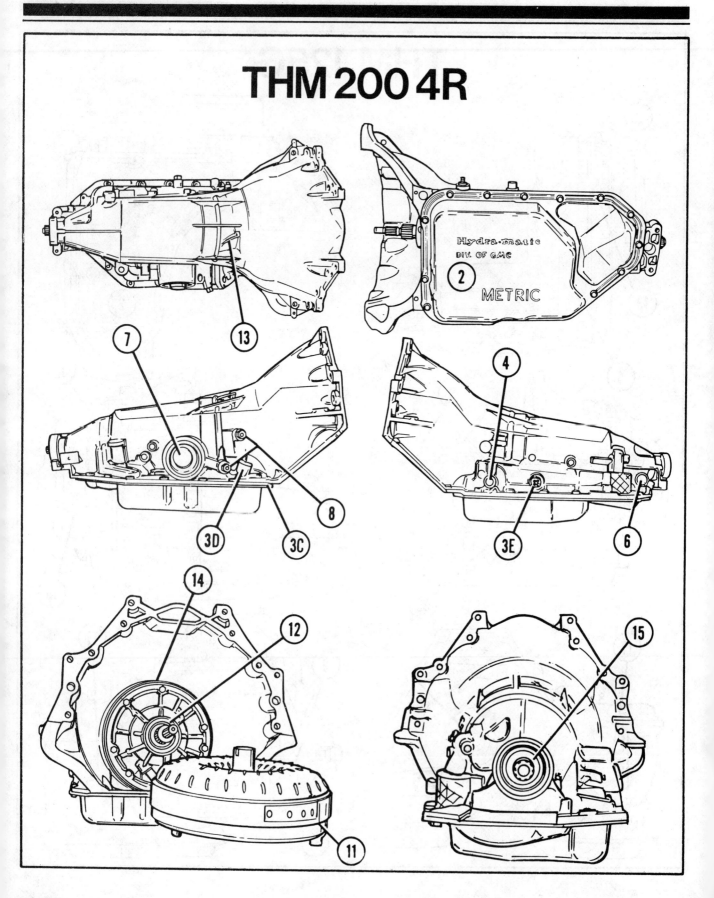

THM 325 4L

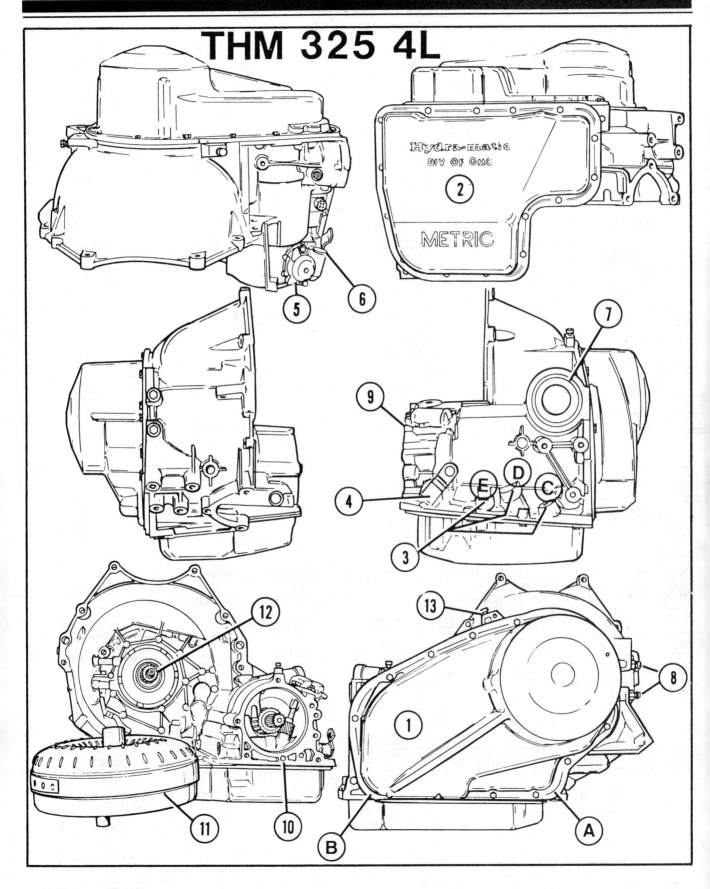

MODIFICATIONS

INTERCHANGEABILITY

The following chart lists the 1984 AOD transmissions that can be used to service the 1983 models and applications listed.

1984 Model	Replaces	1983 Models	1983 Model Applications
		PASSENGER CARS	
PKA-AY18 E4AP-7000-FA	—	PKA-AY16 E3AP-7000-DA	5.0L Ford/Mercury (Passenger Car) Ford Police (Alt/Unique Canada) 3.55 Axle without low gear lockout.
PKA-BB18 E4AP-7000-EA	—	PKA-BB16 E3AP-7000-EA	5.0L Ford Police (Alt/Unique Canada) 3.55 Axle with low gear lockout.
PKA-AU23 E4AP-7000-GA	—	PKA-AU21 E3AP-7000-CA	5.0L Ford Police (50S/Unique Canada) 3.08 Axle without low gear lockout.
PKA-M31 E4VP-7000-DA	—	PKA-M29 E3VP-7000-AB	5.0L Lincoln/Mark VI (50S) 3.08 Axle.
PKA-BC12 E4VP-7000-CA	—	PKA-BC10 E3VP-7000-CA	5.0L Lincoln/Mark VI (50S/Can/Alt) 3.55 Axle.
PKA-K6 E4SP-7000-DA	—	PKA-K4 E3SP-7000-FA	5.0L T-Bird/Cougar (50S/Can) 3.08 Axle, (Alt) 3.27 Axle.
*PKA-CB6 E4DP-7000-HA	—	PKA-CB3 E3DP-7000-EA PKA-BR3 E3SP-7000-BA	3.8L T-Bird/Cougar (50S/Can) 3.08 Axle. 3.8L T-Bird/Cougar (50S/Can) 3.08 Axle.
PKA-BT6 E4SP-7000-CA	—	PKA-BT4 E3SP-7000-EA	3.8L T-Bird/Cougar (Alt), LTD/Marquis (49S/Can/Alt) 3.45 Axle.
		LIGHT TRUCKS	
PKB-A26 E3TP-7000-MA	—	PKB-A24 E3TP-7000-MA	5.0L E100/150 3.50 Axle, F100/150, 3.55 Axle, F250 3.54/3.55/3.73 Axle, E250 3.54/3.55 Axle (50S/Can/Alt).
PKB-E5 E3TP-7000-NA	—	PKB-E4 E3TP-7000-NA	4.9L F100/150 (50S/Can) 3.08 Axle.
PKR-F4 E3TP-7000-PA	—	PKB-F3 E3TP-7000-PA	4.9L E100/150 3.50 Axle, E250 3.54/3.55 Axle (50S/Can/Alt), F100/150 3.55 Axle (40S/Can/Alt).
PKB-G5 E3UP-7000-SA	—	PKB-G4 E3UP-7000-SA	5.0L (50S/Can/Alt) 4.10 Axle.

*Use speedometer driven gear C7SP-17271-B 18T gray when using transmission on a 1983 model vehicle.
Main control valve body interchangeability:
All the 1984 main control valve bodies are pull-ahead 1983 and, therefore, are interchangeable.

OIL LEAKAGE REPAIR PROCEDURE

Key	Location	Leak Cause	Correction
		200C AND 200-4R ONLY	
⑭	Pump Assembly	Pump Bolts Loose	Retorque Bolts and Recheck
		Pump "O" Ring Cut or Damaged	Remove "O" Ring and Replace
		Pump Porosity	Replace Pump
⑮	Rear Extension Seal	Torn or Damaged	Remove Seal and Replace
		125C ONLY	
⑯	Valve Body Cover to Case Assembly	Low Bolt Torque	Retorque Bolt and Recheck
		RTV Bead Broken	Remove Cover and Reseal with RTV Gasket
		Gasket Leak	Remove Gasket and Replace
⑰	Axle Shaft Seals	Seals Damaged	Remove Seals and Replace

OIL LEAKAGE REPAIR PROCEDURE

Key	Location	Leak Cause	Correction
		ALL MODELS	
②	Transmission Oil Pan	Low Bolt Torque Gasket Leak Broken RTV Bead	Retorque and Recheck Bolts Remove and Replace Gasket Remove Cover and Reseal with RTV
③	TV Cable Connector "C"	Connector Cocked and interferring with mount	Remove Connector and Reinstall
		Seal Damaged	Remove and Replace Seal
		Connector Cracked	Remove and Replace Cable
	Fill Tube "D"	Seal Missing or Damaged	Remove Fill Tube and Replace Seal
	Electrical Connector "E"	"O" Ring Missing or Damaged	Remove Connector and Replace Seal
		Connector Missing or Damaged	Remove and Replace Connector
④	Manual Shaft	Seal Assembly Damaged	Remove and Replace Seal
⑤	Governor Cover	"O" Ring Damaged	Remove and Replace "O" Ring
		Low Bolt Torque	Retorque Bolts and Recheck
⑥	Speedo Fitting	Low Bolt Torque	Retorque Bolt and Recheck
		Seal Damaged	Remove and Replace Seal
⑦	Servo Cover	"O" Ring Damaged	Remove and Replace "O" Ring
⑧	Cooler Fittings	Low Fitting Torque Cracked Fitting	Retorque Fitting and Recheck Replace and Retorque Fitting
⑪	Converter Assembly	Hub or Seam Weld Leak	Remove and Replace Converter
⑫	Converter Seal	Seal Damaged	Remove and Replace Seal
⑬	Vent	Leaking	Check for the following and correct as necessary: A. Oil Overfill B. Blocked Drainback hole in sprocket support C. Engine Coolant in oil
		325-4L ONLY	
①	Sprocket Cover	RTV Bead Broken	Remove Cover and Reseal with RTV
		Cracked Cover	Replace Cover
	Corner "A"	Leak	Past Design Case-Drill through and Add Nut, Bolt and Lock Washer
	Corner "B"	Leak	Remove Cover and Reseal with RTV
⑨	Case Face to Final Drive	Low Bolt Torque	Retorque Bolts and Recheck
⑩	Case Face to Final Drive Governor Cover	Plug Loose (Leaking)	Remove Plug and Replace (Loctite Plug)
		Plug High	Drive Plug into Case Below Face Surface